Doppler Ultrasound

Second Edition

Doppler Ultrasound
Physics, Instrumentation and Signal Processing
Second Edition

David H. Evans, PhD, FIPEM, FInstP
University of Leicester, UK

and

W. Norman McDicken, PhD, FIPEM
University of Edinburgh, UK

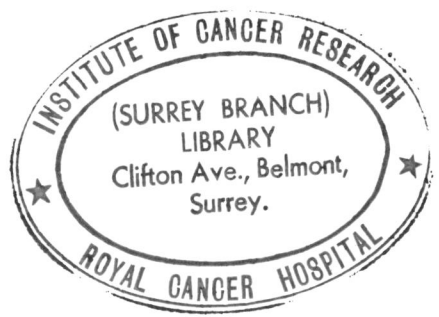

JOHN WILEY & SONS, LTD
Chichester · New York · Weinheim · Brisbane · Singapore · Toronto

First published 1989, as *Doppler Ultrasound: Physics, Instrumentation, and Clinical Applications*,
Second Edition published 2000

Copyright © 1989, 2000 by John Wiley & Sons Ltd,
Baffins Lane, Chichester,
West Sussex PO19 1UD, England

National 01243 779777
International (+44) 1243 779777
e-mail (for orders and customer service enquiries): cs-books@wiley.co.uk
Visit our Home Page on http://www.wiley.co.uk
or http://www.wiley.com

All Rights Reserved. No part of this publication may be reproduced, stored in a retrieval system, or transmitted, in any form or by any means, electronic, mechanical, photocopying, recording, scanning or otherwise, except under the terms of the Copyright, Designs and Patents Act 1988 or under the terms of a licence issued by the Copyright Licensing Agency, 90 Tottenham Court Road, London W1P 9HE, UK, without the permission in writing of the publisher.

Other Wiley Editorial Offices

John Wiley & Sons, Inc., 605 Third Avenue,
New York, NY 10158-0012, USA

WILEY-VCH Verlag GmbH, Pappelallee 3,
D-69469 Weinheim, Germany

Jacaranda Wiley Ltd, 33 Park Road, Milton,
Queensland 4064, Australia

John Wiley & Sons (Asia) Pte Ltd, 2 Clementi Loop #02-01,
Jin Xing Distripark, Singapore 129809

John Wiley & Sons (Canada) Ltd, 22 Worcester Road,
Rexdale, Ontario M9W 1L1, Canada

Library of Congress Cataloging-in-Publication Data

Doppler ultrasound : physics, instrumentation, and signal processing /
  David H. Evans. — 2nd ed.
    p. cm.
    Includes bibliographical references and index.
    ISBN 0-471-97001-8 (cased : alk. paper)
    1. Doppler ultrasonography. 2. Medical physics. I. Evans, D. H., Ph. D.
    [DNLM: 1. Ultrasonography—instrumentation. 2. Ultrasonics. 3. Ultrasonography—methods. WB 289 D692 2000]
  RC78.7.U4D66 2000
  616.07′543—dc21
  DNLM/DLC
  for Library of Congress   99–33585
                                                                 CIP

British Library Cataloguing in Publication Data

A catalogue record for this book is available from the British Library

ISBN 0-471-97001-8

Typeset in 10/12pt Times from the author's disks by Mathematical Composition Setters, Salisbury, Wiltshire
Printed and bound in Great Britain by Antony Rowe Ltd, Chippenham, Wiltshire
This book is printed on acid-free paper responsibly manufactured from sustainable forestry, in which at least two trees are planted for each one used for paper production.

To Christina, Jennifer and William
who were, once again, the main casualties

Contents

Preface xi

Preface to First Edition xiii

Plates xv

1 Introduction
1.1 The Doppler effect 1
1.2 Ultrasound 1
1.3 Doppler ultrasound in medicine 2
1.4 Summary 3
1.5 References 3

2 Blood Flow
2.1 Introduction 5
2.2 Basic concepts and definitions 5
2.3 Steady flow in rigid tubes 6
2.4 Pulsatile flow in rigid tubes 9
2.5 Pulsatile flow in elastic and viscoelastic tubes 14
2.6 The effects of geometric changes 16
2.7 Velocity profiles in human arteries . 20
2.8 Computational fluid dynamics 22
2.9 Summary 24
2.10 Notation 24
2.11 References 25
2.12 Recommended reading 26

3 Physics of Ultrasound Propagation
3.1 Introduction 27
3.2 Ultrasonic waves 27
3.3 Ultrasonic phenomena 34
3.4 Ultrasonic tissue characterisation parameters 39
3.5 The Doppler effect 39
3.6 Summary 40
3.7 Notation 40
3.8 References 41

4 Doppler Systems: A General Overview
4.1 Introduction 43
4.2 Velocity detecting systems 43
4.3 Duplex systems 54
4.4 Profile detecting systems 56
4.5 Velocity imaging systems 58
4.6 Summary 67
4.7 References 67

5 Ultrasonic Transducers, Fields and Beams
5.1 Introduction 71
5.2 Transducers 71
5.3 Ultrasonic beams and fields 84
5.4 Doppler sample volumes 88
5.5 Measurement of fields, zones, beams and sample volumes 90
5.6 Sterilisation of transducers 93
5.7 Damage to transducers 94
5.8 Summary 94
5.9 References 94

6 Signal Detection and Pre-processing: CW and PW Doppler
6.1 Introduction 97
6.2 Continuous wave systems 97
6.3 Pulsed wave systems 105
6.4 Direction detection 107
6.5 Analogue envelope detectors 114
6.6 Complex Doppler systems 116
6.7 Summary 117
6.8 References 117

7 The Doppler Power Spectrum
7.1 Introduction 119
7.2 Blood as an ultrasonic target 120
7.3 Continuous wave Doppler spectra ... 131
7.4 Pulsed wave Doppler spectra 142
7.5 Summary 146
7.6 Notation 146
7.7 References 147

8 Doppler Signal Processors: Theoretical Considerations
- 8.1 Introduction 150
- 8.2 Spectral estimation techniques...... 150
- 8.3 Mean frequency processors 169
- 8.4 Maximum frequency processors 179
- 8.5 Zero-crossing processors (RMS followers) 187
- 8.6 Other signal location estimators 189
- 8.7 Envelope averaging techniques 190
- 8.8 Comparison and choice of envelope extraction techniques.............. 191
- 8.9 Summary....................... 192
- 8.10 Notation 192
- 8.11 References..................... 194

9 Waveform Analysis and Pattern Recognition
- 9.1 Introduction 200
- 9.2 Pattern recognition principles 200
- 9.3 Pre-processing 201
- 9.4 Feature extraction.............. 201
- 9.5 Classification 219
- 9.6 Summary....................... 222
- 9.7 Notation 222
- 9.8 References..................... 223

10 Colour Flow Imaging (CFI) Systems
- 10.1 Introduction 229
- 10.2 Ultrasonic imaging 230
- 10.3 Doppler imaging methods 231
- 10.4 Specialised Doppler imaging methods...................... 233
- 10.5 Features of Doppler imagers....... 237
- 10.6 Performance and use of colour Doppler imaging 239
- 10.7 Artefacts in Doppler techniques 241
- 10.8 Summary....................... 243
- 10.9 References..................... 244

11 Signal Processing for Colour Flow Imaging
- 11.1 Introduction 245
- 11.2 Current narrow-band colour imaging systems................. 245
- 11.3 Other narrow-band (phase shift estimation) methods 256
- 11.4 Wide band (time shift estimation) methods...................... 263
- 11.5 Direction of arrival methods 280
- 11.6 Future directions 282
- 11.7 Summary....................... 282
- 11.8 Notation 282
- 11.9 References..................... 284

12 Volumetric Blood Flow Measurement
- 12.1 Introduction 288
- 12.2 Flow measurement with duplex scanners...................... 288
- 12.3 Flow measurement with multigate systems—1-D profiles............ 293
- 12.4 Flow measurement with multigate systems—2-D profiles............ 296
- 12.5 Flow measurement—attenuation-compensated method.............. 298
- 12.6 Flow measurement—assumed velocity profile method 300
- 12.7 Flow measurement—C-mode Doppler techniques 300
- 12.8 Flow measurement using contrast agents........................ 303
- 12.9 Miscellaneous techniques 303
- 12.10 Flow measurement errors.......... 307
- 12.11 Summary....................... 307
- 12.12 References..................... 307

13 Miscellaneous Doppler Techniques
- 13.1 Introduction 311
- 13.2 Contrast agents 311
- 13.3 Perfusion techniques 317
- 13.4 Emboli detection 319
- 13.5 Transverse Doppler 323
- 13.6 Vector Doppler techniques......... 325
- 13.7 Speckle tracking techniques 333
- 13.8 Decorrelation based techniques..... 337
- 13.9 3-D Doppler imaging 339
- 13.10 Tissue motion imaging 343
- 13.11 Anti-aliasing techniques 347
- 13.12 B-Flow........................ 350
- 13.13 Summary....................... 350
- 13.14 References..................... 350

14 Safety Considerations in Doppler Ultrasound
- 14.1 Introduction 359
- 14.2 Ultrasonic field measurement....... 359
- 14.3 Ultrasonic output from Doppler units......................... 362
- 14.4 Physical effects of ultrasound...... 363
- 14.5 Bioeffects of ultrasound 365

14.6	Contrast agents	367
14.7	Safety standards	369
14.8	Safety statements	374
14.9	Minimising patient exposure	376
14.10	Summary	377
14.11	References	378

Appendix 1 Special Functions Arising from Womersley's Theory 381

Appendix 2 Doppler Test Devices
PR Hoskins and KV Ramnarine

A2.1	Introduction	382
A2.2	Flow phantoms	382
A2.3	Other moving target devices	391
A2.4	Electronic injection devices	395
A2.5	Measurement using test objects	397
A2.6	Which test objects are useful?	401
A2.7	Conclusion	402
A2.8	References	402

Appendix 3 Recording and Reproduction of Doppler Signals and Colour Doppler Images
T Anderson

A3.1	Introduction	405
A3.2	Characteristics of Doppler audio signals	406
A3.3	Recording Doppler audio signals	407
A3.4	Recording colour Doppler images	410
A3.5	Terminology of specifications	414
A3.6	Summary	414
A3.7	References	414
A3.8	Further reading	414

Index 415

Preface

Doppler ultrasound has continued to grow in importance since the publication of the first edition of this book some 10 years ago, to the extent that is now considered an indispensable part of most ultrasound examinations (which have themselves grown tremendously in both number and variety). Not only have many new applications for Doppler ultrasound been identified but ultrasound machines have improved dramatically, which in turn has led to yet further applications. The vast improvements in machines have resulted from many advances, including those in transducer design and manufacture, improved instrumentation and signal processing techniques, and from the rapid increase in computing power available to system designers.

Over the same period of time there have been many excellent text-books published which deal with clinical aspects of Doppler ultrasound (these same books often include a short section on the technical aspects of Doppler ultrasound) but, remarkably, there are still very few, if any, books that deal with technical aspects of Doppler ultrasound in both breadth and depth. There has also been a rapid expansion of the scientific literature relating to technical aspects of Doppler ultrasound in the scientific journals, but this makes it more difficult, rather than easier for the non-specialist to keep abreast of latest developments.

In order to obtain the most from Doppler ultrasound techniques, and in particular to avoid its many pitfalls, it is vital that the user has a firm grasp of its underlying physical principles. The purpose of this book is two-fold; it is to provide a sound theoretical basis for clinical users of the technique, and to provide an up-to-date survey of the many new innovations that have been described as potentially useful for detecting, measuring and imaging blood flow. Thus, it is hoped that this book may be a suitable companion to all Doppler enthusiasts whether their primary discipline is medical, scientific, or technical.

The subject of Doppler ultrasound has moved on considerably since the first edition of this book was published. In particular, there are now so many texts on the clinical aspects of the subject that there seemed little point in including even a superficial survey of clinical applications in this edition. Furthermore, Doppler ultrasound imaging has assumed such an important role that those parts of the book dealing with colour flow imaging have been significantly enlarged, although it is important to bear in mind that many of the general discussions of Doppler signals and the way they are influenced by many factors apply equally to all Doppler ultrasound methods, whether or not they involve imaging. Another change has been in the level of awareness and sophistication amongst clinical users of Doppler ultrasound, and for that reason no attempt has been made in this edition to separate text into 'essential' and 'advanced', but it is hoped that the text should prove equally accessible to Doppler novices as to physicists and engineers whose major interest is in Doppler ultrasound. Although many equations have been included, most of the arguments can be followed without a detailed understanding of these, and the important points contained in equations have been brought out in the discussion.

As for the first edition, notational consistency has proved a problem, because as far as possible we have tried to maintain the 'standard' notation being used by workers in each particular branch of the subject, and unfortunately this leads to a number of clashes where the same symbol is used for different quantities. While we have attempted to use a given symbol for a given quantity throughout the book, the same symbol may indicate different quantities in different chapters. Symbols are generally defined on the first occasion they appear in each chapter, and chapters which contain significant numbers of equations are each provided with a notation list.

Many people have wittingly or unwittingly contributed to the production of this book. These include colleagues throughout the Doppler ultrasound community (both fellow workers in the field and engineers from some of the major ultrasound companies) and our past and present students. These are too numerous to mention by name, and to pick a few would be invidious. We must, however, record our special thanks to Peter Hoskins, Kumar Ramnarine and Tom Anderson for contributing Appendices 2 and 3, and to our secretaries, Mrs Tina Craig and Mrs Irene Craig (no relation!), who have cheerfully coped with the myriad tasks that a book like this begets. We are also indebted to a number of individuals and publishers for allowing us to reproduce figures and tables. We are grateful to GE Ultrasound for their generous contribution towards the production costs of the colour figures.

David Evans
Norman McDicken

Preface to First Edition

Doppler ultrasound has come of age in the last ten years. No longer is it a technique confined to research laboratories and specialist centres, it is now in widespread clinical use, and its use continues to grow at a rapid pace. In concert with its rapid growth a great many books have been published that deal with the clinical aspects of the subject. Such books often contain a brief technical introduction to Doppler ultrasound, but these are invariably superficial in nature, and there are virtually no books that deal primarily with the physics and instrumentation of the method. This is unfortunate because in order to obtain the most from Doppler ultrasound it is important that the user has a firm grasp of the underlying physical principles.

The purpose of this book is to provide an up-to-date introduction to the more technical aspects of Doppler ultrasound. We have attempted to address it to as wide a readership as possible and we hope that it will be suitable as a companion to all Doppler enthusiasts whether their primary discipline be medical, scientific or technical. In order to achieve this we have, wherever possible, avoided the use of complex equations, and leave the more mathematically inclined to pursue these in the cited literature. Most of the included mathematics can be avoided during an initial study of the text.

In order to satisfy the needs of as wide a range of interests as possible we have divided the text into two parts which could be described as 'essential' and 'advanced'. The latter is indicated by a vertical rule in the left-hand margin. We have confined less important technical material and all but the most important mathematics to the advanced sections, and have tried to arrange the separation of the two parts in such a way that the essential material is complete in itself. We hope that this arrangement will appeal to physicists and medical personnel alike and that this structure will tempt *all* readers to dip into the advanced material of special relevance to themselves.

The final part of the book is a short section on clinical applications. It has already been mentioned that there are many books available that deal with this facet of Doppler ultrasound and it is clear that a book such as this cannot hope to compete with these in terms of depth or breadth of coverage. We have included this material since it both illustrates the applications of the physical principles described earlier, and serves as an entry point into the clinical literature for those who have not approached the subject from a medical background.

We have tried to maintain notational consistency throughout the book, but this has not always been easy since, for example, the 'standard notation' of haemodynamics often clashes with that of ultrasonic wave propagation. We have therefore arrived at something of a compromise: we have attempted to use a given symbol for a given quantity throughout the book, but the same symbol may indicate different quantities in different chapters. Symbols are usually defined on the first occasion they appear in each chapter, and Chapters 2, 3, 8 and 10 each contain notation lists.

Many people have contributed to the production of this book directly or indirectly; they are too numerous to mention individually, but to them we offer our heartfelt thanks. We are also indebted to a number of individuals and publishers for allowing us to reproduce tables and illustrative material. We are grateful to Oxford Sonicaid Ltd for their generous support in the production of some of the illustrations.

D.H.E.
W.N.McD.
R.S.
J.P.W.

Plates

PLATE I

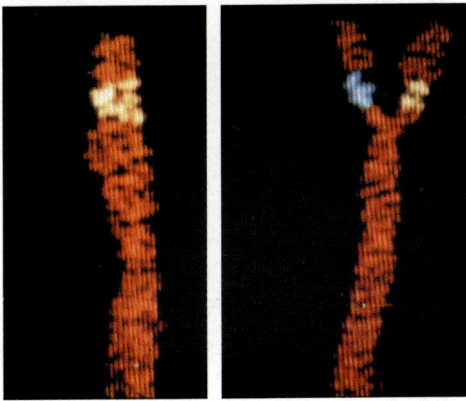

Figure 4.12 Colour-coded Doppler images of diseased carotid bifurcations. The left-hand image is from a patient with a completely occluded right internal carotid, which is therefore not displayed. There was also an atheromatous plaque in the external carotid causing a stenosis, displayed as a moderately increased velocity (yellow) on the Doppler scan. The right-hand image is from a patient with a tight stenosis of the left external carotid artery displayed on the Doppler scan as markedly increased velocity (blue) and two moderate stenoses of the internal carotid, displayed as moderately increased velocity (yellow) (reproduced by permission of Elsevier Science, from Curry and White 1978, *Ultrasound in Medicine and Biology*, © World Federation of Ultrasound in Medicine and Biology)

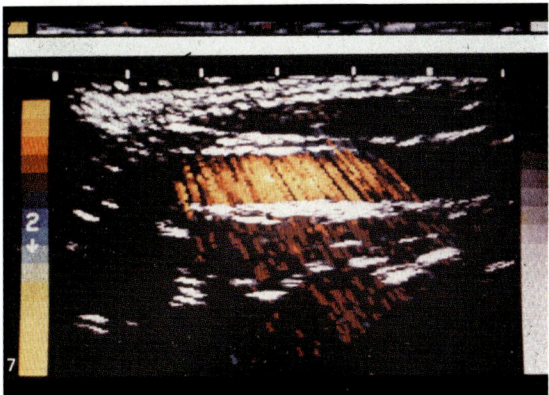

Figure 4.14 Colour-coded Doppler flow map of the common carotid artery obtained from the mid-neck region (reproduced by permission of Elsevier Science, from Eyer et al 1981, *Ultrasound in Medicine and Biology*, © World Federation of Ultrasound in Medicine and Biology)

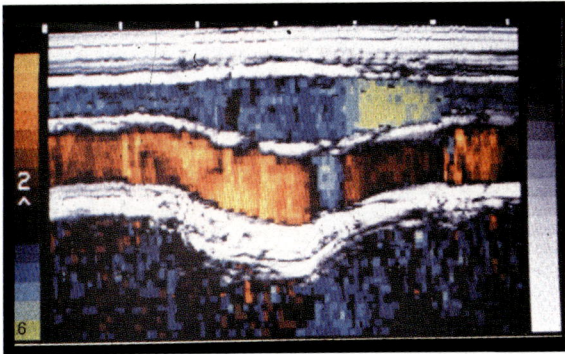

Figure 4.15 Colour-coded Doppler M-mode image of the jugular vein and common carotid artery during a valsalva manoeuvre. Increasing time is defined to be from left to right with the entire horizontal axis covering 3 s (reproduced by permission of Elsevier Science, from Eyer et al 1981, *Ultrasound in Medicine and Biology*, © World Federation of Ultrasound in Medicine and Biology)

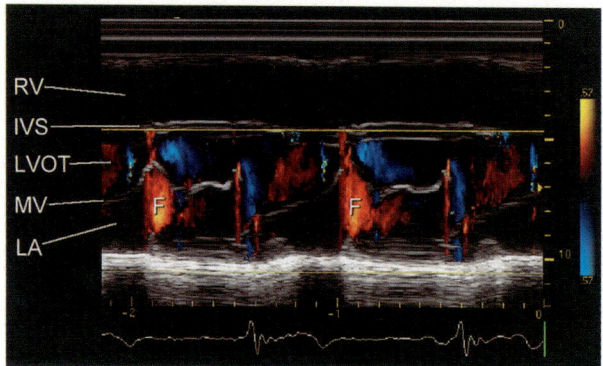

Figure 4.16 Parasternal long axis, M-mode, through the mitral valve leaflets of the heart with colour Doppler overlaid, showing diastolic ventricular filling through the open mitral leaflets and into the left ventricular outflow tract. RV, right ventricle; IVS, intra-ventricular septum; LVOT, left ventricular outflow tract; MV, mitral valve; LA, left atrium. F indicates the ventricular filling phase (image courtesy of GE Ultrasound)

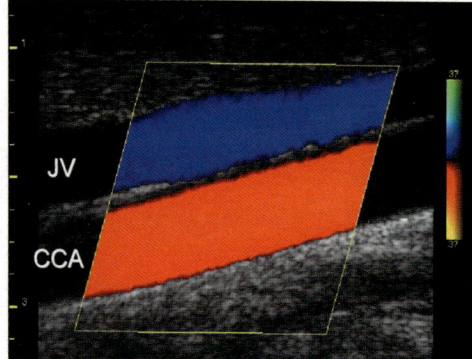

Figure 4.17 Colour Doppler flow image of a common carotid artery (CCA) and jugular vein (JV). The patient's head lies to the left of the scan, his feet to the right. The colour scale on the right of the image indicates that the component of flow detected by the scanner in the CCA is away from the transducer (i.e. moving from right to left). Conversely, the component of flow detected from the JV is towards the transducer (image courtesy of GE Ultrasound)

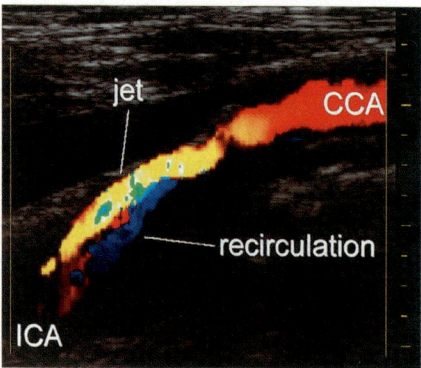

Figure 4.18 Colour Doppler flow image of an internal carotid artery (ICA) stenosis. The patient's head lies to the left of the scan. The colour coding scale is similar to that shown in Fig. 4.17. The colour of the flow changes from red to yellow (and then to green) for two reasons. First, the Doppler angle is increasing as the vessel curves deeper into the tissue, but also because the velocity increases as the same volume of flow has to pass through the stenosis. Once the blood has passed through the stenosis, the jet continues for some distance and gives rise to a region of recirculation (coded in blue because the component of flow is towards the transducer in this region). Note that the region of green in the jet is not reverse flow but aliasing, due to the Doppler shift exceeding half the pulse repetition frequency (image courtesy of GE Ultrasound)

PLATE III

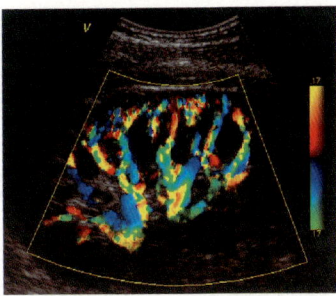

Figure 4.19 Colour Doppler flow map of the renal vasculature. Note that there is considerable aliasing in this image due to an inadequate sampling frequency (image courtesy of GE Ultrasound)

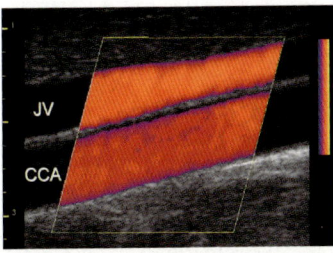

Figure 4.23 Power Doppler image of the same vessels shown in Fig. 4.17. Note the colour scale on the right side of the image, showing that velocity has no influence on the displayed colour. Instead the colour coding depends on the amount of Doppler power detected in each sample volume, with small values being coded blue/purple, and higher power coded orange/yellow. Note also that there is no directional information available from this type of image (image courtesy of GE Ultrasound)

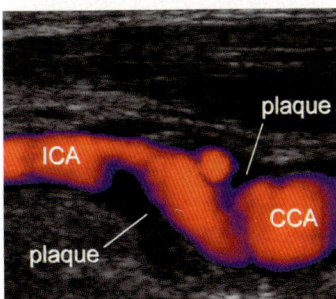

Figure 4.24 Power Doppler image showing narrowing of the internal carotid artery (ICA) lumen due to the presence of atheromatous plaque. Note once again that there is no velocity (speed or direction) information available from this type of image (image courtesy of GE Ultrasound)

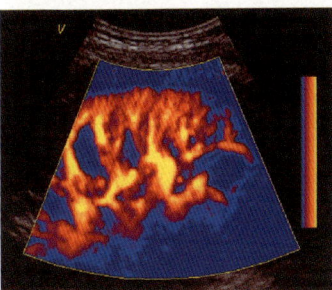

Figure 4.25 Power Doppler image of the same kidney as shown in Fig. 4.19. The pulse repetition frequency is the same as previously, but aliasing has no effect on power mode Doppler because it does not affect the total power of the signal. Note also the additional sensitivity of the power Doppler method, which is particularly noticeable in the renal cortex (image courtesy of GE Ultrasound)

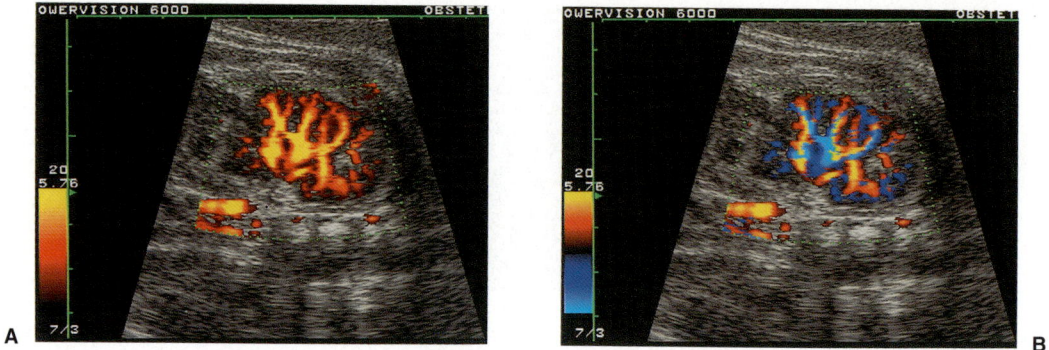

Figure 10.2 Directional colour-coded power Doppler image of fetal kidney vasculature. In this mode, where available, direction of flow information is used to colour-code the power Doppler image: (A) conventional power Doppler; (B) directional colour-coded power Doppler (courtesy of Toshiba)

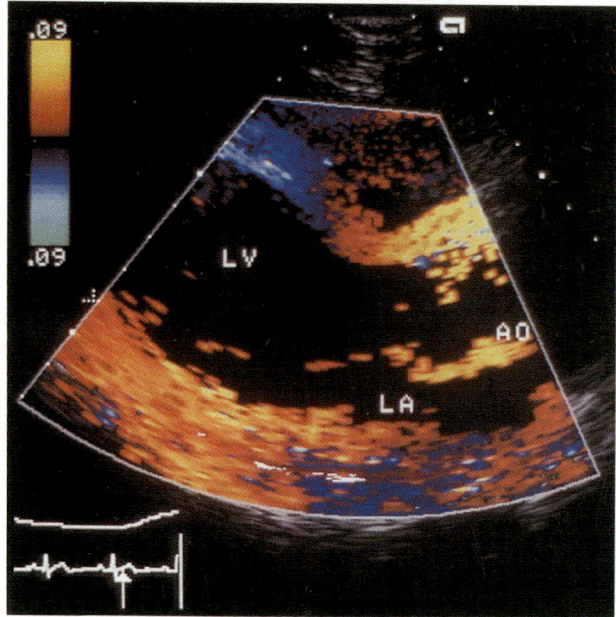

Figure 10.3 A colour Doppler tissue image. The blood flow image has been filtered out and the velocity of the moving tissue has been colour-coded. LV, left ventricle; LA, left atrium; AO, aorta

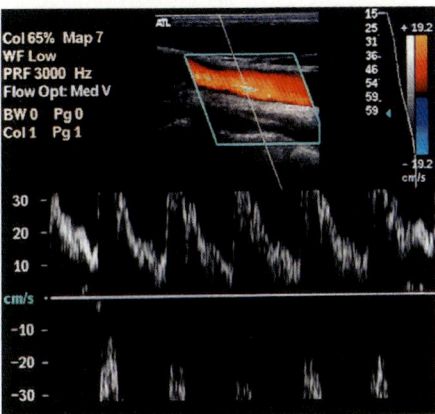

Figure 10.4 A triplex image representing B-mode, colour Doppler velocity image and spectral Doppler in the same display. The aliasing artefact is seen in both the colour Doppler velocity image and in the corresponding spectral Doppler

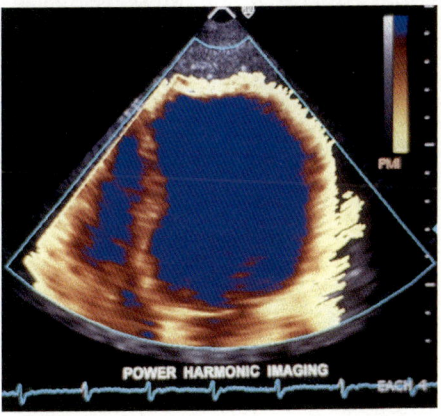

Figure 10.6 Harmonic power Doppler image using microbubble contrast agent to delineate clearly the endocardium boundary (courtesy of ATL)

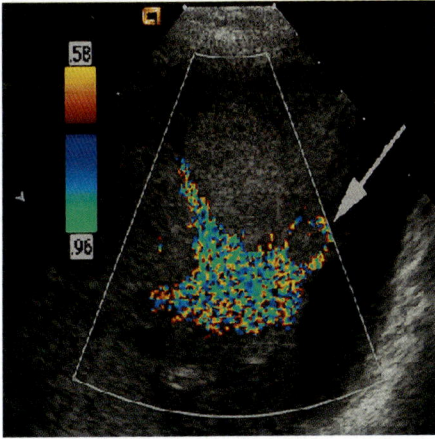

Figure 10.9 Stimulated Acoustic Emission image showing uptake of microbubble agent in liver metastases (courtesy of M. Blomley)

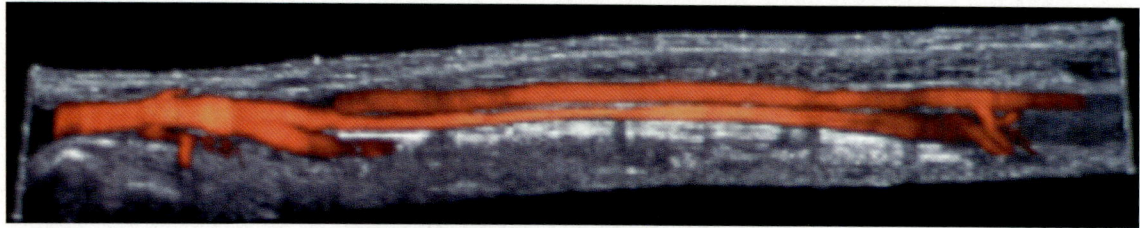

Figure 10.10 Extended field-of-view image. A 38 cm long field-of-view showing femoral artery, vein and main branches (courtesy of Siemens)

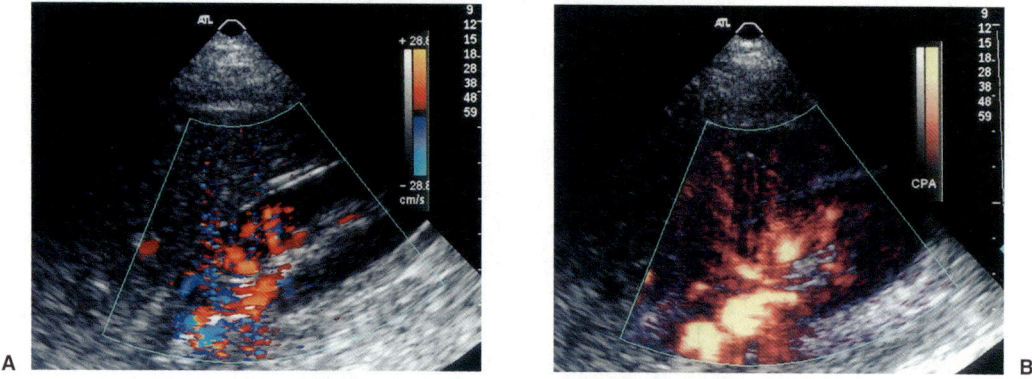

Figure 10.11 Noise in Doppler images: (A) Low-level noise in a colour velocity Doppler image appears randomly over the velocity colour scale; (B) low-level noise in a power Doppler image appears low in the power scale

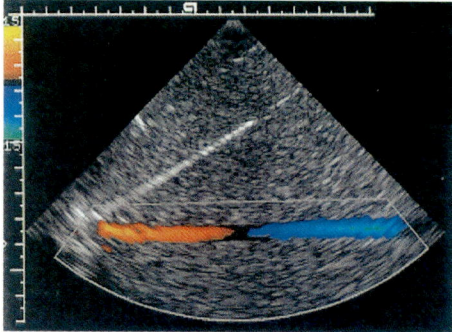

Figure 10.13 Constant flow in a tube depicted as changing colours as the beam/vessel angle alters (courtesy of P. R. Hoskins)

PLATE VII

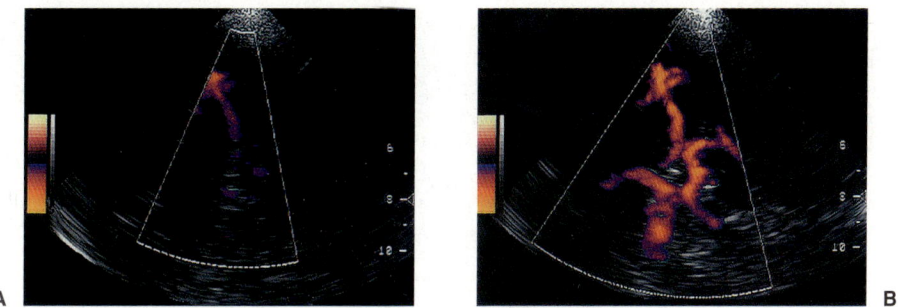

Figure 13.1 Enhanced power Doppler image of the Circle of Willis in an adult using contrast agent, (A) before injection, (B) after injection (courtesy of S. Fontaine)

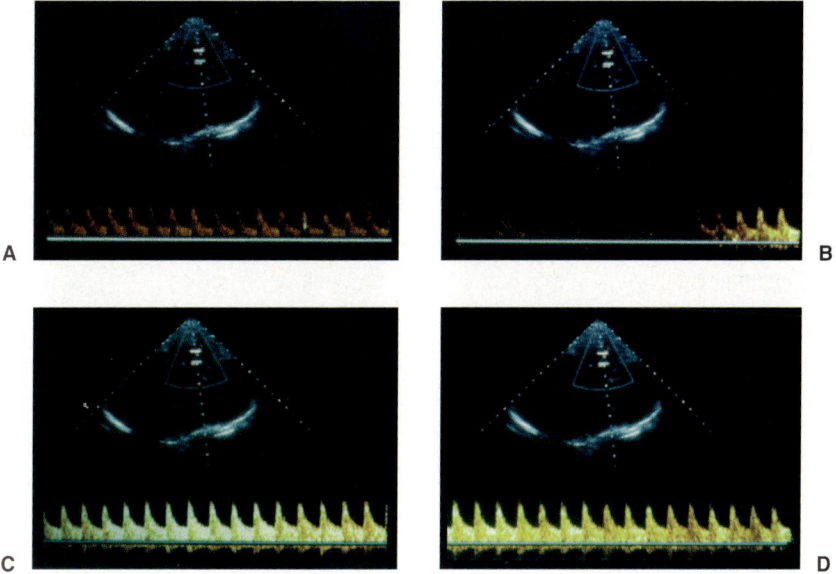

Figure 13.2 Recording of spectral Doppler in cerebral vessel showing increase in power of signal on the arrival of the contrast at the site selected by duplex scanning: (A) no contrast agent; (B) gain reduced to avoid overload, enhancement begins about three-quarters of the way along the sonogram; (C and D) signal power significantly enhanced throughout these recordings (courtesy of P. Allan)

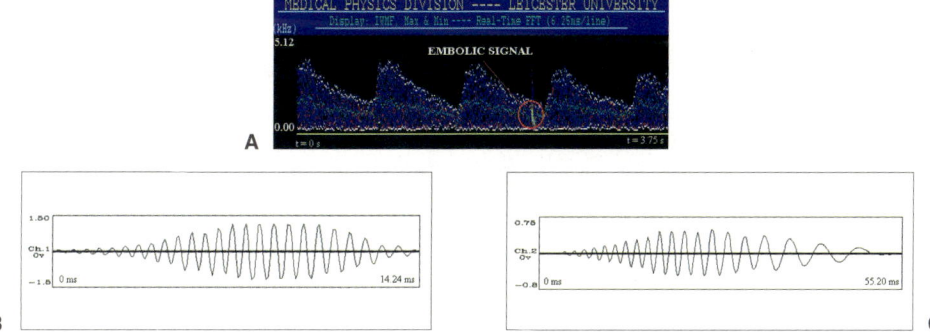

Figure 13.5 Doppler signals generated by emboli in the middle cerebral artery of a patient undergoing carotid artery surgery. (A) Conventional sonogram display. (B) Time domain display, showing amplitude modulation as an embolus passes through the Doppler sample volume. (C) Time domain display containing both amplitude and frequency modulation

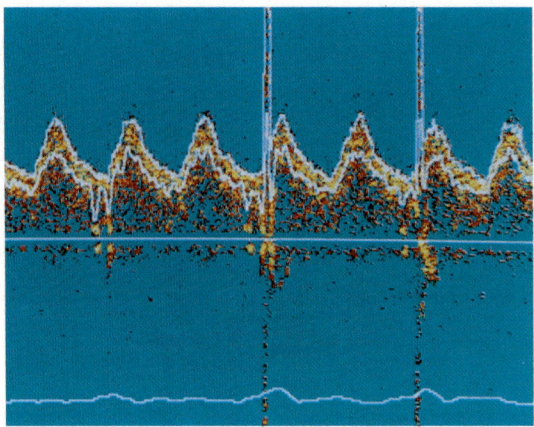

Figure 13.8 Sonogram showing the effect of overload as two emboli with large scattering cross-sections move through the Doppler sample volume

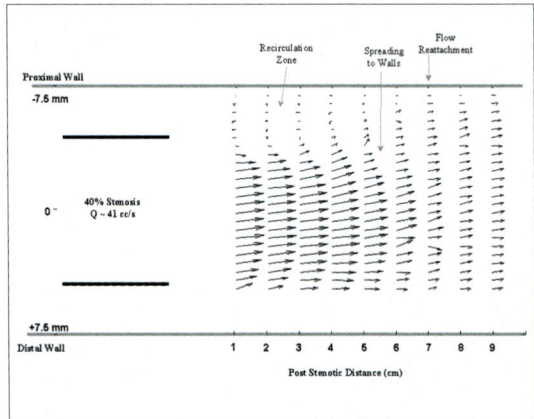

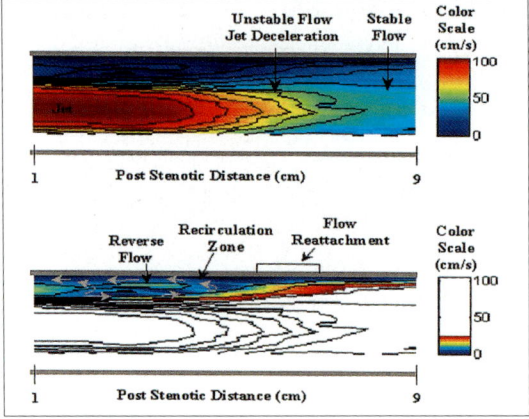

Figure 13.12 Vector Doppler results from an *in vitro* flow rig. The 2-D vectors and corresponding contour maps were obtained by systematically moving the Doppler sample volume to sample all relevant parts of the flow field. In this particular example the steady flow rate through the 15 mm diameter tube containing a 40% stenosis was approximately 2.46 l min^{-1}. In the uppermost diagram the modal velocity vectors are represented by arrows; in the lower two diagrams the magnitude of the velocities are colour-coded (and thus there are no negative values). In the upper colour diagram all the velocities are coded, whilst in the lower colour diagram the colour scale is compressed to highlight the recirculation zone and shear layer (courtesy of B. Dunmire, University of Washington, WA)

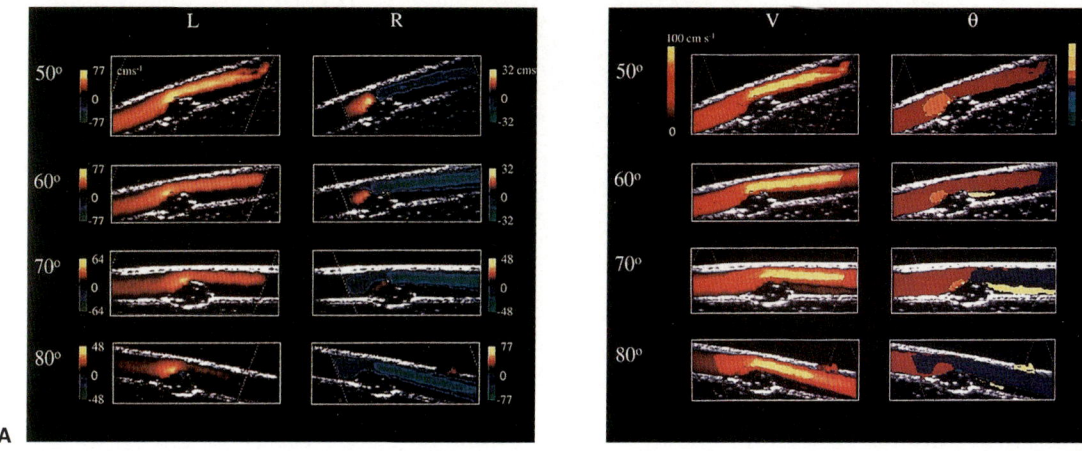

Figure 13.16 Illustration of vector Doppler results obtained by steering a linear array alternatively in two co-planar directions. The images were all obtained from an *in vitro* flow rig containing a 73% stenosis. (A) Colour Doppler images from 'left' (L) and 'right' (R) projections obtained at beam vessel angles of 50°, 60°, 70° and 80°. (B) Colour images of velocity-vector magnitude *V* and vector angle Θ obtained at the same angles as the colour flow images (reproduced by permission of Elsevier Science, from Hoskins 1997, *Ultrasound in Medicine and Biology*, © World Federation of Ultrasound in Medicine and Biology)

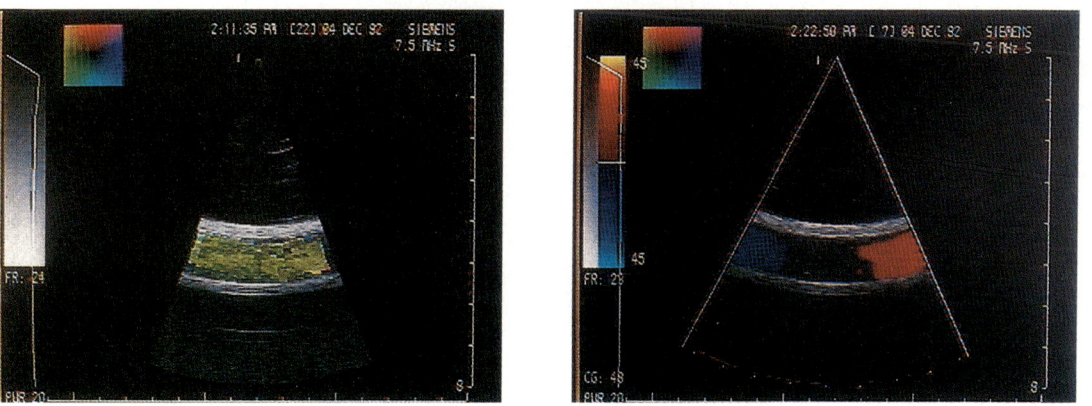

Figure 13.19 Colour flow images of a test phantom obtained using real-time speckle tracking and conventional colour Doppler processing. (A) The vector velocity image indicates flow from right to left via green coloration throughout the phantom tube. (B) In contrast, the Doppler image of the same phantom displays the unidirectional flow with different colours on opposite sides of the image, and does not detect the flow where the Doppler angle is 90° (reproduced by permission of Elsevier Science, from Bohs et al 1993b, *Ultrasound in Medicine and Biology*, © World Federation of Ultrasound in Medicine and Biology)

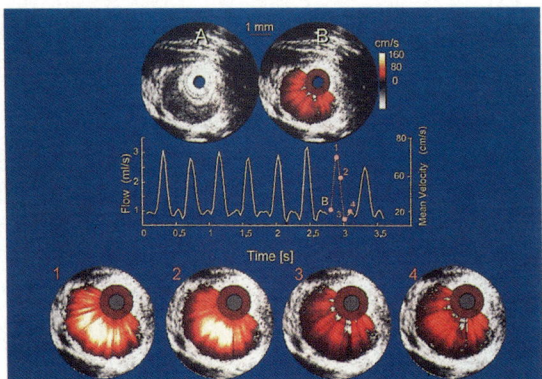

Figure 13.20 IVUS image from the iliac artery of a pig (A), together with five IVUS images with superimposed colour flow maps derived from decorrelation data (B, 1, 2, 3, 4). The phase of the cardiac cycle during which each image was captured is shown on the velocity trace (in the centre of the figure) which has also been derived using decorrelation data (reproduced by permission of Elsevier Science, from Li et al 1998, *Ultrasound in Medicine and Biology,* © World Federation of Ultrasound in Medicine and Biology)

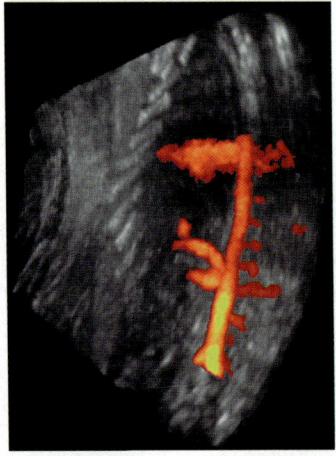

Figure 13.22 3-Scape™ power and B-mode image of fetal aortic vasculature (image courtesy of Siemens)

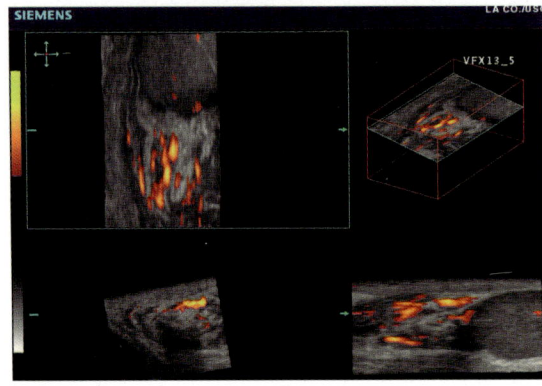

Figure 13.24 Various cuts through the 3-D volume acquired during a 3-Scape™ scan of a testicular variocele (image courtesy of Siemens)

PLATE XI

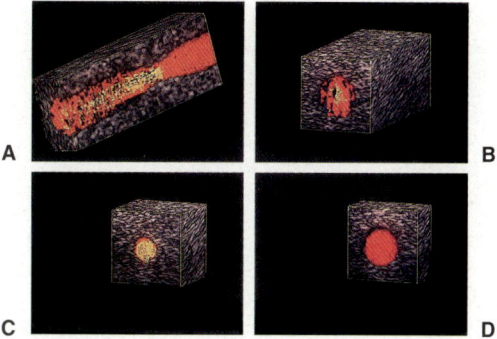

Figure 13.25 Display of 3-D colour flow information using multiplane visualisation with texture mapping. (A) Longitudinal cut through a stenosis phantom showing complex flow pattern after the stenosis. (B, C, D) Transverse cuts through the phantom distal to, at, and proximal to the stenosis, respectively (reproduced by permission of Elsevier Science, from Guo et al 1995, *Ultrasound in Medicine and Biology*, © World Federation of Ultrasound in Medicine and Biology)

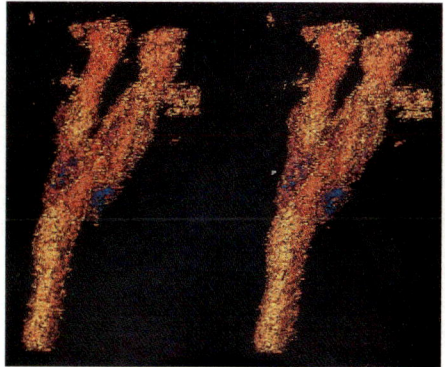

Figure 13.26 3-D image of carotid bifurcation at peak systole. The images are a stereo pair. These were formed by ray-tracing, with the sum of voxels along each ray determining the colours to be rendered on the surfaces (reproduced by permission from Rankin et al 1993, © American Roentgen Ray Society)

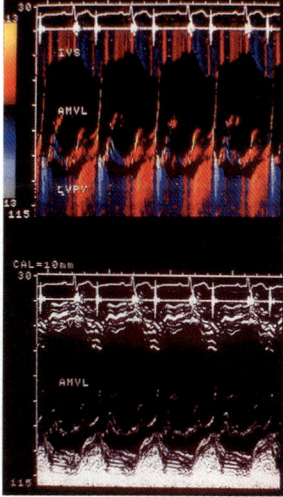

Figure 13.29 A colour Doppler tissue M-mode of a mitral valve leaflet and heart wall, and the corresponding pulse-echo M-mode. IVS, intra-ventricular septum; AMVL, anterior mitral valve leaflet; LVPW, posterior wall of the left ventricle

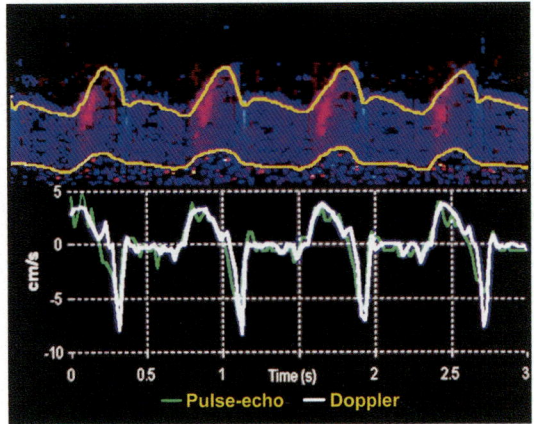

Figure 13.31 (Top) Doppler M-mode of myocardium. (Bottom) Agreement between mean velocity across the myocardium measured from Doppler M-mode and from the changing wall thickness as measured by the pulse-echo M-mode

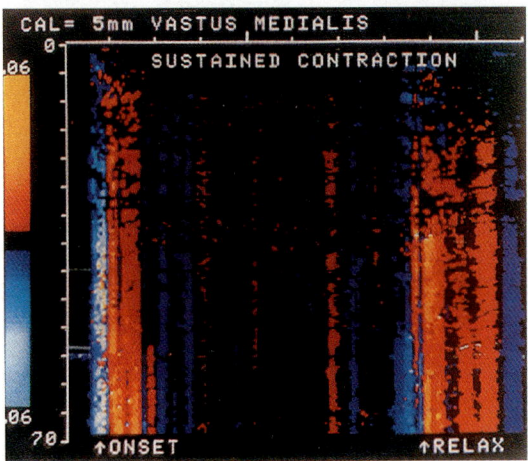

Figure 13.32 Doppler M-mode showing internal velocities in the vastus medialis muscle at different times during sustained contraction and relaxation (courtesy of N. Grubb)

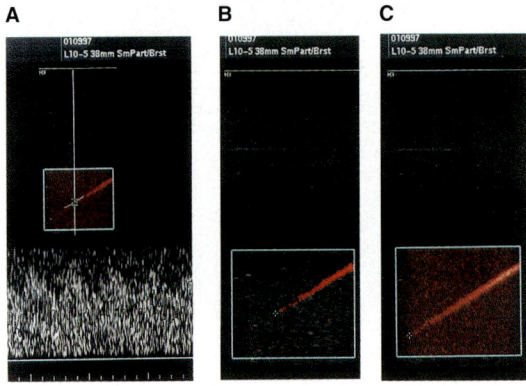

Figure A2.21 Measurements of penetration depth using a flow phantom: (A) spectral Doppler; (B) colour Doppler; (C) power Doppler

1

Introduction

Doppler ultrasound is an important technique for non-invasively detecting and measuring the velocity of moving structures, and particularly blood, within the body. Doppler ultrasound was first introduced into medicine in the latter half of the 1950s, has steadily increased in importance and is now regarded as an indispensable tool in many diagnostic situations. Few diagnostic ultrasound machines are now sold that do not have a Doppler capability. The correct interpretation of the results obtained from Doppler equipment depends to a considerable extent on the understanding of both the physical mechanisms and signal processing methods that result in the Doppler signal. A major aim of this book is to describe these processes and the effects they may have on clinical studies.

Interestingly, it is possible to argue that the only ultrasonic instruments that truly use the Doppler effect to detect motion are simple continuous wave devices, and that pulsed wave systems (including colour flow systems) are not Doppler devices at all in a true sense (Section 4.2.2.5). Strictly this argument may be correct, and certainly devices that use time domain correlation in one or two dimensions do not rely on the Doppler effect for their operation. Despite this, the many ultrasonic techniques which may be used to measure velocities within the body are closely related to each other and have grown out of simple Doppler instruments, and are discussed in this book irrespective of whether or not they can be truly classed as Doppler devices.

The remainder of this chapter consists of a very brief introduction to the Doppler effect and its use in medicine.

1.1 THE DOPPLER EFFECT

When an observer is moving relative to a wave source, the frequency he measures is different from the emitted frequency. If the source and observer are moving towards each other, the observed frequency is higher than the emitted frequency; if they are moving apart the observed frequency is lower. This simple but important assertion is known as the 'Doppler effect' after the Austrian physicist Christian Andreas Doppler (1803–1853), who first postulated it in a paper given before the Royal Bohemian Society of Learning in 1842 (Andrade 1959, White 1982) which was published in the proceedings of that society during the following year (Doppler 1843).

The details of Doppler's famous paper, and the subsequent experimental work of Buys Ballot in 1845 using the Amsterdam–Utrecht railway (originally intended to disprove Doppler's theory) have been described by a number of authors (Jonkman 1980, White 1982, Eden 1986, Eden 1988) and make fascinating reading.

In his original paper Doppler made a number of incorrect observations concerning the velocities of the stars, and therefore it seems ironical that one of the most important applications of the Doppler effect has been for making velocity measurements in astronomy (using the absorption lines present in the light spectra from stars and galaxies). The Doppler effect has also been widely used in terrestrial applications, and it is only relatively recently that is has become of major importance in medicine. Both ultrasound and laser Dopplers have found a place for detecting motion within the body, but the ultrasound technique is much more widely used and is the subject of this book.

1.2 ULTRASOUND

Ultrasound is sound that has a frequency above the audible range of man, that is greater than 20 kHz. In both imaging and Doppler applications in medicine the usual range of frequencies used is between 2 MHz and 10 MHz (although higher frequencies are used in specialist applications). The lower limit is determined by wavelength considerations (the longer the wavelength the poorer the spatial resolution—both axial and transverse), and the upper limit by acceptable power levels (attenuation rises very rapidly with frequency and so a very small proportion of the transmitted power is returned to the transducer at high frequencies).

Both imaging and Doppler devices function by transmitting a beam of ultrasound into the body, and collecting and analysing the returning echoes. Imaging devices are able to calculate the co-ordinates from which echoes originated and these, together with knowledge of the echo amplitudes, allow cross-sectional images of the body to be constructed. Doppler devices may also be able to determine the position of the source of echoes, but their fundamental concern is the frequency of the returning echoes and whether there has been a Doppler shift as a result of interaction with a moving target.

1.3 DOPPLER ULTRASOUND IN MEDICINE

The method of utilising the Doppler principle in medicine varies slightly from the 'classical' Doppler method in that the targets do not themselves spontaneously emit a radiation, and it is therefore necessary to transmit a signal into the body and to observe the changes in frequency that occur when it is reflected or scattered from the targets.

It can be shown that under these conditions (see Section 3.5) there is a shift in the ultrasound frequency, f_d, given by:

$$f_d = f_t - f_r = (2 f_t v \cos\theta)/c \qquad 1.1$$

where f_t and f_r are the transmitted and received ultrasound frequencies, respectively, v the velocity of the target, c the velocity of sound in the medium, and θ the angle between the ultrasound beam and the direction of motion of the target. The velocity, c, and the transmitted frequency, f_t, are known in any given situation, and therefore the velocity of a target can be found from the expression:

$$v = K f_d / \cos\theta \qquad 1.2$$

where K is a known constant given by $c/2f_t$. This equation may be used to monitor changes in velocity, and if the angle θ can be determined, absolute velocity may be calculated.

In practice there is unlikely to be just a single target contributing to the Doppler shift, and furthermore multiple targets are unlikely all to have the same velocity, so the Doppler shift signal will contain not a single frequency, but a spectrum of frequencies. It is this spectrum that must be interpreted if full use is to be made of the Doppler shift signal.

Early Doppler units were continuous wave (CW) non-directional devices which presented the Doppler signal either as an audible signal or as a simple envelope signal related to the instantaneous average frequency, but since then a number of developments have taken place (Table 1.1) which have made the Doppler technique both sophisticated and widely applicable. Kaneko (1986) has published an interesting account of the early history of the development of Doppler ultrasound instruments.

Almost without exception Doppler units are now directional and are able to deal simultaneously with motion both towards and away from the transducer. Most units include real-time spectrum analysis hardware and this has greatly facilitated the interpretation of the Doppler signal. Pulsed Doppler units have virtually replaced CW units and allow the operator to select signals from a particular range by gating the returning signal. Both conventional ultrasound imaging and Doppler imaging have added a new dimension to quantitative Doppler studies by allowing the operator to place the Doppler sample volume at a known point in the body or at a point of particular haemodynamic interest, whilst the Doppler images themselves have proved to be of great value and are now regarded as an essential complement to pulse-echo images.

Doppler techniques are now used in a multitude of clinical applications and it seems that there are few, if any, branches of medicine that do not or will not benefit from such methods. It is hoped that this book will help to improve the understanding of the underlying principles of Doppler measurements.

Table 1.1 Major developments in Doppler ultrasound

Development	Approximate Year	Some early key references*
Doppler effect described	1842	Doppler (1843)
Doppler ultrasound used in medical applications	1957	Satomura (1959) Franklin et al (1961, 1963) Stegall et al (1966) Kaneko et al (1966)
Directional Doppler	1966	McLeod (1967) Cross and Light (1971) Nippa et al (1975)
Pulsed wave Doppler	1967	Wells (1969) Peronneau and Leger (1969) Baker (1970)
Ultrasound contrast agents	1968	Gramiak and Shah (1968) Carroll et al (1980) Keller et al (1987)
Multigate and infinite gate systems	1970/5	Baker (1970) Keller et al (1976) Brandestini (1978) Nowicki and Reid (1981)
Doppler imaging	1971	Mozersky et al (1971) Reid and Spencer (1972) Fish (1975)
Duplex echo-Doppler systems	1974	Barber et al (1974) Phillips et al (1980)
Time-domain processing	1976	Dotti et al (1976) Foster (1985) Bonnefous and Pesque (1986)
Real-time colour flow mapping	1981	Eyer et al (1981) Namekawa et al (1982) Kasai et al (1985)
Harmonic Doppler	1981	Miller (1981) Schrope et al (1992) Schrope and Newhouse (1993)
Power Doppler	1985	Seo et al (1985) Jain et al (1991) Rubin et al (1994)

*Most of these references have been chosen because they are full papers written in English. In many cases the earliest reports in the literature are in less complete and less accessible conference proceedings.

1.4 SUMMARY

Doppler ultrasound is a powerful technique for the non-invasive measurement and imaging of velocities within the body. The Doppler signal, however, is influenced by a variety of factors which need to be appreciated if the maximum benefit is to be gained from the method. This book describes the physical mechanisms involved in the generation of Doppler signals, the signal processing methods that are used to extract velocity information from them, and a wide range of applications of the technique.

1.5 REFERENCES

Andrade EN da C (1959) Doppler and the Doppler effect. Endeavour 18: 14–19

Baker DW (1970) Pulsed ultrasonic Doppler blood-flow sensing. IEEE Trans Sonics Ultrason SU-17:170–185

Barber FE, Baker DW, Nation AWC, Strandness DE, Reid JM (1974) Ultrasonic duplex echo-Doppler scanner. IEEE Trans Biomed Eng 21:109–113

Bonnefous O, Pesque P (1986) Time domain formulation of pulse-Doppler ultrasound and blood velocity estimation by cross correlation. Ultrason Imaging 8:73–85

Brandestini M (1978) Topoflow—a digital full range Doppler velocity meter. IEEE Trans Sonics Ultrason SU-25:287–293

Carroll BA, Turner RJ, Tickner EG, Boyle DB, Young SW (1980) Gelatin encapsulated nitrogen microbubbles as ultrasonic contrast agents. Invest Radiol 15:260–266

Cross G, Light LH (1971) Direction-resolving Doppler instrument with improved rejection of tissue artifacts for transcutaneous aortovelography. J Physiol 217:5–7

Doppler C (1843) Ueber das farbige Licht der Doppelsterne und einiger anderer Gestirne des Himmels. Abhandl. d. Konigl. Bohmischen Gesellschaft der Wissenschaften Sers. 2:465–482

Dotti D, Gatti E, Svelto V, Ugge A, Vidali P (1976) Blood flow measurements by ultrasound correlation techniques. Energia Nucleare 23:571–575

Eden A (1986) The beginnings of Doppler. In: Aaslid R (ed) Transcranial Doppler Sonography, pp. 1–9. Springer-Verlag, Vienna, New York

Eden A (1988) Christian Doppler: Thinker and Benefactor. Christian Doppler Institute, Salzberg

Eyer MK, Brandestini MA, Phillips DJ, Baker DW (1981) Color digital echo/Doppler image presentation. Ultrasound Med Biol 7:21–31

Fish PJ (1975) Multichannel, direction-resolving Doppler angiography. In: Kazner E, de Vlieger M, Muller HR, McCready VR (eds) Ultrasonics in Medicine, pp. 153–159. Excerpta Medica, Amsterdam, Proceedings of the 2nd European Congress on Ultrasonics in Medicine, Munich, 12–16 May 1975

Foster SG (1985) A pulsed ultrasonic flowmeter employing time domain methods. PhD Dissertation, University of Illinois, Urbana-Champaign, IL

Franklin DL, Schlegel W, Rushmer RF (1961) Blood flow measured by Doppler frequency shift of back-scattered ultrasound. Science 134:564–565

Franklin DL, Schlegel WA, Watson NW (1963) Ultrasonic Doppler shift blood flowmeter: circuitry and practical applications. Bio-Med Sci Instrum 1:309–311

Gramiak R, Shah PM (1968) Echocardiography of the aortic root. Invest Radiol 3: 356–366

Jain SP, Fan PH, Philpot EF, Nanda NC, Aggarwal KK, Moos S, Yoganathan AP (1991) Influence of various instrument settings on the flow information derived from the power mode. Ultrasound Med Biol 17:49–54

Jonkman EJ (1980) Doppler research in the nineteenth century. Ultrasound Med Biol 6: 1–5

Kaneko Z (1986) First steps in the development of the Doppler flowmeter. Ultrasound Med Biol 12:187–195

Kaneko Z, Shiraishi J, Omizo H, Kato K, Motomiya M, Izumi T, Okumura T (1966) Analysing blood flow with a sonagraph. Ultrasonics 4:22–23

Kasai C, Namekawa K, Koyano A, Omoto R (1985) Real-time two-dimensional blood flow imaging using an autocorrelation technique. IEEE Trans Sonics Ultrason SU-32:458–464

Keller HM, Meier WE, Anliker M, Kumpe DA (1976) Non-invasive measurement of velocity profiles and blood flow in the common carotid artery by pulsed Doppler ultrasound. Stroke 7:370–377

Keller MW, Feinstein SB, Watson DD (1987) Successful left ventricular opacification following peripheral venous injection of sonicated contrast agent: an experimental evaluation. Am Heart J 114:570–575

McLeod FD (1967) A directional Doppler flowmeter. p. 213. In: Digest of the 7th International Conference on Medical and Biological Engineering, Stockholm

Miller DL (1981) Ultrasonic detection of resonant cavitation bubbles in a flow tube by their second-harmonic emissions. Ultrasonics 19:217–224

Mozersky DJ, Hokanson DE, Baker DW, Sumner DS, Strandness DE (1971) Ultrasonic arteriography. Arch Surg 103:663–667

Namekawa K, Kasai C, Tsukamoto M, Koyano A (1982) Realtime bloodflow imaging system utilizing autocorrelation techniques. In: Lerski RA, Morley P (eds) Ultrasound '82, pp. 203–208. Pergamon Press, New York

Nippa JH, Hokanson DE, Lee DR, Sumner DS, Strandness DE (1975) Phase rotation for separating forward and reverse blood velocity signals. IEEE Trans Sonics Ultrason SU-22:340–346

Nowicki A, Reid JM (1981) An infinite gate pulse Doppler. Ultrasound Med Biol 7:41–50

Peronneau PA, Leger F (1969) Doppler ultrasonic pulsed blood flowmeter. Proc 8th Int Conf Med Biol Eng. 10–11.

Phillips DJ, Powers JE, Eyer MK, Blackshear WM, Bodily KC, Strandness DE, Baker DW (1980) Detection of peripheral vascular disease using the Duplex scanner III. Ultrasound Med Biol 6:205–218

Reid JM, Spencer MP (1972) Ultrasonic technique for imaging blood vessels. Science 176:1235–1236

Rubin JM, Bude RO, Carson PL, Bree RL, Adler RS (1994) Power Doppler US: a potentially useful alternative to mean frequency-based color Doppler US. Radiology 190:853–856

Satomura S (1959) Study of the flow patterns in peripheral arteries by ultrasonics. J Acoust Soc Jap 15:151–158

Schrope BA, Newhouse VL (1993) Second harmonic ultrasonic blood perfusion measurement. Ultrasound Med Biol 19:567–579

Schrope B, Newhouse VL, Uhlendorf V (1992) Simulated capillary blood flow measurement using a nonlinear ultrasonic contrast agent. Ultrason Imaging 14:134–158

Seo Y, Shiki E, Hongo H, Iinuma K (1985) 2-D color flow mapping systems—power mode, In: Proceedings of the 47th Meeting of the Japan Society of Ultrasonics in Medicine, p.481.

Stegall HF, Rushmer RF, Baker DW (1966) A transcutaneous ultrasonic blood-velocity meter. J Appl Physiol 21:707–711

Wells PNT (1969) A range-gated ultrasonic Doppler system. Med Biol Eng 7:641–652

White DN (1982) Johann Christian Doppler and his effect—a brief history. Ultrasound Med Biol 8:583–591

2

Blood Flow

2.1 INTRODUCTION

Although Doppler ultrasound may be used for the study of various types of motion within the body, its major use remains the detection and quantification of flow in the heart, arteries and veins. Doppler signals from these sources contain a great deal of information about flow, but before these signals can be interpreted it is essential to understand the basics of haemodynamics.

Blood flow in arteries is complex. The flow is pulsatile, the blood is an inhomogeneous non-Newtonian fluid, and the tethered viscoelastic arteries branch, curve and taper. The disease process only adds further complication. Despite this, useful insight into the way in which blood flows may be gained from the study of relatively simple models.

In this chapter the major emphasis is on blood flow and blood velocity profiles as these have such an important bearing on the Doppler signal, but flow and pressure are inseparably linked and the latter cannot be completely ignored. Indeed, the detection or measurement of blood velocity may allow pressure information to be derived, as in the case of the quantification of the pressure drop across a stenosed heart valve (Section 2.6.1) or the inference of pressure changes from changes in pulse wave velocity (Section 2.5.1).

2.2 BASIC CONCEPTS AND DEFINITIONS

2.2.1 Viscosity

When one layer of a fluid moves with respect to an adjacent layer, a frictional force arises between them due to viscosity. Put more formally, viscosity is that property of a fluid whereby it offers resistance to shear. Treacle and engine oils are examples of highly viscous fluids, while water and ethanol are examples of liquids with relatively low viscosities.

For most simple fluids (those known as Newtonian fluids) viscosity is independent of shear rate (i.e. the rate at which the adjacent layers slip over each other), but many fluids exhibit a change in viscosity with shear rate. Blood is not a simple fluid; it is essentially a suspension of red blood cells in plasma and exhibits anomalous viscous properties at low shear rates (where it appears to have a finite yield stress) and in small tubes (where the size of the blood cells becomes significant in comparison to the size of the tube). Fortunately neither effect is really significant in blood vessels with diameters of 0.1 mm or more and it is possible to make use of the 'asymptotic' viscosity (the value to which the apparent viscosity tends at high shear rates) in calculations of blood flow in major arteries. The asymptotic viscosity of blood is dependent on both haematocrit and temperature, but is roughly four times as great as that of pure water (i.e. 4 mPa s (0.004 kg m^{-1} s^{-1}), or 4 cP (centipoise—the cgs unit)).

For Newtonian fluids the viscosity, μ, may be written:

$$\mu = \tau \left/ \frac{dv}{dr} \right. \qquad 2.1$$

where τ is shear stress with units of kg m^{-1} s^{-1}, and dv/dr is shear rate with units of reciprocal seconds (s^{-1}).

In addition to the 'absolute' or 'dynamic' viscosity introduced above, the 'kinematic viscosity', which is equal to absolute viscosity divided by density, is often used in equations describing fluid behaviour. The kinematic viscosity of blood is approximately 3.8×10^{-6} m^2 s^{-1} or 0.038 St (stokes—the cgs unit).

The viscous behaviour of blood has been discussed by Nichols and O'Rourke (1990) and reviewed in some detail by Whitmore (1968), McDonald (1974) and Fung (1981).

2.2.2 Laminar, Turbulent and Disturbed Flow

There are two distinctly different types of fluid flow. The first type is known as laminar or streamlined flow, the second as turbulent flow. In laminar flow the fluid particles move along smooth paths in layers (or laminae) with every layer sliding smoothly over its neighbour. Laminar flow becomes unstable at high velocities (see Section 2.3.3) and breaks down into turbulent flow. In turbulent flow the particles follow very irregular and erratic paths, their velocity vectors varying continually both in magnitude and direction. The term 'turbulence' is often rather loosely used in the vascular literature to indicate any non-laminar flow, but vortices and irregular movements of large bodies of fluid which can be traced to some obvious source of disturbance do not constitute turbulence, and are properly referred to as disturbed flow.

2.2.3 Steady Flow

A steady flow is one in which all conditions (such as velocity and pressure) at any point in a stream remain constant with respect to time. Strictly, therefore, turbulent flow may never be described as steady as there are continual fluctuations in both velocity and pressure at every point. However, the definition of steady flow is usually expanded to include flow in which the conditions fluctuate equally on both sides of a constant average value.

2.3 STEADY FLOW IN RIGID TUBES

The steady flow of Newtonian fluids in rigid tubes is well understood and serves as a convenient starting point in the discussion of blood flow in arteries. It is also relevant because pulsatile flow can be considered to be the sum of a steady component and a number of oscillatory components which interact neither with each other nor with the steady component (this assumes that the system is linear, a concept which will be discussed in Section 2.4.1).

2.3.1 Poiseuille Flow

Established steady laminar flow in long cylindrical pipes is sometimes referred to as Poiseuille flow. In this most basic type of flow, the fluid (blood) moves in a series of concentric shells such that the velocity profile across the vessel is parabolic, with the blood in the centre of the vessel moving most rapidly and the blood in contact with the vessel wall not at all.

The velocity of any lamina at radius r from the centre of the vessel may be written:

$$v(r) = v_{max}(1 - r^2/R^2) \qquad 2.2$$

where v_{max} is the velocity of the centre stream and R the radius of the vessel. This type of flow is illustrated in Fig. 2.1a.

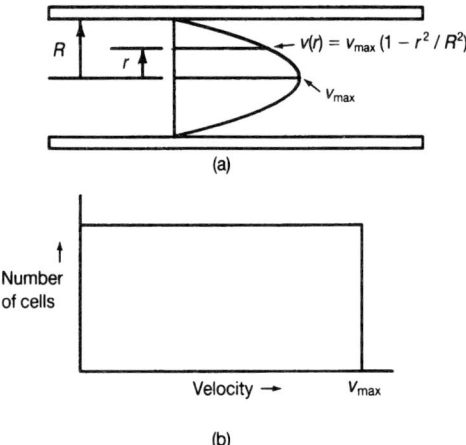

Figure 2.1 (a) Parabolic velocity profile found in steady laminar flow in a long cylindrical pipe. (b) Histogram illustrating the number of cells moving with a given velocity in the profile shown in (a). From this it follows that for this type of profile the maximum velocity is twice the mean velocity

For parabolic flow the shear rate will be given by:

$$\frac{dv}{dr} = -\frac{2rv_{max}}{R^2} \qquad 2.3$$

and the shear rate at the vessel wall by:

$$\left.\frac{dv}{dr}\right|_{r=R} = -\frac{2v_{max}}{R}. \qquad 2.4$$

It follows that the wall shear stress is given by:

$$\tau_{wall} = -\mu\frac{2v_{max}}{R} \qquad 2.5$$

Both high and low wall shear have been cited as major factors leading to the development of atherosclerosis, and ultrasound techniques have been developed for measuring velocity profiles with the specific goal of estimating wall shear rates and hence wall shear stress in human vessels (Hughes and How 1993, Hoeks et al 1995, Forsberg et al 1996).

An interesting feature of parabolic flow is that the number of blood cells with velocities in the range v to $v + \delta v$ is independent of v. This is because, although the shear rate dv/dr is proportional to r, the cross-sectional area of equally spaced laminae also rises in proportion to the distance from the vessel axis, and these two effects exactly cancel. The velocity distribution histogram corresponding to the profile shown in Fig. 2.1a is illustrated in Fig. 2.1.b.

The mean velocity is of special importance because when multiplied by the cross-sectional area of the vessel it yields the volumetric flow, and it follows from the shape of the velocity histogram illustrated in Fig. 2.1b that the mean velocity in a vessel containing a parabolic velocity profile is exactly one-half of the maximum (centre stream) velocity.

This can be shown more formally by integrating the velocity given by eqn. 2.2 over the vessel and dividing by the total area:

$$\bar{v} = 1/\pi R^2 \int_{r=0}^{R} v_{max}(1 - r^2/R^2)2\pi r\, dr \qquad 2.6$$

i.e.

$$\bar{v} = v_{max}/2 \qquad 2.7$$

Using this relationship, and given that the volumetric flow, Q, may be written:

$$Q = \bar{v}\pi R^2 \qquad 2.8$$

then the wall shear stress for parabolic flow may also be written:

$$\left.\frac{dv}{dr}\right|_{r=R} = -\frac{4Q}{\pi R^3} \qquad 2.9$$

The relationship between flow and pressure drop in a section of tube containing steady parabolic lamina flow is particularly simple and is named after Poiseuille in honour of his detailed measurements of flow through capillary tubes, published in 1840. In essence Poiseuille's law states that the pressure drop across a section of tube, Δp, is proportional to the volumetric flow, the fluid viscosity and the length of the tube, and is inversely proportional to the fourth power of the radius. Mathematically it may be expressed as:

$$\Delta p = 8\mu Q L/\pi R^4 \qquad 2.10$$

where μ is the (absolute) viscosity, Q the volumetric flow rate, L the length of the tube, and R its radius.

2.3.2 Entrance Effects and the Inlet Length (Steady Laminar Flow)

When flow enters tube it must travel for some distance before it achieves its steady-state velocity profile. This is because the viscous drag exerted by the walls of the tube can only be transmitted to the central part of the tube by a progressive growth of the region of shear stress or boundary layer. This process is illustrated in Fig. 2.2.

Initially only the fluid directly in contact with the wall is stationary, and a large velocity gradient is present at this point. This gradient becomes reduced as more of the fluid in the core becomes sheared. Since the volumetric flow though any

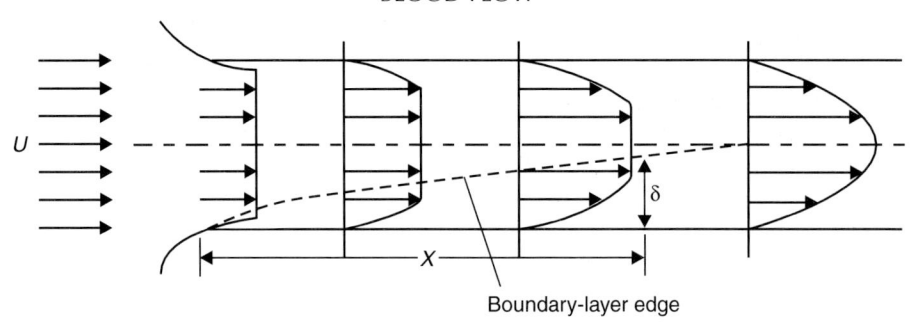

Figure 2.2 The development of the steady-state parabolic velocity profile at the entrance to a tube. The initially flat profile becomes progressively modified as the boundary layer grows with distance from the inlet (reproduced by permission from Caro et al 1978, Oxford University Press)

cross-section of the pipe must be the same, the fluid at the centre of the tube must accelerate to compensate for the deceleration at the periphery. The boundary layer thickness, δ, is initially proportional to the square root of the distance from the entrance of the tube, x, and may be written:

$$\delta \propto (vx/U)^{1/2} \qquad 2.11$$

where v is the kinematic viscosity and U is the free stream velocity (i.e. the velocity of the unsheared central core) (Caro et al 1978).

The distance required for the profile to achieve its final steady-state shape is known as the inlet length, and has been found experimentally to be approximately:

$$X = 0.03D(Re) \qquad 2.12$$

where D is the diameter of the pipe and Re the Reynolds number.

Because of the additional energy required to accelerate the flow in the inlet length, the pressure gradient in this region is much greater than where the flow is fully established.

2.3.3 Turbulent Flow

Laminar flow becomes unstable at high velocities and breaks down into turbulent flow. The point at which the transition between the two flow regimes occurs cannot be predicted exactly, but is largely determined by the Reynolds number, Re, which for a circular pipe is defined as:

$$Re = \bar{v}D/v \qquad 2.13$$

where $\bar{v}$ is the average velocity of flow across the pipe, D the diameter of the pipe, and v the kinematic viscosity. The critical Reynolds number, Re_{crit} (i.e. the Re at which the flow becomes turbulent) can, depending on geometry and the extent of upstream disturbances, be anywhere between about 2000 and 50 000. For values of less than 2000 it its practically impossible for turbulent flow to persist in a straight smooth pipe, and for values of greater than 10 000 the flow is inherently unstable, and the least disturbance will transform it into turbulent flow immediately. The critical Reynolds number is affected by vessel geometry and is greater in converging vessels, and less in curved, diverging and rough vessels, where it may be as low as 1000. For most practical purposes Re_{crit} is taken to be between 2000 and 2500.

Both the average velocity profile and the pressure gradient observed in turbulent flow differ from those in laminar flow (because of the erratic paths followed by individual particles it is necessary to distinguish between the instantaneous velocity profile and the average velocity profile, even in 'steady' turbulent flow). The average velocity profile found in turbulent flow is much flatter than that in steady laminar flow, so that the maximum velocity at the centre of the vessel is only about 20% higher than the average velocity (this compares with a value of 100% for laminar flow). The pressure gradient in a given tube is also much larger, and the pressure drop is proportional not to the flow rate but rather to the flow rate squared.

Fortunately, from the standpoint of the interpretation of Doppler signals, turbulence is rarely seen in the healthy circulation (although it is possible that the flow in the proximal aorta can sometimes be turbulent), and its very presence is usually a clue that an abnormality is present.

2.3.4 Entrance Effects (Turbulent Flow)

The entrance length for turbulent flow is less than that for laminar flow and may be estimated from eqn. 2.14 (Caro et al 1978):

$$X_{\text{turbulent}} = 0.69D(Re)^{1/4} \qquad 2.14$$

2.4 PULSATILE FLOW IN RIGID TUBES

So far we have considered only steady flow. Arterial blood flow is pulsatile and in this section we show how such a flow may be analysed in terms of a steady flow together with a number of superimposed sinusoidal components.

2.4.1 Fourier Analysis

In essence, Fourier's theorem states that any periodic waveform may be broken down into (and resynthesized from) a series of sinusoidal waveforms with frequencies that are integral multiples of the repetition frequency of the original waveform. For any given waveform the amplitudes and phases of the sinusoidal components are unique and can be simply calculated. The way in which a waveform may be reconstructed from its Fourier components is illustrated in Fig. 2.3. The reason why Fourier analysis is so important is that the response of most physical systems to a signal of a single frequency is independent of the presence and magnitude of any other frequency components. Such systems are known as linear systems because a change in the size of the input signal is accompanied by the same proportional change in the output signal.

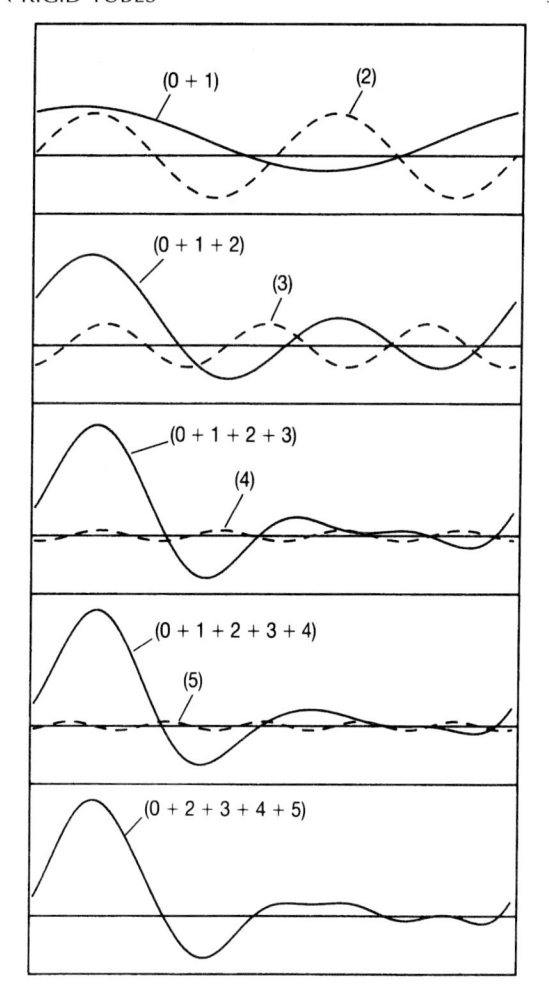

Figure 2.3 Synthesis of a common femoral flow waveform from its Fourier components. The numbers in the brackets refer to the harmonics used to construct each of the waveforms. The coefficients used to synthesise this waveform can be found in the first half of Table 2.2

Thus, for example, if the relationship between the blood pressure gradient and blood flow is a linear one (as we assume in normal peripheral arteries) and we know the relationship between the pressure gradient and flow at any given frequency, we can predict the flow due to a complex pressure gradient in three simple steps. Firstly, the pressure gradient is split into its sinusoidal (Fourier) components; secondly, the flow component (both its amplitude and phase) at each frequency is calculated from the pressure gradient at the same frequency; thirdly, the derived flow components are added together to give the complete flow waveform. In Section 2.4.3 an analogous method will be used to derive the velocity profiles seen in pulsatile flow.

Mathematically, the Fourier series for a periodic function $F(t)$ may be written as:

$$F(t) = F_0/2 + F_1 \cos(\omega t - \phi_1) + F_2 \cos(2\omega t - \phi_2) + \cdots \quad 2.15$$

or

$$F(t) = F_0/2 + \sum_{p=1}^{\infty} F_p \cos(p\omega t - \phi_p) \quad 2.16$$

$F_0/2$ is the amplitude of the mean term (or the zeroth component), and F_p and ϕ_p the amplitudes and phases of the pth components. ω is the angular frequency and is equal to $2\pi f$, where f is the frequency of repetition of the waveform. The coefficients F_p and ϕ_p may be calculated using the following equations and there are now many computer routines which will do this swiftly and accurately:

$$F_p = (a_p^2 + b_p^2)^{1/2} \quad 2.17$$
$$\phi_p = -\tan^{-1}(b_p/a_p) \quad 2.18$$

where

$$a_p = 2/T \int_0^T F(t)\cos(p\omega t)dt \quad 2.19$$

and

$$b_p = 2/T \int_0^T F(t)\sin(p\omega t)dt \quad 2.20$$

Alternatively, these expressions may be written in a discrete form appropriate to the analysis of digital data. If the periodic function is represented as Θ_n, $n = 0, 1, 2 \ldots N-1$, then the coefficients a_p and b_p may be calculated from:

$$a_p = \frac{2}{N} \sum_{n=0}^{N-1} \Theta_n \cos\left(\frac{2\pi pn}{N}\right) \quad 2.21$$

and

$$b_p = \frac{2}{N} \sum_{n=0}^{N-1} \Theta_n \sin\left(\frac{2\pi pn}{N}\right) \quad 2.22$$

2.4.2 Sinusoidal Flow

There are two important aspects of sinusoidal flow to consider. The first is the relationship between pressure gradient and volumetric flow (i.e. the sinusoidal version of Poiseuille's law), and the second the relationship between volumetric flow and velocity profile. For both of these it is necessary to refer to the results obtained by Womersley (1955). Womersley showed that Poiseuille's equation (eqn 2.10) can be modified to describe the relationship between pulsatile flow and pressure, and can be written thus:

$$Q = \frac{M'_{10}}{\alpha^2} \frac{\pi R^4}{\mu L} \Delta P \sin(\omega t - \phi + \varepsilon'_{10}) \quad 2.23$$

where the pressure difference across the section of tube under consideration is $\Delta P \cos(\omega t - \phi)$. M'_{10}/α^2 and ε'_{10} are both functions of the non-dimensional parameter, α:

$$\alpha = R(\omega/v)^{1/2} \quad 2.24$$

(where v is the kinematic viscosity) which characterises kinematic similarities in liquid motion. The definitions of M'_{10} and ε'_{10} are given in Appendix 1 (eqns. A1.1 and A1.2) and plotted in Fig. 2.4. It can be seen from this figure that M'_{10}/α^2 tends to $1/8$ and ε'_{10} tends to $90°$ when ω tends to zero, and therefore eqn. 2.23 reduces to eqn. 2.10 for very low values of α. As α increases (i.e. the frequency increases) and inertial effects tend to dominate viscous effects, the amplitude of flow oscillations in response to given pressure oscillations become smaller, and lag further and further behind the pressure gradient, and the artery effectively acts as a low pass filter. Some typical values of α found in the human circulation are given in Table 2.1.

Perhaps more important than the pressure–flow relationship is the flow-velocity profile rela-

Table 2.1 Some typical values of haemodynamic parameters measured from or calculated for the human circulation

Vessel	Diameter (mm)	Flow (ml min^{-1})	Mean velocity (cm s^{-1})	Reynolds number	Fundamental α	Entrance length (cm)
Ascending aorta	(31)	6400†	18‡	1500‡	21‡	(140)
Abdominal aorta	18*	(2000)	14‡	640‡	12‡	(34)
Common carotid	5.9§	387§	14§	(217)	(4.2)	(3.8)
Renal artery	(6.2)	(725)	40‡	700‡	4‡	(13)
External iliac	8.2*	380¶	(12)	(260)	(5.8)	(5.0)
Superficial femoral	6.4*	150¶	12‡	200‡	4‡	(3.8)
Posterior tibial	3.8*	10¶	3.5¶	(35)	(2.7)	(0.4)

Values in brackets are derived values. Wherever possible measured figures are quoted, and therefore values for the same site are not always totally consistent with each other.
*Callum et al (1983).
†Ganong (1971).
‡Milnor (1982).
§Payen et al (1982).
¶Strandness and Sumner (1975).

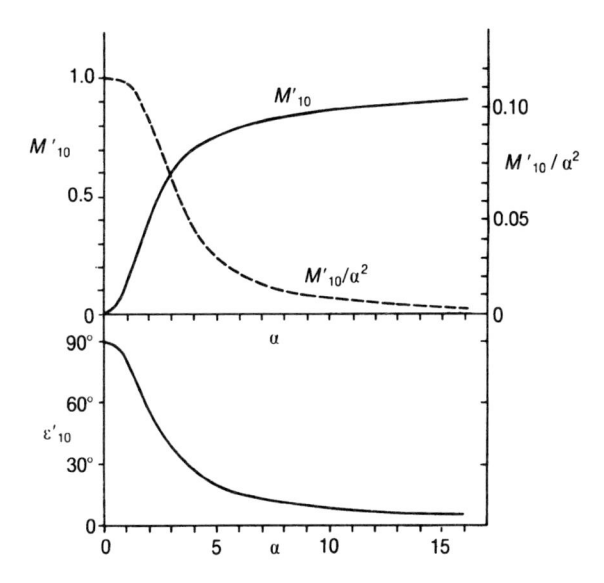

Figure 2.4 Womersley's parameters M'_{10}, M'_{10}/α^2 and ε'_{10} plotted as a function of α (reproduced by permission from Milnor 1982, © 1982, The Williams & Wilkins Co, Baltimore)

$Q \cos(\omega t - \phi)$ is given by:

$$U(y) = \frac{1}{\pi R^2} Q |\psi| \cos(\omega t - \phi + \chi) \qquad 2.25$$

where y is the non-dimensional radial co-ordinate, r/R, and $|\psi|$ and χ are the amplitude and phase of a complex function ψ, which is defined in Appendix 1 (eqn. A1.5.). ψ is a function of both y and α, and its modulus and phase are plotted in Fig. 2.5. For very low values of α the velocity profile is parabolic and all the laminae move with the same phase, but as α becomes large the profile becomes quite blunt and the motion of the outside laminae leads that of those near the centre. This can be seen in Fig. 2.6, which shows the velocity profiles that develop in a tube for a variety of values of α. Note that these shapes differ from those in Fig. 2.5a because of the phase differences across the vessel.

2.4.3 Pulsatile Flow

tionship, since it is the velocity profile that will determine the Doppler signal we detect from the artery. We saw that for steady flow in a long rigid tube, the velocity profile is parabolic (eqn. 2.2). For oscillatory flow, however, the profile is a function of the parameter α. By manipulating Womersley's equations it is possible to show (Evans 1982) that the velocity profile resulting from a sinusoidal flow

We have discussed steady flow and sinusoidal flow and the way in which Fourier's theorem allows us to treat pulsatile flow as a combination of a steady flow and a series of sinusoidal components. These principles are now used to calculate the types of velocity profiles to be found in healthy common femoral and common carotid arteries in man (we

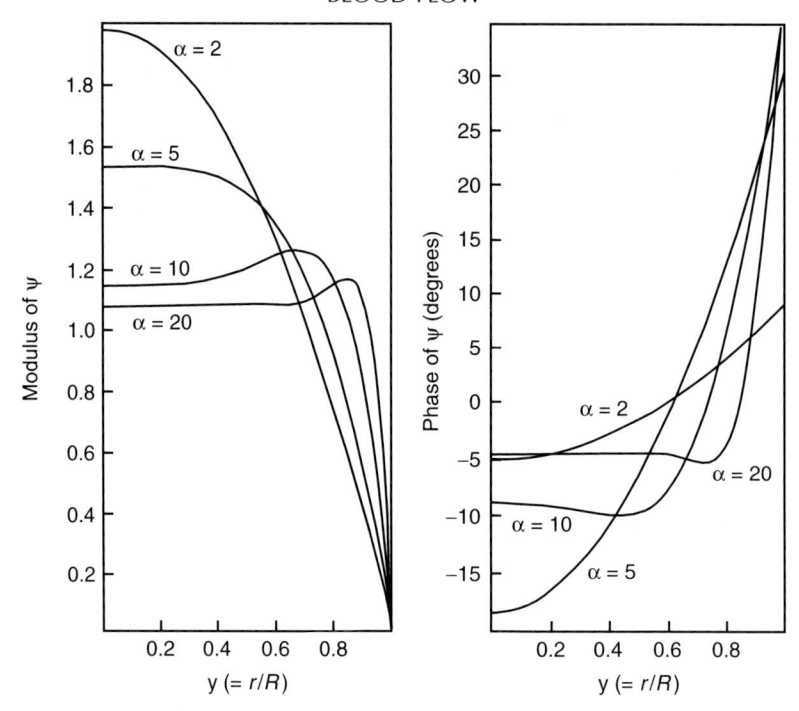

Figure 2.5 The modulus and phase of the function ψ which relates the shape of the velocity profile to sinusoidal laminar flow, plotted for α values of 2, 5, 10 and 20 (reproduced by permission of Elsevier Science, from Evans 1982, *Ultrasound in Medicine and Biology*, © World Federation of Ultrasound in Medicine and Biology)

shall assume that the entrance effects can be ignored).

The starting points for these calculations are the mean velocity waveforms illustrated in Fig. 2.7. These waveforms were actually derived from ultrasound Doppler signals. The first step is to express the velocity signals as a Fourier series:

$$V(t) = V_0 + \sum_{p=1}^{\infty} V_p \cos(p\omega t - \phi_p). \quad 2.26$$

The values of V_0, V_p and ϕ_p, $p = 1$ to 8, which have been calculated from the waveforms in Fig. 2.7, are given in Table 2.2. It is also necessary to find the values of α corresponding to each of the harmonics and these have been calculated from eqn 2.24 and are also to be found in Table 2.2. The shape of the velocity profile at any phase of the cardiac cycle may now be calculated by summing the contributions from the velocity profiles arising from each harmonic, i.e:

$$U(y) = \left\{ 2V_0(1 - y^2) \right. \\ \left. + \sum_{p=1}^{\infty} V_p |\psi|_p \cos(p\omega t - \phi_p + \chi_p) \right\}$$

2.27

In practice, the magnitude of V_p decreases rapidly with increasing p and the summation can be truncated after only a few terms.

The results obtained from this calculation are shown in Fig. 2.8. The velocity profiles that develop in the two arteries are clearly very different from each other; this will have a major influence on the type of Doppler signal that will be obtained from the two arteries and has implications for sampling and signal processing techniques. These velocity profiles will be discussed in Chapter 7 and further information on velocity

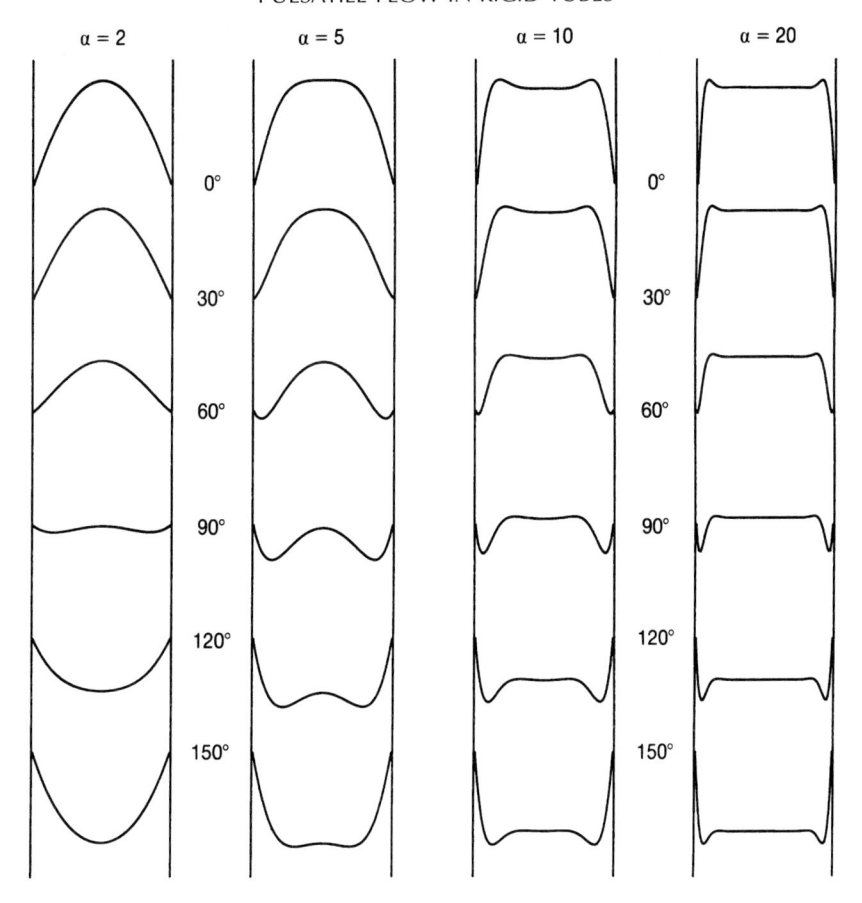

Figure 2.6 The velocity profiles that occur during sinusoidal flow with α values of 2, 5, 10 and 20 plotted at 30° intervals. Each series has been normalised to have the same maximum velocity at 0°, but note that in reality the flow oscillations in response to a given pressure oscillation become smaller as frequency increases. Only half of the cycle has been illustrated as the profiles in the other half have the same shape but are of opposite sign. These shapes differ from those in Fig. 2.5a because of the phase differences across the vessel. The wall velocities are always zero

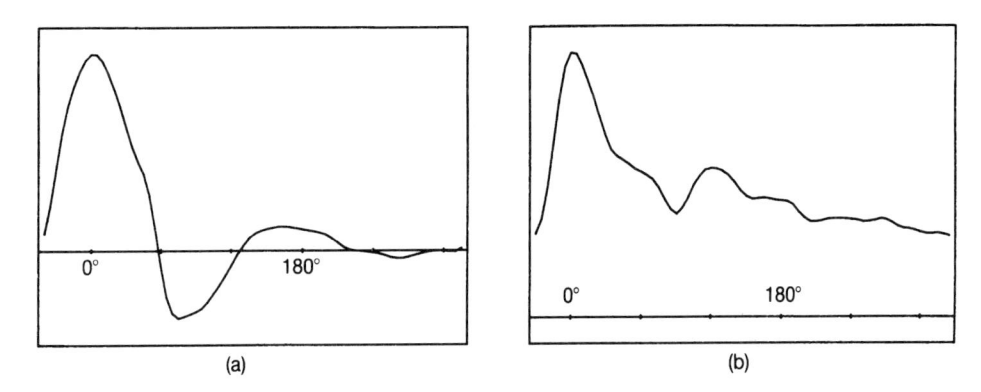

Figure 2.7 Mean velocity waveforms recorded from (a) the common femoral and (b) the common carotid arteries of a normal volunteer. The scale of the y axes is arbitrary. The Fourier components of these waveforms are given in Table 2.2 and the calculated velocity profiles in Fig. 2.8

Table 2.2 Fourier components and corresponding values of the non-dimensional parameter α for flow waveforms recorded from the common femoral and common carotid arteries of a healthy young subject. The values of V_p have been normalised to V_0, and the angle ϕ_p is given in degrees from an arbitrary starting point

	Common femoral (diameter* = 8.4 mm; heart rate† = 62 bpm; viscosity = 0.038 St)					Common carotid (diameter* = 6.0 mm; heart rate† = 62 bpm; viscosity = 0.038 St)			
Harmonic	Frequency	α	V_p	ϕ_p	Harmonic	Frequency	α	V_p	ϕ_p
0	–	–	1.00	–	0	–	–	1.00	–
1	1.03	5.5	1.89	32	1	1.03	3.9	0.33	74
2	2.05	7.7	2.49	85	2	2.05	5.5	0.24	79
3	3.08	9.5	1.28	156	3	3.08	6.8	0.24	121
4	4.10	10.9	0.32	193	4	4.10	7.8	0.12	146
5	5.13	12.2	0.27	133	5	5.13	8.7	0.11	147
6	6.15	13.4	0.32	155	6	6.15	9.6	0.13	179
7	7.18	14.5	0.28	195	7	7.18	10.3	0.06	233
8	8.21	15.5	0.01	310	8	8.21	12.4	0.04	218

* Estimated values.
† These waveforms were recorded from a healthy young subject with a relatively low heart rate.

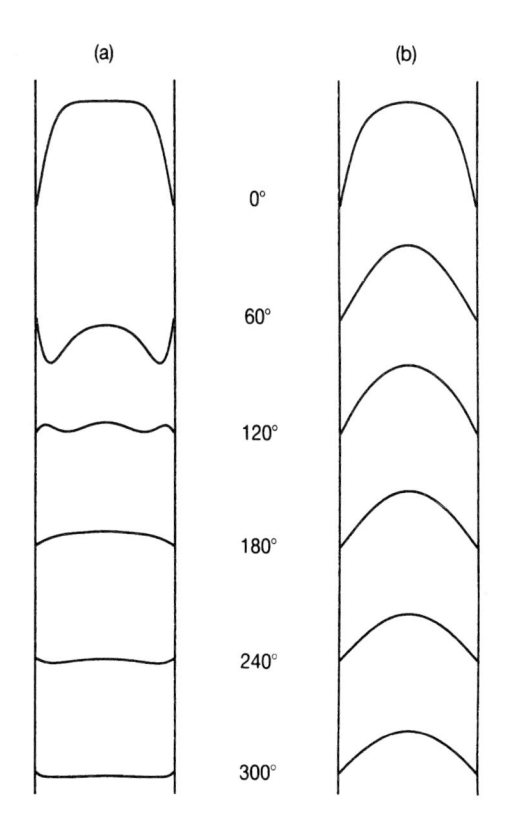

Figure 2.8 Velocity profiles for (a) a common femoral artery and (b) a common carotid artery, calculated from the mean waveforms shown in Fig. 2.7, plotted at 60° intervals. The two sets of profiles have been normalised to have the same maximum velocity at 0°

profiles in human arteries is to be found in Section 2.7.

2.4.4 Entrance Effects in Pulsatile Flow

The mean velocity profile in pulsatile flow develops, at the entrance to a tube, just as it does for steady flow (Section 2.3.2). It will, however, have the pulsatile components of the profile superimposed on it, and these too require some distance to develop fully. In general, however, the pulsatile components become fully established much more rapidly, and the unsteady entrance length X_{unsteady} may be written:

$$X_{\text{unsteady}} \approx 3.4 v_{\text{core}}/\omega \qquad 2.28$$

where v_{core} is the core velocity (Caro et al 1978).

2.5 PULSATILE FLOW IN ELASTIC AND VISCOELASTIC TUBES

A rigid tube model is adequate under most circumstances for predicting the velocity profiles which develop within arteries, and to a certain extent for predicting the relationship between pressure gradient and flow. However, it cannot give any information about arterial pulsations or the velocity of the pulse wave, and therefore it is necessary to develop other models to describe these aspects of haemodynamics.

2.5.1 Pulse Wave Velocity

It is important to distinguish between two types of velocity in the artery. The first is the velocity of the blood itself, the second (the pulse wave velocity, PWV) the velocity with which the blood flow waveform propagates. The distinction is particularly easy to envisage in the flow of an incompressible fluid through a rigid pipe. If a force is applied to the fluid at the entrance to the tube it will move with a finite velocity; however, the fluid at every point along the tube will also start to move with the same velocity at exactly the same time and the PWV will therefore be infinite. The PWV is finite in the case of elastic tubes but is still considerably higher than the blood flow velocity.

A further point to consider is exactly what is meant by PWV. If the pressure (or flow) wave did not alter in shape as it propagated this would be easy; it would simply be the distance travelled by any one feature of the waveform in unit time. In the arterial system, however, the pulse wave changes its shape as it propagates due to the effects of wave reflection and because both attenuation and (phase) velocity are frequency dependent. It has been usual in the past to measure the velocity of the 'foot' of the wave (the point at the end of diastole where the steep rise of the wavefront begins) since this point is easy to identify and is little affected by reflections, but this is a question of expediency.

The equation that predicts the velocity of pulse propagation in a thin-walled elastic tube filled with a liquid that is incompressible and has no viscosity was derived over a century ago and is known as the Moens–Korteweg equation, after two Dutch scientists. The wave velocity c_0 is given by:

$$c_0 = (Eh/\rho D)^{1/2} \qquad 2.29$$

where E is Young's modulus of elasticity in the circumferential direction, h is the wall thickness, ρ is the density of the liquid, and D is the diameter of the tube. There have been a number of more sophisticated equations derived which take account of such parameters as the viscosity of the fluid, but eqn. 2.29 is remarkably good at predicting c_0, and produces theoretical values that are within 15% of those found by experiment (Caro et al 1978).

A point of particular interest from eqn. 2.29 is that the PWV is proportional to the square root of Young's modulus of elasticity. It is possible to measure PWV using a Doppler technique, and thus to follow changes in E (Wright et al 1990). Furthermore, E is affected by blood pressure and therefore it may be possible to monitor blood pressure changes by observing changes in PWV.

In an attempt to remove the confounding effects of blood pressure when using PWV to characterise the elastic properties of the arterial wall, Lehmann et al (1993) have suggested the use of a quantity they call 'the intrinsic distensibility of the arterial wall', which is calculated from PWV, blood density and systolic and diastolic pressure.

2.5.2 Arterial Pulsations

Radial arterial pulsations are of relevance to Doppler studies in several respects, and are in general sources of annoyance. The arterial wall/blood interface is a comparatively strong reflector of ultrasound and gives rise to high-amplitude low-frequency Doppler signals known as 'wall thump'. The changes in diameter during the cardiac cycle also introduce an additional source of error into the ultrasonic measurement of blood flow (Section 12.2.3) and make it difficult to examine the local blood flow close to an arterial wall without some type of tracking device.

The radial dilatation is mathematically difficult to predict exactly because it is affected by the exact characteristics of the vessel wall, but Womersley (1957) showed that, in an elastic tube with very stiff longitudinal tethering, the total excursion of the wall 2ξ can be written:

$$2\xi = R(1 - \sigma^2)(Pe^{i\omega t}/\rho c_0^2) \qquad 2.30$$

where R is the tube radius, σ Poisson's ratio (normally taken to be 0.5), $Pe^{i\omega t}$ the pressure wave, ρ the density of blood, and c_0 the Moens–Korteweg velocity. In this case the dilatation is exactly in phase with the pressure wave, but if the wall has other properties the wall movement may lead the pressure wave (for example if the constraint is less stiff) or lag behind it (if a highly constrained wall also has viscous properties).

2.6 THE EFFECTS OF GEOMETRIC CHANGES

Each curve, branch and constriction in the arterial system will modify both the velocity profile and the local pressure gradient.

2.6.1 Constrictions and Projections

The haemodynamic effects of constrictions and projections are particularly important because a major application of Doppler ultrasound is the identification and quantification of stenotic lesions. Figure 2.9 is a schematic diagram of a simple smooth symmetrical stenosis, which illustrates some of the important characteristics of the flow through a stenosis. The first point to notice is that since the same quantity of blood must flow through the stenosis as through the unstenosed vessel, the local velocity of flow must increase. If the volumetric flow through the vessel is Q, the spatial mean velocities in the unstenosed and stenosed regions are $\bar{v}_0$ and $\bar{v}_s$, respectively, and the cross-sectional areas in the corresponding regions are A_0 and A_s, then:

$$Q = \bar{v}_0 A_0 = \bar{v}_s A_s \qquad 2.31$$

or

$$\frac{\bar{v}_s}{\bar{v}_0} = \frac{A_0}{A_s} \qquad 2.32$$

Thus, the ratio of spatial mean velocities in the stenosed and unstenosed regions of the vessel is the same as the reciprocal of the ratio of their cross-sectional areas. This velocity change forms the basis of the common method of identifying stenoses with Doppler ultrasound by searching for local increases in the Doppler shift frequency from suspicious areas within the vessel (see Sections 9.4.2 and 9.4.9.3).

A second point to notice from Fig. 2.9 is that in order for the blood to pass through the stenosis, the flow stream-lines must converge. This is important because it may change the Doppler angle (the angle between the ultrasound beam and the direction of flow) and lead to a potential misinterpretation of velocity and velocity changes, and hence to stenosis severity. A third point to note is the presence of regions of recirculation and reverse flow distal to the stenosis, which may sometimes be observed on colour flow images.

Wong et al (1991) have published interesting illustrations of flow patterns through stenoses created using finite element computer simulations, whilst Ojha et al (1990) and Hoskins (1997) have published images of flow through model stenoses derived using photochromic visualisation and vector Doppler methods, respectively.

The effects of stenoses on volumetric flow depends not only on the stenosis itself, but also on the characteristics of the vascular bed proximal to and distal to the stenosis, although tight stenoses reduce both mean flow and the pulsatility of the flow waveform. It is a common observation, however, that the arterial lumen needs to be reduced by as much as 80% before the mean resting flow rate is affected (Mann et al 1938, Shipley and Gregg 1944, May et al 1963, Kindt and Youmans 1969, Eklöf and Schwartz 1970), and therefore considerable effort has been directed towards both the detection of the much more significant changes in local velocity mentioned

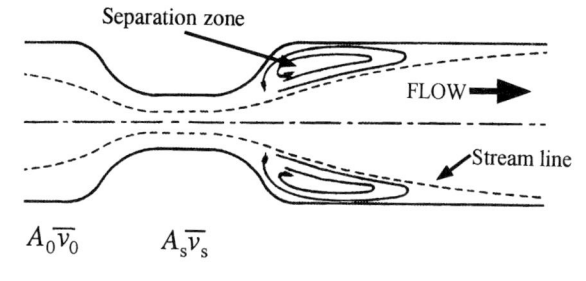

Figure 2.9 Schematic diagram of flow through a simple smooth stenosis, showing convergence of the stream-lines (and hence a velocity increase) and post-stenotic recirculation

above, and changes in the Doppler waveform shape caused by lesser degrees of narrowing (Chapter 9).

Doppler ultrasound may also be used to detect the flow disturbances caused by the encroachment of atheromata into the arterial lumen. If the obstacle projecting into the stream is a small one, small eddies will form but will soon die away. If the obstacle is larger, the eddies will grow in size and, depending on the Reynolds number, may cause a complete breakdown of laminar flow.

For a sharp-edged projection into the lumen of height h, laminar flow will only be maintained if:

$$h < 4r/Re \qquad 2.33$$

where r is the radius of the tube and Re the Reynolds number (Goldstein 1938, cited in McDonald 1974).

Examples of the Doppler signal found distal to atheromatous plaques are to be found in Figs 9.2b and 9.2c. Flow disturbances caused by a stenosis may propagate for some distance but cannot usually be detected beyond 12–16 unstenosed diameters downstream (Clark 1980, Evans et al 1982).

The mechanisms of pressure loss across arterial stenoses is complicated and beyond the scope of this chapter, and the reader is referred to a series of papers by DF Young and his colleagues (Young and Tsai 1973a, 1973b, Young et al 1975, Seeley and Young 1976, Young et al 1977) for an introduction to the problem. In general, however, the losses are made up of three terms and may be written:

$$\Delta P = \Delta P_s + \Delta P_c + \Delta P_e \qquad 2.34$$

where ΔP_s is the pressure drop caused by the resistance of the stenosis (and predicted by Poiseuille's law if the flow is laminar), ΔP_c is a result of energy lost due to the contraction of the stream, and ΔP_e is the result of the post-stenotic expansion. The relative sizes of these three terms depend heavily on the geometry of the stenosis, but one particularly simple case is that of an orifice, where only the final term, ΔP_e, is significant. The equation for the pressure drop across an orifice is of particular relevance because it can be used to estimate the pressure drop across a stenosed heart valve from Doppler velocity measurements.

The starting point for the derivation of the pressure drop is Bernoulli's equation for steady flow, which is an expression of conservation of hydraulic energy (energy per unit volume) in frictionless incompressible fluids, and which may be written:

$$\underset{\text{pressure energy}}{P} + \underset{\text{kinetic energy}}{\tfrac{1}{2}\rho v^2} + \underset{\substack{\text{gravitational} \\ \text{potential} \\ \text{energy}}}{\rho g z} = \text{constant} \qquad 2.35$$

where P is the pressure in the fluid, ρ its density, v the velocity of flow, g the acceleration due to gravity, and z the height relative to a given reference point. Applying this equation to the flow through an orifice, we may write:

$$P_1 + \tfrac{1}{2}\rho v_1^2 = P_2 + \tfrac{1}{2}\rho v_2^2 \qquad 2.36$$

where the subscripts 1 and 2 refer to conditions before and at the orifice. The potential energy term has been omitted because z_1 and z_2 are assumed to be nearly equal. Collecting terms and writing the pressure drop as ΔP leads to:

$$\Delta P = \tfrac{1}{2}\rho(v_2^2 - v_1^2) \qquad 2.37$$

Finally, if the stenosis is tight, then $v_2^2 \gg v_1^2$ and the second term in the bracket may be ignored, and eqn. 2.37 reduces to:

$$\Delta P \approx 4 v_2^2 \qquad 2.38$$

where ΔP is in mmHg and v_2 is in m s^{-1}.

Strictly, P_2 was the pressure within the orifice. However, because of the flow disturbances that occur downstream of a sudden expansion, very little of the kinetic energy gained during the contraction of the stream is converted back to pressure energy, but is dissipated as heat. Thus, eqn. 2.38 may be used to estimate the pressure drop across stenosed heart valves.

For non-steady flow the Bernoulli equation has an additional term due to the energy that is required to accelerate the flow, and eqn. 2.36 should be rewritten:

$$P_1 + \frac{1}{2}\rho v_1^2 = P_2 + \frac{1}{2}\rho v_2^2 + \rho \int_1^2 \frac{dv}{dt} ds \quad 2.39$$

where s is the position along a streamline. The extra term only becomes important for large values of dv/dt, specifically during heart valve opening and closure. In practice this means that there is a lag between velocity and pressure drop and that eqn. 2.38 is not applicable to early systole; despite this, it gives a clinically acceptable estimate of the pressure drop in both mitral and aortic stenoses (Hatle and Angelsen 1982).

It can be seen from eqn. 2.38 that the relationship between flow and pressure drop across a stenosis where exit effects dominate is non-linear. This would not be the case for a stenosis where the pressure drop resulted purely from a Poiseuille type of loss, but in general the relationship between flow and pressure drop is non-linear and may be written in the form:

$$\Delta P = AQ + BQ^2 \quad 2.40$$

where A and B are constants.

2.6.2 Curves

When flow enters a curved section of a tube it experiences a centrifugal force, which is proportional to the velocity squared and inversely proportional to the radius of curvature. The resulting change in flow pattern depends on the velocity profile existing before the curve was entered. If the profile is parabolic, the fluid at the centre of the tube has the highest velocity and will experience the greatest force, and will therefore tend to move towards the outside of the tube. This means that the velocity profile becomes skewed towards this wall, and also that secondary flows are set up, such that individual particles tend to follow helical paths (see Fig. 2.10). If the velocity profile is flat when the curve is entered, the velocity profile becomes skewed in the opposite direction (Fig. 2.11). This is because the pressure at the outside of the bend is larger than that at the inside (otherwise the fluid would not be forced around the bend) and therefore, because the total energy of a streamline cannot change, the kinetic energy (and hence the velocity) of the particles on the inside of the bend must be greater than those of particles on the outside.

It is important to realise that these effects are non-linear, and therefore the behaviour of a complex pulsatile flow waveform cannot be predicted by calculating the effects of the pulsatile components of flow separately and superimposing the answers thus obtained on steady flow profiles.

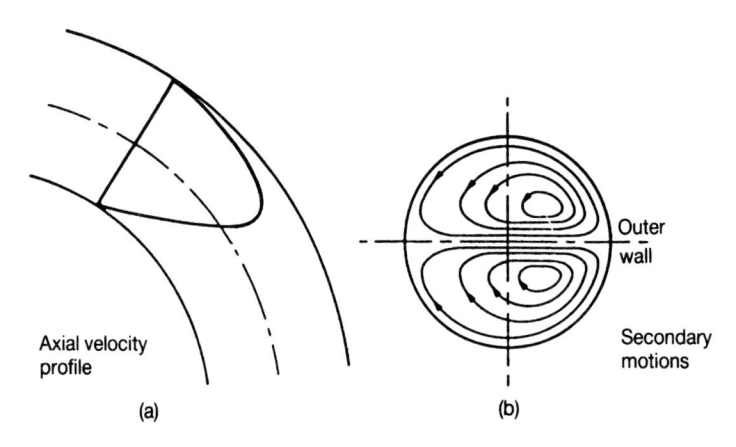

Figure 2.10 (a) Distortion of a parabolic velocity profile resulting from tube curvature. (b) Cross-section of the vessel showing the form of the secondary flow (reproduced by permission from Caro et al 1978, Oxford University Press)

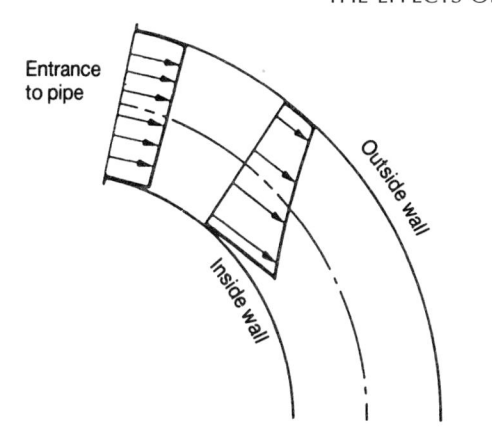

Figure 2.11 Distortion of a flat velocity profile resulting from tube curvature (reproduced by permission from Caro et al 1978, Oxford University Press)

2.6.3 Branches

The arterial system branches repeatedly, and at each junction new velocity profiles must develop, and there may be changes in the mean velocity, Reynolds number and pressure gradient.

McDonald (1974) divided arterial junctions into three main classes: bifurcation junctions, fusion junctions and side branches. Fusion junctions are almost exclusively confined to the venous system, the only arterial example being the fusion of the two vertebral arteries into the basilar artery, but the other two types and various hybrids of the two are found throughout the body.

The effects on flow profile will depend on the exact geometry of the junction. At a pure symmetric bifurcation the flow is split into two and the largest velocities, which were in the centre of the stream of the parent vessel, will then be very close to the inside walls of the daughter vessels, leading to skewed velocity profiles. Brech and Bellhouse (1973) have studied the flow in symmetrical branches using rigid-walled models perfused with steady and pulsatile flows. Velocity distributions downstream of the branches were investigated using a thin film anemometry technique. Wong et al (1991) have published the results of steady flow simulations in a model Y-shaped bifurcation at a Reynolds number of 750.

At the same time as splitting, because each of the two streams must turn a bend, secondary motions of the type described in Section 2.6.2 are set up. These secondary motions sweep some of the fast-moving fluid on the inside of the daughter tubes round to the top and bottom walls, leading to an M-shaped velocity profile in the plane perpendicular to the junction (Caro et al 1978). This situation may be further complicated by the fact that the geometry of bifurcations is commonly non-planar (Caro et al 1996). Hoskins and colleagues (Hoskins et al 1994, Stonebridge et al 1996) have demonstrated the existence of spiral flow distal to bifurcations in the lower limb *in vivo* using ultrasonic Doppler techniques.

Figure 2.12 illustrates the types of profile found two diameters downstream of a symmetric bifurcation. These disturbed profiles gradually return to

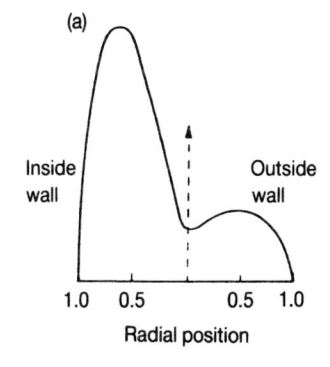

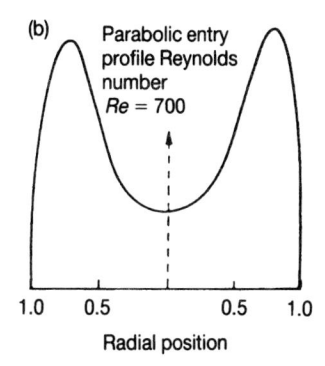

Figure 2.12 Velocity profiles two diameters downstream of a symmetric bifurcation, at a parent tube Reynolds number of 700. (a) Profile in the plane of the junction; (b) profile in the perpendicular plane (reproduced by permission from Schroter and Sudlow 1969, Elsevier Science Publishers BV)

a more normal shape as viscous effects exert themselves, and a steady-state profile is reached at a distance comparable with the usual entrance length (eqn. 2.12).

There are so many configurations for side branches leaving a parent vessel that it is difficult to generalise on the effects of this type of junction. The type of flow found in a junction when a daughter tube branches off a parent tube at right angles is illustrated in Fig. 2.13. There is flow separation both at the inner wall of the daughter tube and on the wall of the main tube opposite and just downstream of the branch. Strong secondary motions are set up in the daughter tube due mainly to the acute bend, and some distance is required before a steady-state profile is established.

In general, the total cross-sectional area of branched vessels exceeds that of the parent vessel and therefore both the mean velocity of flow and the Reynolds number reduce as vessels branch, and this leads to a greater stability in flow.

2.6.4 Tapers

All blood vessels, with the exception of capillaries, show some degree of both geometrical and elastic taper (that is, the elastic modulus of the vessel changes progressively with distance). The major effects of the convergence of individual vessels found in the arterial system is to stabilise the laminar flow, raise the critical Reynolds number, and slightly flatten the velocity profile (Whitmore 1968).

2.7 VELOCITY PROFILES IN HUMAN ARTERIES

Experimental studies of velocity profiles in arteries have been marred by the difficulties of making point measurements in anything but the largest of arteries, and the majority of published results still appear to relate to the profiles found in canine aortas. These results may, to some extent, be extrapolated to man, but for other smaller arteries the profiles that develop must be inferred from the theories outlined earlier in this chapter. Hopefully this is a situation that will be rectified as Doppler techniques become even more sophisticated.

Figure 2.14 illustrates the time-averaged profiles (normalised to the centreline time-averaged velocity) at various positions in the canine aorta. As predicted by the theory discussed in Section 2.3.2., the profile starts off very blunt and slowly becomes more rounded as flow progresses down the aorta. There is also a slight asymmetry across the vessel diameter which is much more marked for the velocity profiles at peak systole (Fig. 2.15). This asymmetry is thought to be a result of both the tight curve of the aortic arch and the consequent lateral pressure gradient (Section 2.6.2), and to be due to some extent to the local distortion caused by the major branches feeding the head and arms.

For the human aorta, both the diameter and the mean Reynolds number are greater than those found in the dog, and therefore the steady-state profile will take longer to develop (eqn. 2.12), and even though the human aorta is longer, the profiles will be less developed at corresponding anatomical points. Substituting approximate values for diameter ($\sim$3.1 cm) and mean Reynolds number ($\sim$1500) into eqn. 2.12 gives an entrance length of approximately 1.4 m, which suggests that the whole of the human aorta must be regarded as an entrance region. Vieli et al (1986) have reported the results of multigate Doppler measurements from

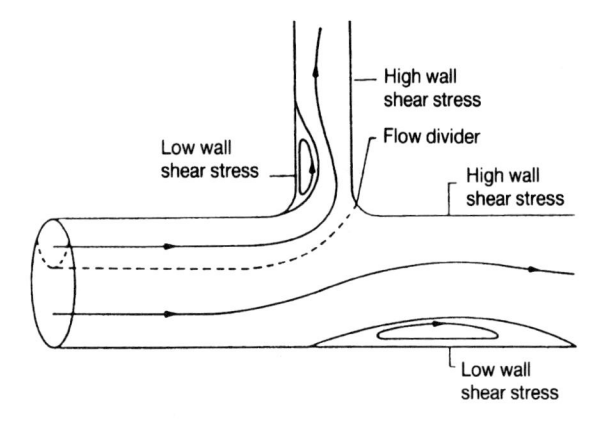

Figure 2.13 Flow at right-angled junction. The dashed line is the surface dividing the fluid which flows down the side branch from that continuing in the main tube; the solid lines are stream lines. Note the closed eddies in the two regions of separated flow. Regions of high and low shear are also indicated (Reproduced by permission from Caro et al 1978, Oxford University Press)

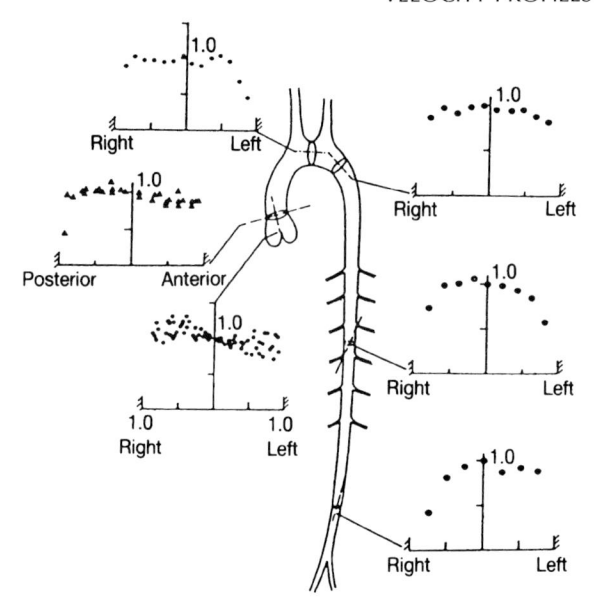

Figure 2.14 Time-averaged velocity profiles (normalised to the centreline time-averaged velocity) measured at various positions in the canine aorta. Note how the profiles become progressively more rounded as the measurement site moves more distally (reproduced by permission from Schultz 1972, © 1972 Academic Press)

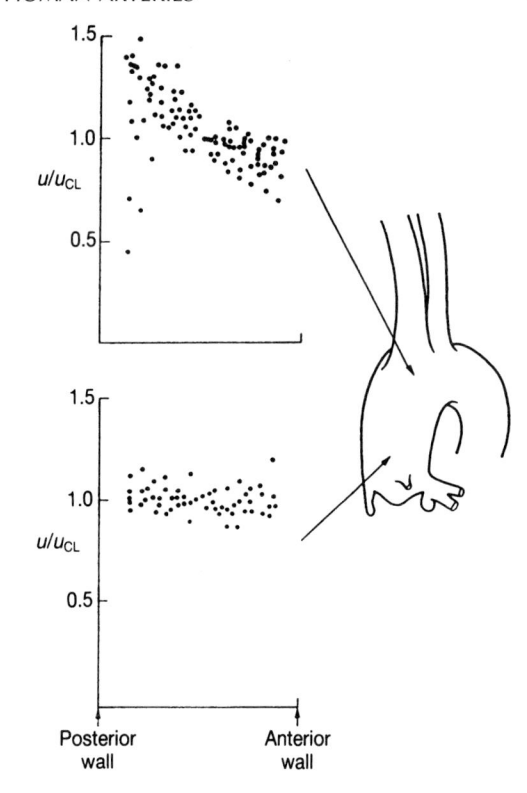

Figure 2.15 Peak systolic velocities across two diameters in the ascending aorta of the dog. Note that the asymmetry across the vessel diameter is much more marked than for the time-averaged profiles (Fig. 2.14) (reproduced by permission from Caro et al 1978, Oxford University Press)

the ascending aorta of healthy humans which demonstrate relatively smooth and flat profiles with considerable skewing. Hessevik at al (1994) have published illustrations of velocity profiles in the human ascending aorta derived using data gathered from an intraluminal Doppler ultrasound probe during open heart surgery together with a model for flow in the ascending aorta. Vieli et al (1989) have published illustrations of abdominal aortic velocity profiles, derived using both MRI and ultrasound techniques (some of which are reproduced in Fig. 2.16) which confirm the blunt nature of human aortic profiles.

As the arterial system branches, both the diameter and the Reynolds number fall and each new junction results in new inlet phenomena. In some arteries (such as the renal arteries) a steady-state mean profile can never be established, but in others (for example, the superficial femoral artery) entrance effects are rapidly dispelled. Some entrance lengths for various human arteries, estimated from arterial dimensions and blood flow values quoted by a number of authors, are given in Table 2.1. In children the smaller values of Re and arterial diameter will result in a much more rapid achievement of steady-state conditions.

Because of the role of flow patterns in atherogenesis, special interest has tended to be focused on the coronary circulation (Batten and Nerem 1982) and particularly the carotid circulation where the flow patterns are relatively complex and Doppler ultrasound is so widely used for diagnostic purposes (Keller et al 1976, Ku et al 1985, Sillesen et al 1988, Tortoli et al 1988, van Merode et al 1988, 1989, Wong et al 1991, Perktold et al 1991, 1994), and it is hoped that these and similar studies will lead to a deeper understanding of flow in human arteries. Figure 2.17, reproduced from Perktold et al (1994), shows the velocity profiles in a modelled human carotid bifurcation, calculated for two geometries and two points in the cardiac cycle.

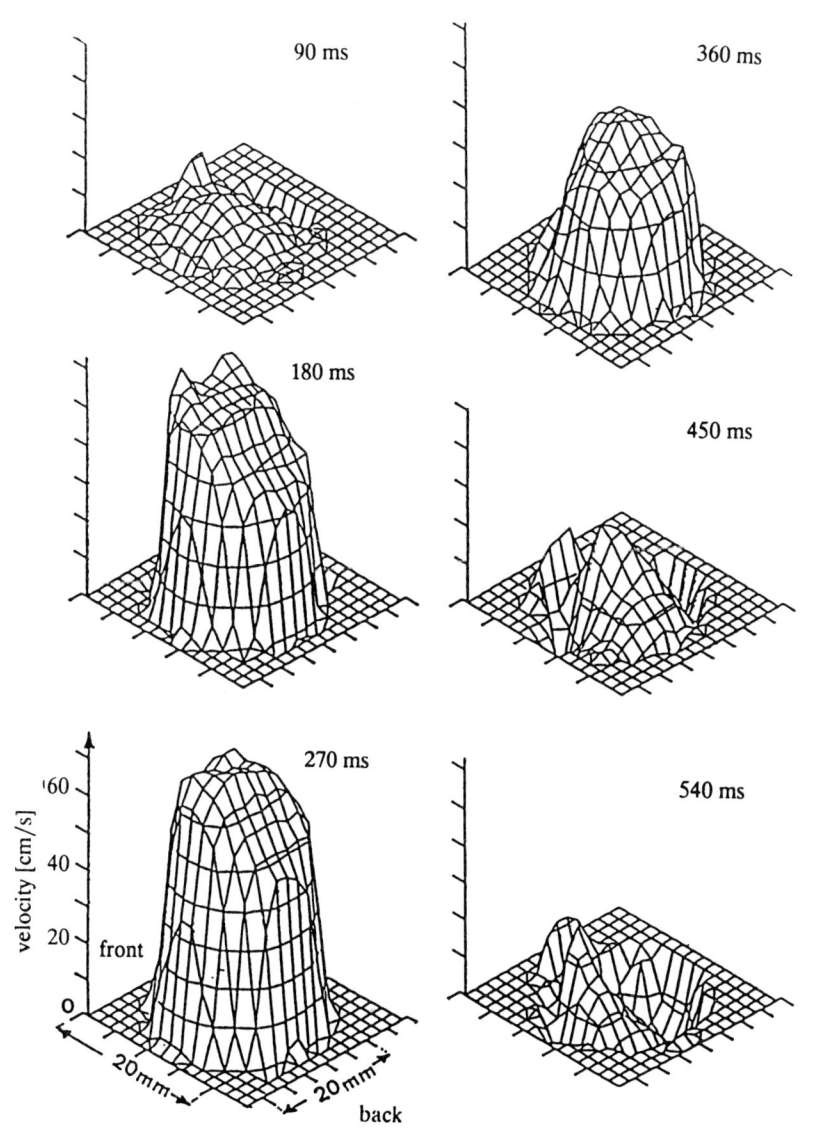

Figure 2.16 Velocity profiles in the normal human abdominal aorta at six points during the cardiac cycle derived using MRI data (reproduced by permission of Elsevier Science, from Vieli et al 1989, *Ultrasound in Medicine and Biology*, © World Federation of Ultrasound in Medicine and Biology)

2.8 COMPUTATIONAL FLUID DYNAMICS

Flow modelling in recent years has increasingly been studied using the numerical simulation approach known as computational fluid dynamics or CFD. This is essentially a predictive method employing the basic partial differential equations for fluid flow (the Navier–Stokes equations). Highly specialised software permits the treatment of 3-D time-dependent (even non-Newtonian) flows in irregular geometries. It can be appreciated that such an approach has not just research but also clinical potential. In fact, CFD code can be combined with additional MRI or ultrasound-based image processing to allow

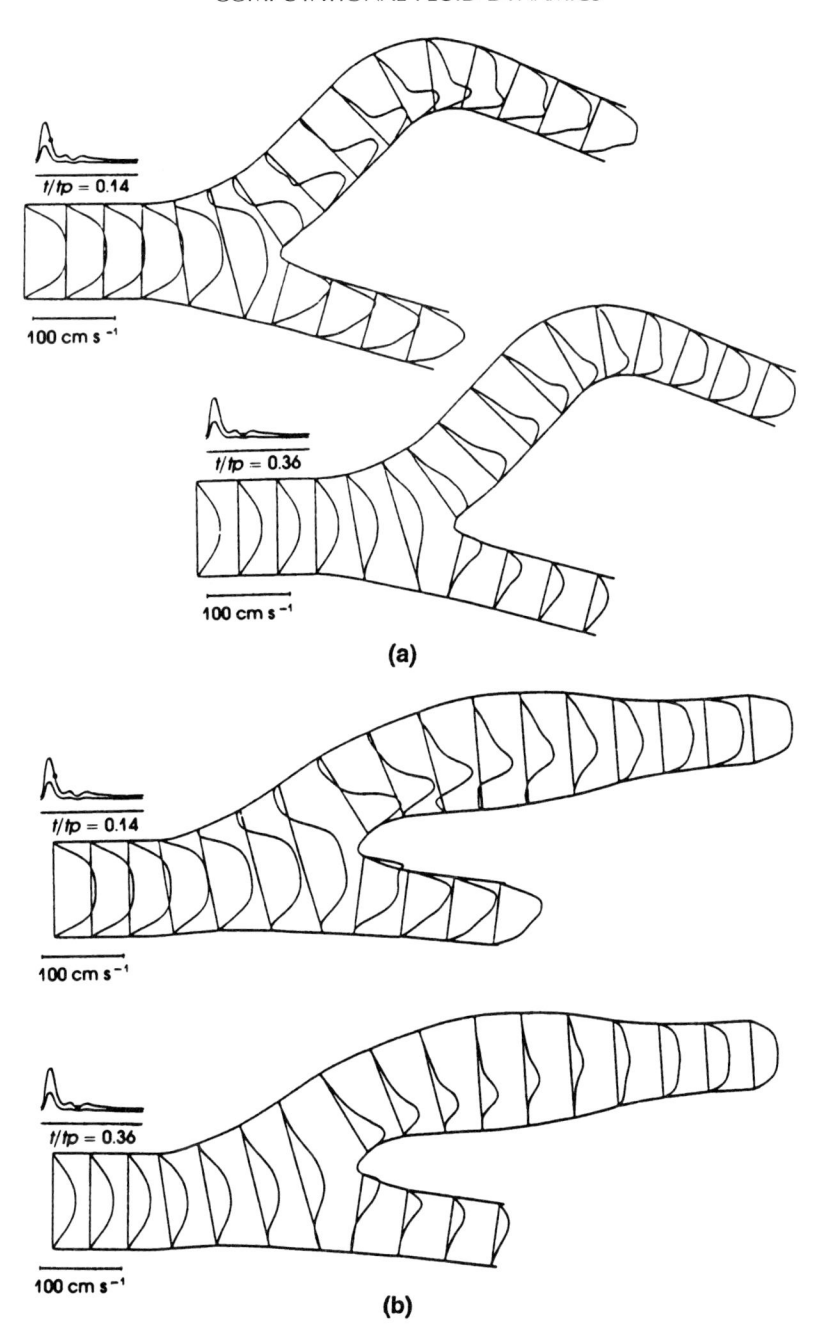

Figure 2.17 Velocity profiles calculated from a mathematical model of the human carotid bifurcation (in the branching plane) for two different geometries, and at two points during the cardiac cycle (systolic deceleration phase, $t/tp = 0.14$; diastolic minimum flow rate, $t/tp = 0.36$) (reproduced by permission from Perktold et al 1994, © IFMBE)

simulations to be made directly from clinical data. Another enhancement, given the compliant behaviour of arterial walls, is to couple the CFD code with a parallel high-level solid mechanics code. This allows the wall–fluid interaction nature of the arterial haemodynamics to be properly addressed. Such methods represent the cutting-edge of CFD research in the area of haemodynamics. A number of commercial CFD codes are available, of which the UK Harwell-based CFX is typical.

2.9 SUMMARY

Blood flow in arteries is complex but, despite this, useful approximations can be made using fairly simple models. The flow in the normal circulation is for the most part both laminar and pulsatile, and may be considered the sum of a steady flow component and a series of oscillatory components. Once established, the mean flow component results in a parabolic profile, whilst the pulsatile components cause oscillating profiles which vary between parabolic and blunt (but with phase differences across the vessel diameter) depending on the value of Womersley's parameter α. At each branch and bifurcation the velocity profiles are disturbed and the return to steady-state 'established' flow requires a finite distance. Each branch and curve also gives rise to secondary flows which also persist for a finite distance. Constrictions of, and projections into, the arterial lumen cause energy losses over and above simple Poiseuille losses and may cause disturbed or even turbulent flow conditions. Computational fluid dynamics has considerable potential for helping to elucidate the complex interaction between flow and complex arterial geometry.

In this chapter it has only been possible to consider those aspects of blood flow that are most relevant to the recording of a Doppler signal. Much more detailed information on these and other facets of blood flow are to be found in the books recommended in Section 2.12.

2.10 NOTATION

A_0	Unstenosed vessel cross-sectional area
A_s	Stenosed vessel cross-sectional area
c_0	Moens–Korteweg velocity
dv/dr	Shear rate
D	Diameter of tube or artery
E	Young's modulus of elasticity
f	Frequency
$F(t)$	A periodic function in time
$F_0/2$	The mean value of $F(t)$
F_p	The amplitude of the pth Fourier component of $F(t)$
g	Acceleration due to gravity
h	Wall thickness
L	Length of a tube
M'_{10}	Amplitude of a special function arising from Womersley's theory (see Appendix 1)
P	Pressure
$Pe^{i\omega t}$	Pressure wave
Q	Volumetric flow
r	Distance from the centre of the tube
R	Tube radius
Re	Reynold's number
s	Position along a streamline
t	Time
U	Free stream velocity
$v(r)$	Velocity of lamina at radius r
$v(y)$	Velocity of lamina at radial co-ordinate r/R
v_{max}	Velocity of central lamina
$\bar{v}$	Spatial mean velocity
$\bar{v}_0$	Spatial mean velocity in unstenosed vessel
$\bar{v}_s$	Spatial mean velocity in stenosed vessel
v_{core}	Core velocity
$V(t)$	Periodic mean velocity signal
V_0	Mean component of $V(t)$
V_p	Amplitude of pth harmonic of $V(t)$
x	Distance from the tube entrance
X	Entrance length
y	Non-dimensional radial co-ordinate, r/R
z	Height above a reference point
α	Womersley's parameter characterising kinematic behaviour
δ	Boundary layer thickness
ΔP	Pressure drop across length of tube
ε'_{10}	Phase of a special function arising from Womersley's theory (see Appendix 1)
Θ_n	The nth sample of a periodic function
μ	Absolute viscosity
υ	Kinematic viscosity
ξ	Wall movement
ρ	Density

σ	Poisson's ratio
τ	Shear stress
ϕ_p	The phase of the pth harmonic of a Fourier series
χ	Phase of a special function relating velocity to flow
$\|\psi\|$	Amplitude of a special function relating velocity to flow
ω	Angular frequency

2.11 REFERENCES

Batten JR, Neren RM (1982) Model study of flow in curved and planar arterial bifurcations. Cardiovasc Res 16:178–186

Brech R, Bellhouse BJ (1973) Flow in branching vessels. Cardiovasc Res 7:593–600

Callum KG, Thomas ML, Browse NL (1983) A definition of arteriomegaly and the size of arteries supplying the lower limbs. Br J Surg 70:524–529

Caro CG, Pedley TJ, Schroter RC, Seed WA (1978) The Mechanics of the Circulation. Oxford University Press, Oxford

Caro CG, Doorly DJ, Tarnawski M, Scott KT, Long Q, Dumoulin CL (1996) Non-planar curvature and branching of arteries and non-planar-type flow. Proc R Soc Lond A 452:185–197

Clark C (1980) The propagation of turbulence produced by a stenosis. J Biomech 13:591–604

Eklöf B, Schwartz SI (1970) Critical stenosis of the carotid artery in the dog. Scand J Clin Lab Invest 25:349–353

Evans DH (1982) Some aspects of the relationship between instantaneous volumetric blood flow and continuous wave Doppler ultrasound recording—III. The calculation of Doppler power spectra from mean velocity waveforms, and the results of processing these spectra with maximum, mean, and RMS frequency processors. Ultrasound Med Biol 8:617–623

Evans DH, Macpherson DS, Asher MJ, Bentley S, Bell PRF (1982) Changes in Doppler ultrasound sonagrams at varying distances from stenoses. Cardiovasc Res 16:631–636

Forsberg F, Morvay Z, Rawool NM, Deane CR, Needleman L (1996) Shear rate estimation in stenotic vessels using a clinical ultrasound scanner. In: Levy M, Schneider SC, McAvoy BR (eds) Proc. 1996 IEEE Ultrasonics Symposium, pp. 1225–1228. IEEE, Piscataway

Fung YC (1981) Biomechanics: Mechanical Properties of Living Tissues. Springer-Verlag, New York

Ganong WF (1971) Review of Medical Physiology. Lange Medical Publications, Los Altos, CA

Goldstein S (1938) Modern Developments in Fluid Dynamics, p. 142. Clarendon Press, Oxford

Hatle L, Angelsen B (1982) Doppler Ultrasound in Cardiology: Physical Principles and Clinical Applications, 2nd edn, pp. 24–26. Lea & Febiger, Philadelphia, PA

Hessevik I, Matre K, Kvitting P, Segadal L (1994) Intraluminal recording of cross-sectional blood velocity distribution of human ascending aorta by ultrasound Doppler technique. Med Biol Eng Comput 32:S171–S177

Hoeks APG, Samijo SK, Brands PJ, Reneman RS (1995) Assessment of wall shear rate in humans: an ultrasound study. J Vasc Invest 1:108–117

Hoskins PR (1997) Peak velocity estimation in arterial stenosis models using colour vector Doppler. Ultrasound Med Biol 23:889–897

Hoskins PR, Fleming A, Stonebridge P, Allan PL, Cameron D (1994) Scan-plane vector maps and secondary flow motions in arteries. Eur J Ultrasound 1:159–169

Hughes PE, How TV (1993) Quantitative measurement of wall shear rate by pulsed Doppler ultrasound. J Med Eng Technol 17:58–64

Keller HM, Meier WE, Anliker M, Kumpe DA (1976) Noninvasive measurement of velocity profiles and blood flow in the common carotid artery by pulsed Doppler ultrasound. Stroke 7:370–377

Kindt GW, Youmans JR (1969) The effect of stricture length on critical arterial stenosis. Surg Gynecol Obstet 128:729–734

Ku DN, Giddens DP, Phillips DJ, Strandness DE (1985) Hemodynamics of the normal human carotid bifurcation: *in vitro* and *in vivo* studies. Ultrasound Med Biol 11:13–26

Lehmann ED, Gosling RG, Parker JR, deSilva T, Taylor MG (1993) A blood pressure independent index of aortic distensibility. Br J Radiol 66:126–131

Mann FC, Herrick JF, Essex HE, Baldes EJ (1938) The effect on the blood flow of decreasing the lumen of a blood vessel. Surgery 4:249–252

May AG, van de Berg L, DeWeese JA, Rob CG (1963) Critical arterial stenosis. Surgery 54:250–259

McDonald DA (1974) Blood Flow in Arteries, 2nd edn. Edward Arnold, London

Milnor WR (1982) Haemodynamics, pp. 135–156. Williams and Wilkins, Baltimore

Nichols WW, O'Rourke MF (1990) McDonald's Blood Flow in Arteries: Theoretical, Experimental and Clinical Principles, 3rd edn. Edward Arnold, London

Ojha M, Cobbold RSC, Johnston KW, Hummel RL (1990) Detailed visualisation of pulsatile flow fields produced by modelled arterial stenoses. J Biomed Eng 12:463–469

Payen DM, Levy BI, Menegalli DJ, Lajat YI, Levenson JA, Nicholas FM (1982) Evaluation of hemispheric blood flow based on noninvasive carotid blood flow measurements using the range-gated Doppler technique. Stroke 13:392–398

Perktold K, Peter RO, Resch M, Langs G (1991) Pulsatile non-Newtonian blood flow in three-dimensional carotid bifurcation models: a numerical study of flow phenomena under different bifurcation angles. J Biomed Eng 13:507–515

Perktold K, Thurner E, Kenner T (1994) Flow and stress characteristics in rigid walled and compliant carotid artery bifurcation models. Med Biol Eng Comput 32:19–26

Schroter RC, Sudlas MF (1969) Flows patterns in models of the human bronchial airways. Resp Physiol 7:341–355

Schultz DL (1972) Pressure and flows in large arteries. In: Bergel DH (ed.), Cardiovascular Fluid Dynamics, Vol. 1, Academic Press, London, p. 291

Seeley BD, Young DF (1976) Effect of geometry on pressure losses across models of arterial stenoses. J Biomech 9:439–448

Shipley RE, Gregg DE (1944) The effect of external constrictions of a blood vessel on blood flow. Am J Physiol 141:289–296

Sillesen H, Neergaard K, Hansen HJB (1988) Quantitative assessment of carotid artery disease by continuous wave and pulsed multigated Doppler—prospective comparison with angiography. Ultrasound Med Biol 14:641–648

Stonebridge PA, Hoskins PR, Allan PL, Belch JFF (1996) Spiral laminar flow *in vivo*. Clin Sci 91:17–21

Strandness DE, Sumner DS (1975) Hemodynamics for Surgeons, pp. 209–289. Grune and Stratton, New York

Tortoli P, Andreuccetti F, Manes G, Atzeni C (1988) Blood flow images by a SAW-based multigate Doppler system. IEEE Trans Ultrason Ferroelec Freq Contr 35:545–551

van Merode T, Hick PJJ, Hoeks APG, Reneman RS (1988) The diagnosis of minor to moderate atherosclerotic lesions in the carotid artery bifurcation by means of spectral broadening combined with the direct detection of flow disturbances using a multi-gate pulsed Doppler system. Ultrasound Med Biol 14:459–464

van Merode T, Lodder J, Smeets FAM, Hoeks APG, Reneman RS (1989) Accurate noninvasive method to diagnose minor atherosclerotic lesions in carotid artery bulb. Stroke 20:1336–1340

Vieli A, Jenni R, Anliker M (1986) Spatial velocity distributions in the ascending aorta of healthy humans and cardiac patients. IEEE Trans Biomed Eng BME-33:28–34

Vieli A, Moser U, Maier S, Meier D, Boesiger P (1989) Velocity profiles in the normal human abdominal aorta: a comparison between ultrasound and magnetic resonance data. Ultrasound Med Biol 15:113–119

Whitmore RL (1968) Rheology of the Circulation. Pergamon, Oxford

Womersley JR (1955) Oscillatory motion of a viscous liquid in a thin-walled elastic tube—1: The linear approximation for long waves. Phil Mag 46:199–221

Womersley JR (1957) Oscillatory flow in arteries: the constrained elastic tube as a model of arterial flow and pulse transmission. Phys Med Biol 2:178–187

Wong PKC, Johnston KW, Ethier CR, Cobbold RSC (1991) Computer simulation of blood flow patterns in arteries of various geometries. J Vasc Surg 14:658–667

Wright JS, Cruickshank JK, Kontis S, Dore C, Gosling RG (1990) Aortic compliance measured by non-invasive Doppler ultrasound: description of a method and its reproducibility. Clin Sci 78:463–468

Young DF, Tsai FY (1973a) Flow characteristics in models of arterial stenoses—I. Steady flow. J Biomech 6:395–410

Young DF, Tsai FY (1973b) Flow characteristics in models of arterial stenoses—II. Unsteady flow. J Biomech 6:547–559

Young DF, Cholvin NR, Roth AC (1975) Pressure drop across artificially induced stenoses in the femoral arteries of dogs. Circ Res 36:735–743

Young DF, Cholvin NR, Kirkeeide RL, Roth AC (1977) Hemodynamics of arterial stenoses at elevated flow rates. Circ Res 41:99–107

2.12 RECOMMENDED READING

Caro CG, Pedley TJ, Schroter RC, Seed WA (1978) The Mechanics of the Circulation, Oxford University Press, Oxford

McDonald DA (1974) Blood Flow in Arteries, 2nd edn. Edward Arnold, London

Milnor WR (1982) Hemodynamics. Williams and Wilkins, Baltimore

Nichols WW, O'Rourke MF (1990) McDonald's Blood Flow in Arteries, 3rd edn. Edward Arnold, London

Whitemore RL (1968) Rheology of the Circulation. Pergamon Press, Oxford

3

Physics of Ultrasound Propagation

3.1 INTRODUCTION

Medical ultrasound and its propagation in tissue have been discussed in several texts (Hill 1986, Kremkau 1998, McDicken 1991, Wells 1977). For an in-depth study of the physics of ultrasound the reader is referred to these publications. For a comprehensive review of the values of acoustic parameters in tissue such as speed of sound, attenuation, etc., the reader is referred to the text by Duck (1990). In this chapter ultrasound physics is considered specifically from the point of view of Doppler techniques.

There is more interest in quantifying Doppler signals than there is in quantifying ultrasound images, and an understanding of wave propagation phenomena assists in the measurement of parameters such as velocity, flow and pressure gradients, and in an evaluation of potential sources of error.

3.2 ULTRASONIC WAVES

Ultrasonic waves, like audible sound waves, are compressional waves produced by the push–pull action of the sources on the propagating medium. These waves are sometimes called 'longitudinal waves', since the oscillatory motion of the particles of the medium is parallel to the direction of propagation. The source is normally a transducer in which the vibrating element is a piece of piezoelectric ceramic or plastic driven by an appropriate voltage signal. In most Doppler techniques the frequencies employed are in the range 2–10 MHz. This range is slightly lower than that in pulse-echo imaging, where 3–20 MHz is commonly used. The Doppler range is lower since, with pulsed Doppler, aliasing problems are more troublesome at high frequencies (Section 4.2.2.3).

Corresponding to the 2–10 MHz range, the wavelengths are on average 0.77–0.15 mm in soft tissue.

Doppler instruments generate either continuous wave (CW) or pulsed wave (PW) ultrasound. In the latter case the pulses vary in length from 2 to 10 cycles depending on the design of the instrument. Typically the spatial peak temporal average intensities (I_{spta}) generated by CW and PW devices are 100 and 1000 mW cm^{-2} respectively. The pressure amplitudes for CW and PW units are around 0.05 and 2.0 MPa, respectively (Section 14.3). It should be noted that there is a wide spread of outputs around these values. Since these output values are of direct concern to the question of safety, they are examined in detail in Chapter 14.

Other types of ultrasonic wave such as shear (transverse) or surface waves are very rarely applied in medical ultrasonics. Shear waves are strongly attenuated in soft tissue.

From the point of view of Doppler techniques, the parameters that describe a wave, i.e. amplitude, frequency and phase, are important, as in pulse-echo imaging. Indeed, frequency and phase are more important for Doppler methods since the velocity of blood is obtained from the shifts in frequency and changes in phase of the scattered waves (Chapter 6). The phenomena discussed below that affect ultrasound waves are also important. This may not be immediately obvious to the investigator unless careful consideration is given to possible distortions of the ultrasound beam in tissue.

3.2.1 Speed of Sound

The speed of sound, c, in soft tissue depends on its bulk modulus K and density ρ_0, and may be written:

$$c \cong (K/\rho_0)^{1/2} \qquad 3.1$$

(The approximation sign is used here because, rigorously, K should be replaced by K', a function of the adiabatic bulk modulus and the shear modulus—see Wells 1969, pp. 3–4). The bulk modulus K is given by applied pressure, p_a, divided by the fractional change in volume $-\Delta V/V$, i.e:

$$K = -p_a V/\Delta V \qquad 3.2$$

In general, the less compressible a material, the higher the speed of sound (Goss et al 1978, Wells 1977). This can be confirmed for materials of interest in medical ultrasound (Table 3.1). From this table it should be noted that the values for soft tissue are closely clustered around the average of 1540 ms^{-1}. The speed of ultrasound in blood is reported to be in the range 1540–1600 ms^{-1}. Range calibration of Doppler instruments for a mixture of soft tissues is therefore simply implemented and beam shapes are not grossly distorted by phenomena such as refraction and scattering at soft tissue boundaries. However, the high velocity in bone does cause problems. To obtain diagnostic information with either B-mode or colour Doppler imaging, it is usually necessary to avoid bone unless it is thin.

In the frequency range 1–20 MHz, the speed of ultrasound is effectively independent of frequency, i.e. dispersion effects need not be taken into account. The effect of small temperature changes can also be ignored.

The speed of sound, c, appears in the denominator of the basic Doppler equation (eqn 1.1). This is the speed of sound in blood. It is usually taken to be 1570 or 1580 ms^{-1}. Note, however, that the use of the value for soft tissue, i.e. $\sim$1540 ms^{-1}, compensates for the error due to refraction at the vessel wall/blood interface (Section 12.2.4) and is therefore to be preferred.

The speed of ultrasound is not of prime importance in Doppler techniques. However, it should be remembered that the sample volume in a duplex system (Section 4.3) may not be positioned exactly at the range indicated in the image if the average velocity for the Doppler beam path is significantly different from 1540 ms^{-1}. For example, an average velocity of 1480 ms^{-1}, perhaps due to fat in the beam path, would result in a sample volume at a depth of 5 cm being displaced 2 mm from the location indicated on the display. This can result in an erroneous signal or a complete loss of signal.

Table 3.1 Speed of ultrasound, and acoustic impedance in some common materials. Data from Wells (1969), Goss et al (1978) and Bamber (1986). The acoustic impedance cannot be calculated where the density of the material is not known

Material	Speed (ms^{-1})	Acoustic impedance (kg m^{-2} s^{-1})
Air (NTP)	330	0.0004×10^6
Amniotic fluid	1510	—
Aqueous humour	1500	1.50×10^6
Blood	1570	1.61×10^6
Bone	3500	7.80×10^6
Brain	1540	1.58×10^6
Cartilage	1660	—
Castor oil	1500	1.43×10^6
CSF	1510	—
Fat	1450	1.38×10^6
Kidney	1560	1.62×10^6
Lens of eye	1620	1.84×10^6
Liver	1550	1.65×10^6
Muscle	1580	1.70×10^6
Perspex	2680	3.20×10^6
Polythene	2000	1.84×10^6
Skin	1600	—
Soft tissue average	1540	1.63×10^6
Tendon	1750	—
Tooth	3600	—
Vitreous humour	1520	1.52×10^6
Water (20 °C)	1480	1.48×10^6

3.2.2 Intensity and Power

As for any form of radiation, the intensity at a point in a transmitted field is the rate of flow of energy through unit area at that point. The intensity is related to the pressure amplitude p_A, of the wave by:

$$I = \tfrac{1}{2}(p_A^2/\rho_0 c) \qquad 3.3$$

and to the particle displacement amplitude, x_A, and the particle velocity amplitude, u_A by the following equation:

$$I = \tfrac{1}{2}\rho_0 c(2\pi f)^2 x_A^2 = \tfrac{1}{2}\rho_0 c u_A^2 \qquad 3.4$$

The points selected for measurement of intensity are commonly the focus of the beam or within 1 or 2 cm of the transducer face. For a continuous wave beam, the intensity may be measured at the spatial peak to give I_{sp}, or it may be averaged across the beam to give the spatial average, I_{sa}. With pulse ultrasound, temporal averaging as well as spatial averaging may be carried out. For example, the spatial peak intensity may be averaged over the duration of a sequence of several pulses to give I_{spta} (intensity—spatial peak, temporal average) or over the pulse length for I_{sppa} (intensity—spatial peak, pulse average). Other quantities of interest are I_{sptp} (intensity—spatial peak, temporal peak) and I_{sata} (intensity—spatial average, temporal average). The relevance of these quantities to biological effects is discussed in Chapter 14. From the machine user's point of view, it is probably sufficient to ascertain I_{sp} or I_{spta} and, if possible, I_{sptp}. The latter is in fact a difficult quantity to measure accurately.

Knowledge of the above quantities allows the outputs of different machines to be compared and their safe usage considered, as is discussed in Chapter 14, where it is also seen that there is interest in pressure amplitude when safety aspects are being considered.

The power of an ultrasonic beam is the rate of flow of energy through the cross-sectional area of the beam. This is a popular quantity to quote since it is relatively easy to measure and is not dependent on identifying the spatial peak.

The transmitted intensity or power of basic Doppler units is often fixed and sensitivity is altered by manipulating the receiver gain. From the point of view of safety, it is actually better to fix the gain at the maximum level and manipulate the power.

Power and intensity controls on ultrasonic units are often labelled using the decibel notation. For example, taking the maximum intensity position of the control as I_0, other values, I, are calibrated relative to it as:

Output intensity in decibels $= 10 \log_{10}(I/I_0)$ 3.5

This method of labelling power and intensity controls is used since it is fairly difficult to measure absolute values in mW or mW cm^{-2} and a separate calibration is required for each transducer. More recently the outputs of machines have been related to possible biological effects, in particular heating and cavitation, and thermal and mechanical indices are displayed (Section 14.7.4).

Ultrasonic intensity is normally measured with a hydrophone, which takes the form of a small probe with a piezoelectric element in it (Section 5.5).

Ultrasonic power is normally measured with a radiation pressure balance (Section 14.2.1). Associated with the flow of energy in a beam there is a flow of momentum. When this flow of momentum is intercepted by an object, a force is experienced by the object which is directly related to the power of the beam, if the whole beam is intercepted (Livett et al 1981). For a beam of power W incident normally on a totally reflecting flat surface:

Radiation force $= 2W/c$ 3.6

The force experienced is 0.135 mg mW^{-1}. Power levels down to 0.1 mW can be measured with such devices. If the object totally absorbs the beam, the force is half that of the reflection case. To measure these small forces, the object is often placed under water on the pan of a sensitive chemical balance (Hill 1970, Kossoff 1965, Rooney 1973). With a small object that only intercepts part of a beam it is possible to measure intensity at a point, but this approach has largely been replaced by the hydrophone one. Portable radiation balances have been designed for easy transport to the ultrasound machines (Farmery and Whittingham 1978), and are commercially available (Fig. 14.1).

In Doppler applications the main interest in intensity and power is in considerations of safety. At present, several pulsed Doppler devices are capable of generating unacceptably high levels of intensity and care must be taken in their application (Chapter 14).

3.2.3 Diffraction and Interference

Diffraction is the term used to describe the spreading out of a wave from its source as it passes through a medium. The way in which a wave spreads out is highly dependent on the shape and size of the source relative to the wavelength of the sound. For instance, the field pattern of a disc-shaped crystal, i.e. its diffraction pattern, may be reasonably cylindrical for a short distance, after which it starts to diverge. The diffraction pattern from a small element of an array transducer is divergent from close to the transducer. Within a diffraction pattern there may be fluctuations in intensity, particularly close to the transducer. Diffraction also occurs beyond an obstacle such as a slit aperture or an array of slits which partially blocks a wavefront. When a beam interacts with less geometrically regular objects, the term 'scattering' is usually applied to describe spreading out of the beam. The narrowness of a beam or the sharpness of a focus is limited by diffraction, indeed they are said to be 'diffraction limited'.

Interference occurs when two or more waves overlap in the same medium. The resultant pressure at any point is obtained by adding algebraically the pressures from each wave at the point. This principle of superposition is widely used to predict ultrasound field shapes using mathematical models. For example, the shape of the transducer field is derived by considering the crystal face to be subdivided into many small parts. The diffraction pattern for each small part is calculated and the effect of them overlapping and interfering gives the resultant ultrasound field shape. The shapes of ultrasonic fields and beams, and the distinction between fields and beams as used in Doppler techniques, are discussed in Section 5.3.

Interference and diffraction are rarely considered in detail since, as techniques stand at present, ultrasound beams and fields are rarely well defined.

3.2.4 Standing Waves and Resonance

Standing waves are formed when two waves of the same frequency travelling in opposite directions interfere. They are most readily seen if the waves have similar amplitude. Examination of the pressure variations show alternate stationary regions of high and low pressure amplitude, i.e. a pattern of nodes and antinodes. When standing waves are present in tissue, the flow of red cells in capillaries can be arrested, i.e. blood stasis occurs (Dyson et al 1974). The cells gather in bands separated by half a wavelength (Fig. 3.1). Further

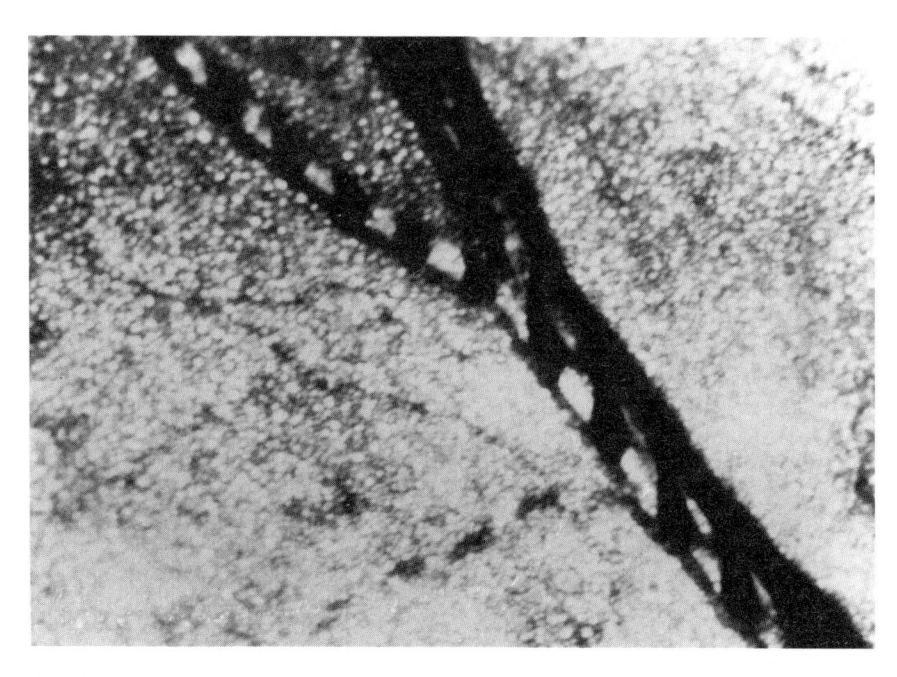

Figure 3.1 Blood stasis due to standing waves in a vein of a chick embryo. Standing waves were generated by 3 MHz continuous wave ultrasound (courtesy of M. Dyson)

analysis of this phenomenon has been carried out by ter Haar and Wyard (1978). Blood stasis is unlikely to occur with diagnostic techniques because standing waves are unlikely to be set up.

Standing wave patterns can result from a sound wave being reflected back and forth between two flat parallel surfaces. If the separation of the surfaces is a whole number of half wavelengths a marked increase in the pressure amplitudes is observed. This is due to the waves adding constructively to give a resonance. A marked resonance, the fundamental, occurs when the separation of the surfaces is half a wavelength. Ceramic piezoelectric transducer elements are made equal in thickness to one-half of the wavelength corresponding to their operating frequency to give efficient generation and detection of ultrasound. Continuous wave Doppler transducers have little or no damping and resonate at their operating frequency. Pulsed wave Doppler transducers have some damping, which gives them a wider frequency response. In general, when the size of a structure has a special numerical relationship to the wavelength, a resonant vibration will occur. One interesting possibility is that small bubbles in blood could be driven by ultrasound to resonate at a frequency which is related to blood pressure (Fairbank and Scully 1977). Microbubble contrast agents of a size which allows passage through the lungs have resonant frequencies in the low MHz range similar to that used in diagnostic ultrasound (Section 13.2).

Standing waves and interference are of interest to the designer of transducers. They are two of the phenomena that contribute to the overall performance of a transducer and provide some insight into its complexity of operation. Given this complexity and scope for variation it is always worth thoroughly assessing the performance of each transducer in clinical use.

3.2.5 Non-linear Propagation

The discussion so far has assumed that the waveforms, pulsed or continuous, retain their shape as they pass through material. This is true for very low amplitude waves in a non-attenuating medium such as water. For waves of larger amplitude, i.e. finite amplitude, the speed of sound is higher in the portions of tissue influenced at any instant by the positive half cycles of pressure than it is in the portions experiencing the negative half cycles of pressure. The speed is changed because the stiffness of the medium changes with pressure. The positive half cycles of the waveform therefore catch up on the negative half cycles, resulting in distortion of the waveform (Fig. 3.2). At increasing depth in the propagating medium, the distortion increases until the waveform becomes a sawtooth shape (Fig. 3.3). When sharp discontinuities appear in the waveform, it is called a shock wave. Figure 3.4 illustrates distorted pulses from a commercial Doppler imaging unit and a transcranial Doppler unit (both measured in water).

For a wave of finite amplitude travelling in an isotropic medium, the relationship between pres-

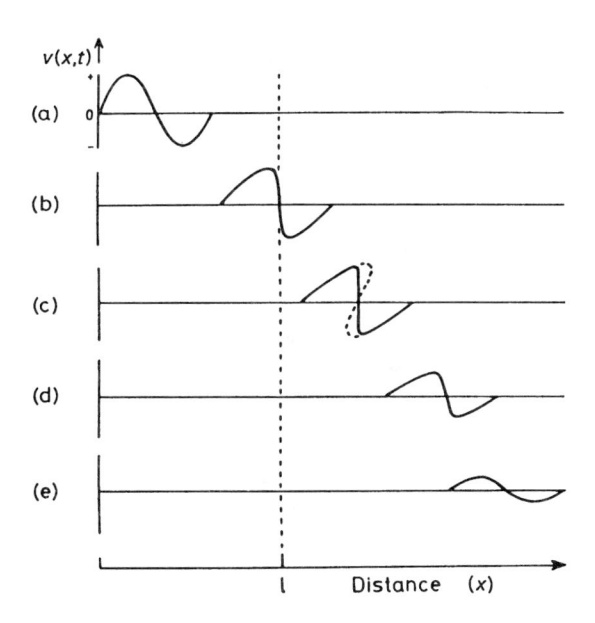

Figure 3.2 The evolution of a sinusoidal wave launched at distance $x = 0$ and time $t = 0$ in a non-linear, weakly and dispersively attenuating medium. Time, and therefore distance, increases from (a) to (e). Each section shows a single cycle representation of the velocity amplitude waveform $v(t)$ at some relative distance from the source. (a) Sine wave is launched; (b) after accumulative distortion a shock front forms; (c) shock front increases in strength; (d) high frequency components attenuated (stabilisation region); (e) far from source wave returns to a (attenuated) sine wave (reproduced by permission from Bamber 1986, © CR Hill/Ellis Horwood Ltd)

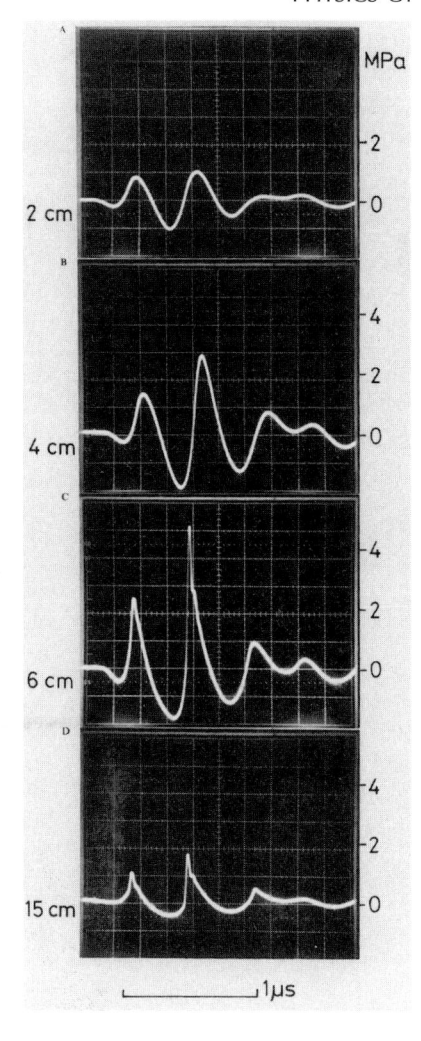

Figure 3.3 Distortion of a pulse of 2 MHz ultrasound due to non-linear propagation in water. Pulse shapes are recorded at 2, 4, 6, and 15 cm ranges (reproduced by permission from Duck and Starritt 1984, © 1984, British Institute of Radiology)

sure and density can be expressed by the Taylor expansion:

$$p = p_0 + A\left(\frac{\rho - \rho_0}{\rho_0}\right) + \frac{B}{2}\left(\frac{\rho - \rho_0}{\rho_0}\right)^2 + \cdots \quad 3.7$$

where

$$A = \rho_0\left[\left(\frac{\partial p}{\partial \rho}\right)_s\right]_{\rho=\rho_0} = \rho_0 c_0^2 \quad 3.8$$

and

$$B = \rho_0^2\left[\left(\frac{\partial^2 p}{\partial \rho^2}\right)_s\right]_{\rho=\rho_0} \quad 3.9$$

p and ρ are instantaneous pressure and density, p_0 and ρ_0 are ambient pressure and density, and c_0 is the velocity of sound for low amplitude waves. The suffix s denotes constant entropy.

The ratio of B/A is an important parameter in non-linear acoustics, and can be shown to be given by:

$$\frac{B}{A} = 2\rho_0 c_0\left[\left(\frac{\partial c}{\partial p}\right)_T\right]_{\rho=\rho_0}$$

$$+ \left(\frac{2c_0 T\xi}{C_p}\right)\left[\left(\frac{\partial c}{\partial T}\right)_p\right]_{\rho=\rho_0} \quad 3.10$$

where the suffixes T and p denote constant temperature and pressure, ξ is the volume coefficient of thermal expansion and C_p is the specific heat at constant pressure (Bjørnø 1976, Beyer 1960). The ratio B/A is known as the parameter of non-linearity of the medium and can be obtained by measurement of the second harmonic generated in the wave under controlled conditions. The larger the value of B/A for a medium, the earlier the shock discontinuity appears in the waveform as it propagates from the transducer.

The first term in eqn. 3.10 (which is the dominant part of the expression) can be interpreted as expressing the relative increase in the phase velocity due to changes in pressure, and the second as the increase due to changes in temperature (Beyer 1960).

The speed of sound, c, and the particle velocity, u, at any point, y, are related by:

$$c(y) = c_0 + (1 + B/2A)u(y) \quad 3.11$$

B/A is a property of the medium. For liquids and soft tissues it takes values in the range 2–13 (Bamber 1986). It does not appear to be simply related to physical properties, being influenced by

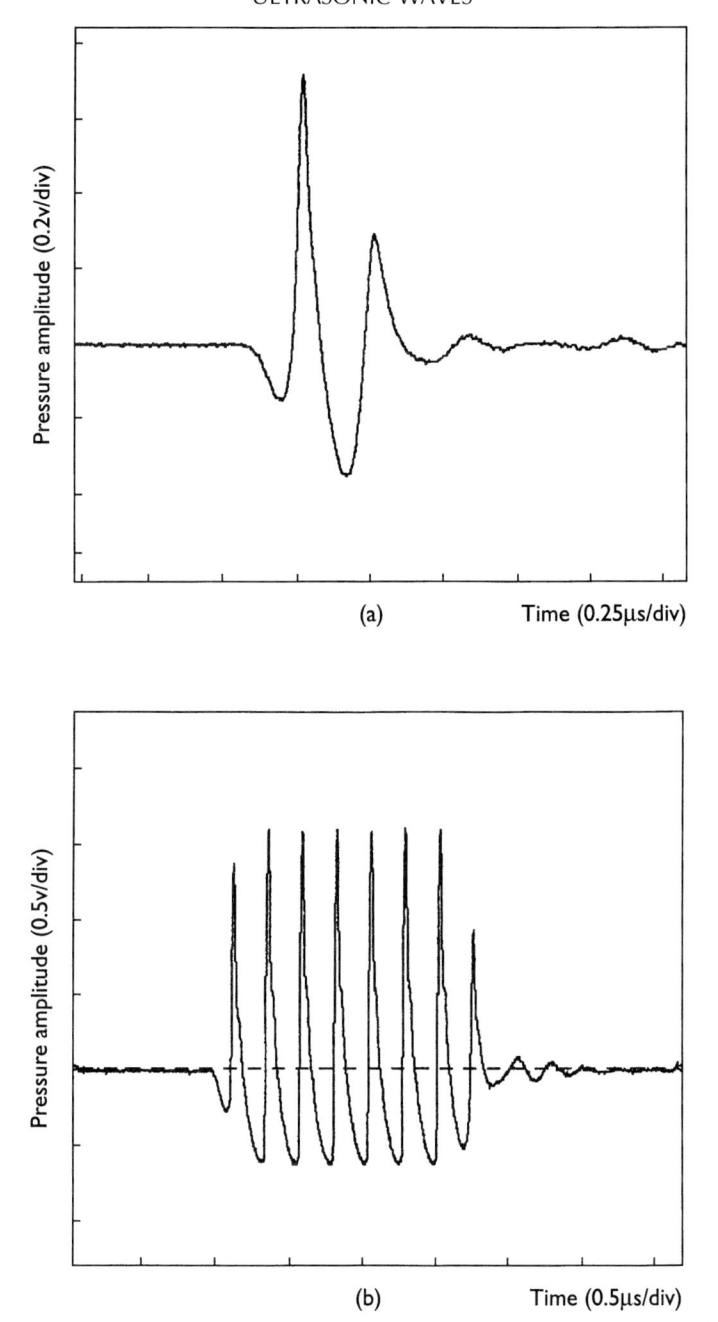

Figure 3.4 (a) Pulse waveform from a Doppler imaging unit showing non-linear distortion. (b) Pulse waveform from a transcranial Doppler unit showing non-linear distortion (courtesy of S. Pye)

factors such as concentration of macromolecules, solute–solvent interactions and tissue structure. Temperature and pressure do not strongly influence values of B/A. Table 3.2 shows some typical values for B/A.

The significance of non-linear propagation in medical ultrasonics has been appreciated for some time (Cartensen et al 1980, Muir and Cartensen 1980, Law et al 1985, Bacon 1984, Duck and Starritt 1984). From the theory it is evident that non-linear propagation will be more readily observed for high-amplitude waves and also for high-frequency waves, since a smaller distance moved by the positive half cycle relative to the negative half cycle causes significant distortion at high frequency. Shock wave formation occurs more readily in water than in tissue, since the high frequency components introduced to the waveform in water are attenuated less than in tissue (see Section 3.3.3). It is therefore of particular relevance in techniques that involve waterbaths or any liquid path along which the beam passes. The presence of shock wave formation at diagnostic intensity levels in absorbing tissue (liver and calf muscle) has been demonstrated (Duck et al 1986). In this work it was shown that for a 2.5 MHz pulse of initial amplitude 0.58 MPa, the second harmonic increased by about 10 dB in the first 5 cm of transmission. Modelling of second harmonic frequency component production in a 2 MHz beam has shown that the 4 MHz second harmonic beam has a narrower width and lower sidelobes than the 2 MHz fundamental beam, and that the harmonic signal levels are of sufficient magnitude to permit use in tissue imaging (Christopher 1997). The technique has also been used in underwater acoustic imaging (Muir 1980). Non-linear distortion is influential in the conditions pertaining in diagnostic ultrasound and has been exploited to improve the clarity of images. The introduction of harmonic frequencies as a result of the distortion of the propagating pulse occurs most strongly along the central axis of the beam, where the pressure amplitude is greatest. The high harmonic frequencies are filtered out to provide the echo signals for image construction. Less harmonic frequency production occurs in the weaker beam sidelobes or in spurious multiple reflection echoes. In the case of absorbing media, the high-frequency components generated by this distortion are eventually absorbed and a smooth waveform of reduced amplitude results. Harmonic tissue imaging has been exploited commercially, e.g. in native tissue imaging by the Acuson Corporation. To date, second harmonic tissue imaging has only been used to improve B-mode images of tissue structures but not Doppler blood flow images, presumably because the signal from blood is too low. It could, however, be used for Doppler tissue imaging where the echo signals are 20–40 dB stronger than those from blood.

The following points can be made in relation to non-linear propagation:

1. From the limited data available, soft tissue, blood and other liquids appear to have similar properties.
2. Due to non-linear propagation, the frequency components of the continuous or pulsed carrier wave vary with depth. In the light of the extensive work with Doppler methods in test phantoms, errors due to non-linear distortion are probably relatively small (Li et al 1993).
3. Non-linear distortion is more severe with PW devices than with CW devices, since larger amplitude waves are employed in the former case.
4. Increased pressure amplitudes in the form of shock waves need to be considered from a hazard standpoint (Chapter 14).

3.3 ULTRASONIC PHENOMENA

3.3.1 Scattering

When an ultrasonic wave travelling through a medium strikes a discontinuity of dimensions

Table 3.2 Values of the non-linearity parameter. Data from Bamber (1986)

Material	B/A
Water	5.0
Blood	6.3
Liver	7.8
Fat	11.1

similar to or less than a wavelength, some of the energy of the wave is scattered in many directions. Scattering is the process of central importance in diagnostic ultrasonics, since it provides most of the signals for both echo imaging and Doppler techniques. The discontinuities may be changes in density or compressibility or both. Detailed information is not available on the nature of the discontinuities in soft tissue but they are usually considered to be a random distribution of closely packed scattering centres. These centres are modelled as discrete centres or as fluctuations in a continuum (Chivers 1977). Computer models of a wave interacting with such structures can predict a scattered wave with properties like those observed in practice. For example, statistical fluctuations in simulated echo signals are similar to those that give rise to the speckle pattern in images (Burkhardt 1978, Abbot and Thurstone 1979, Morrison et al 1980). The red cells in blood, singly or in groups, act as scattering centres which produce the signals used in Doppler techniques.

The ratio of the total power, S, scattered by a target (or by unit volume of material) to the incident intensity, I, is called the total scattering cross-section, σ_t, of the target or the volume of material:

$$\sigma_t = S/I \qquad 3.12$$

This ratio is used to compare the scattering powers of different structures.

The differential scattering cross-section, $\sigma(\phi)$, is the ratio of the power scattered into unit solid angle located in a particular direction in space, $S(\phi)$, to the incident intensity, I:

$$\sigma(\phi) = S(\phi)/I. \qquad 3.13$$

Of particular interest is the differential cross-section at 180° to the direction of the incident ultrasound. This is known as the backscatter cross-section and determines the size of the signal back to the transducer in imaging and Doppler techniques.

Theoretical treatment of scattering in a material is possible if sufficient simplifications are made. Dickinson (1986) considers scattering of a plane wave at centres in an inhomogeneous medium. The medium is considered to exhibit no absorption, small fluctuations in density and compressibility, and no multiple reflection. This results in an expression for the intensity of the scattered wave at a distance, s, which may be written in the form:

$$I(s, \phi) \propto \frac{k^4}{s^2} [g_1(k, \beta) + g_2(k, \rho_0)\cos \phi]^2 \qquad 3.14$$

where $k = 2\pi/\lambda$ and ϕ is the angle between the incident and scattered waves. $g_1(k, \beta)$ and $g_2(k, \rho_0)$ are functions that depend on ultrasound frequency, and on the distribution of compressibility and density (respectively) in the target volume.

The following conclusions can be drawn from this idealised case:

1. The scattered wave is spherical with modifications due to the terms in brackets, which depend on the distribution of compressibility and density.
2. If the fluctuations in compressibility and density are small and $g_1(k, \beta)$ and $g_2(k, \rho_0)$ are constant over a significant frequency range, the scattered intensity has a fourth power dependence on frequency due to k^4, i.e. there is Rayleigh scattering.
3. The scattered intensity is dependent on the angle of observation, ϕ.

The above analysis is reassuring in that it confirms that our understanding of scattering from limited experimental measurements in soft tissue and blood is reasonably correct. However, attempts to apply theory to scattering emphasise the complexity of the situation. Usually the model has to be simplified to such an extent that the results are little more than confirmation of intuitive reasoning. The development of techniques involving scattering proceed more by experimental investigation than by theoretical modelling. On the other hand, realistic computer modelling is becoming more viable.

Measurements of scattering cross-sections in the literature are fairly scarce and quite variable. This is due in part to the difficulties in making such measurements. For example, there are problems in separating the forward scattered ultrasound from the original beam and the physical size of the beam or side lobes restricts the angular range of measurement. Estimates of the contribution of scattering to the total attenuation in tissue range up to 50%. The contribution is obviously dependent on the tissue and the frequency of the sound, but most investigators report the scattering contribution to be less than 10%. Scattering increases with the amount of fat and collagen but decreases with the water content of tissue. One exception to the paucity of scattering data is for blood, which has been investigated quite thoroughly.

Shung et al (1976) give values for the scattering cross-section at 5, 8.5 and 15 MHz (Table 3.3) and show that the values are in agreement with theoretical ones calculated with a formula by Morse and Ingard (1968). The angular distribution of scattering in blood has been shown to be asymmetric, the back-scattered cross-section being about 6 dB greater than that at forward angles (Shung et al 1977). Scattering from blood is considered in more detail in Section 7.2.

From the point of view of Doppler techniques, the study of scattering is important since it improves our understanding of CW and PW instruments. In day-to-day usage of instruments, however, the operator need not be concerned with scattering except to note that the signal from blood is very much weaker than that from soft tissue. The sample volume for soft tissue is therefore much larger than that for blood (Section 5.4). Wall thump filters are normally included in Doppler devices to reduce low frequency signals from moving tissue.

3.3.2 Reflection and Refraction

Reflection is a special case of scattering which occurs at smooth surfaces on which the irregularities are very much smaller than a wavelength. The physical property of tissue which is of importance in the reflection process is the acoustic impedance, z, of the material. This is the ratio of the wave pressure over the particle velocity, i.e. p_A/u_A. For a plane wave in weakly absorbing medium the acoustic impedance may be written:

$$z = \rho_0 c \qquad 3.15$$

In the S.I. system of units, acoustic impedance has the units of kg m^{-2} s^{-1}, or Rayl. For soft tissue, acoustic impedances are in the range $1.3–1.9 \times 10^6$ Rayl. The acoustic impedance of blood is 1.6×10^6 Rayl. The higher the density or stiffness of a material, the higher is its acoustic impedance. Table 3.1 lists some commonly encountered values of acoustic impedance.

Consideration of reflection at perpendicular incidence to a flat boundary between two media of impedances $\rho_1 c_1$ and $\rho_2 c_2$ provides some appreciation of the magnitude of echo amplitudes (Fig. 3.5a). The values in Table 3.4 were calculated using:

$$a_r = a_i \times \left(\frac{\rho_2 c_2 - \rho_1 c_1}{\rho_2 c_2 + \rho_1 c_1} \right) \qquad 3.16$$

Table 3.3 Approximate scattering cross-section values for blood with a haematocrit of 40% over a range of frequency. Data extracted from Shung et al (1976)

Frequency (MHz)	Scattering cross-section (cm^2)
5.0	2×10^{-14}
8.5	1.4×10^{-13}
15	1.3×10^{-12}

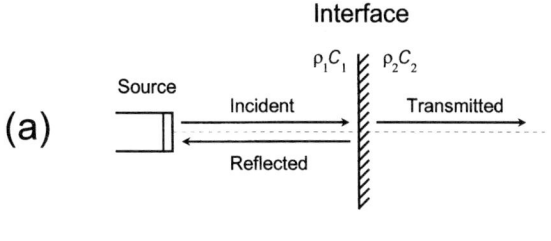

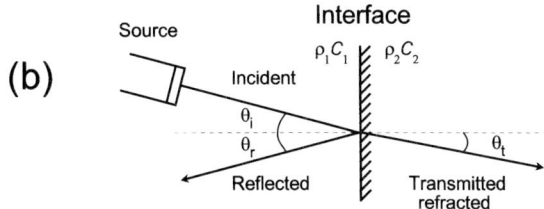

Figure 3.5 (a) Reflection of ultrasound at a plane boundary (perpendicular incidence). (b) Reflection and refraction at a plane boundary (non-perpendicular incidence)

ULTRASONIC PHENOMENA

Table 3.4 The ratio of the reflected wave to the incident wave amplitude and the percentage energy reflected for perpendicular incidence (energy is proportional to amplitude squared). Data from McDicken (1991)

Reflecting interface	Ratio of reflected to incident wave amplitude	Percentage energy reflected
Fat/muscle	0.10	1.08
Fat/kidney	0.08	0.64
Muscle/blood	0.03	0.07
Bone/fat	0.70	48.91
Bone/muscle	0.64	41.23
Lens/aqueous humour	0.10	1.04
Soft tissue/water	0.05	0.23
Soft tissue/air	0.9995	99.9
Soft tissue/PZT crystal	0.89	80
Soft tissue/polyvinylidene difluoride	0.47	0.22
Soft tissue/castor oil	0.07	0.43

where a_r is the reflected amplitude and a_i the incident amplitude. Substantial errors may exist in these values since there is a spread in reported values for the velocity of sound in tissues.

When a wavefront strikes a smooth surface at an oblique angle of incidence, it is reflected at an equal and opposite angle (Fig. 3.5b). In this case the amplitude of the echo is given by a more complex formula (Kinsler et al 1982):

$$a_r = a_i \times \left(\frac{\rho_2 c_2 \cos \theta_i - \rho_1 c_1 \cos \theta_t}{\rho_2 c_2 \cos \theta_i + \rho_1 c_1 \cos \theta_t} \right) \quad 3.17$$

where θ_i is the angle of incidence and θ_t is the angle of transmission.

Refraction is the deviation of a beam when it crosses a boundary between two media in which the speeds of sound are different. The resultant angle of propagation is given by the familiar Snell's Law:

$$\frac{\sin \theta_i}{\sin \theta_t} = \frac{c_1}{c_2} = \mu \quad 3.18$$

where μ is the refractive index, and θ_t is the angle of refraction.

Table 3.5 shows the deviation experienced by a beam for increasing angles of incidence at a muscle/blood interface. Although deviations of a few degrees can occur at soft tissue boundaries, experience in ultrasonic imaging indicates that blood vessels can often be clearly depicted, so refraction is not a severe problem.

From the point of view of Doppler techniques, reflection is primarily of interest in the visualisation of blood vessel walls and cardiac structures when duplex and flow imaging systems are being used. Since reflected pulses have a finite length, the accuracy with which dimensions such as vessel diameter can be measured is limited. This in turn significantly affects the accuracy of measurement of blood flow by several techniques. Refraction distorts an ultrasonic beam as it crosses the muscle/blood interface of a vessel. Greater deviations are to be expected when a wall is more rigid due to the presence of plaque. Distortion of ultrasonic beams at curved surfaces has been modelled and shown to be significant (LaFollette and Ziskin 1986). However, the case of refraction at a blood vessel has not been treated in detail. Distortion of a beam at a vessel wall can be expected to alter the intensity pattern in the vessel and will add to the error in measurement of mean velocity (Section 7.3.2).

3.3.3 Absorption and Attenuation

Absorption is the process of conversion of wave motion energy into heat. For ultrasonic

Table 3.5 Deviation experienced by an ultrasound beam passing through a muscle/blood interface at different angles of incidence

Angle of incidence (degrees)	Deviation (degrees)
0	0
10	0.07
20	0.13
30	0.21
40	0.31
50	0.43
60	0.62
70	0.97
80	1.87
90	6.44

propagation at low megahertz frequencies through biological materials, the largest contribution to absorption is that due to relaxation mechanisms (Wells 1975). In these mechanisms the stress imposed by the wave on the medium is relaxed by the flow of energy to various energy states of the tissue. For example, thermal relaxation is the flow of wave energy to internal molecular energy and structural relaxation is flow to changed structural states of different energy. When the stress is released, some of the energy flows back to the wave energy but it is out of phase with the original wave and the resultant amplitude is reduced. Each relaxation mechanism is most effective at one particular frequency, the relaxation frequency. As might be expected in tissue, many such relaxation processes are present and the effect of individual ones cannot be observed (Bamber 1986). Isolated processes can only be studied in simple liquids. Absorption increases rapidly with frequency in tissue over the frequency range used in diagnostic techniques.

In practice, absorption on its own is rarely of interest, since other processes are simultaneously contributing to the total attenuation of the wave. These processes are scattering, reflection, refraction, non-linear propagation and beam divergence. Since many of these factors are strongly frequency dependent, it is not surprising that attenuation rises rapidly over the frequency range 1–20 MHz. Table 3.6 shows the attenuation coefficients for a variety of tissues. It is common practice to quote attenuation coefficients in dB cm^{-1} when imaging or Doppler techniques are being considered.

Intensity (or power) attenuation coefficient

$$= \frac{10}{x} \log_{10}\left(\frac{I}{I_0}\right) \quad 3.19$$

Amplitude attenuation coefficient

$$= \frac{20}{x} \log_{10}\left(\frac{p}{p_0}\right) \quad 3.20$$

where x is the thickness of the tissue layer. These coefficients have the same numerical value.

From Table 3.6 it can be seen that the attenuation in blood is low. The decrease in the scattered signal from blood across a 1 cm vessel is 0.8 dB at 2 MHz and 2.0 dB at 5 MHz. This source of error is usually neglected in calculations of mean velocity. It can be allowed for by computation or, in the case of PW Doppler, by attenuation compensation (TGC) in the receiving amplifier. No detailed data exist for attenuation in plaque in its various forms. However, solid plaque attenuates ultrasound strongly owing to high reflection and refraction and therefore casts shadows in the region behind it.

The frequency dependence of attenuation results in the high frequency components of an ultrasonic pulse being preferentially reduced in amplitude relative to the low frequency components. The resulting distortion of the ultrasonic pulse is rarely taken into account in present-day practice. It has been evaluated and shown to reduce the central frequency of the pulse bandwidth by amounts which can be significant (Holland et al 1984) (Section 8.3.5). With high amplitude pulses, non-linear propagation has the opposite effect on the frequency bandwidth by generating harmonics (Section 7.4.2).

Absorption and attenuation are very important in Doppler techniques, and their influence on any examination must be considered. A frequency of ultrasound is selected which will provide the required penetration. Typically, Doppler devices operating at around 7 MHz are used for superficial vessels, while 2 MHz is employed for deep ones. The high attenuation of bone and gas severely limits access to the adult brain and thorax. When Doppler signals are to be detected

Table 3.6 Attenuation coefficients for common tissues in dB cm^{-1} at 1 MHz. The value at a higher frequency may be obtained approximately by multiplying by the frequency in MHz (note, however, that for water the value should be multiplied by the square of frequency)

Tissue	Attenuation coefficient (dB cm^{-1})
Blood	0.2
Muscle	1.5
Liver	0.7
Brain (adult)	0.8
Brain (infant)	0.3
Bone	10.0
Fat	0.6
Water	0.002
Soft tissue (average)	0.7
Castor oil	1.0

at a series of ranges, as in colour flow imaging, attenuation compensation may be supplied by electronic circuitry which applies increasing gain as signals are received from increasing depth (TGC).

3.4 ULTRASONIC TISSUE CHARACTERISATION PARAMETERS

Parameters related to ultrasonic phenomena are often used in attempts to characterise tissue for diagnostic and therapeutic purposes. Examples of parameters extracted from echo signals are speed of sound, attenuation coefficient, scattering coefficient and the frequency dependence of these quantities (Hill 1986). Attempts to characterise tissue are often thwarted by the degradation of the ultrasound beam in the intervening non-uniform tissues, such as fat and muscle, between the transducer and the site of interest. However, there is continuing interest in the characterisation of blood vessel walls, atheromatous plaque and thrombi due to the importance of vascular disease, good access to superficial vessels and the fact that, with intravascular scanners, these tissues can be interrogated via a short beam path through blood (Wilson et al 1994, Bridal et al 1997, Spencer et al 1997). The method used to obtain values of parameters must be carefully examined, as a significant amount of processing of the echo signals is usually involved.

3.5 THE DOPPLER EFFECT

The Doppler effect is the change in the observed frequency of a wave due to motion. This motion may be of the source or the observer (Fig. 3.6). When the observer moves towards the source, the increased frequency, f_r, due to passing more wave cycles per second, is given by:

$$f_r = f_t \frac{c + v}{c} \qquad 3.21$$

where f_t is the transmitted frequency and v is the velocity of the observer.

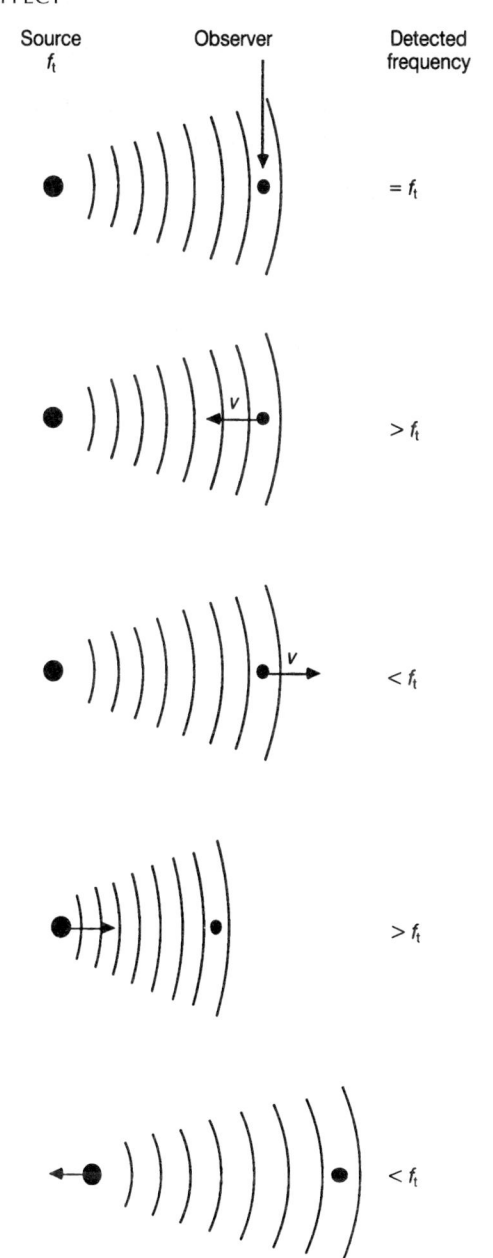

Figure 3.6 The Doppler effect due to motion of the source or observer. The detected frequency is increased or decreased depending on the direction of motion

If the velocity of the observer is at an angle θ to the direction of the wave propagation, v is replaced by the component of the velocity v in the wave direction, $v \cos \theta$:

$$f_r = f_t \frac{c + v \cos \theta}{c} \qquad 3.22$$

If the observer is at rest and the source moves with velocity v in the direction of wave travel, the wavelengths are compressed. The resulting observed frequency is:

$$f_r = f_t \frac{c}{c - v}. \qquad 3.23$$

Taking angle into account:

$$f_r = f_t \frac{c}{c - v \cos \theta} \qquad 3.24$$

Both of these motions which give rise to changes in the observed frequency are in fact slightly different effects, since in the first the wave is not altered and in the second it is compressed.

In medical ultrasonic applications, an ultrasonic beam is backscattered from moving blood cells or tissue. Both of the above effects combine to give the resultant Doppler shift in frequency. The observed frequency is then given by:

$$f_r = f_t \frac{c + v \cos \theta}{c} \cdot \frac{c}{c - v \cos \theta} \qquad 3.25$$

$$= f_t \frac{c + v \cos \theta}{c - v \cos \theta} \qquad 3.26$$

The Doppler shift, f_d $(= f_r - f_t)$, is therefore given by:

$$f_d = f_t \frac{c + v \cos \theta}{c - v \cos \theta} - f_t \qquad 3.27$$

since $c \gg v$:

$$f_d = \frac{2 f_t v \cos \theta}{c} \qquad 3.28$$

Table 3.7 Doppler shifts for the range of velocities and ultrasonic frequencies encountered in practice. An angle of insonation of 45° has been assumed

Speed (cm s^{-1})	Ultrasonic frequency (MHz)			
	2	3	5	10
1	18 Hz	27 Hz	46 Hz	92 Hz
10	183 Hz	275 Hz	459 Hz	918 Hz
50	915 Hz	1.37 kHz	2.29 kHz	4.59 kHz
100	1.83 kHz	2.75 kHz	4.59 kHz	9.18 kHz
500	9.15 kHz	13.75 kHz	22.95 kHz	45.90 kHz

It is instructive to consider some numerical evaluations of this formula for values encountered in medical ultrasonics, as presented in Table 3.7. The Doppler shift is seen to be around one part in 1000. The Doppler shift frequencies are also in the audible range.

3.6 SUMMARY

In this chapter some basic physics encountered in the use of ultrasonic techniques has been briefly described. Further detail can be obtained in other texts. Points of particular relevance to Doppler techniques have been emphasised. The properties of ultrasonic waves have been discussed with particular attention being paid to intensity and power. The ultrasonic phenomena which occur in tissue, such as scattering and attenuation, have been considered in some detail since they strongly influence the magnitude of signals received from tissue. Finally, the Doppler effect has been explained.

3.7 NOTATION

a_i	Incident amplitude of ultrasonic wave
a_r	Reflected amplitude of ultrasonic wave
B/A	Non-linearity parameter
c	Velocity of sound in tissue
c_0	Velocity of low amplitude sound waves
C_p	Specific heat at constant pressure
f	Frequency
f_r	Received or observed frequency
f_t	Transmitted or source frequency

$g_1(k, \beta)$	Function appearing in the scattering equation
$g_2(k, \rho)$	Function appearing in the scattering equation
I	Intensity
I_0	Reference intensity
k	Wave number ($=2\pi/\lambda$)
K	Bulk modulus
p	Instantaneous pressure
p_a	Applied pressure
p_A	Acoustic pressure amplitude
p_0	Ambient or reference pressure
s	Entropy
S	Total scattered power
$S(\phi)$	Power scattered into a unit solid angle
T	Temperature
u	Particle velocity
u_A	Particle velocity amplitude
v	Velocity of a source, observer or target
W	Power
x	Thickness of a tissue layer
x_A	Displacement amplitude
z	Acoustic impedance
β	Adiabatic compressibility
$\Delta V/V$	Fractional change in volume
θ	Angle between a velocity and direction of wave propagation
θ_i	Angle of incidence at surface
θ_r	Angle of reflection
θ_t	Angle of transmission/refraction
λ	Wavelength
μ	Refractive index
ξ	Volume coefficient of thermal expansion
ρ	Instantaneous density
ρ_0	Ambient density
σ_t	Total scattering cross-section
$\sigma(\phi)$	Differential scattering cross-section
ϕ	Scattering angle

3.8 REFERENCES

Abbott JG, Thurstone FL (1979) Acoustic speckle: theory and experimental analysis. Ultrason Imaging 1:303–324

Bacon DR (1984) Finite amplitude distortion of the pulsed fields used in diagnostic ultrasound. Ultrasound Med Biol 10:189–195

Bamber JC (1986) Attenuation and absorption. In: Hill CR (ed) Physical Principles of Medical Ultrasonics, pp. 118–199. Ellis Horwood, Chichester

Beyer RT (1960) Parameter of nonlinearity in fluids. J Acoust Soc Am 32:719–721

Bjørnø L (1976) Nonlinear acoustics. In: Stevens RWB, Leventhall HG (eds) Acoustic and Vibration Progress, vol 2, pp. 101–198. Chapman and Hall, London

Bridal SL, Fornès P, Bruneval P, Berger G (1997) Parametric (integrated backscatter and attenuation) images constructed using backscattered radio frequency signals (25–56 MHz) from human aortae in vitro. Ultrasound Med Biol 23:215–229

Burckhardt CB (1978) Speckle in ultrasound B-mode scans. IEEE Trans Sonics Ultrason SU-25:1–6

Cartensen EL, Law WK, McKay ND, Muir TG (1980) Demonstration of nonlinear acoustical effects at biomedical frequencies and intensities. Ultrasound Med Biol 6:359–368

Chivers RC (1977) The scattering of ultrasound by human tissues—some theoretical models. Ultrasound Med Biol 3:1–13

Christopher T (1997) Finite amplitude distortion-based inhomogeneous pulse echo ultrasonic imaging. IEEE Trans Ultrason Ferroelec Freq Contr 44:125–139

Dickinson RJ (1986) Reflection and scattering. In: Hill CR (ed) Physical Principles of Medical Ultrasonics, pp. 225–260. Ellis Horwood, Chichester

Duck FA (1990) Acoustic Properties of Tissue: A Comprehensive Reference Book. Academic Press, London

Duck FA, Starritt HC (1984) Acoustic shock generation by ultrasonic imaging equipment. Br J Radiol 57:231–240

Duck FA, Starritt HC, Hawkins AJ (1986) Observations of finite-amplitude distortion in tissue. Proc Inst Acoustics 18:71–77

Dyson M, Pond JB, Woodward B, Broadbent J (1974) The production of blood cell stasis and endothelial damage in blood vessels of chick embryos treated with ultrasound in a stationary wave. Ultrasound Med Biol 1:133–148

Fairbank WM, Scully MO (1977) A new noninvasive technique for cardiac pressure measurement: resonant scattering of ultrasound from bubbles. IEEE Trans Biomed Eng BME-24:107–110

Farmery MJ, Whittingham TA (1978) A portable radiation-force balance for use with diagnostic ultrasonic equipment. Ultrasound Med Biol 3:373–379

Goss SA, Johnston RL, Dunn F (1978) Comprehensive compilation of empirical ultrasonic properties of mammalian tissues. J Acoust Soc Am 64:423–457

Hill CR (1970) Calibration of ultrasonic beams for biomedical applications. Phys Med Biol 15:241–248

Hill CR (ed) (1986) Physical Principles of Medical Ultrasonics. Ellis Horwood, Chichester

Holland SK, Orphanoudakis SC, Jaffe CC (1984) Frequency-dependent attenuation effects in pulsed Doppler ultrasound: experimental results. IEEE Trans Biomed Eng BME-31:626–631

Kinsler LE, Frey AR, Coppens AB, Sanders JV (1982) Fundamentals of Acoustics, 3rd edn. Wiley, New York

Kossoff G (1965) Balance technique for the measurement of very low ultrasonic power outputs. J Acoust Soc Am 38:880–881

Kremkau FW (1998) Diagnostic Ultrasound: Principles and Instrumentation, 5th edn. Saunders, Philadelphia

LaFollette PS, Ziskin MC (1986) Geometric and intensity distortion in echography. Ultrasound Med Biol 12:953–963

Law WK, Frizzell LA, Dunn D (1985) Determination of the nonlinearity parameter B/A of biological media. Ultrasound Med Biol 11:307–318

Li S, McDicken WN, Hoskins PR (1993) Nonlinear propagation in Doppler ultrasound. Ultrasound Med Biol 19:359–364

Livett AJ, Emery EW, Leeman SJ (1981) Acoustic radiation pressure. J Sound Vibration 76:1–11

McDicken WN (1991) Diagnostic Ultrasonics: Principles and Use of Instruments, 3rd edn. Churchill Livingstone, London

Morrison D, McDicken WN, Wild R (1980) Statistical variations in echo amplitudes and their effect on interpretation of grey-scale images. Appl Radiol 9:109–112

Morse PM, Ingard KU (1968) Theoretical Acoustics. McGraw-Hill, New York

Muir TG (1980) Non-linear effects in acoustical imaging. In: Proceedings of the 9th International Symposium on Acoustic Imaging, pp. 93–109. Plenum Press, New York

Muir TG, Cartensen EL (1980) Prediction of nonlinear acoustic effects at biomedical frequencies and intensities. Ultrasound Med Biol 6:345–357

Rooney JA (1973) Determination of acoustic power outputs in the microwatt–milliwatt range. Ultrasound Med Biol 1:13–16

Shung KK, Sigelmann RA, Reid JM (1976) Scattering of ultrasound by blood. IEEE Trans Biomed Eng 23:460–467

Shung KK, Sigelmann RA, Reid JM (1977) Angular dependence of scattering of ultrasound from blood. IEEE Trans Biomed Eng BME-24:325–331

Spencer T, Ramo MP, Salter DM, Anderson T, Kearney PP, Sutherland GR, Fox KAA, McDicken WN (1997) Characterisation of atherosclerotic plaque by spectral analysis of intravascular ultrasound: an *in vitro* methodology. Ultrasound Med Biol 23:191–203

ter Haar G, Wyard SJ (1978) Blood cell banding in ultrasonic standing wave fields: a physical analysis. Ultrasound Med Biol 4:111–123

Wells PNT (1969) Physical Principles of Ultrasonic Diagnosis. Academic Press, London

Wells PNT (1975) Absorption and dispersion of ultrasound in biological tissue. Ultrasound Med Biol 1:369–376

Wells PNT (1977) Biomedical Ultrasonics. Academic Press, London

Wilson LS, Neale ML, Talhami HE, Appleberg M (1994) Preliminary results from attenuation-slope mapping of plaque using intravascular ultrasound. Ultrasound Med Biol 20:529–542

4

Doppler Systems: A General Overview

4.1 INTRODUCTION

Developments in Doppler technology have led to a vast increase in the number of non-invasive blood velocity investigations carried out in all areas of medicine. There are now very many types of Doppler ultrasound device available commercially for detecting, measuring and imaging blood flow and other movements within the body. There are also various types of apparatus which have been used in the past, or which are currently used only in research laboratories. The purpose of this chapter is to provide an overview of Doppler devices, the most important of which will be described and analysed in further detail in subsequent chapters.

The majority of Doppler systems may, for the sake of convenience of description, be classified into one of the following groups:

1. Velocity detecting systems.
2. Duplex systems.
3. Profile detecting systems.
4. Velocity imaging systems.

The descriptions given in this chapter are based on a systems approach, and the reader will be referred to the relevant sections of other chapters for more details of individual systems and of transducers, signal processing methods and other aspects of their operation.

4.2 VELOCITY DETECTING SYSTEMS

The simplest Doppler units are stand-alone systems that produce an output signal related to the velocity of the targets in a single sample volume, be it large or small. The transducers for such systems are not connected to any type of position-sensing gantry, and are often hand-held. Such systems may be very basic and produce a non-directional audio output, or may be quite sophisticated, producing directional signals sampled from predetermined depths in the tissue; they may also derive various types of information from the Doppler signal and output one or more Doppler envelope signals.

Because these systems are not combined with an imaging facility, the angle of insonation, θ, between the ultrasound beam and the direction of blood flow is unknown except at some favourable anatomical locations, and therefore their outputs cannot be calibrated in absolute terms. They can, however, be used for detecting the presence or absence of flow (and perhaps its direction), monitoring changes in flow and for recording the shape of the flow waveform, which may contain considerable information about the integrity of the cardiovascular system. They may also be used to make velocity measurements where the angle θ can be reliably estimated (particularly if θ is small, because the $\cos\theta$ term in eqn. 1.1 is then close to unity and changes very slowly with angle). Examples of suitable sites for such velocity measurements are the aortic arch insonated through the suprasternal notch, and the middle cerebral artery insonated through the temporal bone.

4.2.1 Continuous Wave Velocity Detecting Systems

Most Doppler systems may be broadly categorised as either continuous wave (CW) or pulsed wave (PW). CW systems are the simpler of the two, but have some advantages over PW systems, and are therefore still widely used and may even be found in some otherwise sophisticated instruments. CW units both transmit and received ultrasound continuously, and because of this it is usual for them to employ two separate transducer crystals, although these are usually housed in the same probe (see Fig. 5.4). Because transmission and reception are continuous, CW systems have no depth resolution, except in the sense that signals originating from close to the transducer experience less attenuation than those from distant targets (and hence are stronger). Also, since the transmitting and receiving crystals are separate, the distance between the two crystals and the angle at which they are set with respect to each other, allows control of the overlap of the transmitted field and reception zone and hence the macroscopic Doppler sample volume (see Section 5.4).

4.2.1.1 CW Systems—Basic Principles

A block circuit diagram of a simple non-directional CW Doppler unit is shown in Fig. 4.1. The master oscillator usually produces a frequency of between 2 and 10 MHz, but higher and lower frequencies have been used. The frequency chosen depends on the depth of interest, since ultrasound attenuation is highly dependent on frequency (Section 3.3.3); for superficial vessels, frequencies of around 8 MHz are used, and for deep vessels, frequencies as low as 2 MHz. The oscillations are amplified by the transmitting amplifier and the output used to drive the transmitting crystal. The crystal converts the electrical energy into acoustic energy (Section 5.2), which is then propagated as a longitudinal (compression) wave into the body. The ultrasound energy is reflected and scattered by both moving and stationary particles within the ultrasound beam, and a small portion finds its way back to the receiving crystal, which re-converts the acoustic energy into electrical energy. This small signal is then amplified by a radio frequency amplifier and mixed with a reference signal from the master oscillator. The process of mixing produces both the sum of the transmitted and received frequencies, and the required difference frequency or Doppler shift frequency. The low pass filter removes all signals outside the audio range, and this leaves only the Doppler difference frequency, which is high-pass filtered to remove high-amplitude low-frequency signals from stationary or nearly stationary targets, and then amplified and used to drive a loudspeaker or headphones, or sent for further processing. The process of recovering the Doppler audio signal from the Doppler-shifted ultrasound (radio frequency) signal is known as demodulation.

Simple non-directional Doppler units are of value for detecting the movement of blood, for example to check whether an artery or graft is patent, or for measuring blood pressure in conjunction with a sphygmomanometer, but otherwise their value is limited.

Simple units of the type described above are unable to distinguish between motion towards and away from the transducer, and most modern instruments have additional circuitry that allows this distinction to be made. There are a number of ways of realising this (Sections 6.2.5 and 6.4), but the most widespread technique is that of quadrature phase detection, whereby the received signal is mixed with two quadrature reference signals (signals separated by a 90° phase shift) from the master oscillator (see Fig. 6.8). This results in two audio frequency signals, both containing the

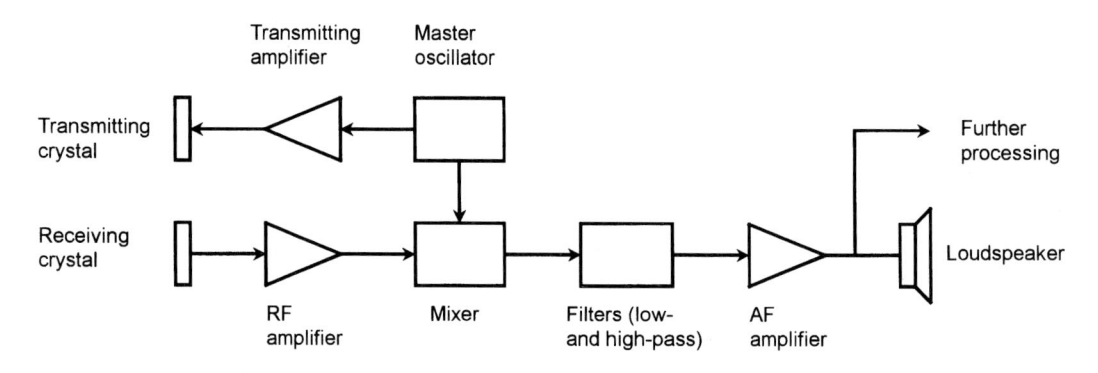

Figure 4.1 Block diagram of a simple non-directional continuous wave Doppler system

Doppler information, but shifted by ±90° to each other, depending on whether flow is towards or away from the probe. Further electronic processing is still required to unscramble the directional information into two channels. The technique most commonly used for this is one of phase domain processing (Section 6.4.2), which results in separate forward and reverse flow signals which may be listened to through stereo headphones, recorded on audio tape (Appendix 3) or sent for further processing.

Directional information is important clinically for several reasons; so that arterial and venous flows may be separated, because the direction of flow in an artery may change completely (for example when steal occurs); and because in peripheral arteries the direction of flow may reverse once or more during each cardiac cycle. It is also of great value in studies of the heart, where the flow patterns are complex.

Although with experience an operator may derive considerable information by listening to the audio signal, more objective methods of analysis are desirable. Further processing of the Doppler signal allows the Doppler shift frequencies to be presented in terms of a time-varying trace, so that written records may be made and the time-varying aspects of the blood velocity patterns analysed. The earliest attempts at producing such traces were based on the zero-crossing technique (Franklin et al 1961). The zero-crossing detector is a crude form of frequency processor, which functions by counting the number of times a signal crosses its own mean value in a give time. Under 'ideal' conditions this produces an output that is proportional to the root mean square frequency of the input signal. In practice, the zero-crossing detector suffers from a variety of problems (Section 8.5) and is highly susceptible to noise. Its performance is particularly poor when a wide range of Doppler shift frequencies are present, as is the case with CW studies where both slow-moving blood near the arterial wall and fast-moving blood in the vessel centre are insonated simultaneously. Despite these drawbacks, zero-crossing detectors are still incorporated into some low-cost Doppler instruments to provide a chart recorder output. These outputs are best avoided.

There are a now a range of methods available for obtaining a pictorial record of the Doppler shift signal, of which the best and most commonly used is real-time spectral analysis (Section 8.2). The output of spectral analysers is usually represented as a sonogram (Fig. 4.2). In this type of display the horizontal axis represents time (t), the vertical axis frequency (f), and the intensity at co-ordinates (t, f), the power of the signal at that frequency and at that point in time (Kaneko, 1966). Both the time-varying maximum frequency and mean frequency envelopes may also be extracted from the output of the analyser. The maximum frequency envelope, or outline of the Doppler spectrum versus time, is the most commonly used parameter for Doppler waveform analysis (Chapter 9), whilst the intensity-weighted mean frequency envelope is most commonly used for computing blood flow velocity and volumetric flow (Chapter 12). Furthermore, the combination of the mean and maximum frequency gives the user some information about the instantaneous flow profiles in the vessel (for instance, for parabolic flow, the maximum frequency or velocity should be approximately twice that of the mean frequency, whereas for plug flow the two should be approximately the same).

The use of a spectral analyser with simple CW Doppler units significantly increases their cost, and therefore less expensive analysis techniques have been developed specifically to extract the time-varying maximum frequency (Section 8.4) and the time-varying mean frequency envelopes (Section 8.3) from the Doppler signal.

4.2.2 Pulsed Wave Velocity Detecting Systems

Because CW Dopplers transmit ultrasound continuously, they usually provide no information about the range at which movement is occurring. Whilst this may not be a problem when high-frequency ultrasound is used to study superficial vessels (particularly since the maximum range is limited anyway by rapid attenuation), it can cause considerable problems in studying deep structures, particularly the heart and vascular organs such as the brain. Even for superficial vessels it is

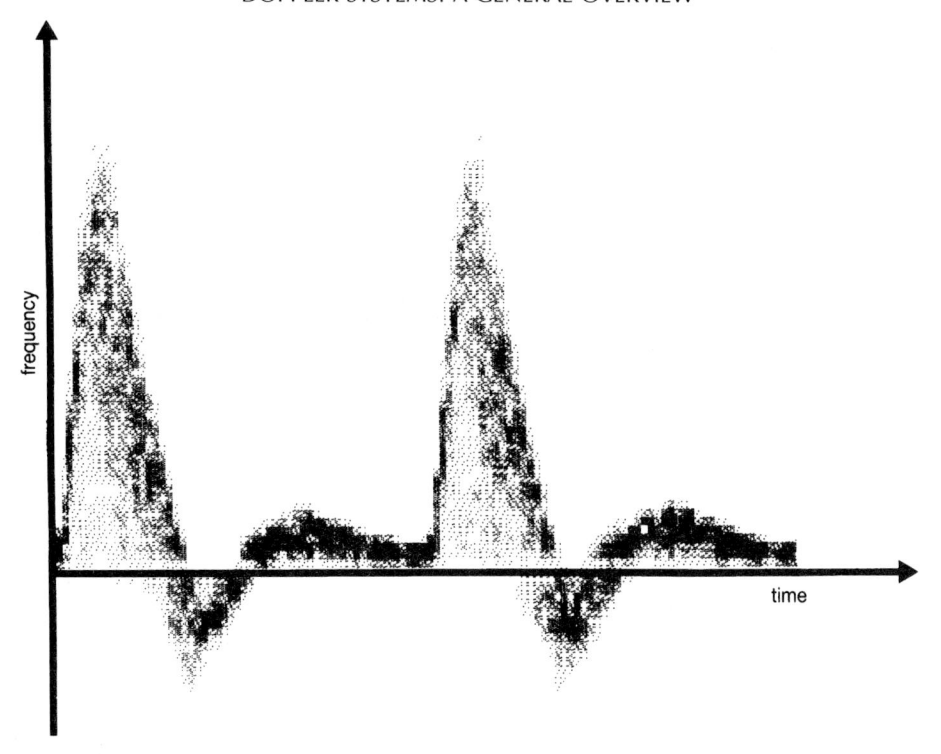

Figure 4.2 Sonogram of the Doppler signal recorded from a common femoral artery. The horizontal axis represents time, the vertical axis the Doppler shift frequency, and the intensity of each pixel the power of the signal at the corresponding frequency and time. Flow towards the transducer appears above the time axis, flow away below it

sometimes difficult to separate the signals from arteries and veins (for example in the popliteal fossa) with CW Doppler. Pulsed Doppler systems overcome these problems by transmitting short bursts of ultrasound at regular intervals, and receiving only for a short period of time following an operator-adjustable delay. The length of the delay determines approximately the range from which signals are gathered (Wells 1969, Peronneau and Leger 1969, Baker 1970).

PW units tend to be used in two distinct fashions. Either the sample volume is made sufficiently large to encompass an entire blood vessel (or other regions containing movement), or sufficiently small, such that just a small part of velocity field is probed. In the former case, the range discrimination is used primarily to reject signals from other nearby structures; in the latter, the high spatial resolution is used to extract information about flow or movement in a specific area.

PW Doppler units differ from CW units in respect of both their electronics and their transducers. Additional circuitry is necessary to gate the transmitted and received signals at appropriate times, and to sample and hold the demodulated signals, but the demodulation process itself is similar. Simple PW transducers need contain only a single piezoelectric element which is used both to transmit and later to receive ultrasound, and must be damped so that short pulses may be transmitted and received. Further details of PW electronics and PW transducers may be found in Sections 6.3 and 5.2.2 respectively.

PW systems have one major drawback; they are only able to detect velocities unambiguously up to a finite maximum which is related to the depth of examination (Section 4.2.2.3). The maximum Doppler shift frequency a pulsed Doppler unit is able to detect is half the pulse repetition frequency (PRF). As the depth of the region of interest increases, the PRF must be decreased to allow the pulses sufficient time for the round journey, and thus for vessels deep within the body only lower velocities may be detected. The maximum velocity problem is particularly severe when it is necessary to make measurements on high-velocity jets in the heart, and many cardiac instruments are equipped with both a PW and a CW option.

4.2.2.1 PW Systems—Basic Operation

A block diagram of a simple non-directional PW Doppler unit is shown in Fig. 4.3. For PW operation the signal from the master oscillator is gated under the control of a PRF generator. The length of time the transmission gate remains open depends on the required length of the sample volume (see next section) but is usually sufficient to permit the passage of a number of complete cycles from the oscillator. The resulting RF pulse is amplified and used to drive the transducer, which transmits a burst of ultrasound into the tissue (the shape and length of the burst is determined both by the exciting pulse and the transducer characteristics —see Section 5.2.1). Ultrasonic echoes returning from the tissue are converted to electrical signals by the same transducer, and these signals are amplified and mixed with a reference signal from the master oscillator before being low-pass filtered to remove RF components, and fed to a receiver gate which is opened during each transmission cycle after an operator-determined delay, to admit signals to a sample and hold circuit. It is the delay between transmission and the opening of the receiver gate that determines the region from which the signals are gathered, whilst the time for which the gate is left open, together with the length of the transmitted pulse, determines the sample volume length (see next section). The output from the sample and hold circuit is filtered to remove both the sampling frequency and unwanted low frequency components, then amplified and sent for further processing and display, as for CW devices. In practice, PW systems, like all but the simplest CW devices, include circuitry to detect the direction of motion, and this is virtually always based on quadrature demodulation techniques, as introduced in Section 4.2.1.1 and described in more detail in Chapter 6.

4.2.2.2 PW Systems—Principal Sample Volume

PW systems emit short bursts of ultrasound several thousand times every second, usually at regular intervals. After each pulse has been transmitted, there is a delay before one or more gates in the receiving circuit are opened for a short period of time to admit signals returning from a small volume of tissue. The delay between transmission and opening the gate may be altered by the operator to determine the depth from which the signals are gathered, whilst the time for which the gate is left open (and to some extent the receiver bandwidth), taken together with the length of the transmitted pulse, determines the length of the sample volume. Specifically, the distance from the transducer to the beginning of the range cell, Z_1, will be given by:

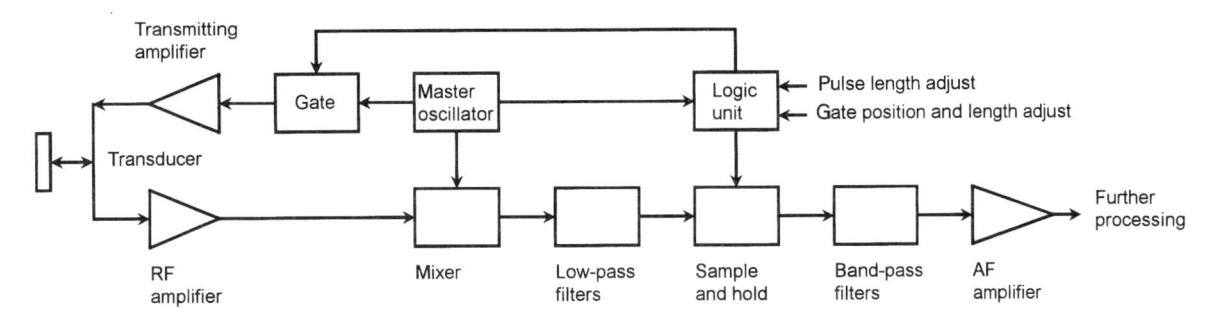

$$Z_1 = c(t_d - t_p)/2 \qquad 4.1$$

Figure 4.3 Block diagram of a simple non-directional pulsed wave Doppler system. The logic unit controls the PRF, the ultrasonic pulse length, and the receive gate position and length

where c is the velocity of the ultrasound in tissue, t_p the pulse length, and t_d the time delay between the start of transmission and the moment at which the receiver gate opens. The distance from the transducer to the end of the range cell, Z_2, will be given by:

$$Z_2 = c(t_d + t_g)/2 \qquad 4.2$$

where t_g is the period for which the gate is open. The length of the range cell may therefore be written:

$$Z_r = Z_2 - Z_1 = c(t_g + t_p)/2 \qquad 4.3$$

It should be noted that the range cell is not equally sensitive throughout, and its effective distribution will depend on the shape of the ultrasound pulse, the relative lengths of the transmitted pulse and the sample gate, and the bandwidth of the receiving electronics. Figure 4.4 shows the one-dimensional distribution of the received energy for a rectangular pulse of length t_p and a rectangular sample gate of length t_g for different t_p to t_g ratios, derived simply by convolving the receive gate shape with the pulse shape for different ranges. This can equally be interpreted as the axial sensitivity of the sample volume. Peronneau et al (1974) implicitly made the suggestion that setting $t_p \approx t_g$ gives the best sensitivity for a given axial resolution, whilst Kristoffersen (1986) made the point that selecting $t_p \neq t_g$ provides more equal axial weighting of the sample volume, which may be desirable in blood flow measurements based on estimation of the mean Doppler shift over the vessel cross-section. Hoeks et al (1994b) have also made the point that, for values of $t_g > t_p$, the intensity of the Doppler signal is no longer preserved because of the increasing chance that scattered signals with opposite phase are contained within the observation window. Loupas et al (1995) showed by extensive simulations, that for the autocorrelation method of estimating mean frequency (see 4.5.3.1), matching the range gate to the pulse length represented the optimum choice.

In reality, the ultrasound sample volume is not one-dimensional, as the ultrasound beam will have a finite width. Also, the ultrasound beam will not be perpendicular to the direction of flow (otherwise the Doppler shift frequency will be nominally zero) and therefore the width of the beam will also be important in determining the weighting of the received energy across a blood vessel, particularly if the beam width is large compared with the sample volume length. Furthermore, due to the properties of the ultrasound transducer and the limited bandwidth of the receiving electronics, the ideal rectangular pulse and rectangular receiving gate shapes will not be attained in practice. Peronneau et al (1974) have given a qualitative description of the effects of three-dimensional sampling of blood vessels by finite ultrasound beams.

4.2.2.3 PW Systems—Maximum Velocity Limit

With CW Doppler there is no practical limit on the maximum velocity that can be measured. This is

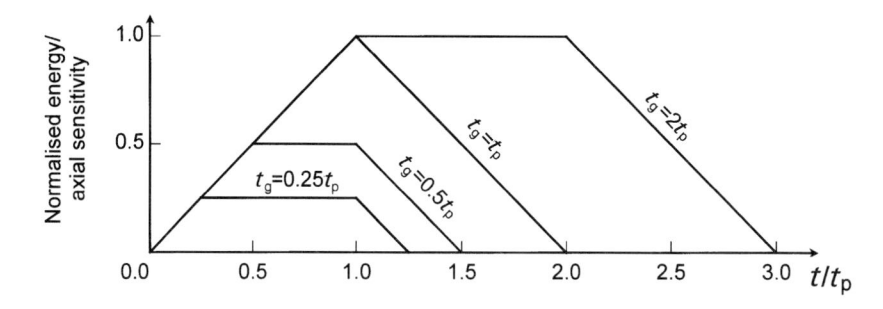

Figure 4.4 Axial sensitivity of a PW sample volume for a rectangular pulse of length t_p, and a rectangular receive gate of length t_g. The abscissa is scaled in terms of pulse length and can be converted to distance by multiplying by $(t_p c)/2$ where c is the velocity of ultrasound. The sensitivity, plotted along the ordinate, is normalised to the maximum attained using a value of $t_g \geqslant t_p$

not so with PW Doppler because of the finite sampling rates employed. In order to extract a Doppler shift from the ultrasound signal, the velocimeter compares the phase relationships between a signal from a reference oscillator and each successive returning ultrasound pulse. The maximum phase change that can be observed between two pulses is limited to the range $-\pi$ to $+\pi$ radians (since angular measurements repeat themselves every 2π radians), and therefore if the target moves a distance of more than $\lambda/4$ between samples (equivalent to a round-trip distance of $\lambda/2$ for the ultrasound pulse) its velocity may be interpreted incorrectly. This limitation is simply an expression of the sampling theorem (Shannon 1949, Jerri 1977), which states that it is necessary to sample a signal at at least twice the highest frequency present in the signal to avoid ambiguity. Mathematically this may be stated as:

$$f_d(\max) = f_s/2 \qquad 4.4$$

where $f_d(\max)$ is the maximum Doppler shift that can be unambiguously detected, and f_s is the pulse repetition frequency or sampling frequency. The critical frequency $f_s/2$ is commonly known as the Nyquist frequency (in honour of Nyquist's celebrated work on telegraph transmission theory—Nyquist, 1928). The maximum velocity that may be unambiguously detected, $V_{\max}$, is found by substituting eqn. 1.1 into eqn. 4.4, i.e.,

$$V_{\max} = f_s c/(4 f_t \cos\theta). \qquad 4.5$$

The effect of aliasing in a directional PW Doppler system (the incorrect interpretation of frequencies above the Nyquist limit) is illustrated in Fig. 4.5a. As soon as the Doppler shift frequency exceeds $f_s/2$, it is interpreted as being $-f_s/2$ and the top of the sonogram appears at the bottom of the reverse channel. Although the Doppler shift is normally interpreted as being between $-f_s/2$ and $+f_s/2$, it may be interpreted as any other range of frequencies provided that the total range is only f_s, for example from $-f_s/4$ to $3f_s/4$ or from 0 to f_s. In the case of the sonogram shown in Fig. 4.5a, a shift of the range to $-0.1 f_s$ to $+0.9 f_s$ results in the sonogram shown in Fig. 4.5b, which is in this case the correct interpretation of the signal. Some analysers have the ability to shift sonograms

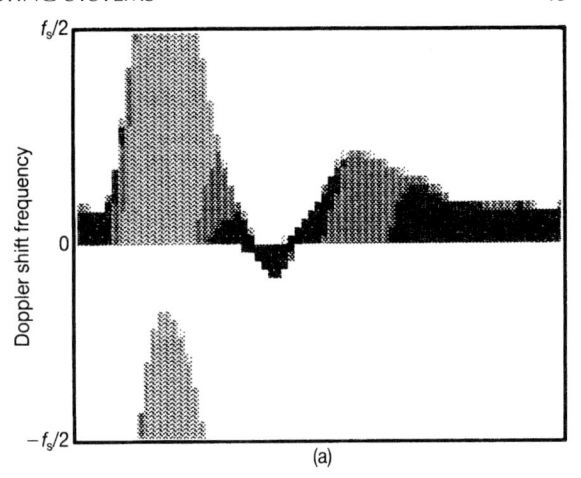

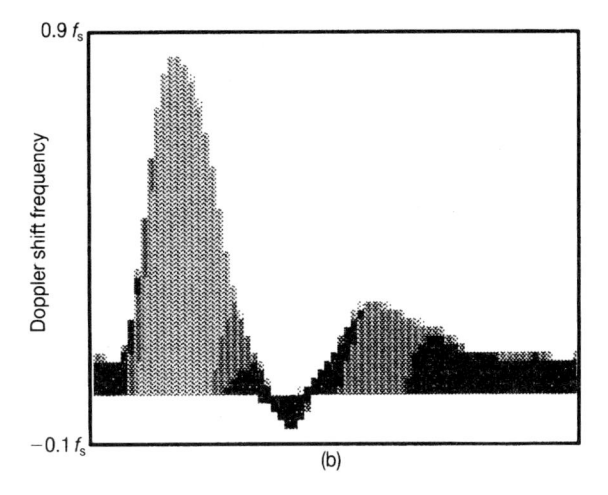

Figure 4.5 Effect of transmitted pulse repetition frequency aliasing on a sonogram. (b) Correction of aliasing by changing the interpretation of the information derived in (a)

graphically in this way; they cannot change the permissible range of frequencies, only the way in which they map onto the frequency scale.

4.2.2.4 PW Systems—Range Ambiguity

In the last but one section it was stated that the position of the range gate in a PW system is determined by the time delay between transmission and the commencement of signal acquisition. In fact, there is a degree of range ambiguity since the signals arriving at the transducer at a given time may be echoes from the latest transmission pulse, the previous pulse, or even earlier pulses. Signals are therefore collected from ranges located around Z_n given by:

$$Z_n = (c/2)(t_d + nt_s) \qquad 4.6$$

where t_s is the time between subsequent pulse transmissions and n is zero or any non-negative integer. In practice, because of attenuation, the signals returning from deeper tissue are weaker than signals from the more superficial tissues, and if f_s is low they may be negligible. If f_s is high, two or more significant gates may exist and it is for this reason that f_s cannot be increased at will to overcome the maximum velocity limit discussed in the last section. Another way of viewing this is that, as f_s is increased, the PW case tends to the CW case, where there is no maximum velocity limit but neither is there any depth resolution.

If a flowmeter is to use pulses that return during the same transmission cycle and thus avoid ambiguity, the maximum range Z_{max} at which it can operate is given by:

$$Z_{max} = c/2f_s \qquad 4.7$$

If, in addition, there is to be no velocity ambiguity, then eqn. 4.5 must be satisfied, and therefore there is a maximum range-velocity limit given by:

$$Z_{max} V_{max} = c^2/8f_t \cos\theta. \qquad 4.8$$

In practice, some range ambiguity is not usually a serious problem with single gate PW systems, except possibly in the heart, because provided that the sites of the range gates are known care can be taken to ensure that only one of them encompasses a vessel.

Kremkau (1990) has made the point that in practice V_{max} is effectively independent of transmitted frequency, because at higher frequencies attenuation decreases the range from which significant echoes can be detected. By assuming a velocity of sound in tissue of 1540 ms^{-1}, he was able to write an expression for the maximum practical maximum depth in terms of attenuation coefficient and dynamic range, i.e:

$$V_{max} = \frac{6000\gamma}{DR \cos\theta} \qquad 4.9$$

where γ is the attenuation coefficient in dB.cm^{-1}MHz^{-1}, DR is the dynamic range in dB, and θ is the usual Doppler angle. This does not affect the validity of eqn. 4.8.

Some commercial PW systems that do not incorporate a CW facility attempt to overcome the problem of measuring high velocities by deliberately increasing their PRF beyond the maximum required to prevent range ambiguity. Doubling the PRF will cause the device to monitor not only velocities at the depth of interest, but also velocities at a depth which is in between the probe and the depth of interest. Continuing to increase the PRF will add more equally spaced sample volumes between the probe and initial depth of interest, but this often does not present a problem as the higher velocities usually only occur at one point in the beam. Methods of attempting to circumvent aliasing are discussed in Chapter 13 (Section 13.11).

4.2.2.5 PW Systems—Difference between the CW and PW Doppler Shift Frequency

Although the demodulated signals from CW and PW Doppler devices are usually treated and processed in the same way, there are fundamental differences between them which really only become important when considering the effects (or rather the perhaps unexpected lack of effects) on the Doppler signal of such phenomena as frequency-dependent attenuation. Indeed, the surprising fact is that, whilst CW devices measure the 'classical' Doppler shift, PW devices do not, and indeed could not, because the (wide-band) spectrum of a short pulse of ultrasound changes so significantly as it propagates through tissue and is scattered by blood. It is possible to debate whether or not PW Doppler devices should even be called Doppler devices (and this of course applies equally to more complicated devices which use pulses of ultrasound, such as colour flow systems), but this argument is really one of semantics because the demodulated signals from CW and PW devices have virtually the same characteristics (leaving aside issues such as aliasing).

In CW systems, the returning ultrasound signal, which is either a slightly expanded or slightly compressed version of the transmitted signal, due to the motion of the targets, is continuously multiplied by signals from the master oscillator, and the result (after low pass filtering) is the difference between the transmitted (monochromatic) frequency and the received frequency. In PW systems a series of bursts are transmitted and sampled at the fixed pulse repetition frequency, and there are two distinct mechanisms affecting the received signal (Wilhjelm and Pedersen 1993a, Thomas and Leeman 1995). The first is that of a 'classical' Doppler shift (as for CW) where, because of the motion of the target, each received pulse is either an expanded or compressed version of the transmitted pulse. The second is that because the targets in the sample volume (which have a particular 'ultrasonic signature') have either moved towards or away from the transducer, consecutive received signals undergo a progressive time shift with respect to the time of transmission. This time shift gives rise to a progressive change in the phase relationship between the ultrasound signature and the master oscillator, which is exactly what the demodulator detects. Curiously enough, it can be argued that the 'classical' Doppler shift on the ultrasound pulse actually decreases the precision of velocity estimates using a PW system. The important thing to note is that the frequency shift detected due to the changing phase relationship is virtually identical to that detected by a classical CW system, and any differences can be ignored for most practical purposes. Consequently, the basic Doppler equation (eqn. 1.1) applies equally in the case of both CW and PW systems.

Although simple single-gate Doppler instruments work by detecting the rate of change of *phase-shift* between returning pulses and the master oscillator, an alternative strategy, which is used in some multigate instruments, is to measure the pulse to pulse *time-shift* between echoes from the same target (characterised by their ultrasonic signature) using cross-correlation techniques (Bonnefous and Pesque 1986, Foster et al 1990). Such techniques are described in more detail in Section 11.4.

The differences between CW and PW signals are discussed further in Sections 7.3 and 7.4.

4.2.3 Coded Signal Velocity Detecting Systems

Although most Doppler devices may be thought of as either CW or PW, several workers have proposed devices which code the transmitted signal in a more complex way than the simple amplitude modulation used by standard PW systems. Such devices may be conveniently described as coded signal systems. There are two main rationales for the use of such systems, the first related to improved spatial resolution, the second related to overcoming range–velocity ambiguity.

It has already been shown that the spatial resolution of a standard PW system is related to the length of the transmitted pulse (eqn. 4.3), and it follows from this equation that to achieve very short range cells, or very good spatial resolution, it is necessary to use very short pulses. The signal-to-noise ratio (SNR) of a Doppler system is, however, related to the average transmitted power and therefore, if the pulses are made shorter in order to maintain SNR it is necessary to increase their amplitude (the pulse repetition frequency cannot be arbitrarily increased because of the need to avoid range ambiguity by only having one pulse in flight at a given time). In general this is undesirable, since the higher the peak pressures in the tissue become, the greater the potential there is for biological damage. It can, however, be shown that for a more general Doppler system it is the signal bandwidth rather than the signal duration that determines the range resolution. In a standard PW system that uses sine wave bursts, these two quantities are of course intimately connected (such that their product is approximately unity), but by using other coding techniques it is possible to considerably increase the duration of the transmitted signal and yet still achieve a high bandwidth. This leads to a reduced peak to average power ratio, and therefore for the same average transmitted power, a potentially better range resolution for the same SNR.

The problem of range–velocity ambiguity has been discussed in Section 4.2.2.4. Basically, in order to avoid range ambiguity it is necessary that only one ultrasound pulse is in flight at any given moment, which means that the maximum sampling frequency that a PW system is able to employ is limited. This limitation arises because each transmitted pulse is identical, and therefore echoes from different transmitted pulses cannot be distinguished. If, however, an appropriate coding system is used, then echoes from different pulses or different sections of the transmitted signal can, in principle, be distinguished, and the range–velocity problem can be ameliorated or even completely overcome.

The two major classes of coded signal Doppler systems, that is those that use random signals and those that use frequency-modulated signals, are briefly discussed in Sections 4.2.3.1 and 4.2.3.2.

4.2.3.1 Random and Pseudo-random Signal Doppler Systems

Following an early report by Waag et al (1972), a number of Doppler systems which use random or pseudo-random signals have been described in the literature, although to the authors' knowledge none have been produced commercially. Figure 4.6 is a schematic diagram of a basic random signal Doppler system as described by Bendick and Newhouse (1974). A wide-band noise source (which may be continuous or intermittent) is used to drive a transmitting transducer which is directed so as to insonate a region of interest within the body. A separate receiving transducer continuously receives the returning signals, which are then correlated with a delayed version of the transmitted signal. Since the signal is random, in general there will be no correlation between the received signal and the time delayed reference, except where the delay experienced by the ultrasonic echoes as they travel through the tissues is equal to that introduced in the reference channel, and therefore it is only signals from the range thus defined that will produce an output.

A major problem with the simple technique described above is that the uncorrelated clutter signal due to reflections at ranges other than that of interest can be very large compared to the signal (particularly if the sample volume is deep within the body) and will contribute additional noise to the receiver input. To avoid this, it is necessary either to transmit bursts of random noise and keep the PRF low so as to minimise the number of pulses that exist simultaneously between the transducer and region of interest, or to arrange the transmitter field and receiver sensitive zone only to overlap around the depth of interest (Bendick and Newhouse 1974, Jethwa et al 1975). Both of these solutions, of course, to some degree reduce the putative advantages of random signal coding. In an attempt to overcome these problems, Jethwa et al (1975) introduced the idea of random pulse staggering, which it was claimed would lead to 'considerable improvement over the conventional pulsed Doppler ultrasonic system in the measurement of maximum unambiguous blood velocities in vessels at all ranges from the face of the transducer'.

Further work on similar systems has been reported by Cathignol et al (1980) and Cathignol (1986), who have investigated systems employing pseudo-random codes, and sine waves phase modulated by a pseudo-random code, which they claim to give superior results to those from the random noise flowmeter.

The fact that little has been written about random signal devices in the recent past seems to suggest that the problems of making them function

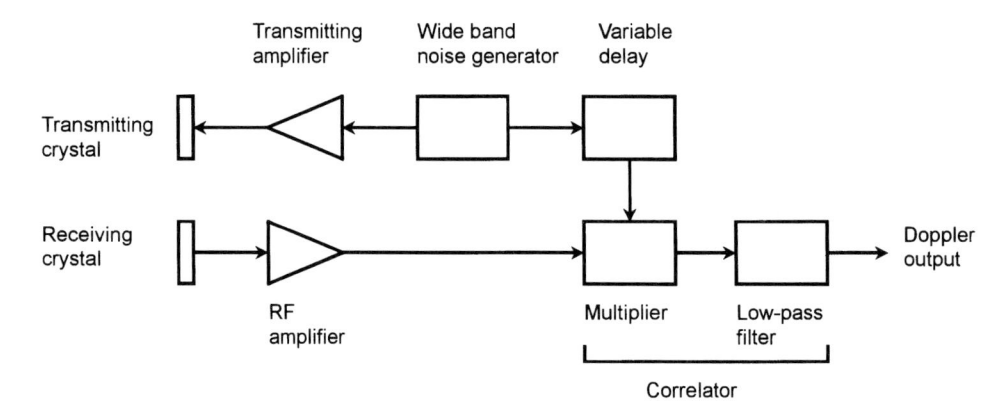

Figure 4.6 Schematic diagram of a basic random signal Doppler system

adequately in the clinical environment are difficult to overcome.

4.2.3.2 Frequency-modulated Doppler Systems

Although frequency-modulated (FM) systems had previously been discussed at conferences (Kemper et al 1971), the first papers on the subject appear to have been those of McCarty and Woodcock (1974, 1975). In these papers the authors described the use of an FM Doppler system to measure the diameter of tubes containing flow and combined this information with mean velocity estimates made with the Doppler unit operating in conventional CW mode.

Figure 4.7 is a schematic diagram of the FM system used by McCarty and Woodcock. The output from a saw-tooth generator is applied to a voltage controlled oscillator, which produces a frequency-modulated constant voltage sine wave output. The output from the oscillator is amplified and transmitted, and the returning echoes are amplified and mixed with a reference signal also taken from the oscillator and finally low-pass filtered to remove all but the difference frequency. Even if echoes return from a stationary target, then because of the time taken for the ultrasound pulse to propagate, the frequency of the reference signal will be higher than the received signal by an amount determined by the rate of change of the oscillator output frequency and the delay between transmission and reception, and consequently the range of any such target can be calculated. If the target is moving with respect to the transducer there will be an additional (positive or negative) shift in the frequency of the returning signal due to the Doppler effect, which will, at least at first sight, be difficult to disentangle from a frequency shift related to range, and indeed Atkinson and Woodcock (1982) later wrote that 'Using a simple sawtooth sweep modulation, it is not possible to differentiate a stationary target at one range from a moving target with accompanying Doppler shift, at a different range'.

The system used by McCarty and Woodcock was further analysed by Bertram (1979), who compared theoretically and experimentally the spatial resolution of an FM system and an equivalent pulse-echo system (i.e. one using the same transducer) and concluded that the FM system was inferior to the pulse-echo system, both in respect of range measurement precision and multiple-target resolution.

Much more recently interest in FM systems has been re-awakened by the publication of a pair of papers by Wilhjelm and Pedersen (1993a, 1993b), which suggest that it might after all be possible to separate out the effects of range and velocity on signals from an FM system and thereby measure velocity profiles. The concept of the system described by Wilhjelm and Pedersen is illustrated in Fig. 4.8. It differs from that described by McCarty and Woodcock in that rather than continuously transmitting an FM signal, it transmits a series of chirps (i.e. linear frequency-modulated tones) in an manner analogous to a PW system transmitting a series of tone bursts, although the duration of the chirps is much longer than the bursts used for PW excitation. Also the

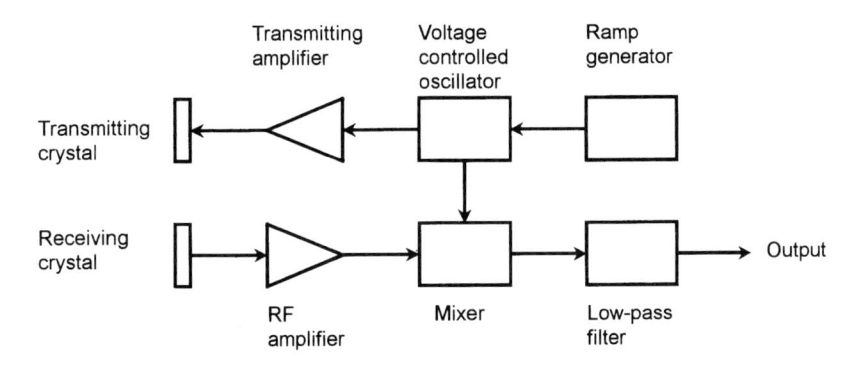

Figure 4.7 Schematic diagram of FM 'flowmeter' described by McCarty and Woodcock (1974, 1975)

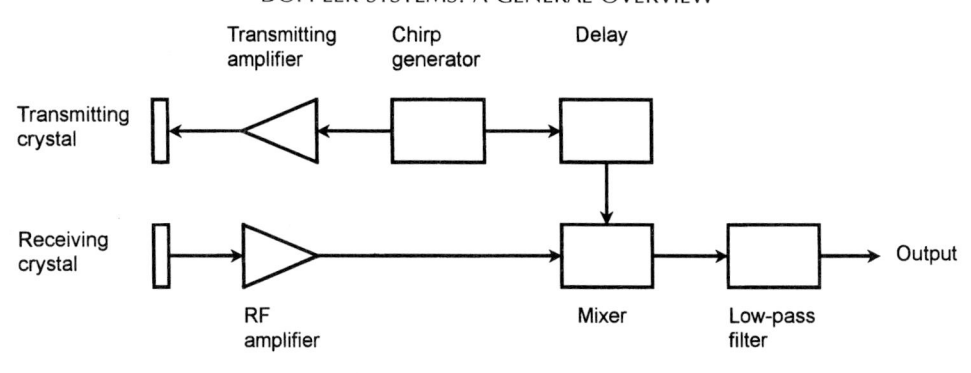

Figure 4.8 Concept of the FM Doppler system described by Wilhjelm and Pedersen (1993a,b)

reference signal from the sweep generator is delayed by a time corresponding to the range from which it is required to sample the signal before it is correlated with the returning signals. Wilhjelm and Pedersen were able to show that (for velocities and depths in the physiological range) using a single chirp it is theoretically possible to derive the approximate range of a target from the initial instantaneous frequency of the chirp, and the velocity from the sweep rate of the demodulated signal. Unfortunately, as for conventional PW systems, the Doppler shift is hidden by the spectral 'smearing' due to the finite duration of the signal, and therefore a different approach is necessary to extract the required velocity information in practice. It transpires, however, that in a manner analogous to that used in conventional PW Doppler systems, it is possible to obtain the velocity information by comparing the centre or instantaneous frequencies of adjacent pulses, and that this can be achieved using either phase shift measurement or time shift-based measurement techniques. Wilhjelm and Pedersen concluded that the FM systems they described feature the advantage of long coded excitation signals in allowing operation at a low peak transmitted power relative to PW Doppler systems, but that they exhibit range ambiguities in the form of range 'side-lobes' as well as a minor range cell displacement, proportional to velocity, due to the strong coupling between velocity and range information in the received signal. They also emphasised the advantages of the time shift measurement approach in comparison to the phase shift approach in terms of avoiding aliasing at higher velocities (in an exact analogy with conventional PW systems).

In follow-ups to their original papers, Wilhjelm and Pedersen (1995, 1996) have evaluated the comparative performance of FM and PW Doppler systems by means of numerical simulations and measurements of actual flow profiles, and have found the two systems to 'have very similar levels of performance'. It will be interesting to see if such systems prove to be of practical value.

4.3 DUPLEX SYSTEMS

Duplex scanners are devices that combine a pulse-echo B-scan system (McDicken 1991b) and a Doppler system so that the Doppler shift signal can be recorded from known anatomical locations (Barber et al 1974, Phillips et al 1980). There are a number of ways of combining the two modalities, but they all share certain characteristics; the permissible directions for obtaining Doppler information all lie within the scan plane of the pulse-echo imager, and the direction of the Doppler beam at any instant is indicated by a cursor superimposed on the image. If the Doppler is PW, the position and usually the extent of the sample gate are also indicated (Fig. 12.1).

Early duplex systems combined mechanical sector scanners for imaging with a separate Doppler transducer, which might or might not have been offset from the imaging transducer (Fig. 4.9), but nearly all duplex systems now use the same array transducers for both imaging and

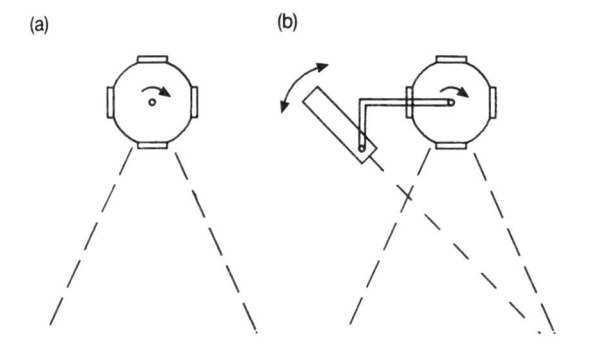

Figure 4.9 Schematic diagram of two early types of duplex transducer. (a) Spinning wheel with four separate crystals, one of which is a PW Doppler crystal. When switched into Doppler mode the wheel stops with the Doppler crystal pointing in the selected direction. Once this has occurred only the range and not the direction of the sample volume can be changed. (b) Spinning wheel imaging transducer with offset Doppler probe. With this type of system the wheel usually stops spinning during Doppler recordings, but both angle and range of the Doppler sample volume may be changed at any time. Most contemporary duplex systems use linear or phased array transducers for both pulse–echo and Doppler acquisition (see Chapter 5)

Doppler measurements (and, as will be seen later, for colour flow imaging) (see Fig. 5.5). There are clear advantages to using the same transducer for both imaging and Doppler purposes, including ease of construction and operation, avoidance of alignment problems and beam agility considerations, but this arrangement also has a number of drawbacks which stem from the compromises necessary in order to use the same elements for two purposes.

Firstly, it should be noted that in order to achieve good axial resolution with a pulse-echo system, it is necessary to use very short pulses, which can only be generated by the use of heavily damped transducer elements. Such elements are unfortunately very inefficient and in particular rather insensitive, which necessitates transmitting more ultrasound power into the tissue than would otherwise be the case. This is not a problem with imaging pulse regimens, but with Doppler, where the transmitted pulses are longer, where the ultrasound beam is static, and where the pulse repetition frequency has to be relatively high in order to adequately sample the Doppler shift frequency, this can lead to very high temporal average intensities and the potential for biological damage (the reader is referred to Chapter 14 for an in-depth discussion of safety issues relating to Doppler ultrasound). A second area of compromise with dual purpose transducers is that of the out-of-plane width of the ultrasound beam. For imaging purposes it is desirable to produce as narrow a beam as possible to produce the best resolution; on the other hand, in Doppler applications it is often advantageous to insonate an entire blood vessel, especially if measurements of mean velocity or flow are required. Ultrasound scanners are essentially sold on their ability to produce excellent images rather than to make accurate velocity measurements, and therefore ultrasound beams are often too narrow for the latter purpose. As two-dimensional array technology becomes available, there should be no reason why the out-of-plane focusing should not be different for imaging and Doppler pulses, which will mitigate this problem. Finally, when the same transducer is used for both imaging and Doppler measurements there is the issue of insonation angles. When making Doppler measurements on a blood vessel, it is often important also to be able to simultaneously produce a good image of the vessel. Unfortunately, using the same transducer, the best angle for Doppler measurements, i.e. a small angle with respect to the direction of flow, is incompatible with the best angle for imaging the vessel walls, i.e. 90°, and so once again a compromise is required (actually this is inevitable even if separate transducers are used, but the use of the same transducer can considerably exacerbate the problem). The reader is referred to Chapter 5 for more details of modern duplex transducer systems.

To operate a duplex system, the sonographer first finds the blood vessel or portion of the heart of interest using the imaging facility, and then places the Doppler sample volume at the required anatomical location using the on-screen cursor as a guide. The scanner is then switched to duplex mode to make the required measurements or recordings. With array systems it is generally possible to adjust the way in which time is divided between the imaging and Doppler modes. Thus, the operator can freeze the image and simply adjust the Doppler sample location based on the frozen image, or can choose to update the image at regular intervals at the same time as acquiring Doppler information. It is not possible to simultaneously acquire imaging and Doppler information, but from the operator's point of view it can appear as though the image is being updated in real-time while the Doppler information is also being acquired. Clearly, because of the finite velocity of ultrasound, the degree to which this is possible is influenced by factors such as the field-of-view depth, the number of scan lines, the frame rate and the acceptability of interpolating the Doppler data.

4.3.1 Velocity Measurement with Duplex Scanners

The measurement of absolute velocity using the Doppler technique depends on knowing the angle of insonation of the Doppler beam relative to the axis of flow (eqn. 1.1). Using a duplex system it is possible to image the vessel walls and hence estimate the flow axis. Duplex scanners incorporate a special angle-measuring cursor which can be rotated about the Doppler beam cursor, and which produces a direct readout of the appropriate angle (Fig. 12.1). It is usually also possible to use this angle information to convert the system output automatically from Doppler shift frequency to velocity.

4.3.2 Flow Measurement with Duplex Scanners

The volumetric flow through a vessel is equal to the product of the average velocity of flow over the cardiac cycle and the vessel cross-sectional area. Both these quantities may be measured using a duplex scanner. The former is calculated from the instantaneous mean velocities, obtained as described above, averaged over the cardiac cycle. The vessel area is usually calculated by measuring the vessel diameter from the B-scan image, and assuming the vessel cross-section to be circular. Details of both the method and potential sources of error in duplex flow measurements are given in Section 12.2.

4.4 PROFILE DETECTING SYSTEMS

In every blood vessel there is a spatial variation of velocity across the vessel which varies with time (Section 2.7) and is referred to as the velocity profile. The shape of this velocity profile can be useful in calculating volume flow, and can also be used as an indicator for the presence of arterial disease. For instance, if a plaque reduces the lumen of the vessel, the velocity profile will be distorted. To date, the use of velocity profile detecting systems has been largely restricted to research laboratories. However, as we will see later (Section 4.5.3), by colour-coding the velocities in the vessel an image may be built up which provides the user with a much simpler display that can be more easily interpreted in a clinical situation.

Using a simple pulsed Doppler system, it is possible to obtain a measure of the velocity profile in a vessel by sequentially moving the sample volume from one side of the vessel to the other and noting the velocity information at each point of measurement (Peronneau et al 1972, Histand et al 1973). This technique is time-consuming and clumsy and does not allow the simultaneous measurement of velocities across the vessel.

An alternative to using a single-gate PW Doppler is to have a multichannel Doppler system where the information from a number of range gates is processed in parallel. Velocity profiles may then be reconstructed by combining the signals from each of the range gates (Baker 1970, McLeod 1974, Keller et al 1976). The major disadvantage of this approach is that each individual range gate requires its own separate circuitry, and therefore the electronics required can become both bulky and costly. Furthermore, each of the channels must be accurately tuned and matched to prevent artefactual results and this can be both time-consuming and tedious.

A much more elegant solution is to use a single serial processing system that behaves like a multi-gate system, and a number of instruments capable of this have been described in the literature. Analogue 'infinite' gate systems based on delay line cancellation and phase detection (Grandchamp 1975, Brandestini 1975, Nowicki and Reid 1981) showed early promise, but have been superseded by serial digital processing systems (Brandestini 1978, Brandestini and Forster 1978, Hoeks et al 1981, Hoeks et al 1984, Reneman et al 1986, Tortoli et al 1988, Takabatake et al 1990, Tortoli and Guidi 1995, Tortoli et al 1996, Tortoli et al 1997), which comprise a standard pulsed Doppler 'front end' followed by analogue-to-digital converters and digital circuitry. These latter systems are very versatile and their outputs may be processed to display either the velocity versus time traces for pre-selected gates (Fig. 4.10a) or, alternatively, the velocity profiles in the vessel at pre-selected times (Fig. 4.10b). These systems are described further in Section 6.6.1. Ahn and colleagues (1988) have suggested that it is possible to implement a multigate instrument using direct 'second order' sampling of the RF signal, which results in a fully digital system.

The concept of real-time two-dimensional (2-D) colour flow imaging is introduced in Section 4.5.3. The techniques used for colour flow mapping, which require velocity to be estimated at many points in the tissue, can also be used to measure flow profiles along one line, but it should be noted that some of the algorithms employed by such devices are optimised for speed rather than accuracy. Foster et al (1990) have analysed the errors involved in using time domain correlation techniques, which might be used in colour flow mapping, to derive one-dimensional (1-D) velocity profiles across a vessel, and have obtained encouraging results.

In a slightly different approach to profile measurement, Moser and colleagues (1992a,b, 1995) have described the use of a C-mode Doppler system, which is able to measure the 2-D velocity profile in a vessel at 10–30 ms intervals. This system is based on a 2-D array transducer which is placed more or less perpendicular to the axis of flow (Fig. 12.12). This is similar to one of the approaches suggested by Hottinger and Meindl (1975), and is discussed further in Section 12.7 in the context of volumetric flow measurement.

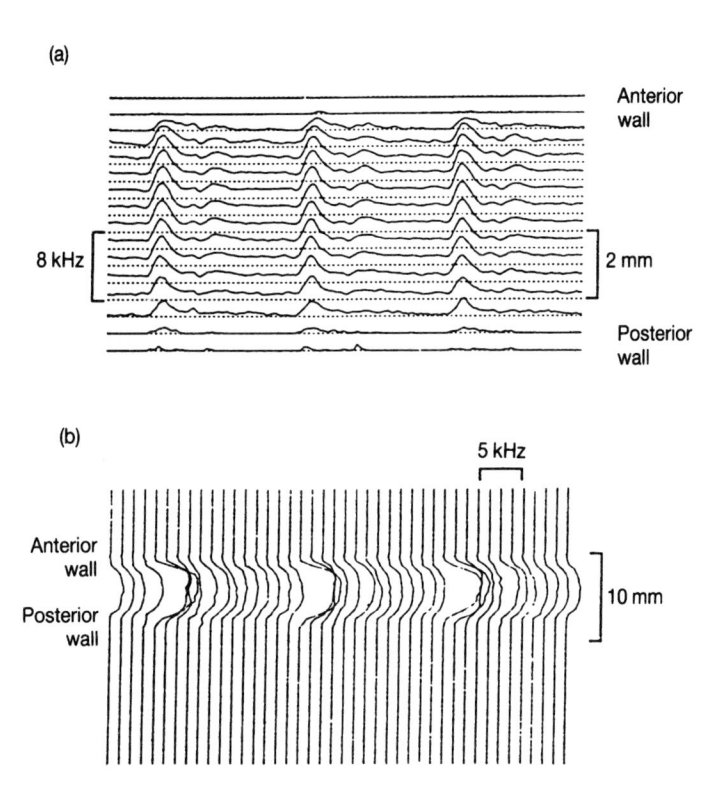

Figure 4.10 (a) The instantaneous velocity waveforms as recorded simultaneously from various gates along the ultrasound beam in the common carotid artery of a healthy subject using the system described by Hoeks et al (1981). (b) The axial velocity profiles at discrete time intervals during the cardiac cycle, synthesised from the velocity waveforms in (a) (reproduced by permission of Elsevier Science, from Reneman et al 1986, *Ultrasound in Medicine and Biology*, © World Federation of Ultrasound in Medicine and Biology)

4.5 VELOCITY IMAGING SYSTEMS

If the sample volume of a Doppler system is moved around inside the body, there will be some positions, corresponding to the locations of blood vessels, from which strong Doppler signals will be received. At other points, where large vessels are absent, there will be an absence of Doppler signals. Therefore, if the position of the Doppler sample volume is mapped onto a 2-D display which is intensity-modulated in response to either the amplitude or the frequency of the Doppler shift signal, blood flow images may be produced. This is exactly analogous to ultrasonic pulse-echo imaging except that the Doppler signal rather than the size of the reflected ultrasound pulse is used to build the image.

4.5.1 Continuous Wave Doppler Imaging

The simplest type of Doppler imaging system consists of a focused CW transducer connected to a position-sensing arm in such a way that the probe can be moved in two dimensions over the skin surface (Reid and Spencer 1972, Spencer et al 1974). The image on the screen is built up by multiple manual sweeps of the transducer over the vessel of interest. The position of the 'active' pixel of the image mimics the motion of the transducer and is intensified if a sufficiently large Doppler signal is present. This results in a plan view of the vessels (Fig. 4.11), which is the same projection as that obtained during conventional arteriography. The relatively rapid attenuation of ultrasound by tissue limits the depth from which information may be obtained to a few centimetres. In pulse-echo terms, this type of projection is equivalent to C-scanning.

CW imaging systems may be improved considerably by colour-coding the Doppler shift information (Curry and White, 1978) (Fig. 4.12: see Plate I). Usually, a temperature scale is used that progresses from dull red for low forward velocities to white for high forward velocities (Chan and Pizer 1976). Flow in the opposite direction may also be imaged, using a blue temperature scale. The colour chosen for each

Figure 4.11 C-scan image of a carotid bifurcation built up using a raster scanning motion over the skin surface

pixel is that corresponding to the peak frequency obtained from the sample during the cardiac cycle and this, of course, means that the probe must dwell at each measurement point for at least one complete heart beat. Colour-coded systems are particularly useful because a reduction in vessel lumen shows up both as a reduction in image size and as a local increase in blood velocity. Because the ultrasound beam is of a finite size, the increase in blood velocity through relatively mild stenoses may be much easier to detect than the change in the diameter of the flow channel.

The long scanning time limitation of CW Doppler may be overcome by using a linear array transducer to sweep rapidly across the vessel and to sample each scan line several times during the cardiac cycle. Arenson et al (1982) have taken this one step further with an ingenious real-time two-dimensional 'long pulse' CW imaging system which combines a stepped linear array of Doppler crystals with a rotating mirror. In a different approach to speeding up CW imaging, Cathignol et al (1988) have described a technique using several transducer pairs, each with a unique emission frequency, and each having a separate CW receiver.

4.5.2 Pulsed Wave Doppler Imaging

Because they have no depth resolution, CW imaging systems can only provide a single view of the artery under examination, i.e. its projection on the skin surface. If, however, a PW Doppler system is used, it is possible to produce an image from any selected plane (Hokanson et al 1971, Mozersky et al 1971, Fish 1972), (Fig. 4.13). In order to produce a scan whose plane is perpendicular to the body surface, a pulsed Doppler probe is moved in just one dimension (x-axis) across the skin; the other dimension of scan (z-axis) being produced by altering the range of the Doppler sample gate. With the probe in its initial position the delay between the pulse transmission and the sample gate opening is increased periodically (and usually automatically) in small steps corresponding to perhaps 0.5 mm of tissue. The probe is then moved in the x-scan direction in small increments and the vertical stepping of the z-axis is repeated for each value of x until the entire target vessel has been explored and mapped.

Because the sample volume must remain at each point for at least one cardiac cycle, this procedure can be extremely time-consuming and a complete scan may take around 15 min. This is both wasteful of time and practically difficult because the part the body being scanned must be kept completely immobile for this period. To overcome these difficulties, Fish (1975) devised a multi-channel instrument which allowed the flow to be simultaneously interrogated at 30 sites along the ultrasound beam. Further developments of the same system allow the user to display velocity profiles at various points in the cardiac cycle (Section 4.4) and to measure volumetric flow (Section 12.3) (Fish 1977, 1981). A further increase in scanning speed was achieved by Pourcelot (1979), who used a linear array transducer to sweep the pulsed ultrasound beam from side to side, but the images produced by this prototype system exhibited very poor resolution.

4.5.3 Real-time Two-dimensional Colour Flow Imaging

To a large extent the images of vessels produced by Doppler flow mapping systems and ultrasound pulse-echo systems are complementary. The former provides an image of the flow channel, but not of the vascular wall; the latter an image of the healthy vessel wall but not necessarily of the flow channel, as many non-calcified lesions mimic blood with their low acoustic impedance. Thus, the combination of these two types of image is potentially of great value. The first step towards producing such

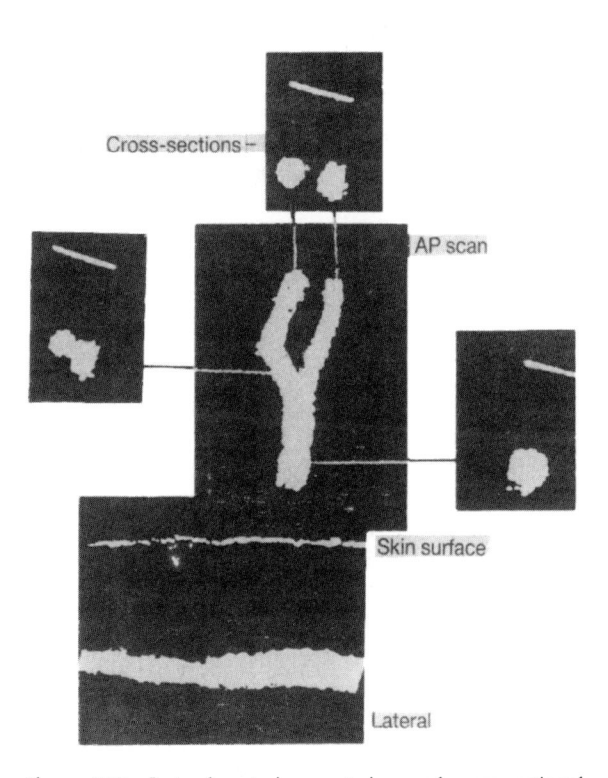

Figure 4.13 Lateral, anterior–posterior, and cross-sectional projections of a carotid bifurcation, produced by a PW Doppler imaging system

images was to combine a multigate Doppler imaging system with a pulse-echo B-scan system (Eyer et al 1981). The pulse-echo information was used to produce a monochrome image, and this was then supplemented with flow information encoded by means of a colour scale (Fig. 4.14: see Plate I). As with previous multigate systems, the slow spatial sweep of the Doppler beam over the region of interest took 20–30 s, since it was necessary to dwell for a complete cardiac cycle with the Doppler transducer in each new sampling direction. (This particular device used a sector rather than a linear scanning action.) Nevertheless, the system seemed capable of producing clinically useful images.

Although the system described above could not produce real-time 2-D Doppler images, like other multigate systems it could be used to produce real-time 1-D Doppler images. Eyer and his colleagues made use of this facility to colour-code time–motion scans of arteries and veins and so to elucidate changes of flow in one section of the artery with time (Fig. 4.15: see Plate I) (Time–motion or M-mode pulse-echo scans are commonly used in cardiac studies; a pulse-echo transducer is fixed in a given direction and the ranges of significant interfaces are plotted on a chart as a function of time; McDicken 1991a). One area where Doppler colour coding of the M-mode scan is of particular value is for examining the motion of blood through the heart (Fig. 4.16: see Plate II). Flow can be coded using a red or blue scale, depending on its direction, and this makes it easy to recognise abnormalities such as retrograde flow in regurgitant valvular disease.

Perhaps the greatest single advance in Doppler velocity imaging was the introduction of real-time colour flow imaging (CFI) systems which combine conventional real-time pulse echo B-scanning with real-time 2-D Doppler imaging (Namekawa et al 1982a, 1982b, Omoto et al 1984, Kasai et al 1985). As with the earlier, slower systems, the pulse-echo information is used to produce an ordinary grey-scale image, whilst the Doppler shift information from all or part of the scan area is superimposed as a colour image (the type of information written to each individual pixel on the display will depend on a priority encoder that uses various pieces of information, including the Doppler power returning from the corresponding sample volume, to determine whether to suppress the grey-scale information and display the Doppler information, or vice versa). A number of examples of images from commercially available CFI systems can be found in Figs 4.17–4.19 (see Plates II–III).

Initially, CFI appeared to have its greatest value in cardiac applications, but it is now used routinely for sites throughout the body. It is valuable for gaining a rapid assessment of the anatomy and pathology of blood vessels and for guiding the placement of the single gate pulsed-wave Doppler sample volume necessary for quantitative assessment of blood flow. It is particularly useful for finding specific regions of flow (for example, regions of increased velocity in a stenosis) and for delineating the extent of specific regions of flow (for example, a jet in a cardiac chamber). It has the advantage that it makes it easier to relate specific regions of flow to their associated structures (for example, a stenosis), and that flow effects at more than one site can be observed simultaneously. Because the velocity information is gathered from a relatively extended region, it is also possible to 'track' flow at a particular location, even when it is moving relative to the ultrasonic transducer (as for example in a moving heart chamber). Modern CFI systems are extremely flexible, and provide the user with a wide range of facilities. Details of CFI systems can be found in Chapter 10.

Before discussing the operation of CFI instrumentation, it is instructive to compare the requirements for the Doppler processing in CFI systems with those for conventional pulsed wave and multi-gate Doppler systems. The fundamental difference between the two stems from the fact that in CFI, in order to produce real-time images made up of a substantial number of image lines, only a very limited number of ultrasound pulses are available for the calculation of each velocity estimate. In the case of conventional Doppler systems, the sample time is limited only by the stationarity of the Doppler signal, which is of the order of 10 ms; this means that for a target at a depth of 50 mm (allowing a pulse repetition frequency of up to 15 kHz) approximately 150 samples are available for each spectral estimation. By contrast, if the same target were studied with a CFI system using 64 scan lines with a frame rate of 15 frames per second, the maximum time available during each frame for interrogating the target would be 1 ms, giving at most 15 samples for analysis. This has important consequences both for the design of the high pass filters required to reject stationary and near stationary echoes, and also for the design of the frequency estimation algorithms.

Several algorithms have been proposed for extracting the information necessary to form colour flow images from the ultrasonic signal, and most operate by estimating the mean velocity and direction of motion (i.e. towards or away from the transducer) within each sample volume of interest. Details of CFI signal processing methods are given in Chapter 11, but an overview of the two methods that have found their way into commercial machines (one based on phase demodulation, and one on time domain cross-correlation) can be found in the next two sections. An alternative approach to using the Doppler shift information extracted from the ultrasonic echoes is to measure and display the total Doppler power from each sample volume rather than its mean Doppler shift frequency, and this technique is introduced in Section 4.5.3.3.

4.5.3.1 CFI Systems—Basic Operation

A much simplified block diagram of the main components of a phase domain (PD) CFI system is shown in Fig. 4.20. The transmitter stage is similar to that of a simple PW Doppler system. The output from an oscillator is gated and after amplification is used to drive the ultrasound transducer. In principle, any type of transducer capable of real-time scanning can be used, although linear and phased array transducers are now much more widely used than the once popular mechanical sector scanners (see Chapter 5 for a detailed discussion of transducers for CFI). The same transducer is used for reception, and after amplification the received signals are processed along one of three different signal paths—note, however, that different pulses are used for each of the three modalities—pulsed Doppler requires a large number of echoes from one sample volume, colour Doppler information is derived from at least three and usually no more than 16 pulses transmitted along the same imaging vector, and grey-scale imaging requires only the echoes from

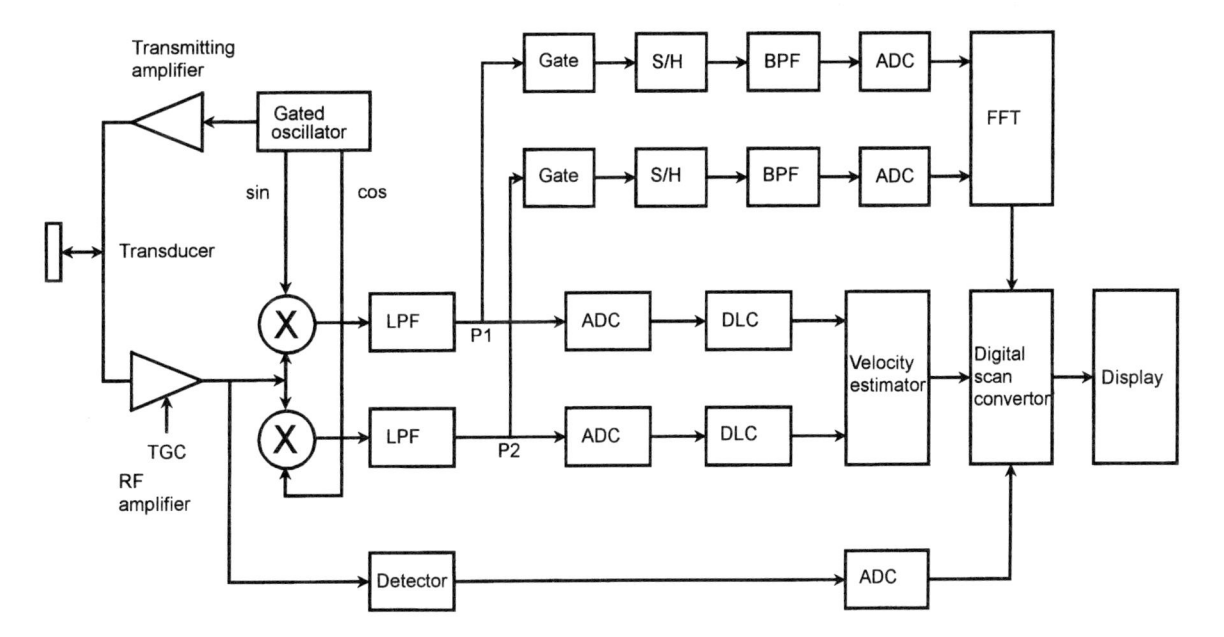

Figure 4.20 Block schematic of a simple phase domain colour flow imaging system. The top processing path is for the PW Doppler information, the middle path for the colour vector, and the lower path for the pulse–echo imaging vector. TGC, time gain compensation; LPF, low-pass filter; ADC, analogue to digital converter; DLC, delay line canceller; S/H, sample and hold; BPF, band-pass filter; FFT, fast Fourier transform. Timing control lines are omitted for simplicity

one transmitted pulse per scan line, which for good spatial resolution will usually be shorter than those used for the Doppler measurements.

The conventional Doppler signal is obtained in exactly in the same way as for a stand-alone PW system. (The PW section of Fig. 4.20 shows the two quadrature paths necessary to preserve directional information, one of which was omitted from Fig. 4.3 for the sake of simplicity. More details of directional separation for CW and PW systems can be found in Section 6.4.) The outputs of the band-pass filters are digitised and Fourier-transformed, and the resulting spectra are available to the scan converter for display in the form of sonograms.

The grey-scale image is formed exactly as in conventional pulse-echo imaging. The input is taken directly from the receiver output (or in some cases after the quadrature demodulators), and the signal amplitude at each point is suitably processed to form the image vector.

The information for the colour display is, like the simple PW Doppler, obtained from the output of the phase-quadrature demodulator, but is digitised, filtered by delay line cancellers (DLCs) and then used to calculate the velocity (and other derived quantities) at each point along the current colour vector.

The DLCs (also known as stationary echo cancellers or fixed target cancellers) serve the same purpose in a CFI system as the 'wall thump' filters in a conventional CW or PW Doppler system, that is to reject the very high amplitude echoes from stationary and near stationary targets. In essence, DLCs work by subtracting each successive echo signal from its predecessor. Signals arising from stationary objects will be unchanged and thus cancelled, whilst signals from moving objects will change and thus be preserved. This methodology is widely used in radar systems, and can be adapted to produce different frequency characteristics by allowing different degrees of feedback, and cascading one or more stages. Unfortunately, whatever techniques are used, DLCs are unable to approach the sharpness of the filters used in conventional systems, and much of the art in building a good CFI system is in optimising the stationary and near-stationary echo cancellation. Filters for CFI applications are discussed in greater depth in Section 11.2.2.

Once stationary echo cancellation has taken place, the filtered range-phase signals are sent to the 'velocity estimator'. In CFI systems the velocity estimator does not do a full spectral analysis of the Doppler signal from each sample volume, but rather estimates parameters such as mean frequency, and possibly some measure of spectral spread. There are a number of reasons why this is so, including processing speed and volume of data, but the most important are related to the short time segment and the small number of samples available for analysis.

Firstly, when CFI was introduced, real-time fast Fourier transformation (FFT) of the information from each sample volume was beyond the then state of the art in terms of processing speed. In fact, modern hardware would be capable of the required speeds, especially if parallel processing were used, but as will be seen, FFT is not an ideal tool for short segments of Doppler data, and more sophisticated spectral analysis methods demand more processing power.

Secondly, the FFT, the classical method of obtaining signal spectra, has a frequency resolution that is limited to the reciprocal of the data segment analysed, and has an extremely large variance unless some averaging can be performed (Section 8.2.2). Using the example of timing given in Section 4.5.3, the best possible spectral resolution for the 1 ms data segment would be 1 kHz, and any windowing would degrade this. This limitation can to some extent be overcome with interleaved signal processing (Section 10.5.3), but this does not increase the total number of samples available for processing, merely their timing. Modern spectral estimation techniques, such as autoregressive spectral analysis, have considerable advantages over FFT for the analysis of short data segments, and have shown promise as methods of analysing CW and PW Doppler signals (Section 8.2.4), but require even more computational power than FFTs and cannot overcome the intrinsic noisiness of the Doppler data.

A third reason for not performing spectral analysis, and one that will remain even when processing speeds increase further and better spectral analysis techniques can be applied in real time, is simply the problem of displaying the vast quantity of data that would result if spectral analysis were performed on the data from each individual sample volume. In practice, since each pixel on the display represents a small sample volume within the tissue, all the operator usually requires from the CFI system is an indication of mean or maximum velocity in that sample volume, and some indication of spectral spread which may give some information about disturbed flow. If the operator requires more detailed information about the flow in any particular region of the tissue, he or she can define a region of interest and use the PW mode to interrogate it in more detail.

A number of phase domain algorithms have been suggested for frequency estimation in CFI systems, but the autocorrelator was the first (Namekawa et al 1982a,b) and is still the most widely used. The basis of the autocorrelation technique can be intuitively appreciated by considering Fig. 4.21, which shows the phase, relative to a master clock, of a signal vector during two successive samples, $(i-1)$ and (i). The angular frequency, ω, of the vector is defined as its rate of change of phase, or:

$$\omega = \frac{d\Phi}{dt} \approx \frac{\Phi_i - \Phi_{i-1}}{T} \qquad 4.10$$

where T is the pulse repetition interval.

The tangent of the required phase difference, $\Phi_i - \Phi_{i-1}$, may be written in terms of the ratio of the sine and cosine of the phase difference, i.e:

$$\tan(\Phi_i - \Phi_{i-1})$$
$$= \frac{\sin(\Phi_i - \Phi_{i-1})}{\cos(\Phi_i - \Phi_{i-1})}$$
$$= \frac{\sin\Phi_i \cos\Phi_{i-1} - \cos\Phi_i \sin\Phi_{i-1}}{\cos\Phi_i \cos\Phi_{i-1} - \sin\Phi_i \sin\Phi_{i-1}}. \qquad 4.11$$

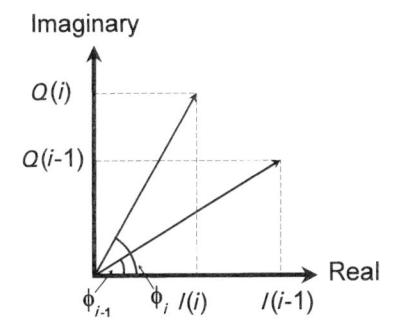

Figure 4.21 Position of a rotating signal vector during two successive samples $(i-1)$ and (i), showing the in-phase components $I(i-1)$ and $I(i)$, and quadrature components $Q(i-1)$ and $Q(i)$

If the sine and cosine terms are now expressed as the in-phase (I) and quadrature (Q) magnitudes of the vectors, and an average frequency calculated by summing over a number of pulse pairs (the justification for this is given in Chapter 11), then from eqn. 4.11, the mean angular frequency can written as:

$$\bar{\omega} = \frac{1}{T} \tan^{-1}\left\{\frac{\sum_{i=1}^{n} Q(i)I(i-1) - I(i)Q(i-1)}{\sum_{i=1}^{n} I(i)I(i-1) + Q(i)Q(i-1)}\right\}.$$

4.12

This algorithm, together with several other possible approaches to mean frequency estimation in CFI systems, is explored in more detail in Chapter 11.

The final common destination for all signals in the CFI system is the colour display, although of course the PW Doppler signal may also be presented as an audible output. The colour vector and image vector information are used to create the real-time colour flow image, whilst the conventional Doppler information derived from a selected sample volume is displayed in the form of a sonogram. It is usually possible either to display the colour image and the spectral Doppler sonogram separately, or to display both simultaneously, although the time sharing necessary to achieve this inevitably reduces the image frame rate.

The colour flow image consists of both anatomical and velocity information; where the Doppler power is insignificant, then the pulse-echo information is used to produce a conventional grey-scale image of the tissue; where the Doppler power is significant, then the grey-scale image is suppressed and the Doppler information is written in a colour-coded format (see Section 11.2.5 for further details).

Several schemes have been used to encode the colour vector information, although most are based on the use of blue shades to represent blood moving away from the transducer and red shades to represent blood moving towards the transducer. The earliest systems (Kasai et al 1985) encoded the magnitude of the mean velocity in terms of the intensity of the two colours, with bright blue and bright red representing high velocities away from and towards the transducer respectively. The variance of the velocity estimate (indicating the bandwidth of the Doppler signal) was also displayed by adding green to the display, so that as bandwidth increased (due to disturbed flow) the red shades tended to yellow and the blue shades tended to cyan. Another commonly used approach to colour display is to encode the velocity information in terms of colour hue rather than intensity, although in this case intensity may also vary at the same time. Whatever the system chosen, the display should include a colour scale showing the range of colours being used, and the operator must not forget that CFI systems, like conventional Doppler systems, can only measure the component of velocity along the ultrasound beam. Naturally, even such sophisticated phase domain systems are limited by the same maximum velocity constraints as simple PW systems (Section 4.2.2) and care is necessary to ensure that high-velocity flow patterns are correctly interpreted.

Most modern machines have a variety of display options from which the operator can choose, together with an array of features which will alert him to possible aliasing, and allow him to perform many post-processing operations such as tagging selected velocities, and acquiring time averaged colour images. The operator will also have control of many other aspects of the scanner operation, such as the size of the flow mapping area (which will often be smaller than the grey-scale image to allow a reasonably high frame rate for the colour information). The reader is referred to Chapter 10 for more details of CFI controls.

4.5.3.2 CFI Systems—Time Domain Correlation Methods

The time domain (TD) approach to CFI is quite different to that of the phase domain autocorrelation method described in the last section, in that it works directly on the RF A-lines rather than the demodulated signals, and that it calculates the movement of the targets by detecting the time shift between successive range gated echoes by finding the maximum of their cross-correlation function, rather than calculating the phase shift between successive samples (Dotti et al 1976, Bonnefous et al 1986, Bonnefous and Pesque 1986, Foster et al 1990, Embree and O'Brian 1990). Figure 4.22, which should be compared with Fig. 4.20, is a block diagram of the major components of a TD system. The PW Doppler vector path is identical to that in a PD system and is therefore merely indicated as an input to the digital scan converter (although this, too, could also be based on a time-domain approach). The colour vector path is, however, completely different. The RF A-lines are digitised immediately after the receiving amplifier, and the rest of the system is entirely digital (note that some modern phase domain-based systems

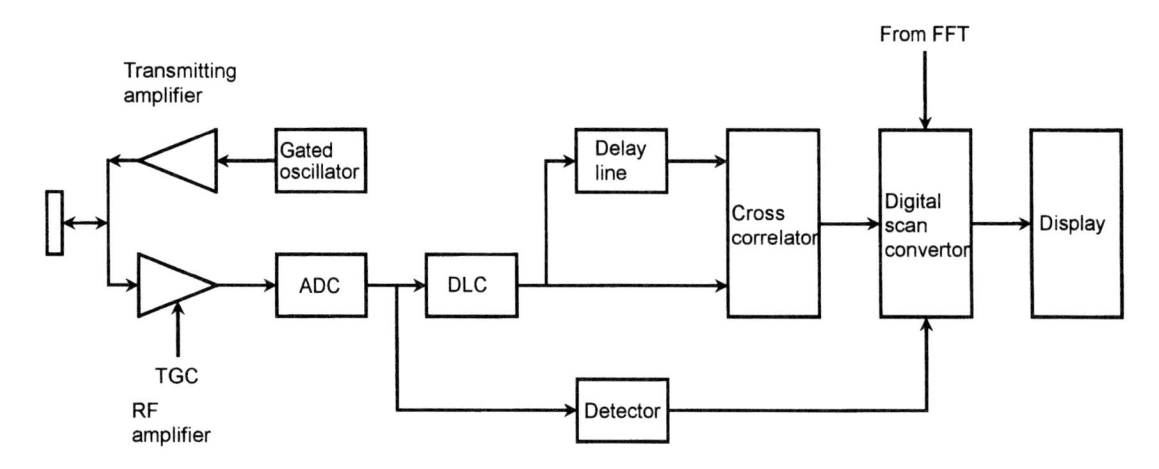

Figure 4.22 Block schematic of a time-domain colour-flow imaging system. Abbreviations are the same as those in Fig. 4.20

also digitise the RF signal right at the front end, although this is not strictly necessary for the autocorrelation algorithm). The signal is passed through a fixed echo canceller (DLC) to remove echoes from stationary structures and then on to the cross-correlation unit, which consists of a further delay line that introduces a delay of exactly one pulse repetition period, and an algorithm for calculating the cross-correlation between the direct and delayed signals for a series of small negative and positive time shifts. The processor finds the maximum of the cross-correlation function, and the velocity is calculated from the apparent displacement of the target that has occurred within the single pulse repetition period.

The image vector path is virtually the same as in Fig. 4.20, except that as the RF signal has to be digitised for the CFI vector, the information needed for the image vector is also available in a digital form right at the beginning of the processing chain.

The TD method of obtaining colour vector information is technically more demanding than the PD method because of the requirement for a much higher digitisation rate, and particularly because of the very high number of numerical operations necessary to perform the cross-correlations (Section 11.4.4), but is claimed to have significant advantages in terms of accuracy, noise immunity and lack of susceptibility to aliasing.

In terms of accuracy, the argument in favour of TD methods is that in order to obtain good spatial resolution it is necessary to use broad-band (short-pulse) excitation of the transducer, and that the shape of the resulting ultrasound pulses will be significantly modified by frequency-dependent attenuation and scattering in the tissues (Newhouse et al 1977, Holland et al 1984, Round et al 1987). This pulse shaping will have a significant effect on the mean frequency of the received pulses, and will thus bias the measured Doppler shift. The TD method, on the other hand, only measures time delays and therefore will not be affected by the filtering effects of the tissues (Bonnefous and Pesque 1986, Suorsa and O'Brien 1988). The effects of frequency-dependent attenuation and scattering on mean frequency estimation are discussed in detail in Section 8.3.5.

The reason for the superior noise immunity of TD systems is claimed to be that PD systems are susceptible to the fluctuations of the local mean frequency of the ultrasonic A-line, induced by the random interference between the echoes from the many scatterers within the tissues, whilst since TD systems are dependent only on times of flight and not frequency, they are not (Bonnefous and Pesque 1986). Practical TD systems do appear to produce estimates with less variance, which means that fewer individual estimates need to be averaged to obtain each colour vector. This in turn means that TD systems can work with higher frame rates and/or a greater line density, and that less interpolation is needed to produce acceptable image quality. More recently, however, Torp et al (1993) and Hoeks et al (1994a) have suggested that the PD method does not have an intrinsically higher variance than the TD method and that the variance can be reduced by suitable averaging (see Section 11.4.5.1).

The third advantage claimed for TD techniques is that since they measure time rather than phase they are not susceptible to aliasing. This may be the case when there is a strong signal and little noise, because the cross-correlation will have an unambiguous global maximum, however because of the limited bandwidth of the signal, and its consequent periodic nature, it seems likely that, under low SNR conditions and when successive A-lines are derived from slightly different scatterer paths, there may be ambiguity about which correlation maximum to use, and thus the appearance of aliasing (Section 11.4.5.3).

Although at least one commercial machine has used TD methods for a number of years (based on a simple one-bit correlation), most CFI machines are still based on PD methods, which, given the vast investment manufacturers have put into development, suggests that at present at least any gain to be made using TD methods are not cost-effective. Furthermore, recently developed phase-shift estimation algorithms may well prove superior to the cross-correlation technique and be much easier to implement (Section 11.3).

4.5.3.3 CFI Systems—Doppler Power Mapping

An alternative to coding and displaying the Doppler shift frequency measured from each sample volume is to measure and display the total Doppler power from each sample volume, which is governed primarily by the volume of moving blood rather than its velocity. This was a facility available on many early machines (Seo et al 1985, Jain 1991, Evans 1993), but was not properly exploited until Rubin and colleagues (1994) demonstrated that, by careful optimisation of the Doppler gain (increasing it), sacrifice of directional information, total suppression of the grey-scale information and suitable averaging, it was possible to produce acceptable images of flow in much smaller vessels than had previously been the case. Figures 4.23–4.25 (see Plate III) are examples of images formed using the power Doppler mode of a commercial scanner.

Much of the advantage of power Doppler stems from the achievable increase in Doppler gain, but the virtual independence of the displayed image from measured velocity is also a contributory factor. In any Doppler system, as the sensitivity is increased to detect weak signals, random noise from various sources will influence the detection algorithms. With a conventional velocity map (VM) this leads to a noisy 'mottled display' that changes rapidly with time because each short segment of noise analysed by the system from each sample volume may have virtually any mean frequency (which may also be interpreted as corresponding to a positive or negative flow direction). With a power map (PM), however, it is the power of the noise from each sample volume which is calculated and displayed, and because the total power of a noise segment is very much less variable than its mean frequency (and must be positive), the display is relatively uniform. The result of this is that small signals close to the noise level are difficult to detect using conventional VMs, but should stand out on the PM because they significantly increase the total power.

Because Doppler power, unlike Doppler frequency, is (more or less) angle-independent, this also helps to improve the appearance of tortuous vessels or networks of small vessels. Indeed, because of the effects of intrinsic spectral broadening (Section 7.3.4) it is possible to detect Doppler power even when the transducer is at right-angles to a blood vessel, and even though the mean Doppler shift frequency is very close to zero (due to the presence of both forward and reverse components), and thus image segments of a vessel using a PM which would not be possible with a VM. Note, however, that wall thump filters will inevitably reduce the total power returning from a vessel in such an orientation.

A third advantage of using a PM is that aliasing due to high velocities will, in general, not affect the display and may make it easier to interpret. A final advantage is that vessels appear to have smoother edges in a PM than a VM. This is because if at the edge of a vessel only a proportion of a sample volume contains significant flow, the PM will show proportionally less power but the VM will either show the mean velocity of the significant flow or, if there is insufficient Doppler power, the colour will be suppressed by the priority encoder and the grey-scale information written instead. This means that with VMs the pixels along the vessel edge will tend to assume one of two binary states, depending on the exact location of each sample volume, and this leads to an apparently jagged edge.

PM CFI clearly has a number of advantages over VM CFI, particularly for imaging small vessels. It must not be forgotten, however, that these advantages are only gained at the expense of suppressing all information about velocity (including direction), which in many circumstances is clinically vital. Fortunately, the difference between implementing PM and VM CFI lies purely in the software, and thus the operator is able to switch rapidly from one mode to the other to utilise the advantages of both techniques.

4.5.4 Velocity Imaging Systems — Emerging Technologies

Doppler velocity imaging has made tremendous strides over the past decade, and continues to be the subject of active research in many laboratories around the world. Considerable effort has been put into designing new algorithms for clutter rejection and velocity estimation for CFI systems, and some of these are described in Chapter 11. An obvious extension to 2-D CFI is 3-D CFI (Picot et al 1993, Guo et al 1995, Ferrara et al 1996, Guo and Fenster 1996, Ritchie et al 1996), and this too has been an area of active research, and is discussed further in Chapter 13 (Section 13.9). A major limitation of all the methods discussed in this chapter is that the velocity calculated from the Doppler equation is merely the component of the true velocity vector in the direction of the transducer, and several groups are currently working on vector Doppler techniques designed to overcome this problem, usually by interrogating each sample volume from different angles, so that the true velocity vectors can be found (Overbeck et al 1992, Hoskins et al 1994, Beach et al 1996, Giarrè et al 1996). These techniques are destined to become extremely important, and a summary of progress in this area to date can also be found in Chapter 13 (Section 13.6). Two further aspects of velocity imaging systems also described in Chapter 13 are attempts that have been made to overcome the aliasing problem experienced with phase domain algorithms (Section 13.11) and the use of 2nd harmonic Doppler imaging (in conjunction with ultrasonic contrast agents) to greatly improve signal-to-noise ratio (Schrope and Newhouse 1993, Chang et al 1995, Chang et al 1996) (Section 13.2).

Doppler imaging has improved dramatically over the last decade; the next decade holds similar promise.

4.6 SUMMARY

Doppler systems have advanced from simple flow detectors to their present state of sophistication in a matter of only 40 years, with progress over the last 10 having been particularly rapid. There is now a wide range of Doppler instruments available, each suited to a particular task and all useful in their own way. The modern colour flow scanner is capable of producing the most remarkable images of flowing blood within the body using totally non-invasive techniques.

The purpose of this chapter has been to give the reader a general overview of the Doppler systems in common use, together with others that have been mooted in the past. Details of systems which are still at a development stage, and more detailed explanations of the physics and instrumentation of Doppler ultrasound systems, follow throughout this book.

4.7 REFERENCES

Ahn YB, Kim YG, Park SB (1988) New multigate pulsed Doppler system using second-order sampling. ElecLetters 24:1091–1093

Arenson JW, Cobbold RSC, Johnston KW (1982) Real-time two-dimensional blood flow imaging using a Doppler ultrasound array. In: Ash EA, Hill CR (eds) Acoustical Imaging, vol 12, pp. 529–538. Plenum Press, New York

Atkinson P, Woodcock JP (1982) Doppler Ultrasound and Its Use in Clinical Measurement, p. 46. Academic Press, London

Baker DW (1970) Pulsed ultrasonic Doppler blood-flow sensing. IEEE Trans Sonics Ultrason SU-17:170–185

Barber FE, Baker DW, Nation AWC, Strandness DE, Reid JM (1974) Ultrasonic duplex echo-Doppler scanner. IEEE Trans Biomed Eng 21:109–113

Beach KW, Dunmire B, Overbeck JR, Waters D, Billeter M, Labs K-H, Strandness DE (1996) Vector Doppler systems for arterial studies. Part I: Theory. J Vasc Invest 2:155–165

Bendick PJ, Newhouse VL (1974) Ultrasonic random-signal flow measurement system. J Acoust Soc Am 56:860–865

Bertram CD (1979) Distance resolution with the FM–CW ultrasonic echo-ranging system. Ultrasound Med Biol 5:61–67

Bonnefous O, Pesque P (1986) Time domain formulation of pulse-Doppler ultrasound and blood velocity estimation by cross correlation. Ultrason Imaging 8:73–85

Bonnefous O, Pesque P, Bernard X (1986) A new velocity estimator for colour flow mapping. In: Proceedings of the 1986 Ultrasonics Symposium, pp. 855–860. IEEE, Piscataway, NJ

Brandestini M (1975) Application of the phase detection principle in a transcutaneous velocity profile meter. In: Kazner E, de Vlieger M, Muller HR, McCready VR (eds) Ultrasonics in Medicine, pp. 144–152. Proceedings 2nd European Congress on Ultrasonics in Medicine, Munich 12–16th May, 1975. Excerpta Medica, Amsterdam

Brandestini M (1978) Topoflow—a digital full range Doppler velocity meter. IEEE Trans Sonics Ultrason SU-25:287–293

Brandestini MA, Forster FK (1978) Blood flow imaging using a discrete-time frequency meter. In: Proceedings of the 1978 IEEE Ultrasonics Symposium, pp. 348–352. IEEE, Piscataway, NJ

Cathignol DJ (1986) Signal-to-clutter ratio in pseudo-random Doppler flowmeter. Ultrason Imaging 8:272–284

Cathignol DJ, Fourcade C, Chapelon J-Y (1980) Transcutaneous blood flow measurements using pseudorandom noise Doppler system. IEEE Trans Biomed Eng BME-27:30–36

Cathignol DJ, Birer A, Unterreiner R (1988) Multifrequency Doppler flowmeter: a way to rapid CW Doppler imaging. IEEE Trans Biomed Eng 35:416–419

Chan FH, Pizer SM (1976) An ultrasonogram display system using a natural color scale. J Clin Ultrasound 4:335–338

Chang PH, Shung KK, Wu S-J, Levene HB (1995) Second harmonic imaging and harmonic Doppler measurements with Albunex. IEEE Trans Ultrason Ferroelec Freq Contr 42:1020–1027

Chang PH, Shung KK, Levene HB (1996) Quantitative measurements of second harmonic Doppler using ultrasound contrast agents. Ultrasound Med Biol 22:1205–1214

Curry GR, White DN (1978) Colour coded ultrasonic differential velocity arterial scanner (echoflow). Ultrasound Med Biol 4:27–35

Dotti D, Gatti E, Svelto V, Ugge A, Vidali P (1976) Blood flow measurements by ultrasound correlation techniques. Energia Nucleare 23:571–575

Embree PM, O'Brien WD (1990) Volumetric blood flow via time-domain correlation: experimental verification. IEEE Trans Ultrason Ferroelec Freq Contr UFFC-37:176–189

Evans DH (1993) Techniques for color-flow imaging. In: Wells PNT (ed) Advances in Ultrasound Techniques and Instrumentation—Clinics in Diagnostic Ultrasound, vol 28, pp. 87–107. Churchill Livingstone, New York

Eyer MK, Brandestini MA, Phillips DJ, Baker DW (1981) Color digital echo/Doppler image presentation. Ultrasound Med Biol 7:21–31

Ferrara KW, Zagar B, Sokil-Melgar J, Algazi VR (1996) High resolution 3D color flow mapping: applied to the assessment of breast vasculature. Ultrasound Med Biol 22:293–304

Fish P (1972) Visualising blood vessels by ultrasound. In: Roberts C (ed) Blood Flow Measurement, pp. 29–32. Sector Publishing, London

Fish PJ (1975) Multichannel, direction-resolving Doppler angiography. In: Kazner E, de Vlieger M, Muller HR, McCready VR (eds) Ultrasonics in Medicine, pp. 153–159. Proceedings 2nd European Congress on Ultrasonics in Medicine, Munich, 12–16th May 1975. Excerpta Medica, Amsterdam

Fish PJ (1977) Recent progress in the field of Doppler devices: reprint from Excerpta Medica International Congress Series No.436; Recent Advances in Ultrasound Diagnosis. Proceedings of the International Symposium on Recent Advances in Ultrasound Diagnosis, Dubrovnik, October 1977

Fish PJ (1981) A method of transcutaneous blood flow measurement—accuracy considerations: reprint from Recent Advances in Ultrasound Diagnosis 3. Proceedings of the 4th European Congress on Ultrasonics in Medicine, Dubrovnik, May 1981

Foster SG, Embree PM, O'Brien WD (1990) Flow velocity profile via time-domain correlation: error analysis and computer simulation. IEEE Trans Ultrason Ferroelec Freq Contr UFFC-37:162–175

Franklin DL, Schlegel W, Rushmer RF (1961) Blood flow measured by Doppler frequency shift of back-scattered ultrasound. Science 134:564–565

Giarrè M, Dousse B, Meister JJ (1996) Velocity vector reconstruction for color flow Doppler: experimental evaluation for a new geometrical method. Ultrasound Med Biol 22:75–88

Grandchamp PA (1975) A novel pulsed directional Doppler velocimeter: the phase detection profilometer. In: Kazner E, de Vlieger M, Muller HR, McCready VR (eds) Ultrasonics in Medicine, pp. 137–143. Proceedings of the 2nd European Congress on Ultrasonics in Medicine, 12–16th May 1975. Excerpta Medica, Amsterdam

Guo Z, Fenster A (1996) Three-dimensional power Doppler imaging: a phantom study to quantify vessel stenosis. Ultrasound Med Biol 22:1059–1069

Guo Z, Moreau M, Rickey DW, Picot PA, Fenster A (1995) Quantitative investigation of in vitro flow using three-dimensional colour Doppler ultrasound. Ultrasound Med Biol 21:807–816

Histand MB, Miller CW, McLeod FD (1973) Transcutaneous measurement of blood velocity profiles and flow. Cardiovasc Res 7:703–712

Hoeks APG, Reneman RS, Peronneau PA (1981) A multigate pulsed Doppler system with serial data processing. IEEE Trans Sonics Ultrason SU-28:242–247

Hoeks APG, Peeters HHPM, Ruissen CJ, Reneman RS (1984) A novel frequency estimator for sampled Doppler signals. IEEE Trans Biomed Eng BME-31:212–220

Hoeks APG, Brands PJ, Arts TGJ, Reneman RS (1994a) Subsample volume processing of Doppler ultrasound signals. Ultrasound Med Biol 20:953–965

Hoeks APG, Brands PJ, Reneman RS (1994b) Effect of sample window length on the correlation between RF signal and pulsed Doppler signal intensity. Ultrasound Med Biol 20:35–40

Hokanson DE, Mozersky DJ, Sumner DS, Strandness DE (1971) Ultrasonic arteriography: a new approach to arterial visualisation. Biomed Eng 6:420

Holland SK, Orphanoudakis SC, Jaffe CC (1984) Frequency-dependent attenuation effects in pulsed Doppler ultrasound: experimental results. IEEE Trans Biomed Eng BME-31:626–631

Hoskins PR, Fleming A, Stonebridge P, Allan PL, Cameron D (1994) Scan-plane vector maps and secondary flow motions in arteries. Eur J Ultrasound 1:159–169

Hottinger CF, Meindl JD (1975) Unambiguous measurement of volume flow using ultrasound. Proc IEEE 63:984–985

Jain SP, Fan PH, Philpot EF, Nanda NC, Aggarwal KK, Moos S, Yoganathan AP (1991) Influence of various instrument settings on the flow information derived from the power mode. Ultrasound Med Biol 17:49–54

Jerri AJ (1977) The Shannon sampling theorem—its various extensions and applications: a tutorial review. Proc IEEE 65:1565–1596

Jethwa CP, Kaveh M, Cooper GR, Saggio F (1975) Blood flow measurements using ultrasonic pulsed random signal Doppler system. IEEE Trans Sonics Ultrason SU-22:1–11

Kaneko Z, Shiraishi J, Omizo H, Kato K, Motomiya M, Izumi T, Okumura T (1966) Analysing blood flow with a sonagraph. Ultrasonics 4:22–23

Kasai C, Namekawa K, Koyano A, Omoto R (1985) Real-time two-dimensional blood flow imaging using an autocorrelation technique. IEEE Trans Sonics Ultrason SU-32:458–464

Keller HM, Meier WE, Anliker M, Kumpe DA (1976) Non-invasive measurement of velocity profiles and blood flow in the common carotid artery by pulsed Doppler ultrasound. Stroke 7:370–377

Kemper WS, Franklin D, Vatner SF (1971) Practical implementation of the FM ultrasonic flowmeter. In: Proceedings of the 24th Annual Conference on Engineering in Medicine and Biology, p. 269. Las Vegas, NV

Kremkau F (1990) Calculation of V_{max} incorporating dynamic range and attenuation. Ultrasound Med Biol 16:L522

Kristoffersen K (1986) Optimal receiver filtering in pulsed Doppler ultrasound blood velocity measurements. IEEE Trans Ultrason Ferroelec Freq Contr 33:51–58

Loupas T, Powers JT, Gill RW (1995) An axial velocity estimator for ultrasound blood flow imaging, based on a full evaluation of the Doppler equation by means of a two-dimensional autocorrelation approach. IEEE Trans Ultrason Ferroelec Freq Contr 42:672–688

McCarty K, Woodcock JP (1974) The ultrasonic Doppler shift flowmeter—a new development. Biomed Eng 9:336–341

McCarty K, Woodcock JP (1975) Frequency modulated ultrasonic Doppler flowmeter. Med Biol Eng 13:59–64

McDicken WN (1991a) M-scan instruments, performance and use. In: Diagnostic Ultrasonics—Principles and Use of Instruments, 3rd edn, pp. 187–195. Churchill Livingston, Edinburgh

McDicken WN (1991b) Real-time B-scan instruments. In: Diagnostic Ultrasonics—Principles and Use of Instruments, 3rd edn, pp. 123–137. Churchill Livingston, Edinburgh

McLeod FD (1974) Multichannel pulse Doppler techniques. In: Reneman RS (ed) Cardiovascular Applications of Ultrasound, pp. 85–107. North-Holland, Amsterdam

Moser U, Anliker M, Schumacher P, Vieli A, Pinter P (1992a) C-Mode Doppler: a quantitative ultrasonic real-time procedure for the measurement of 2-D velocity fields and volume flow in large vessels. Proceedings of Eurodop '92 Conference, pp. 63–64. British Medical Ultrasound Society, London

Moser U, Vieli A, Schumacher P, Pinter P, Basler S, Anliker M (1992b) Ein Doppler-Ultraschall-Gerät zur Bestimmung des Blut-Volumenflusses. Ultraschall in Med 13:77–79

Moser U, Schumacher PM, Anliker M (1995) Benefits and limitations of the C-mode Doppler procedure. Acoustical Imaging 21:509–514

Mozersky DJ, Hokanson DE, Baker DW, Sumner DS, Strandness DE (1971) Ultrasonic arteriography. Arch Surg 103:663–667

Namekawa K, Kasai C, Tsukamoto M, Koyano A (1982) Imaging of blood flow using autocorrelation. Ultrasound Med Biol 8 (Suppl. 1): 138

Namekawa K, Kasai C, Tsukamoto M, Koyano A (1982b) Realtime bloodflow imaging system utilizing auto-correlation techniques. In: Lerski RA, Morley P (eds) Ultrasound '82, pp. 203–208. Pergamon Press, New York

Newhouse VL, Ehrenwald AR, Johnson GF (1977) The effect of Rayleigh scattering and frequency dependent absorption on the output spectrum of Doppler blood flowmeters. In: White D, Brown RE (eds) Ultrasound in Medicine, vol 3B, pp. 1181–1191. Plenum Press, New York

Nowicki A, Reid JM (1981) An infinite gate pulse Doppler. Ultrasound Med Biol 7:41–50

Nyquist H (1928) Certain topics in telegraph transmission theory. AIEE Trans 47:617–644

Omoto R, Yokote Y, Takamoto S, Kyo S, Ueda K, Asano H, Namekawa K, Kasai C, Kondo Y, Koyano A (1984) The development of real-time two-dimensional Doppler echocardiography and its clinical significance in acquired valvular heart disease. Jap Heart J 25:325–340

Overbeck JR, Beach KW, Strandness DE (1992) Vector Doppler: accurate measurement of blood velocity in two dimensions. Ultrasound Med Biol 18:19–31

Peronneau PA, Leger F (1969) Doppler ultrasonic pulsed blood flowmeter. Proceedings of the 8th International Conference on Medical and Biological Engineering, pp. 10–11.

Peronneau P, Xhaard M, Nowicki A, Pellet M, Delouche P, Hinglais J (1972) Pulsed Doppler ultrasonic flowmeter and flow pattern analysis. In: Roberts VC (ed) Blood Flow Measurement, pp. 24–28. Sector, London

Peronneau PA, Bournat JP, Bugnon A, Barbet A, Xhaard M (1974) Theoretical and practical aspects of pulsed Doppler flowmetry: real-time application to the measure of instantaneous velocity profiles *in vitro* and *in vivo*. In: Reneman RS (ed) Cardiovascular Applications of Ultrasound, pp. 66–84. North-Holland, Amsterdam

Phillips DJ, Powers JE, Eyer MK, Blackshear WM, Bodily KC, Strandness DE, Baker DW (1980) Detection of peripheral vascular disease using the Duplex scanner III. Ultrasound Med Biol 6:205–218

Picot PA, Rickey DW, Mitchell R, Rankin RN, Fenster A (1993) Three-dimensional colour Doppler imaging. Ultrasound Med Biol 19:95–104

Pourcelot LG (1979) Real-time blood flow imaging. In: Lancee CT (ed) Echocardiology, pp. 421–429. Martinus Nijhoff, The Hague

Reid JM, Spencer MP (1972) Ultrasonic technique for imaging blood vessels. Science 176:1235–1236

Reneman RS, Van Merode T, Hick P, Hoeks APG (1986) Cardiovascular applications of multi-gate pulsed doppler systems. Ultrasound Med Biol 12:357–370

Ritchie CJ, Edwards WS, Mack LA, Cyr DR, Kim Y (1996) Three-dimensional ultrasonic angiography using power-mode Doppler. Ultrasound Med Biol 22:277–286

Round WH, Bates RHT (1987) Modification of spectra of pulses from ultrasonic transducers by scatterers in non-attenuating and in attenuating media. Ultrason Imaging 9:18–28

Rubin JM, Bude RO, Carson PL, Bree RL, Adler RS (1994) Power Doppler US: a potentially useful alternative to mean frequency-based color Doppler US. Radiology 190:853–856

Schrope BA, Newhouse VL (1993) Second harmonic ultrasonic blood perfusion measurement. Ultrasound Med Biol 19:567–579

Seo Y, Shiki E, Hongo H, Iinuma K (1985) 2-D color flow mapping systems—power mode. In: Proceedings of the 47th Meeting of the Japan Society of Ultrasonics in Medicine, p. 481

Shannon CE (1949) Communications in the presence of noise. Proc Inst Radio Engineers 37:10–21

Spencer MP, Reid JM, Davis DL, Paulson PS (1974) Cervical carotid imaging with a continuous-wave Doppler flowmeter. Stroke 5:145–154

Suorsa VT, O'Brien WD (1988) Effects of ultrasonic attenuation on the accuracy of the blood flow measurement technique utilizing time domain correlation. In: Proceedings of the 1988 IEEE Ultrasonics Symposium, pp. 989–994. IEEE, Piscataway, NJ

Takabatake H, Sasayama S, Asanoi H, Tanaka I, Ando M, Kihara Y, Fujita M (1990) A new two-dimensional color flow mapping system for the visualization of flow velocity profile. J Cardiovasc Technol 9:173–180

Thomas N, Leeman S (1995) The attenuation effect in pulsed Doppler flowmeters. Acoustical Imaging 21:543–552

Torp H, Lai XM, Kristoffersen K (1993) Comparison between cross-correlation and auto-correlation technique in color flow imaging. In: Proceedings of the 1993 IEEE Ultrasonics Symposium, pp. 1039–1042. IEEE, Piscataway, NJ

Tortoli P, Guidi F (1995) Detection of low velocities by a novel ultrasound multigate system. In: Levy M, Schneider SC, McAvoy BR (eds) Proceedings of the 1995 IEEE Ultrasonics Symposium, pp. 1523–1526. IEEE, Piscataway, NJ

Tortoli P, Andreuccetti F, Manes G, Atzeni C (1988) Blood flow images by a SAW-based multigate Doppler system. IEEE Trans Ultrason Ferroelec Freq Contr 35:545–551

Tortoli P, Guidi F, Guidi G, Atzeni C (1996) Spectral velocity profiles for detailed ultrasound flow analysis. IEEE Trans Ultrason Ferroelec Freq Contr 43:654–659

Tortoli P, Guidi G, Berti P, Guidi F, Righi D (1997) An FFT-based flow profiler for high-resolution *in vivo* investigations. Ultrasound Med Biol 23:899–910

Waag RC, Myklebust JB, Rhoads WL, Gramiak R (1972) Instrumentation for non-invasive cardiac chamber flow rate measurement. In: Proceedings of the 1972 IEEE Ultrasonics Symposium, pp. 74–77. IEEE, Piscataway, NJ

Wells PNT (1969) A range-gated ultrasonic Doppler system. Med Biol Eng 7:641–652

Wilhjelm JE, Pedersen PC (1993a) Target velocity estimation with FM and PW echo ranging Doppler systems—Part 1: Signal analysis. IEEE Trans Ultrason Ferroelec Freq Contr 40:366–372

Wilhjelm JE, Pedersen PC (1993b) Target velocity estimation with FM and PW echo ranging Doppler systems—Part II: Systems analysis. IEEE Trans Ultrason Ferroelec Freq Contr 40:373–380

Wilhjelm JE, Pedersen PC (1995) Comparison between PW Doppler system and enhanced FM Doppler system. In: Levy M, Schneider SC, McAvoy BR (eds) Proceedings of the 1995 IEEE Ultrasonics Symposium, pp. 1549–1552. IEEE, Piscataway, NJ

Wilhjelm JE, Pedersen PC (1996) Analytical and experimental comparisons between the frequency-modulated frequency-shift measurement and the pulsed-wave time-shift measurement Doppler systems. J Acoust Soc Am 100:3957–3970

5

Ultrasonic Transducers, Fields and Beams

5.1 INTRODUCTION

Several general discussions of ultrasonic transducers, fields and beams for medical imaging are available (Kinsler et al 1982, Hill 1986, Wells 1977, Mortimer 1982, Shung and Zipparo 1996, Whittingham 1997). In this chapter, transducers will be discussed from the point of view of Doppler techniques. State-of-the-art transducers as supplied by manufacturers are surprisingly similar; it is as though the field has moved toward the same solution for transducer design with the presently available materials. As with ultrasonic pulse-echo imaging, the performance of the transducer is of paramount importance in a Doppler system. The significance of a poor transducer may not be as immediately obvious in Doppler techniques as in B-mode imaging, where degradation of resolution is more readily appreciated, but this is changing as Doppler images improve. In the Doppler case, poor transducers may result in additional flow signals from neighbouring vessels or noise in the Doppler signal and sonogram due to low sensitivity. Transducers designed for B-mode imaging are highly damped to generate short pulses and focused to provide a narrow beam. Such transducers are also used for Doppler imaging and in duplex Doppler techniques, particularly in the case of array transducers. Some transducers, designed purely for Doppler methods, have little or no damping since they are designed to generate long pulses or continuous wave ultrasound. These Doppler transducers are therefore more sensitive than those which are also used for imaging.

The Doppler beam is the region in front of the transducer in which blood flow or tissue motion may be detected. The characteristics of the beam are of importance in the operation of any Doppler system. The shape of the beam is determined by a combination of the shape of the zone into which ultrasound is transmitted (the ultrasonic field) and the reception zone from which ultrasonic signals may be received. For a single element transducer used for both transmission and reception of pulses, the ultrasonic field and the reception zone have the same shape and are orientated in the same direction (Fig. 5.1a). With a dual-element transducer the shape of the field and reception zone are often the same but are orientated to converge in front of the transducer (Fig. 5.1b). The ultrasonic beam is then the region of overlap. In the case of multi-element arrays which employ electronic focusing, the transmission field and the reception zone have different shapes and again the beam is the region of overlap (Fig. 5.1c).

5.2 TRANSDUCERS

An ultrasonic transducer converts electrical energy into acoustic energy during transmission when its active element is excited by a voltage or current signal. Conversely, during reception, the acoustic energy of the returned ultrasound is converted into an electrical signal. The active elements of transducers used in diagnostic medicine depend on the piezoelectric effect for their operation. Initially, transducer elements were made from naturally occurring materials such as quartz. Later, piezoelectric ceramics, for example lead zirconate titanate (PZT), were introduced and are at present still the most commonly used material. Piezoelectric polymer elements are now slowly being introduced, e.g. polyvinylidene difluoride (PVDF) (Chen et al 1978, Chen 1983, Woodward 1977, Lancée et al 1985, Hunt et al 1983, Sherar and Foster 1989). These have acoustic properties closer to tissue than the solid ceramics, therefore the passage of ultrasound across the transducer/tissue interface is more efficient. The design of ultrasonic transducers has been described in the scientific literature (Bainton

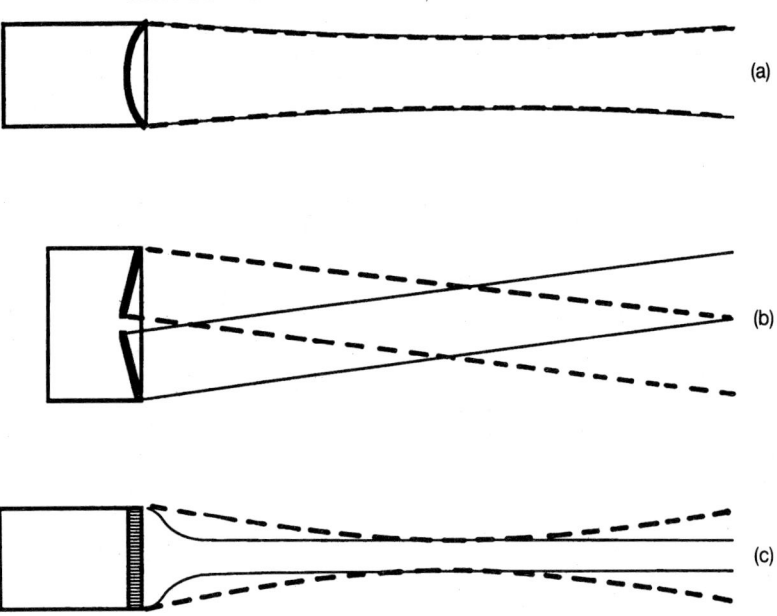

Figure 5.1 The transmission fields (dashed lines) and reception zones (solid lines) of (a) single element, (b) dual-element and (c) multi-element transducers

and Silk 1980, McKeighen 1983, Hunt et al 1983, Kino 1987); however, design details are often not completely available as they are of commercial value. The user or purchaser of transducers is therefore advised to assess them thoroughly in clinical tests. Nominally identical transducers do not always have identical performance, since small variations in materials or structure can have a large influence on the final performance. Custom-made transducers for special projects can often have a disappointing performance, since they are usually made in small numbers and do not benefit from improvements over a period of development. Transducers found on commercial equipment have normally benefited from several years of development. In the following subsections the types of transducer encountered in equipment will be discussed, with particular emphasis being placed on their influence on Doppler techniques. Table 5.1 lists the types of transducer employed for each Doppler application. Research workers and routine clinical users involved with Doppler techniques are not required to know great detail of the structure and function of transducers but some knowledge of them can point the way to new applications.

The piezoelectric effect, featured by the material of a transducer element, results from the internal alteration in relative positions of the ions in the material and hence the charge distribution across it when a pressure is applied across the element. This altered charge distribution is the basis of detection of ultrasonic pressure waves. Conversely, the thickness of the element can be altered when an applied voltage moves the relative position of the ions and hence alters the thickness of the element. This is the basis of ultrasound generation. Several materials are potentially useful for diagnostic transducers. The suitability of a material depends on a number of factors, in particular the piezoelectric pressure constant and the electric field distortion constant commonly denoted by g and d, respectively (Shung and Zappiro 1996). The pressure constant, or receiving constant g, is the open circuit electric field per unit of applied stress and has units of volt metre/Newton. The electric field distortion constant, or the transmission constant d, is the strain produced per unit of applied electric field when the external stress is zero and has units of metre/volt. The electromechanical coupling coefficient, k, is defined as the ratio of the stored mechanical energy to the total

Table 5.1 Transducer types for various Doppler applications

Application	Scan type	Transducer
Cardiac	Sector (90°)	Phased array
		Mechanical scanner
		Transoesophageal
		PW or CW single beam
Superficial	Linear	Linear array
		Mechanical waterbath scanner
Paediatric	Sector (90°) or linear	Phased array
		Curved array
		Mechanical scanner
		Mechanical waterbath scanner
Abdominal	Sector (90°) or linear	Phased array
		Curved array
		Mechanical scanner
Obstetric	Sector (90°) or linear	Phased array
		Linear array
		Curved array
		Mechanical scanner
		PW or CW single beam
Ophthalmological	Sector (90°) or linear	Phased array
		Linear array
		Mechanical scanner
Internal	Sector (90°, 360°) or linear	Phased array
		Linear array
		Curved array
		Mechanical scanner
Intravascular	Sector (360°) or linear	Phased array
		Linear array
		Mechanical scanner
		PW or CW single beam
Transcranial	Sector (90°)	Phased array
		PW single beam

Table 5.2 Properties of common piezoelectric materials (after Shung and Zipparo 1996 by permission, © 1996 IEEE)

Property	PVDF	Quartz (X-cut)	PZT-5H
d (10^{-12} m/V)	15	2.31	583
g (10^{-2} Vm/N)	14	5.78	1.91
k (unitless)	0.11	0.14	0.55
ε^T (10^{-11} F/m)	9.7	3.98	3010
c (m/s)	2070	5740	3970
ρ (kg/m³)	1760	2650	7450
curie temp (°C)	100	573	190

ρ, density; c, speed of sound; d, transmission constant; g, reception constant; k, electro-mechanical coupling constant; ε^T, free dielectric constant

stored energy and gives an indication of the efficiency of conversion between electrical and mechanical energy for each material. However, the situation is complex, for instance the parameters quoted g, d and k depend on the direction through the material being considered. In addition, internal losses due to heating and the generation of unwanted modes of vibration are significant. Table 5.2 allows comparison of ceramic PZT and polymer PVDF materials relative to quartz. It can be seen that PZT ceramics are much superior to quartz when the combined transmission/reception performance is considered and that the polymer PVDF has an intermediate performance. The choice between PZT and PVDF elements is often down to other practical considerations. For example, although PVDF has an acoustic impedance better matched to tissue or water than PZT, it is usually only available in thin sheets and is therefore normally considered only for high frequency use, i.e. above 15 MHz (Armitage et al 1995, Hatfield et al 1994). On the other hand, PVDF is flexible and can be wrapped round catheters. PZT can be cast or machined and, although brittle, it has been used to high frequencies in some applications, e.g. 50 MHz (Lockwood et al 1992, 1996). Although PVDF has some interesting properties, PZT transducer elements are still by far the most common.

5.2.1 Basic Transducer Structure and Function

Before considering the more complex transducers often found in Doppler machines, it is instructive to look at a basic one as illustrated in Fig. 5.2. Each face of a piezoelectric element is coated with a thin metallic layer which acts as an electrode. Wire connections to these electrodes allow the excitation signal to be applied and the echo signal to be passed to the receiver amplifier. When a ceramic element is used, a single or multilayered waveplate is attached to the front face of the transducer to reduce the mismatch of acoustic impedance at the transducer/tissue interface and hence give an increased transfer of acoustic energy across the interface. Waveplate matching is most

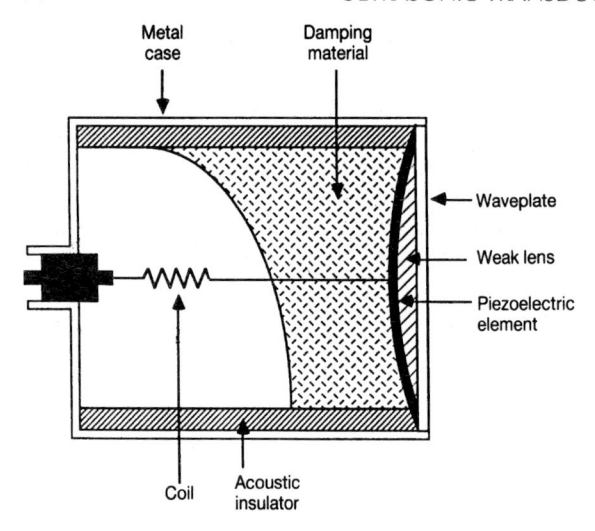

Figure 5.2 A single-element damped transducer for the generation and detection of pulsed wave ultrasound

effective for CW Doppler transducers, since the thickness of the plate can be made equal to one-quarter of the wavelength of the ultrasound. In the case of PW Doppler, the pulse has a frequency bandwidth and hence a spread of wavelengths which the waveplate structure cannot accommodate exactly. However, waveplating is still a worthwhile feature of pulsed transducers (Kossoff 1966, Kino 1987). To reduce the vibrations of the active element after the excitation pulse has ceased, some damping material, e.g. metal-loaded epoxy, is bonded to its back surface. Ideally, less damping is employed than in transducers designed for imaging, but it is common practice to use the same transducer for both Doppler and imaging. The piezoelectric element and its attachments are supported on the end of a tube of acoustic insulator which is housed in a metal cylinder. The metal cylinder is earthed to act as a screen to prevent pick-up of electromagnetic signals. This is important in the light of the weak echo signals to be detected by the transducer. The front electrode is also earthed to complete the screened case round the element. The back electrode is linked to the transmitter and receiver of the instrument. An induction coil is sometimes placed across the capacitance of the element in conjunction with an electrical matching network to optimise power transfer during transmission and reception. To assist with preserving the signal-to-noise ratio, a pre-amplifier chip may be located very close to the piezoelectric element. Finally, the whole assembly may be inserted into a plastic case, which is sealed to keep out water or coupling liquid. It is perfectly feasible for researchers to manufacture basic transducers quite cheaply since the materials are robust and do not contaminate easily.

The design of ultrasonic transducers is no longer a black art. The electrical and mechanical performance of a transducer can be modelled using equivalent circuits. Commonly used models are that of Mason, and that of Krimholtz, Leedom and Matthaei (the KLM model), (Silk 1984). These are one-dimensional models which can be used to predict the transmitted pulse shape. Their main attraction is that they can provide quick solutions. A more complete model of the action of a transducer over its front face and other internal planes is obtained using Finite Element Analysis packages, for example the ANSYS (Swanson Analysis Systems, Canonsburg, USA) or PZFLEX (Weidlinger Associates, Los Altos, USA) packages. Calculations can be carried out in either the time or the frequency domain. For calculation of beam shapes, the front surface is subdivided into a large array of small sources which transmit spherical waves which superimpose according to Huygens principle. A computational package such as MATLAB is used to calculate the field pressure amplitude at each point in front of the transducer. Another option for calculating ultrasonic fields and wave propagation is 'WAVE 2000' (CyberLogic Inc., New York), a stand-alone computer package designed specifically for computational ultrasonics.

There is considerable interest in the frequency bandwidth response of a transducer from a number of points of view. Apart from the normal relationship of interest between transducer bandwidth and the possible bandwidth of the transmitted pulse, a wideband transducer permits additional flexibility in the generation and reception of specific frequencies. For example, the transducer can be operated in a number of different wavebands. Typically, it may transmit and receive at a selected central frequency, say 3 or 6 MHz, without changing the transducer. Alternatively, it may transmit at 3 MHz and receive at 6 MHz; this technique is employed in harmonic imaging.

The percentage bandwith of a transducer may be defined as:

$$\Delta B(\%) = \frac{f_2 - f_1}{f_0} \times 100\% \qquad 5.1$$

where f_1 and f_2 are the lower and upper frequencies at which the amplitude response falls by 3 dB and f_0 is the central frequency. Manufacturers have now increased the bandwidth of medical transducers to as high as 140%.

The Q-factor for a transducer is defined as:

$$Q = \frac{f_0}{f_2 - f_1} \qquad 5.2$$

For CW Doppler devices a high Q transducer is best, whereas for pulsed ultrasound the Q factor is designed to be low, e.g. between 1 and 3. The Q of a transducer is altered by electrical and/or mechanical matching.

The shape of the ultrasonic pulse transmitted from a transducer depends on the bandwidth of the transducer and the shape of the electrical excitation pulse applied to it. In basic Doppler units, the transmission may be a continuous single frequency wave or a pulse containing 5–10 cycles. Such units are narrow band and therefore low noise in operation. In pulsed mode, the excitation varies from short delta function spikes to Gaussian shaped pulses, depending on the bandwidth and dynamic range the manufacturer wishes to achieve. On reception the echoes are passed through low noise preamplifiers and in most state-of-the-art units are then immediately digitised for subsequent B-mode imaging and Doppler signal processing. Usually no compensation for attenuation of echo amplitude with tissue depth (TGC) is employed in Doppler, since it is desired to use all signals detected above the noise level of the electronics.

The introduction of wideband transducers and the corresponding signal processing techniques for pulse-echo imaging, colour Doppler imaging (Chapter 11) and harmonic contrast imaging (Chapter 13) represent one of the major areas of advance in recent years. Increase in bandwidth over that of older transducers has been achieved by the use of improved matching layers and composite piezoelectric elements. Two or more matching layers are commonly found on transducers. They lower the reflection that occurs at the transducer/tissue interface and hence produce short broadband pulses by reducing the ringing of the element. To produce a composite material, grooves (kerfs) are cut in the PZT and filled with a suitable polymer (Gururaja et al 1985, Hayward and Hossack 1990, Hayward et al 1995, Hall et al 1993, Bui et al 1988). The addition of low density attenuating polymer reduces the acoustic impedance of the PZT/polymer composite, enabling ultrasonic energy to pass more readily between the transducer and tissue and so giving broadband characteristics to the transducer. It also reduces the energy in radial modes of vibration in the transducer, contributing further to the efficiency of the device. In arrays the polymer is between the individual elements of the array, whereas in transducers which act as single elements the grooves are cut to form PZT pillars in the polymer (Fig. 5.3). Cross-talk between array elements is also lowered by the attenuation in the polymer.

5.2.2 Single-element Transducers

A single element transducer, as found in PW Doppler devices, is shown schematically in Fig. 5.2. Focusing at a fixed range is achieved by using a lens (Tarnoczy 1965) or a concave piezoelectric element. In the latter case, to provide a flat face on the transducer and hence easier coupling to the skin, a weak convex lens is attached to the front of the concave element. The combined effect of the concave element and the convex lens is designed to give focusing at the desired range.

In stand-alone pulsed Doppler systems, the transducer is excited with an electrical signal of frequency and duration corresponding to that of the required ultrasonic pulse. Typically the pulse is of length 5–10 cycles. In Doppler imaging systems a short pulsed voltage is applied and the length of the ultrasonic pulse is determined by the ringing of the active element. This is the same technique as utilised in pulse-echo imaging. It is not the best method for generating low-noise Doppler signals but it is very convenient and gives good spatial resolution. The central frequency of pulsed Doppler transducers varies from 1 MHz for the

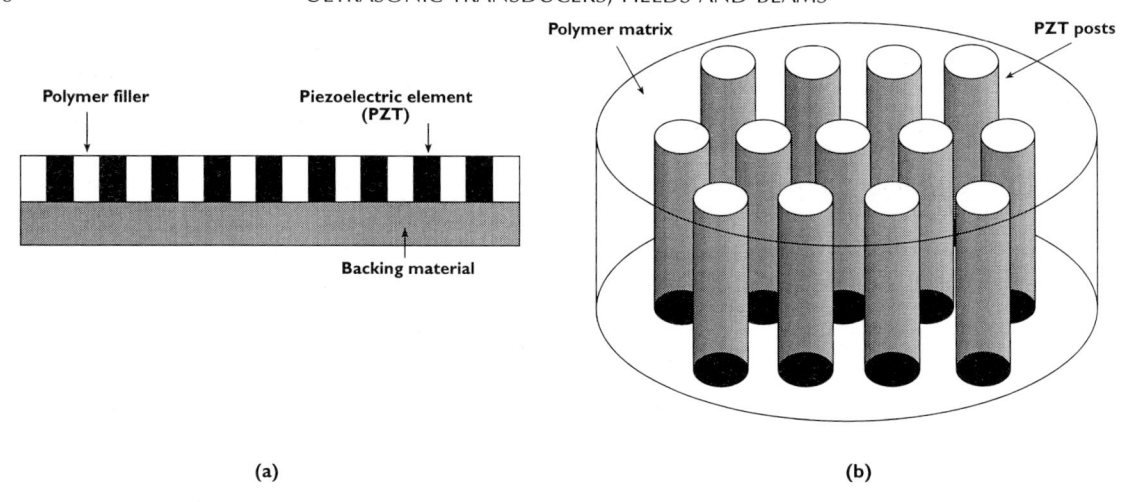

Figure 5.3 Piezoelectric composite transducers. (a) slotted version for linear and phased arrays, (b) pillar version for single element transducers

examination of deep vessels to 10 MHz for superficial ones. Specialised units operate at higher frequencies, e.g. 30 MHz in a Doppler catheter (Doucette et al 1992).

5.2.3 Double-element Transducers

A double-element transducer, as used in CW Doppler devices, is shown schematically in Fig. 5.4. Its structure is similar to that of the PW unit except that no damping material is bonded to the back surface of the piezoelectric element. The two elements are fixed with their faces at an angle to each other. In operation, one element generates a transmission field which overlaps the reception zone of the other element. When the elements are placed side by side in this manner, they are usually D-shaped or rectangular. Another variation is one central disc element surrounded by an annular element. The operating frequencies of CW Doppler transducers cover the range from 2 MHz for fetal hearts and deep vessels to 10 MHz for vessels near the skin.

In Doppler fetal monitoring systems, similar transducers find application in the generation of ultrasound to detect fetal heart activity. To provide a wide beam which accommodates some fetal movement, several elements are used in the transmission and reception modes.

5.2.4 Array Transducers

Array transducers that consist of a large number of separate elements are now widely used in both imaging and Doppler applications. They have many important features which are listed in this section, and in the following section on digital beam forming.

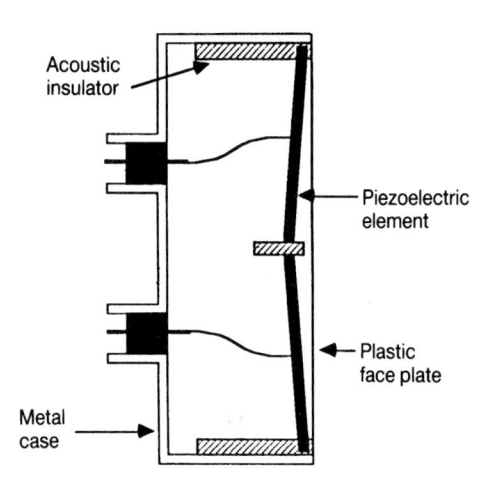

Figure 5.4 A dual-element undamped transducer for the generation and detection of continuous wave ultrasound

1. No moving parts are required to sweep the ultrasound beam direction.

2. The position of the focus can be electronically controlled and hence moved very rapidly.
3. They can be made fairly small.
4. Fortunately, the large number of piezoelectric elements in them is very compatible with computer technology. This makes for great flexibility in terms of the way in which beams are formed and moved around.
5. The lack of moving parts suits them to the detection of slow flow by Doppler methods.
6. Since there is more than one element in each transducer, both PW and CW Doppler methods can be employed.
7. The flexibility of beam production makes them well suited to advanced Doppler techniques such as vector Doppler and 3-D Doppler.

The disadvantages are:

1. Focusing for one-dimensional linear and phased arrays is only in the in-plane direction; the out-of-plane focusing depends on the addition of a cylindrical lens attached to the front of the array. Annular arrays produce true focusing around the beam axis but they are required to be moved mechanically to sweep the beam direction. Two-dimensional arrays are being developed to give true focusing without a lens in front of the array or the need to move the transducer mechanically to scan the beam.
2. It is difficult to produce very high frequency array transducers.

Commonly encountered array structures are shown in Fig. 5.5. Each is discussed in subsequent sections of this chapter; however, before that it is worth studying the ways in which modern arrays are digitally controlled.

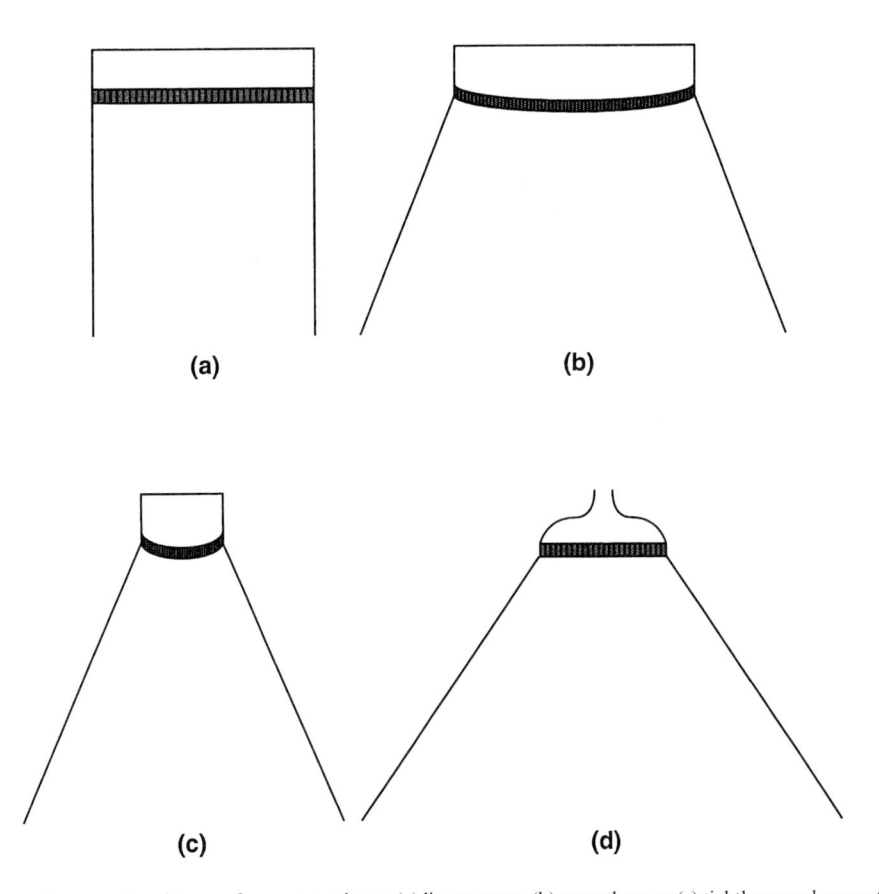

Figure 5.5 Commonly encountered types of array transducer: (a) linear array; (b) curved array; (c) tightly curved array; (d) phased array

5.2.4.1 Digital Beam Forming

It was noted above that array transducers with large numbers of piezoelectric elements are well matched to digital computing methods (Steinberg 1992). This is true of both transmission and reception. On transmission, the size of the excitation signal to each element, the timing of the excitation of each element and the number of elements excited can all be controlled and altered rapidly. On reception, the echo signal at each element can be digitised accurately and rapidly to preserve amplitude and phase information. The gain of each receiver, the delay applied to each signal and the number of elements in use can also be controlled. The complete signal from each element over the tissue range of interest is stored for future combination with the signals from other elements. For example, a 5 MHz echo train of bandwidth 2–8 MHz may be digitised at 32 MHz to an accuracy of 10 bits over a duration of 200 µs. Digital control allows the following features to be included in beam forming:

1. Stepped focusing on transmission—by altering the relative timing of the excitation of the elements (Fig. 5.6).
2. Beam stepping on transmission—by using different elements across a linear array.
3. Beam steering on transmission—by altering the relative timing of the excitation of elements (Fig. 5.7).
4. Generation of several simultaneous transmission beams—by using elements in different parts of a linear array or timing the excitation of elements of a phased array to produce a multi-lobe field.
5. Apodisation on transmission—by weighting the size of the excitation signals applied to each element to improve the beam shape and in particular reduce side and grating lobes.
6. Variable aperture on transmission—by varying the number of elements used for each focal range and hence to optimise the sharpness of the focus at each depth.
7. Dynamic focusing on reception—by very rapidly changing the delay pattern applied to the echo signals from elements of an array, the

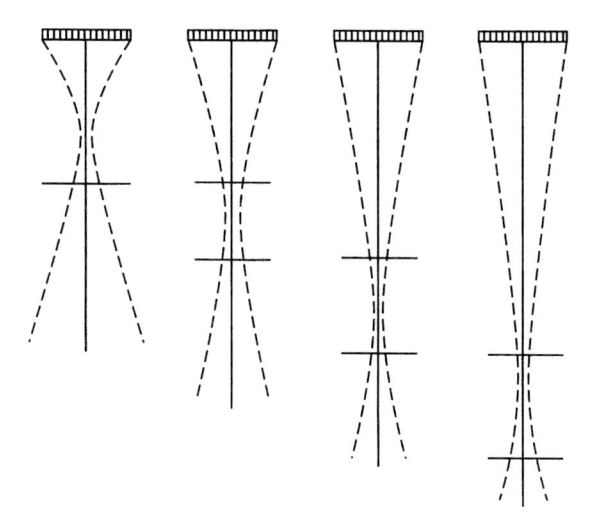

Figure 5.6 Stepped focusing on transmission with array transducers. Each of the four focal zones requires a separate pulse transmission cycle

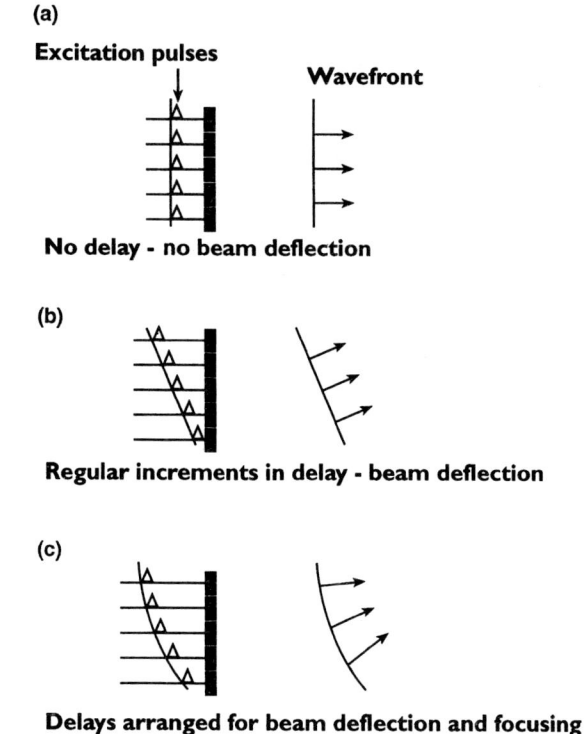

Figure 5.7 Beam steering on transmission with linear and phased arrays: (a) no delay—no beam deflection; (b) regular increment in delay—beam deflection; (c) delays arranged for beam deflection and focusing

array can be focused to coincide at each instant with the depth of the target which produced the echoes. By this means the focus is swept along the axis of the beam to give an extended focus. This dynamic focusing ability is probably the main attraction of array transducers.
8. Beam stepping on reception—by using different elements across a linear array.
9. Beam steering on reception—by altering the relative delays, or phase changes, applied to the received echoes.
10. Reception of several reception zones—just as the array can be made sensitive to one beam direction by using a particular set of signal delays, it can be made sensitive to several beam directions simultaneously by processing each echo train in parallel, using several sets of delay patterns. This multi-beam reception mode is normally used in combination with a wide beam transmission mode to ensure that energy is first propagated in all of the directions of interest.
11. Apodisation on reception—by weighting the gain applied to each echo signal to improve the beam shape and in particular reduce the side and grating lobes.
12. Variable aperture on reception—by varying the number of elements used for each focal range and hence to optimise the sharpness of the focus.

On reception, for complete computer control the echo signals are digitised as close to each transducer as possible and with an accuracy which preserves both amplitude and phase information. It is worth noting that this digital beam forming requires large numbers of replicated driver circuits for independent control of each element to be achieved. Some of these techniques can be accomplished by analogue means but the versatility of the digital approach makes it the obvious choice now that the appropriate digital circuitry is available.

5.2.4.2 *Linear Array Transducers*

Linear array transducers are widely used for imaging (Vogel et al 1979, Kino and DeSilets 1979). In many duplex and Doppler flow map systems the linear array performs both the echo imaging and the Doppler function (Fig. 5.8a) (Berson et al 1987). Occasionally an independent Doppler transducer is linked to the end of the

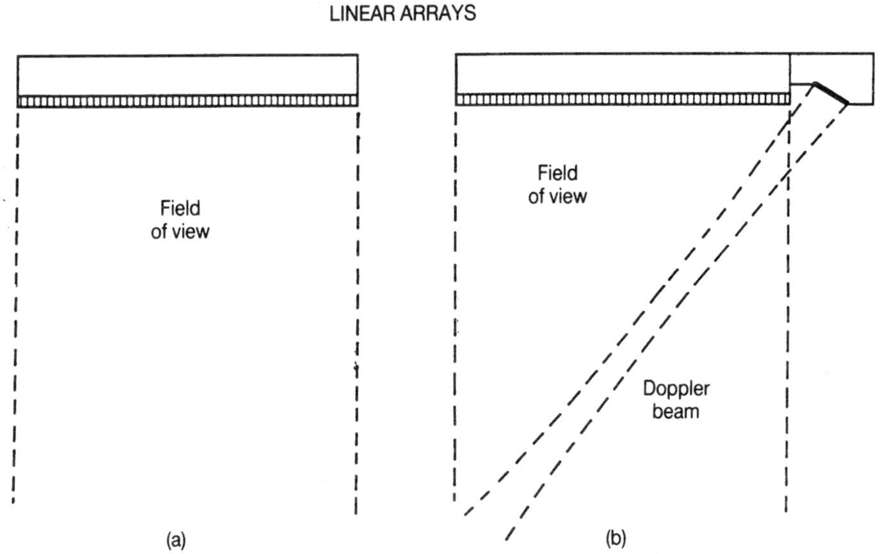

LINEAR ARRAYS

Figure 5.8 (a) A linear array in which the same transducer is employed for imaging and Doppler modes. The Doppler mode beam may be emitted at an angle across the field-of-view using electronic beam steering. (b) a linear array with a separate Doppler transducer

array and directs the Doppler beam into the field-of-view of the imager (Fig. 5.8.b). When the array elements are used to generate the Doppler beam, it can be focused or defocused at the selected blood vessel. By changing the active group of elements, the point of origin of the beam can be moved along the array and it may also be possible to alter the angle of the beam to the array by employing beam steering techniques. These two manipulations introduce flexibility into the application of linear arrays for Doppler, but the need to maintain contact between the flat surface of the transducer and the skin can be restricting.

Linear arrays typically contain up to 256 piezoelectric elements whose operation is under computer control. There is great flexibility in the selection of the elements in use at any one time, the number ranging from 8 to 128. When 128 elements are employed, sharp focusing may be obtained deep within the body due to the large size of the source, i.e. the large aperture. However, if too wide an aperture is used ultrasound converges into the sample volume over too large a range of angles, giving rise to spectral broadening. Since there are several elements available at any time, the CW mode as well as the PW mode can be catered for with this type of transducer.

5.2.4.3 Curved Array Transducers

Introducing curvature to linear arrays has proven to be popular in imaging since the field of view is increased and contact with the skin surface is easier in some situations (Fig. 5.5b). Sharp curvature converts linear arrays into sector scanners. The smaller transducer size, or 'footprint', allows imaging and Doppler measurements to be made where access is limited (Fig. 5.5c) (Ishiyama et al 1985).

5.2.4.4 Phased Array Transducers

A modern phased array typically consists of 128 thin linear elements and has a length of 1–2 cm (Fig. 5.5d) (von Ramm and Thurstone 1976, Somer 1968). The same transducer functions in both imaging and Doppler modes. As for linear and curved arrays, both CW and PW Doppler techniques can be accommodated by phased arrays. The images produced by phased arrays have always been slightly inferior to those from linear arrays, since it is difficult to steer a beam without a significant amount of energy being transmitted in directions other than that of the main beam. The same problem occurs for Doppler beams from phased arrays so that moving structures well off the main beam axis may contribute spurious signals to the Doppler output from the site of interest.

Phased array transducers find widespread application in duplex and Doppler flow mapping instruments designed for cardiac examinations, since their small size and sector field of view permit access avoiding bone and lung. Small size is of particular advantage in transoesophageal transducers where it is even possible to mount two arrays in biplanar transducers or move one to give omnidirectional scanning.

5.2.4.5 2-D Array Transducers

Just as 1-D phased arrays can produce 2-D scans, 2-D phased arrays can produce 3-D scans (Fig. 5.9). 2-D scanner arrays are technologically difficult to produce because of the large number of connections to be made to the piezoelectric elements and the corresponding increase in electronics (Lockwood and Foster 1996, Smith et al 1992). If a 1-D array has 128 elements, a 2-D array of similar quality would require 16384. Another problem with 3-D scanning is that the

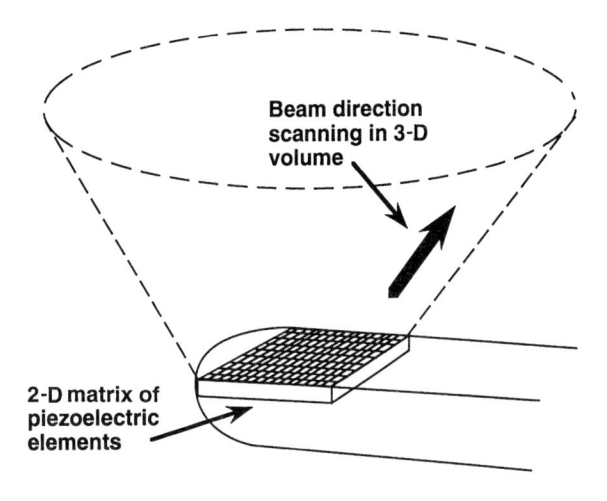

Figure 5.9 A two-dimensional phased array for three-dimensional scanning

echo-collection time from a volume of tissue is relatively long and it is difficult to perform scanning in real time. Present-day 2-D scanners giving, say, 64 frames per second make use of all the echo collection time available. Truly real-time 3-D scanning awaits the development of 2-D arrays which can generate many simultaneous beams. At the time of writing one commercially available scanner (the Volumetrics Medical Imaging Model 1) uses almost 2000 crystal elements to create a circular aperture and generate 4096 lines in each volume, and can acquire up to 45 volumes per second.

Apart from 3-D imaging, there is one other use for 2-D arrays, namely to reduce the thickness of the slice scanned by a linear array using electronic focusing. In addition to the main line of elements in a linear array, a few parallel arrays are laid alongside it. The whole array then has typical dimensions of 256×5 and is commonly referred to as a 1.5-D array. This type of electronic focusing is successful because the requirements for focusing along the central axis are much less demanding than for beam steering (Shung and Zipparo 1996).

5.2.4.6 Annular Array Transducers

The electronic focusing of both linear and phased arrays is only in the direction of the plane of scan, i.e. perpendicular to the length of the elements. As noted above, to provide focusing in the out-of-plane direction, 2-D arrays containing a large number of elements are required. However, annular arrays using a smaller number of elements can achieve symmetrical electronic focusing about the axis of the transducer (Fig. 5.10). Theoretical modelling of the beam shape from an annular array indicates that at least five ring elements are necessary to generate a good quality focused beam (Melton and Thurstone 1978, Parks et al 1979, Arditi et al 1981, Dietz et al 1979, Pye 1983). The disadvantage of annular arrays is that they are required to be moved mechanically to perform a scan.

In the quantitative measurement of blood flow it may be desirable to insonate the vessel of interest with an uniform ultrasonic beam (Chapter 12). It has been shown that a uniform beam may be generated with an annular array composed of just two elements (Evans et al 1986).

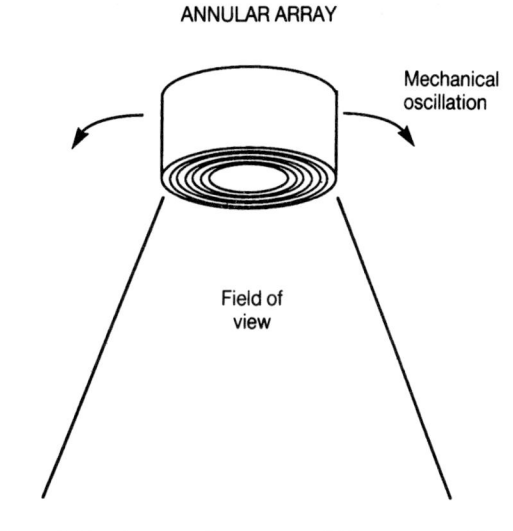

Figure 5.10 An annular array which can provide true electronic focusing about the central axis of the beam

5.2.5 Internal Transducers

Simple single- and double-element transducers are used to examine exposed vessels during surgery or in the orifices of the body (Duck et al 1974). However, more information is obtained if the transducer is incorporated into a real-time scanner for internal application. As for external application, both duplex and flow mapping Doppler modes are valuable. At present, commercially available internal scanners are based on rotating transducers, linear arrays and phased arrays (Fig. 5.11). The frequency of operation of internal Doppler units is normally in the range 5–10 MHz. When arrays or double-element transducers are used both CW and PW Doppler operations are possible.

A well-established use of internal transducers is transoesophageal imaging of the heart using single plane, biplane and multiplane devices (Souquet et al 1982, Krebs et al 1996). Although these techniques cause some discomfort to the patient, they have the great attraction of access to the heart without attenuation by bone or lung. Very complete pulse-echo and Doppler images can be obtained.

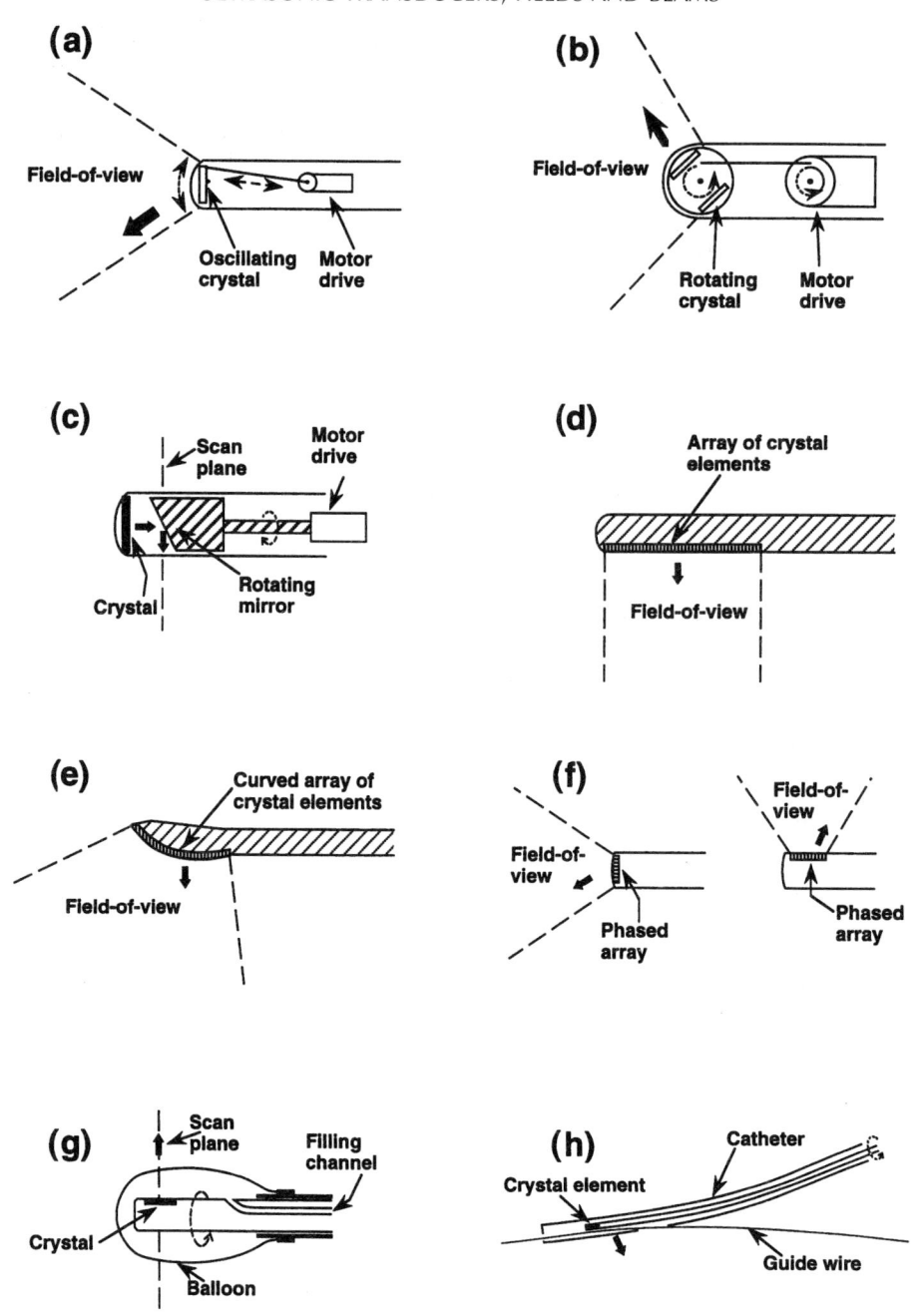

Figure 5.11 Types of transducer for internal scanning: (a) oscillating transducer, (b) rotating transducer, (c) moving mirror transducer, (d) linear array, (e) curved array, (f) phased array, (g) water-bath transducer, (h) catheter transducer (reproduced by permission of Martin Dunitz Ltd, London, from Lees WR and Lyons EA (eds) 1996, Invasive Ultrasound)

5.2.6 Intravascular Ultrasound Transducers (IVUS)

Ultrasound transducers can be miniaturised to allow them to be incorporated into catheters of diameter of around 2 mm or less. The operating frequency is typically 30 MHz but this is now increasing to cover a range up to 100 MHz. As usual the frequency depends on the tissue penetration required. In their present design IVUS scanners normally direct the ultrasound beam perpendicularly to the axis of the catheter to examine the artery wall (Fig. 5.12). The initial commercially available devices employed mechanical beam scanning through the 360° field-of-view and now electronic arrays are also on the market. The beam directions are not ideal for Doppler techniques since they are essentially at 90° to the direction of blood flow. One technique has been developed to detect blood flow by looking at the changes in the blood pulse-echo speckle pattern using correlation methods (Li et al 1998) (see Section 13.8). IVUS imaging catheters with more forward-looking beams are also being investigated (Lee and Benkeser 1991, Evans et al 1994, Ng et al 1994).

Simple Doppler catheters with forward-looking transducers, commonly known as 'Doppler wires', are widely used to check blood velocities in diseased arteries before and after treatment. Typically they operate in PW mode at 15 MHz and are 0.5 mm in diameter (Doucette et al 1992, Hartley 1989, Moraes and Evans 1995, Roth et al 1993).

Doppler catheter devices can also be combined with optical endoscopes by inserting them through the biopsy channel (Martin et al 1985, Beckley et al 1982). Having observed a site of interest via the endoscope, the Doppler transducer is placed on it to detect blood flow, for example in vessels prior to surgery or in varices and ulcerations prior to

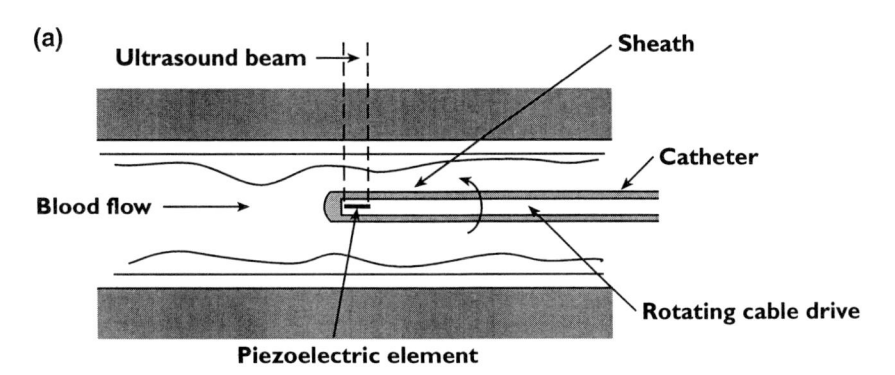

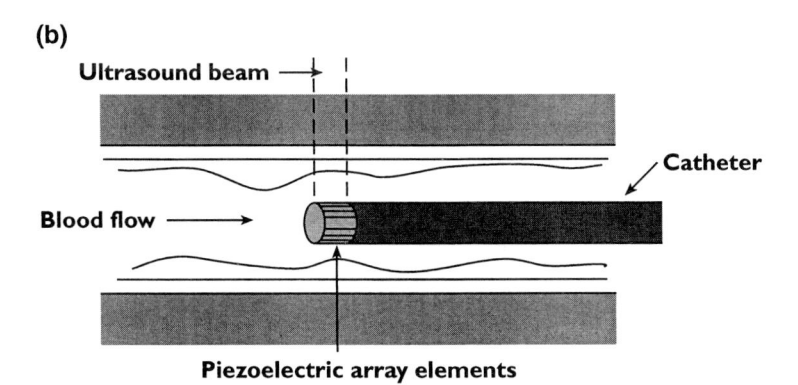

Figure 5.12 Intravascular ultrasound (IVUS) imaging catheters, (a) mechanical scanner, (b) phased array. Field of view is 360° for both

therapy. Such Doppler transducers typically operate at around 10 MHz and are required to be able to pass down a channel of 2.5 mm diameter. The field of view is 360° and is generated to be at right angles to the catheter axis by using a cylindrical piezoelectric element. Smaller fields of view are made by deactivating part of the element, which improves the localisation of flowing blood.

5.3 ULTRASONIC BEAMS AND FIELDS

The wavelength of ultrasound in soft tissue for all commercially used frequencies is less than 1 mm. As a result of these short wavelengths, highly directional fields and beams can be generated using transducers with an active element of dimensions around 1 cm or less. The small size of the transducers has significant consequences for the convenience and versatility of diagnostic ultrasonics. The high directionality of ultrasonic beams is of paramount importance, since virtually all imaging and Doppler techniques are dependent on it. In current practice of Doppler ultrasound, these beam shapes and dimensions are rarely well known to the user. In addition array transducers may have significant grating lobes (Whittingham 1997). With time these uncertainties should be resolved to some extent. Improved knowledge of beam properties will benefit clinical applications, since the source of Doppler signals will then be identified more precisely.

5.3.1 Continuous Wave Fields and Beams

The transmitted field from a flat disc element oscillating uniformly at a well-defined frequency in a thickness mode of vibration is well documented, since it is often used as an introduction to the ultrasound fields employed in medical imaging. Complete analytical solutions are available for the pressure amplitude distribution along the transducer axis, and also off-axis in the far field (Kinsler et al 1982). The pressure amplitude along the transducer axis is given by:

$$p(z) = p_0 \sin\left(\frac{\pi}{\lambda}[a^2 + z^2]^{1/2} - z\right) \quad 5.3$$

where p_0 is the maximum pressure amplitude, λ the ultrasound wavelength, z is the distance along the axis, and a is the radius of the disc.

Figure 5.13a shows the plot of the intensity distribution along the transducer axis. It can be seen that there are rapid fluctuations out to a distance of a^2/λ from the transducer face. This region is known as the near or Fresnel field. The last axial maximum occurs at the end of the near field, beyond which the intensity falls as the inverse square of distance. This region is known as the far or Fraunhofer field. Fluctuations also occur in the direction perpendicular to the axis (Fig. 5.13b). In the near field they are rapid, whereas in the far field they are slow.

The reception zone is identical in shape to the transmission field, since reception is the converse of transmission for a transducer in which there is no electronic focusing. The resultant ultrasonic beam, which depends on the combination of the transmission field and the reception zone, exhibits even greater fluctuations than these two individual components. It will be recalled that the beam sensitivity at each point is obtained by multiplying the transmission field and the reception sensitivity for that point.

The fluctuations in the beam are of significance in the clinical application of CW Doppler instruments. Vessels that fall within the near field are not uniformly insonated. The measured mean velocity will be in error and the maximum velocity may also be inaccurate if the high velocity coincides with a minimum in the beam sensitivity. One obvious way to ensure more uniform insonation of the moving cells is to arrange for the vessel to fall within the far field. Table 5.3 is a list of near field lengths for some typical piezoelectric element dimensions and operating ultrasonic frequencies.

5.3.2 Pulsed Wave Fields and Beams

Theoretical calculation can also be made to determine the intensity distribution from transducers operating in the pulsed mode. The field may be found by summing the contributions from the frequency components of the pulse or by using the impulse response approach (Duck 1981). These calculations have been performed for circular,

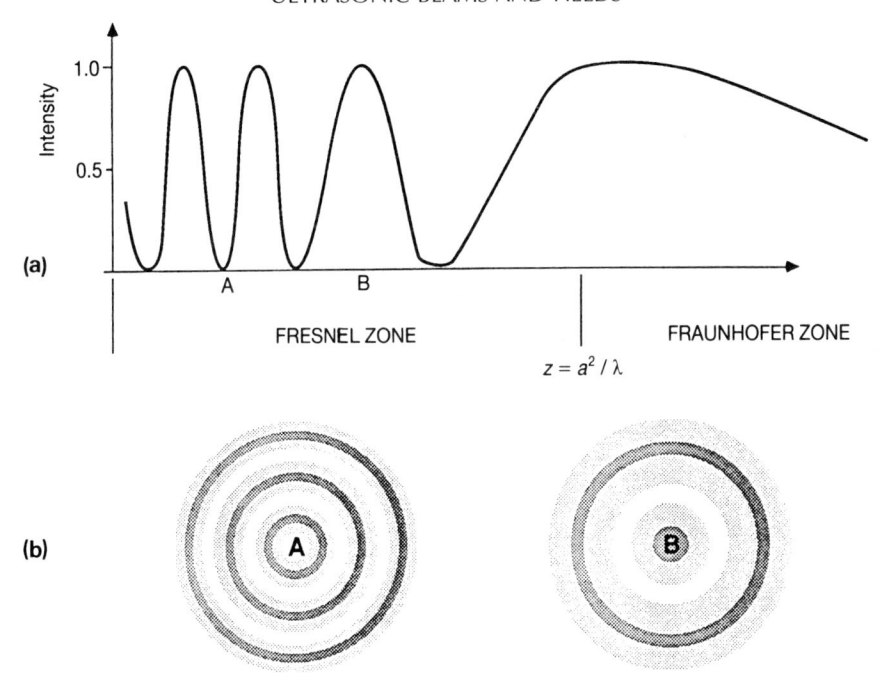

Figure 5.13 (a) Intensity variation along the axis of a continuous wave field. (b) Intensity variations represented by the grey scale, of a continuous wave field perpendicular to the axis at points A and B

Table 5.3 Near field lengths for some typical transducer element diameters and related operating frequencies

Frequency (MHz)	Diameter (mm)	Near field length (mm)
1	20	65
2	15	73
5	8	52
10	5	42
15	3	22
20	2	16
30	1	5

annular, rectangular and bowl-shaped sources (Cahill and Baker 1997, Duerinckx 1981, Jensen and Svendson 1992, Baker et al 1995).

As noted earlier, numerical calculations on a computer are commonly employed to predict the transmitted field from transducers. This approach is particularly appropriate in the case of transducers of complex shape, e.g. annular arrays. Good agreement is obtained between computed fields and those measured experimentally (Beaver 1974, Pye 1983).

An example of a pulsed wave field calculated numerically is shown in Fig. 5.14. It is seen that the fluctuations in the field are much smaller than in the CW case. This is due to the minima and maxima of the pressure distribution for each frequency component of the pulse occurring at different locations and hence smoothing the field.

5.3.3 Multi-beam Systems

To increase the frame rate for the observation of rapidly changing events in B-mode and Doppler imaging, several beams may be generated simultaneously. This is achieved with linear and phased arrays, which generate a wide transmission field and receive simultaneously on, say, four reception zones (Fig. 5.15). For each direction of the reception zone, the signals from each element are combined with the appropriate delays to give sensitivity along that direction. This has been made much more feasible by the introduction of completely digital beam formers in which the

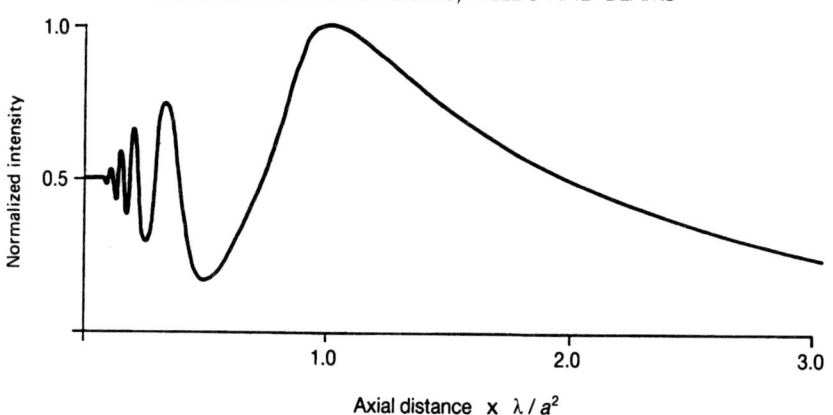

Figure 5.14 Intensity variation along the axis of a pulsed wave field. This pattern corresponds to a three cycle pulse. Fluctuations perpendicular to the beam are smaller than in the continuous wave case (courtesy of D Pye)

signal from each transducer element is digitised prior to any processing. Frame rates of 120 s^{-1} have been achieved for B-mode and 40 s^{-1} for Doppler imaging. The high frame rates make colour Doppler images appear to change more smoothly, as distinct from the disjointed presentation of frames at lower frame rates (e.g. 15 s^{-1}) of earlier scanners.

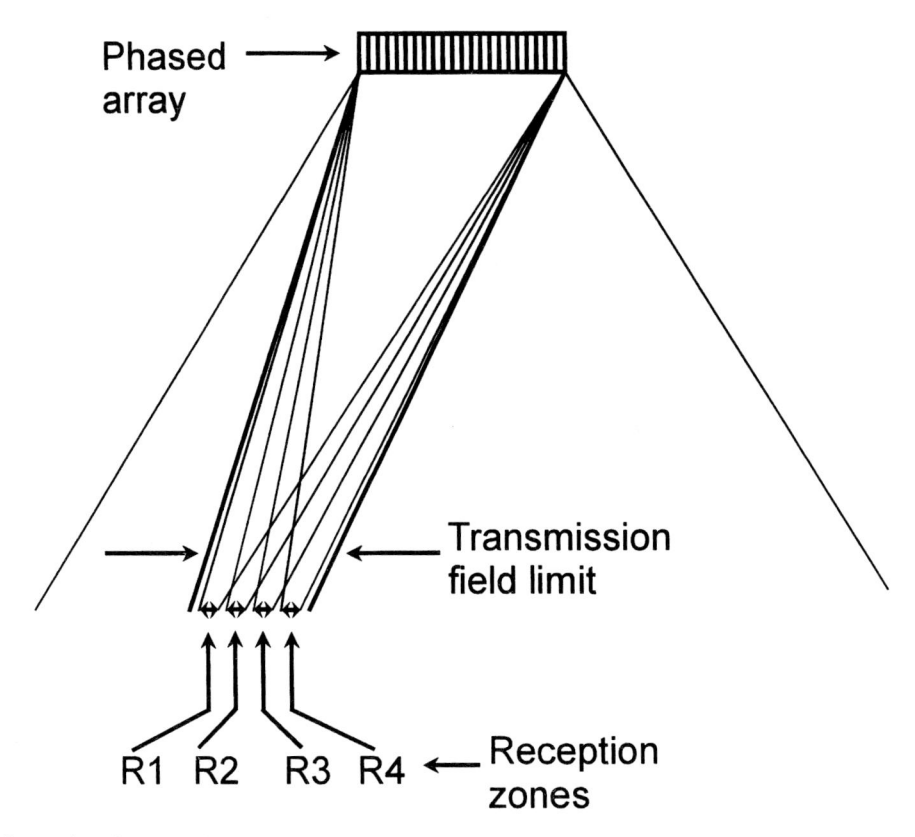

Figure 5.15 Multi-beam phased array. Wide field on transmission, four simultaneous focused zones (R1, R2, R3, R4) on reception

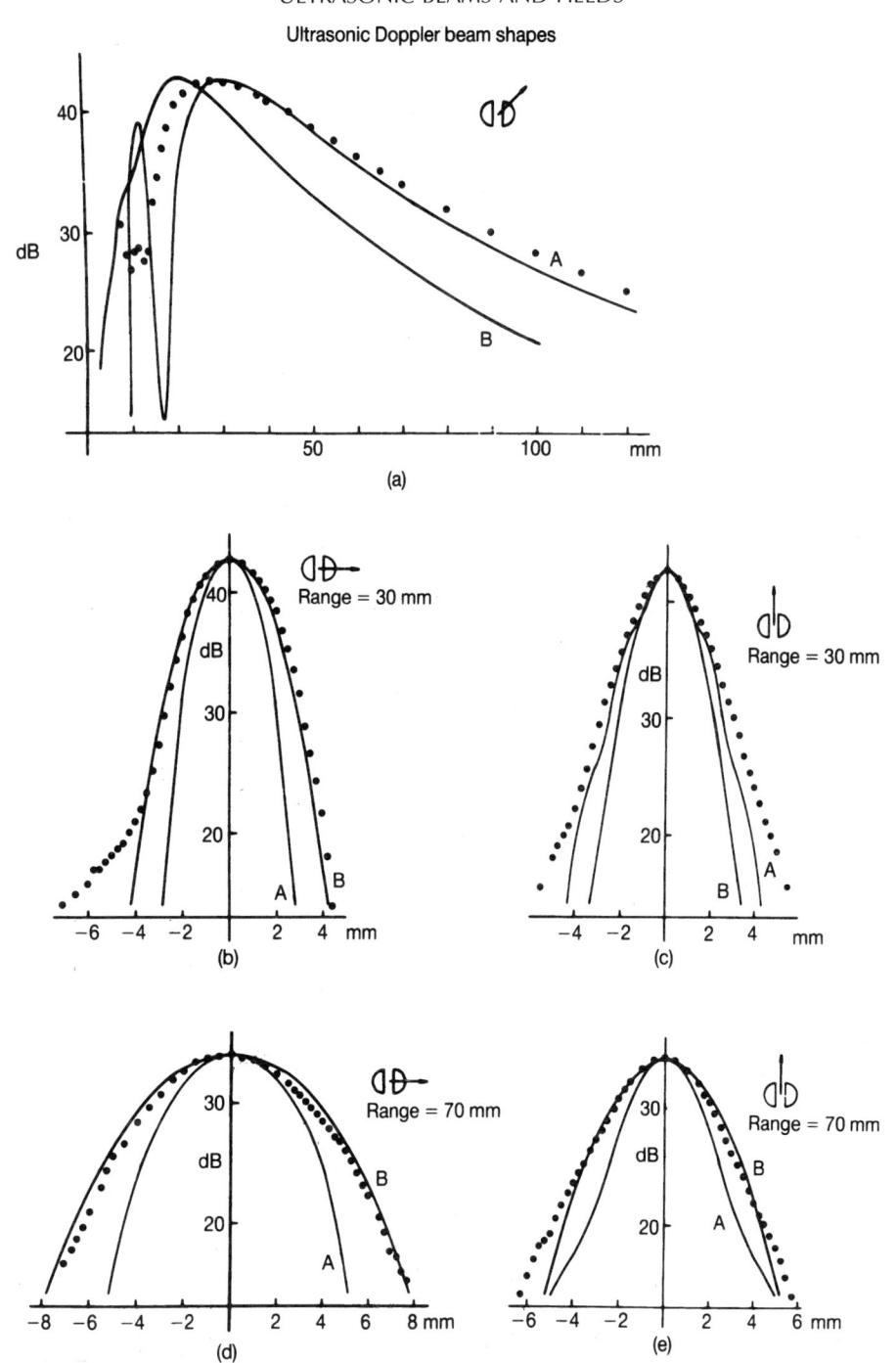

Figure 5.16 Echo amplitude plots from a dual-element Doppler transducer along the transducer axis and perpendicular to it at two different ranges (30 mm and 70 mm). The transducer operated at 3.9 MHz. The points represent experimental results, the lines the predictions of two theoretical models (reproduced by permission of Elsevier Science, from Evans and Parton 1981, *Ultrasound in Medicine and Biology*, © World Federation of Ultrasound in Medicine and Biology)

5.4 DOPPLER SAMPLE VOLUMES

In the continuous wave case the sample volume is identical to the whole of the Doppler beam. Note that in a CW system the sample volume position is usually but not necessarily fixed. It can be moved by altering the positions of the transmitted field or reception zone. This type of device has been made using two single-element transducers (McHugh et al 1981) and commercially with a single-element transducer linked to a phased array.

For PW systems the length of sample volume depends on the combination, i.e. convolution, of the transmitted pulse length and the length of the gated range (see Section 4.2.2.2). The width of the sample volume is determined by the width of the beam at the position of the range gate. It is also worth noting that a change in the sensitivity of the beam of the Doppler unit will alter the effective length of the transmitted pulse and the effective beam width. The sample volume is therefore not a fixed quantity but depends to some extent on the settings of the sensitivity controls. Likewise, the width of the beam in a Doppler flow mapping system also depends on the sensitivity settings.

5.4.1 CW Sample Volumes

Few detailed plots of the sample volume of CW beams have been published (Evans and Parton 1981, Wells 1970, Eriksson et al 1994, Douville et al 1983). Figure 5.16 illustrates some results of Evans and Parton. These distributions were obtained by using one element to transmit a long pulse with the other acting as an echo receiver, which is equivalent to CW operation from a beam-plotting point of view. The reflecting target was a small ball-bearing which was systematically moved in the region in front of the transducer. In this approach it is not the magnitude of a Doppler signal that is plotted but the echo amplitude from the ball-bearing. The distributions are therefore similar but not identical to those that would be obtained using moving blood cells as the target. These results give an appreciation of the size of a CW beam sample volume, in particular its extended range along the transducer axis.

The above plots were found to agree best with those of a model of transducer action in which the active element is considered to be restricted at its edge, resulting in some apodisation. This is a realistic situation, since the means of mounting an element can restrict its action at the edge.

5.4.2 PW Sample Volumes

Pulsed wave sample volumes have not been studied in extensive detail. Figures 5.17 and 5.18 present the results of Walker et al (1982) and Hoeks et al (1984).

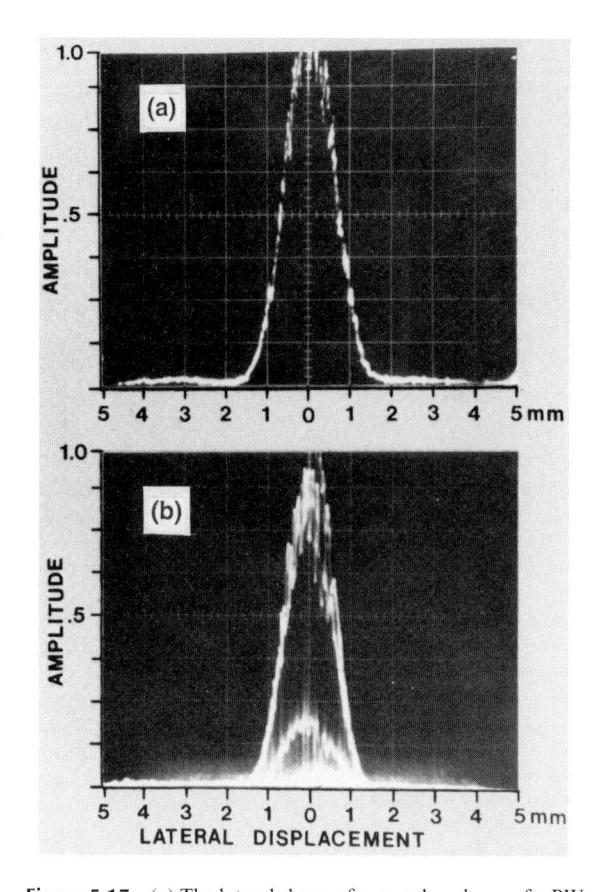

Figure 5.17 (a) The lateral shape of a sample volume of a PW beam obtained by examining the Doppler signal from a moving string. (b) The lateral shape of a sample volume of a PW beam obtained by examining the echoes from a stationary string. The transducer was a 5 MHz annular array (reproduced by permission of the publisher Churchill Livingstone, from Walker et al 1982)

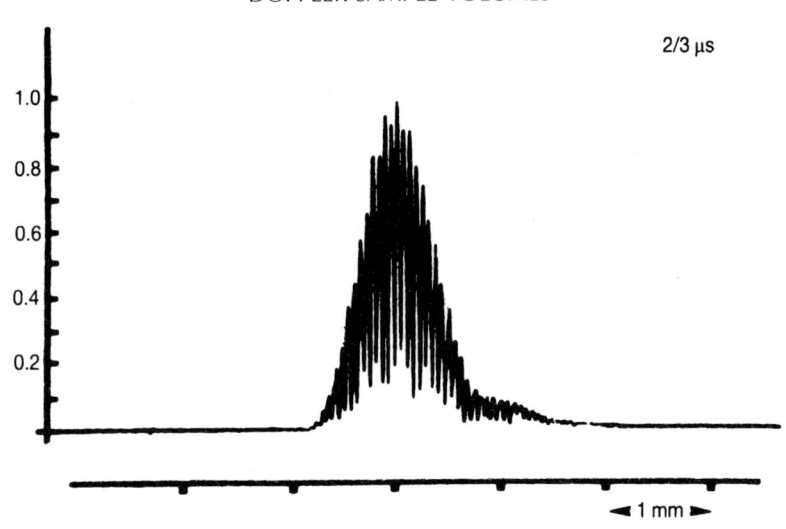

Figure 5.18 The axial sample volume shape of a PW beam at 6.1 MHz (emission duration 2/3 μs). The fine structure in the trace is due to the measurement technique. (reproduced by permission of Elsevier Science, from Hoeks et al 1984, *Ultrasound in Medicine and Biology*, © World Federation of Ultrasound in Medicine and Biology)

Sample volumes for transcranial pulsed Doppler have been measured by Arnolds et al (1989).

The shape of the sample volume of a PW beam is often described as a 'teardrop' (Fig. 5.19). This arises from the combination of the 3-D shape of the transmitted pulse and gated range (see Section 4.2.2.2).

The lack of results relating to sample volumes is due in part to the lack of development of convenient and accurate methods for measuring sample volumes. There is also some debate as to whether sample volume shape is best measured by a small moving point reflector or by a line of moving reflectors that mimic a stream of moving cells.

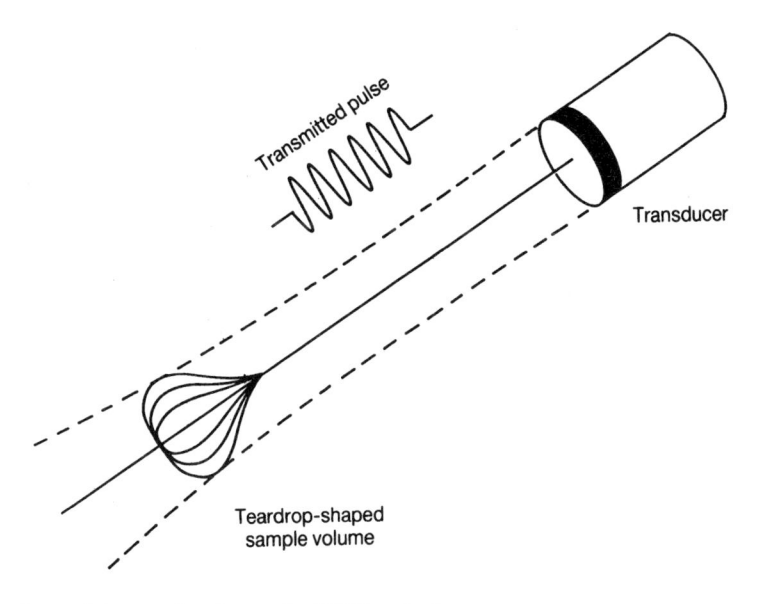

Figure 5.19 Teardrop shape of the sample volume of a PW Doppler beam

5.4.3 Resolution in Colour Doppler Images

Although Doppler imaging technology has matured and now produces good quality images, little has been reported with regard to assessing image quality (Hoskins and McDicken 1997). One attempt to assess spatial resolution in images uses a grid pattern of slits on a tube through which flows blood-mimicking fluid. The ultrasound is blocked by an air barrier, except at the slits, where it can detect the flow in the tube. Making use of a series of grid patterns, each of which has a different slit separation, allows the identification of the smallest slit pattern for which each slit can be imaged without merging of the blood flow signals. This gives a measure of the smallest separation of flow structures that the scanner can register in the display, i.e. a measure of the spatial resolution as measured by this test procedure. Although diffraction at the slits may affect the accuracy of the result, this approach does allow comparison of machines operating at the same ultrasonic frequency (Li et al 1997). Another approach to spatial resolution measurement (Phillips et al 1997) is to use an oscillating thin film test object (see Section A2.3.3).

5.5 MEASUREMENT OF FIELDS, ZONES, BEAMS AND SAMPLE VOLUMES

From the points of view of sensitivity and imaging performance, it is the reception zone sensitivity pattern and the distribution of the pressure amplitude in the transmitted field which are of interest. They are discussed below. The intensity pattern of the transmitted field and the output power are of interest for bioeffects and are discussed in Chapter 14. In Chapter 3 the definitions of pressure amplitude, intensity, power, transmission and reception zones are presented.

Measurement of the ultrasonic field (transmission zone) of a transducer is undertaken by moving a small detector systematically in the region in front of the transmitting element. The transducer face and the detector (hydrophone) are normally immersed in water (Chivers et al 1986, Harris 1988). The hydrophone motion is usually restricted to a plane containing the axis of the field and to a few planes perpendicular to the axis (Fig. 5.20). The latter scans by the hydrophone check on the symmetry of the field, which cannot be presumed.

A typical hydrophone-detecting element is a piece of piezoelectric material of diameter 0.2–1 mm, although devices have been made as small

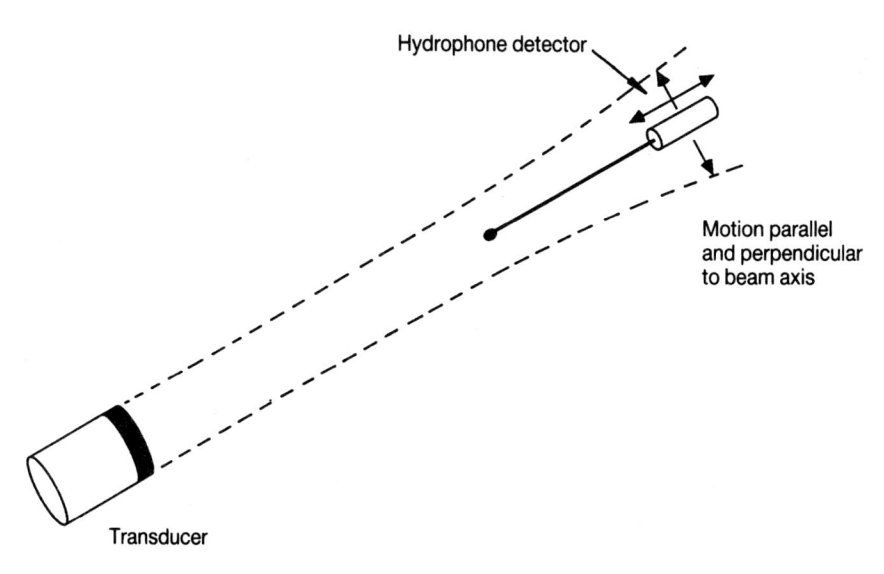

Figure 5.20 Schematic diagram of a hydrophone field plotting system

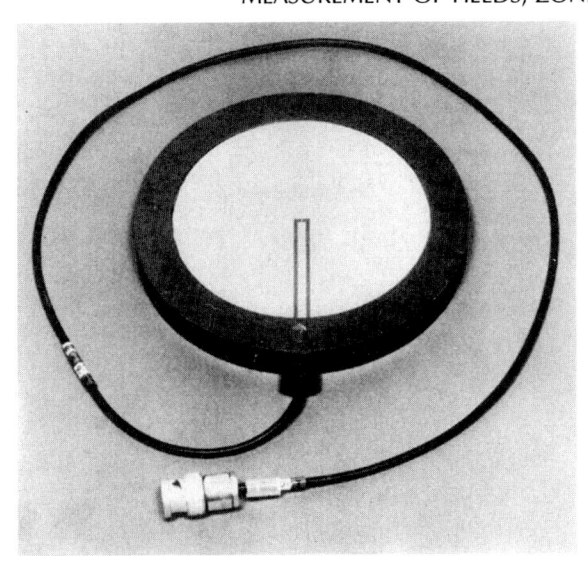

Figure 5.21 A membrane hydrophone. The active element is a small disc of diameter less that 1 mm at the centre (courtesy of R Preston)

devices, a useful plot of the field shape can be obtained. The development of plastic (PVDF) piezoelectric material has allowed hydrophones to be constructed in the form of a thin membrane (Fig. 5.21). Small PZT and PVDF elements can be mounted on the tip of a needle (Fig. 5.22). These hydrophones are designed to disturb the field very little during the measurement (Preston et al 1983, Shotton et al 1980, Lewin 1981, Bacon 1981). For example, the membrane may be 0.05 mm thick and 10 cm in diameter with a 0.5 mm diameter active element at its centre, or the needle may have an element of 0.1 mm diameter on its tip. For rapid beam plotting, a 1-D or 2-D array of active elements can be incorporated into the membrane (Preston et al 1985, Precision Acoustics, Dorchester, UK). PVDF hydrophones are stable, have a known frequency response and are easy to use, requiring only an amplifier and a fast oscilloscope. The needle approach lends itself to the manufacture of very small devices. The method used to measure the response of the hydrophones should be carefully checked to ensure that the specific frequencies of interest are part of the calibration process and that they have not just been included by interpolation (Fay et al 1994).

as 0.1 mm in diameter. For detailed plotting of an ultrasonic field of frequency in the range 1–10 MHz, i.e. of wavelength 1.5–0.15 mm, the latter are required. However, even with the larger

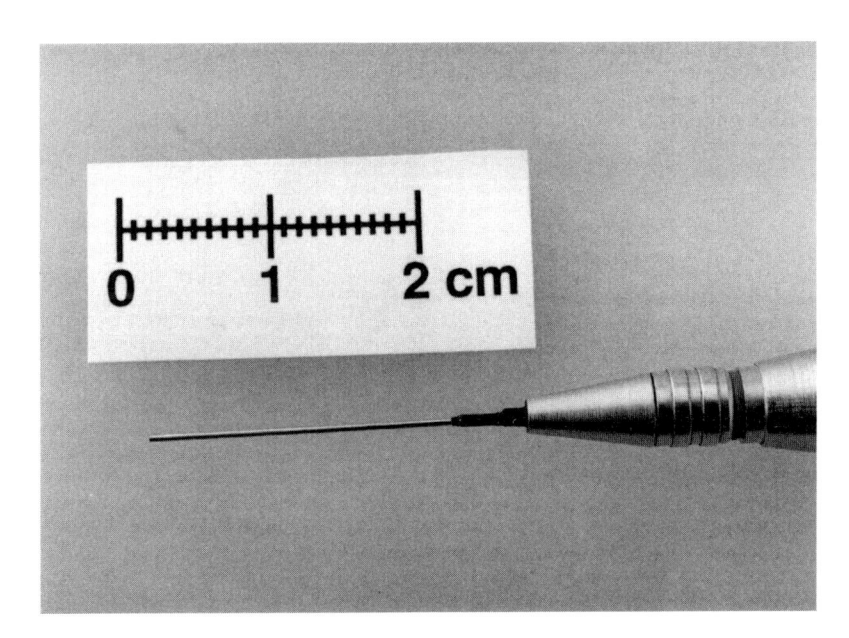

Figure 5.22 PVDF needle hydrophones. The 9 μm thick sensor element has an active diameter of 0.2 mm (supplied by Precision Acoustics Ltd, Dorchester, UK)

The signal from a hydrophone at any point is proportional to the wave pressure at that point. Values of the pressure amplitude are recorded and displayed numerically or as shades of grey in a 2-D contoured image of the section through the beam (Fig. 5.23) (Whittingham and Roberts 1986). Note that in most ultrasonic pulses as used in diagnosis the positive pressure amplitude at a point is usually different from the negative. The intensity at any point is calculated using the square of the pressure amplitude. This latter procedure is not strictly valid in the near field of the transducer, where the pressure and particle velocity are not in phase. Current practice is therefore to characterise the beam in terms of pressure rather than intensity. It should also be remembered that the beam characteristics measured in water are different from those in tissue, where dispersive absorption is present and non-linear propagation is less significant. Nevertheless, a great deal can be deduced about a beam in tissue from a plot obtained in water.

The plotting of a reception zone in isolation would require a point transmitter to be moved in a regular fashion in front of the transducer. Another approach would be to calculate the shape of the reception zone from the transmitted field shape and the beam shape. Reception zone shapes are rarely determined in practice.

An approximation to a Doppler beam shape may be obtained by operating the transducer in a pulse echo mode in which the reflecting target is a small ball-bearing, wire or tube (Arnolds et al 1989). As the target is systematically moved in front of the transducer, the echo is detected after amplification in the receiver of the pulsed Doppler unit. A long ultrasonic pulse of 10 cycles or more simulates the operation of a CW unit. This approach is reasonable when the properties of the transducer are under study. However, if the influence of the whole system on the beam shape is to be known, a small moving target that generates a Doppler shift signal is required. Test objects for this purpose based on moving string, ball-bearings and liquid jets, are described in detail in Appendix 2.

An elegant way of visualising an ultrasonic field is known as the 'schlieren' method (Willard 1947, Wells 1977, Follett 1986). It depends on the fact that pressure fluctuations in an ultrasonic field cause fluctuations in the refractive index of water. In a schlieren system, a wide beam of light is directed through a water tank and brought to a sharp focus using a lens or mirror. When no ultrasound is transmitted into the tank, a stop placed at the focus blocks the light beam. However, when ultrasound is present the light is refracted in the region where the refractive index is varying and hence by-passes the stop. The optical

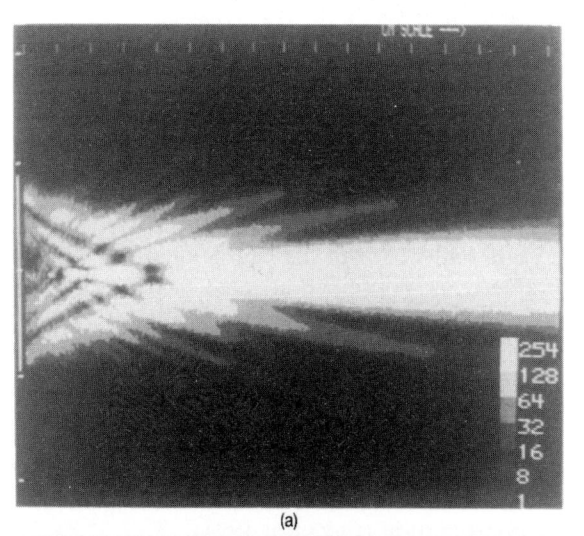

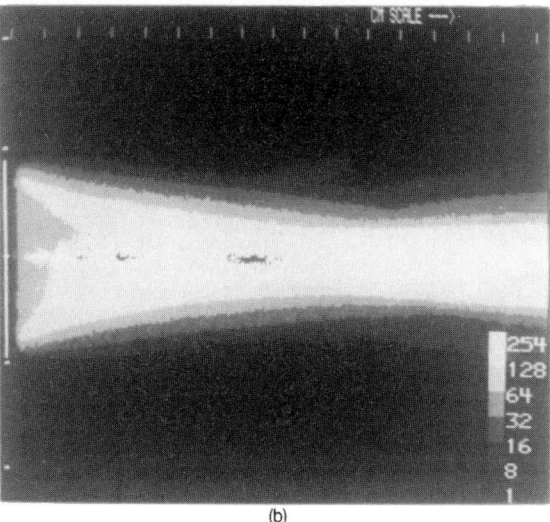

Figure 5.23 (a) Grey-tone hydrophone plot of the pressure amplitudes in the field of a CW transducer. (b) Grey-tone hydrophone plot of the pressure amplitudes in the field of a PW transducer (reproduced by permission of the Institute of Physics and Engineering in Medicine, from Whittingham and Roberts 1986)

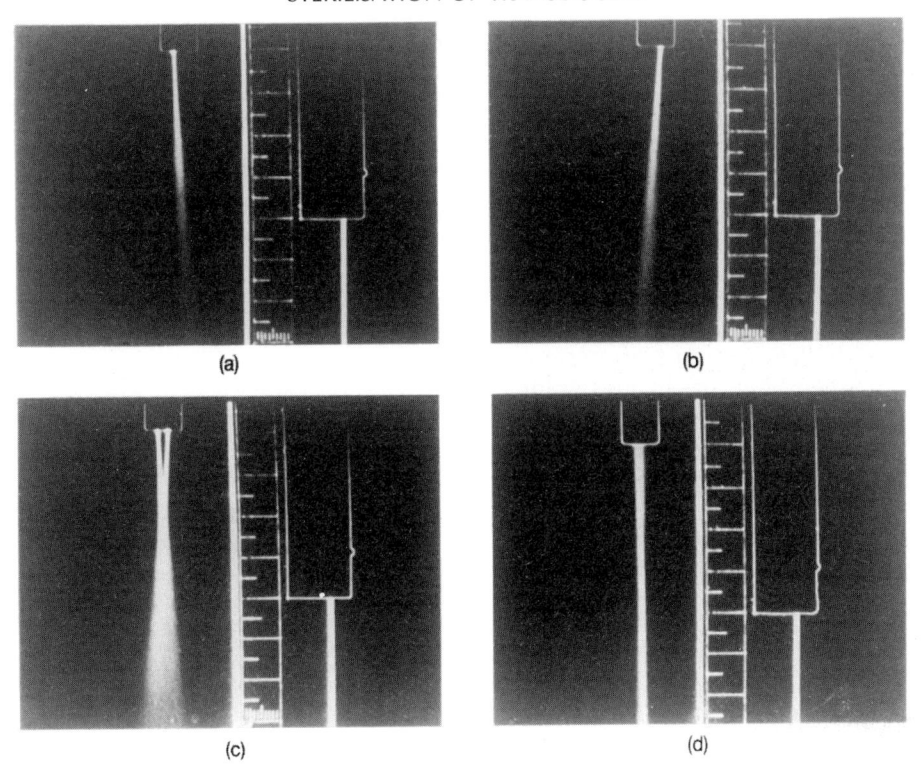

Figure 5.24 Schlieren photographs of ultrasound transmitted into water from the crystals of a continuous wave Doppler transducer. The transducer is a 10 MHz device with rectangular crystals of dimension 5 mm × 2 mm and separated by 0.75 mm. The scale shows 1 cm and 0.5 cm divisions. (a) One crystal transmitting; (b) the other crystal transmitting; (c) a double exposure obtained by driving each crystal in turn to demonstrate overlap of transmission and reception zones. If the crystals are driven simultaneously a strong inference pattern is obtained. (d) One crystal transmitting with the transducer rotated 90° about its axis relative to the position corresponding to the top pictures. The lack of circular symmetry in the field is illustrated (courtesy of DH Follett)

system then forms an image of the ultrasound field. This image can be viewed directly, projected on to a screen or photographed. The technique functions for both continuous and pulsed wave ultrasound. In the pulsed case, the optical beam is supplied by a stroboscope, which is synchronised to the ultrasonic pulses. At low and medium diagnostic intensity levels, the schlieren image may be qualitatively interpreted as exhibiting higher brightness at regions of higher intensity. Quantitative interpretation is much more difficult due to the complex 3-D structure of the ultrasonic field. The value of a schlieren system is that it can provide a quick appreciation of field characteristics, e.g. focal region, asymmetry or pulse dimension (Fig. 5.24). Schlieren systems are commercially available.

5.6 STERILISATION OF TRANSDUCERS

Transducers are often very expensive items and care is required in their sterilisation. Steam sterilisation in an autoclave is most effective but will probably damage the transducer or the bonding of its case. The most practical approach is gas sterilisation with ethylene oxide, but this normally takes several days. A degree of sterilisation can be achieved with anti-bacterial and anti-viral liquids. If there is doubt about sterility, professional advice should be sought, but apart from autoclaving no guarantees of complete sterility are usually forthcoming. Re-use of catheter transducers is not usually permitted due to the risk of infection.

5.7 DAMAGE TO TRANSDUCERS

Dropping transducers is the most common cause of damage. Smaller impacts can often damage the windows of mechanical transducers. Replacement of the windows is usually an expensive repair. The life of mechanical scanners can be greatly extended by ensuring that they are not moving when they are not in use. A few machines automatically switch off the drive when the transducer is replaced in its holder. Although electronic array transducers are more robust than mechanical scanning devices, when they are damaged it may not be possible to repair them.

Care should be taken in selecting the cleaning liquid for use on a probe, as some are known to attack plastic and rubber. It is probably best to use water. The manufacturer's instructions should also be followed.

5.8 SUMMARY

In this chapter the transmitted field was distinguished from the reception zone and the ultrasonic beam was seen to depend on their combined effect. The types of transducers found in simple, duplex and flow mapping systems were then described and their suitability for different applications indicated. The ultrasonic fields, beams and sample volumes associated with CW and PW equipment were discussed. Finally, care of the transducers was considered.

5.9 REFERENCES

Arditi M, Foster FS, Hunt JW (1981) Transient fields of concave annular arrays. Ultrason Imaging 3:37–61

Armitage AD, Scales NR, Hicks PJ, Payne PA, Chen OX, Hatfield JV (1995) An integrated array transducer receiver for ultrasound imaging. Sensors and Actuators: A—Physical 47:542–546

Arnolds BJ, Kunz D, von Reutern G-M (1989) Spatial resolution of transcranial pulsed Doppler technique *in vitro* evaluation of the sensitivity distribution of the sample volume. Ultrasound Med Biol 15:729–735

Bacon DR (1981) Characteristics of a PVDF membrane hydrophone for use in the range 1–100 MHz. IEEE Trans Sonics Ultrason SU-29:18–25

Bainton KF, Silk MG (1980) Some factors which affect the performance of ultrasonic transducers. Br J Non-destructive Testing January:15–20

Baker AC, Berg AM, Sahin A, Tjotta JN (1995) The nonlinear pressure field of plane rectangular apertures: experimental and theoretical results. J Acoust Soc Am 97:3510–3517

Beaver WL (1974) Sonic nearfields of a pulsed piston radiator. J Acoust Soc Am 56:1043–1048

Beckley DE, Casebow MP, Pettengell KE (1982) The use of a Doppler ultrasound probe for localizing blood flow during upper gastrointestinal endoscopy. Endoscopy 14:146–147

Berson M, Roncin A, Arbeille P, Patat F, Pourcelot L (1987) A linear array system for deep vessel explorations. Ultrasound Med Biol 13:267–274

Bui T, Chan HLW, Unsworth J (1988) Specific acoustic impedances of piezoelectric ceramic and polymer composites used in medical applications. J Acoust Soc Am 83:2416–2421

Cahill MD, Baker AC (1997) Increased off-axis energy deposition due to diffraction and nonlinear propagation of ultrasound from rectangular sources. J Acoust Soc Am 102:199–203

Chen WH (1983) Analysis of the high-field losses of polyvinylidene fluoride transducers. IEEE Trans Sonics Ultrason (suppl 30): 238–249

Chen WH, Shaw HJ, Weinstein DG, Zitelli LT (1978) PVG2 transducers for NDE. Proc IEEE Ultrasonics Symp:780–783

Chivers RC, Adach J, Filmore PR (1986) Ultrasonic Doppler probe assessment by scanned hydrophone. Clin Phys Physiol Meas 7:161–178

Dietz DR, Parks SI, Linzer M (1979) Expanding-aperture annular array. Ultrason Imaging 1:56–75

Doucette JW, Corl D, Payne HM, Flynn AE, Goto M, Nassi M, Segal J (1992) Validation of a Doppler guide wire for intravascular measurement of coronary artery flow velocity. Circulation 85:1899–1911

Douville Y, Arenson JW, Johnston KW, Cobbold RSC, Kassam M (1983) Critical evaluation of continuous-wave Doppler probes for carotid studies. J Clin Ultrasound 11:83–90

Duck FA (1981) The pulsed ultrasonic field. In: Moores BM, Parker RP, Pullan BR (eds) Physical Aspects of Medical Imaging, pp. 97–108. Wiley, Chichester

Duck FA, Hodson CJ, Tomlin PJ (1974) Esophageal Doppler probe for aortic flow velocity monitoring. Ultrasound Med Biol 1:233–241

Duerinckx AJ (1981) Modelling wavefronts from acoustic phased arrays by computer. IEEE Trans Biomed Eng BME-28:221–234

Eriksson R, Persson HW, Dymling SO, Lindstrom K (1994) Overall sensitivity patterns in the near field of CW Doppler transducers for perfusion measurements. Eur J Ultrasound 1:191–196

Evans DH, Parton L (1981) The directional characteristics of some ultrasonic Doppler blood-flow probes. Ultrasound Med Biol 7:51–62

Evans JL, Ng KH, Vonesh MJ, Kramer BL, Meyers SN, Mills TA, Kane BJ, Aldrich WN, Jang YT, Yock PG, Rold MD,

Roth SI, McPherson DD (1994) Arterial imaging with a new forward-viewing intravascular ultrasound catheter. I—Initial studies. Circulation 89:712–717

Evans JM, Skidmore R, Wells PNT (1986) A new technique to measure blood flow using Doppler ultrasound. In: Evans JA (ed) Physics in Medical Ultrasound, pp. 141–144. Institute of Physical Sciences in Medicine, London

Fay B, Ludwig G, Lankjaer C, Lewin PA (1994) Frequency response of PVDF needle-type hydrophones. Ultrasound Med Biol 20:361–366

Follett DH (1986) Light diffraction by ultrasound as evidence of finite amplitude distortion. Proc Inst Acoustics 8:55–62

Gururaja TR, Schulze WA, Cross LE, Newham RE (1985) Piezoelectric composite-materials for ultrasonic transducer applications. 2—Evaluation of ultrasonic medical applications. IEEE Trans Sonics Ultrason 32:499–513

Hall DDN, Hayward G, Gorfu Y (1993) Theoretical and experimental evaluation of a two-dimensional composite matrix array. IEEE Trans Ultrason Ferroelec Freq Contr 40:704–709

Harris GR (1988) Hydrophone measurements in diagnostic ultrasound fields. IEEE Trans Ultrason Ferroelec Freq Contr 35:87–101

Hartley CJ (1989) Review of intracoronary Doppler catheters. Int J Cardiac Imag 4:159–168

Hatfield JV, Scales NR, Armitage AD, Hicks PJ, Chen OX, Payne PA (1994) An integrated multielement array transducer for ultrasound imaging. Sensors and Actuators: A—Physical 41:167–173

Hayward G, Hossack JA (1990) Unidimensional modeling of 1–3 composite transducers. J Acoust Soc Am 88:599–608

Hayward G, Bennet J, Hamilton R (1995) A theoretical study on the influence of some constituent material properties on the behaviour of 1–3 connectivity composite materials. J Acoust Soc Am 98:2187–2196

Hill CR (1986) The generation and structure of acoustic fields. In: Hill CR (ed) Physical Principles of Medical Ultrasonics, pp. 68–92. Ellis Horwood, Chichester

Hoeks APG, Ruissen CJ, Hick P, Reneman RS (1984) Methods to evaluate the sample volume of pulsed Doppler systems. Ultrasound Med Biol 10:427–434

Hoskins PR, McDicken WN (1997) Colour ultrasound imaging of blood flow and tissue motion. Br J Radiol 70:878–890

Hunt JW, Arditi M, Foster FS (1983) Ultrasound transducers for pulse-echo medical imaging. IEEE Trans Biomed Eng 30:453–481

Ishiyami K, Yangagawa T, Sato T, Yano M, Yoshikawa N (1985) Development of small radius convex scanning system with view angle of 98 degrees. In: Gill RW, Dadd MJ (eds) Proceedings of the 4th Meeting of World Federation for Ultrasound in Medicine and Biology, p. 542. Pergamon Press, Sydney

Jensen JA, Svendson NB (1992) Calculation of pressure fields from arbitrarily shaped, apodized, and excited ultrasound transducers. IEEE Trans Ultrason Ferroelec Freq Contr 39:262–267

Kino GS (1987) Acoustic Waves: Devices, Imaging, and Analog Signal Processing. Prentice-Hall, Englewood Cliffs, NJ

Kino GS, DeSilets CS (1979) Design of slotted transducers with matched backing. Ultrason Imaging 1:189–209

Kinsler LE, Frey AR, Coppens AB, Sanders JV (1982) Fundamentals of Acoustics, 3rd edn. Wiley, New York

Kossoff G (1966) The effects of backing and matching on the performance of piezoelectric ceramic transducers. IEEE Trans Sonics Ultrason SU-13:20–31

Krebs W, Klues HG, Steinert S, Sivarajan M, Job FP, Flachskampf FA, Franke A, Reineke T, Hanrath P (1996) Left-ventricular volume calculations using a multiplanar transesophageal echoprobe—*in vivo* validation and comparison with biplane angiography. Eur Heart J 17:1279–1288

Lancée CT, Souquet J, Ohigashi H, Bom N (1985) Ferroelectric ceramics versus polymer piezoelectric materials. Ultrasonics 23:138–142

Lee CK, Benkeser PJ (1991) Investigation of a forward looking IVUS imaging transducer. In: Proceedings of the 1991 IEEE Ultrasonics Symposium, pp. 691–694. IEEE, Piscataway, NJ

Lees WR, Lyons EA (eds) (1996) Invasive Ultrasound. Martin Dunitz, London

Lewin PA (1981) Miniature piezoelectric polymer ultrasonic hydrophone probes. Ultrasonics 19:213–219

Li S, Hoskins PR, McDicken WN (1997) Rapid measurement of the spatial resolution of colour flow scanners. Ultrasound Med Biol 23:591–596

Li W, van der Steen AFW, Lancée CT, Céspedes I, Bom N (1998) Blood flow imaging and volume flow quantitation with intravascular ultrasound. Ultrasound Med Biol 24:203–214

Lockwood GR, Foster FS (1996) Optimizing the radiation-pattern of sparse periodic 2-dimensional arrays. IEEE Trans Ultrason Ferroelec Freq Contr 43:15–19

Lockwood GR, Ryan LR, Gotlieb AI, Lonn E, Hunt JW, Liu P, Foster FS (1992) *In vitro* high resolution intravascular imaging in muscular and elastic arteries. J Am Coll Cardiol 20:153–160

Lockwood GR, Turnbull DH, Christopher DA, Foster FS (1996) Beyond 30 MHz: applications of high-frequency ultrasound imaging. IEEE Eng Med Biol 15 (6):60–71

Martin RW, Gilbert DA, Silverstein FE, Deltenre M, Tygat G, Gange RK, Myers J (1985) An endoscopic Doppler probe for assessing intestinal vasculature. Ultrasound Med Biol 11:61–69

McHugh R, McDicken WN, Thompson P, Boddy K (1981) Blood flow detection by an intersecting zone ultrasonic Doppler unit. Ultrasound Med Biol 7:371–375

McKeighen RE (1983) Basic transducer physics and design. Seminars in Ultrasound 4:50–59

Melton HE, Thurstone FL (1978) Annular array design and logarithmic processing for ultrasonic imaging. Ultrasound Med Biol 4:1–12

Moraes R, Evans DH (1995) Effects of nonuniform insonation by catheter-tipped Doppler transducers on velocity estimation. Ultrasound Med Biol 21:779–791

Mortimer AJ (1982) Physical characteristics of ultrasound. In: Repacholi MH, Benwell DA (eds) Essentials of Medical Ultrasound, pp. 1–34. Humana, Clifton, NJ

Ng KH, Evans JL, Vonesh MJ, Meyers SN, Mills TA, Kane BJ, Aldrich BS, Jang YT, Yock PG, Rold MD, Roth SI,

McPherson DD (1994) Arterial imaging with a new forward-viewing intravascular ultrasound catheter. II—Three dimensional reconstruction and display of data. Circulation 89:718–723

Parks SI, Linzer M, Shawker TH (1979) Further development and clinical evaluation of the expanding aperture annular array system. Ultrason Imaging 1:378–383

Phillips D, McAleavey S, Parker KJ (1997) Color Doppler spatial resolution measurements with an oscillating thin film test object. In: Schneider SC, Levy M, McAvoy BR (eds) Proceedings of the 1997 IEEE Ultrasonics Symposium, pp. 1517–1520. IEEE, Piscataway, NJ

Preston RC, Bacon DR, Livett AJ, Rajendran K (1983) PVDF membrane hydrophone performance properties and their relevance to the measurement of the acoustic output of medical ultrasonic equipment. J Phys E: Sci Instrum 16:786–796

Preston RC, Taylor GEM, Thompson RC, Zeqiri B, Livett AJ (1985) The performance of the ultrasound beam calibrator BECA2. p. 537. In: Gill RW, Dadd MJ (eds) Proceedings of the 4th Meeting of World Federation for Ultrasound in Medicine and Biology, Pergamon Press, Sydney

Pye DW (1983) Medical ultrasonics: dynamic focusing in diagnostic imaging. PhD Dissertation, University of Edinburgh

Roth T, Erbel R, Brennecke R, Rupprecht HJ, Meyer J, Seelen WV (1993) Variations in acoustical beam properties of intracoronary Doppler catheters. Catheterization and Cardiovas Diag 30:257–263

Sherar MD, Foster FS (1989) The design and fabrication of high frequency poly(vinylidene fluoride) transducers. Ultrason Imaging 11:75–94

Shotton KC, Bacon DR, Quilliam RM (1980) A PVDF membrane hydrophone for operation in the range 0.5 MHz to 15 MHz. Ultrasonics 18:123–126

Shung KK, Zipparo M (1996) Ultrasonic transducers and arrays. IEEE Eng Med Biol 15 (6):20–30

Silk MG (1984) Ultrasonics Transducers for Non-destructive Testing. Adam Hilger, Bristol

Smith SW, Trahey GE, von Ramm OT (1992) Two-dimensional arrays for medical ultrasound. Ultrason Imaging 14:213–233

Somer JC (1968) Electronic sector scanning for ultrasonic diagnosis. Ultrasonics 6:153–159

Souquet J, Hanrath P, Zitelli L, Kremer P, Langenstein BA, Schulter M (1982) Trans-esophageal phased-array for imaging the heart. IEEE Trans Biomed Eng 29:707–712

Steinberg BD (1992) Digital beamforming in ultrasound. IEEE Trans Ultrason Ferroelec Freq Contr 39:716–721

Tarnoczy T (1965) Sound focusing lenses and waveguides. Ultrasonics 3:115–127

Vogel J, Bom N, Ridder J, Lancée C (1979) Transducer design considerations in dynamic focusing. Ultrasound Med Biol 5:187–193

von Ramm OT, Thurstone FL (1976) Cardiac imaging using a phased array ultrasound system. 1. System design. Circulation 53:258–262

Walker AR, Phillips DJ, Powers JE (1982) Evaluating Doppler devices using a moving string test target. J Clin Ultrasound 10:25–30

Wells PNT (1970) The directivities of some ultrasonic Doppler probes. Med Biol Eng 8:241–256

Wells PNT (1977) Biomedical Ultrasonics. Academic Press, London

Whittingham TA (1997) New and future developments in ultrasonic imaging. Br J Radiol 70:S119-S132

Whittingham TA, Roberts TJ (1986) Practical beamshapes as visualised by the Newcastle beam plotting system. In: Evans JA (ed) Report 47—Physics in Medical Ultrasound, pp. 67–77. Institute of Physical Sciences in Medicine, London

Willard GW (1947) Ultrasound waves made visible. Bell Lab Res 25:194–200

Woodward B (1977) The stability of polyvinylidene fluoride as an underwater transducer material. Acoustica 38:264–268

6

Signal Detection and Pre-processing: CW and PW Doppler

6.1 INTRODUCTION

The physical principles of both continuous wave (CW) and pulsed wave (PW) Doppler systems have been outlined in Chapter 4. The practical implementation of these principles will be outlined in this chapter, which is intended primarily for those who have an interest in understanding the detailed workings of such devices.

6.2 CONTINUOUS WAVE SYSTEMS

Figure 6.1 is a block diagram of a typical CW Doppler system. The master oscillator operates at a constant frequency and drives the transmitting crystal of the probe via a transmitting amplifier. The returning ultrasound signal, containing echoes from both stationary and moving targets, is fed to the radio frequency (RF) amplifier from the receiving crystal. This amplified signal is then demodulated and filtered to produce audio frequency signals (usually in phase quadrature) whose frequencies and amplitudes provide information about motion within the ultrasound beam. Although CW systems usually operate in this way, Tortoli et al (1994) have also demonstrated the feasibility of using a single RF channel 'under sampling' technique.

6.2.1 The Master Oscillator

The master oscillator is the heart of the Doppler system, and provides both the operating reference frequency, which is amplified by the transmitter circuit, and reference signals for the demodulation stage. It is important that the oscillator has high short-term stability so that artificial Doppler shift noise is not generated. Nowadays the necessary signals are provided by purpose-built integrated circuit crystal oscillator chips.

6.2.2 The Transmitter

The function of the transmitter or transmitting amplifier is to drive the transmitting crystal with the correct time-varying voltage in order to produce the required acoustic signal. For continuous wave devices the voltage applied across the crystal is usually in the range of 2–5 V peak to peak. For a 50 Ω matched system the transmitter output stage has to be capable of delivering 50 mA. For single-frequency CW systems the transmitter is usually a transformer coupled output stage. A typical example of such a circuit is illustrated in Fig. 6.2. For multifrequency systems, where the frequencies range from 2 to 10 MHz, the output stage has to be wideband and is usually in the form illustrated in Fig. 6.3. These wideband transmitter systems do not excite the crystal with a sine wave, but with a square wave; the higher harmonics do not prove a problem as they are filtered by the narrow bandwidth of the piezoelectric crystal.

6.2.3 The Probe

The CW Doppler probe consists of two crystals, one acting as a transmitter while the other acts as a receiver. The most usual configuration is known as a 'split D' and is produced by splitting a circular transducer. It is usual to angle each crystal so that a region of sensitivity is defined by the overlap of the transmission field of one transducer and the

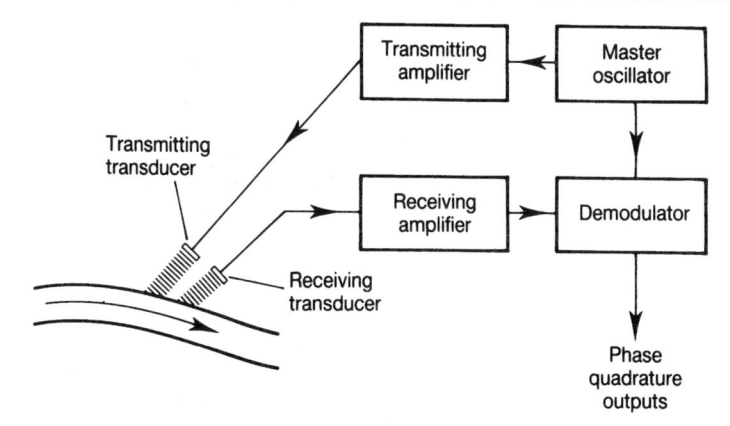

Figure 6.1 Block diagram of a continuous wave Doppler system

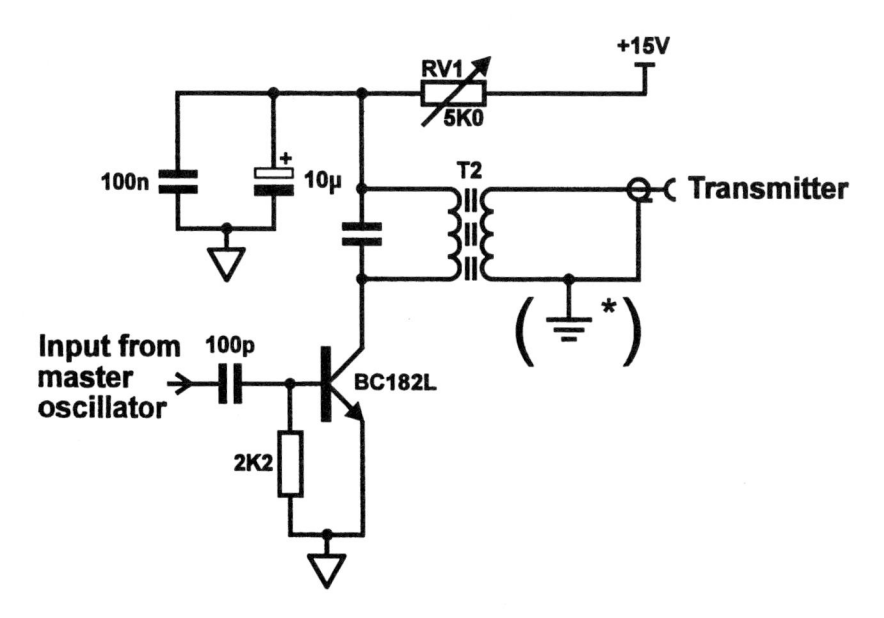

Figure 6.2 Transmitter stage for a single-frequency continuous wave Doppler velocimeter. The output transformer is tuned to the frequency of operation and the variable resistor RV1 provides adjustment for the overall output voltage. *Isolation of the secondary winding may be used to achieve a 'floating' patient applied part (i.e. the probe)

reception field of the other. The piezoelectric crystals are attached to coaxial cables, which in turn are connected to the transmitter and receiver. The crystals are housed and separated by an acoustic insulator that reduces crosstalk between the transmitter and receiver. Excessive crosstalk will overload the RF amplifier and demodulator stages in the Doppler device leading to degrading of the directional separation of forward and reverse flow signals. Further details of Doppler transducers and the acoustic fields they produce may be found in Chapter 5. Follett (1993) has discussed methods of ensuring good electrical isolation of Doppler transducers without undue signal loss, with particular reference to catheter transducers.

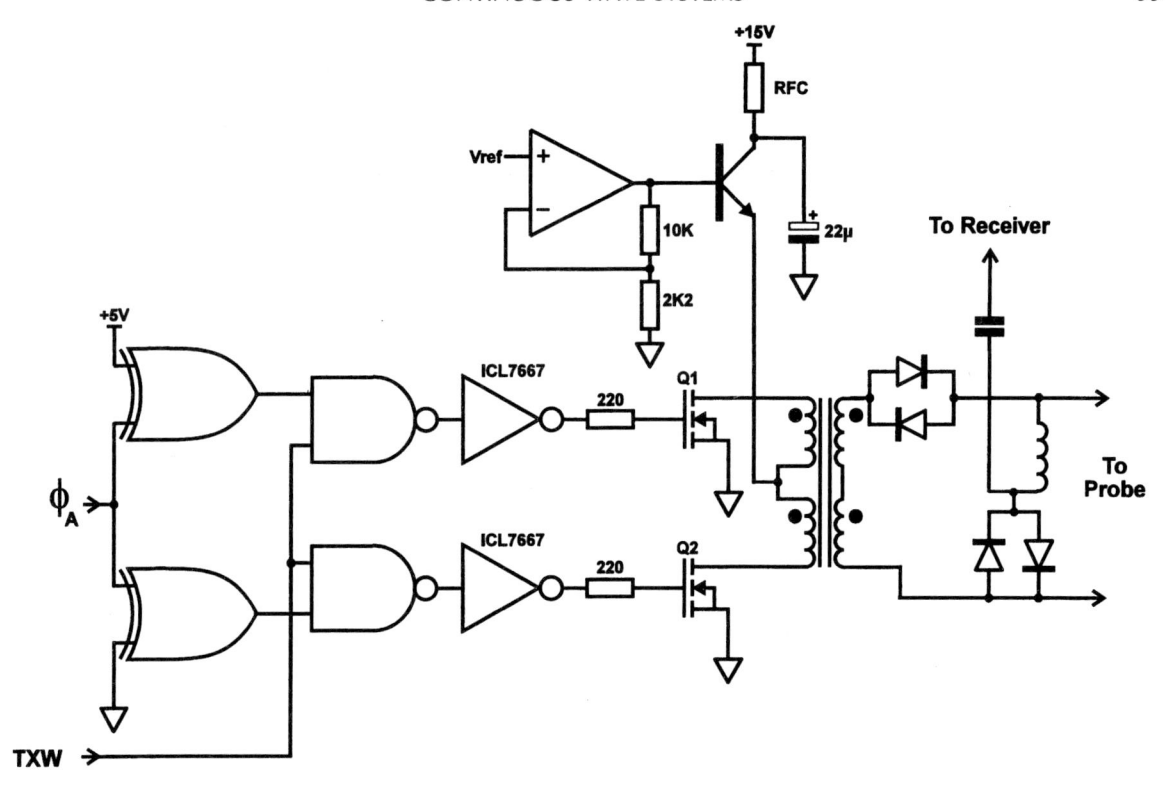

Figure 6.3 Circuit diagram of a wideband transmitter system for continuous and pulsed wave applications. ϕ_A, operating clock frequency; TXW, transmission burst width (PW operation only); Vref, output voltage control. Q_1, Q_2 are MOSFET devices

6.2.3.1 Probe Matching

Most continuous wave Doppler pencil probes do not contain electrical matching networks. Consequently, there can be variation of the acoustic output power from one probe to another. Furthermore, the variation in output impedance of the receiving transducer will cause variations in matching to the receiving amplifier, which will in turn produce variations in signal-to-noise ratio. Changing probes can sometimes give the user the impression that one probe is more sensitive than another and although this may be true in certain circumstances, the apparent increase in sensitivity is mainly due to the increase in acoustic output power where there is a drop in the impedance of the transmitting crystal. Ideally the transmitting crystal and the receiving crystal should be matched to the connecting cable which is either 75 or 50 Ω.

In order to obtain an acoustic match between the piezoelectric crystal and the body, a matching layer of epoxy resin of quarter-wavelength thickness can be used. However, for the higher frequencies of operation this thickness becomes practically too small and sometimes three-quarter wave matching is used instead. In practice, inexpensive CW probes forego this rigour and the thickness of epoxy resin is not well controlled, giving rise to considerable variation of sensitivity from one probe to another.

6.2.4 The Receiver

The function of the receiver is to amplify the small voltages produced by the receiving crystal, and for Doppler shifted signals themselves this is of the order of 10–100 µV. However, leakage from the transmitter to receiver crystal constitutes a received signal approaching 10 mV. A typical receiver amplifier stage is illustrated in Fig. 6.4, where

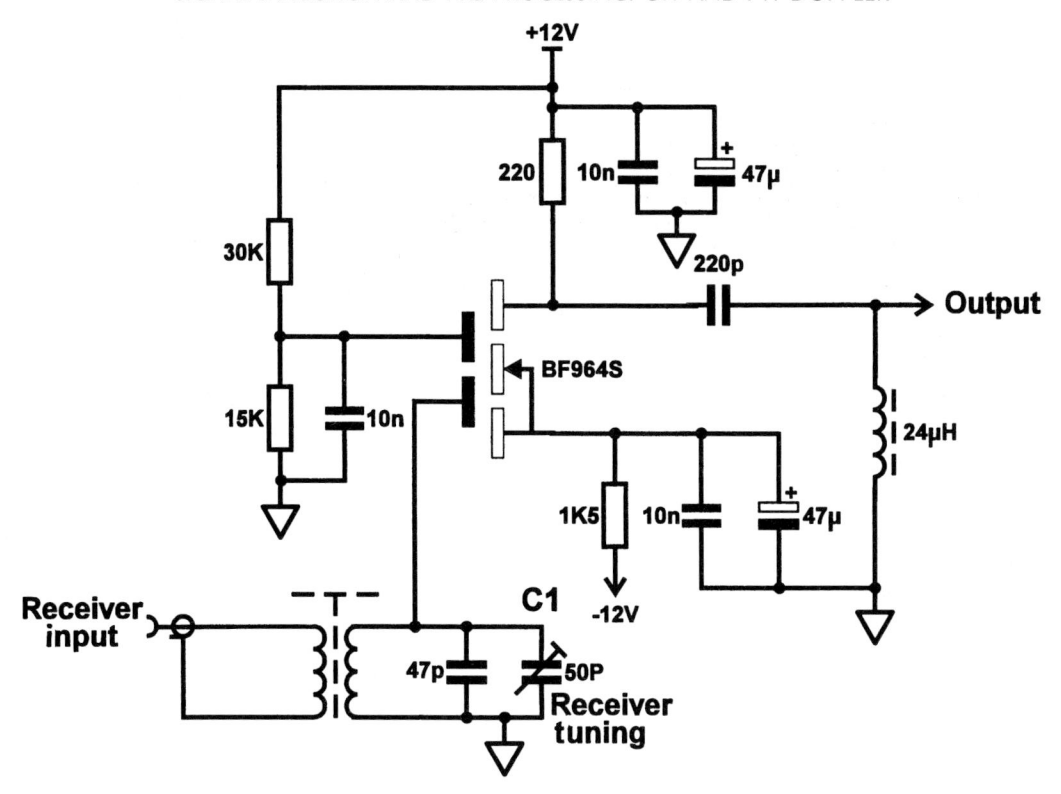

Figure 6.4 Circuit diagram of a continuous wave receiver. The variable capacitor C1 is tuned for optimum sensitivity at the frequency of operation (2 MHz)

transformer matching has been employed in order to optimise signal-to-noise characteristics of the FET amplifier. Follett (1991) has discussed sources of electromagnetic RF interference in Doppler equipment, and methods of reducing it, in some detail.

The receiver must have a sensitivity of better than 1 µV RMS and a dynamic range capability in excess of 80 dB. For single-frequency CW devices, the RF amplifier need only have a limited bandwidth in the region of 200 kHz; however, for wideband systems the receiver bandwidth needs to be of the order of the range of operating frequency, usually 2–10 MHz. A typical receiver circuit for a wideband system is illustrated in Fig. 6.5.

6.2.5 The Demodulator

The purpose of the demodulator is to remove the carrier frequency and provide an output consisting of the Doppler sidebands. All but the simplest Doppler units now use coherent demodulation techniques (i.e. the reference signal for demodulation is taken from the master oscillator rather than the clutter signals that arise from stationary or near stationary tissue). Straightforward coherent demodulation does not provide directional information because both upper and lower Doppler sidebands (corresponding to flow towards and away from the probe) are shifted into the same region of the baseband, and therefore alternative techniques must be used. Three such solutions that have been successfully employed are single sideband detection, heterodyne detection and quadrature phase detection, and each of these has been discussed by Coghlan and Taylor (1976) and Atkinson and Woodcock (1982).

Single sideband (SSB) detection (see Fig. 6.6) is achieved by directly filtering the returning RF signal with low and high pass filters to remove the

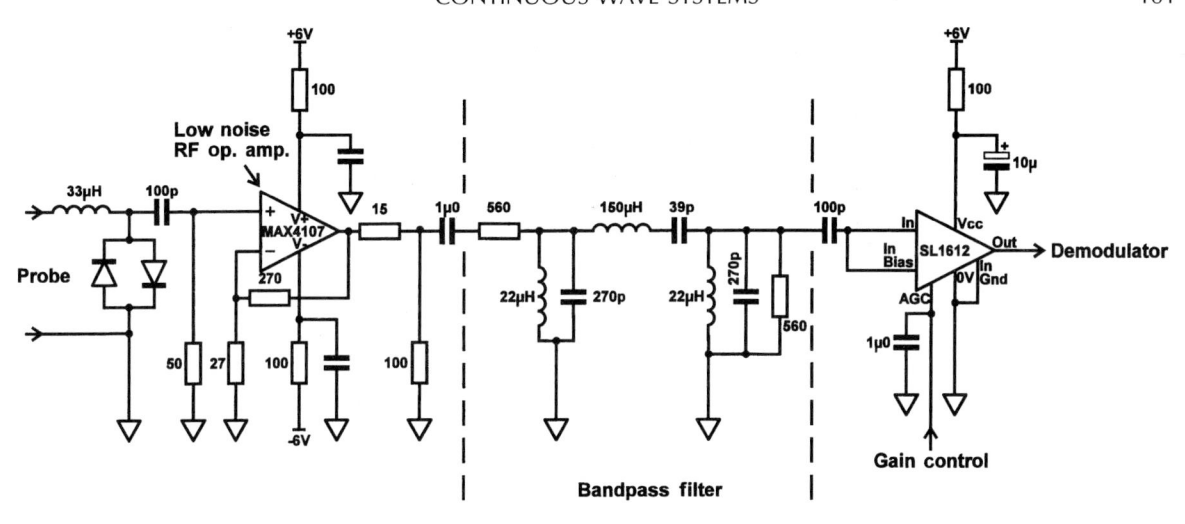

Figure 6.5 Circuit diagram of a wideband continuous or pulsed wave Doppler receiver. The 33 μH coil and the diode combination are protection from the transmission burst for pulsed Doppler applications. The bandpass filter illustrated is for a 2 MHz system. The final amplification stage is a voltage-controlled amplifier

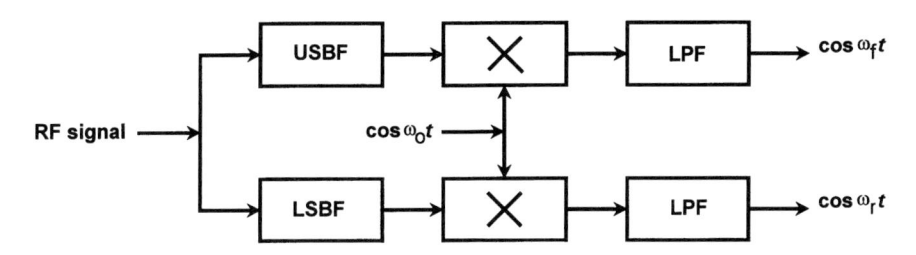

Figure 6.6 Single sideband detection system. USBF, upper sideband filter; LSBF, lower side band filter; LPF, low pass filter

upper and lower sidebands, respectively. The two channels are then independently coherently demodulated to produce separate audio signals, one composed of flow away from the probe, the other of flow towards the probe. A major problem with this method is that the filters must be exceedingly sharp (with a Q-factor approaching 10^6) if proper channel separation is to be obtained. The only practical method of achieving this is to use multistage crystal filters. A practical SSB detection system, which employs frequency conversation prior to the single sideband filter, has been described by de Jong et al (1975).

Heterodyne demodulation is achieved by coherently demodulating the returning ultrasound signal with a signal that is of a slightly lower frequency than that of the master oscillator, but which is derived from it by mixing its output with that of a heterodyne oscillator (Fig. 6.7).

For the sake of simplicity, consider an ultrasound signal that is composed of three distinct components: the carrier and two signals resulting from motion towards and away from the probe, respectively. Such a signal may be written:

$$S(t) = A_0 \cos(\omega_0 t + \phi_0) + A_f \cos(\omega_0 t + \omega_f t + \phi_f) \\ + A_r \cos(\omega_0 t - \omega_r t + \phi_r) \qquad 6.1$$

where A, ω and ϕ refer to the amplitude, angular frequency and phase of each signal, and the subscripts 0, f and r to the carrier, forward and reverse signals. This composite signal is coherently demodulated with a difference signal, $H(t)$, derived

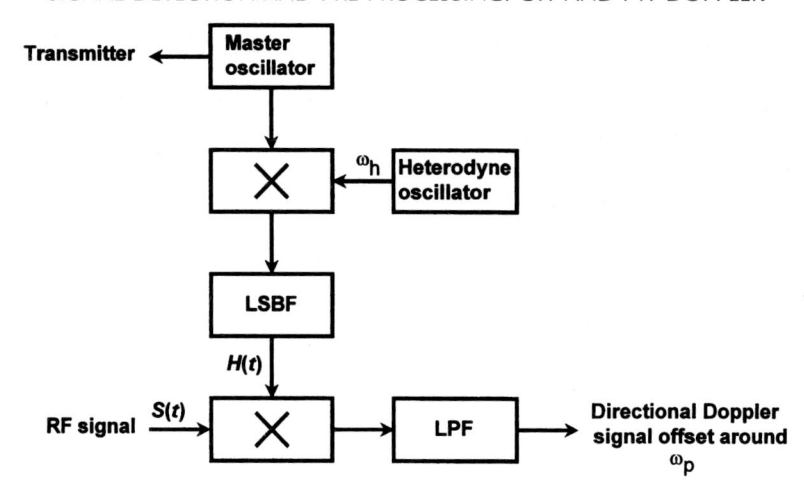

SIGNAL DETECTION AND PRE-PROCESSING: CW AND PW DOPPLER

Figure 6.7 Heterodyne detection system. LSBF, lower side band filter; LPF, low pass filter

by mixing the output of the master oscillator with that of the heterodyne oscillator, which may be written:

$$H(t) = \cos(\omega_0 t - \omega_h t) \qquad 6.2$$

where ω_h is the angular frequency of the heterodyne oscillator (note that there is a second component of angular frequency, $\omega_0 + \omega_h$, which must be removed before demodulation takes place).

The product of $S(t)$ and $H(t)$ is given by:

$$\begin{aligned}S(t) \cdot H(t) = \tfrac{1}{2}\{&A_0 \cos(\omega_h t + \phi_0) \\ +& A_0 \cos(2\omega_0 t - \omega_h t + \phi_0) \\ +& A_f \cos(\omega_h t + \omega_f t + \phi_f) \\ +& A_f \cos(2\omega_0 t + \omega_f t - \omega_h t + \phi_f) \\ +& A_r \cos(\omega_h t - \omega_r t + \phi_r) \\ +& A_r \cos(2\omega_0 t - \omega_r t - \omega_h t + \phi_r)\} \end{aligned}$$
$$6.3$$

Eliminating all frequencies of the order of ω_0 by low pass filtering leaves:

$$\tfrac{1}{2}\{A_0 \cos(\omega_h t + \phi_0) + A_f \cos[(\omega_h + \omega_f)t + \phi_f] \\ + A_r \cos[(\omega_h - \omega_r)t + \phi_r]\} \qquad 6.4$$

Provided that the heterodyne frequency is higher than the highest Doppler difference frequency received from targets receding from the Doppler transducer, the reverse and forward Doppler

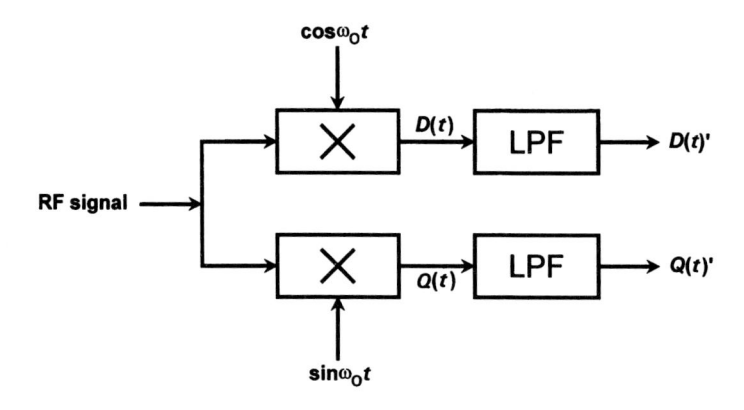

Figure 6.8 Quadrature phase detection system. LPF, low pass filter

components are completely separated and situated on either side of the heterodyne frequency ω_h. The large clutter component that occurs at the heterodyne frequency (the first term in eqn. 6.4) must be removed by a very sharp notch filter centred at that frequency. This type of output is particularly suited to spectral analysis, since both the forward and reverse flow signals can be analysed using a single analysis channel. A practical heterodyne demodulation system has been described by Light (1970).

The third and most widely used type of directional demodulation is quadrature phase detection. The object of this type of demodulation is to preserve the real and imaginary Doppler difference components, and this is achieved by coherently demodulating the returning Doppler signal both with the master oscillator and with a signal derived from the master oscillator but shifted in phase by 90° (Fig. 6.8).

Consider first the effect of multiplying the returning ultrasound signal given by eqn. 6.1 by $\cos(\omega_0 t)$; the result is a 'direct' signal $D(t)$ given by:

$$D(t) = \tfrac{1}{2}A_0[\cos(\phi_0) + \cos(2\omega_0 t + \phi_0)]$$
$$+ \tfrac{1}{2}A_f[\cos(\omega_f t + \phi_f)$$
$$+ \cos(2\omega_0 t + \omega_f t + \phi_f)]$$
$$+ \tfrac{1}{2}A_r[\cos(\omega_r t - \phi_r)$$
$$+ \cos(2\omega_0 t + \omega_r t + \phi_r)] \quad 6.5$$

Filtering out the DC component and terms of order $2\omega_0$ leads to:

$$D(t)' = \tfrac{1}{2}A_f \cos(\omega_f t + \phi_f)$$
$$+ \tfrac{1}{2}A_r \cos(\omega_r t - \phi_r) \quad 6.6$$

Multiplying the signal given by eqn. 6.1 by $\sin(\omega_0 t)$ leads in a similar way to a filtered 'quadrature' signal, $Q(t)'$, given by:

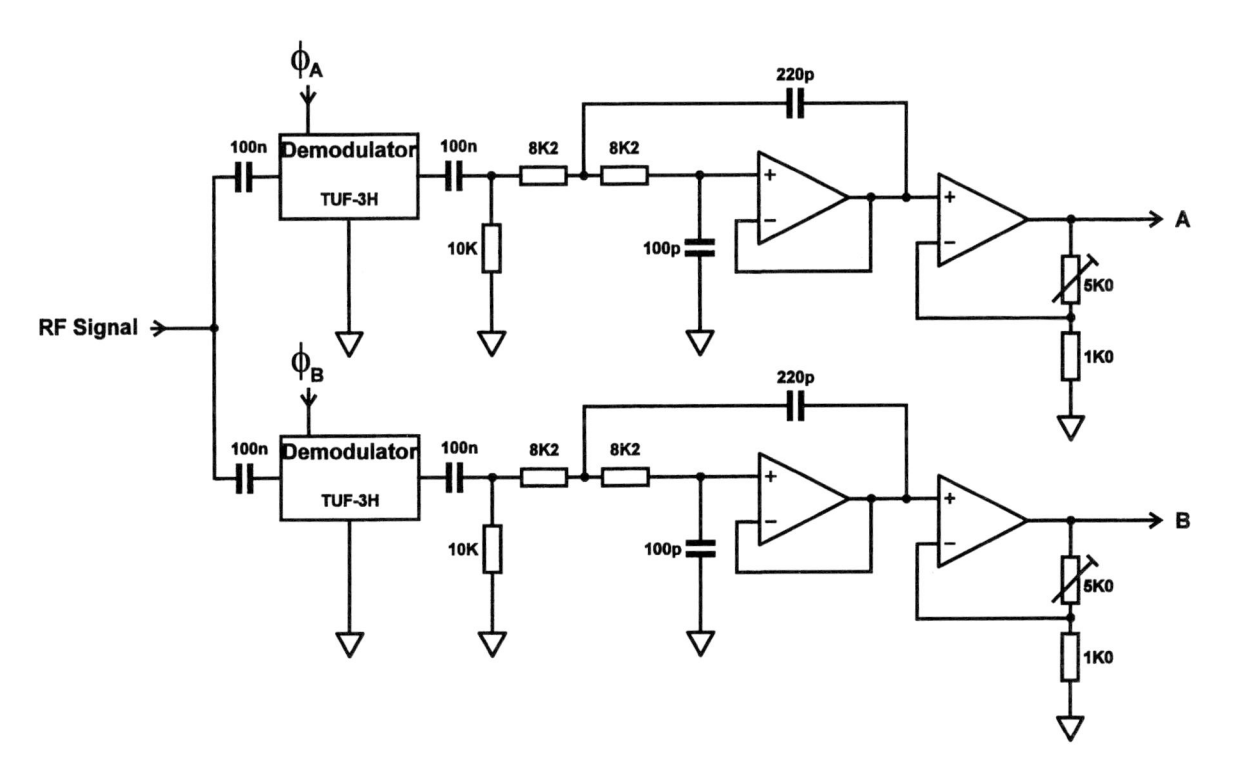

$$Q(t)' = -\tfrac{1}{2}A_f \sin(\omega_f t + \phi_f)$$
$$+ \tfrac{1}{2}A_r \sin(\omega_r t - \phi_r) \quad 6.7$$

Figure 6.9 Circuit diagram for a practical phase quadrature demodulation system. ϕ_A, master clock, ϕ_B master clock phase shifted by 90°; A, B, phase quadrature outputs

or

$$Q(t)' = \tfrac{1}{2}A_f \cos(\omega_f t + \phi_f + \pi/2) + \tfrac{1}{2}A_r \cos(\omega_r t - \phi_r - \pi/2) \qquad 6.8$$

It can be seen by comparing eqn. 6.6 with eqn. 6.8 that for a signal resulting from flow which is solely towards the probe the direct signal lags the quadrature signal by 90°, whilst for flow that is solely away from the probe the direct signal leads the quadrature signal by 90°. However, before forward and reverse flow can be completely separated, some type of further processing must occur (see Section 6.4).

Practically, demodulation is achieved using commercially available demodulation integrated circuits. Figure 6.9 illustrates such a circuit for phase quadrature demodulation. Because of the transmitter receiver breakthrough, the demodulator needs to have a large dynamic range so that it can detect low-level signals due to moving blood. Acoustic breakthrough is particularly severe at low frequencies, and with 2 MHz systems such as those used in cardiology it is usually necessary to filter

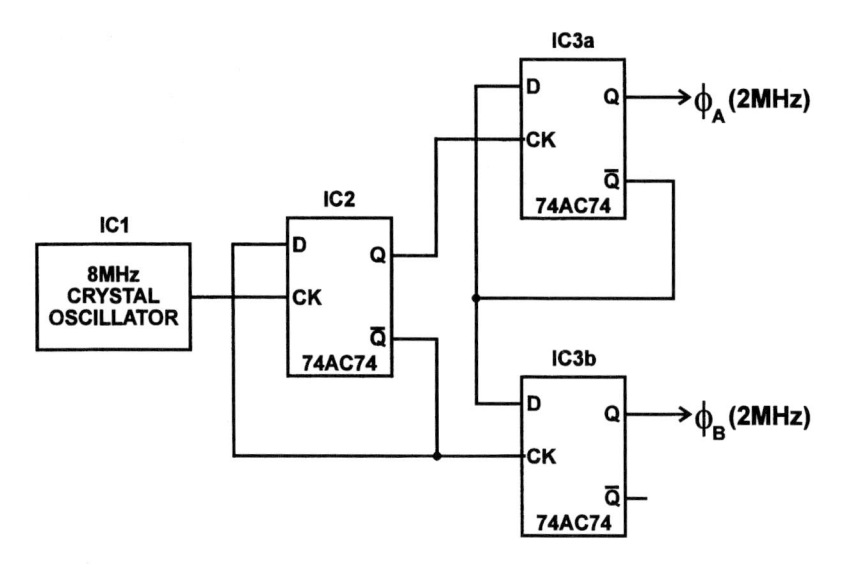

Figure 6.10 Logic system required for obtaining phase quadrature reference signals. Note that the master oscillator is set to four times the operating frequency of the Doppler system

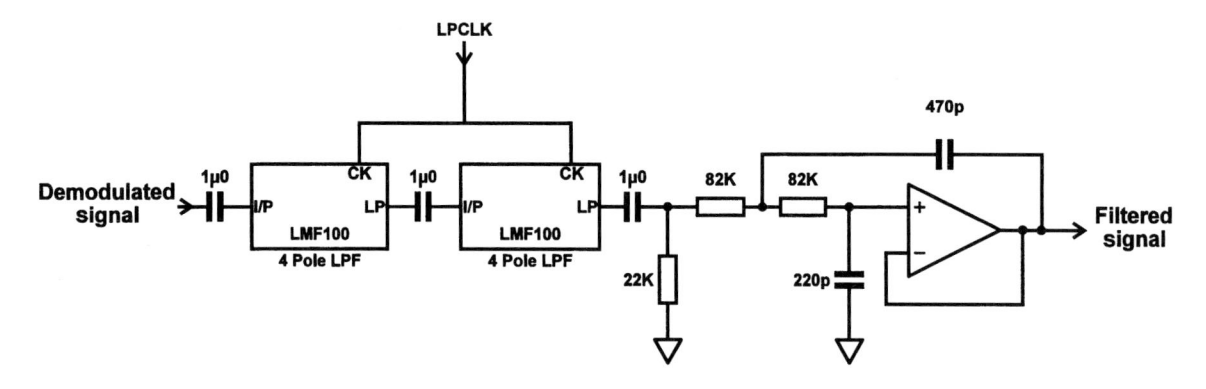

Figure 6.11 Example of a switched filter technique to provide a variable eight-pole low pass filter. The cut-off frequency is set by the LPCLK clock frequency. Similar configurations may be used for high pass filter applications

the receiver signal with a crystal notch filter before demodulation can be successfully achieved.

It is important for phase quadrature demodulation that the two clock signals are truly shifted by 90°. This is best achieved using a logic technique such as that illustrated in Fig. 6.10.

6.2.6 Filters

The demodulator output signals are filtered so that the high-frequency noise is removed and the low-frequency signals due to wall motion are also removed. For systems that have fixed filtering, standard operational amplifier filtering techniques are used. However, for variable filtering, switched filter techniques are now usually employed. An example of a switched filter circuit which employs custom-built integrated circuits is illustrated in Fig. 6.11.

6.3 PULSED WAVE SYSTEMS

Pulsed Doppler systems are used to obtain Doppler information at a specific range from the face of the transducer. Figure 6.12 is a block diagram of a typical pulsed Doppler system. The main difference between a CW and a pulsed system is that the transducer is excited with bursts of pulses instead of being continuously excited. The burst of ultrasound travels into the body, where it is Doppler shifted by moving structures along the sound path. Returning echoes from both stationary and moving targets are received by the same transducer. This process is then repeated for the next burst of ultrasound. The Doppler shift information is extracted by the demodulation process in exactly the same way as it is for the CW systems. The demodulation signal contains the range–phase information, i.e. for each burst of ultrasound the difference in phase between the reference signal (provided by the master oscillator) and the received echo at the specified range. The output from the demodulator is sampled at a specified point in time relative to the onset of the transmission pulse. The time of sampling defines the depth of interest and is chosen to correspond to the time it takes for sound to travel from the transducer to the depth of interest and back again. The sampling is achieved using a sample and hold amplifier which is updated after every transmission burst. The output from the sampler is filtered to remove the sampling frequency, thereby providing the Doppler shift frequency information at the range of interest. The physical principles of PW Doppler are discussed further in Section 4.2.2.

6.3.1 The Transmitter

Pulsed Doppler probes are usually excited with peak-to-peak voltages of between 20 and 100 V and with a burst length ranging between 1 and 10 μs. The main constraint on the transmitter is that it has to have a sufficient bandwidth to cope with such a signal. Also, because a single element transducer is normally permanently connected to the receiver input, the output noise from the transmitter must be kept to an absolute minimum. An on–off ratio of greater than 150 dB is usually adequate (see Fig. 6.3).

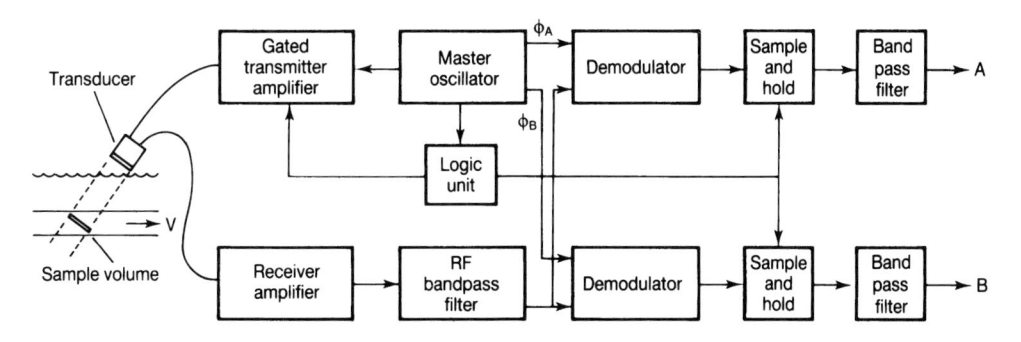

Figure 6.12 Block diagram of a pulsed Doppler system

6.3.2 The Probe

As the length of the transmission burst is very small compared with the reception time, a single crystal transducer can be used which operates firstly as a transmitter of ultrasound, and then as a receiver. As the reception path is identical to the transmission path, the problem of alignment is avoided. The probe is electrically matched to the input impedance of the receiver and is acoustically backed and matched so that it provides an ultrasound burst that matches the electrical excitation voltage produced by the transmitter, i.e. the bandwidth of the probe has to be as large as the bandwidth of the transmitted signal. See Chapter 5 for further details.

6.3.3 The Receiver

The receiver amplifier receives signals from the probe. The signals from stationary targets and the small signals from moving blood are amplified equally. Therefore, the receiver requires a large dynamic range in order to avoid saturation from vessel wall signals. Furthermore, the receiver needs to be protected from the transmission burst, as its input is directly connected to the transmitter. This is most commonly achieved using a protection network of the type shown in Fig. 6.5, which blocks the larger signals from the transmitter but allows the passage of small signals from the reflected and back-scattered ultrasound. The receiver must also have the capability to recover quickly from saturation and requires a bandwidth that is inversely proportional to the axial resolution of the system.

6.3.4 The Demodulator

The phase quadrature demodulation system used by most CW systems may also be incorporated in the pulsed Doppler system where the requirements are similar. The required bandwidth is inversely proportional to the axial resolution of the system.

6.3.5 The Sample-and-hold Amplifier

The sample-and-hold circuit samples the output from the demodulator stage, and it is common practice to integrate over the sample period and then hold this integrated value until the next sampling period. A suitable circuit is illustrated in Fig. 6.13. The sampling duration, together with the transmitted pulse length, sets the range over which velocity information is gathered (see Section 4.2.2.2). For optimum performance (i.e. the best sensitivity for a given axial resolution), the length of the transmission burst is usually set to the same length as the sampling or receiver gate length (Peronneau et al 1974)—but see Section 4.2.2.2. It is desirable to reduce the transmitted output voltage when the transmitted pulse length is increased so that the transmitted ultrasound power may be kept constant. Most pulsed Doppler systems allow the user to vary the sample volume length, to match the vessel diameter. The required bandwidth of the system is inversely proportional to the gate length and so an

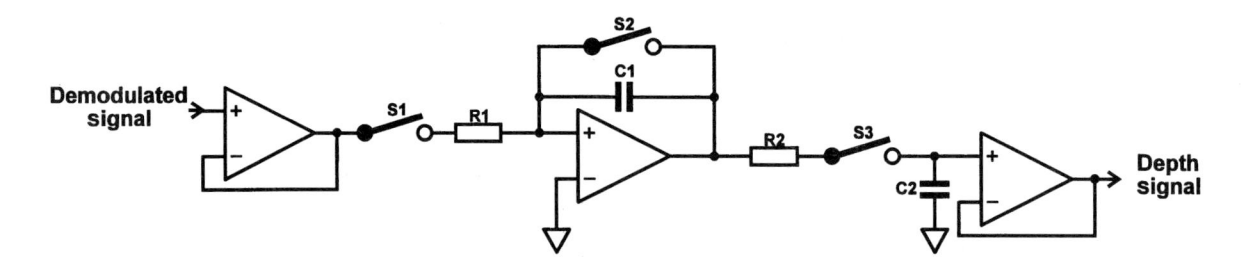

Figure 6.13 Pulsed Doppler integrating sample and hold system. Switch S1 samples the signal for a period defined by the gate length. During this time the signal is integrated. The final integrated voltage is then sampled and held via S3/C2. The integrator is then reset by switch S2 and the sequence restarted on the next pulse repetition cycle

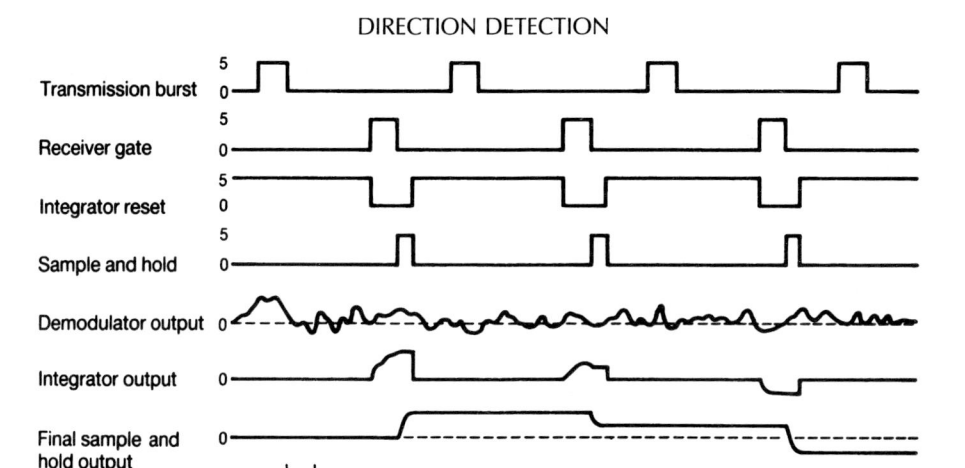

Figure 6.14 Timing diagram illustrating the sequence of events during the sampling of a Doppler signal in a PW system

improvement in signal-to-noise ratio can be achieved by increasing gate length.

6.3.6 Filters

The output from the sample-and-hold circuit contains not only the Doppler shift frequencies but also the sampling frequency which has to be removed. In order that the full range of Doppler shift frequencies can be utilised, i.e. up to half the pulse repetition frequency (PRF), a very sharp (at least 8 pole) low pass filter is used to eliminate the sampling frequency without degrading the Doppler signal. This filter must also be variable if a variable PRF system is used and this is best achieved using switched capacitor filters (see Fig. 6.11). As well as the low pass filter which removes the PRF, a high pass filter is also employed to remove the high-energy low-frequency signals that result from moving structures such as the heart and vessel walls. The high pass filter or 'wall' filter is usually variable over a range of 100–800 Hz, and is best implemented using switched capacitor filtering techniques.

6.3.7 Control Logic

The control logic provides the necessary timing signals to implement the transmitter burst, demodulator phase quadrature reference signals, and to initiate the sample-and-hold amplifiers at the appropriate time relative to the transmission burst. Figure 6.14 shows a typical timing diagram for a pulsed Doppler system.

6.4 DIRECTION DETECTION

If, as is now usual, the output from the Doppler system is in phase quadrature form, further processing is necessary to completely separate forward and reverse flow signals. There are at least four methods of achieving this, each of which is described in the following four sections.

6.4.1 Switched Channel Processing

The simplest method of separating the forward and reverse components present in quadrature signals, switched channel processing, was first described by McLeod (1967) and is illustrated in Fig. 6.15. A logic unit is used to determine which of the two quadrature signals (described by eqns 6.6 and 6.8) is in the lead and which is lagging. Depending on this relationship, the output of one of the quadrature channels (the one chosen is unimportant) is switched to either the forward flow or the reverse flow channel. This simple method works correctly only when the

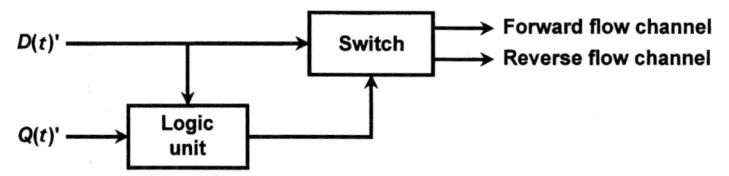

Figure 6.15 Switched channel processor for use with quadrature phase detected signals

flow is unidirectional, because if simultaneous forward and reverse signals are present, the phase relationship between the direct and quadrature signals is indeterminate. Furthermore, the method is susceptible to switching artefacts, particularly in the presence of high-amplitude low-frequency components such as vessel wall thump (Coghlan and Taylor 1976). This method has been used extensively in simple CW systems, but is not recommended for other purposes.

6.4.2 Phasing-filter Technique

Equations 6.6 and 6.8 can be solved simultaneously to extract the forward and reverse components, and this is in essence the method by which the phasing-filter technique is able to separate the two flow channels. The basis of the method by which this is achieved is illustrated in Fig. 6.16. Both the direct and quadrature channels are phase-shifted by 90° and added to the other

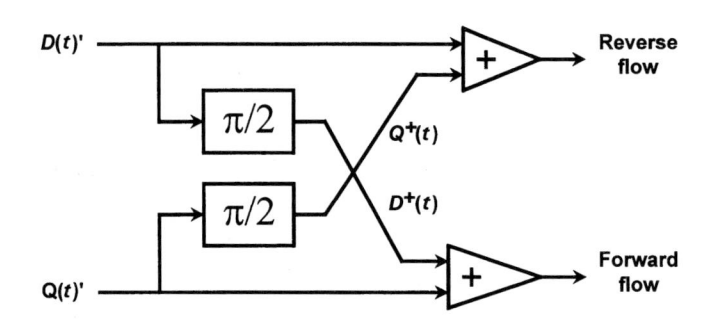

Figure 6.16 Phasing filter processor for use with quadrature phase detected signals

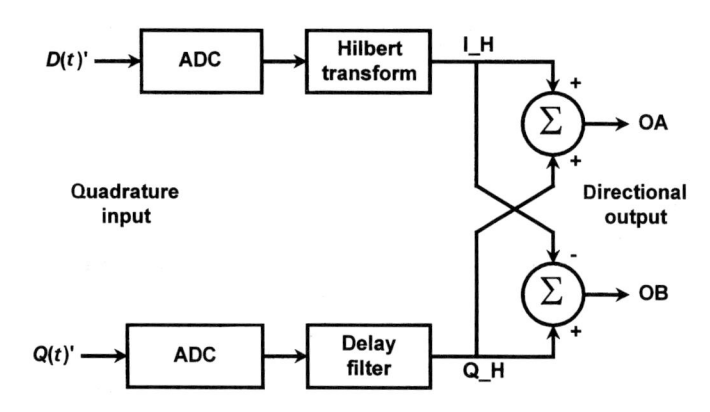

Figure 6.17 Phasing filter implementation described by Aydin and Evans (1994)

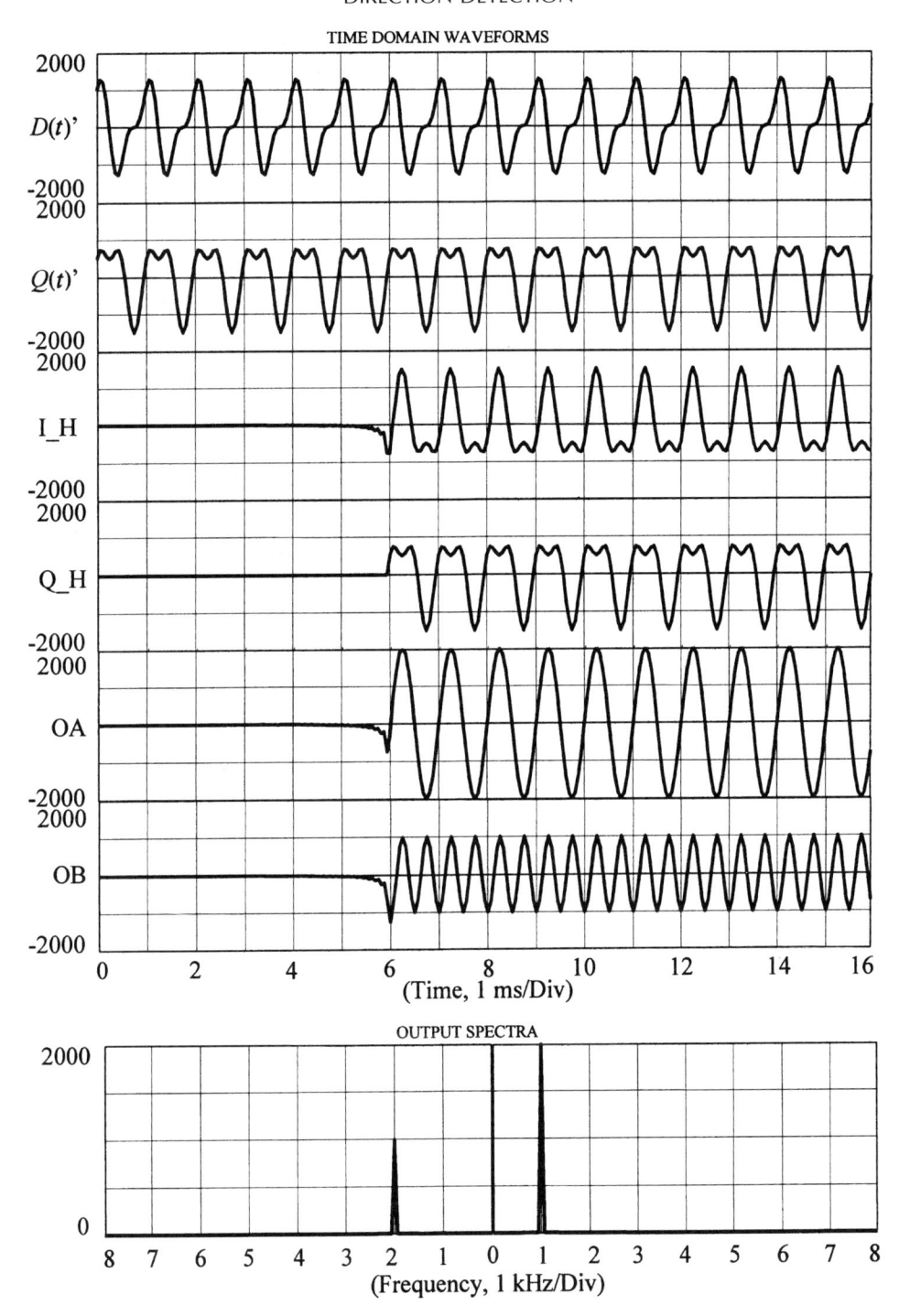

Figure 6.18 Time domain displays at each stage of the processing chain shown in Fig. 6.17, and the final output spectrum (see text for details of input signal) (reproduced by permission from Aydin and Evans 1994, © IFMBE)

(unshifted) channel, and this results in two completely separate flow channels.

Phase-shifting the 'direct' signals given by eqn. 6.6 leads to:

$$D^+(t) = -\tfrac{1}{2}A_\text{f} \sin(\omega_\text{f} t + \phi_\text{f})$$
$$\quad\quad\quad - \tfrac{1}{2}A_\text{r} \sin(\omega_\text{r} t - \phi_\text{r}) \quad\quad 6.9$$

Adding this signal to the unshifted 'quadrature' signal (eqn. 6.7) eliminates the reverse flow component and gives:

$$F(t) = -A_\text{f} \sin(\omega_\text{f} t + \phi_\text{f}) \quad\quad 6.10$$

Similarly, shifting the 'quadrature' signal by 90° gives:

$$Q^+(t) = -\tfrac{1}{2}A_\text{f} \cos(\omega_\text{f} t + \phi_\text{f})$$
$$\quad\quad\quad + \tfrac{1}{2}A_\text{r} \cos(\omega_\text{r} t - \phi_\text{r}) \quad\quad 6.11$$

which when added to the unshifted direct channel (eqn. 6.6) eliminates the forward component and yields:

$$R(t) = A_\text{r} \cos(\omega_\text{r} t - \phi_\text{r}) \quad\quad 6.12$$

There are a number of equivalent methods of achieving separation using the phasing-filter method (Nippa et al 1975, Coghlan and Taylor 1976), and Coghlan and Taylor (1978) have given details of a possible analogue implementation. A modern practical digital implementation described by Aydin and Evans (1994) is shown in Fig. 6.17. The implementation is based on a wide-band digital Hilbert transform (HT), which produces a 90° phase shift in the 'direct' input, and an all-pass delay filter which simply serves to introduce a delay identical to that produced by the HT filter. For this design, the forward channel is derived as for eqn. 6.10, but the reverse channel is derived by simply subtracting eqn. 6.9 from eqn. 6.7, i.e.

$$R(t) = A_\text{r} \sin(\omega_\text{r} t - \phi_\text{r}). \quad\quad 6.13$$

Figure 6.18 shows the results of a simulation study carried out by Aydin and Evans, which illustrates the signals present at each node in Fig. 6.17 when the Doppler shift signal contains a 1 kHz 'forward' component of peak-to-peak amplitude of 4000 units, and a 2 kHz 'reverse' component of amplitude 2000 units. In another study, Moraes and Evans (1996), using a digital implementation of the circuit shown in Fig. 6.16, were able to show that it is possible to compensate for phase imbalance in the original quadrature demodulation process, by adding the phase imbalance to the $\pi/2$ radians already produced by the symmetrical phase shift networks. This can easily be achieved in practice using all-pass IIR filters.

6.4.3 Weaver Receiver Technique

A third type of processing, the Weaver receiver method, is illustrated in Fig. 6.19. Both the direct and quadrature signals $D'(t)$ and $Q'(t)$, are mixed with quadrature signals from a pilot oscillator. This results in the forward and reverse flow components being separated on either side of the pilot frequency (ω_p).

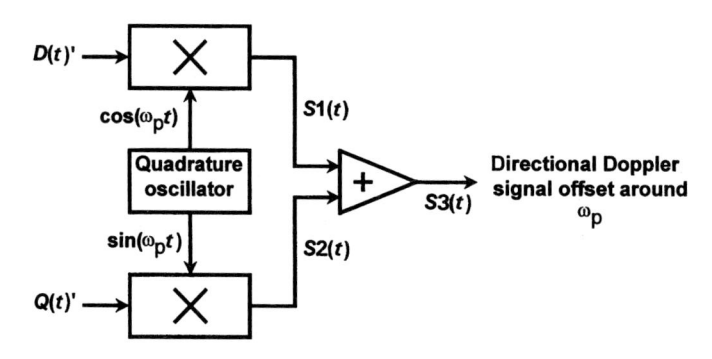

Figure 6.19 Weaver receiver processor for use with quadrature phase detected signals

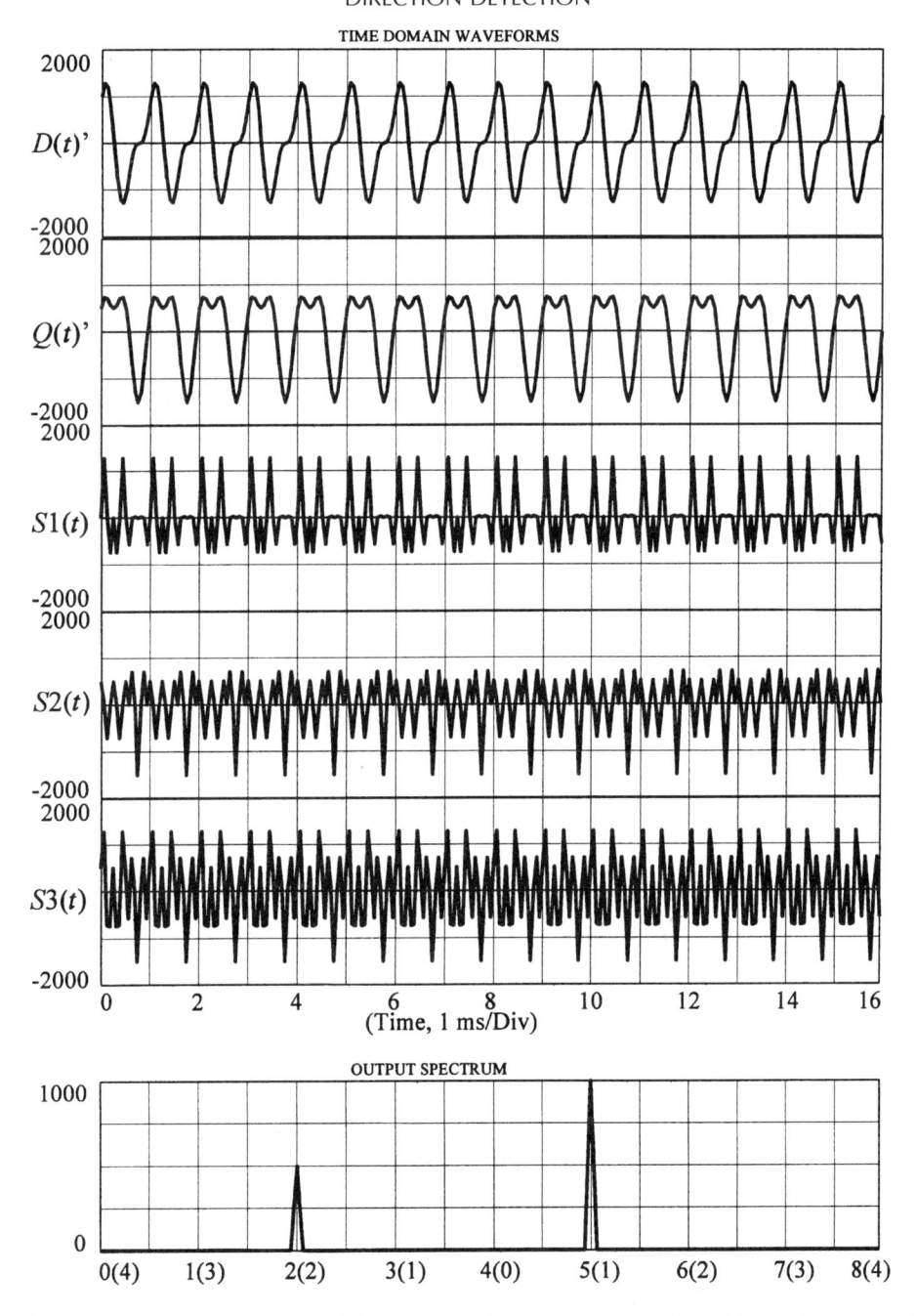

Figure 6.20 Time domain displays at each stage of the processing chain shown in Fig. 6.19, and the final output spectrum (see text for details of input signal) (reproduced by permission from Aydin and Evans 1994, © IFMBE)

Multiplying eqn. 6.6 by $A_p \cos(\omega_p t)$ where A_p and ω_p are the amplitude and angular frequency of the signals from the pilot oscillator gives:

$$S1(t) = \tfrac{1}{2}A_p\{A_f \cos(\omega_f t + \phi_f)\cos(\omega_p t) + A_r \cos(\omega_r t - \phi_r)\cos(\omega_p t)\} \quad 6.14$$

which can be expanded to give:

$$S1(t) = \tfrac{1}{4}A_p\{A_f[\cos(\omega_p t - \omega_f t - \phi_f) + \cos(\omega_p t + \omega_f t + \phi_f)] + A_r[\cos(\omega_p t - \omega_r t + \phi_r) + \cos(\omega_p t + \omega_r t - \phi_r)]\} \quad 6.15$$

Multiplying eqn. 6.8 by $A_p \sin(\omega_p t)$ results in:

$$S2(t) = \tfrac{1}{4}A_p\{A_f[-\cos(\omega_p t - \omega_f t - \phi_f) + \cos(\omega_p t + \omega_f t + \phi_f)] + A_r[\cos(\omega_p t - \omega_r t + \phi_r) - \cos(\omega_p t + \omega_r t - \phi_r)]\} \quad 6.16$$

Finally, adding eqns 6.15 and 6.16 results in:

$$S3(t) = S1(t) + S2(t)$$
$$= \tfrac{1}{2}A_p\{A_f \cos[(\omega_p + \omega_f)t + \phi_f] + A_r \cos[(\omega_p - \omega_r)t + \phi_r]\} \quad 6.17$$

Practical aspects of analogue implementation of this type of circuit have been discussed by Coghlan and Taylor (1976, 1978), and Aydin and Evans (1994) have described a digital implementation. Figure 6.20 shows the results of a simulation carried out using this implementation, which shows the signals at each node of Fig. 6.19, with the same inputs used for the simulation described in the last section. As can be seen from eqn. 6.17, the Weaver receiver technique produces a single output where the forward and reverse components are disposed on either side of a pilot frequency. This type of output is particularly suited to spectrum analysis, since only one analysis channel is necessary to display flow in both directions, and the position of the base-line corresponding to zero velocity can be simply moved by altering the pilot frequency; however, separate forward and reverse channels are more suited to some types of analogue envelope extraction and can be listened to on stereo head-phones. In an attempt to overcome this limitation, Aydin et al (1994) described an 'extended Weaver receiver technique' which provides the standard Weaver receiver output, together with two completely separated outputs representing forward and reverse flow. As for the phasing-filter separation technique, Moraes and Evans (1996) have shown that it is possible to compensate for phase imbalance in the original quadrature demodulation process using the Weaver receiver method. In this instance the correction is achieved by adjusting the phase relationship between the two pilot frequency signals, which are nominally separated by $\pi/2$ radians.

6.4.4 Complex Fourier Transform

A fourth method of obtaining directional information from quadrature signals, and one suitable only for digital implementation, is that of complex Fourier transformation (CFT). This method, illustrated in Fig. 6.21, is remarkably simple since because of the symmetry properties of the CFT (Brigham, 1974), if the complex input signal is in quadrature, the forward and reverse components appear on different sides of the dual-sided output spectrum (for a complex time signal, if the real part is even and the imaginary part odd, then the CFT is real; if the real part is odd and the imaginary part even, then the CFT is real). The use of the CFT for directional separation has been described by Aydin and Evans (1994) and the results of their simulation is shown in Fig. 6.22.

The major drawback of the CFT method is that the separation is purely in the frequency domain,

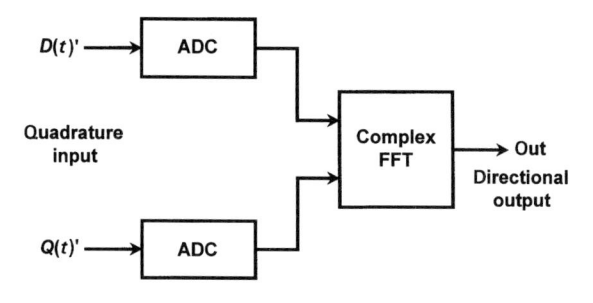

Figure 6.21 Complex Fourier transform processing of quadrature phase detected signals

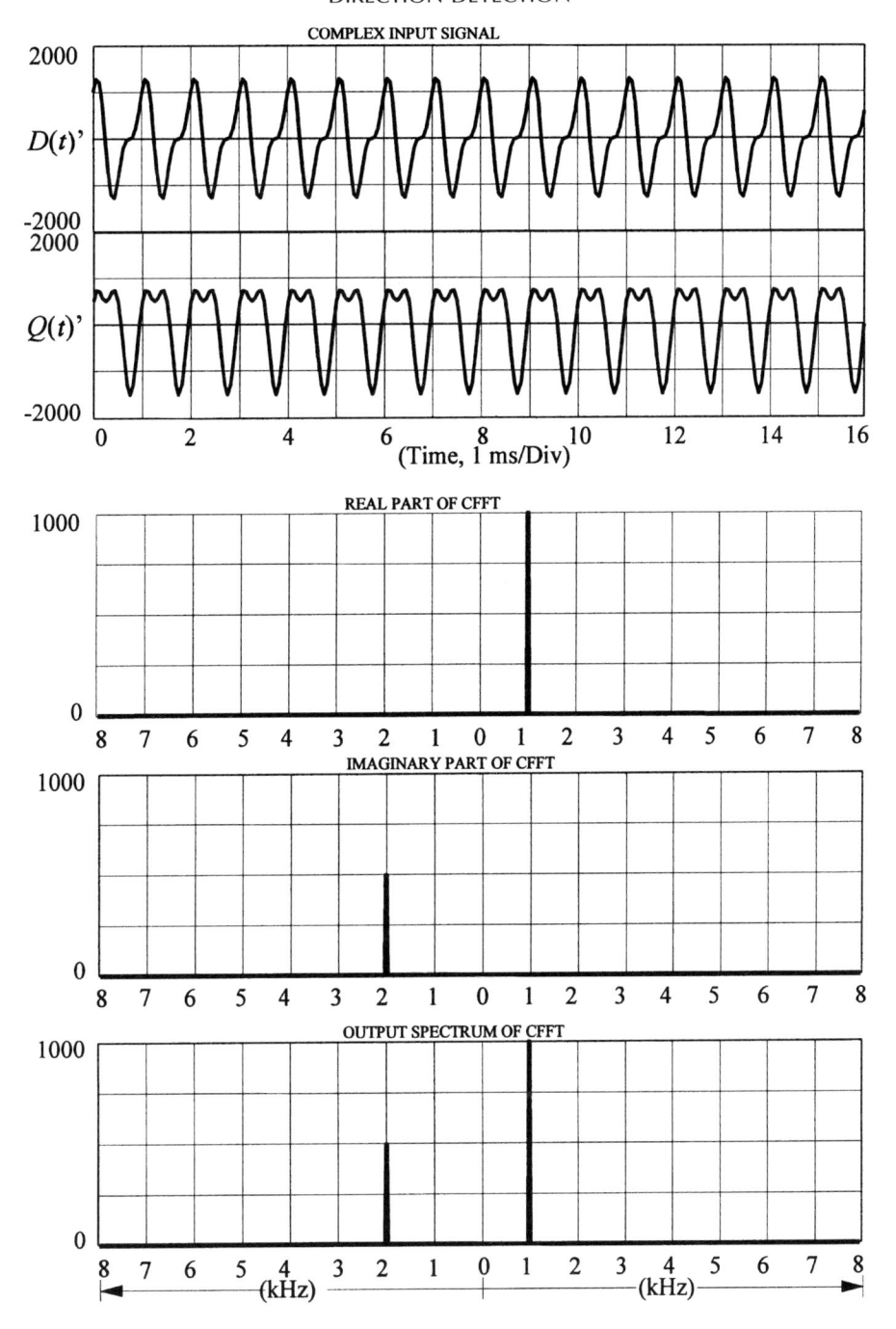

Figure 6.22 Real and imaginary output spectra from system shown in Fig. 6.21 (see text for details of input signal) (reproduced by permission from Aydin and Evans 1994, © IFMBE)

and therefore there is no time domain output suitable for recording or further analysis (and it should be noted that the input quadrature signals cannot easily be recorded since neither analogue (Smallwood 1985) not digital audio tape recorders (Bush and Evans 1993) are capable of preserving the precise phase relationship of the quadrature pair necessary to permit the subsequent separation of forward and reverse signals). Aydin et al (1994) have shown that it is possible to obtain separated directional outputs from the CFT by taking inverse FFTs of the spectra derived from the CFT, but whilst this is quite possible, it rather detracts from the simple elegance of the CFT as a direction separating method.

6.5 ANALOGUE ENVELOPE DETECTORS

Many of the Doppler processing methods described later in this book (see Chapters 8, 9 and 12) rely on an envelope signal (such as the maximum or mean frequency waveform) derived from the Doppler audio signal. The best method of deriving such a signal is probably to Fourier transform the Doppler signal and digitally calculate the envelope signal, but analogue envelope detectors are still in widespread use and are therefore briefly discussed below.

6.5.1 Intensity Weighted Mean Frequency Processors

Under ideal conditions the output of the intensity-weighted mean frequency follower is proportional to volumetric flow (see Chapter 8) and has thus gained widespread popularity. Various methods have been proposed for deriving the mean frequency envelope, but the best and most popular are those of Arts and Roevros (1972) and Roevros (1974) and the so-called 'root f' followers of de Jong et al (1975) and Gerzberg and Meindl (1977, 1980). A phase-lock loop method has been described by Sainz et al (1976) but the details of the performance of this method with wideband signals are not clear (Angelsen 1981). Block circuit diagrams of both the Arts and Roevros and 'root f' followers are shown in Fig. 6.23, and details of the operation of the former are given below. This particular method has the advantage of operating directly on the phase quadrature signals.

Consider two Doppler signals of frequency ω_1 and ω_2 having amplitudes of A and B, respectively. Referring to eqns 6.6 and 6.7 shows that they will result in two quadrature signals given by:

$$D(t)' = A\cos(\omega_1 t + \phi_1) + B\cos(\omega_2 t + \phi_2) \quad 6.18a$$

$$Q(t)' = -A\sin(\omega_1 t + \phi_1) - B\sin(\omega_2 t + \phi_2) \quad 6.18b$$

Figure 6.23a shows the basic sequence of operations which leads to 'numerator' and 'denominator' signals which are then divided to form the mean. Consider first the numerator. Differentiating the quadrature signal $Q(t)'$ leads to:

$$\begin{aligned} dQ'/dt = &-\omega_1 A\cos(\omega_1 t + \phi_1) \\ &-\omega_2 B\cos(\omega_2 t + \phi_2) \end{aligned} \quad 6.19$$

Multiplying this by the direct signal $D(t)'$ gives:

$$\mathrm{Num}(t) = -\tfrac{1}{2}\omega_1 A^2 - \tfrac{1}{2}\omega_2 B^2 \quad 6.20a$$

$$\left.\begin{aligned} &-\tfrac{1}{2}\omega_1 A^2\cos(2\omega_1 t + 2\phi_1) \\ &-\tfrac{1}{2}\omega_2 B^2\cos(2\omega_2 t + 2\phi_2) \end{aligned}\right\} \quad 6.20b$$

$$\left.\begin{aligned} &-\tfrac{1}{2}(\omega_1+\omega_2)AB\{\cos[(\omega_1-\omega_2)t \\ &+\phi_1-\phi_2]+\cos[(\omega_1+\omega_2)t+\phi_1 \\ &+\phi_2]\} \end{aligned}\right\} \quad 6.20c$$

This signal is then low pass filtered to leave only the DC and quasi-DC components. This removes the terms labelled 6.20b and, provided ω_1 and ω_2 are not too similar, the terms labelled 6.20c, to leave:

$$\mathrm{Num}(t) = -\tfrac{1}{2}\omega_1 A^2 - \tfrac{1}{2}\omega_2 B^2 \quad 6.21$$

In the case of a wideband signal from a Doppler unit, some difference components will leak through and these all contribute to the noise in the output signal. Fortunately this noise is not of sufficient size to invalidate the use of the processor.

The denominator signal is formed by squaring and low pass filtering the direct signal $D(t)'$.

ANALOGUE ENVELOPE DETECTORS

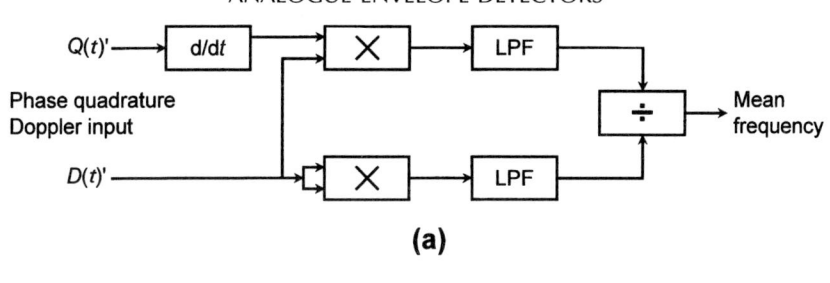

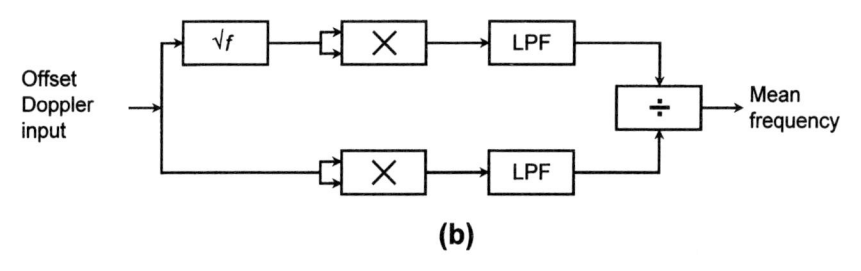

Figure 6.23 (a) The Arts and Roevros circuit for calculating the intensity weighted mean frequency from phase quadrature inputs. (b) The root f circuit for calculating the mean frequency from offset Doppler inputs. LPF, low pass filter

Squaring eqn 6.18a leads to:

$$\text{Denom}(t) = \tfrac{1}{2} A^2 + \tfrac{1}{2} B^2 \quad \text{6.22a}$$

$$\left. \begin{array}{l} + \tfrac{1}{2} A^2 \cos(2\omega_1 t + 2\phi_1) \\ + \tfrac{1}{2} B^2 \cos(2\omega_2 t + 2\phi_2) \end{array} \right\} \quad \text{6.22b}$$

$$\left. \begin{array}{l} + AB\{[\cos(\omega_1 - \omega_2)t + \phi_1 - \phi_2] \\ + [\cos(\omega_1 + \omega_2)t + \phi_1 + \phi_2]\} \end{array} \right\} \quad \text{6.22c}$$

Filtering as before leads to:

$$\text{Denom}(t) = \tfrac{1}{2} A^2 + \tfrac{1}{2} B^2 \quad \text{6.23}$$

Once again for a wideband signal some of the difference components will leak through and contribute to the noise. Finally, dividing the numerator by the denominator signal leads to:

$$\frac{\text{Num}(t)}{\text{Denom}(t)} = \frac{-(\omega_1 A^2 + \omega_2 B^2)}{A^2 + B^2} \quad \text{6.24}$$

which is the intensity-weighted mean of the two original frequencies. A more complete proof that the circuit gives the true intensity-weighted mean is given by Arts and Roevros (1972). Evans et al (1987) have published a practical implementation of the Arts and Roevros circuit, whilst Kassam et al (1982) have published details of an implementation of the 'root f' follower.

6.5.2 Other Analogue Frequency Processors

A number of other types of analogue processors have been described, including the widespread but unsatisfactory zero-crossing detector (Franklin et al 1961, Flax et al 1970, Lunt 1975) which was at one time popular because of its simplicity and low cost, and the maximum frequency follower (Sainz et al 1976, Skidmore and Follett 1978, Nowicki et al 1985) which is more usually implemented using digital techniques (Gibbons et al 1981, D'Alessio 1985, Mo et al 1988, Marasek and Nowicki 1994, Routh et al 1994). Theoretical aspects of the performance of these and a number of other 'signal location estimators' are considered in Chapter 8.

6.6 COMPLEX DOPPLER SYSTEMS

Several types of complex Doppler systems have been described in Chapter 4, including multigate profile detecting systems and real-time colour flow mapping systems. Most of these systems share the same basic front-end design as simple PW systems, and it is only after quadrature phase detection that they differ in the way in which they handle the resulting range–phase signals. Details of the signal processing in colour flow imaging systems are to be found in Chapter 11, whilst an example of a multi-gate device intended for measuring velocity profiles rather than real-time two dimensional flow imaging is discussed in the next section.

6.6.1 Multigate Pulsed Doppler Systems

Multigate pulsed Doppler systems may use either parallel or serial processing. If parallel processing is chosen, separate circuitry will be necessary for each gate and the resulting electronics will rapidly become both bulky and costly as the number of channels increases (Section 4.4). A much better solution is to use serial digital signal processing (Brandestini 1978, Hoeks et al 1981, Reneman et al 1986).

The arrangement of the serial processing system described by Hoeks et al is illustrated in Fig. 6.24. As far as the analogue-to-digital converters (ADCs), the system is the same as a standard single gate PW Doppler system employing quadrature phase detection (see Fig. 6.8); from this point on until the display the system is purely digital. Because of the large dynamic range of the quadrature signals (due in particular to the unwanted but large low-frequency signals from tissue interfaces) the ADCs are required to be 12-bit devices. The output of each of the ADCs is filtered by a pair of first-order high pass filters (HPFs) in order to reject the low-frequency clutter signal. The basic design of these filters is as illustrated in Fig. 6.25; the high pass output at a given time is the difference between the input and a running average of input signals from the same range. The number of memory locations in the filter determines the number of gates that can be used, and the multiplication factor K (which controls the proportion of the difference signal contributing to the running average) the filter cut-off frequency. The filtered quadrature signals are fed into a pair of zero-crossing detectors (ZCDs) where each is compared with a threshold to produce a single bit of information (above or below threshold). The output from the comparators is stored in a memory and, if different from the previous processing cycle (indicating a zero-crossing), a signal of plus or minus one, depending on the detected direction of flow, is put on the input of the final low-pass filter (LPF). This filter is of the same design as those used to high pass filter the input (Fig. 6.25) with a cut-off frequency

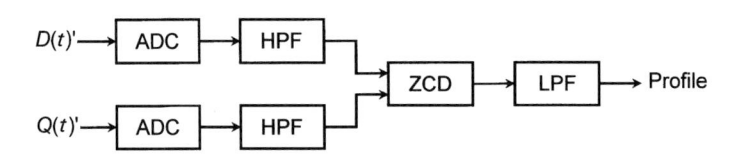

Figure 6.24 Block diagram of a multigate pulsed Doppler system with serial data processing. The input signals are taken from a circuit such as that illustrated in Fig. 6.8. All abbreviations are given in the text

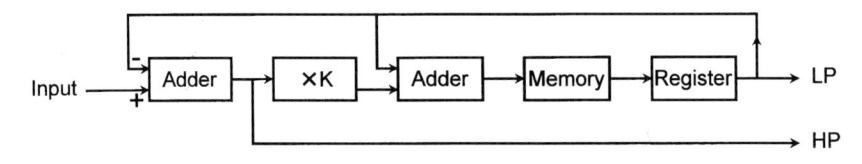

Figure 6.25 Block diagram of the first-order high pass/low pass filters used by Hoeks et al (1981)

Table 6.1 Summary of demodulation techniques

Method	Type of output	Comments
Non-coherent	Single channel	No direction resolving capability; forward and reverse flow treated identically
Single sideband	Dual channel	Channel A contains forward flow, channel B contains reverse flow
Heterodyne	Single channel	Heterodyne frequency corresponds to zero velocity, lower frequencies to reverse flow and higher frequencies to forward
Quadrature detection and switched channel processing	Dual channel	Only one channel switched in at any one time; confused by simultaneous forward and reverse flow
Quadrature detection and phasing filter processing	Dual channel	Channel A contains forward flow, channel B contains reverse flow
Quadrature detection and Weaver receiver processing	Single channel	Heterodyne frequency corresponds to zero velocity, lower frequencies to reverse flow and higher frequencies to forward flow
Quadrature detection and complex Fourier transform processing	Frequency domain only	Further processing required to produce separated audio outputs

of either 6 or 12 Hz. During each processing cycle the instantaneous velocity distribution along the ultrasonic beam appears at the output of the low pass filter, and this digital signal may be further processed to display either the velocity versus time traces for preselected gates (Fig. 4.10a) or, alternatively, the velocity profile in the vessels at preselected times (Fig. 4.10b). In addition most multigate systems, including the one described here, allow the user to select any channel for both audio presentation and full spectral analysis.

6.7 SUMMARY

In this chapter the basic building blocks from which all pulsed and continuous wave Doppler units are constructed have been examined. A number of Doppler demodulation techniques have been described and their attributes are summarised in Table 6.1. Simple analogue envelope detectors have been discussed. Most advanced colour flow mapping systems share the same basic front-end design as simple PW units, and the processing methods they employ are described in Chapter 11.

6.8 REFERENCES

Angelsen BAJ (1981) Instantaneous frequency, mean frequency, and variance of mean frequency estimators for ultrasonic blood velocity Doppler signals. IEEE Trans Biomed Eng BME-28:733–741

Arts MGJ, Roevros JMJG (1972) On the instantaneous measurement of bloodflow by ultrasonic means. Med Biol Eng 10:23–34

Atkinson P, Woodcock JP (1982) Doppler Ultrasound and Its Use in Clinical Measurement. Academic Press, London

Aydin N, Evans DH (1994) Implementation of directional Doppler techniques using a digital signal processor. Med Biol Eng Comput 32:S157–S164

Aydin N, Fan L, Evans DH (1994) Quadrature to directional format conversion of Doppler signals using digital methods. Physiol Meas 15:181–199

Brandestini M (1978) Topoflow—a digital full range Doppler velocity meter. IEEE Trans Sonics Ultrason SU-25:287–293

Brigham EO (1974) The Fast Fourier Transform, pp. 11–49. Prentice-Hall, Englewood Cliffs, NJ

Bush G, Evans DH (1993) Digital audio tape as a method of storing Doppler ultrasound signals. Physiol Meas 14:381–386

Coghlan BA, Taylor MG (1976) Directional Doppler techniques for detection of blood velocities. Ultrasound Med Biol 2:181–188

Coghlan BA, Taylor MG (1978) On methods for preprocessing direction Doppler signals to allow display of directional blood-velocity waveforms by spectrum analysers. Med Biol Eng Comput 16:549–553

D'Alessio T (1985) 'Objective' algorithm for maximum frequency estimation in Doppler spectral analysers. Med Biol Eng Comput 23:63–68

de Jong DA, Megens PHA, De Vlieger M, Thon H, Holland WPJ (1975) A directional quantifying Doppler system for measurement of transport velocity of blood. Ultrasonics 13:138–141

Evans JM, Beard JD, Skidmore R, Horrocks M (1987) An analogue mean frequency estimator for the quantitative measurement of blood flow by Doppler ultrasound. Clin Phys Physiol Meas 8:309–315

Flax SW, Webster JG, Updike SJ (1970) Statistical evaluation of the Doppler ultrasonic blood flowmeter. Bio-Med Sci Inst 7:201–222

Follett DH (1991) Electromagnetic radiofrequency interference with Doppler equipment. Phys Med Biol 36:1443–1455

Follett DH (1993) Transducer electrical isolators without critical adjustments. Phys Med Biol 38:1675–1682

Franklin DL, Schlegel W, Rushmer RF (1961) Blood flow measured by Doppler frequency shift of back-scattered ultrasound. Science 134:564–565

Gerzberg L, Meindl JD (1977) Mean frequency estimator with applications in ultrasonic Doppler flowmeters. In: White D, Brown RE (eds) Ultrasound in Medicine, vol 3B, pp. 1173–1180. Plenum, New York

Gerzberg L, Meindl JD (1980) The root f power-spectrum centroid detector: system considerations, implementation, and performance. Ultrason Imaging 2:262–289

Gibbons DT, Evans DH, Barrie WW, Cosgriff PS (1981) Real-time calculation of ultrasonic pulsatility index. Med Biol Eng Comput 19:28–34

Hoeks APG, Reneman RS, Peronneau PA (1981) A multigate pulsed Doppler system with serial data processing. IEEE Trans Sonics Ultrason SU-28:242–247

Kassam MS, Cobbold RSC, Johnston KW, Graham CM (1982) Method for estimating the Doppler mean velocity waveform. Ultrasound Med Biol 8:537–544

Light LH (1970) A recording spectrograph for analysing Doppler blood velocity signals in real time. J Physiol 207:42–44

Lunt MJ (1975) Accuracy and limitations of the ultrasonic Doppler blood velocimeter and zero crossing detector. Ultrasound Med Biol 2:1–10

Marasek K, Nowicki A (1994) Comparison of the performance of three maximum Doppler frequency estimators coupled with different spectral estimation methods. Ultrasound Med Biol 20:629–638

McLeod FD (1967) A Directional Doppler Flowmeter. In: Digest of the 7th International Conference on Medical and Biological Engineering, Stockholm, p. 213

Mo LYL, Yun LCM, Cobbold RSC (1988) Comparison of four digital maximum frequency estimators for Doppler ultrasound. Ultrasound Med Biol 14:355–363

Moraes R, Evans DH (1996) Compensation for phase and amplitude imbalance in quadrature Doppler signals. Ultrasound Med Biol 22:129–137

Nippa JH, Hokanson DE, Lee DR, Sumner DS, Strandness DE (1975) Phase rotation for separating forward and reverse blood velocity signals. IEEE Trans Sonics Ultrason SU-22:340–346

Nowicki A, Karlowicz P, Piechocki M, Secomski W (1985) Method for the measurement of the maximum Doppler frequency. Ultrasound Med Biol 11:479–486

Peronneau PA, Bournat JP, Bugnon A, Barbet A, Xhaard M (1974) Theoretical and practical aspects of pulsed Doppler flowmetry: real-time application to the measure of instantaneous velocity profiles *in vitro* and *in vivo*. In: Reneman RS (ed) Cardiovascular Applications of Ultrasound, pp. 66–84. Proceedings of an International Symposium held at Janssen Pharmaceutica, Beerse, Belgium, 1973. North-Holland, Amsterdam

Reneman RS, Van Merode T, Hick P, Hoeks APG (1986) Cardiovascular applications of multi-gate pulsed doppler systems. Ultrasound Med Biol 12:357–370

Roevros JMJG (1974) Analogue processing of C.W. Doppler flowmeter signals to determine average frequency shift momentaneously without the use of a wave analyser. In: Reneman RS (ed) Cardiovascular Applications of Ultrasound, pp. 43–54. Proceedings of an International Symposium held at Janssen Pharmaceutica, Beerse, Belgium, 1973. North-Holland, Amsterdam

Routh HF, Powrie CW, Bellevue RB, Peterson R (1994) Continuous display of peak and mean blood flow velocities. USA Patent No. 5287753, issued 22 Feb 1994 (to Advanced Technology Laboratories, Inc., Bothell, Washington, DC)

Sainz A, Roberts VC, Pinardi G (1976) Phase-locked loop techniques applied to ultrasonic Doppler signal processing. Ultrasonics 19:128–132

Skidmore R, Follett DH (1978) Maximum frequency follower for the processing of ultrasonic Doppler shift signals. Ultrasound Med Biol 4:145–147

Smallwood RH (1985) Recording Doppler blood flow signals on magnetic tape. Clin Phys Physiol Meas 6:357–359

Tortoli P, Bessi L, Guidi F (1994) Bidirectional Doppler signal analysis based on a single RF sampling channel. IEEE Trans Ultrason Ferroelec Freq Contr 41:1–3

7

The Doppler Power Spectrum

7.1 INTRODUCTION

In this chapter we consider the origin of the power spectrum of signals from CW and PW Doppler instruments. The simple Doppler equation for a single target passing through an infinitely wide ultrasound beam has already been presented in Chapters 1 and 3. The Doppler shift frequency f_d is given by:

$$f_d = f_t - f_r = (2f_t v \cos\theta)/c \qquad 7.1$$

where f_t and f_r are the transmitted and received ultrasound frequencies, v the velocity of the target, c the velocity of sound in tissue and θ the angle between the ultrasound beam and the direction of motion of the target. When ultrasound is used to interrogate the flow within a blood vessel, there are of course numerous targets in the ultrasound field with a range of velocities, and the Doppler shift signal therefore contains not just a single frequency, as shown by eqn. 7.1, but rather a spectrum of frequencies which varies in shape as the velocity distribution within the vessel changes with time.

The Doppler shift frequency is proportional to velocity, and under ideal uniform sampling conditions the power in a particular frequency band of the Doppler spectrum is proportional to the

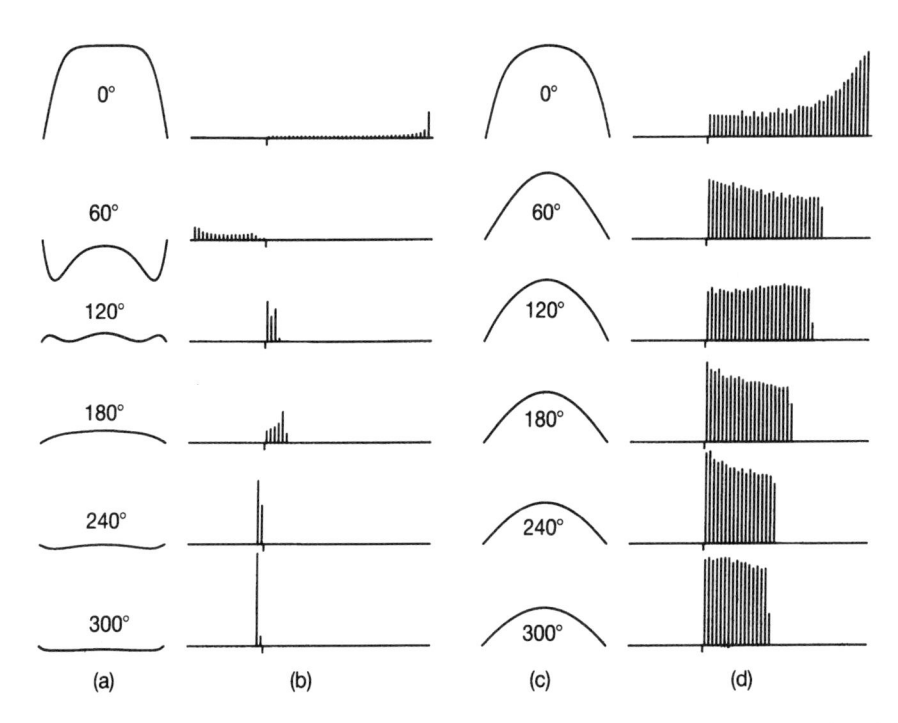

Figure 7.1 Series of velocity profiles for a common femoral artery (a) and common carotid artery (c), together with corresponding velocity distribution histograms (b) and (d). The peak of forward flow has been arbitrarily called 0° and the maximum velocities have been scaled to have the same amplitude

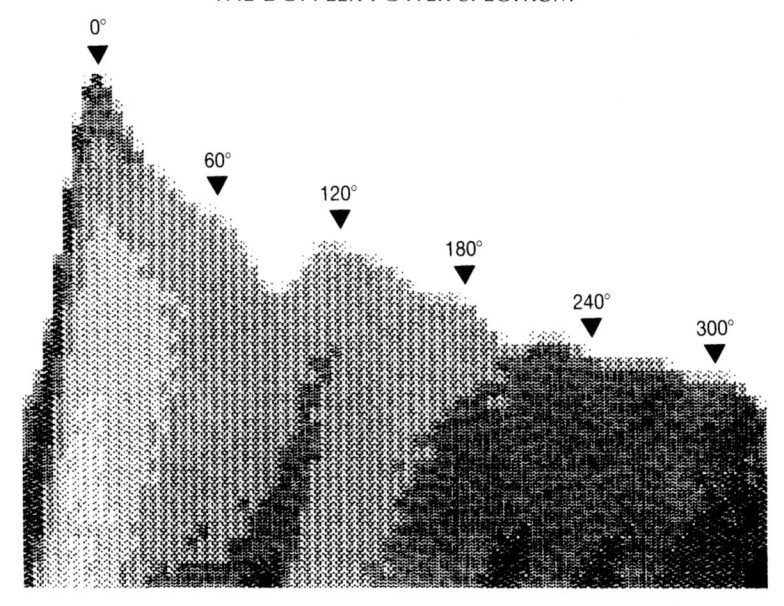

Figure 7.2 Sonogram of the Doppler signal from the common carotid artery whose velocity profiles are given in Figure 7.1c. Time is represented along the horizontal axis, Doppler shift frequency along the vertical axis, and spectral amplitude by the blackness of the paper. The time slices corresponding to the histograms in Figure 7.1d are marked with small arrowheads

volume of blood moving with velocities that produce frequencies in that band, and therefore the Doppler power spectrum should have the same shape as a velocity distribution plot for the flow in the vessel. The velocity distributions corresponding to a variety of realistic velocity profiles (as derived for Fig. 2.8) are shown in Fig. 7.1. The spectra corresponding to flat velocity profiles have much of the power concentrated in a relatively small range of frequencies, whilst those corresponding to parabolic profiles are almost flat. In turbulent flow, the velocities of the targets in the ultrasound field fluctuate rapidly with time (Section 2.3.3) and this causes a broadening of the spectrum that would otherwise be obtained from flow with the same temporal average velocity profile. Gross haemodynamic disturbances, such as large vortices, may cause irregular spectra with large isolated forward and reverse components.

The variation in the shape of the Doppler power spectrum as a function of time is usually presented in the form of a sonogram (Fig. 7.2). In this type of display, time is plotted along the horizontal axis, frequency along the vertical axis, and the power at a particular frequency and time as the intensity of the corresponding pixel. Thus, a single line on the sonogram corresponds to a single power spectrum, much in the same way as a line on an ultrasound B-scan corresponds to a single A-scan.

There are a number of factors that distort the power spectra and which may limit the accuracy with which the velocity distribution in a vessel can be determined. In this chapter we consider the source of scattering of the ultrasound and the way in which the Doppler power spectrum is influenced by various physical and electronic mechanisms.

7.2 BLOOD AS AN ULTRASONIC TARGET

Despite its appearances, blood is not a homogeneous liquid, but a suspension of cells and other particles in a clear straw-coloured fluid called plasma. Because of this microscopic structure, ultrasound is scattered by blood and Doppler shift measurements of blood velocity are possible.

The so-called formed elements of blood consist of the erythrocytes (red blood cells), the leukocytes (white blood cells), and platelets. The relative size and concentrations of these components are summarised in Table 7.1. It is generally believed that scattering of ultrasound by blood is almost entirely due to erythrocytes because they are much more numerous than the slightly larger leukocytes and significantly larger than the platelets. Reid et al (1969) found that at 5 MHz the scattering cross-section of platelets is approximately 10^{-3} times that of the erythrocytes and that at normal concentrations their contribution is undetectable. Van der Heiden et al (1995) reported that at 30 MHz platelets 'did not contribute to the backscatter power'.

The behaviour of erythrocytes as targets is dictated by their size, acoustic properties, concentration, the way in which they are packed, and the acoustic properties of the embedding media, the plasma.

Erythrocytes are flexible biconcave discs with a diameter of 7.2 μm and a thickness of 2.2 μm, and their concentration is such that they normally occupy between about 36% and 54% of the total blood volume (the haematocrit is thus said to be between 36% and 54%). In general, at ultrasound frequencies used for blood velocity measurements, the diameter of the erythrocytes is much smaller than an ultrasound wavelength and therefore, in low (unphysiological) concentrations, they act as a random distribution of point targets. In the high concentrations found in normal whole blood, however, they cannot be treated as having a purely random distribution because their positions are no longer independent of each other; Shung et al (1976) pointed out that when the haematocrit is 45% the average distance between two red cells is only about 10% of their diameter. This significantly complicates the behaviour of blood as a scattering target.

The acoustic properties of the blood constituents that influence the scattering of ultrasound are their densities and adiabatic compressibilities, and these are summarised in Table 7.2 (Urick 1947). The attenuation of ultrasound by blood is relatively low with a value of approximately 0.15 dBcm^{-1} MHz$^{-1.2}$ (Narayana et al, 1984), and the contribution of scattering towards this is negligible for frequencies of less than 15 MHz (Shung et al 1976).

The scattering of waves by particles that are small in comparison to the wavelength was first studied by Lord Rayleigh in 1872 (Strutt 1873), and such scattering is therefore usually referred to as 'Rayleigh scattering'. Two of its important characteristics are that the shape of the scatterers is unimportant, and that the scattered power is proportional to the fourth power of frequency (and the sixth power of diameter). As already mentioned, the behaviour of a concentrated ensemble of particles is rather different from that of individual scatterers, but Shung et al (1976) showed both theoretically and experimentally that the scattering of ultrasound by blood is proportional to the fourth power of frequency between 5 MHz and 15 MHz, whilst van der Heiden et al (1995) found the power law to have an exponent of 3.3 at 30 MHz for rouleau-suppressed (see Section 7.2.1) blood. At high frequencies (35–65 MHz), Lockwood et al (1991) found the exponent to be much lower (1.3–1.4) and suggested that this was because, at the higher frequencies, the scattering can no longer be regarded as Rayleigh. The rapid increase in scattering at more normal ultrasound frequencies has the practical implication that the performance of a Doppler system falls off with frequency less rapidly than that of a pulse–echo system, because the increase in scattered power with frequency partially offsets the increased attenuation of ultrasound by the intervening tissue.

Table 7.1 The sizes and concentrations of the major formed elements of blood

	Concentration (particles/mm^3)	Dimensions (μm)	Total of blood volume (%)
Erythrocytes	5×10^6	7.2 × 2.2	45
Leukocytes	8×10^3	9–25	~0.8
Platelets	2.5×10^5	2–3	~0.2

Table 7.2 The density and compressibility of the major components of blood. Data based on Urick (1947)

	Density (kg m^{-3}) (ρ)	Adiabatic compressibility (m^2 N^{-1}) (β)
Erythrocytes	1.091×10^3	3.41×10^{-10}
Plasma	1.021×10^3	4.09×10^{-10}

7.2.1 Models of Ultrasound Scattering by Blood

There have been numerous models of ultrasonic scattering of blood proposed and the reader is referred to Mo (1991), Shung and Thieme (1993) and Cloutier and Qin (1997) for a detailed discussion of these. The models are generally based on the Born approximation (i.e. there is assumed to be no multiple scattering) and usually assume plane wave insonation of stationary or laminar flow. There have been two basic modelling approaches used, the particle approach and the continuum approach, although Mo and Cobbold (Mo 1991, Mo and Cobbold 1992) have described a 'unified approach', which has since been extended by Bascom and Cobbold (1995). The particle approach, exemplified by the studies of Atkinson and Berry (1974), Shung et al (1976), Mo and Cobbold (1986a), Routh et al (1987) and Zhang et al (1994), models the blood as being composed of individual red cells, generally moving with different velocities. The continuum approach, exemplified by the studies of Angelsen (1980) and Shung and Kuo (1994) treats the blood as an isotropic continuum, with the source of scattering being fluctuations in the compressibility and mass density of the continuum.

A particularly enlightening formulation of the problem is that of Mo and Cobbold (1986b) in which they treat the blood as being composed of small aggregates of red blood cells or 'rouleaux'. They argue that erythrocyte aggregation can occur even under normal blood flow conditions, and that the extent of the aggregation varies dramatically with mean shear rate (although under the high shear rates of normal arterial blood flow, the rouleaux tend to be small, consisting of singlets, doublets or triplets). The consequence of this is that it is necessary to treat blood as a suspension of scatterers whose effective volume is a random variable, the probability density function of which is dependent on shear rate. From this starting point they showed that the backscattering coefficient for blood, Φ_{bs}, can be written:

$$\Phi_{bs} = \sigma_b \frac{HW}{V_c} \overline{m}(1 + \sigma_m^2/\overline{m}^2) \qquad 7.2$$

where σ_b is the backscattering cross-section of a single erythrocyte, H is the haematocrit, V_c the average volume of one erythrocyte, m a measure of the size of the aggregate scattering unit (with $\overline{m}$ and σ_m being the mean and standard deviation of the scattering size distribution), and W the packing factor associated with a particular aggregate size distribution.

The approximate functional dependency of W on H is known for the packing of identical hard spheres, aligned cylinders and slabs (Shung 1982), but since an assumption of the Mo and Cobbold model was that the scattering units are of variable size and shape, they chose to leave it as a parameter. They also make the point that as the orientation and aggregation of erythrocytes is shear-dependent, in the general case W is dependent both on H and the prevailing flow conditions. Finally, Mo and Cobbold make the point that several prior models of blood scattering can be treated as special cases of their aggregate model with $\sigma_m = 0$, $\overline{m} = 1$ (or larger integer) and W an explicit function of H.

The backscattering cross-section of a single erythrocyte may be written:

$$\sigma_b = \frac{V_c^2 \pi^2}{\lambda^4} \left[\frac{\beta_e - \beta_p}{\beta_p} + \frac{3\rho_e - 3\rho_p}{2\rho_e + \rho_p} \right]^2 \qquad 7.3$$

where λ is the ultrasound wavelength, and β_e, β_p, ρ_e and ρ_p are the adiabatic compressibility and density of the erythrocytes and plasma (see Table 7.2). Hence, eqn. 7.2 may be written:

$$\Phi_{bs} = \frac{V_c \pi^2}{\lambda^4} \left[\frac{\beta_e - \beta_p}{\beta_p} + \frac{3\rho_e - 3\rho_p}{2\rho_e + \rho_p} \right]^2$$

$$\times HW\overline{m}(1 + \sigma_m^2/\overline{m}^2) \qquad 7.4$$

Note that Mo and Cobbold (1986b) use a slightly simplified term within the square bracket, which they argue is more appropriate. The expression used in eqn. 7.3 and 7.4 is strictly applicable only for spherical objects, but is used here to maintain consistency with other derivations.

7.2.2 Effect of Haematocrit on Ultrasound Backscatter

It is now well established that the backscattering coefficient of saline suspensions of erythrocytes initially increases with haematocrit, and reaches a maximum at between 15% and 20% (depending on flow conditions) and decreases thereafter (Shung et al 1984, Mo et al 1994). Figure 7.3 is reproduced from the work of Shung et al 1984, and illustrates this behaviour for a bovine erythrocyte suspension under uniform flow conditions. The theoretical line is derived from the Percus–Yevick packing theory for pair-correlated hard spheres (Twersky 1978), which allows the packing factor W to be expressed in terms of haematocrit, i.e:

$$W = \frac{(1-H)^4}{(1+2H)^2} \qquad 7.5$$

From eqn. 7.2, provided that the probability density function of the aggregate size does not vary (as is implicitly assumed in most theories of ultrasound scattering by blood), then the backscatter coefficient may be written:

$$\Phi_{bs} \propto \frac{H(1-H)^4}{(1+2H)^2} \qquad 7.6$$

which is identical in form to the theoretical curve shown in Fig. 7.3.

More recently a new theoretical model, which uses a fractional packing dimension to represent the way in which the erythrocytes are packed, has been proposed by Bascom and Cobbold (1995). Their expression for packing factor may be written:

$$W = \frac{(1-H)^{p+1}}{(1+H[p-1])^{p-1}} \qquad 7.7$$

where p is the fractional packing dimension. In this model $p = 1$ corresponds to the packing of uniform slabs, $p = 2$ to the packing of infinite cylinders, and $p = 3$ to the packing of spheres. It is also possible for p to take on values of greater than 3 under special circumstances described by Bascom and Cobbold. For $p = 3$ (spherical packing), eqn. 7.7

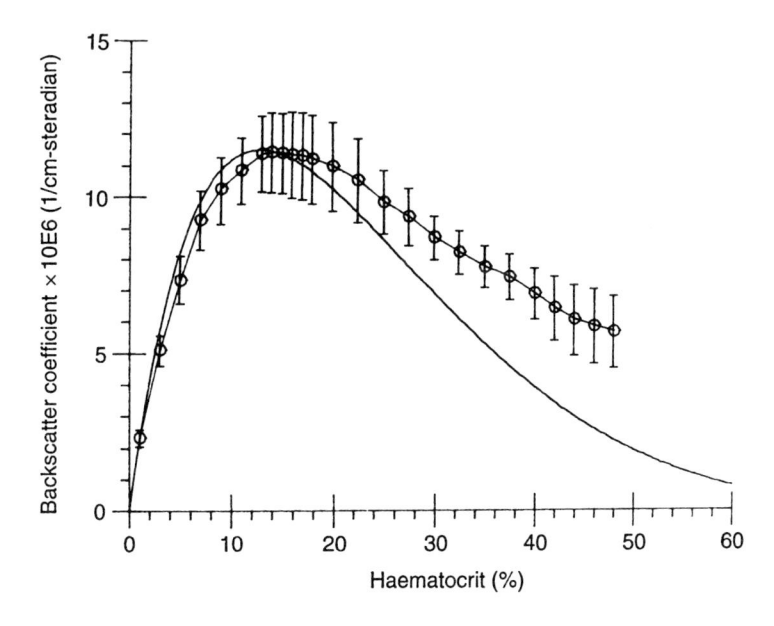

Figure 7.3 Ultrasonic backscatter coefficient from a bovine erythrocyte suspension under uniform flow conditions plotted as a function of haematocrit. The circles are the experimental points and the solid line is a theoretical curve (see text for details) (reproduced by permission from Shung et al 1984, © 1984 Acoustical Society of America).

reduces to eqn. 7.5. Bascom and Cobbold found that the best fit of eqn. 7.7 for porcine erythrocytes suspended in a saline solution (no aggregation) was obtained with a value of $p = 2.54$, which implies a scatterer with a geometry between a sphere and a cylinder.

It is worth noting that frequently cited evidence that the maximum backscatter occurs at a haematocrit of between 20% and 30% (Shung et al 1976) is now thought to be erroneous due to the presence of turbulent flow in the measurement chamber used in that study (Shung et al 1984). Also contrary to a commonly held belief, there is now known to be no second peak in the backscattered power versus haematocrit curve at very high values of haematocrit (Mo et al 1994).

7.2.3 Angular Dependence of Ultrasound Scattering by Blood

In general Doppler studies are performed using either a single transducer that both transmits and receives ultrasound or, in the case of continuous wave Doppler, two transducers located close together, and therefore scattering of ultrasound in directions other than 180° to the incident wave are of little relevance. There are, however, occasions when the receiver and transmitter may be separated, in which case the angular dependence of scattering becomes important. Note that this is a separate issue from any angular dependence of backscatter from flowing blood, which is addressed in Section 7.2.4.1.

The angular dependence of scattering of ultrasound by blood has been addressed by Shung et al (1977), who compared experimental measurements with two theoretical descriptions of scattering, one given by Rschevkin (1963) and Morse and Ingard (1968), and another due to Ahuja (1972). The two theories lead to similar results, but that of Ahuja is more complete in that it recognises that the embedding medium is not frictionless and that therefore shear waves are generated by the oscillating particles. The measurements of Shung et al (1977) agreed well with both theories, but particularly closely with that of Ahuja. Because the Rschevkin model is much simpler and performs reasonably well, its essential results are reproduced here, and the reader is referred to the paper of Shung et al (1977, eqn. 6) for the more exact theoretical description. The Rschevkin formula for the power received by a transducer $P(\phi)$ may be written:

$$P(\phi) \propto \left(\frac{\beta_e - \beta_p}{\beta_p} + \frac{3\rho_e - 3\rho_p}{2\rho_e + \rho_p} \cos\phi \right)^2 \quad 7.8$$

where ϕ is the scattering angle. This function is plotted in the form of a polar diagram in Fig. 7.4. It may be seen that the scattered power is at a minimum when ϕ is 0° and a maximum when ϕ is 180°.

7.2.4 Effect of Flow Conditions on Ultrasound Backscatter

It is widely acknowledged that the scattering properties of blood are altered by flow conditions, and there is now an extensive literature on the subject, much of which has been published in the fairly recent past. The results of *in vitro* studies of ultrasound backscatter from blood are heavily dependent on experimental conditions and the type of blood or blood substitute used, and

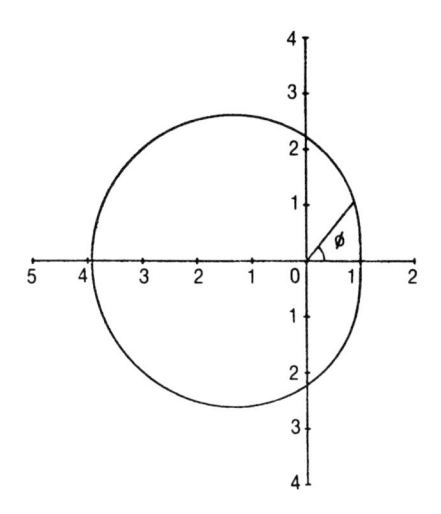

Figure 7.4 Normalised polar diagram of the scattered power from blood as a function of the scattering angle ϕ, calculated using eqn. 7.8. Note that the backscattered power is about 6 dB greater than at forward angles

caution should be exercised in extrapolating them to the *in vivo* situation.

At low shear rates there is a tendency for erythrocytes to aggregate into multi-cellular clumps, known as rouleaux, which can lead to a dramatic increase in the backscattered power from the blood. On the other hand, turbulent conditions are also known to increase backscattered power, although the mechanism for this is perhaps less well understood.

7.2.4.1 Low Blood Shear Rates and Erythrocyte Aggregation

If the shear rates in flowing blood are particularly low, then the erythrocytes aggregate into multicellular clumps or rouleaux, and this leads to an increase in backscattering of ultrasound. The results of this are frequently observed in pulse-echo ultrasound, where normally anechoic blood become echogenic in large veins (shear rate is proportional to velocity and inversely proportional to vessel radius—see Section 2.3.1), particularly if some degree of obstruction is present (Machi et al 1983, Sigel et al 1983). The question therefore arises as to what extent the backscatter coefficient of blood can be regarded as independent of flow conditions in other situations, and this problem has been studied by many authors and led Mo and Cobbold (1986b) to introduce a factor relating the degree of red cell aggregation into their model of ultrasonic backscatter [the term $\bar{m}(1 + \sigma_m^2/\bar{m}^2)$ in equations 7.2 and 7.4].

Shung et al (1992) worked *in vitro* with porcine and bovine whole blood, and found the backscattered power from the former to be extremely sensitive to shear rate (even at relatively high shear rate) over the physiological range of haematocrits. They made the point, however, that red cell aggregation tendency in porcine blood is greater than in human blood, and that human blood might be expected to behave more like bovine blood at shear rates of $50\ s^{-1}$ or greater, where red cell aggregation is believed to be minimal.

More recently, Cloutier and colleagues (1996) have made measurements of the variation of backscattered power as a function of shear rate at 10 MHz using porcine whole blood. Their measurements were made from different sample volumes from 1.27 cm diameter tubes over a range of mean shear rates ranging between 6 and $74\ s^{-1}$, and they concluded that there is a rapid reduction in power between samples subject to shear rates of between 1 and $5\ s^{-1}$, a transition region between 5 and $10\ s^{-1}$, and a very slow reduction beyond $10\ s^{-1}$. A further paper from the same group (Allard et al 1996) reported an angular dependence (in relation to flow direction) of backscattered power under erythrocyte aggregation conditions. For porcine blood the anisotropy was about 5 dB for shear rates between 17 and $51\ s^{-1}$, whilst for lower $(8.5\ s^{-1})$ and higher $(102\ s^{-1})$ rates the anisotropy was reduced to approximately 2 dB. Their results for blood with a haematocrit of 40% are reproduced in Fig. 7.5, and clearly show both the shear rate and angular dependence of backscattered Doppler power in porcine blood. No such angular dependence was obtained for saline

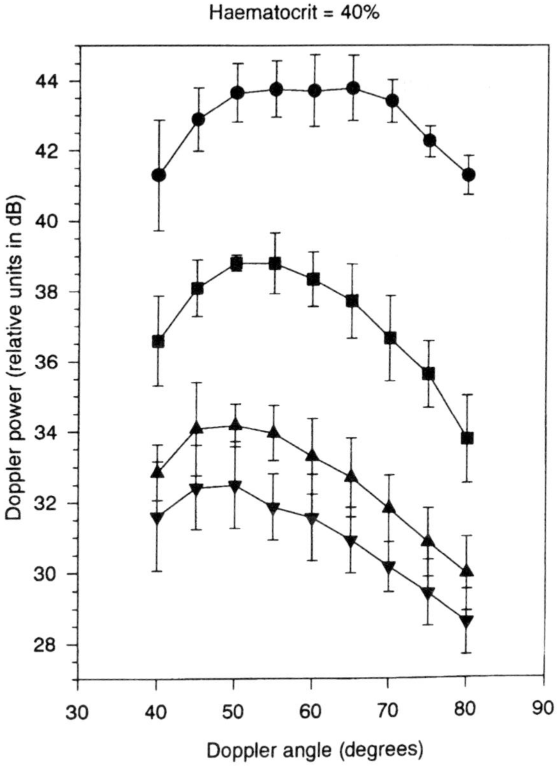

Figure 7.5 Doppler power as a function of shear rate, and the angle of insonation for porcine whole blood (haematocrit 40%). Shear rates: circle, $8.5\ s^{-1}$; square, $17\ s^{-1}$; triangle, $25\ s^{-1}$; inverted triangle, $51\ s^{-1}$ (reproduced by permission from Allard et al 1996, © 1996 IEEE).

suspensions of calf erythrocytes, which supports their hypothesis that the explanation for the anisotropic effects is the structure of red cell aggregates.

Van der Heiden et al (1995) have reported the effect of shear rate on both the magnitude and spectral slope of ultrasound backscatter from whole human blood, and from both 'rouleau-enhanced' and 'rouleau-suppressed' human blood, at 30 MHz. They reported that the backscatter from rouleau-suppressed blood showed no shear rate dependence and had a spectral slope of 3.3. At high shear rates (>80 s^{-1} for integrated backscatter power and >11 s^{-1} for spectral slope), whole blood and rouleau-enhanced blood tended to the results for rouleau-suppressed blood. At low shear rates and increasing rouleau size, the backscattered power was found to increase by 11 dB and the spectral slope to decrease to from 3.3 to 1. Figure 7.6, reproduced from van der Heiden et al, shows their results for the shear-rate dependence of backscattered power.

7.2.4.2 High Reynolds Numbers and Turbulent Flow

It has been known for many years that turbulence may have a significant effect on the scattering of ultrasound by blood. Shung et al (1976) reported that for haematocrits above 50%, turbulence could affect the scattering of ultrasound by blood, and it was later shown by Shung and his colleagues (1984) that, contrary to their original belief, turbulent flow can also cause an increase in ultrasound backscatter from erythrocyte suspensions with haematocrits of down to 10%. Angelsen (1980) also recognised the increase in power due to turbulence whilst developing his continuum model of ultrasound scattering from blood, and ascribed it to local accelerations in the velocity field causing a separation between the plasma and erythrocytes due to their different mass densities, and the resulting increase in fluctuations of local cell concentrations enhancing backscatter.

A further study of the increase in backscattered power due to turbulence was reported by Shung et al (1992). Controlled turbulence was introduced into porcine erythrocyte suspensions of differing haematocrits and the power of the resulting Doppler spectra compared with the power measured under laminar flow conditions. Figure 7.7 is reproduced from Shung et al (1992), and clearly shows an increase in backscattered power of the order of 100% for turbulent flow. Bascom and colleagues (1993), using human blood cells in

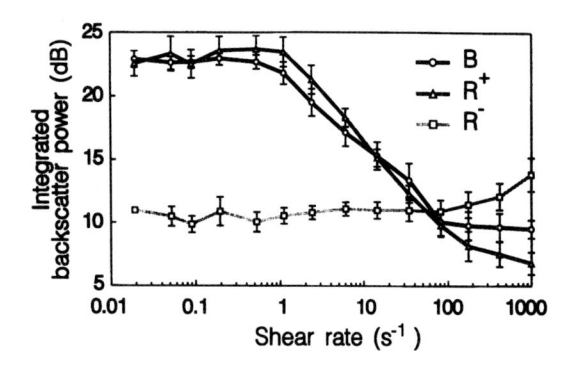

Figure 7.6 Average shear-rate dependency of integrated backscatter power (mean ± SEM) of the three main subsamples of blood used by van der Heiden et al. B, whole blood; R$^+$, rouleau-enhanced blood; R$^-$, rouleau-suppressed blood (reproduced by permission of Elsevier Science, from van der Heiden et al (1995), *Ultrasound in Medicine and Biology*, © World Federation of Ultrasound in Medicine and Biology)

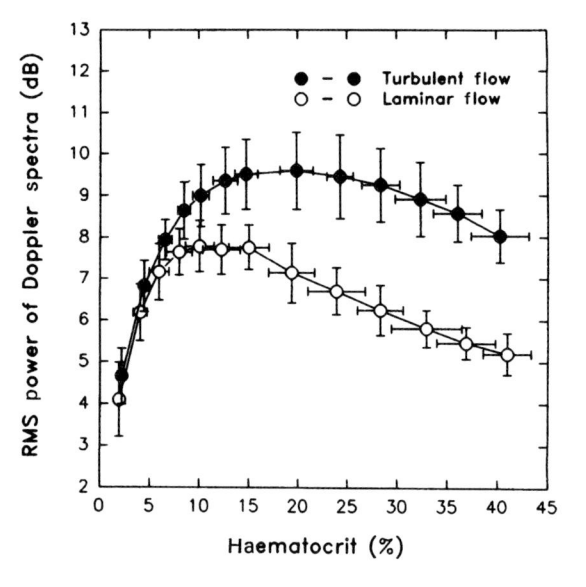

Figure 7.7 The RMS (root mean square) values of Doppler spectra for porcine erythrocyte suspensions under laminar and turbulent flow conditions as a function of haematocrit (reproduced by permission from Shung et al 1992, © 1992 IEEE)

saline (haematocrit = 42%), measured the Doppler signal from steady flow through an asymmetrical stenosis model and showed that the maximum backscattered power (equivalent to a 60–65% increase) was obtained from the turbulent region distal to the stenosis. Figure 7.8, which summarises the findings of that study, is taken from a subsequent theoretical paper from the same group (Bascom and Cobbold, 1996). In a further study, Bascom et al (1997) combined the use of their asymmetric stenosis model with pulsatile flow and with a photochromic technique for visualising flow profiles. They reported that the onset of turbulence for both steady and pulsatile flow increased the backscattered Doppler power, and that the location of the peak Doppler power coincided with the region of maximum turbulence observed using their visualisation technique. Wu et al (1998) made *in situ* measurements of Doppler power and flow turbulence intensity to quantify the relationship between these two quantities. Controlled levels of flow turbulence were generated in porcine red blood cell suspensions and measured using constant-temperature anemometry. These measurements showed that at a fixed haematocrit Doppler power increases in a complex and non-linear way with turbulent intensity.

Finally, Lockwood et al (1991) studied backscatter of ultrasound from blood from 35–65 MHz (which is outside the range for which Rayleigh scattering can be expected) and found an increase of 60% as the velocity of flow in their rig was increased from 18 to 36 cm s^{-1}. They suggested that this was due to a transition to turbulent flow.

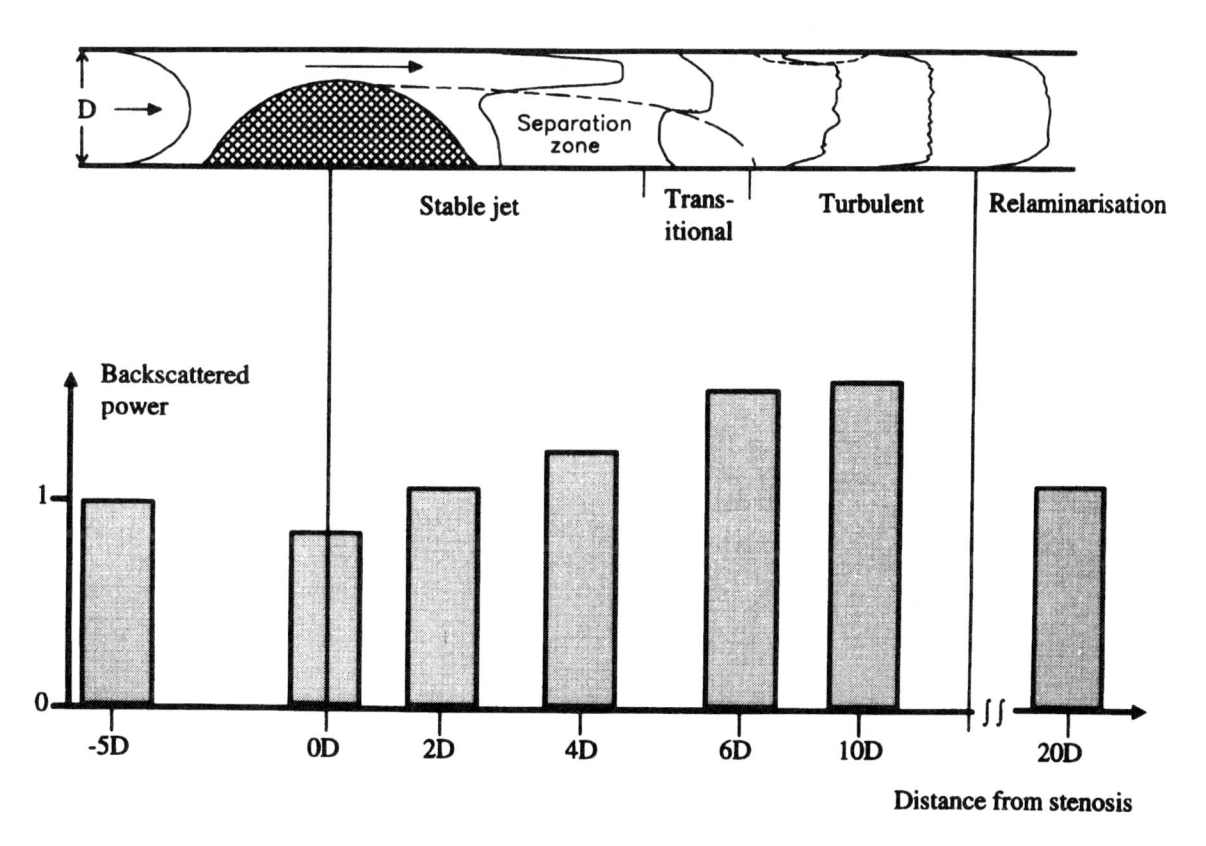

Figure 7.8 Illustration of how the backscattered Doppler power from a 5 MHz CW system changed as the insonation site was moved relative to a 70% asymmetric stenosis. The results were obtained under steady flow conditions at a Reynolds number of 545 using a 40% suspension of human erythrocytes in saline. The backscattered Doppler power was determined by averaging 200 samples of the output of a true RMS meter over 2 min (reproduced by permission from Bascom and Cobbold 1996, © 1996 IEEE)

7.2.4.3 Cyclic Changes in Backscattered Power in Pulsatile Flow

So far we have considered only steady flow conditions but there is, perhaps not surprisingly, evidence that there are cyclic changes in backscattered power during the cardiac cycle as a result of varying shear rates.

Some of the earliest reports of cyclic changes in Doppler power over the cardiac cycle appear to be those of Thompson et al (1985, 1986), who documented an increase in the power from fetal umbilical arteries at the time of peak systole, even after allowances had been made for the effects of high-pass filtering and cyclic arterial dilation. One possible explanation advanced for this behaviour was 'some effect involving the scatterers, such as rouleaux formation'.

De Kroon et al (1991) observed cyclic changes in the echo density of ultrasound images of blood in human iliac arteries recorded *in vivo* with a 30 MHz intravascular imaging device. The averaged echo density was shown to be consistently higher at end-diastole than at end-systole, although the method of measurement did not allow quantification of these changes. De Kroon et al attributed these changes to changes in the state of erythrocyte aggregation being induced by variations in the shear rate. They discussed the transit time between different aggregation states (characterised by the so-called 'aggregation half-time') and cited various published values of half-time constant ranging from 10 s down to 3.5 s, as well as 'short-time constants' of 0.5–1.0 s. Most of these constants seem relatively long in terms of the cardiac cycle, but De Kroon et al suggested the existence of faster changes than have previously been reported, together with the possibility of a higher sensitivity of high frequency ultrasound for detecting such changes, due to the scattering being outside the Rayleigh region.

Cloutier and Shung (1993a) performed experiments at 10 MHz with porcine erythrocytes suspended in saline solutions under both laminar and turbulent flow conditions. Under laminar conditions, and using a haematocrit of 40% they found no variations in Doppler power, but when turbulence was induced in the flow model the power increased during systole. A maximum was observed early after peak systole, and a decrease was obtained in diastole during deceleration of flow. Figure 7.9 is reproduced from this work. Cloutier and Shung attributed the cyclic variations observed at high flow velocities and in the presence of turbulence to cyclic changes in the correlation among particles.

Cloutier and Shung (1993b) circulated porcine whole blood specimens through a flow model with

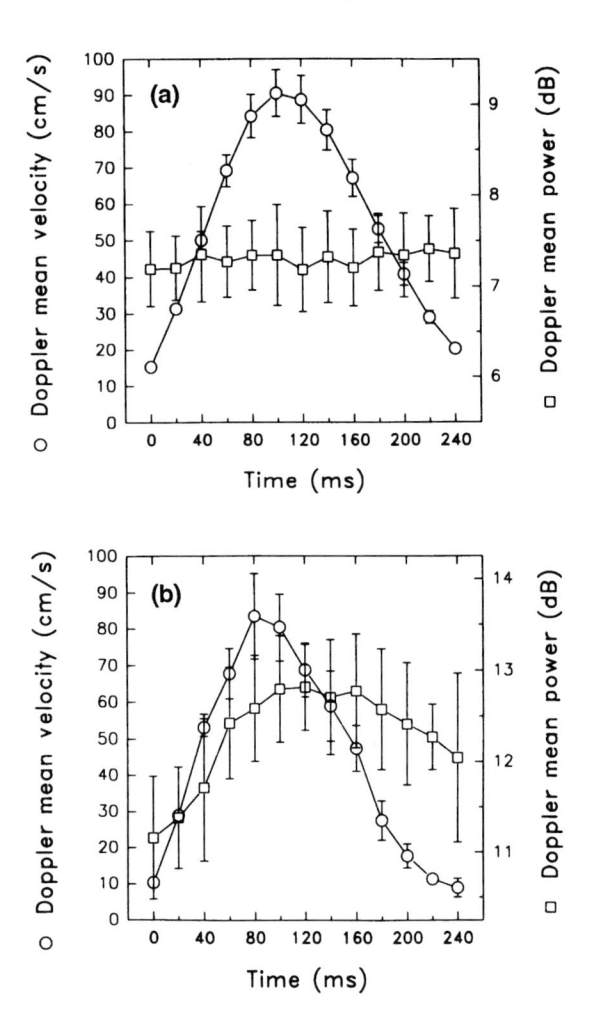

Figure 7.9 Mean velocity within the Doppler sample volume of porcine erythrocyte suspensions at a haematocrit of 40% and mean power as a function of the timing within systolic and early diastolic periods; (a) corresponds to laminar flow experiments, while (b) represents measurements performed in turbulent flow. Both experiments were carried out at a mean velocity (over the entire cycle) of 11 cm s^{-1}. Each data point represents the mean ±SD computed over five experiments. The mean spectra were averaged over 200 cycles (reproduced by permission from Cloutier and Shung 1993a, © 1993 IEEE)

a tube diameter of 0.476 cm at different mean velocities and pulsation rates. At cycle rates of 70 bpm for mean velocities of 13 cm s^{-1} and 63 cm s^{-1}, no cyclic variation of Doppler power was observed, 'suggesting the absence of rouleaux build-up and rouleaux disruption'. At cycle rates of 20 bpm and mean velocities of 11 cm s^{-1} and 38 cm s^{-1}, however, statistically significant cyclic variations were measured. Cloutier and Shung suggested that 'aggregate size enlargement, rouleaux orientation with the flow field and the effect of shear stress on rouleaux disruption [were] possible causes for the observed cyclic variation of the Doppler power within the flow cycle at a pulsation of 20 bpm'.

In a further paper, Cloutier and his colleagues (1995) reported on changes in backscattered power downstream of concentric and eccentric stenoses of between 47% and 91% area reduction under pulsatile flow conditions. As with the work of Bascom et al (1993) on steady flow through stenoses, they found an increase in power downstream of the narrowing, with a maximal power at approximately 10 diameters beyond the stenosis. The increase in power was loosely related to the percentage area reduction, with values ranging from approximately 30% to 100% for stenoses of between 75% and 90%. In addition to the increase in power, a cyclic variation of the backscattered intensity was observed within the flow cycle downstream of severe stenoses, but upstream of all stenoses included in the study, and downstream of the 47% and 52% area reduction stenoses, the power was constant during the acceleration and deceleration phases of the cycle. As a result of this, Cloutier and his colleagues reiterated the belief that the correlation among the scattering particles under disturbed flow is a main determinant affecting the intensity of the backscattered signal.

Finally, Wu and Shung (1996) have reported results from an *in vitro* pulsatile flow rig containing porcine whole blood, with which they studied cyclic variations of Doppler power as a function of the flow cycle, radial position in the vessel, and vessel compliance. At low shear rates they observed cyclic changes in the gross power (6 dB at 30 bpm) but could not consistently observe cyclic variations when the stroke rate was greater than 56 bpm. At the low stroke rates the peak of the Doppler power from the whole blood flowing near the centre stream coincided with the peak of flow velocity. However, it began to lead the velocity peak as the measurement site was moved away from the centre stream. This behaviour was attributed to the fact that the shear rate changes at different radial locations in the vessel have different phases. Their inability to detect cyclic changes in power for stroke rates of greater than 56 bpm was attributed to the fact that the time between consecutive strokes was not sufficiently long for disrupted cell aggregates to reform, and they cited studies by Schmid-Schöbein (1968), Shehada et al (1994), Shiga et al (1983) and Skalak (1984) suggesting that the aggregation time of erythrocytes ranges from a few seconds to minutes, depending on the experimental conditions and aggregate sizes.

7.2.4.4 Practical Implications

In some circumstances it is assumed that the power scattered by a small volume of blood is independent of its location. From the preceding discussion it is clear that this may not always be the case, and therefore the question arises as to the practical implications of possible non-uniformities in backscattered power.

The most common use of Doppler ultrasound is to estimate blood flow velocity (often to facilitate the production of velocity images), although with the advent of 'power Doppler' it is also being used as a method of imaging the 'quantity' of blood perfusing a specific region of tissue.

In the case of velocity measurement, inhomogeneities in back-scattered power are not important, provided that the sample volume is not large compared with the scale of variation. Thus, for example, if the backscatter cross-section of blood varies across a large vessel due to shear rate changes, it is still possible correctly to reconstruct a velocity profile across the vessel by interrogating a number of sample volumes across the vessel and then to use this information to estimate the mean velocity of flow through the vessel. What would not be acceptable, however, if there were large inhomogeneities, would be to calculate the mean velocity using one large sample volume encompassing the whole vessel, because velocities from regions of flow with the highest scattering cross-section would be weighted too heavily, compared with those with low scattering cross-sections. Thus, in general non-homogeneities in scattering cross-section are probably of little concern in colour flow mapping, in multigate systems and in pulsed Doppler systems using a small sample volume, but may need to be considered when reviewing the accuracy of measurements made with continuous wave Doppler and long-gate pulsed Doppler. The same principle applies in terms of temporal variation, as long as the sample length is short compared with changes in backscatter, then individual velocity estimates should remain valid.

Unfortunately, it is not entirely clear under what circumstances the backscatter cross-section of blood varies significantly across *in vivo* blood vessels, although the evidence to date would seem to suggest that for the reasonably high shear rates (i.e. $\geqslant 50$ s^{-1}) and cardiac cycle rates (i.e. $\geqslant 60$ bpm) associated with normal arteries in man the problem is not one of major proportions. The same cannot be said for veins, and particularly partially obstructed veins, where the shear rates may be quite low and vary across the vessel, or for stenosed arteries where turbulence may significantly alter the scattering cross-section of the blood in very localised areas.

With regard to power Doppler imaging, since it is the returning Doppler power that is used directly to form the image, then any alteration in scattering cross-section will influence the image directly, which may lead to false interpretations if the user is not aware of the factors that affect the ultrasound backscattered from blood. Particular caution must be exercised when interpreting the results from turbulence distal to stenoses (although Wu et al, 1995, have made the point that increases in intensity in Doppler power imaging may aid in the detection of stenoses and the resulting flow disturbances) and from veins or any other area where there is sluggish flow.

7.2.5 Statistical Properties of Ultrasound Scattered by Blood

Previous sections have dealt with only the time-averaged value of the scattered power, but it is well known that the power returning from blood fluctuates as a function of time. In fact, it is generally accepted that the Doppler signal can be regarded as a Gaussian random process (Angelsen 1980, Mo and Cobbold 1986a, Mo and Cobbold 1992) which is band-limited, i.e. non-white Gaussian 'noise'. This follows from noting the uniformly random phases of the signal returning from a large number of uncorrelated scattering units (erythrocytes and erythrocyte aggregates) randomly distributed in the same small target area, and considering the superposition of these signals.

As noted by Mo and Cobbold (1992), this has led to two approaches to the simulation of such signals, that of summing many sinusoidal components with random amplitude and phase shifts (Mo and Cobbold 1986a, van Leeuwen et al 1986, Mo and Cobbold 1989), which mimics the genesis of such signals, and that of passing white noise through a filter whose coefficients are determined by the desired power spectral density function (Sheldon and Duggen 1987, Kristoffersen and Angelsen 1988, Moraes et al 1995), which models the result of the process. The statistical properties of the Doppler signal has implications for the estimation of the Doppler spectrum (and therefore the estimation of the velocity distribution in the sample volume) and is discussed in Section 8.2. For details of statistical models of the Doppler signal, the reader is referred to the relevant literature, but the models of Mo and Cobbold (1986a and 1989) are briefly described in the following two paragraphs as interesting examples of such models.

According to the model of Mo and Cobbold (1986a), the backscattered Doppler signal $x(t)$, which is regarded as a band-limited, wide sense stationary Gaussian random process, can be represented over a stationary time interval T by a sum of sinusoids having discrete frequencies f_i ($i = 1, ..., M$), i.e.:

$$x(t) = \lim_{M \to \infty} \sum_{i=1}^{M} a_i \cos(2\pi f_i t + \xi_i) \qquad 7.9$$

where

$$f_i = (i - 0.5)\Delta f \qquad 7.10$$

and

$$a_i = \sqrt{2 S_x(f_i) \Delta f y_i} \qquad 7.11$$

where $S_x(f_i)$ is the power spectral density function defined over the frequency range $[0, f_{max}]$ and $\Delta f = f_{max}/M$. The ξ_i's and y_i's are independent random variables that account for the random nature of the signal, with each ξ_i being uniformly distributed over $[0, 2\pi]$, and each y_i being a χ-squared random variable with two degrees of freedom (χ_2^2). Since the a_i's are square roots of χ_2^2, they are Rayleigh distributed. In the limit as $M \to \infty$, the model becomes an exact representation of the in-phase component of a narrow-band Gaussian process. In the specific simulation described in their 1986b paper, Mo and Cobbold used the deterministic computer model of Bascom et al (1986), which incorporates the spectral broadening effects of circular transducer apertures, to obtain an estimate of a theoretical power spectral density $S_x(f_i)$ and thus generate values of a_i and finally $x(t)$.

In their subsequent paper, Mo and Cobbold (1989) generalised their model to simulate a non-stationary signal by varying the a_i coefficients in eqn. 7.9, by allowing the power spectral density function to vary with time, so that:

$$a_i(t) = \sqrt{2S(f_i, t)\Delta f y_i}. \qquad 7.12$$

From a physical viewpoint, $S(f_i, t)$ is determined by the velocity profile, whereas the χ-squared random fluctuations in each frequency bin is a result of the interference of backscattered wavelets from the corresponding volume of blood. In this instance, Mo and Cobbold used a simple empirical approach to derive $S(f_i, t)$ from clinical data.

Fontaine et al (1997) have recently presented results which suggest that the Doppler signal from the jet of a severe stenosis may not be considered to be a Gaussian random process.

7.3 CONTINUOUS WAVE DOPPLER SPECTRA

There are two distinct types of Doppler shift velocimeter (see Chapter 4), continuous wave (CW) and pulsed wave (PW). CW units, which are the subject of this section, both transmit and receive continuously; PW units transmit short burst of ultrasound at regular intervals and switch to a receive mode during the inter-burst period. The Doppler spectra resulting from these two types of instrument may differ significantly and they are therefore treated separately.

As stated in the introduction to this chapter, the spectrum of Doppler shift frequencies will under 'ideal' circumstances, have the same shape as the velocity distribution plot for the vessel of interest. In the sections that follow (7.3.1–7.3.6), a number of factors that distort the spectra from CW instruments are discussed.

7.3.1 Non-uniform Target Distribution

There is some evidence that, at least for steady flow and high shear rates *in vitro*, red blood cells in blood flowing in a tube are more concentrated in the centre stream than near the wall (Aarts et al 1988). In addition, it is known that both shear rate and turbulence influence the scattering cross-section of flowing blood (see Section 7.2). In general, therefore, it cannot necessarily be assumed that the 'ultrasonic targets' are completely uniformly distributed in flowing blood. The question of the implications of possible non-uniformities of backscattered power has already been considered in Section 7.2.4.4, where it was concluded that for the reasonably high shear rates and cardiac cycle rates associated with flow in normal arteries in man, the evidence suggests that the non-uniformities are probably not of major consequence, but this issue requires further investigation.

7.3.2 Non-uniform Insonation

Uniform insonation of blood vessels (particularly larger ones) is not easy to achieve. The spatial variation of sensitivity of a CW transducer depends on both the transmitting and receiving crystals and their relative positions and orientations. Evans and Parton (1981) and Douville et al (1983) have published studies of the directional characteristics of a number of twin-crystal CW transducers, and both groups noted a significant difference between probes of apparently identical manufacture. The near field, particularly close to the transducer, may be quite complex. In the far field there is usually, but not always, a single principal maximum at each range, and although side lobes may be present they are not of great significance. The width and rate of decay of the main lobe is highly dependent on transducer geometry.

If the sensitivity of the probe is not uniform across the diameter of a vessel from which measurements are to be made, some parts of the vessel will be preferentially sampled, and the frequencies corresponding to the velocities in the most sensitive part of the beam over-represented in the Doppler power spectrum.

Ultrasound transducers are usually manipulated to give the 'best' signal, and the axis of the ultrasound beam is likely to pass through, or close to, the centre of the vessel. If the vessel is small, the ultrasound beam may be sufficiently flat across the vessel for no undue distortion of the Doppler spectrum to occur, but if the vessel is large it is likely that the beam will effectively insonate only the middle portion of the vessel and the signal from the central streamlines will be over-represented. A number of authors have performed calculations on the effects of partial sampling of the blood vessel. Powalowski et al (1975) and Evans (1982a) considered models in which rectangular beams of ultrasound of varying size passed through the centre of vessels containing parabolic velocity profiles. Cobbold et al (1983) considered rectangular and Gaussian beams, both on-axis and off-axis, and also made some allowances for tissue attenuation. Evans (1982b) demonstrated the effect of using a narrow beam to insonate vessels containing complex velocity profiles, and later (Evans 1985) discussed the effect of non-uniform insonation on the measurement of mean velocity in vessels containing complex profiles. Bascom and Cobbold (1990) used a theoretical model to show how the Doppler spectrum from various axisymmetric velocity profiles is affected by beam misalignment and incomplete insonation, and derived a closed-form expression for the power spectral density received by an on-axis transducer with a Gaussian beam profile. Aldis and Thompson (1992) described a volume integral method for the calculation of Doppler power spectra for arbitrary beams and obtained results for uniform rectangular and circular insonating beams, and for non-uniform beams with Gaussian, jinc and sinc profiles. In a follow-up paper, the same authors (Thompson and Aldis, 1996) considered the effects of a cylindrical refracting interface on the CW Doppler spectrum (the relevance of this being the curved acoustic impedance interfaces found *in vitro* in flow phantoms and *in vivo* in vascular disease). Thompson and Aldis showed that even for an initially uniform beam, such interfaces lead to non-uniform insonation—a source of error in flow phantoms, which Thompson and colleagues first highlighted in an earlier paper on measurement artefacts (Thompson et al 1990). More recently, Tortoli et al (1997) have shown that for flow phantoms made of plastic materials, where both shear and longitudinal waves are supported, these effects can be even greater. Since the propagation speeds, and therefore the refractive angles and propagation paths, for the two modes are different, the different contributions can add together in such a way that the spectrum is significantly distorted. This problem should not occur *in vivo* as soft tissues do not support shear waves.

In the simple case of a rectangular beam passing through the centre of a blood vessel, it can be shown that the fraction, F, of the lamina at radius r which is intersected by an ultrasound beam of width w is given by (Evans 1982a):

$$F = (2/\pi)\sin^{-1}(w/2r) \quad \text{for } w < 2r \quad 7.13$$

If the radial distribution of velocities in the vessel is known, eqn. 7.13 may be written in terms of velocity rather than radius. A particularly interesting case is that of a parabolic velocity profile which, under conditions of uniform insonation, yields a flat power spectrum (Section 2.3.1). Such a profile may be written:

$$v(r) = v_{\max}(1 - r^2/R^2) \quad 7.14$$

where $v_{\max}$ is the maximum velocity found at the centre of the vessel of radius R.

Substituting for r in eqn. 7.13 leads to:

$$F = (2/\pi)\sin^{-1}[(w/2R)(1 - v/v_{\max})^{1/2}]$$
$$\text{for } w < 2R(1 - v/v_{\max})^{1/2} \quad 7.15$$

A graph of this function (Fig. 7.10) shows the way in which the Doppler power spectrum from a

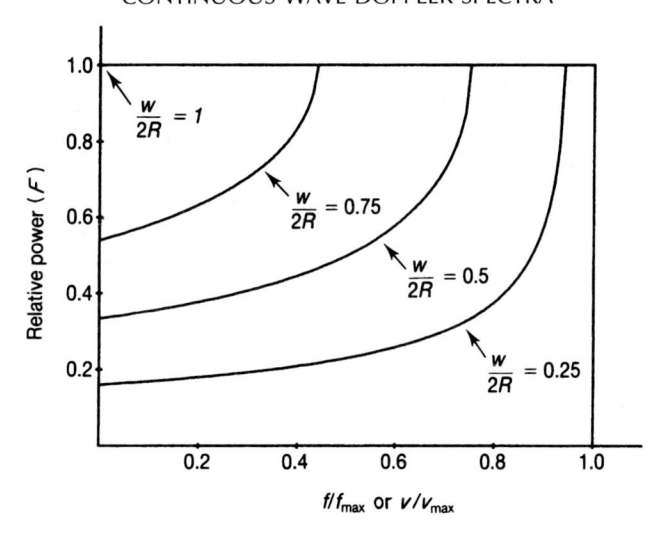

Figure 7.10 Power spectra resulting from insonating a vessel containing steady laminar flow with ultrasound beams of different relative widths

parabolic velocity profile is distorted by incomplete vessel sampling.

By applying eqn. 7.15 to the two halves of the vessel separately it is also possible to deduce the effect of using an ultrasound beam which is displaced to one side. Alternatively, the same results may be calculated numerically, using a simple model such as that suggested by Cobbold et al (1983). Although such an approach is unnecessarily complicated for such simple situations, it allows the investigation of the effects of more complex ultrasonic beam profiles and the differential attenuation between blood and soft tissue. The results obtained by Cobbold et al for a parabolic velocity profile interrogated by a square ultrasound beam are shown in Fig. 7.11.

Velocity profiles found in arteries vary both from vessel to vessel and throughout the cardiac

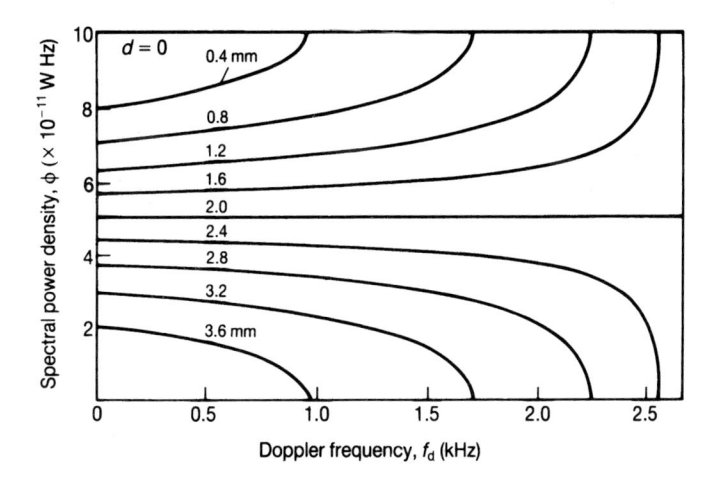

Figure 7.11 Spectral density graphs for a parabolic velocity profile interrogated by a square ultrasound beam with a width equal to the vessel diameter (4 mm). The results are plotted for successive 0.4 mm displacements of the beam central axis (reproduced by permission from Cobbold et al 1983, © 1983 IEEE)

cycle, and therefore the effects of partial sampling may also vary widely. The results of incomplete sampling of some complex velocity profiles (those shown in Fig. 7.1) are illustrated in Fig. 7.12.

It is by now clear that non-uniform vessel sampling may severely distort the shape of the Doppler spectrum; this may significantly affect spectral broadening indices (Section 9.4.5), the output of Doppler signal processors (Chapter 8) and the measurement of volumetric flow (Chapter 12).

7.3.3 Attenuation

Ultrasound waves are attenuated by a variety of mechanisms as they propagate through the body (Section 3.3.3). Attenuation rates in soft tissue (approximately $0.8 \, \text{dB} \, \text{MHz}^{-1} \, \text{cm}^{-1}$) are much greater than those in blood (approximately $0.2 \, \text{dB} \, \text{MHz}^{-1} \, \text{cm}^{-1}$) and therefore echoes returning from different parts of a blood vessel may experience different amounts of attenuation if they traverse different acoustic pathways. In particular, signals from the centre of the vessel will be stronger than those from its lateral edges, where the ultrasound traverses more soft tissue and less blood, and this has much the same effect as reducing the effective width of the ultrasound beam and exacerbates the effects of non-uniform insonation discussed in the last section. The effect becomes greater as the Doppler angle, θ, decreases and the path lengths in tissue and blood become more disparate, and is more pronounced at higher frequencies where the rates of attenuation are greater. Cobbold et al (1983) have documented this effect for a number of specific cases, and the results of one of their studies is reproduced in Fig. 7.13. This figure is directly comparable with Fig. 7.11; it differs only in that attenuation due to both blood and soft tissue have been included in the calculations. The angle θ was taken to be $60°$ and the transmitted frequency 8 MHz.

7.3.4 Intrinsic Spectral Broadening (ISB)

Equation 7.1 relates the Doppler shift frequency from a single target passing through an infinitely

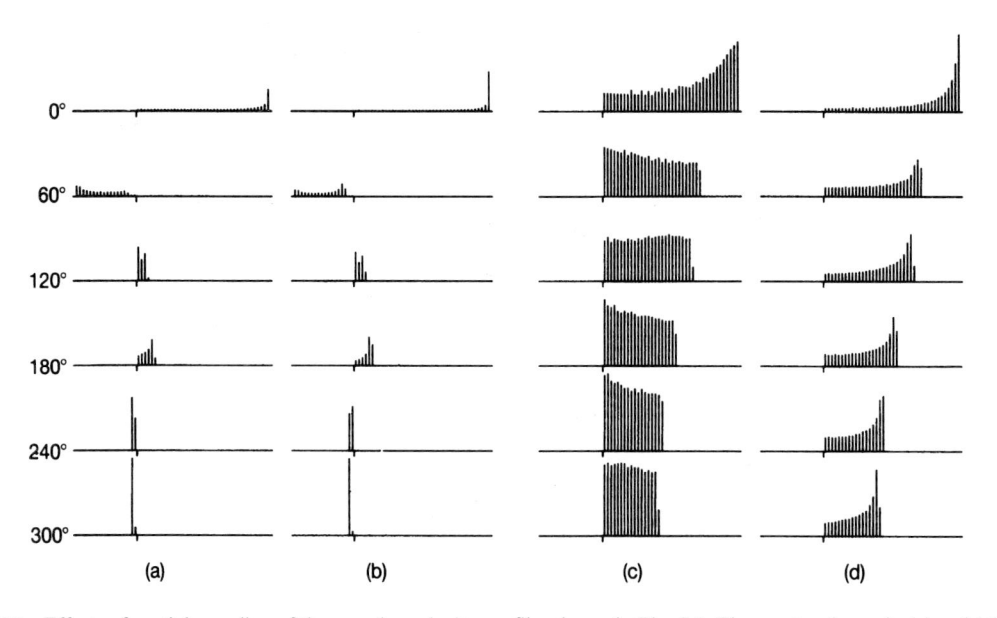

Figure 7.12 Effects of partial sampling of the complex velocity profiles shown in Fig. 7.1. The spectra shown in (a) and (c) are those resulting from uniform sampling and are identical in shape to the velocity histograms shown in Fig. 7.1b and d. The spectra shown in (b) and (d) are those resulting from interrogating the vessel with a uniform beam that is only 25% of the width of the vessel. Each series of spectra has been normalised in terms of both maximum frequency and maximum power, but the magnitude relationships within each series have been maintained

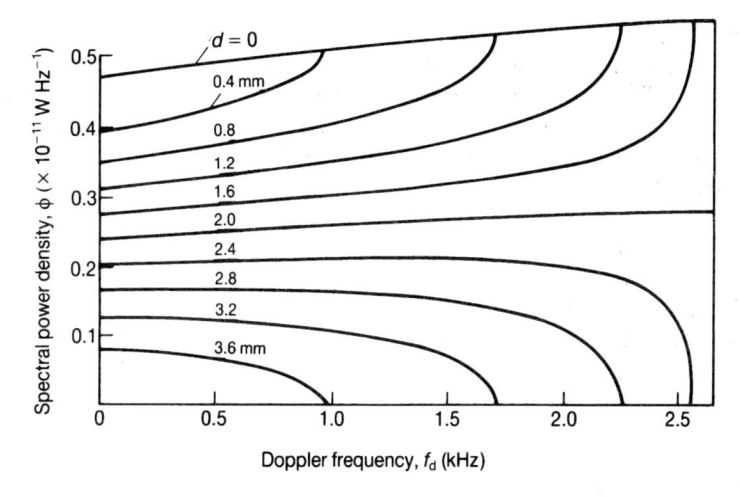

Figure 7.13 Spectral density graphs for a parabolic velocity profile interrogated by a square ultrasound beam with a width equal to the vessel diameter. These results differ from those shown in Fig. 7.11 because attenuation effects have been considered (the ultrasound frequency is 8 MHz and the assumed values of attenuation are 6.4 dB cm^{-1} in tissue and 1.44 dB cm^{-1} in blood) (reproduced by permission from Cobbold et al 1983, © 1983 IEEE).

wide plane ultrasound beam to its velocity. In practice, the ultrasound beam has only a finite width and is non-planar and, for reasons that will be explained subsequently, even a single target produces a spectrum of Doppler shift frequencies rather than a single frequency. When there are many targets passing through a finite ultrasound beam each contributes a spectrum of Doppler shift frequencies to the overall spectrum, which will therefore be broader and more smeared than would otherwise have been the case. This type of spectral broadening, which is due to the properties of the measurement system rather than the nature of the system being measured, is referred to as intrinsic spectral broadening (ISB).

ISB can be explained either in terms of the range of angles available to the incident and back-scattered radiation as the target traverses the ultrasound beam (so-called geometrical broadening) or in terms of the amplitude modulation caused by the finite transit time of the target through the ultrasound beam (so-called transit-time broadening). The equivalence of these two mechanisms was not appreciated in early publications and this has led to some confusion. Green (1964) identified what he considered to be three independent mechanisms of spectral spreading, unrelated to velocity gradients, in Doppler shift fluid flowmeters. The first, that due to Brownian motion of the individual scatterers, he showed to be negligible, but he found both geometrical and transit-time broadening to be significant. Griffith et al (1976) and Newhouse et al (1976) studied transit-time broadening and Newhouse et al (1977c) geometrical broadening, but the latter group of workers (Newhouse et al 1977b, 1980) were subsequently able to show the two effects to be equivalent. It is interesting that the same mistakes concerning these two apparently separate sources of spectral broadening were made by a number of workers in the laser Doppler field until the equivalence of the two effects was pointed out by Angus and his colleagues (Angus et al 1971, Edwards et al 1971). Although the two explanations for ISB result in the same conclusions, both are described here for completeness and because each may have advantages in terms of intuitive understanding and computation under certain circumstances.

7.3.4.1 Geometrical Explanation

The Doppler shift frequency, f_d, has been shown to be proportional to the cosine of the angle θ between the ultrasound beam and the direction of flow. In practice, because of the finite size of the transducer, the velocity vector of each target in an ultrasound field subtends not a single angle but a range of angles, and therefore each target contributes a range of Doppler frequencies to the overall Doppler spectrum. This is illustrated for the case of a focused beam in Fig. 7.14, where only two edge rays are shown. Newhouse et al (1977c, 1980) have shown that the magnitude of the spectral broadening may under some circumstances be calculated simply by considering the 'extreme angle rays', i.e. the range of angles over which back-scattered ultrasound is received by the transducer from the sample volume (which because of the finite size of the focus, will be slightly greater than that shown in the figure).

7.3.4.2 Transit Time Explanation

Each individual target that traverses an ultrasound beam scatters ultrasound for a limited period of time, i.e. for as long as it takes to cross from one edge of the beam to the other. Even whilst the target is within the beam, the intensity of ultrasound returning to the transducer will change as a result of the non-homogeneous ultrasound field. Although the spectrum of a continuous sine wave contains only a single frequency, the spectrum of a sinusoidal burst, or 'amplitude-modulated' sine wave, contains a whole spectrum of frequencies, and therefore even a single target must result in a spectrum of frequencies being received by the transducer. The wider and more homogeneous the ultrasound beam, the longer the returning ultrasound pulse from each target and the narrower the frequency spectrum.

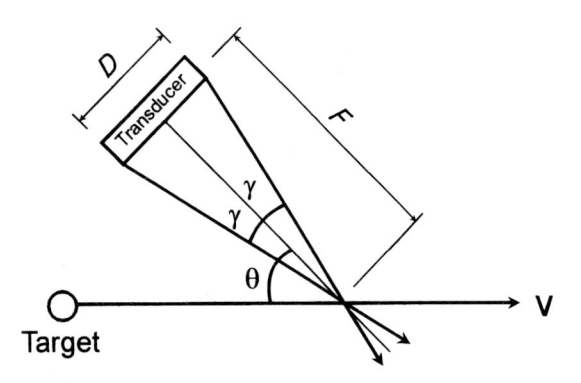

Figure 7.14 Schematic diagram of a target moving through the waist of a focused ultrasound beam. The velocity vector (**V**) subtends a range of angles at the transducer face, and therefore produces a range of Doppler shift frequencies

The frequency representation of a sinusoidal burst is a spectrum with a relative width ($\delta f/f$) given approximately by the reciprocal of the number of oscillations in the burst (the exact relationship depends on the shape of the modulating function and the definition of spectral width), i.e:

$$\delta f/f \approx 1/N_B \qquad 7.16$$

where N_B is the number of oscillations in the received burst. Note that f is the ultrasound frequency and not the Doppler shift frequency and therefore eqn. 7.16 may be rewritten:

$$\delta f = 1/T_t \qquad 7.17$$

where T_t is the transit time of the target through the Doppler sample volume.

So far the explanation has involved only a single target. In practice there are many targets within the ultrasound beam at any given time, but since the 'ultrasonic signature' of the target (which is determined by the instantaneous spatial distribution of the erythrocytes within the sample volume) is constantly changing as blood continuously enters through one edge of the sample volume and leaves from another, the net effect is approximately the same as for individual targets. For this reason the coherent signal processing time cannot exceed the average transit time across the sample volume, and is for practical purposes restricted to less (roughly half the transit time) because the backscattered signal changes significantly before an individual cell crosses the entire beam.

Early calculations (e.g. Newhouse et al 1977c) suggested that in the near field (Fresnel zone) transit-time spectral broadening was much smaller in magnitude than geometrical spectral broadening, and this strengthened the view that the two effects were independent. In fact, the broadening calculated by the transit-time approach was erroneously small because no account was taken of the strong modulation of the ultrasound signal in the extremely complicated near field.

7.3.4.3 Extent and Shape of ISB

Several authors have studied theoretical aspects of spectral broadening (Green 1964, Griffith et al 1976, Newhouse et al 1976, 1977b,c, 1980, 1987,

Censor et al 1988, Kim and Park 1989, Newhouse and Reid 1990, McArdle and Newhouse 1996). In each case the assumed transducer geometries and beam shapes were idealised and much simpler than those produced by practical CW Doppler units (Evans and Parton 1981, Douville et al 1983). The only way to calculate the effects of real CW Doppler transducers with their twin-crystal arrangement would be numerically; nevertheless the formulae that have been derived in analytical studies provide a good insight into the magnitude of ISB effects. Bascom et al (1986) have published some numerical calculations on the effect of ISB on the received Doppler spectra from flat and parabolic velocity profiles interrogated by square and circular transducers, but these too assumed the transmitting and receiving crystals to be coincident.

The most important formulae given in the literature can be derived very easily by making a few simplifying assumptions, and the analyses for three such cases are given below.

The first example is for uniform velocity flow through the waist of a focused ultrasound beam, a situation which has been discussed in detail by Newhouse et al (1980) and Censor et al (1988). The relevant geometry is given in Fig. 7.14 where, for the sake of simplicity, it has been assumed that the focus is at a single point. The Doppler shift frequencies f_1 and f_2 corresponding to the extreme rays may be written:

$$f_1 = \frac{2 f_t v \cos(\theta - \gamma)}{c} \qquad 7.18$$

and

$$f_2 = \frac{2 f_t v \cos(\theta + \gamma)}{c} \qquad 7.19$$

where f_t is the transmitted ultrasound frequency, v is the velocity of the target, θ the angle between the axis of the transducer and the flow direction, γ is half the angle subtended by the transducer at the target, and c is the velocity of ultrasound.

The bandwidth of the Doppler signal, B_d, may be therefore written:

$$B_d = f_1 - f_2 = \frac{2 f_t v}{c} \{\cos(\theta - \gamma) - \cos(\theta + \gamma)\} \qquad 7.20$$

or, expanding the cosine terms and cancelling:

$$B_d = \frac{2 f_t}{c} \cdot 2 \sin \gamma \cdot v \sin \theta \qquad 7.21$$

But $\sin \gamma \approx D/2F$, where D is the active Doppler aperture width, and F the transducer focal length, and therefore eqn. 7.21 may be rewritten:

$$B_d \approx \frac{2 f_t}{c} \frac{D}{F} v \sin \theta \qquad 7.22$$

which is the expression derived much more rigorously by Censor et al 1988.

The relative broadening B_d/f_d may be calculated by substituting eqn. 7.1 into eqn. 7.22, i.e:

$$\frac{B_d}{f_d} \approx \frac{D}{F} \tan \theta \qquad 7.23$$

which shows that the relative Doppler bandwidth for targets travelling with a single velocity is velocity-independent.

Note also that this expression is a close approximation to that derived and experimentally verified by Newhouse et al (1980) for ISB at the waist of a focused ultrasound beam, i.e:

$$\frac{B_d}{f_d} = \Delta\tau \cdot \tan \theta \qquad 7.24$$

where $\Delta\tau$ ($=2\gamma$) is the angle subtended by the transducer face at the focal spot.

Although eqn. 7.22 has been derived here for a specific position in the ultrasound field, it has been theoretically predicted (Newhouse and Reid 1990) that whilst the spectral width is dependent on the angle θ and the transducer geometry, it is in fact independent of the flow line location in the sound field. In other words, eqn. 7.22 should be equally valid for flow crossing the beam at positions other than at the focus. The 'Doppler bandwidth invariance' theorem has been tested *in vitro* by Tortoli et al (1992), who used a string phantom to measure the Doppler bandwidth at various positions in the field of two different focused transducers at Doppler angles of 70° and

90° and found little change at ranges from as little as 0.2 focal lengths up to more than twice the focal length. Since 1992 eqn. 7.22 has appeared regularly in the literature in one guise or another, particularly in relation to 'transverse Doppler' (Section 13.5), and to correcting maximum velocity estimates (Section 8.4.5), and has become the *de facto* standard equation for calculating the magnitude of ISB for focused beams.

The second case to be analysed is that of spectral broadening in the far field of a non-focused Doppler transducer. The simplified geometry for this case is shown in Fig. 7.15. It can be seen that using the same arguments as for the focused transducer, the bandwidth of the Doppler signal may now be written:

$$B_d = \frac{2f_t}{c} \cdot 2 \sin \chi \cdot v \sin \theta \qquad 7.25$$

where χ is half the angle of divergence of the main lobe. For a plane disc transducer, this angle is given by $\chi = \arcsin(1.22\lambda/D)$ (Wells 1977), where λ is the ultrasound wavelength and as before D is the active Doppler transducer width. Substituting this value of χ into 7.25 leads to:

$$B_d = \frac{4.88}{D} v \sin \theta \qquad 7.26$$

Substituting eqn. 7.1 into eqn. 7.26, the relative broadening may be written:

$$\frac{B_d}{f_d} = 2.44 \left(\frac{\lambda}{D} \right) \tan \theta \qquad 7.27$$

which is approximately the expression derived theoretically and verified experimentally by Newhouse et al (1980) for ISB in the far field of an unfocused transducer.

The third and final case to be analysed is that for transit time broadening due to uniform velocity flow through a parallel ultrasound beam with the same diameter as the transducer, and the following analysis is taken from Griffith et al (1976). Referring to Fig. 7.16, the bandwidth of the Doppler signal, B_d, returning from scatterers all moving with the same velocity will be equal to the width of the ultrasound spectrum δf, as defined by eqn. 7.17, i.e:

$$B_d = 1/T_t \qquad 7.28$$

Figure 7.15 Schematic diagram of a target moving through the far field of a non-focused ultrasound beam. In this part of the field the beam diverges, and therefore the velocity vector subtends a range of angles depending on its position, producing a range of Doppler shift frequencies

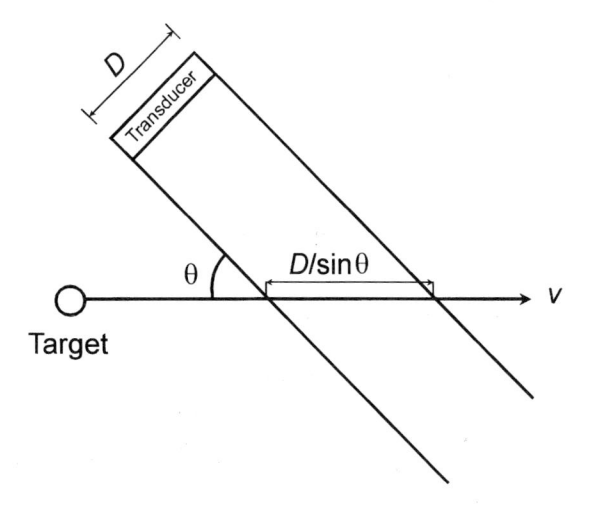

Figure 7.16 Schematic diagram of a target crossing a parallel ultrasound beam. An alternative way of looking at ISB is to consider the finite time which a single scatterer spends in the beam

Griffith et al (1976) showed that the effective transit time T_t is given by the following expression:

$$T_t = \frac{D}{2} \frac{1}{\sin\theta} \frac{1}{v} \frac{1}{2} \qquad 7.29$$

where v is the target velocity, and θ the usual Doppler angle. Griffith et al used the factor $D/2$ rather than D as the beamwidth diameter as they argued that it is a better approximation because little power is received from the fringes of the acoustic beam. They also introduced the factor $1/2$ to account for the continual entering and leaving of red cells from the sample volume—the argument being that the backscattered signal changes considerably by the time one cell moves half way across the beam.

Combining 7.28 and 7.29 leads to:

$$B_d = 4v \sin\theta / D. \qquad 7.30$$

Now from eqn. 7.1 the velocity v may be written in terms of the Doppler shift frequency:

$$v = \frac{cf_d}{2\cos\theta f_t} = \frac{\lambda f_d}{2\cos\theta} \qquad 7.31$$

which, when substituted into eqn. 7.30, gives:

$$\frac{B_d}{f_d} = 2\left(\frac{\lambda}{D}\right) \tan\theta \qquad 7.32$$

which is virtually identical to eqn. 7.27. Equations 7.27 and 7.32 have of course been derived under rather different conditions, and their almost exact agreement is somewhat coincidental, but this does serve to demonstrate that the geometrical and transit time descriptions of ISB can lead to similar results and conclusions.

There are two important facts to be gleaned about ISB from the preceding mathematics. The first is that for CW Doppler, ISB is proportional to the tangent of the angle between the ultrasound beam and the flow axis, and therefore the smaller the angle, the smaller is the effect of ISB (this is intuitively obvious, since the smaller the angle, the larger the sample length). The second observation is that ISB can be significant in CW applications. For example, for a 3 MHz transducer ($\lambda = 0.5$ mm) of diameter 10 mm and a beam to vessel angle of 45°, the spectral broadening is approximately 10%. Equation 7.27 has been evaluated and plotted in Fig. 7.17 for a range of values of λ/D likely to be encountered in practice, and it may be seen that particularly at large angles, B_d/f_d is quite appreciable.

So far we have considered the extent of, but not the shape of, the broadening function, which is approximately triangular in form. The reason for this can be appreciated by considering the geometry shown in Fig. 7.14. Following the argument of Willink and Evans (1996), the highest Doppler shift frequencies observed at the transducer result from energy transmitted and received by the lower left-most region of the transducer, and the lowest frequencies from the energy transmitted and received by the upper right region. Middle frequencies are favoured because they arise not only from energy transmitted and received by the middle part of the transducer, but also from energy transmitted by the lower left and received on the upper right, and vice versa, and this results in an approximately triangular form. More rigorous analysis (Newhouse et al 1987, Censor et al 1988) leads to the conclusion that the broadening function due to a circular aperture transducer has a slightly rounded triangular form. This particular geometry is not strictly applicable to the CW Doppler situation but is indicative of the type of broadening to be expected.

The practical effect of spectral broadening on a complex Doppler spectrum is to blur its shape, and in particular to smooth out sharp changes. The effect of different degrees of ISB on a number of power spectra is illustrated in Fig. 7.18. Guidi et al (1995) have described a method for calculating the shape of Doppler spectra based on the summation of individual broadened flow-line spectra.

The importance of ISB depends on the use to be made of the spectral information. Even large degrees of ISB will not influence the measurement of mean velocity, since the broadening is more or less symmetrical, but the maximum velocity becomes difficult to define precisely and may well be overestimated if too low a threshold value is chosen (see Fig. 7.19) and can potentially lead to misdiagnosis if the effect is not corrected for or at

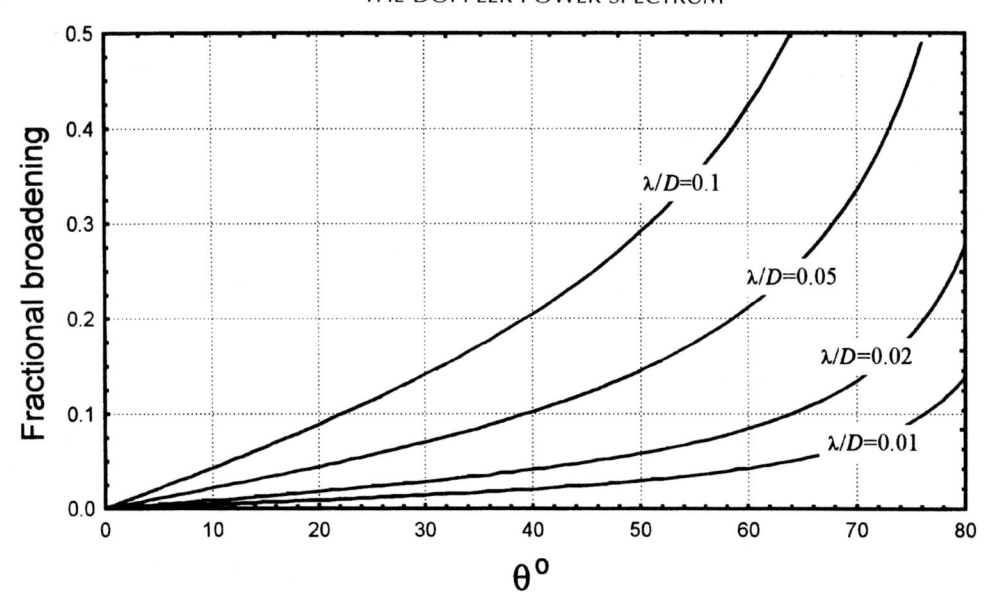

Figure 7.17 The fractional broadening in CW applications calculated from eqn. 7.27. Each curve represents the effect for a given ratio between the ultrasound wavelength and the transducer diameter

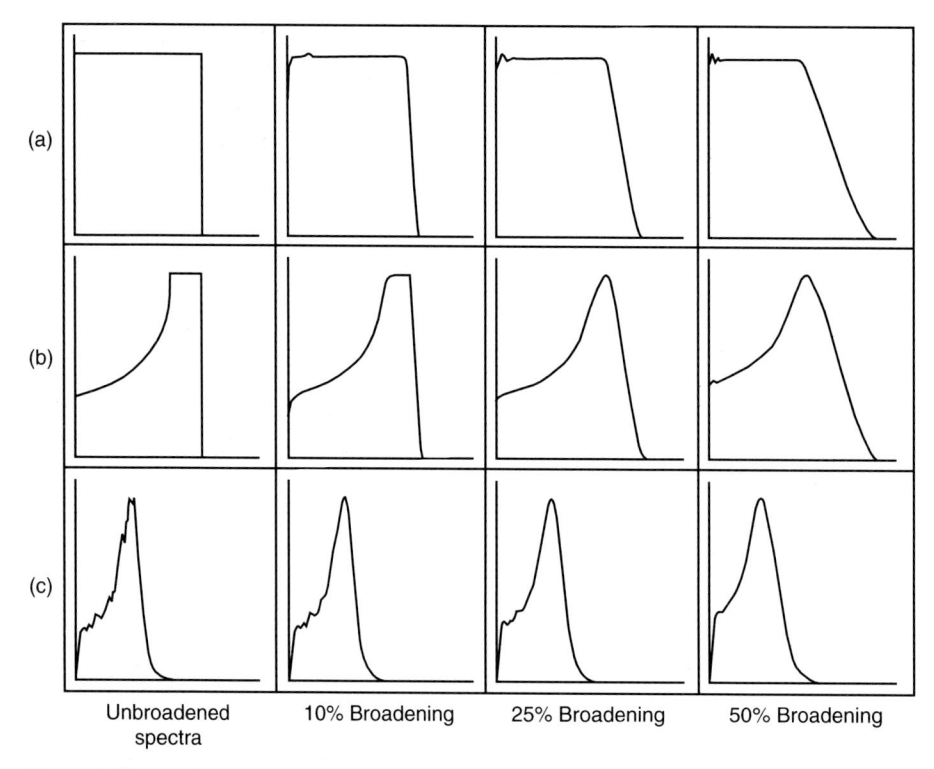

Figure 7.18 Effect of different degrees of spectral broadening on three different power spectra: (a) the spectrum resulting from uniform insonation of parabolic flow; (b) the spectrum from partially sampled uniform flow, cf. Figure 7.10; (c) the averaged spectrum recorded from a common carotid artery at peak systole

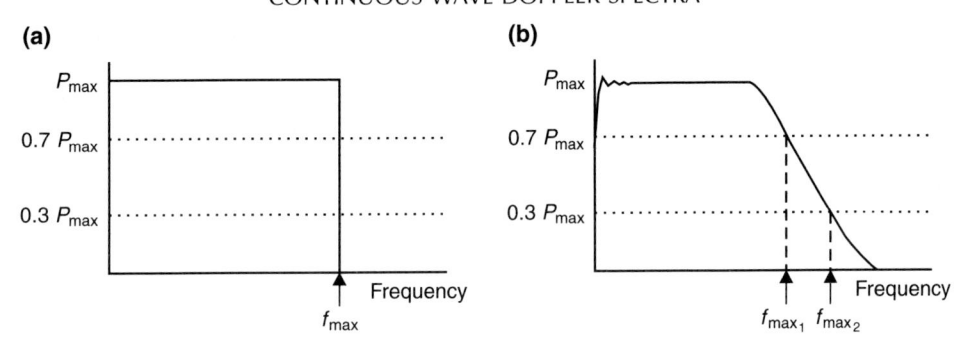

Figure 7.19 Effect of intrinsic spectral broadening on the output of a maximum frequency follower: (a) the Doppler spectrum from parabolic flow in the absence of ISB. Changes in threshold have no effect on the derived maximum frequency; (b) the spectrum from parabolic flow with ISB. The derived maximum frequency is influenced by threshold level

least appreciated (Daigle et al 1990, Thrush and Evans 1995, Winkler and Wu 1995, Hoskins 1996). This issue is discussed further in Section 8.4.5. Information derived from the spectral shape itself must clearly be treated with caution if there is a possibility of significant ISB.

Whilst ISB is generally undesirable, as it is a source of error in maximum velocity measurements and limits velocity resolution, it can be turned to advantage in situations where the classical orientation of the Doppler beam to a vessel is not possible (Newhouse et al 1987) and may also have applications in identifying the Doppler angle and allowing the vectorial evaluation of blood velocities at specific sites (Tortoli et al 1993, 1994). Furthermore the technique of Doppler power imaging relies on ISB to permit imaging of vessels that are perpendicular to the Doppler beam. The deliberate use of very high Doppler angles for 'transverse Doppler' studies is discussed further in Section 13.5; Doppler power imaging is discussed in Chapters 10 and 11.

7.3.5 Filtering

CW Doppler is incapable of precise range resolution and therefore, in addition to the Doppler shifted signal arising from blood flowing through the ultrasonic field, there is inevitably a component of the Doppler signal that arises from other moving structures, such as the vessel wall, and indeed from the effects of muscle vibrations in the patient and even the operator (Heimdal and Torp 1997). Fortunately, most of these signals are of low frequency, but they may have amplitudes which are considerably greater than those from blood. It is therefore necessary to incorporate high pass filters into a Doppler unit, and in addition to removing the unwanted clutter signal they will remove the signal returning from blood, which has a low velocity. This leads to a 'gap' in the power spectrum around the frequencies that represent zero flow, which may be particularly troublesome during diastole, and when recordings of very low flow velocities (such as those observed from fetuses and low birthweight babies) are to be made. The cut-off frequency of the high pass filters (often called wall thump filters) is usually user-adjustable and they should always be set as low as is compatible with recording a 'noise-free' spectrum. Unfortunately, it is not always possible to reduce the filter frequency adequately, either because the equipment manufacturer has not allowed sufficient adjustment, or because low-frequency interference signals are so strong that they prevent the use of a very low frequency cut-off. The effects of filters on the performance of frequency followers is discussed in Chapter 8.

7.3.6 Spectral Analysis Limitations

In order to extract the power spectrum from a Doppler signal, it must be transformed into the frequency domain. Because of the nature of the Doppler signal, such a procedure can only yield an estimate of the true underlying spectrum, and the reader is referred to Section 8.2 for a detailed discussion of this problem.

The most popular method of spectral estimation is Fourier transformation, using a fast Fourier transform (FFT) algorithm. This method leads to estimates that may appear very noisy due to the high spectral variance of FFT estimation on random signals, and may result in significant algorithm related spectral broadening, if the signal is highly non-stationary. Two examples of the high spectral variance associated with the FFT can be found in Figs 8.1a,c, which are individual spectral estimates of the Doppler signal from steady laminar flow in a tube and from a human carotid artery at peak systole, respectively.

7.4 PULSED WAVE DOPPLER SPECTRA

PW Doppler is dealt with separately from CW Doppler in this chapter because the Doppler spectra resulting from the two types of instrument may differ considerably. PW units may be roughly classified into two (overlapping) groups; those that use long gate times sufficient to interrogate an entire vessel, and those (including multigate instruments) that use very short gates so that the flow in only a small part of the vessel is interrogated. Many of the characteristics of these two types of operation are similar but they differ in other respects, so that the long gate PW operations may actually be more like CW operations than short-gate PW operations from the standpoint of the spectra they produce. Wherever possible it will be made clear if, and how, short and long-gate operations differ. Details of the operation of PW devices and of the generation and demodulation of PW signals are given in Chapters 4, 5 and 6.

7.4.1 Non-uniform Insonation

Whilst CW transducers often use separate elements for transmitting and receiving, it is usual in PW applications to use a single crystal or group of crystals for both purposes. Hence, the field shape produced by a PW unit differs from that of a CW unit both radially (due to crystal geometry) and longitudinally (due to finite pulse length).

As with CW Doppler, it is essential that the field from a long-gate PW Doppler unit is substantially uniform over the entire lumen of the vessel if the Doppler power spectrum is to be representative of the velocity distribution in the vessel, and so, in addition to the ultrasound beam being sufficiently wide to encompass the whole vessel, the gate must be open for a sufficiently long period of time. Short-gate Doppler units do not insonate the vessel uniformly and would ideally reflect the velocity distribution of the erythrocytes in the small sample volume they interrogate. Unfortunately spectral broadening mechanisms can become particularly important in such instances (see Section 7.4.3) and considerable care must be taken with the interpretation of such spectra.

7.4.2 Non-linear Propagation

Pulsed wave systems use higher peak pressures than CW systems, both because they only transmit for a small percentage of the time (and therefore, to achieve an adequate signal-to-noise ratio, the short bursts they use must have greater amplitude) and because PW transducers are less efficient than CW transducers. It has been shown for the pressure amplitudes used by commercial PW Doppler (and pulse-echo) machines that this not only leads to significant non-linear propagation under some circumstances, but can even lead to acoustic shock generation (Bacon 1984, Duck and Starritt 1984, Duck et al 1985)—see also Section 3.2.5 for further discussion of non-linear effects in ultrasound propagation. Because non-linear propagation leads to changes in the spectral content of ultrasound pulses, this will in turn influence the Doppler spectrum. However, studies that have examined this potential source of error have shown that non-linear effects to make little (Thomas et al 1989, Thomas and Leeman 1991) or no difference (Li et al 1993) to measured Doppler shift frequencies.

7.4.3 Intrinsic Spectral Broadening

In CW applications, ISB is determined by the length of the bursts of ultrasound reflected by targets as they traverse the ultrasound beam (Section 7.3.4.2). In PW applications, if the sample volume length (SVL) is sufficiently short and the Doppler angle sufficiently small, then the transit time is determined by the SVL rather than the ultrasound beam width. The two possible situations are illustrated in Fig. 7.20. For the situation depicted in Fig. 7.20a the beam width determines the transit time of the target, whilst in Fig. 7.20b it is the Doppler SVL that determines the transit time. Simple geometrical considerations show that the SVL limits the transit time if:

$$\theta < \tan^{-1}(w/Z_r) \qquad 7.33$$

where θ is the usual Doppler angle, w the ultrasound beam width, and Z_r the SVL.

An expression for the relative broadening in case (a) has already been derived in Section 7.3.4.3 (equation 7.32). A simple analysis of case (b) follows.

Referring to Fig. 7.20b, the effective transit time of the blood through the sample volume may be written:

$$T_t \approx \frac{1}{2} \frac{Z_r}{\cos\theta} \frac{1}{v} \qquad 7.34$$

where θ is the usual Doppler angle, and v the velocity. The factor 1/2 has been introduced to account for the continual entering and leaving of blood cells from the sample volume, as in equation 7.29.

The sample volume length may be written in terms the length of the transmitted pulse length t_p, and the receiving gate length t_g (equation 4.3):

$$Z_r = c(t_g + t_p)/2 \qquad 7.35$$

where c is the velocity of ultrasound.

Substituting 7.35 into 7.34 leads to:

$$T_t \approx \frac{1}{4} \frac{c(t_g + t_p)}{\cos\theta} \frac{1}{v} \qquad 7.36$$

Now, substituting 7.36 into 7.17 to get an expression for spectral width leads to:

$$B_d \approx \frac{4v\cos\theta}{c(t_g + t_p)} \qquad 7.37$$

Finally, dividing 7.37 by eqn. 7.1 leads to an expression for relative broadening:

$$\frac{B_d}{f_d} \approx \frac{2}{(t_g + t_p)f_t} \qquad 7.38$$

(actually, this expression can be derived much more directly from equation 7.16).

For the situation where $t_g \approx t_p$, as is frequently the case in practice (Section 4.2.2.2), then equation 7.38 may be rewritten:

$$\frac{B_d}{f_d} \approx \frac{1}{t_p f_t}. \qquad 7.39$$

More detailed analysis of SVL limited spectral broadening shows that for the case where $t_g \ll t_p$, if there is uniform flow within the sample volume, the Doppler spectrum will be a replica of the square of the spectrum of the transmitted ultrasound pulse, but scaled down in frequency by a factor of f_d/f_t (Newhouse et al 1976) (Fig. 7.21).

Equation 7.38 indicates that considerable spectral broadening occurs where very short SVLs are

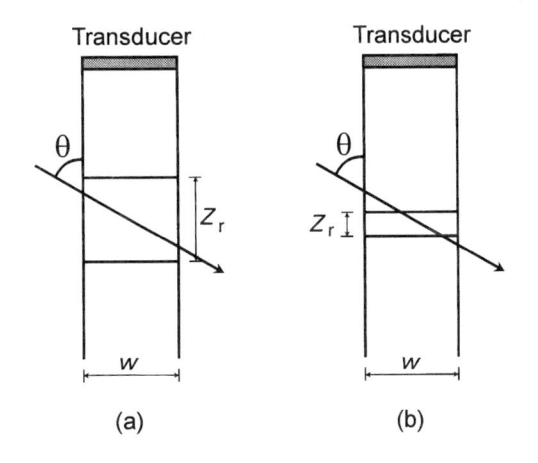

Figure 7.20 For CW Doppler systems and PW systems with relatively long sample volume lengths, the transit time of the target through the sample volume, and hence the ISB, is determined solely by beam width (a). For PW systems with very short gate lengths or where the Doppler angle is particularly small, the transit time, and hence the ISB, is determined by the sample volume length (b)

The use of pulses with large bandwidths creates additional problems in that, as both scattering and attenuation are frequency-dependent, the shape of an ultrasound pulse, and hence its Doppler replica, can be considerably distorted as it travels through the body and is scattered by the blood cells. This is discussed further in the next two sections.

7.4.4 Frequency-dependent Scattering

Scattering of ultrasound by blood is proportional to the fourth power of frequency (Shung et al 1976) and therefore the higher frequency components of a pulse of ultrasound are more strongly scattered than the lower frequencies. This leads to an increase in the centre frequency of the pulse (Round and Bates 1987) which, depending on the Doppler estimator employed, may in turn result in an overestimate of the Doppler shift frequency (Newhouse et al 1977a, Forsberg 1989, Embree and O'Brien 1990, Ferrara et al 1992). Fortunately, even for estimators that are sensitive to this source of error, it only becomes significant for extremely high bandwidth pulses. It will be seen in Section 8.3.5, where this source of error is discussed in the context of the mean frequency processor, that even for a pulse with a centre-frequency-to-bandwidth ratio of three, the error in measuring the modal Doppler frequency from uniform flow is less than 4%.

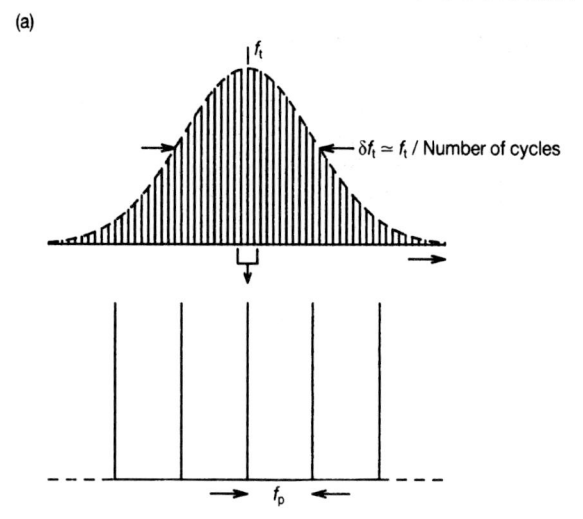

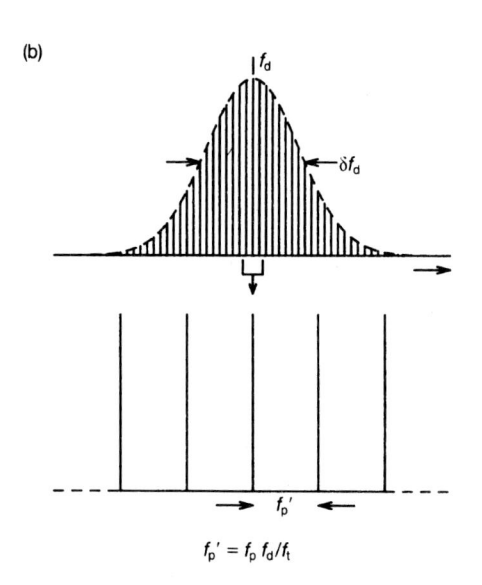

Figure 7.21 Spectrum of ultrasound power transmitted by a PW Doppler unit, and (b) the corresponding audio frequency replica derived by demodulation. In each case the lower part of the diagram represents an expanded version of the centre of the upper diagram

used, and it is difficult to distinguish between broadening caused by a true distribution of velocities in the sample volume and the intrinsic broadening. For this reason it is necessary to exercise caution in the interpretation of reports concerning the ability to resolve spectral broadening due to turbulence using high spatial resolution measurements.

7.4.5 Attenuation

As in the case of CW Doppler (Section 7.3.3), attenuation may effectively modify the shape of a PW Doppler beam, but this will normally only be of importance with long-gate Doppler systems where the objective is to insonate the vessel uniformly.

For short-gate Dopplers it is the frequency-dependent nature of attenuation, rather than attenuation itself, that is important in distorting the Doppler power spectrum. Attenuation of ultrasound in tissue increases with increasing frequency, and therefore the higher frequency components of the transmitted ultrasound pulse are attenuated more rapidly than the lower frequencies, which leads to a decrease in the centre frequency of the pulse (Ophir and Jaeger 1982). As with frequency-dependent scattering, this can lead to an error in the demodulated audio frequency signal, but in this instance, as it is the lower frequencies that are relatively accentuated, the mean frequency may be underestimated (Newhouse et al 1977a, Forsberg 1989, Embree and O'Brien 1990, Ferrara et al 1992).

It is more difficult to generalise over the effects of frequency-dependent attenuation than frequency-dependent scattering, because the former is not only determined by the Doppler frequency estimator used and the pulse bandwidth, but also by the attenuation coefficient of the tissue through which the pulse travels (and its frequency dependency) and by the depth of the target. In general the errors can be somewhat larger than those introduced by frequency-dependent scattering, but are not sufficient to cause practical difficulties unless the pulse bandwidth is large and/or the target is deep within the body. Further discussion of these issues can be found in Section 8.3.5, where some typical error values are given.

To some extent any increase in the mean Doppler shift frequency due to scattering is partially compensated for by the decrease due to frequency-dependent attenuation. Reid and Klepper (1984) have suggested that the operating frequency of a Doppler unit could be chosen to minimise the net error over a finite range if the attenuation of the tissue were known, and that the same choice would maximise the average echo power.

7.4.6 Filtering

The effects of filtering on the Doppler spectrum have been discussed in Section 7.3.5, where it was explained that high pass filters are necessary to remove high-amplitude low-frequency signals arising from tissue movement and especially 'vessel wall thump'. Exactly the same considerations apply to long-gate PW Doppler, but for short-gate PW the sample may be placed entirely within the blood vessel of interest, and therefore many of the low-frequency signals will be rejected by the range gating. In these circumstances it is possible to reduce the filtering requirements.

7.4.7 Spectral Analysis Limitations

The spectral estimation of Doppler signals from PW systems is subject to the same limitations as for signals from CW systems. There is, however, one extra constraint to be considered, particularly in relation to the FFT, and that is that if there is

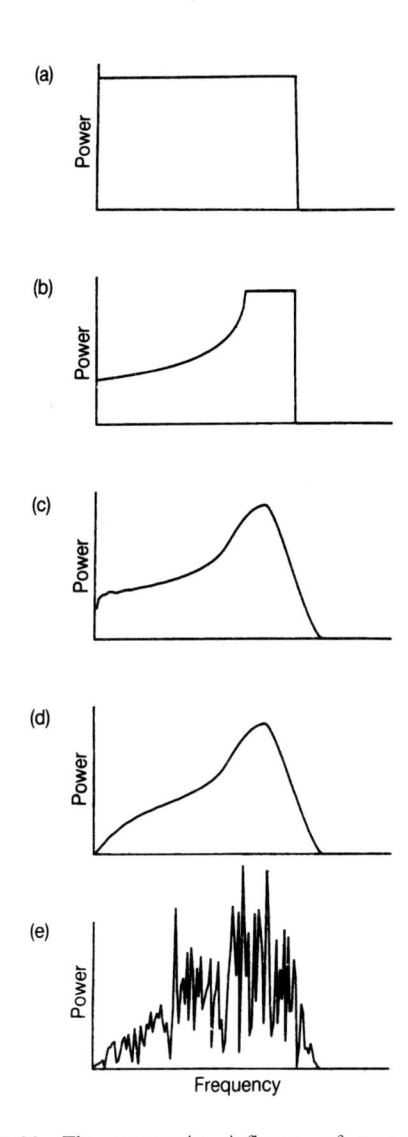

Figure 7.22 The progressive influence of a number of distorting mechanisms on the idealised Doppler power spectrum from steady laminar flow; (a) idealised spectrum; (b) the effect of partial sampling of the vessel—Section 7.3.2; (c) the effect of spectral broadening—Section 7.3.4; (d) the effect of high pass filtering—Section 7.3.5; (e) the effect of spectral estimate variance—Section 7.3.6

excessive ISB due to the use of a very short gate length, then the spectral resolution may be limited by the ISB rather than the length of the stationary data sample available for analysis. In extreme cases where the reciprocal of the intrinsic broadening width is much less than the stationarity period, then since spectral resolution cannot be improved by using the entire stationary data section, it is better to segment it and use Bartlett's averaging procedure to reduce the spectral variance (Section 8.2.2).

7.5 SUMMARY

Under ideal uniform sampling conditions the Doppler power spectrum would have the same shape as a histogram of the velocity distribution of the erythrocytes within the Doppler sample volume. In practice, this shape is modified by a number of mechanisms, including non-uniform sampling, attenuation, intrinsic spectral broadening, filtering and the limitations of spectral estimation techniques. Figure 7.22 illustrates the way in which the ideal spectrum from steady laminar flow might be progressively influenced by these mechanisms.

There are significant differences between CW and PW Doppler, due to both the effects of sampling and the wide signal bandwidths used by some of the latter type of device. In the case of short-gate PW Doppler, these differences may profoundly alter the shape of the Doppler power spectrum.

7.6 NOTATION

B_d	Doppler signal bandwidth
c	Velocity of ultrasound in tissue
D	Active Doppler aperture width
f	Centre frequency of an ultrasound burst
f_d	Doppler shift frequency
f_i	Sinusoidal contributions to the Doppler signal
f_r	Received frequency
f_t	Transmitted frequency
f_1 and f_2	Doppler shift frequencies corresponding to two extreme rays from a transducer
F	Weighting factor due to non-uniform insonation, or the focal length of a transducer
H	Haematocrit
m	Measure of size of aggregate scattering unit
$\bar{m}$	Mean of scattering size distribution
N_B	Number of oscillations in a received burst
p	Fractional packing dimension
$P(\phi)$	Power as a function of a scattering angle
r	Radial co-ordinate
R	Radius of a blood vessel
$S_x(f_i)$	Power spectral density function
$S(f_i, t)$	Time varying power spectral density function
t	Time
t_g	Time for which a gate is open
t_p	Length of an ultrasonic pulse
T_t	Transit time of target through Doppler sample volume
v	Velocity of a target
v_{max}	Maximum velocity in a blood vessel
$v(r)$	Velocity profile
V_c	Average volume for one erythrocyte
w	Width of an ultrasound beam
W	Packing factor
$x(t)$	Backscattered Doppler signal
Z_r	Length of Doppler range cell
β_e	Adiabatic compressibility of erythrocyte
β_p	Adiabatic compressibility of plasma
γ	Half-angle subtended by transducer at target
δf	Spectral width of an ultrasound burst
$\Delta \tau$	Angle subtended by a transducer at its focal spot
θ	Angle between ultrasound beam and blood flow
λ	Wavelength of ultrasound
ρ_e	Density of erythrocytes
ρ_p	Density of plasma
σ_b	Backscattering cross-section for single erythrocyte
σ_m	Standard deviation of scattering size distribution
ϕ	Scattering angle
Φ_{bs}	Backscattering coefficient for blood
χ	Half angle of ultrasound beam divergence

7.7 REFERENCES

Aarts PAMM (1988) Blood platelets are concentrated near the wall and red blood cells in the centre of flowing blood. Arteriosclerosis 8:819–824

Ahuja AS (1972) Effect of particle viscosity on propagation of sound in suspensions and emulsions. J Acoust Soc Am 51:182–191

Aldis GK, Thompson RS (1992) Calculation of Doppler spectral power density functions. IEEE Trans Biomed Eng 39:1022–1031

Allard L, Cloutier G, Durand L-G (1996) Effect of the insonation angle on the Doppler backscattered power under red blood cell aggregation conditions. IEEE Trans Ultrason Ferroelec Freq Contr 43:211–219

Angelsen BAJ (1980) A theoretical study of the scattering of ultrasound from blood. IEEE Trans Biomed Eng BME-27:61–67

Angus JC, Edwards RV, Dunning JW (1971) Signal broadening in the laser Doppler velocimeter. AIChE J 17:1509–1510

Atkinson P, Berry MV (1974) Random noise in ultrasonic echoes diffracted by blood. J Phys A: Math Nucl Gen 7:1293–1302

Bacon DR (1984) Finite amplitude distortion of the pulsed fields used in diagnostic ultrasound. Ultrasound Med Biol 10:189–195

Bascom PAJ, Cobbold RSC (1990) Effects of transducer beam geometry and flow velocity profile on the Doppler power spectrum: a theoretical study. Ultrasound Med Biol 16:279–295

Bascom PAJ, Cobbold RSC (1995) On a fractile packing approach for understanding ultrasonic backscattering from blood. J Acoust Soc Am 98:3040–3049

Bascom PAJ, Cobbold RSC (1996) Origin of the Doppler ultrasound spectrum from blood. IEEE Trans Biomed Eng 43:562–571

Bascom PAJ, Cobbold RSC, Roelofs BHM (1986) Influence of spectral broadening on continuous wave Doppler ultrasound spectra: a geometric approach. Ultrasound Med Biol 12:387–395

Bascom PAJ, Cobbold RSC, Routh HF, Johnston KW (1993) On the Doppler signal from a steady flow asymmetrical stenosis model: effects of turbulence. Ultrasound Med Biol 19:197–210

Bascom PAJ, Johnston KW, Cobbold RSC, Ojha M (1997) Relation of the flow field distal to a moderate stenosis to the Doppler power. Ultrasound Med Biol 23:25–39

Censor D, Newhouse VL, Vontz T, Ortega HV (1988) Theory of ultrasound Doppler-spectra velocimetry for arbitrary beam and flow configurations. IEEE Trans Biomed Eng 35:740–751

Cloutier G, Qin Z (1997) Ultrasound backscattering from non-aggregating and aggregating erythrocytes—a review. Biorheology 34:443–470

Cloutier G, Shung KK (1993a) Cyclic variation of the power of ultrasonic signals backscattered by polystyrene microspheres and porcine erythrocyte suspensions. IEEE Trans Biomed Eng 40:953–962

Cloutier G, Shung KK (1993b) Study of red cell aggregation in pulsatile flow from ultrasonic Doppler power measurements. Biorheology 30:433–461

Cloutier G, Allard L, Durand L-G (1995) Changes in ultrasonic Doppler backscattered power downstream of concentric and eccentric stenoses under pulsatile flow. Ultrasound Med Biol 21:59–70

Cloutier G, Qin Z, Durand L-G, Teh BG (1996) Power Doppler ultrasound evaluation of the shear rate and shear stress dependences of red blood cell aggregation. IEEE Trans Biomed Eng 43:441–450

Cobbold RSC, Veltink PH, Johnston KW (1983) Influence of beam profile and degree of insonation on the CW Doppler ultrasound spectrum and mean velocity. IEEE Trans Sonics Ultrason SU-30:364–371

Daigle RJ, Stavros AT, Lee RM (1990) Overestimation of velocity and frequency values by multielement linear array Dopplers. J Vasc Tech 14:206–213

de Kroon MGM, Slager CL, Gussenhoven WJ, Serruys PW, Roelandt JRTC, Bom N (1991) Cyclic changes of blood echogenicity in high-frequency ultrasound. Ultrasound Med Biol 17:723–728

Douville Y, Arenson JW, Johnston KW, Cobbold RSC, Kassam M (1983) Critical evaluation of continuous-wave Doppler probes for carotid studies. J Clin Ultrasound 11:83–90

Duck FA, Starritt HC (1984) Acoustic shock generation by ultrasonic imaging equipment. Br J Radiol 57:231–240

Duck FA, Starritt HC, Aindow JD, Perkins MA, Hawkins AJ (1985) The output of pulse-echo ultrasound equipment: a survey of powers, pressures and intensities. Br J Radiol 58:989–1001

Edwards RV, Angus JC, French MJ, Dunning JW (1971) Spectral analysis of the signal from the laser Doppler flowmeter: time-independent systems. J Appl Phys 42:837–850

Embree PM, O'Brien WD (1990) Pulsed Doppler accuracy assessment due to frequency-dependent attenuation and Rayleigh scattering error sources. IEEE Trans Biomed Eng 37:322–326

Evans DH (1982a) Some aspects of the relationship between instantaneous volumetric blood flow and continuous wave Doppler ultrasound recordings—I. The effect of ultrasonic beam width on the output of maximum, mean and RMS frequency processors. Ultrasound Med Biol 8:605–609

Evans DH (1982b) Some aspects of the relationship between instantaneous volumetric blood flow and continuous wave Doppler ultrasound recording—III. The calculation of Doppler power spectra from mean velocity waveforms, and the results of processing these spectra with maximum, mean, and RMS frequency processors. Ultrasound Med Biol 8:617–623

Evans DH (1985) On the measurement of the mean velocity of blood flow over the cardiac cycle using Doppler ultrasound. Ultrasound Med Biol 11:735–741

Evans DH, Parton L (1981) The directional characteristics of some ultrasonic Doppler blood-flow probes. Ultrasound Med Biol 7:51–62

Ferrara KW, Algazi R, Liu J (1992) The effect of frequency dependent scattering and attenuation on the estimation of blood velocity using ultrasound. IEEE Trans Ultrason Ferroelec Freq Contr 39:754–767

Fontaine I, Cloutier G, Allard L (1997) Non-Gaussian statistical property of the ultrasonic Doppler signal downstream of a severe stenosis. Ultrasound Med Biol 23:41–45

Forsberg F (1989) An assessment of artefacts in Doppler blood flow velocity measurement. PhD Dissertation, Technical University of Denmark, Lyngby.

Green PS (1964) Spectral broadening of acoustic reverberation in Doppler-shift fluid flowmeters. J Acoust Soc Am 36:1383–1390

Griffith JM, Brody WR, Goodman L (1976) Resolution performance of Doppler ultrasound flowmeters. J Acoust Soc Am 60:607–610

Guidi G, Newhouse VL, Tortoli P (1995) Doppler spectrum shape analysis based on the summation of flow-line spectra. IEEE Trans Ultrason Ferroelec Freq Contr 42:907–915

Heimdal A, Torp H (1997) Ultrasound Doppler measurements of low velocity blood flow: Limitations due to clutter signals from vibrating muscles. IEEE Trans Ultrason Ferroelec Freq Contr 44:873–881

Hoskins PR (1996) Accuracy of maximum velocity estimates made using Doppler ultrasound systems. Br J Radiol 69:172–177

Kim YM, Park SB (1989) Modeling of Doppler signal considering sample volume and field distribution. Ultrason Imaging 11:175–196

Kristoffersen K, Angelsen BAJ (1988) A time-shared ultrasound Doppler measurement and 2-D imaging system. IEEE Trans Biomed Eng BME-35:285–295

Li S, McDicken WN, Hoskins PR (1993) Nonlinear propagation in Doppler ultrasound. Ultrasound Med Biol 19:359–364

Lockwood GR, Ryan LK, Hunt JW, Foster FS (1991) Measurement of the ultrasonic properties of vascular tissues and blood from 35–65 MHz. Ultrasound Med Biol 17:653–666

Machi J, Sigel B, Beitler JC, Coelho JCU, Justin JR (1983) Relation of in vivo blood flow to ultrasound echogenicity. J Clin Ultrasound 11:3–10

McArdle A, Newhouse VL (1996) Doppler bandwidth dependence on beam to flow angle. J Acoust Soc Am 99:1767–1778

Mo LYL (1991) A unifying approach to modelling the backscattered Doppler ultrasound from blood. PhD Dissertation, University of Toronto.

Mo LYL, Cobbold RSC (1986a) Speckle in continuous wave Doppler ultrasound spectra: a simulation study. IEEE Trans Ultrason Ferroelec Freq Contr UFFC-33:747–753

Mo LYL, Cobbold RSC (1986b) A stochastic model of the backscattered Doppler ultrasound from blood. IEEE Trans Biomed Eng BME-33:20–27

Mo LYL, Cobbold RSC (1989) A non-stationary signal simulation model for continuous and pulsed Doppler ultrasound. IEEE Trans Ultrason Ferroelec Freq Contr 36:522–530

Mo LYL, Cobbold RSC (1992) A unified approach to modeling the backscattered Doppler ultrasound from blood. IEEE Trans Biomed Eng 39:450–461

Mo LYL, Kuo I-Y, Shung KK, Ceresne L, Cobbold RSC (1994) Ultrasound scattering from blood with hematocrits up to 100%. IEEE Trans Biomed Eng 41:91–95

Moraes R, Aydin N, Evans DH (1995) The performance of three maximum frequency envelope detection algorithms for Doppler signals. J Vasc Invest 1:126–134

Morse PM, Ingard KU (1968) Theoretical Acoustics, p. 427. McGraw-Hill, New York

Narayana PA, Ophir J, Maklad NF (1984) The attenuation of ultrasound in biological fluids. J Acoust Soc Am 76:1–4

Newhouse VL, Reid JM (1990) Invariance of Doppler bandwidth with flow axis displacement. In: McAvoy BR (ed) IEEE Ultrasonics Symp, vol 3, pp. 1533–1536. IEEE, Piscataway

Newhouse VL, Bendick PJ, Varner LW (1976) Analysis of transit time effects on Doppler flow measurement. IEEE Trans Biomed Eng BME-23:381–387

Newhouse VL, Ehrenwald AR, Johnson GF (1977a) The effect of Rayleigh scattering and frequency dependent absorption on the output spectrum of Doppler blood flowmeters. In: White D, Brown RE (eds) Ultrasound in Medicine, vol 3B, pp. 1181–1191. Plenum, New York

Newhouse VL, Johnson FG, Furgason ES (1977b) Transit time and geometrical broadening of Doppler ultrasound spectra. Proceedings of the 30th Annual Conference on Engineering in Medicine and Biology, p. 353

Newhouse VL, Varner LW, Bendick PJ (1977c) Geometrical spectrum broadening in ultrasonic Doppler systems. IEEE Trans Biomed Eng BME-24:478–481

Newhouse VL, Furgason ES, Johnson GF, Wolf DA (1980) The dependence of ultrasound Doppler bandwidth on beam geometry. IEEE Trans Sonics Ultrason SU-27:50–59

Newhouse VL, Censor D, Vontz T, Cisneros JA, Goldberg BB (1987) Ultrasound Doppler probing of flows transverse with respect to beam axis. IEEE Trans Biomed Eng 34:779–789

Ophir J, Jaeger P (1982) Spectral shifts of ultrasonic propagation through media with nonlinear dispersive attenuation. Ultrason Imaging 4:282–289

Powalowski T, Borodzinski K, Nowicki A (1975) Effect of ultrasonic beam width on blood flow estimation by means of CW Doppler flowmeter. Scripta Medica Univ Brno 48:97–103

Reid JM, Klepper J (1984) Frequency errors in pulse Doppler systems (Abstract). Ultrason Imaging 6:212

Reid JM, Sigelmann RA, Nasser MG, Baker DW (1969) The scattering of ultrasound by human blood. Session 10-7. In: Proceedings of the 8th International Conference on Medical and Biological Engineering, Chicago, 1969

Round WH, Bates RHT (1987) Modification of spectra of pulses from ultrasonic transducers by scatterers in non-attenuating and in attenuating media. Ultrason Imaging 9:18–28

Routh HF, Gough W, Williams RP (1987) One-dimensional computer simulation of a wave incident on randomly distributed inhomogeneities with reference to the scattering of ultrasound by blood. Med Biol Eng Comput 25:667–671

Rschevkin SN (1963) A course of Lectures on the Theory of Sound, p. 374. Pergamon, Oxford

Schmid-Schonbein H, Gaehtgens P, Hirsch H (1968) On the

shear rate dependence of red cell aggregation *in vitro*. J Clin Invest 47:1447–1453

Shehada REN, Cobbold RSC, Mo LYL (1994) Aggregation effects in whole blood: influence of time and shear rate measured using ultrasound. Biorheology 31:115–135

Sheldon CD, Duggen TC (1987) Low-cost Doppler signal simulator. Med Biol Eng Comput 25:226–228

Shiga T, Imaizuni K, Harada N, Sekiya M (1983) Kinetics of rouleaux formation using TV image analyzer. I. Human erythrocytes. Am J Physiol 245:252–258

Shung KK (1982) On the ultrasound scattering from blood as a function of hematocrit. IEEE Trans Sonics Ultrason SU-29:327–331

Shung KK, Kuo IY (1994) Analysis of ultrasonic scattering in blood via a continuum approach. Ultrasound Med Biol 20:623–627

Shung KK, Thieme GA (eds) (1993) Ultrasonic Scattering in Biological Tissues. CRC Press, London

Shung KK, Sigelmann RA, Reid JM (1976) Scattering of ultrasound by blood. IEEE Trans Biomed Eng 23:460–467

Shung KK, Sigelmann RA, Reid JM (1977) Angular dependence of scattering of ultrasound from blood. IEEE Trans Biomed Eng BME-24:325–331

Shung KK, Yuan YW, Fei DY, Tarbell JM (1984) Effect of flow disturbance on ultrasonic backscatter from blood. J Acoust Soc Am 75:1265–1272

Shung KK, Cloutier G, Lim CC (1992) The effects of hematocrit, shear rate, and turbulence on ultrasonic Doppler spectrum from blood. IEEE Trans Biomed Eng 39:462–469

Sigel B, Machi J, Beitler JC, Justin JR (1983) Red cell aggregation as a cause of blood-flow echogenicity. Radiology 148:799–802

Skalak R (1984) Aggregation and disaggregation of red blood cells. Biorheology 21:463–476

Strutt JW (1873) Investigation of the disturbance produced by a spherical obstacle on the waves of sound. Proc London Math Soc 4:253–283

Thomas N, Leeman S (1991) Blood velocity estimation in medical imaging—artefacts and errors. In: Ultrasonics International '91, pp. 95–98. Butterworth-Heinemann, Oxford

Thomas N, Deane I, Leeman S (1989) Artefacts in pulsed Doppler measurement. In: Proceedings of the Ultrasonics International '89 Conference, pp. 1179–1185

Thompson RS, Aldis GK (1996) Effect of a cylindrical refracting interface on ultrasound intensity and the CW Doppler spectrum. IEEE Trans Biomed Eng 43:451–459

Thompson RS, Trudinger BJ, Cook CM (1985) Doppler ultrasound waveforms in the fetal umbilical artery: quantitative analysis technique. Ultrasound Med Biol 11:707–718

Thompson RS, Trudinger BJ, Cook CM (1986) A comparison of Doppler ultrasound waveform indices in the umbilical artery. II. Indices derived from the mean velocity and the first moment waveforms. Ultrasound Med Biol 12:845–854

Thompson RS, Aldis GK, Linnett IW (1990) Doppler ultrasound spectral power density distribution: measurement artefacts in steady flow. Med Biol Eng Comput 28:60–66

Thrush AJ, Evans DH (1995) Intrinsic spectral broadening: a potential cause of misdiagnosis of carotid artery disease. J Vasc Invest 1:187–192

Tortoli P, Guidi G, Mariotti V, Newhouse VL (1992) Experimental proof of Doppler bandwidth invariance. IEEE Trans Ultrason Ferroelec Freq Contr 39:196–203

Tortoli P, Guidi G, Pignoli P (1993) Transverse Doppler spectral analysis for a correct interpretation of flow sonograms. Ultrasound Med Biol 19:115–121

Tortoli P, Guidi G, Guidi F, Atzeni C (1994) A review of experimental transverse Doppler studies. IEEE Trans Ultrason Ferroelec Freq Contr 41:84–89

Tortoli P, Berti P, Guidi F, Thompson RS, Aldis GK (1997) Flow imaging with pulsed Doppler ultrasound: Refraction artefacts and dual mode propagation. In: Schneider SC, Levy M, McAvoy BR (eds) Proceedings of the 1997 IEEE Ultrasonics Symposium, pp. 1269–1272. IEEE, Piscataway

Twersky V (1978) Acoustic bulk parameters in distributions of pair-correlated scatterers. J Acoust Soc Am 64:1710–1719

Urick RJ (1947) A sound velocity method for determining the compressibility of finely divided substances. J Appl Phys 18:983–987

van der Heiden MS, de Kroon MGM, Bom N, Borts C (1995) Ultrasound backscatter at 30 MHz from human blood: influence of rouleau size affected by blood modification and shear rate. Ultrasound Med Biol 21:817–826

van Leeuwen GH, Hoeks APG, Reneman RS (1986) Simulation of real-time frequency estimators for pulsed Doppler systems. Ultrason Imaging 8:252–271

Wells PNT (1977) Biomedical Ultrasonics, p. 29. Academic Press, London

Willink RD, Evans DH (1996) The effect of geometrical spectral broadening on the estimation of mean blood velocity using wide and narrow ultrasound beams. IEEE Trans Biomed Eng 43:238–248

Winkler A, Wu J (1995) Correction of intrinsic spectral broadening errors in Doppler peak velocity measurements made with phased sector and linear array transducers. Ultrasound Med Biol 21:1029–1035

Wu SJ, Shung KK (1996) Cyclic variation of Doppler power from whole blood under pulsatile flow. Ultrasound Med Biol 22:883–894

Wu S-j, Reyner J, Shung KK, Routh HF (1995) A study of the feasibility of using power level for detection of turbulence and vessel differentiation in Doppler power imaging. In: Levy M, Schneider SC, McAvoy BR (eds) Proc. 1995 IEEE Ultrasonics Symposium, pp. 1527–1530. IEEE, Piscataway

Wu SJ, Shung KK, Brasseur JG (1998) *In situ* measurements of Doppler power vs. flow turbulence intensity in red cell suspensions. Ultrasound Med Biol 24:1009–1021

Zhang J, Rose JL, Shung KK (1994) A computer model for simulating ultrasonic scattering in biological tissues with high scatterer concentration. Ultrasound Med Biol 20:903–913

8

Doppler Signal Processors: Theoretical Considerations

8.1 INTRODUCTION

The Doppler shift signal that results from the demodulation process contains a wealth of information about blood flow occurring within the sample volume of the Doppler velocimeter. The most complete way to display this information is to perform a full spectral analysis and present results in the form of a sonogram (see for example Fig. 7.2). Whilst such a display may be of great value for assessing the general quality of the signal and for making qualitative statements about the disease state, it contains so much information that some feature reduction must usually take place before quantitative statements may be made. It is therefore usual to extract some kind of envelope signals from the full Doppler shift spectrum, either using analogue processing methods on the raw Doppler signals or using digital methods to process the Fourier-transformed Doppler data. The most useful envelope detectors are probably the mean frequency and maximum frequency processors, but a number of others have been used and may even have some advantages. It has already been shown in Chapter 7 that the underlying Doppler power spectrum is influenced by a number of factors that are unrelated to the velocity distribution in the target blood vessel, and the same must necessarily be true of envelope signals derived from the Doppler spectrum.

In this chapter we first discuss methods for estimating the spectral content of Doppler signals, and then compare the theoretical merits of a number of envelope processors and consider the effects that various distorting factors may have on their output.

8.2 SPECTRAL ESTIMATION TECHNIQUES

Spectral analysis of the Doppler signal has in the past been performed using a variety of equipment, including parallel filter analysers and swept filter analysers of both the analogue and digital time-compression varieties (Atkinson and Woodcock, 1982), but virtually all new instruments now use the fast Fourier transform (FFT) method. There has been much discussion in the literature of alternative spectral estimation techniques for use with Doppler signals, but so far these have only been used in specialist applications.

8.2.1 The Time Interval Histogram

The time interval histogram (TIH) analyser was at one time widely used as a method of estimating the spectral content of Doppler signals (Daigle and Baker 1977, Baker et al 1977). The basic idea behind this method is that the time delay, τ_i, between two adjacent zero-crossings of the Doppler signal at $t = t_i$ and $t = t_{i+1}$ is measured and used to form a frequency estimate, $\hat{f}_i$, given by:

$$\hat{f}_i = \frac{1}{2\tau_i} = \frac{1}{2(t_i - t_{i-1})} \qquad 8.1$$

A sonogram-like display is then formed by plotting t_i along the horizontal axis and $\hat{f}_i$ along the vertical axis of a screen or chart. If the signal is narrow band, then the zero-crossings will occur at regular intervals and all the dots will cluster around its central frequency, but if the signal is wide band, then the zero crossings will occur at irregular intervals and the dots will be widely scattered around the central frequency.

The exact relationship between the TIH and the spectrum of the ultrasonic Doppler signal was explored in a series of papers published in the early 1980s (Angelsen 1980a, Angelsen 1981, Burckhardt 1981, Angelsen and Kristoffersen 1983) which showed that the TIH gives only an estimate of the average frequency and width of the spectrum, with no detailed information about its shape. For this reason the TIH method was rapidly superseded when real-time FFT analysers became readily available.

8.2.2 The Fourier Transform Analyser

The concept of Fourier analysis has already been introduced in Chapter 2 (2.4.1) in the context of finding the spectral content of arterial flow waveforms. The Fourier transform may also be used to estimate the power spectrum of the Doppler signal using the so-called periodogram approach. The periodogram, which was first introduced by Schuster at the end of the nineteenth century (Schuster 1898, 1899) in an attempt to find 'hidden periodicities' in records of the variation of sun spot numbers, takes the form of a histogram whose bin widths correspond to the minimum resolvable frequency interval in the Fourier transform operation (i.e. the reciprocal of the data length), and whose heights or areas correspond to the estimate of power within each interval.

The periodogram estimate of the power spectrum of a time series $x_0, ..., x_{N-1}$ is given by the squared modulus of the discrete Fourier transform (DFT) of the data and may be written (Kay and Marple 1981):

$$\hat{P}_m = \left| \frac{1}{N} \sum_{n=0}^{N-1} x_n \exp(-j2\pi mn/N) \right|^2 \quad 8.2$$

Note that in this formulation it is the height of each bin, rather than its area, that is equal to the estimate of the power for each frequency bin.

The development that made the periodogram approach to spectral estimation so popular in many engineering fields was the development of the fast Fourier transform (FFT) (Cooley and Tukey 1965), which provides an extremely rapid method of calculating the DFT of a data series of length $N = 2^p$ where p is any positive integer (approximately $2N \log_2 N$ arithmetic operations are required to evaluate all N DFT coefficients using the FFT approach, in comparison to the N^2 operations required for the straightforward approach—Cochran et al 1967). The details of the implementation of the FFT are beyond the scope of this book, and have been discussed in detail elsewhere (Cochran et al 1967, Rabiner and Gold 1975, Brigham 1974). Schlindwein et al (1988) have published details of an FFT analyser for Doppler signals based on a personal computer and digital signal processing chip.

Despite its popularity, the periodogram of a stochastic signal has a large variance associated with it and individual estimates may appear very noisy. An example of this is provided by Fig. 8.1a, which is a single estimate of the power spectrum from a test rig containing steady flow.

Oppenheim and Schafer (1975) have shown that for a band-limited Gaussian random process (to which a Doppler signal may be approximated—Angelsen 1980b, Mo and Cobbold 1992) the

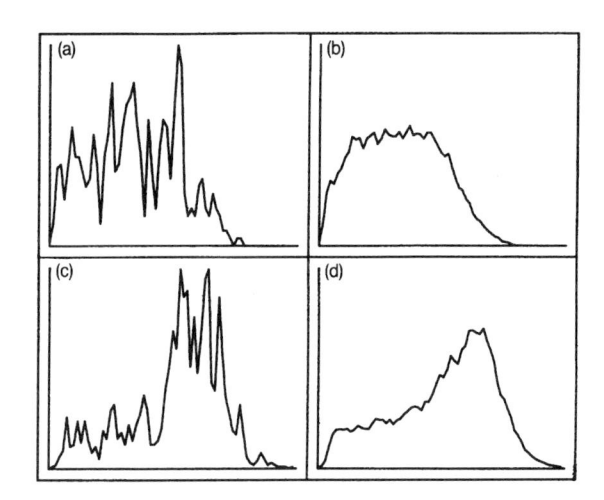

Figure 8.1 Individual and averaged periodogram estimates: (a) individual spectral estimate for steady flow in a tube; (b) Bartlett estimate of power spectrum for the same flow using 64 individual estimates; (c, d) individual and averaged spectral estimates taken from the human carotid artery at peak systole

variance of the spectral estimate may be written:

$$\text{var}[\hat{P}(\omega)] = P(\omega)^2 \left[1 + \left(\frac{\sin \omega N}{N \sin \omega}\right)^2\right] \quad 8.3$$

where $\hat{P}(\omega)$ is the estimated power spectral density (PSD) as a function of ω, the angular frequency ($=2\pi f$); $P(\omega)$ is the true underlying PSD; and N is the number of points in the Fourier transform. Even for large values of N, the variance remains finite and tends to the square of the spectrum, and therefore the estimated PSD will always fluctuate wildly about the true spectrum value. For large N we may write:

$$\text{var}[\hat{P}(w)] \approx P(w)^2 \quad 8.4$$

A more detailed treatment (Priestley 1981) gives the result that for a zero-mean purely random Gaussian process the estimates, other than for the highest and lowest bins, are independently distributed and follow the form of a chi-squared distribution with two degrees of freedom, and each has a variance equal to the square of its mean. For the extreme bins, corresponding to the constant component and the frequency component at the Nyquist frequency, the means are unchanged but the variance is doubled, and follow a chi-squared distribution with one degree of freedom.

It is possible to reduce the variance of the estimated PSD by averaging over a number of independent estimates, a method usually known as Bartlett's procedure (Oppenheim and Schafer 1975). In this case the variance may be written:

$$\text{var}[\hat{P}(\omega)] = \frac{1}{K} P(\omega)^2 \left[1 + \left(\frac{\sin \omega N'}{N' \sin \omega}\right)^2\right] \quad 8.5$$

where K is the number of estimates averaged and N' the number of points in each Fourier transform. For large values of N' this reduces to:

$$\text{var}[\hat{P}(\omega)] \approx P(\omega)^2/K \quad 8.6$$

Thus, the spectral variance of the Bartlett estimate is inversely proportional to the number of raw estimates that are averaged together. Figure 8.1b is a Bartlett estimate of the PSD from the same rig at Fig. 8.1a, with 64 successive individual estimates having been averaged, and shows a dramatic reduction in spectral variance. The spectral resolution of the Fourier transform (i.e. the width of each spectral component) is determined solely by the length of the data sample used in the transform, T_a, and may be written:

$$\Delta f_a = 1/T_a \quad 8.7$$

For a given length of data, T_x, it is possible to trade off spectral resolution against spectral variance. The entire sample may be used for a single estimate, in which case the resolution will be given by $1/T_x$ and the spectral variance by $P(\omega)^2$, or the sample may be split into K segments in which case the resolution will be given by K/T_x and the variance by $P(\omega)^2/K$.

It is an implicit assumption of the Fourier transform that the data are stationary, that is to say their statistical properties do not change over the sample period. This must be the case, otherwise the PSD will vary during the sample and so it will not be possible to estimate properties such as (underlying) spectral width with any accuracy. It is the stationarity condition that limits the length of data T_x that may be used to produce a spectral estimate of a Doppler signal (whether using a signal transform of the entire segment or averaging several transforms). Arterial Doppler signals may not be considered stationary for periods of greater than about 10–20 ms and sometimes less, and therefore it is impossible to obtain a frequency resolution of better than 50–100 Hz, even if the Doppler shift signal has a low frequency. The resolution will become even lower if Bartlett's procedure is used to reduce spectral variance, but it is possible to maintain the resolution and decrease the variance by averaging the transforms of the corresponding portions of signals from a number of heartbeats. This is possible because a longer data segment has been analysed without violating the stationarity condition. Figure 8.1c shows a single spectral estimate taken from a human carotid artery at peak systole, whilst Fig. 8.1d is an average of 64 estimates taken from 64 consecutive heartbeats. Non-stationarity broadening is discussed further in Section 8.2.2.1.

In the preceding discussion it has been implicitly assumed that the data window is rectangular. In practice, spectral analysers usually multiply the data by a non-rectangular weighting function, W_n, in order to reduce spectral leakage to form a 'modified' periodogram estimate given by:

$$\hat{P}'_m = \left| \frac{1}{N} \sum_{n=0}^{N-1} x_n W_n \exp(-j2\pi mn/N) \right|^2 \qquad 8.8$$

The details of windowing are not important in the context of this discussion (see Harris 1978 for a comprehensive survey of data windows), but the use of such windows does slightly modify the theory given above. Windowing reduces the effective data length, and therefore the spectral resolution, without decreasing spectral variance. The reduction in the effective data length does, however, mean that it is possible to average spectral estimates from overlapping segments in order to reduce spectral variance, and this is a technique employed in some spectral analysers. This method was first described by Welch (1967), who showed that by using a window similar to a Hanning window, a reduction in the spectral variance of 11/18 could be achieved using a 50% data overlap.

A possible alternative approach to variance reduction has been suggested by Wells (1988) who noted that if several ultrasonic frequencies were transmitted simultaneously, the uncorrelated Doppler spectra could be averaged to reduce random fluctuations caused by changes in backscattering from the blood. Loupas and Gill (1994) examined a variation of this approach which they called 'multifrequency' Doppler, in which they combined Doppler spectra corresponding to the whole range of frequencies in the transmitted bandwidth, derived using a two-dimensional discrete Fourier transform method (Wilson 1991) (Section 8.2.3.1).

8.2.2.1 Non-stationarity Broadening of the Estimated Doppler Spectrum

The effect of non-stationarity of the Doppler signal on the estimated Doppler spectrum has been addressed by a number of authors. Kikkawa et al (1987) studied the problem theoretically and concluded that for Doppler signals from the ascending aorta, the non-stationary broadening during the acceleration phase could be 'very large'. Fish (1991) derived a theoretical expression for the total broadening occurring when Gaussian windows are used in conjunction with the Fourier transform, and then derived results for other window shapes numerically. Guo et al (1993a) studied the stationarity of Doppler signals from the left ventricular outflow tract. They found that 82% of segments of 10 ms or less could be regarded as stationary, whilst less than 75% of segments with a duration of 40 ms could be regarded as such, and concluded that a 10 ms window is a good choice but that a shorter time segment would be preferable. Cloutier et al (1993) used a pulsatile laminar flow model to study various aspects of Doppler spectra, and concluded that for a window of 2.5 ms, broadening due to spectral leakage dominates broadening resulting from non-stationarity, but that for windows of 5 and 10 ms duration the window broadening becomes less significant.

The expression derived by Fish for the root mean square (RMS) width, σ_m, of the total broadening function due to non-stationarity and windowing with a Gaussian function, is:

$$\sigma_m = \{\sigma_{wb}^2 + \sigma_{fm}^2\}^{1/2} \qquad 8.9$$

where σ_{wb}^2 and σ_{fm}^2 are, respectively, the mean square width of the broadening due to the window, and the mean square width due to the frequency variation given by:

$$\sigma_{wb}^2 = 1/8\pi^2 \sigma_t^2 \qquad 8.10$$

and

$$\sigma_{fm}^2 = \sigma_t^2 \beta^2 / 2 \qquad 8.11$$

where σ_t is the RMS width of the Gaussian window (in ms), and β the rate of modulation (in kHz/ms).

Equation 8.9 is plotted in Fig. 8.2, where it can be seen that there is an optimal window length, depending on the rate of modulation, that provides the best spectral resolution. For shorter windows, broadening due to the window function increases rapidly, whilst for longer windows, non-stationary

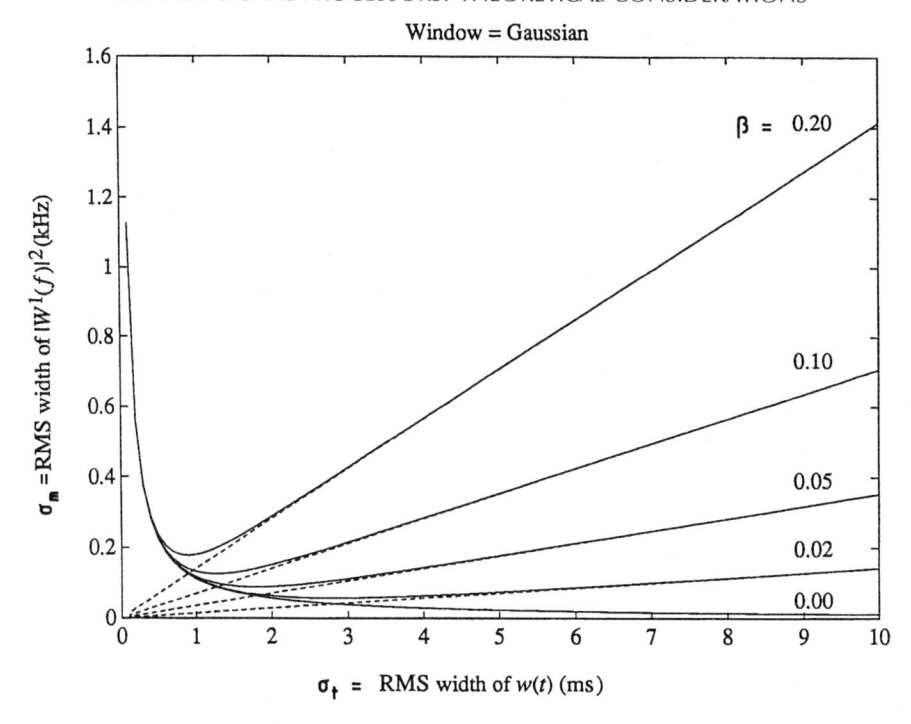

Figure 8.2 The RMS width (σ_m) of the broadening function (due to both the window function and the non-stationarity of the signal), plotted against the RMS width (σ_t) of a Gaussian window for a range of frequency modulation rates (β) (reproduced by permission of Elsevier Science, from Fish 1991, *Ultrasound in Medicine and Biology*, © World Federation of Ultrasound in Medicine and Biology)

broadening becomes progressively more important. Figure 8.3, reproduced from Fish (1991), shows similar results for the Hanning window, which is frequently used in FFT analysers. Wang and Fish (1997) have suggested a method of correcting for non-stationarity and window broadening.

8.2.2.2 Fourier Transform Analyser— Summary

The important properties of the Fourier transform analyser may be summarised as follows:

1. The best spectral resolution obtainable from a transform is given by the reciprocal of the data segment length used. If, for example, the transform is carried out on a 5 ms data segment, the resolution can be no better than 200 Hz. In practice the data are usually 'windowed' to prevent spectral leakage and this reduces the effective data length, and hence spectral resolution, further. The maximum data segment length that may be legitimately used is determined by the stationary period of the signal (the period for which none of its statistics changes significantly).

2. The maximum frequency component that can be detected by analysers operating on a real signal is half the sampling frequency of the analyser. Frequencies above this value must be removed at the input by an analogue low pass filter, otherwise they will be 'aliased' down into the working frequency range (i.e. they are misinterpreted as falling in the analysis range). In the case of analysers operating on complex signals (i.e. using two quadrature components) the frequency of neither signal must exceed the sampling rate.

3. The Fourier transform of a random signal is merely an estimate of the true underlying spectrum and in particular has a large variance.

SPECTRAL ESTIMATION TECHNIQUES

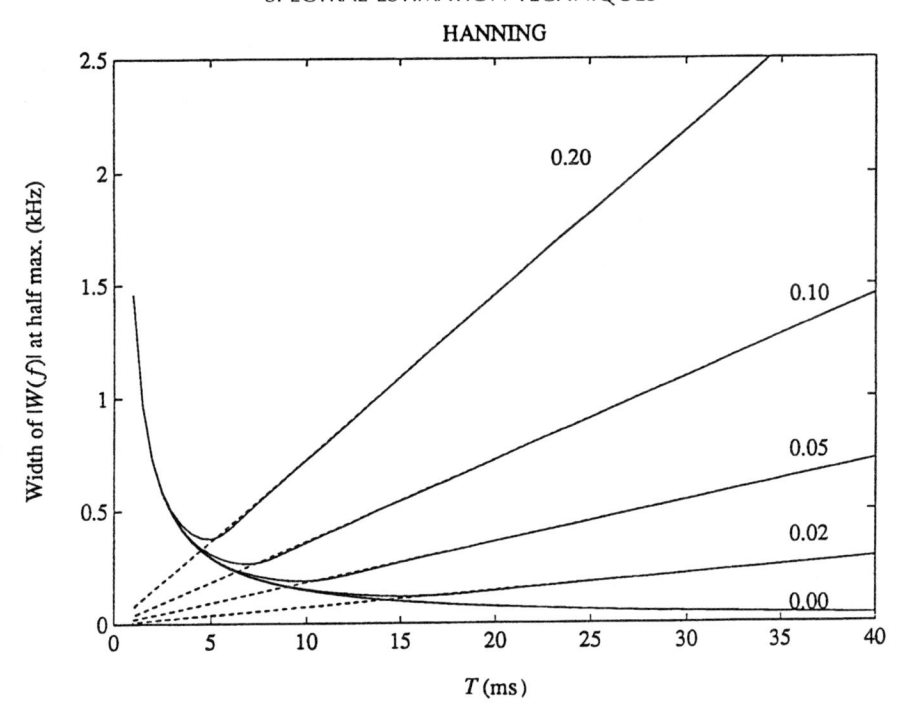

HANNING

Figure 8.3 The width at half-maximum of the broadening function, plotted against the duration (T) of a Hanning window for a range of frequency modulation rates (β) (reproduced by permission of Elsevier Science, from Fish 1991, *Ultrasound in Medicine and Biology*, © World Federation of Ultrasound in Medicine and Biology)

This variance may be reduced by averaging a number of independent estimates using Bartlett's procedure.

Most of the limitations discussed above are common to both analogue and digital implementations of a periodogram or a modified periodogram; however, analogue systems do not suffer from aliasing problems. Despite the problems with the periodogram method, it is by far the most popular approach to spectral estimation of Doppler signals, and sonograms (Section 7.1) produced using FFT analysers are to be found throughout this book (see for example Fig. 9.2).

8.2.3 Two-dimensional Fourier Transform-based Methods

Traditionally the Doppler signal obtained from a single direction through the tissue is viewed as a one-dimensional (1-D) signal containing a series of in-phase and quadrature samples (the real and imaginary parts of the demodulated signal) which vary with time. In the case of pulsed wave Doppler techniques, however, either the RF signal or indeed the demodulated quadrature components may be viewed as two-dimensional (2-D) signals, where the two dimensions are depth (sometimes called fast time) and line number (or slow time), (see Fig. 8.4). This concept leads to a number of different ways of processing the Doppler signal of which some have great potential in the context of colour flow mapping, and these will be discussed in Chapter 11. However, at least two techniques have been described which have potential for improving the spectral estimation of pulsed wave Doppler signals in general and so are considered in this chapter. These two specific techniques, 'multi-frequency Doppler' and 'velocity-matched spectral analysis' are dealt with in Sections 8.2.3.1 and 8.2.3.2, respectively, but we will first introduce the concept of the 2-D Fourier transform as applied to Doppler signals.

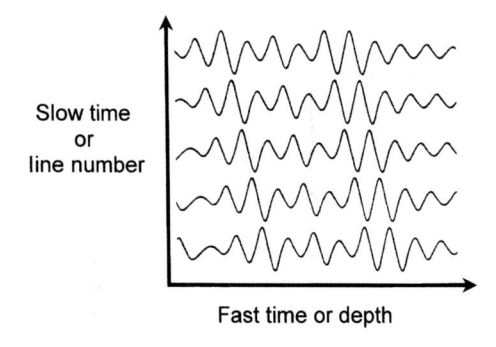

Figure 8.4 Representation of the received RF signal in a pulsed Doppler unit (from a single direction) as a two-dimensional array, where one dimension corresponds to depth in the tissue or 'fast time' and the second dimension corresponds to the a-line number or 'slow time'

The 2-D discrete Fourier transform is well known in image processing and other 2-D signal processing applications, and the 2-D periodogram corresponding to the conventional 1-D version given by eqn. 8.2 may be written (Lim 1990):

$$\hat{P}(m_1, m_2) = \left| \frac{1}{N_1 N_2} \sum_{n_1=0}^{N_1-1} \sum_{n_2=0}^{N_2-1} x(n_1, n_2) \right.$$
$$\times \exp(-j2\pi m_1 n_1/N_1)$$
$$\left. \times \exp(-j2\pi m_2 n_2/N_2) \right|^2 \quad 8.12$$

where the symbols are the same as in eqn. 8.2, except that $\hat{P}$ and x are now 2-D variables.

The value of the 2-D Fourier transform as a descriptor of Doppler signals was first demonstrated by Mayo and Embree (1990) and Wilson (1991), who showed that the axes of the 2-D Fourier transform of a signal such as that shown in Fig. 8.4 correspond to a slow frequency (the Doppler shift frequency) and a fast frequency (the received pulse radio frequency) (see Fig. 8.5). In an idealised case in which the transmitted signal is a pure sine wave which is reflected from a line of particles all moving with the same velocity, then the 2-D spectrum would be a delta function at (f_t, f_d) where f_t and f_d are the transmitted and Doppler shift frequencies, respectively (Fig. 8.5a). In practice, pulsed wave systems necessarily transmit a range of frequencies and in this case, as shown by Wilson (1991), constant velocity Doppler shifts are mapped as radial lines passing through the origin with a slope proportional to the velocity of the moving medium (Fig. 8.5b). Spectral broadening due to the finite sample volume length (Section 7.4.3) causes a frequency spread in the RF direction, and this leads to an effective broadening in the Doppler frequency direction, even if the observation time is large (Fig. 8.5c). Likewise, spectral broadening associated with a finite observation time (Section 8.2.2) will lead to frequency spread in the Doppler frequency direction and thus also in the RF direction, even for large sample volume lengths (Wilson 1991). The net effect of these mechanisms is that echoes of wide band pulses which are scattered from moving targets are mapped as generally elliptical shapes in 2-D Fourier space (Fig. 8.5d). As pointed out by Mayo and Embree (1990), the angle between the major axis of the ellipse and the RF axis is a high-quality measure of the velocity of the scattering medium (note that $\tan \delta = 2v/c$, where v is the scatterer velocity and c the ultrasound velocity). The effect of aliasing on the 2-D Fourier transform is particularly interesting, in that the aliased replicas do not form radial lines passing through the origin, and this property is the basis of use of 'velocity-matched spectral analysis', which is intended to improve the analysis of aliased signals (see Section 8.2.3.2). The reader is referred to Wilson's paper for further discussion of the 2-D Fourier representation of these and other classical Doppler phenomena, whilst Loupas and Gill (1994) provides a useful summary of the statistics of the 2-D transform.

The relationship between the conventional 1-D Fourier transform and the 2-D transform has been addressed by a number of authors, including Mayo and Embree (1990), Wilson (1991), Jones (1993) and Loupas and Gill (1994), who show that conventional narrow band processing is equivalent to an evaluation of the 2-D Fourier transform along the line given by $f_{RF} = f_c$, where f_c is the centre frequency of the transmitted ultrasound pulse. This is illustrated in Fig. 8.6. More formally, the conventional 1-D Fourier spectrum $P_{1D}(f_d)$ may be written:

$$P_{1D}(f_d) = P_{2D}(f_{RF} = f_c, f_d) \quad 8.13$$

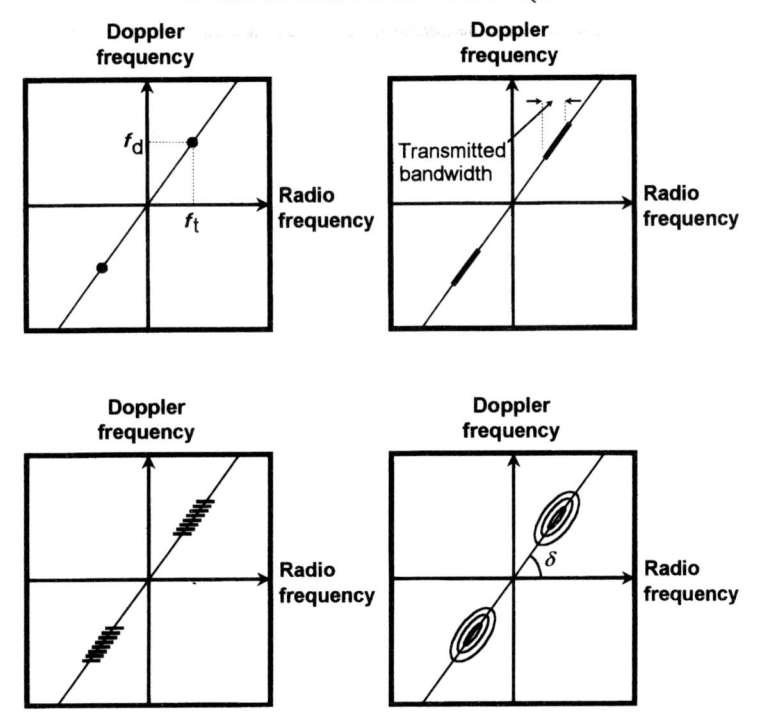

Figure 8.5 2-D Fourier transform representations of various Doppler signals. (a) Idealised case where the transmitted signal is a pure sine wave which is reflected from a line of particles all moving with the same velocity. (b) As for (a), but for a pulsed Doppler system with a finite band width. (c) The effect of spectral broadening due to a finite sample volume length on (b). (d) General representation of a pulsed wave Doppler signal from a line of particles all moving with a single velocity, which takes into account the bandwidth of the transmitted signal and spectral broadening in the Doppler and RF directions

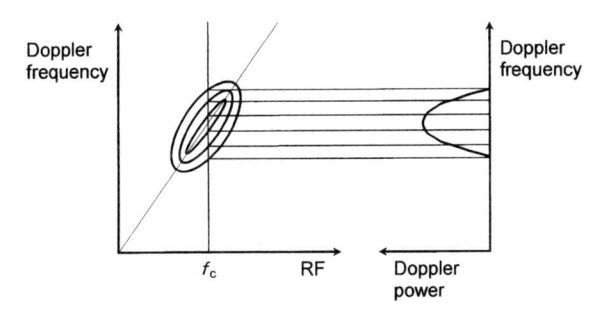

Figure 8.6 The conventional 1-D Fourier transform of a Doppler signal is equivalent to an evaluation of the 2-D transform along the line given by $f_{RF} = f_c$

Note that in the idealised case, where the transmitted signal is a pure sine wave, all the information about the Doppler signal is contained in the 1-D spectrum, but that for all other cases additional information is contained in the 2-D spectrum. This information is utilised by the techniques described in the next two sections.

8.2.3.1 Multifrequency Doppler

It was shown in Section 8.2.2 that the variance of the estimated power spectral density of a Doppler signal found using the periodogram approach is very considerable. One possible way to reduce this variance would be to average the uncorrelated Doppler spectra from several ultrasound frequencies transmitted simultaneously (Wells, 1988). The 'multifrequency Doppler' approach, suggested by Loupas and Gill (1994), is a variation on this method, where the full information contained in the back-scattered RF echoes (corresponding to the whole range of frequencies in the transmitted band-width) is used by combining properly scaled Doppler spectra corresponding to each transmitted frequency component. These spectra are obtained by taking a series of slices through the 2-D Fourier transform at different values of f_{RF} from $f_c - BW/2$ to $f_c + BW/2$, where BW is the RF bandwidth of the transmitted pulse.

Modifying slightly the notation adopted by Loupas and Gill, the multifrequency Doppler power spectral estimate, $\hat{P}_{MF}(f_d)$, may be written:

$$\hat{P}_{MF}(f_d) = \frac{\int_{f_c - BW/2}^{f_c + BW/2} P(f_{RF}, f_d f_{RF}/f_c) df_{RF}}{\int_{f_c - BW/2}^{f_c + BW/2} |U(f_{RF} - f_c)|^2 df_{RF}} \quad 8.14$$

where $U(f)$ is the Fourier transform of the pulse's complex envelope. Note that the contribution to the integral in the numerator at $f_{RF} = f_c$ is the standard 1-D Fourier spectrum (cf. eqn. 8.13), and that contributions to the spectrum from RF frequencies of greater than f_c will be compressed, whilst those from RF frequencies of less than f_c will be expanded so that all the components due to a given velocity contribute to the multifrequency Doppler spectrum at the same value of f_d. An alternative way of viewing eqn. 8.14 is as an integral of the 2-D spectral values along radial lines of the frequency plane, and in this sense eqn. 8.14 is very similar to the formula originally proposed by Mayo and Embree (Loupas and Gill, 1994).

The integral in the denominator of eqn. 8.14 is included simply to normalise the results in terms of the pulse's power in the integration band.

Loupas and Gill (1994) analyse the multifrequency Doppler technique in some detail and show, both by theoretical argument and by simulation, that multifrequency sonograms have consistently higher spectral SNR than the conventional single RF frequency approach.

8.2.3.2 Velocity-Matched Spectrum Analysis

In 1995 Torp and Kristoffersen proposed a new method for suppressing velocity ambiguity in pulsed-wave Doppler which they named 'velocity-matched spectrum analysis'. It will transpire that, although their motivation and approach were very different from those of Loupas and Gill (1994), the final results are effectively identical.

The starting point for the velocity-matched (VM) method is the observation that by treating the Doppler signal as a 2-D signal (Fig. 8.4), and by simultaneously processing several data samples from a range of depths, the movement of the scatterers along the ultrasonic beam can be tracked from pulse to pulse. The advantage of this is that the correlation length of the signal component from targets with a particular velocity will increase when that velocity matches the 'expected' velocity. This principle is illustrated in Fig. 8.7, where it can be seen that when processing is carried out along line number 1 (which corresponds to the velocity of the single target in this example), the correlation length is at a maximum, but that when processing is carried out along line number 2, the correlation length is much shorter. If, however, there were targets moving with a velocity corresponding to the gradient of line 2, then clearly the correlation length for these targets would be at a maximum along this line. Obviously there is a unique gradient corresponding to any given velocity, and for an extended target like blood the Doppler power corresponding to that velocity can be calculated by taking only samples along a line with that unique gradient.

Hence, in the VM method, rather than forming the power spectrum by taking samples from a constant depth, the samples for estimating the power at each Doppler frequency are gathered

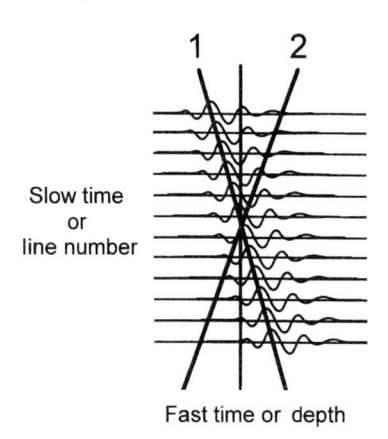

Figure 8.7 The correlation length is at a maximum for a given velocity component if samples are gathered along a skewed line which tracks that particular velocity (line 1)

along skewed lines, such that the correlation length is maximised for each velocity (Fig. 8.8). Using the notation of Torp and Kristoffersen, the VM spectrum can be written:

$$\hat{P}(v) = \left| \sum_k w(k) x(t_0 + k\delta, k_0 + k) \exp(j\omega_0 k\delta) \right|^2 \qquad 8.15$$

where $w(k)$ denotes a window function, $x(t, k)$ is the complex base-band signal envelope where t is the elapsed time from pulse transmission k, and $\delta = 2Tv/c$, where T is the pulse repetition period, v the target velocity, and c the ultrasound velocity. This can be contrasted with the conventional spectrum obtained from samples from a constant depth which may, using the same notation, be written:

$$\hat{P}_{\text{conv.}}(v) = \left| \sum_k w(k) x(t_0, k_0 + k) \exp(j\omega_0 k\delta) \right|^2 \qquad 8.16$$

Turning now to the interpretation of the above calculation in terms of the 2-D Fourier transform, it transpires that the VM method is equivalent to integrating the 2-D power spectrum along lines through the origin for a number of pre-defined velocity values (Torp and Kristoffersen, 1995), which is, of course, exactly the same principle used by Mayo and Embree (1990) and Wilson (1991) for improving mean velocity estimation and by Loupas and Gill (1994) for improving spectral estimation.

The particularly interesting aspect of Torp and Kristoffersen's paper is that they were able to show theoretically (and experimentally for laminar flow with velocities of up to 1.2 ms^{-1}) that the VM method can help suppress the effects of aliasing, and that the method is also immune to the effects of frequency-dependent attenuation (Section 8.3.5). The reason why aliasing is partially suppressed is because VM Doppler makes use of information from all the frequencies present in the transmitted pulse, and whilst lines on the 2-D Fourier transform corresponding to the 'true' Doppler shift from a single velocity can be extrapolated to pass through the origin, aliased replicas do not. This is illustrated in Fig. 8.9. It can be seen that the aliased replica maintains the same gradient because, whilst it is displaced in the 'Doppler shift direction', it is not displaced in the 'RF direction' and therefore no longer passes through the origin. Note that with the conventional Doppler spectrum, formed by taking a single slice through the 2-D transform, it is impossible to distinguish the aliased replica from the original; with the VM Doppler spectrum, however, which is formed by integrating the information along a series of radial slices, the aliased replica is smeared out over a range of velocities, whilst the power from the true Doppler shift is concentrated at the correct frequency.

An additional reason why VM Doppler improves the appearance of aliased spectra is that the high pass filters do not remove all the power at multiples of the sampling frequency. The reason for this can also be seen by referring to

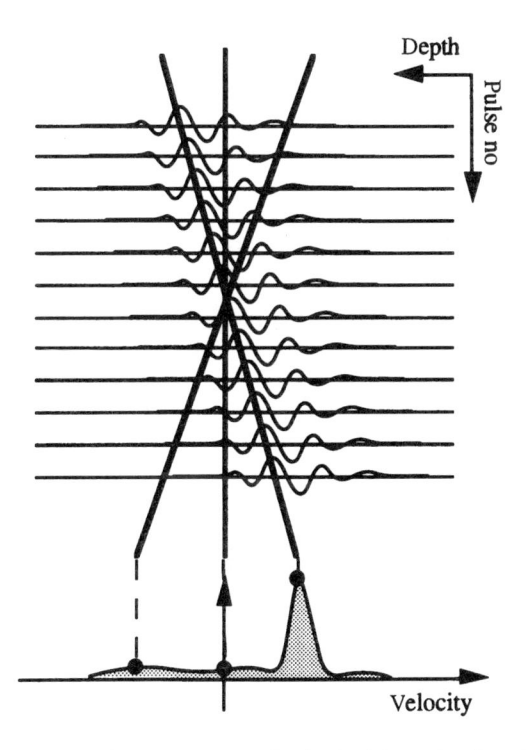

Figure 8.8 The VM spectrum is obtained by estimating the power at each Doppler frequency from skewed lines through the 2-D RF data, such that the correlation length is maximised for each velocity (reproduced by permission of Elsevier Science, from Torp and Kristoffersen 1995, *Ultrasound in Medicine and Biology*, © World Federation of Ultrasound in Medicine and Biology)

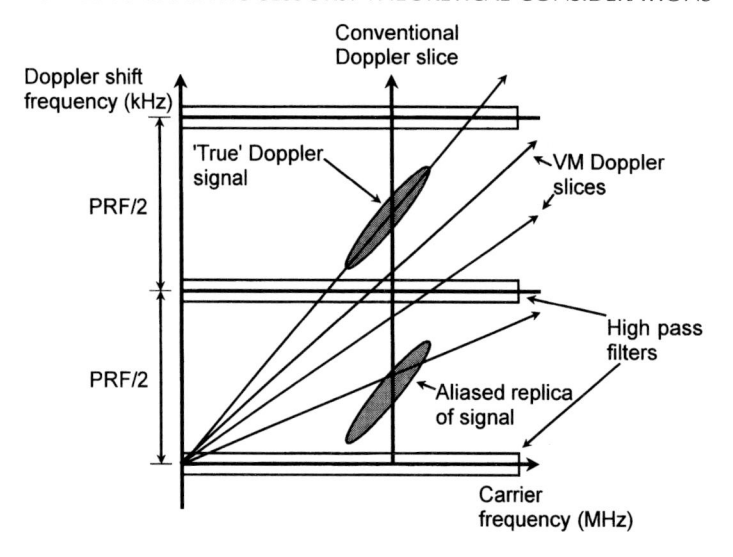

Figure 8.9 Diagram to illustrate the reasons why VM spectral analysis partially overcomes aliasing. See text for details

Fig. 8.9; it can be seen that with VM Doppler, because of the slope of the Doppler signal ellipse, even if part of the signal falls within the filter stop band, other parts of the signal will be outside this band. Figure 8.10, reproduced from Torp and Kristoffersen (1995), compares conventional spectral analysis and VM spectral analysis of the Doppler signal from a subclavian artery where the peak velocity is 4.5 times the Nyquist limit. Whilst the method cannot completely remove aliased spectra, the true Doppler spectrum can easily be seen and its outline extracted by manual if not automatic means.

VM Doppler is immune from the effects of frequency-dependent attenuation because any downshift in the pulse centre frequency simply moves the radial lines, representing the power from each velocity component in 2-D Fourier space, towards the origin and does not change their gradients. A similar argument holds for the effects of frequency-dependent scattering, although in this case the components move away from the origin.

8.2.4 Parametric Spectral Estimation Techniques (AR and ARMA)

The limitations of the periodogram approach to spectral estimation discussed in the previous section has led investigators to explore alternative methods of spectral estimation for Doppler signals. By far the most studied of these approaches are the parametric spectral estimation techniques which have become so popular in many engineering fields over the last 20 or so years. An excellent introduction to these methods is provided by Kay and Marple (1981), and an interesting historical reference by Yule (1927). Vaitkus and Cobbold (1988) have reviewed parametric spectral estimation techniques in the context of Doppler ultrasound signals. The major advantage of these techniques is that they permit accurate estimation of frequency content from a much shorter segment of data and therefore the trade-off between temporal and spectral resolution is not as limiting as for Fourier techniques, and signals which are rapidly changing are more easily interpreted.

As pointed out by Kay and Marple (1981), a useful approach to spectral estimation in general is to consider it in the context of modelling. Use of *a priori* information or assumptions about the data to be analysed permits the selection of an appropriate model for the underlying process, which may then allow a better spectral estimate based on that model. Viewed in this way, spectrum analysis becomes a three-step procedure:

1. Select a time series model.

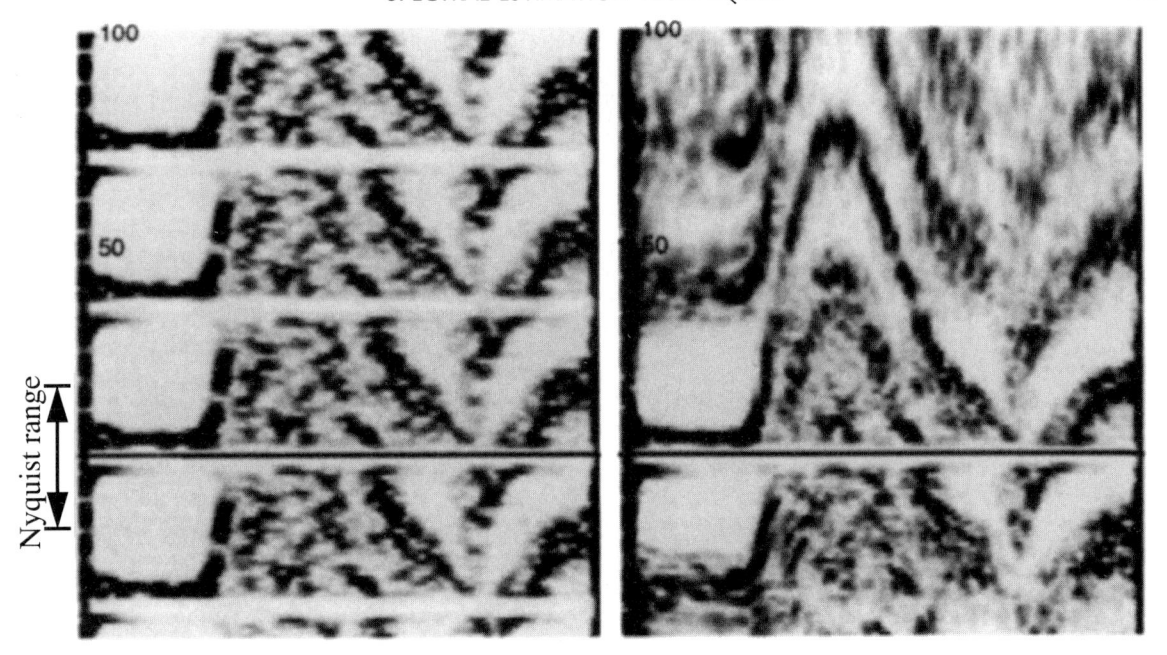

Figure 8.10 VM spectral analysis (right) compared to conventional spectral analysis (left) applied to subclavian artery blood flow with a peak velocity of 4.5 times the Nyquist limit. The velocity is labelled in cm s^{-1} (reproduced by permission of Elsevier Science, from Torp and Kristoffersen 1995, *Ultrasound in Medicine and Biology*, © World Federation of Ultrasound in Medicine and Biology)

2. Estimate the parameters of the assumed model from the available data samples.
3. Obtain the spectral estimate by substituting the estimated model parameters into the theoretical power spectral density (PSD) implied by the model.

In these terms the underlying reason for the shortcomings of the periodogram approach to spectral estimation of Doppler signals is that the model of the data implicitly assumed (i.e. a set of harmonically related complex sinusoids) is inappropriate for a Gaussian random process. A better model for such a stochastic process is known to be a rational transfer model, in which it is assumed that the input driving sequence $u(n)$ and the output signal $x(n)$ are related by the linear difference equation:

$$x(n) = -\sum_{k=1}^{p} a_k x(n-k) + \sum_{k=0}^{q} b_k u(n-k) \quad 8.17$$

This model is the well-known autoregressive moving average model of order p, q (ARMA(p, q)), where a_k and b_k denote, respectively, the autoregressive (AR) and moving average (MA) coefficients. If the input driving signals for such a model can be assumed to be a white noise process of zero mean and variance σ^2, then the PSD of $x(n)$ can be written (Kay and Marple 1981, Marple 1987):

$$P_{\text{ARMA}}(f) = \sigma^2 \Delta t \left| \frac{1 + \sum_{k=1}^{q} b_k \exp(-j2\pi f k \Delta t)}{1 + \sum_{k=1}^{p} a_k \exp(-j2\pi f k \Delta t)} \right|^2$$

8.18

where Δt is the sampling period.

Two special cases of ARMA spectral analysis are of particular interest. When all values of a_k ($k = 1$ to q) are equal to zero, then:

$$x(n) = \sum_{k=0}^{q} b_k u(n-k) \quad 8.19$$

and the process is strictly a moving average of order q (sometimes called an all-zero) model, and the PSD is given by

$$P_{MA}(f) = \sigma^2 \Delta t \left| 1 + \sum_{k=1}^{q} b_k \exp(-j2\pi f k \Delta t) \right|^2 \quad 8.20$$

When all the values of b_k except $b_0 = 1$ are equal to zero, then:

$$x(n) = -\sum_{k=1}^{p} a_k x(n-k) + u(n) \quad 8.21$$

and the process is strictly an autoregression of order p. With this model, sometimes termed an 'all pole model', the present value of the process is expressed as a weighted sum of past values plus a noise term. The corresponding PSD is given by:

$$P_{AR}(f) = \frac{\sigma^2 \Delta t}{\left| 1 + \sum_{k=1}^{p} a_k \exp(-j2\pi f k \Delta t) \right|^2} \quad 8.22$$

The important Wold decomposition theorem relates the ARMA, MA and AR models and states that any stationary ARMA or MA process of finite variance can be approximated by an AR model of higher (possibly infinite) order (Kay and Marple 1981). Since the estimation of parameters for an AR model results in linear equations, it has considerable computational advantages over ARMA and MA parameter estimation techniques, and by far the largest proportion of research effort on rational transform modelling in both the general signal processing field and in Doppler ultrasound spectral estimation has been concerned with the AR model.

In order to calculate the AR spectral estimate of a Doppler ultrasound signal, it is of course necessary to calculate the p AR coefficients together with σ^2. There are a number of approaches to this problem (which may result in different spectra), including methods which estimate the coefficients from the autocorrelation function of the data (such as the Yule–Walker method) and methods which directly estimate the AR coefficients from the raw data (such as the Burg maximum entropy algorithm). The details of these various methods are beyond the scope of this book, and the reader is referred to the literature for further information (Kay and Marple 1981, Vaitkus and Cobbold 1988, Schlindwein and Evans 1992). It is relevant to note, however, that the estimation of the coefficients and the subsequent calculation of the AR spectral estimates is relatively efficient in computational terms and can be achieved in real time for Doppler ultrasound signals with relatively moderate hardware requirements (Schlindwein and Evans 1989). Ruano et al (1993) have published details of parallel computing implementations of an AR spectral estimator for Doppler ultrasound signals. Jensen and colleagues (Jensen et al 1996, Stetson and Jensen 1997) have described an AR approach to Doppler spectral estimation using a lattice filter method. This approach has the theoretical advantage of being able to adapt more rapidly to non-stationary signals than the more classical implementations.

A further important estimator which is closely related to the AR estimator is the so-called maximum likelihood (ML), minimum variance, or Capon method. In this method the PSD is estimated by effectively measuring the power out of a set of narrow-band filters (Kay and Marple 1981, Lacoss 1971). The difference between the ML spectral estimate and the periodogram estimate is that the shapes of the narrow-band filters are, in general, different for each frequency in the former case, whilst they are fixed in the latter. Once again the details of the approach will not be given here and the reader is referred to Kay and Marple (1981) and Vaitkus and Cobbold (1988) for further information. In practice, the ML estimator exhibits more resolution than the periodogram and less than the AR estimator (Lacoss 1971) but less variance than either (Kay and Marple 1981, Baggeroer 1976).

The ML and AR estimators have been related analytically as follows (Kay and Marple 1981, Burg 1972):

$$\frac{1}{\hat{P}_{ML}(f)} = \frac{1}{p} \sum_{m=1}^{p} \frac{1}{\hat{P}_{AR(m)}(f)} \quad 8.23$$

where $\hat{P}_{AR(m)}(f)$ is the estimate of the AR PSD for the mth order model and $\hat{P}_{ML}(f)$ is the estimate of the ML PSD, both based on a known autocorrelation matrix of order p.

The use of AR spectral estimation for analysing Doppler flow signals appears to have been introduced by D'Luna and Newhouse (1981), who were interested in the detection of vortices as a method of detecting mild stenoses. They used an *in vitro* model to investigate the possibility of such a method, and because of the rapid changes in Doppler shift frequency as the vortices crossed the ultrasound beam, they needed to obtain spectral estimates over relatively short time intervals. This precluded the use of Fourier analysis because of the inherent trade-off between time and frequency resolution and so they turned to the AR technique, which overcomes this limitation to some extent. Kitney and Giddens (1986) also used AR techniques in an attempt to identify specific structures in the Doppler signals, such as those resulting from shed vortices, and mentioned the potential for detecting mild and moderate degrees of stenosis. The following year, Kaluzynski (1987) compared the merits of FFT and AR spectral estimation of Doppler signals and found the latter method to give better results when analysing data from small sample volumes.

Since that time there have been a steady stream of papers on AR (and ARMA) techniques applied to Doppler spectral analysis, most of which have compared FFT- and AR-based estimates of the Doppler spectrum in different circumstances and with different problems in mind. Vaitkus and colleagues followed their previously mentioned review of spectral estimation techniques (Vaitkus et al 1988) with a comparative study of FFT, ARMA, AR, MA and ML models on simulated Doppler signals, and also gave examples of the application of the first three of these estimates to CW Doppler signals recorded from a normal internal carotid artery (Vaitkus et al 1988) (see Fig. 8.11). Kaluzynski and colleagues published a series of papers (Kaluzynski and Tedgui 1989, Kaluzynski 1989b, Kaluzynski and Palko 1993) in which they compared various Doppler spectral indices (mean frequency, maximum frequency, spectral broadening indices, coefficient of skewness and indices of turbulence) derived using different estimation methods (including FFT, AR, and ML) under different analysis conditions. Other papers which have tackled the effect of using different spectral estimators with Doppler signals include: Markou and Ku (1991), who compared the performance of FFT, AR and ML estimators for measuring fluid velocity under well-defined flow conditions *in vitro*; David et al (1991), who compared the relative merits of the same methods for the estimation of flow velocity and Doppler spectra *in vitro*; Fan and Evans (1994a), who studied the differences in the power structures of FFT and AR estimates of simulated Doppler signals; Guo et al (1994a), who compared the use of FFT and AR (and the Bessel distribution—see Section 8.2.5) in a pattern recognition system designed to classify lower limb arterial stenoses; Marasek and Nowicki (1994), who compared the performance of FFT, AR and ARMA for maximum frequency estimation of Doppler spectra; Giovannelli et al (1996), who compared a Bayesian approach to determining the AR coefficients with the more usual least squares method on simulated Doppler signals; and Wang and Fish (1996), who compared FFT and AR (and Wigner–Ville and Choi–Williams distributions—see Section 8.2.5) for velocity waveform, turbulence and vortex measurement in a simulation study.

Another important area of study has been that of order selection for AR models of Doppler spectra, since model order can have a profound influence on estimated spectral shape. Kaluzynski (1989a) investigated AR order selection in signals recorded from the carotid arteries of dogs and concluded that there was a major problem in choosing the best order, as existing objective criteria are not applicable due to the nature of the Doppler signal. Schlindwein and Evans (1990) tested four established model order estimation procedures on various human arterial waveforms and on simulated data derived from arterial waveforms, and showed that none worked well for short segments of Doppler signals, but that overestimating the AR model order introduced less error than underestimating it, and suggested the use of an empirical fixed order even though 'true' model order will change throughout the cardiac cycle. Guo et al (1993b) investigated the effect of AR model order on spectral estimates of cardiac Doppler signals taken from 20 patients and, contrary to the two previous studies, found

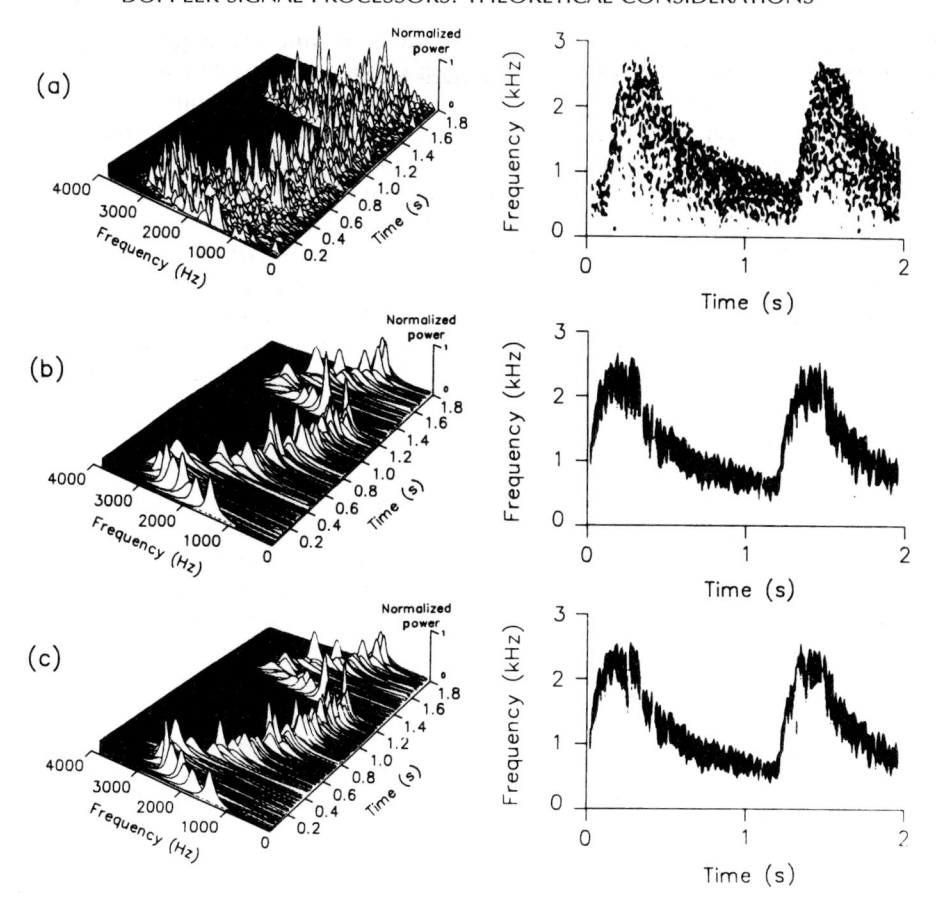

Figure 8.11 Comparison of (a) FFT, (b) ARMA, and (c) AR methods for a CW Doppler signal from a normal internal carotid artery for a 10 ms data window. Both the 3-D and the standard 2-D sonograms are shown (reproduced by permission of Elsevier Science, from Vaitkus et al 1988, *Ultrasound in Medicine and Biology*, © World Federation of Ultrasound in Medicine and Biology)

objective model order determination to be of value, possibly due to their use of many more samples for the estimate. Fort et al (1995) have suggested a new automatic order selection method for use with Doppler signals, which they report to overcome the problems associated with model order selection from short signal segments.

Both Talhami and Kitney (1988) and Forsberg (1991) have specifically examined the use of ARMA modelling for spectral estimation of ultrasound Doppler signals. The former authors used a Kalman filter (based on an ARMA model) to attempt to track oscillations in the modal Doppler frequency during the deceleration phase of pulsatile flow from patients with stenoses in the common carotid artery—apparently with some success. The latter author examined the singular value decomposition ARMA approach, but concluded that it could not be considered suitable for real-time spectral estimation of Doppler ultrasound signals because of the problems of choosing the model order and the requirement to alter algorithm parameters throughout the cardiac cycle.

Parametric spectral estimation techniques have some clear theoretical advantages over the more conventional Fourier transform approach to Doppler spectral estimation. Specifically, they have better stability for short segments of signal and they have better spectral resolution for a given temporal resolution (or vice versa). They are, however, slightly more complicated to implement and are intuitively more difficult to understand. There also remains the serious problem of model order determination, which can significantly alter the spectral estimate—in order to gain advantages from ARMA-based methods, they must be applicable to short signal segments, but a number of workers have shown that AR and ARMA model order estimation with Doppler signals is poor with short data segments, which clearly limits their usefulness. Nevertheless, several workers have recommended their use in specific circumstances, and whilst they are unlikely to replace FFT methods for general purpose Doppler spectral analysis, they are likely to have a role to play in the future where more detailed information is required about Doppler spectra. At present, however, they seem not to have progressed beyond the research laboratory.

8.2.5 Cohen Joint Time–Frequency Distributions

Despite the improved temporal resolution that parametric spectral estimation techniques achieve when compared with conventional methods, there is still a requirement for the signal to be stationary within the analysis time interval. This has led to interest in the use of joint time–frequency distributions for the analysis of rapidly changing Doppler signals. These distributions describe how the frequency content of a signal is changing in time by mapping it into a function of two variables, time and frequency, instead of first dividing the signal into stationary segments for which individual PSDs are then estimated.

Many time–frequency distributions have been described in the past but it was Cohen (1966) who realised that an infinite number of such distributions could be generated using the following equation:

$$C(t, \omega, \phi) = \frac{1}{2\pi} \iiint \exp[j(\xi\mu - \tau\omega - \xi t)] \cdot \phi(\xi, \tau)$$
$$\cdot x\left(\mu + \frac{\tau}{2}\right) \cdot x^*\left(\mu - \frac{\tau}{2}\right) d\mu \, d\tau \, d\xi$$

8.24

where t is time, ω is angular frequency, $x(\mu)$ is the time signal, $x^*(\mu)$ its complex conjugate, ξ and τ are, respectively, frequency and time lags, and $\phi(\xi, \tau)$ is the kernel defining the distribution. Probably the best known Cohen class distributions are the Wigner–Ville and Choi–Williams distributions, and both these have been used to analyse Doppler ultrasound signals. It should be noted that the spectrogram or modified periodogram estimate can also be shown to be a member of Cohen's class (Cohen 1989), although it fails to fulfil many of the desired properties of joint time–frequency distributions (Guo et al 1994b). An excellent review of time–frequency distributions has been given by Cohen (1989).

The concept of the Wigner or Wigner–Ville distribution (WVD) was first introduced by Wigner (1932) in the context of quantum mechanics, and is becoming an important tool in radar, sonar and speech processing. It is obtained from eqn. 8.24 by setting the kernel $\phi(\xi, \tau)$ equal to unity for all values of ξ and τ, which leads to (Cohen 1989):

$$W_x(t, \omega) = \int_{-\infty}^{\infty} x\left(t + \frac{\tau}{2}\right) x^*\left(\tau - \frac{\tau}{2}\right) e^{-j\omega\tau} d\tau$$

8.25

For computational purposes it is necessary to window the signal $x(t)$, which leads to a function of t and ω that resembles, but in general is not, a Wigner distribution, and Claasen and Mecklenbräuker (1980a) introduced the term 'pseudo-Wigner distribution' (PWD) for such a function. Much has been written about the WVD and PWD and for further details of these distributions the reader is referred to Cohen (1989), Claasen and Mecklenbräuker (1980a,b,c), Martin and Flandrin (1985), and Jones and Parks (1992).

A number of workers have also used the PWD to analyse Doppler ultrasound signals. Kitney and Talhami (1987) developed a 'zoom Wigner' transform which they applied to Doppler data taken from a stenosed carotid artery. Kaluzynski (1989b) compared the PWD with Fourier, AR and ML spectral estimators as a means of quantifying turbulence and showed it to be inferior to other methods in the presence of wide-band Doppler signals from steady flow. Oung and Reid (1990) applied a modified PWD to signals from the carotid artery of a healthy subject and suggested it might have advantages in characterising non-stationarities from disturbed flow and vortices. Fan and Evans (1994b,c) developed a method of extracting instantaneous mean frequency information from Doppler signals using the PWD, and went on to develop an analyser to implement the technique in practice. Workers from the same group (Smith et al 1994) used this analyser to process signals from blood-borne emboli, and showed the potential of the Wigner method for analysing extremely short events occurring within the Doppler signal. Zeira et al (1994) studied the PWD as a means of analysing simulated signals, and in more recent reports concerning simulated Doppler signals, Fish and colleagues have compared the merits of several frequency estimators including the PWD for vortex detection (Wang and Fish 1996) and for reducing non-stationary broadening (Cardoso et al 1996).

The PWD has extremely good temporal and frequency resolution (determined only by the sampling frequency of the signal being analysed, and the length of the Fourier transform used to derive the PWD, respectively), but also has the major drawback that it produces time-varying cross-terms between signal components (Cohen 1989, Jeong and Williams 1992, Loughlin et al 1993, Fan and Evans 1994b). For this reason Choi and Williams (1989) devised a Cohen class distribution that to a large extent reduces the spurious cross-terms from multi-component signals. The kernel for the Choi–Williams distribution (CWD) is given by:

$$\phi(\xi, \tau) = \exp(-\xi^2 \tau^2 / \sigma) \qquad 8.26$$

where σ is a parameter of the CWD, and the resulting distribution by:

$$CW_x(t, \omega) =$$

$$\int_{-\infty}^{\infty} e^{-j\omega\tau} \left[\int_{-\infty}^{\infty} \frac{1}{\sqrt{4\pi\tau^2/\sigma}} \exp\left(-\frac{(\mu-t)^2}{4\tau^2/\sigma}\right) \right.$$

$$\left. \cdot x\left(\mu + \frac{\tau}{2}\right) \cdot x^*\left(\mu - \frac{\tau}{2}\right) d\mu \right] d\tau$$

$$8.27$$

Unfortunately, although the CWD has smaller interference terms than the WVD, it also has a poorer frequency resolution, and requires a longer computation time.

Guo et al (1994b), Fan (1994), Wang and Fish (1996) and Cardoso et al (1996) have all used the CWD to analyse Doppler ultrasound signals, with some success, but the middle two reports draw attention to the difficulties of choosing the most appropriate value of the parameter σ, the optimal value of which depends on the rate at which the signal amplitude and frequency are changing.

As has been pointed out, there are an infinite number of Cohen class distributions, and Guo et al (1994c) have proposed a kernel based on the first kind of Bessel function, and given by:

$$\phi(\xi, \tau) = \frac{J_1(2\pi\alpha\xi\tau)}{\pi\alpha\xi\tau} \qquad 8.28$$

where J_1 is the first kind of Bessel function of order one, and α (> 0) a scaling factor. Guo and colleagues have compared the Bessel distribution function (BDF) with the CWD and with AR modelling for the analysis of blood flow signals from the femoral artery, and suggested that the BDF is slightly better than either of the other two methods. They have also used the BDF as part of a pattern recognition process designed to classify lower limb arterial stenosis, with some success (Guo et al 1994a). Cardoso et al (1996) included the BDF as one of the time–frequency estimators in their study of non-stationary broadening reduction. Cicilioni et al (1998) compared the CWD, the BDF and the cone-kernel distribution (CKD) (Zhao et al 1990) on simulated carotid Doppler signals and suggested that, of the three, the CKD had the best performance.

A further interesting development has been the introduction of the so-called 'adaptive-Q' distribution (AQD) (Forsberg and Oung 1995, 1996, 1999), which employs a time- and frequency-varying kernel. Forsberg and Oung (1995) found it to perform better than either the Choi–Williams or the Bessel distribution in tests with synthetic Doppler signals, but to be more computationally demanding. In their later papers (Forsberg and Oung, 1996, 1999) they demonstrated a better trade-off between reducing variance and preserving resolution with the AQD compared with the short-term Fourier transform for *in vivo* data gathered from the aorta of a rabbit.

The Cohen joint time–frequency distributions are much more complicated than the periodogram or even the AR approach to spectral estimation, and are only likely to be widely adopted for Doppler signal analysis where they have very clear advantages over the more conventional methods. Their major advantage is that they have excellent frequency and temporal resolutions, and therefore any role they have is likely to be in the analysis of events of a rapidly changing nature such as turbulence, or the propagation of vortices or emboli. The improvement in resolution when compared to other techniques is only achieved at a price, and all Cohen distributions are susceptible to interference between different frequency components, which lessens their value for the analysis of multi-component or wide-band signals. At present the most suitable application for the PWD and CWD appears to be for the analysis of embolic phenomena, which are both very short-lived and, due to the discrete nature of embolic particles, result in very narrow band Doppler signals.

8.2.6 Sonogram Post-processing

Strictly speaking a sonogram is merely a graphical representation of a 2-D array of power estimates, where one dimension is time and the other frequency (Section 4.2.1.1). The term 'sonogram post-processing' is, however, used here for convenience to indicate manipulation of the data used to form the sonogram (or any other representation of evolving Doppler spectra) rather than the sonogram itself. Sonogram post-processing is also taken to mean that both the input and output of the process are data in the form of a 2-D array of power estimates; thus, envelope extraction techniques, which are discussed in Sections 8.3–8.6, are specifically excluded.

All or most sonogram post-processing techniques described in the literature have been developed in the context of overcoming the limitations of the periodogram approach to spectral estimation, but this is hardly surprising since the overwhelming majority of sonograms are formed in this way, and although some specific issues addressed by post-processing (such as speckle reduction) may not be so relevant to sonograms estimated using other techniques, this does not preclude post-processing of sonograms derived by other methods.

There have been two basic approaches to sonogram post-processing, that of ensemble averaging, and that of 2-D filtering. The value of averaging individual periodogram estimates has already been discussed in Section 8.2.2, where it was seen that spectral variance could be reduced either by averaging spectral estimates adjacent to each other within the cardiac cycle (providing the stationarity condition is met) or by averaging spectra from the equivalent time slice from a number of cardiac cycles. Ensemble averaging of the whole sonogram simply takes this process one step further, and has been reported by a number of authors.

The University of Washington group, in their studies of computer-based pattern recognition of carotid artery Doppler signals for disease classification (Greene et al 1982, Knox et al 1982, Langlois et al 1983, 1984), averaged sonogram data from 20 heartbeats (chosen from 25 or more candidate beats) by synchronising them with respect to the ECG R-wave, and then extracted a series of envelope signals (mode, -3 dB, -9 dB contours) which were used as features in the pattern recognition process. The same group used a similar technique as part of their study of flow patterns in the carotid arteries of young normal subjects (Phillips et al 1983), but on that occasion used data from fewer (16) cardiac cycles in the averaging process. Poots and colleagues (1986), in their comparison of CW Doppler ultrasound spectra with the spectra derived from a flow visualisation model, implemented a noise threshold and then averaged the sonograms from 15 successive pump cycles, which, as expected, proved successful in reducing statistical fluctuations in the sonogram (one advantage of ensemble averaging sonograms from *in vitro* rigs is, of course, that all the cycle lengths can be made identical). Hutchison and Karpinski (1985, 1988), in their work on flow patterns in post-stenotic fields in animal models, measured spectral broadening from averaged sonograms as a method of quantifying turbulence. They first produced an ensemble average of the sonogram data from 15 cardiac cycles in which the ECG R–R interval did not differ by more than 10%, and then, in a similar way to the University of Washington group, extracted three envelope signals corresponding to the mode frequency and the two −8 dB contours on either side of the mode, spectral broadening then being defined as the frequency difference between the two −8 dB contours. Cloutier et al (1990), in a study of 'Doppler spectral envelope area' of sonograms recorded from the left ventricular outflow tract (LVOT) of patients with stenosed aortic valves, used an ensemble average of data from five cardiac cycles, synchronised on the ECG R-wave, to reduce the variance of their Doppler data. The same workers went on to study the improvement in amplitude variability of data obtained from the LVOT by averaging data from differing numbers of cardiac cycles (Cloutier et al 1992). They reported that amplitude variability reduced exponentially with the number of cycles used, that the reduction was more rapid for the diastolic phase than the systolic phase, and that no significant improvement in amplitude variability reduction could be obtained by using more than 20 cardiac cycles. In a similar study of sonograms recorded from the lower limb arteries, the same group (Allard et al 1992) concluded once again that amplitude variability reduces exponentially with the number of cycles averaged together, but, contrary to their findings with cardiac data, there is little to be gained by using more than ten cycles with these waveforms. Hoskins and his colleagues have used ensemble averaging of sonograms derived from a flow rig to derive 'gold standard' sonograms for evaluating speckle reduction filters (Hoskins et al 1990), and 'gold standard' maximum frequency envelopes for evaluating the effect of three physical parameters on the maximum frequency envelopes and pulsatility index (Hoskins et al 1991b). Finally, Loupas et al (1995) have used ensemble averaging of spectral data from between six and nine cardiac cycles in their sophisticated computer analysis of the early diastolic notch in sonograms recorded from uterine arteries (see Section 9.3 for more details).

The second approach to sonogram post-processing, that of two-dimensional filtering, appears to have been much less popular. Heringa et al (1988), in their study of computer processing of cardiac Doppler signals, used independent filtering in the frequency and time dimensions. They first attempted to reduce noise and compensate for the wall thump filters by averaging the background noise level in regions of the sonogram where there appeared to be no signal, subtracting this value from all elements of the sonogram (setting negative bins to zero), and applying an inverse wall filter. They then applied a five-point window (1/16, 4/16, 6/16, 4/16, 1/16) in the frequency dimension, followed by a three-point window (1/3, 1/3, 1/3) in the time dimension. They concluded that these operations produced smoothed sonograms from which parameters such as pulsatility index could be successfully calculated without human intervention.

In a much more sophisticated study, Hoskins and colleagues (1990) compared the ability of three different 2-D filters applied to sonogram data from a flow rig to produce the same results as the 'gold standard' ensemble averaged sonogram (as judged from noise reduction, and bias and distortion of the maximum frequency envelope). The three filters investigated were the double window modified trimmed mean (DWMTM) filter, Lee's filter, and a directional filter—the latter two in combination with different controlling parameters. The DWMTM filter, first described by Lee and Kassam (1985), combines the properties of a median filter (which performs well in suppressing spikes and preserving edges) with the properties of a mean filter (which performs poorly in these respects, but much better in terms of noise reduction) and is implemented by first calculating the median value inside a relatively small window and then, using a second larger window, calculating a running mean of values that are within a specified limit from the median.

Lee's filter (Lee 1980), which like the directional filter is adaptive, is described by:

$$O(x, y) = m(x, y) + C(I(x, y) - m(x, y)) \quad 8.29$$

where $I(x, y)$ and $O(x, y)$ are the input and output of the filters respectively, $m(x, y)$ is the local mean of terms inside the filter window, and C is a controlling parameter which determines the amount of smoothing at each point and depends on a local image measure.

The directional filter used was the smoothing stage of a two-stage directional filter described by Loupas (1988) for combined smoothing and sharpening of ultrasonic B-scan images. This filter is described by:

$$O(x,y) = \frac{\sum_{d=1}^{8} m(d) \times [C(d)]^k}{\sum_{d=1}^{8} [C(d)]^k} \qquad 8.30$$

where $m(d)$ is the mean value of elements in a window aligned along the direction d, $C(d)$ is the local image measure along the same direction, and k is a parameter which determines the relative contributions of the directional means $m(d)$ to the final output.

The two adaptive filters are controlled by local image content characterised by C and $C(d)$, respectively, so that maximum smoothing is performed if a pixel belongs to a uniform area corrupted by noise, whilst the amount of smoothing is reduced when the pixel belongs to part of a resolvable structure which must be retained. There are a number of image measures that can be used for this purpose, of which Hoskins et al (1990) tested measures of local statistics, slope, and edge presence. Overall they concluded that the best performance was offered by the directional filter whose action was controlled by the combination of the local edge content and the slope of the least-squares-fit line passing through the data points along each particular direction, and that the technique was useful for improving the performance of maximum frequency followers and the calculation of quantitative information based on the Doppler spectrum such as spectral broadening indices. Figure 8.12 shows the effect of the optimal directional filter on sonograms from an internal carotid artery.

8.3 MEAN FREQUENCY PROCESSORS

Perhaps the most obvious envelope signal to extract from a Doppler audio signal is the intensity-weighted mean frequency (IWMF), $\bar{f}(t)$, defined by:

$$\bar{f}(t) = \frac{\int_f P(f) f \, df}{\int_f P(f) \, df} \qquad 8.31$$

Here $P(f)$ is the Doppler power spectrum. Under ideal conditions this derived frequency is proportional to the mean velocity of the flow in the target blood vessel. If the angle between the ultrasound beam and the blood vessel is known, the output from a mean frequency follower may therefore be calibrated directly in terms of flow velocity; if, in addition, the cross-sectional area of the vessel is known, the output may be calibrated in terms of volumetric flow.

Several methods of deriving the mean frequency using analogue techniques have been described, of which the best known are the single-correlation

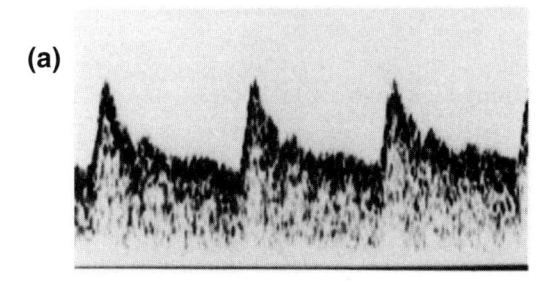

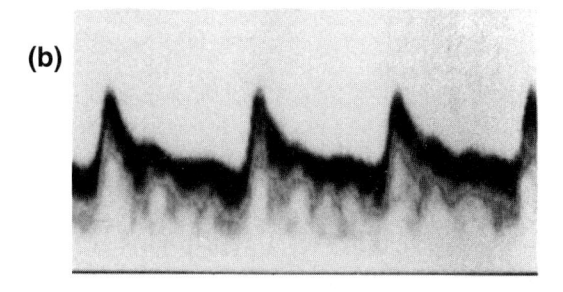

Figure 8.12 Internal carotid artery sonograms (a) before filtering and (b) after filtering with a directional filter (reproduced by permission of Elsevier Science, from Hoskins et al 1990, *Ultrasound in Medicine and Biology*, © World Federation of Ultrasound in Medicine and Biology)

centroid detector of Arts and Roevros (Arts and Roevros 1972, Roevros 1974) and the $\sqrt{f}$ power-spectrum centroid detector (de Jong et al 1975, Gerzberg and Meindl 1977, 1980a, 1980b), and these analogue techniques are discussed in Chapter 6 (Section 6.5.1). An alternative approach, and that which is now most used, is to evaluate eqn. 8.31 numerically using a real-time spectrum analyser and a microcomputer (see, for example, Kontis 1987, Schlindwein et al 1988). In this case eqn. 8.31 may be rewritten in terms of the 'intensity-weighted mean bin number' (*IWMB*) and the estimated spectral power, p_i, in each analyser bin, i.e:

$$IWMB = \frac{\sum_{i=IMIN}^{IMAX} ip_i}{\sum_{i=IMIN}^{IMAX} p_i} \qquad 8.32$$

where *IMIN* and *IMAX* are the lowest and highest bin numbers, respectively (*IMIN* will usually be zero if there is no reverse flow, but may take negative values otherwise). Because the values of p_i are merely estimates of the expected spectral power in the *i*th frequency bins (due to the stochastic nature of the Doppler signal) *IWMB* is in turn only an estimate of the expected value of the intensity-weighted bin number, $\langle IWMB \rangle$. The effects this has are discussed in Section 8.3.2.

8.3.1 Effect of Non-uniform Insonation

The effect of non-uniform insonation on the Doppler power spectrum was discussed in Section 7.3.2, and illustrated for a number of specific situations in Figs 7.10–7.13. Clearly, such dramatic changes will have a considerable effect on the output of mean frequency followers, and most importantly on the estimation of mean velocity. The magnitude of these errors has been investigated for only a few simple idealised geometries (Evans 1982, Cobbold et al 1983, Evans 1985b), but the results show that use of an ultrasound beam which is much narrower than a blood vessel containing a parabolic velocity profile may cause an overestimation of mean velocity of up to 33%.

Interestingly, the error in the time-averaged mean velocity over the cardiac cycle is dependent only upon the shape of the *mean* component of the velocity profile and the way in which this is sampled by the ultrasound beam (Evans 1985b). The shape of the mean component depends on a number of factors (see Evans 1985b and Chapter 2), but will be parabolic at sites that are sufficiently far from curves, branches, bifurcations and other sources of haemodynamic disturbance. Figure 8.13 illustrates the error in the measurement of mean velocity in a vessel containing parabolic flow, using a rectangular ultrasound beam of various sizes with various degrees of offset from the vessel centre.

Most modern duplex and colour flow scanners use the same transducer for both pulse-echo imaging and Doppler measurements, and since in general these transducers are optimised to produce the best images, they use narrow ultrasound beams which can lead to large errors in the calculated intensity-weighted mean frequency. In an attempt to overcome this problem, Willink and Evans (1994, 1995b) have suggested a new alternative estimator which is optimised for narrow ultrasound beams and which they have named the 'position and intensity weighted mean frequency' (PIWMF).

For unidirectional flow, position and intensity-weighted mean bin number (*PIWMB*) can be written:

$$PIWMB = \frac{\sum_{i=0}^{IMAX} ip_i \left(2 \sum_{j=i+1}^{IMAX} p_j + p_i \right)}{\left(\sum_{i=0}^{IMAX} p_i \right)^2} \qquad 8.33$$

The derivation of this estimator is based on a number of assumptions, viz: (1) each half of the velocity profile is a monotonic increasing function of distance from the vessel wall; (2) the Doppler beam dimensions are constant along the sample volume; (3) the sample volume can be thought of as being infinitesimally thin so that all scatterers a given distance along the sample volume can be thought of as having the same velocity; (4) the length, and position in depth, of the sample

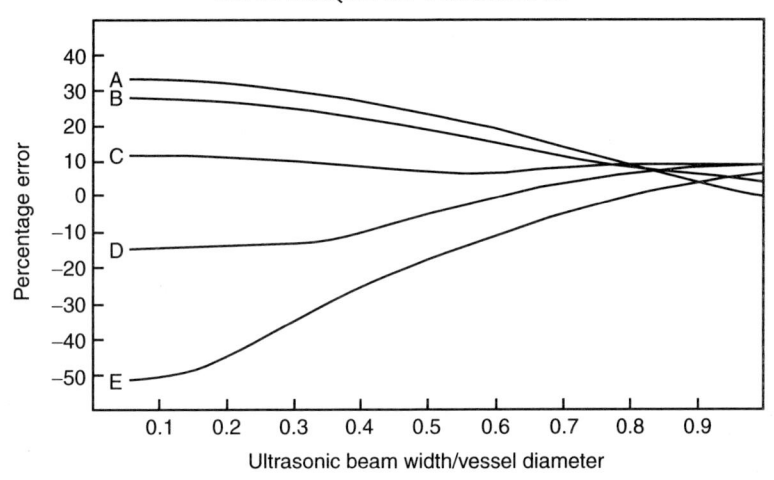

Figure 8.13 The error that results if a mean frequency processor is used with an ultrasound beam that is narrower than the blood vessel. Curve A has been calculated for a centralised beam, curves B–E represent the behaviour of the offset beams. The values of the offsets (in fractions of the vessel radius) are: B, $R/5$; C, $2R/5$; D, $3R/5$; E, $4R/5$ (reproduced by permission of Elsevier Science, from Evans 1985b, *Ultrasound in Medicine and Biology*, © World Federation of Ultrasound in Medicine and Biology)

volume are sufficient for it to extend over a diameter of the cross-section; (5) the sample volume is positioned through the centre of the vessel; (6) the velocity profile can be regarded as constant with axial position throughout the angled sample volume; and (7) the concentration of red blood cells (i.e. scatterers) throughout the sample volume is constant.

Assumptions (4), (6) and (7) are also implicitly made in the definition of IWMF, whilst the effects of deviations from (3) and (5) are relatively small (even if a rectangular beam insonates 40% of a vessel containing parabolic flow, the error in PIWMF is less than 5%, whilst a lateral displacement of a narrow beam by 10% and 20% of the vessel radius leads to errors of 1% and 4%, respectively). Errors in both IWMF and PIWMF (and hence flow measurements derived using these estimates) as a function of beam width to vessel size ratio are illustrated in Fig. 8.14a (for rectangular beams) and Fig. 8.14b (for Gaussian beams), where it can be seen that PIWMF incurs smaller errors than IWMF (under ideal conditions) for beamwidth to vessel diameter ratios of less than 0.65, and σ/R ratios of less than 0.4 (see captions for more details). It remains to be seen whether the departures from assumptions (1) and (2) in real ultrasound beams and velocity profiles outweigh the advantages gained by relaxing the uniform insonation assumption.

For bi-directional flow, assumption (1) can be modified such that the flow profile is considered to be axi-symmetric and made up of a central monotonic section inside a region of reverse flow that is symmetric in itself. The *PIWMB* can then be written:

$$PIWMB_{\text{rev}} = \frac{\sum_{i=1}^{IMAX} ip_i \left(2 \sum_{j=i+1}^{IMAX} p_j + p_i \right) + \left(\sum_{i=IMIN}^{0} p_i + 2 \sum_{i=1}^{IMAX} p_i + \frac{p_0}{3} \right) \sum_{i=IMIN}^{-1} ip_i}{\left(\sum_{i=IMIN}^{IMAX} p_i \right)^2}$$

8.34

which reduces to 8.33 when no reverse flow is present. Willink and Evans (1995b) have shown that for theoretical velocity profiles calculated using Womersley's theory (Womersley 1957) the new estimator is considerably better than

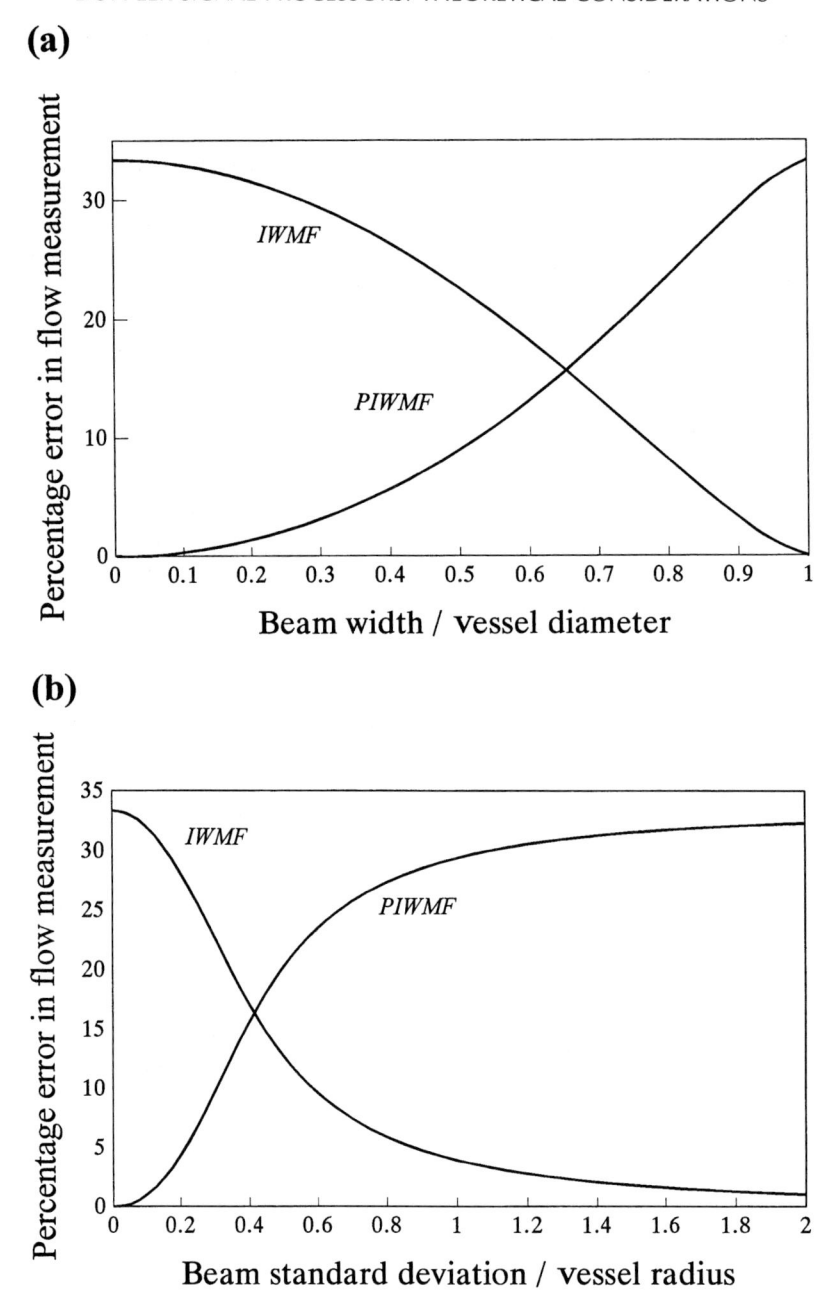

Figure 8.14 (a) Errors in flow and velocity measurements incurred by using intensity-weighted mean frequency (IWMF), and position and intensity-weighted mean frequency (PIWMF) estimators with a parabolic flow profile and a rectangular beam shape of variable width. (b) Errors in flow and velocity measurements incurred by using IWMF and PIWMF with a parabolic flow profile and a Gaussian beam shape of variable width (defined in terms of beam standard deviation σ)

IWMF when used with narrow ultrasound beams. Willink (1999) has discussed the effects of asymmetric velocity profiles on the new estimator.

8.3.2 Statistical Bias and Variance in Mean Frequency Estimators

Estimation of mean blood flow velocity is based on the principle that the expected power of the Doppler signal in a given frequency range is proportional to the number of scatterers in the sample volume with velocities in the corresponding range. However, the stochastic nature of the Doppler signal dictates that the observed values of power in a given frequency range are random estimates of such expected values, dependent on the spatial distribution of the relevant scatterers. Since the power in any given frequency range is a random quantity, so too are the mean velocity estimates derived from the power spectrum, and as such will have a variance associated with them, and may be subject to bias.

The statistical properties of mean frequency estimators have been explored by Gerzberg and Meindl (1980a), Angelsen (1981) and more recently by Willink and Evans (1995a). These last authors examined the statistics of both the conventional intensity-weighted mean frequency, and the so-called position and intensity-weighted mean frequency (Section 8.3.1), derived approximate expressions for their biases and variances when used on uniform Doppler power spectra, and evaluated numerically their magnitudes when used on more complex spectra. The results showed that for velocity profiles between parabolic and plug forms, the biases in both IWMF and PIWMF due to the random nature of the spectral estimates are small (< 1%) in the presence of typical amounts of spectral broadening. For uniform spectra the relative uncertainty (defined as the ratio of standard deviation to expected value) in the observed value of the intensity weighted mean bin number was shown to be given by the expression:

$$r.u.(IWMB) \approx 1/\sqrt{3M} \qquad 8.35$$

where M is the number of bins occupied by the spectrum. Similarly the relative uncertainty (r.u.) in PIWMV was shown to be:

$$r.u.(PIWMB) \approx 2/\sqrt{5M} \qquad 8.36$$

Willink and Evans also gave a summary expression for the relative uncertainty in the estimate of the mean velocity, $\hat{v}$, due to the stochastic nature of the Doppler signal, irrespective of whether IWMF or PIWMF is used:

$$r.u.(\hat{\bar{v}}) = \sqrt{\frac{c}{24 f_t \Delta \tau \cos\theta}} \cdot \sqrt{\frac{\Delta v}{\bar{v}}} \qquad 8.37$$

where c is the velocity of ultrasound, f_t the transmitted ultrasound frequency, $\Delta\tau$, the duration of the segment of signal analysed, θ the usual Doppler angle, Δv the range of velocities present in the sample volume, and $\bar{v}$ the expected value of mean velocity. Using this estimate they showed, with typical parameters, that the relative uncertainty of a single velocity measurement might easily be greater than 10%. However, when mean velocity measurements or flow measurements are made over one or several cardiac cycles, precision is greatly increased, improving with the square root of duration, and if the signal duration analysed is greater than 1 s, a relative uncertainty of less than 1% might be expected. Thus, the random nature of the Doppler signal leads to a minimal error in average velocity measurements or flow measurements, compared to typical errors from other sources.

8.3.3 Effect of Attenuation

The differential rates of attenuation in soft tissue and blood have the effect of slightly reducing the effective width of an ultrasound beam (Section 7.3.3) and therefore of accentuating the higher velocities at the centre of the vessel. Cobbold et al (1983) have published diagrams illustrating the effect of attenuation on the error in mean velocity measurements when square and Gaussian beams interact with a parabolic velocity profile (Fig. 8.15).

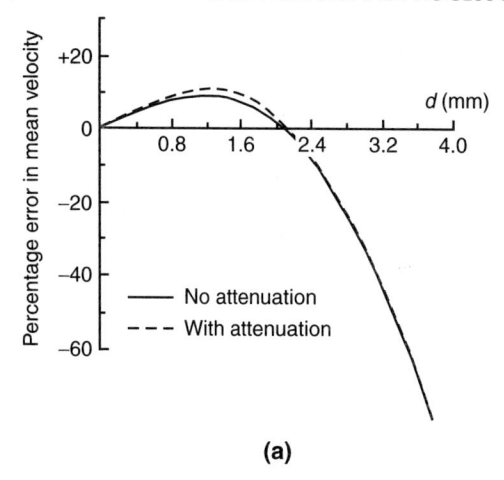

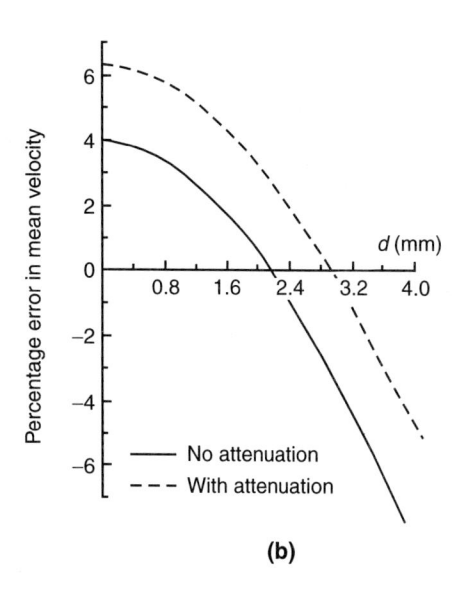

Figure 8.15 The effect of attenuation on the error in mean velocity calculated from the power density spectrum, for a parabolic velocity profile, for various beam displacements from the centre; (a) represents the results from a square beam of 4 mm width insonating a vessel of the same diameter; (b) represents the results from a Gaussian beam with a standard deviation of 2 mm insonating the same vessel as in (a) (reproduced by permission from Cobbold et al 1983, © 1983, IEEE)

8.3.4 Effect of Intrinsic Spectral Broadening

Intrinsic spectral broadening (Sections 7.3.4 and 7.4.3) *per se* does not affect the mean frequency of the Doppler spectrum because the broadening is more or less symmetrical and therefore leads to an increase in variance but not the mean of the spectrum.

8.3.5 Effect of Frequency-dependent Attenuation and Scattering

As has been pointed out in Sections 7.4.4 and 7.4.5, both attenuation and scattering of ultrasound are frequency-dependent mechanisms. For CW Doppler systems this is irrelevant in terms of mean frequency estimation because the transmitted frequency is monochromatic. The same is not, however, true of PW Doppler systems, where the transmitted signal may have a short duration and therefore a relatively large bandwidth. Since attenuation increases with frequency, the mean frequency of an ultrasonic pulse decreases as it propagates through tissue, whilst since scattering increases with frequency the mean frequency of an ultrasonic pulse increases when it is scattered by blood. As will be seen shortly, the frequency shifts involved can be quite large and for ultrasonic pulses containing only a few cycles can easily be much larger than the frequency shift due to the Doppler effect! Fortunately, as has been explained in Section 4.2.2.5, PW Doppler systems actually function by determining the displacement the target undergoes between successive interrogating ultrasound pulses (either by measuring a phase or time difference), rather than the Doppler shift on the pulse itself, and therefore they are relatively insensitive to changes in the spectrum of the transmitted ultrasound pulse. Several authors, including Newhouse et al (1977), Holland et al (1984), Forsberg (1989), Thomas et al (1989), Light (1990), Embree and O'Brien (1990), Ferrara et al (1992), and Thomas and Leeman (1995), have addressed the problem of frequency-dependent effects in relation to their effect on the measured Doppler shift frequency, and there has been a certain amount of controversy with regard to the magnitude of such effects, which appear to be quite difficult to quantify experimentally, and indeed with regard to whether such effects exist at all (see, for example, Light 1990). Fortunately a clearer theoretical position is beginning to emerge, but there is still room for good experimental work in this area to confirm theoretical predictions.

Before discussing changes in the centre frequency of ultrasound pulses due to frequency-dependent mechanisms, we will examine the effect changes in the pulse centre frequency have on the measured demodulated frequency and hence the mean velocity. It will emerge that this depends on the frequency estimation strategy employed.

For the classical PW Doppler estimator the proportional error in the measured mean velocity is the same as the proportional change in the pulse centre frequency, i.e:

$$\frac{\hat{\bar{v}}}{\bar{v}} = \frac{f'_c}{f_c} \qquad 8.38$$

where $\hat{\bar{v}}$ is the estimate of the true mean velocity, $\bar{v}$, f_c the centre frequency of the transmitted pulse, and f'_c the centre frequency of the received pulse (Embree and O'Brien 1990, Ferrara et al 1992). Intuitively this seems sensible, because the classical PW Doppler estimator functions by deriving the relative phase shifts of the returning signals with respect to a master oscillator in each inter-pulse interval. Therefore, if for example the centre frequency of the returning pulse is increased by $x\%$, then the apparent rate of phase change will also increase by $x\%$. Likewise, the autocorrelator estimator (Namekawa et al 1982, Kasai et al 1985), which is widely used in colour flow mapping systems and also functions by determining inter-pulse phase-shifts, suffers from exactly the same error (Ferrara et al 1992). On the other hand, estimators that function by deriving the relative *time* shifts of the returning signals in the inter-pulse interval, such as the cross-correlation estimator, are unaffected by changes in the centre frequency of the returning ultrasound pulses (Bonnefous and Pesque 1986, Ferrara et al 1992). Once again, this seems intuitively sensible. Ferrara et al (1992) have explicitly shown that this is true for their wideband maximum likelihood estimator (Ferrara and Algazi 1991a, 1991b).

Having established the relationship between changes in the centre frequency of the ultrasonic pulse, and the potential errors in velocity estimation, we now explicitly examine the effect of frequency-dependent Rayleigh scattering (i.e. where scattering is proportional to the fourth power of frequency) and attenuation on the pulse centre frequency. Newhouse et al (1977), Round and Bates (1987) and Forsberg (1989) have all derived expressions for the shift in the centre frequency of a Gaussian modulated pulse caused by frequency-dependent scattering. All the results are of the same form, and those of Newhouse and Forsberg are numerically identical. Forsberg's expression for the increase in the centre frequency $\Delta f_{c(scat)}$ may be written:

$$\Delta f_{c(scat)} = \frac{f_c}{2} \left[\left\{ 1 + \frac{1}{\ln 2} \left(\frac{B_0}{f_c} \right)^2 \right\}^{1/2} - 1 \right] \qquad 8.39$$

where f_c and B_0 are respectively the centre frequency and (-6 dB) bandwidth of the pulse before scattering takes place. The bandwidth-to-centre-frequency ratio of a pulse can be approximated by the reciprocal of the number of (significant) cycles in the pulse, and thus from eqn. 8.39 the percentage increase in the centre frequency of the pulse due to Rayleigh scattering may be written:

$$\%\Delta f_{c(scat)} \approx 50 \times \left[\left\{ 1 + \frac{1}{N^2} \frac{1}{\ln 2} \right\}^{1/2} - 1 \right] \qquad 8.40$$

where N is the length of the transmitted pulse in cycles. This function is plotted in Fig. 8.16, where it can be seen that scattering is only a significant cause of frequency shift for very short pulses, and that the error is less than 1% for pulses containing six cycles or more.

The effect of attenuation on the centre frequency of a pulse has been investigated by several authors by theoretical analysis (Newhouse et al 1977, Ophir and Jaeger 1982), numerical simulation (Forsberg 1989, Embree and O'Brien 1990, Ferrara et al 1992) and experimental measurements (Ophir and Jaeger 1982, Holland et al 1984, Thomas 1995). A particularly useful formulation is that of Ophir and Jaeger (1982), who investigated the relationship between the centre frequency downshift and transmitted bandwidth of pulses with Gaussian amplitude spectra propagating

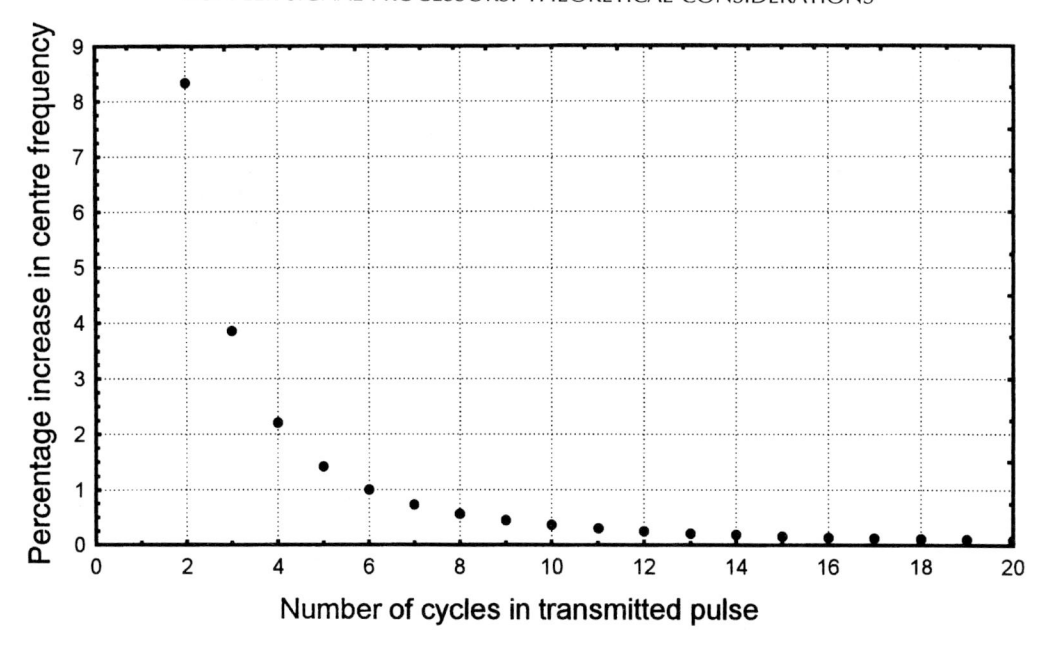

Figure 8.16 The percentage increase in the centre frequency of an ultrasound pulse due to Rayleigh scattering from blood, plotted against the number of cycles in the transmitted pulse (eqn. 8.40)

through lossy media with power law frequency dependence of attenuation. They also went on to confirm their theoretical predictions with experimental measurements. The main result of their theoretical analysis was to show that for the frequency-dependent filter function:

$$|H(f)| = \exp[-2\alpha_0 f^n Z] \qquad 8.41$$

where n is the exponent of frequency dependence (typically $1 \leqslant n \leqslant 2$ for tissue), Z the tissue propagation distance, and α_0 the amplitude attenuation frequency-dependent coefficient of the medium (in Nepers cm^{-1} MHz^{-n}), the equation relating the downshifted centre frequency $f_{c(atten)}$ to the transmitted centre frequency f_c can be written:

$$2n\alpha_0 Z \sigma^2 f_{c(atten)}^{n-1} + f_{c(atten)} - f_c = 0 \qquad 8.42$$

where σ^2 is the variance of the transmitted spectrum.

This equation cannot be solved in a closed form except for some simple cases where n takes a small integer value, but can easily be solved numerically for other values. For the case of linear increase in attenuation with frequency ($n = 1$), which is a reasonable approximation for soft tissue, eqn. 8.42 reduces to:

$$\Delta f_{c(atten)} = -2\alpha_0 Z \sigma^2 \qquad 8.43$$

where $\Delta f_{c(atten)}$ is the decrease in the pulse centre frequency due to attenuation.

The -6 dB bandwidth of a Gaussian spectrum, B_0, is given by $B_0 = 2.36\sigma$ and thus substituting for σ in eqn. 8.43 leads to:

$$\Delta f_{c(atten)} = -2\alpha_0 Z \left(\frac{B_0^2}{5.57}\right). \qquad 8.44$$

Writing the bandwidth to centre frequency of the transmitted pulse as the reciprocal of the number of cycles in the pulse (N), as before, leads to:

$$\Delta f_{c(atten)} \approx -0.36\alpha_0 Z \frac{f_c^2}{N^2} \qquad 8.45$$

This can also be expressed as a percentage change in pulse frequency, i.e:

$$\%\Delta f_{c(atten)} \approx \frac{36\alpha_0 Z f_c}{N^2} \qquad 8.46$$

or

$$\%\Delta f_{c(atten)} \approx \frac{4.14\alpha_{0(dB)} Z f_c}{N^2} \qquad 8.47$$

where $\alpha_{0(dB)}$ has the units of dB cm^{-1} MHz^{-1}, Z is in cm, and f_c in MHz.

The percentage decrease in pulse centre frequency, calculated from eqn. 8.47, has been plotted against the number of cycles in the transmitted pulse for various values of $K = \alpha_{0(dB)} Z f_c$, in Fig. 8.17. It can be seen that the frequency shift only becomes significant when the transmitted pulse is short and the centre frequency and/or target depth is large. In practice the frequency shift due to frequency-dependent scattering partially compensates for the shift due to frequency-dependent attenuation, and Fig. 8.18 has been generated by combining the values from Figs 8.16 and 8.17. For small values of K scattering dominates, whilst for large values of K attenuation dominates irrespective of pulse length. It is interesting to note that for a value of K around 8, the two effects more or less cancel, and thus the operating frequency of a Doppler unit could in principle be chosen to minimise the frequency-dependent errors over a finite range if attenuation were known (Reid and Klepper 1984).

Although frequency-dependent attenuation and scattering are probably the most important mechanisms that affect the spectrum of short ultrasound pulses, it is worth noting that there are other sources of error in this respect, including the Doppler shift on the ultrasound pulse (!), the frequency response of the ultrasound transducer and receiving electronics, and non-linear ultrasound propagation. Recalling that PW Doppler devices operate by determining the displacement that a target undergoes between successive pulses, the Doppler shift on the pulse due to target movement actually introduces an error into phase

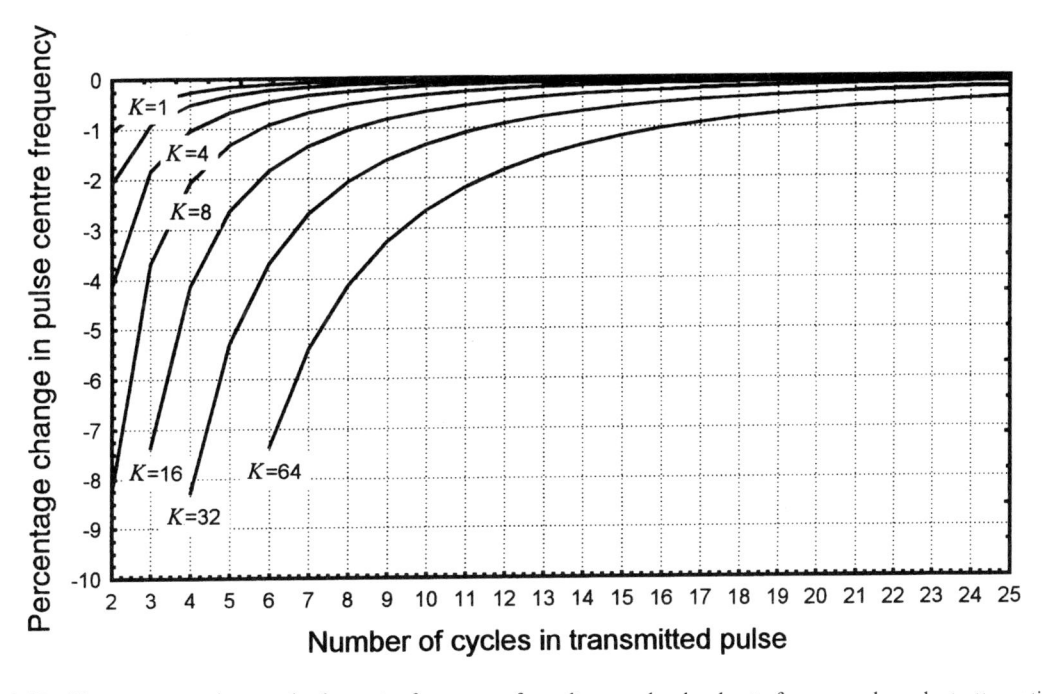

Figure 8.17 The percentage decrease in the centre frequency of an ultrasound pulse due to frequency-dependent attenuation, for various values of K (the product of attenuation coefficient, range and pulse centre frequency) plotted against the number of cycles in the transmitted pulse (eqn. 8.47)

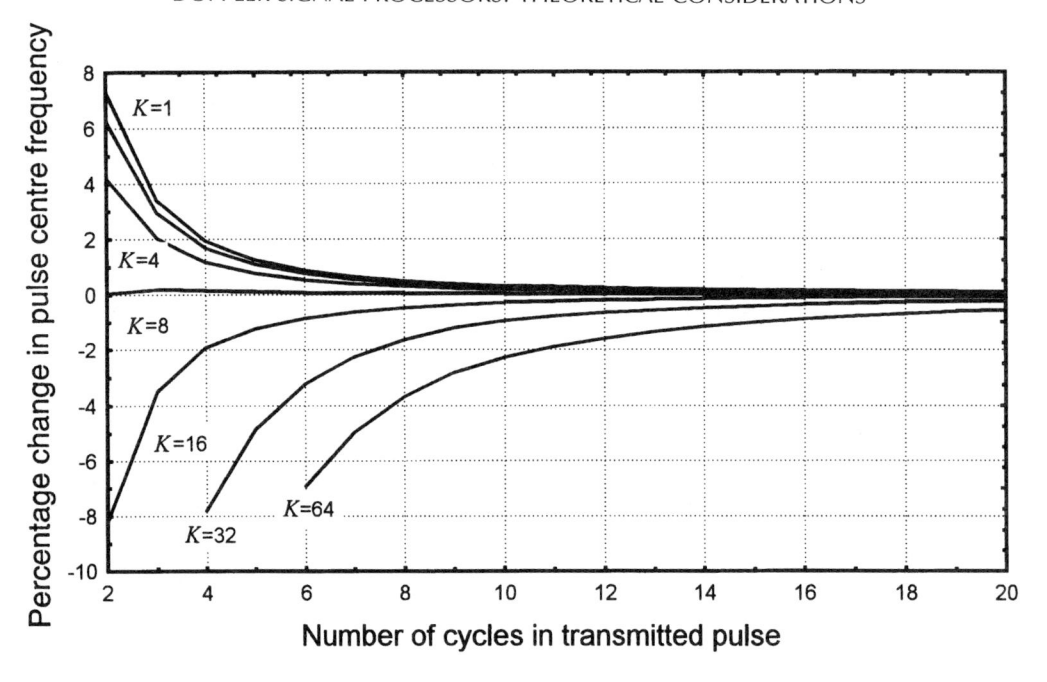

Figure 8.18 Percentage change in the centre frequency of an ultrasound pulse due to scattering and attenuation, for various values of K (see Fig. 8.17) plotted against the number of cycles in the transmitted pulse. This graph combines the results of Figs 8.16 and 8.17

shift-based velocity estimates, albeit small. In general, the filtering effect of the transducer is likely to shift the mean value of the spectrum towards the transmitted frequency, and thus reduce any errors. The effect of non-linear propagation fortunately appears to be negligible (Section 7.4.2).

To summarise this section, frequency-dependent attenuation and scattering only become significant sources of error in mean frequency estimation when very short pulses of ultrasound are used, and only then when the 'Doppler shift frequency' is measured using a phase domain- rather than a time domain-based algorithm. These errors are most likely to become important in multigate techniques (including colour flow imaging) where the ultrasound pulses are necessarily short, but could to some extent be compensated for if desired.

8.3.6 Effect of Filters

As has been explained in Section 7.3.5, Doppler devices contain high pass filters designed to reject high-amplitude low-frequency signals arising from stationary and near stationary tissue. In addition to rejecting unwanted noise, they also reject low frequency signals, and this has the effect of biasing the mean frequency estimate in an upwards direction. Gill (1979) has studied the error that this introduces and has shown that it is influenced by both the cut-off frequency of the filter, f_{HP}, and the velocity profile.

For velocity profiles of the form:

$$v = v_{max}[1 - (r/R)^\beta] \qquad 8.48$$

Gill shows the error to be given by:

$$f_{error} = 2f_{HP}/(2 + \beta) \qquad 8.49$$

For a parabolic profile, β has a value of 2; for blunt profiles a value of greater than 2; and for plug flow β tends to infinity.

Equation 8.49 must be applied to the instantaneous velocity profile and not the time-averaged velocity profile, and therefore it is not easy to correct for this error when pulsatile flow is present if the profile changes dramatically during the cardiac cycle. Willink and Evans (1996) showed that eqn. 8.49 applies equally to both IWMF and PIWMF.

8.3.7 Effect of Signal-to-noise Ratio

The signal-to-noise ratio of the Doppler signal is often not as great as might be hoped. The mean frequency processor has no way of distinguishing between signal and noise within the acceptable band for Doppler signals, and therefore noise will bias the output.

Gerzberg and Meindl (1977) have shown that in the presence of band-limited stationary noise, the output of a mean frequency follower is given by:

$$\bar{f}' = \bar{f}_d/(1 + 1/S) + \bar{f}_n/(1 + S) \qquad 8.50$$

where $\bar{f}_d$ is the mean Doppler shift frequency, $\bar{f}_n$ the centroid of the noise spectrum and S the signal-to-noise power ratio. The effects of various levels of additive white noise on the output of a mean follower (calculated from eqn. 8.50) are shown in Fig. 8.19; these results have been experimentally verified by Gerzberg and Meindl.

If the characteristics of the noise are known, it is theoretically possible to eliminate the effects of this error (Gerzberg and Meindl 1977, 1980b, Gill 1979). Willink and Evans (1996) compared the effects of noise on the IWMF and PIWMF and found the latter to be more sensitive to high frequency noise than the IWMF, but less sensitive to low frequency noise.

8.4 MAXIMUM FREQUENCY PROCESSORS

A second popular envelope detector is the maximum frequency processor. This method was first used to derive a single-valued waveform from sonograms produced off-line with Kay spectrographs, simply by drawing around the darkened area of the sonogram (Gosling et al 1969). When the signal is unidirectional, the basic definition of maximum is self-evident; however, if there is bi-directional flow and particularly simultaneous forward and reverse flow, then a 'composite maximum frequency' must be defined. The most usual method of producing a composite maximum frequency envelope is to follow both the absolute maximum frequency and the absolute minimum frequency and to sum these two envelopes (absolute maximum is not allowed to fall below zero, nor absolute minimum to exceed zero). Thus, when only forward flow is present, the composite maximum corresponds to the highest forward velocity, when only reverse flow is present to the highest reverse velocity, and when both are present to the difference between maximum forward and maximum reverse velocity.

8.4.1 Methods of Extracting the Maximum Frequency Envelope

Many methods of extracting the maximum frequency envelope of the Doppler signal have been described. Early methods relied on the use of analogue processing methods (Sainz et al 1976, Skidmore and Follet 1978, Nowicki et al 1985), but with the widespread availability of real-time spectrum analysers and microcomputers, digital methods have become all but ubiquitous.

In principle the identification of the instantaneous maximum frequency present in a Doppler spectrum may appear simple, since this frequency

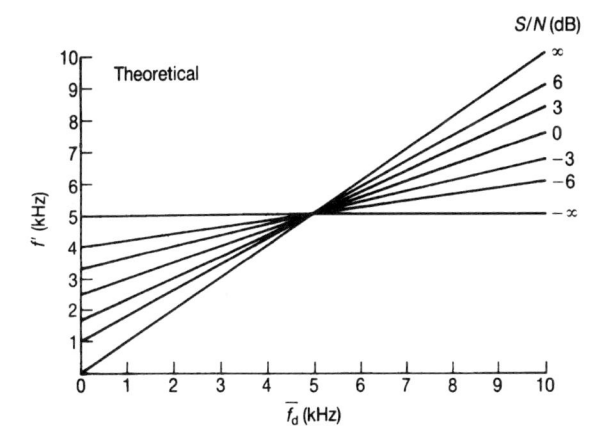

Figure 8.19 Predicted centroid $\bar{f}'$ of a signal $\bar{f}_d$ accompanied by band-limited stationary noise whose centroid is located at 5 kHz (reproduced by permission from Gerzberg and Meindl 1977, Plenum Publishing Corp)

should correspond to a well-defined maximum velocity present in the blood flow being interrogated. In practice this apparently simple task is confounded by two major influences: intrinsic spectral broadening (Sections 7.3.4 and 7.4.3) and noise. The difficulties are illustrated by Fig. 8.20, which shows typical spectra estimated from a short segment of Doppler signal obtained from a carotid artery using both a rectangular window and a Hanning window (8.2.2). The spectrum contains information about both the signal and the noise, and the task of the maximum frequency detector is to find the transition between the two. Unfortunately, because of the effect of intrinsic spectral broadening the maximum frequency is poorly defined and the transition from signal plus noise to noise is not at all obvious. Furthermore, as the flow rate changes during the cardiac cycle, the Doppler power will spread over a different bandwidth so that the signal-to-noise ratios at particular frequencies are constantly changing. A number of approaches have been used in attempts to solve these difficulties, and several of them are described below.

Each method to be described determines values of maximum frequency based on one or more spectral estimates of the Doppler signal and although most have been implemented on FFT-based systems, they can of course be implemented with other estimators. Kaluzynski and Palko (1993) have reported the results of using a number of combinations of spectral and maximum frequency estimators.

8.4.1.1 Simple Threshold Method (STM)

Perhaps the simplest method of determining the transition between signal and noise is that proposed by Gibbons et al (1981) in an early attempt to automate maximum frequency extraction from the output of a spectrum analyser. The method employs a user-determined threshold to superimpose the maximum frequency envelope on a sonogram. In its simplest implementation the algorithm simply examines the power in each frequency bin (starting from the bin representing the highest analysed frequency and working down) until one is found that exceeds the threshold, and that bin is then flagged as the maximum frequency and marked on the sonogram in some way. Slight variations of the algorithm involve searching for more than one bin that exceeds the threshold level. The threshold itself is set subjectively by the operator on the basis of the level that provides a stable envelope that follows the sonogram outline reasonably faithfully. Having found the threshold that produces the best performance, this is kept constant within a particular segment of the signal. An algorithm which seeks to automate the threshold selection has been described by Evans et al (1989a). Despite the simplicity of this method it can be remarkably effective.

8.4.1.2 Percentile Method (PM)

Another simple method of determining a maximum frequency envelope is to calculate the total power for each spectral estimate, and to determine the frequency below which a given percentage of the total power lies. Values of the order of 75–95% of total power have been used in the past, and this method can be successful when the signal-to-noise ratio is high (in which case higher percentage values can be used).

8.4.1.3 Threshold-crossing Method (TCM)

This method suggested by D'Alessio (1985) determines a new threshold for each spectral sweep. The method is based on the assumption that the noise superimposed on the signal can be modelled as a

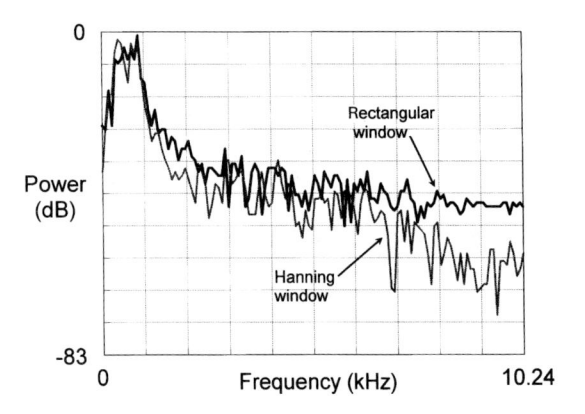

Figure 8.20 Typical spectra estimated from a short segment of a Doppler signal obtained from a carotid artery using both a rectangular window and a Hanning window (reproduced by permission of Churchill Livingstone from Moraes et al 1995)

white Gaussian process, i.e. with a constant spectral density over the analysed bandwidth, and that the tail of the Doppler spectrum can be used to determine that noise level. A threshold level, L_T, is determined based on the estimated noise level, the assumed statistics of the noise, the required specificity of the process and the search algorithm to be used. The magnitude of each bin of the spectrum (scanning from the upper to the lower frequency bins) is compared with the threshold, and when in a sequence of r successive bins there are at least m bins which exceed the threshold, then the required maximum frequency is taken to be the highest bin frequency in that sequence. In practice r and m usually take the values of $(2, 2)$ or $(3, 6)$, and in the former case the appropriate threshold was shown by D'Alessio to be:

$$L_T = -N_0 \ln(1 - r_{\min}) \qquad 8.51$$

where N_0 is the spectral density of the white Gaussian noise, and $r_{\min}$ the minimum specificity setting. As will be explained in the next section, the threshold-crossing method requires modification to function correctly with windowed spectral estimates.

8.4.1.4 Modified Threshold-crossing Method (MTCM)

An assumption of the threshold-crossing method is that no explicit window function is applied to the Doppler signal before the spectral estimation is performed (i.e. a rectangular data window is used). However, rectangular windows introduce large sidelobes (Harris, 1978) and therefore are seldom used in practice. When other windows are applied to the sampled Doppler signal there is a smoothing of the noise present in the tail of the spectrum, and so the noise no longer has a constant spectral density. This phenomenon is illustrated in Fig. 8.20, where it can be seen that the noise level is fairly constant in the spectrum obtained with a rectangular window, but steadily decreases with frequency in the spectrum obtained with a Hanning window. In order to overcome this problem, Mo et al (1988) proposed a modified threshold-crossing method which is the same as D'Alessio's threshold-crossing method with $r = m = 2$, but with a threshold level L'_T set empirically to some multiple of N', the average noise level estimated from the tail of the windowed spectral estimate. Mo et al recommended a threshold level of several times N' to avoid producing large positive bias in the estimated maximum frequency, at the expense of generally producing a small negative bias in the estimated maximum frequency, and in their simulation studies used a value of $8N'$ for SNRs between 5 dB and 13 dB, and $11N'$ for SNRs between 13 dB and 20 dB. It should be noted that for both the TCM and the MTCM the noise in the tail of the spectrum can also be affected by the presence of any anti-aliasing filters. Also, it is assumed that there is no signal in the tail of the spectrum where the noise is estimated, and this can cause difficulties with fixed-length FFT systems because the Doppler spectrum must be confined to the lower part of the spectrum. This implies poorer spectral resolution that will of itself decrease the accuracy of the maximum velocity estimation.

A further modification of the MTCM was suggested by Cloutier et al (1990) whilst studying the Doppler spectral envelope area in patients with aortic stenosis. The method essentially depended on finding a dominant amplitude peak in each spectrum, determining a pair of minimum and maximum frequency values around this peak using a threshold level, and then repeating the process to find a second dominant amplitude peak entirely separate from the first. The pair of minimum and maximum values associated with the peak with the greater bandwidth were then chosen as the final estimates of the minimum and maximum frequencies of the spectrum. The rationale behind this approach is not quite clear, but was presumably to produce both maximum and minimum frequency envelopes which are at least partially immune to the effects of spurious noise spikes, and the high-amplitude low-frequency vibrations which appear to be associated with Doppler recordings from stenosed aortic valves.

8.4.1.5 Hybrid Method (HM)

Mo and his colleagues (1988) suggested a new hybrid method of maximum frequency detection which combines features of both the percentile method and the threshold crossing method. It is similar to the percentile method in that the maximum frequency is defined as the frequency below which a certain percentage of the power lies, but the percentage is determined from the estimated noise power density, and the proportion of the frequency range that is occupied by the signal (which in turn depends on the maximum frequency!). The rationale for this is that if the signal occupies only a small part of the frequency range, the percentage of the total noise power lying above the maximum frequency will be large, whilst if the signal occupies a large part of the frequency range, most of the noise power will actually lie below the maximum frequency. This is particularly important when the signal-to-noise ratio is low.

The implementation of the hybrid method is illustrated in Fig. 8.21(a). The curve $\phi(f)$ is the integrated Doppler power spectrum, defined as:

$$\phi(f) = \int_{f_L}^{f} \hat{P}(f)\, df \qquad 8.52$$

where $\hat{P}(f)$ is the estimated power spectrum of the Doppler signal (including noise and any other distorting influences such as filters) and f_L is some lower frequency below which the noise can no longer be assumed to be white Gaussian (because of the presence of wall thump filters and possible $1/f$ noise).

Note that the curve $\phi(f)$ is relatively steep in the low frequency region, where both the Doppler signal and the noise contribute to the power spectrum, and relatively flat in the high frequency region, where only the noise is present. In this diagram the boundary between the two parts of the curve, the 'knee', represents the transition from signal plus noise to noise alone, that is the maximum frequency of the Doppler signal. The line labelled A, passes through the maximum of the curve $\phi(f)$ with a slope, s, determined from the estimated noise power density $\hat{N}_0$ (determined from the tail of the power spectrum). The hybrid maximum frequency estimate is given by the lowest frequency at which line A crosses $\phi(f)$.

Care is needed in the choice of slope s. If the noise were truly white Gaussian noise, then s could be set to equal $\hat{N}_0$, but any underestimation of N_0, or departure from whiteness (as, for example, will occur with windowing before spectral estimation) could potentially lead to a large positive bias. Because of

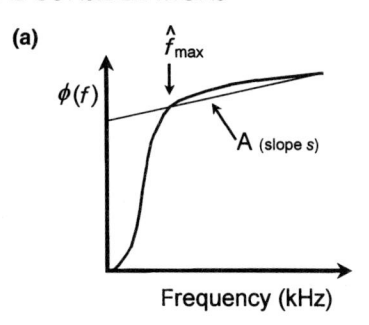

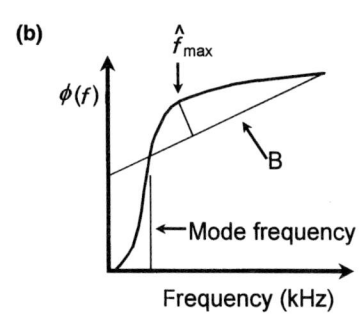

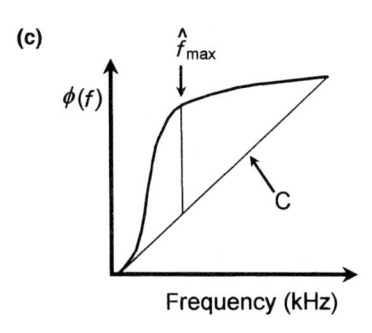

Figure 8.21 Diagrammatic representation of various maximum frequency followers. (a) Hybrid method. The line A passes through the maximum of the integrated Doppler spectrum [$\phi(f)$] with a slope determined from the estimated noise power density in the tail of the power spectrum. (b) Geometric method. The line B passes through the maximum of $\phi(f)$ and the point on $\phi(f)$ corresponding to the mode frequency. (c) Modified geometric method. The line C passes through the maximum and minimum of $\phi(f)$

this, Mo et al recommended a slope greater than $\hat{N}_0$, and in their simulations used values of $3\hat{N}_0$ for signals with SNRs in the range 5–13 dB, and $4.5\hat{N}_0$ for signals with SNRs in the range 13–20 dB.

8.4.1.6 Geometric Method (GM)

The 'geometric method' was introduced by Marasek and Nowicki (1994) as part of their work on the coupling of maximum frequency estimators with different spectral estimation methods. The method was intended to find the 'knee' of the integrated Doppler power spectrum $\phi(f)$ by simple geometric means. The method is illustrated in Fig. 8.21(b). The first step is to construct a straight line (line B) connecting the maximum of the curve $\phi(f)$ and the point on $\phi(f)$ corresponding to the frequency bin in the raw Doppler spectrum with the maximum value of power, i.e. f_{mode}. The maximum frequency is then taken to correspond to the point on $\phi(f)$ where the perpendicular distance between line B and $\phi(f)$ is maximum.

8.4.1.7 10% Max (10% MM)

Marasek and Nowicki (1994) also used a method which defined the maximum frequency as the frequency at which spectral power decreased to 10% of its maximum value, but showed it only to be at all reliable when applied to smoothed spectral estimates (derived in their case using AR and ARMA modelling).

8.4.1.8 Modified Geometric Method (MGM)

A modified version of the geometric method was introduced by Moraes and colleagues (1995) in order to overcome some shortcomings they detected in the geometric method of Marasek and Nowicki. The method is illustrated in Fig. 8.21(c). Three modifications were made in all. The first was simply to introduce an empirical power threshold so that, if there was insignificant signal power present, the maximum frequency was set to zero rather than producing a spurious output. The other two modifications were to replace line B with a straight line connecting the maximum of the curve $\phi(f)$ to the minimum of curve $\phi(f)$ (line C), and to define the maximum frequency as corresponding to the maximum *vertical* distance between $\phi(f)$ and line C. Line C was preferred to line B because the latter depends on f_{mode}, whose position can fluctuate quite markedly due to the variance of the spectral estimator and also be influenced by noise spikes. The vertical distance was preferred to the perpendicular distance because the former, unlike the latter, is independent of system gain.

8.4.1.9 Adaptive Threshold Method (ATM)

Routh et al (1994) have described a method of detecting the maximum frequency envelope using a threshold value that is updated every cardiac cycle depending on the estimated signal-to-noise ratio of the previous cardiac cycle. This method may be thought of as an automatic version of the simple threshold method. During each cardiac cycle the signal-to-noise ratio is estimated from a subset of Doppler spectra. For each individual spectrum the average signal value and average noise value are found by separately summing all power values that are greater than or less than a threshold value, T_i. The average signal and noise values for the whole subset of spectra are then separately averaged to determine average signal and average noise values, and from these a new value of threshold T_{i+1}, calculated as:

$$T_{i+1} = \frac{\text{average noise } (i)}{k \times \text{average noise } (i)} \quad 8.53$$

where k is a constant which is dependent on scroll rate.

In order to initiate the procedure the first value of threshold, T_1, is taken to be a given fraction (-3 dB is suggested) of the peak intensity of each sweep. The algorithm described by Routh et al also incorporates a number of measures to ensure that each sweep really contains valid signal, otherwise no peak determination is made, or the peak is based on a previous estimate. These measures are based on comparing the peak intensity of the sweep with the average noise level of the previous cardiac cycle.

8.4.1.10 Comparison of Extractor Performance

There have been a number of studies comparing the efficacy of two or more of the methods of extracting a maximum frequency envelope. Mo et al (1988) compared the percentile method, the threshold-crossing method, the modified threshold-crossing method and the hybrid method, and found the performance of the various estimators to depend not only on the SNR and bandwidth, but also on the shape of the power spectrum. The PM did not appear to be very reliable for SNR ⩽ 20 dB because it does not adapt to noise level and signal bandwidth. The TCM performed well at SNRs below 13 dB, but at higher SNRs tended to have greater variance than the MTCM and HM. Overall they concluded that the MTCM and HM had the best and most consistent performances over the range of conditions tested.

Marasek and Nowicki (1994) compared the performance of the 10% max method, the modified threshold-crossing method and the geometric method (in conjunction with three spectral estimation techniques) and, like Mo et al (1988), showed the performance of the various estimates to be dependent on SNR, signal bandwidth and spectral shape. They found that it was possible to determine reliable values for the maximum frequency envelope even for very low SNRs (> 3 dB) and that regardless of spectral shape and the spectral estimator used, better estimates of maximum frequency were obtained with narrowband as opposed to wideband signals. They concluded that the 10% MM was only reliable for use with smoothed spectra and that overall the GM performed best, but that the MTCM produced similar results to the GM for SNRs > 6 dB.

Moraes et al (1995) compared the simple threshold method, the modified threshold-crossing method and the modified geometric method. These authors concluded that the STM was the most useful for off-line applications, where several values can be tried for achieving the 'best' maximum frequency envelope, and whilst the MTCM has a simpler and faster algorithm than the MGM, the latter is more suitable for general purpose software, being totally automatic and very robust.

More recently Rickey and Fenster (1996) have extensively tested a commercial implementation of the adaptive threshold method and found it to be of value in a clinical setting.

8.4.2 Relationship to Mean Frequency

The relationship between the instantaneous mean and maximum velocities in an artery depends entirely on the shape of the velocity profile. For plug flow the mean and maximum velocities are equal, whilst for parabolic flow the maximum velocity is exactly twice the mean. In pulsatile flow the velocity profiles may change quite dramatically during the cardiac cycle and thus the ratio of maximum to mean frequency in the Doppler signal will also vary widely. Despite this, the shapes of maximum and mean frequency envelopes derived from the same signals are often superficially similar (Evans and Macpherson 1982).

Under certain flow conditions (such as those found in vessels supplying low-impedance vascular beds) the maximum velocity in the vessel is in its centre throughout the cardiac cycle. If this is the case, the ratio of the time-averaged maximum Doppler frequency to the time-averaged mean Doppler frequency is the same as the ratio of the maximum velocity to the mean velocity in the time-averaged mean component of the velocity profile (Evans 1985b). This means that if the shape of the mean velocity profile is known, then the time-averaged mean velocity can be calculated from the time-averaged maximum velocity. In the case where the mean velocity profile is parabolic, as it will be at sufficient distances from geometry changes, then the time-averaged maximum velocity will be twice the time-averaged mean velocity (Evans 1985b). This has been confirmed to be the case in neonatal cerebral arteries (Evans et al 1989b), and has also been demonstrated in a computer controlled flow rig by Li and colleagues (1993).

8.4.3 Effect of Non-uniform Insonation

One of the great advantages of the maximum frequency processor is that its output is not significantly influenced by ultrasonic beam shape or the ratio of the beam width to vessel diameter, provided that some part of the ultrasound beam passes through the part of the vessel containing the maximum velocity (Evans 1982).

8.4.4 Effect of Attenuation

Since ultrasonic beam shape is not critical for the correct functioning of the maximum frequency follower, differential attenuation between blood and tissue is of little consequence.

8.4.5 Effect of Intrinsic Spectral Broadening (ISB)

The difficulties that ISB presents to maximum frequency followers has already been mentioned in the context of developing algorithms to estimate maximum frequency (8.4.1). This difficulty occurs because, in the presence of significant ISB, the Doppler power spectrum falls off gradually rather than there being a well-defined maximum frequency corresponding to a well-defined maximum velocity. A second and in many ways more serious difficulty is that because maximum frequency algorithms necessarily seek the point in the Doppler power spectrum at which the signal just rises above the noise, in the presence of significant ISB, they extract a frequency value that corresponds to the maximum velocity *plus* an increment due to spectral broadening. Substitution of this maximum frequency into the classical Doppler equation therefore produces an overestimate of the true maximum velocity.

Ignoring the effects of ISB, the maximum Doppler shift frequency $f_{d(max)}$ in a sample volume may be calculated simply by substituting the maximum velocity v_{max} into eqn. 1.1, i.e:

$$f_{d(max)} = \frac{2v_{max}f_t \cos\theta}{c} \qquad 8.54$$

where f_t, θ and c are as defined in eqn. 1.1.

For high values of θ, such that the transit time of flow is determined by the beam lateral edges rather than the ends of the range cell, then the Doppler bandwidth B_d due to a target moving with a velocity, v, through the field of a focused transducer, may be written:

$$B_d \approx \frac{2f_t}{c} \frac{D}{F} v \sin\theta \qquad 8.55$$

where D is the active Doppler aperture width of the transducer and F is the focal length of the transducer (Censor et al 1988; see also 7.3.4). Since ISB is more or less symmetrical, the maximum Doppler shift frequency detected in the presence of significant ISB, $f'_{d(max)}$, will be given by:

$$f'_{d(max)} = f_{d(max)} + B_d/2 \qquad 8.56$$

or, substituting eqns 8.54 and 8.55 into eqn. 8.56:

$$f'_{d(max)} = \frac{2v_{max}f_t}{c}\left\{\cos\theta + \frac{D}{2F}\sin\theta\right\} \qquad 8.57$$

Modern array transducers often use wide apertures, and may be used at insonation angles of 60° or greater, and this can easily lead to overestimates of maximum velocity as high as 40% or even more, as shown by Fig. 8.22.

Several authors have recognised this type of error to be significant in either their own work or that of others. Daigle et al (1990) described the overestimation of frequency by linear array Dopplers. Hoskins at al (1991a) studied the errors in maximum velocity determination with a commercial string phantom and ascribed the overestimates of peak frequency in their own work, and that of Tessler et al (1990), to ISB. Eicke et al (1995) used a string phantom to examine the accuracy with which several ultrasound instruments could estimate maximum velocity using a range of different insonation angles. They found that a PW system using a single-crystal transducer provided acceptably accurate estimates at various angles, but that all the duplex instruments (from three different manufacturers) that used linear array transducers consistently produced an overestimate of the true velocity, which increased with increasing angles of insonation. Thrush and Evans (1995) warned of the possibility of misdiagnosis of carotid artery disease because of variable overestimation of peak frequency, depending on machine design and angle of insonation. Tortoli et al (1995), Winkler and Wu (1995) and Chabria and Newhouse (1997) suggested corrections to improve estimates of maximum velocity, based on eqn. 8.57, although Winkler and Wu had difficulties in making adequate corrections for angles of less than 70°.

8.4.6 Effect of Frequency-dependent Attenuation and Scattering

The effects of frequency-dependent attenuation and scattering on Doppler spectra have been discussed in Section 8.3.5, where it was seen that for very short pulses these mechanisms could significantly change the centre frequency of the

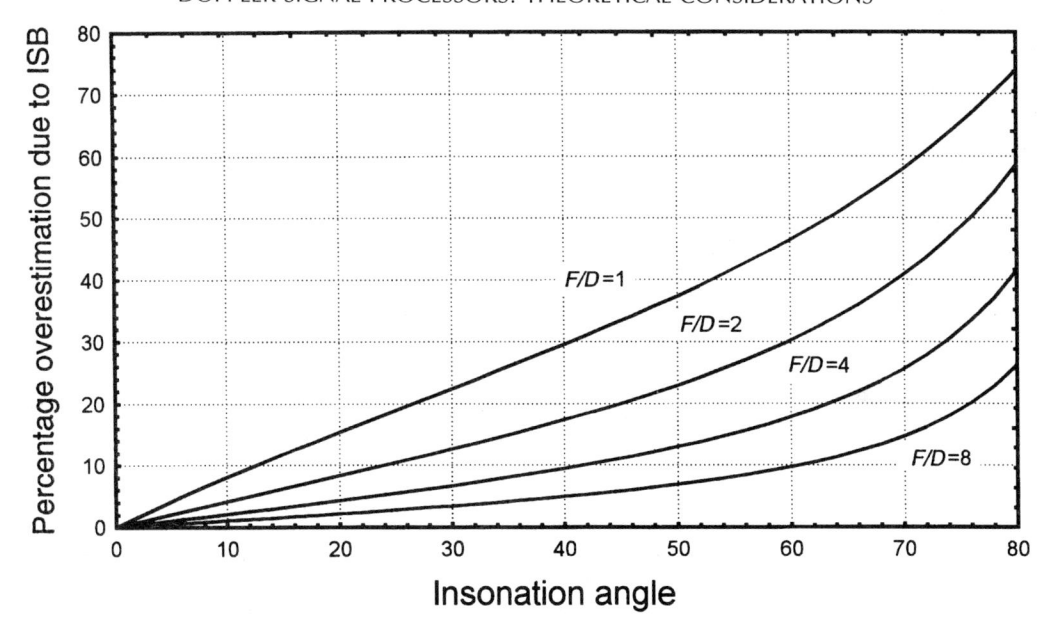

Figure 8.22 Percentage overestimation of maximum frequency due to ISB as a function of insonation angle, plotted for various values of transducer focal length to active aperture width ratio (eqn. 8.57)

pulse and hence both the mean and maximum Doppler frequency. The same mechanisms have very little effect on pulse bandwidth (see for example Embree and O'Brien 1990), and therefore any extra minor effect from this source will be totally overshadowed by the ISB of the Doppler signal due to the use of short pulses.

8.4.7 Effects of Filters

Provided the maximum frequency in the Doppler signal does not fall below the cut-off frequency of the high pass (wall thump) filter, the maximum frequency follower is totally unaffected by its action. If the maximum frequency does fall below this value, the follower output will drop to zero. This problem can only be alleviated by using the lowest practical filter cut-off frequency for each application.

8.4.8 Effect of Signal-to-noise Ratio

A major advantage of maximum frequency followers is their noise immunity, but full advantage of this virtue can only be realised if the maximum frequency envelope is superimposed on a real-time sonogram (Fig. 8.23). Using this method it is possible for an operator to monitor the performance of the follower, and if necessary change its threshold level to ensure that it follows the required signal. A practical method of implementing this has been described by Prytherch and Evans (1985). Experience has shown that the threshold level becomes critical only when there is a very poor signal-to-noise ratio, and that even then it is usually possible by careful adjustment to 'pull out' the required signal. Figure 8.23 also illustrates that the system can separate even the signals from two vessels that have been simultaneously insonated. The threshold has been set to extract the higher frequency signals but, by suitable manipulation of the analyser gain and threshold level, the maximum frequency of the second vessel could be followed. This could not be achieved using any of the other followers described in this chapter. Other sources of noise, such as venous flow and vessel wall thump, are easily recognised on the sonogram and are usually easy to eliminate.

Without simultaneous sonographic display the maximum follower is better at coping with

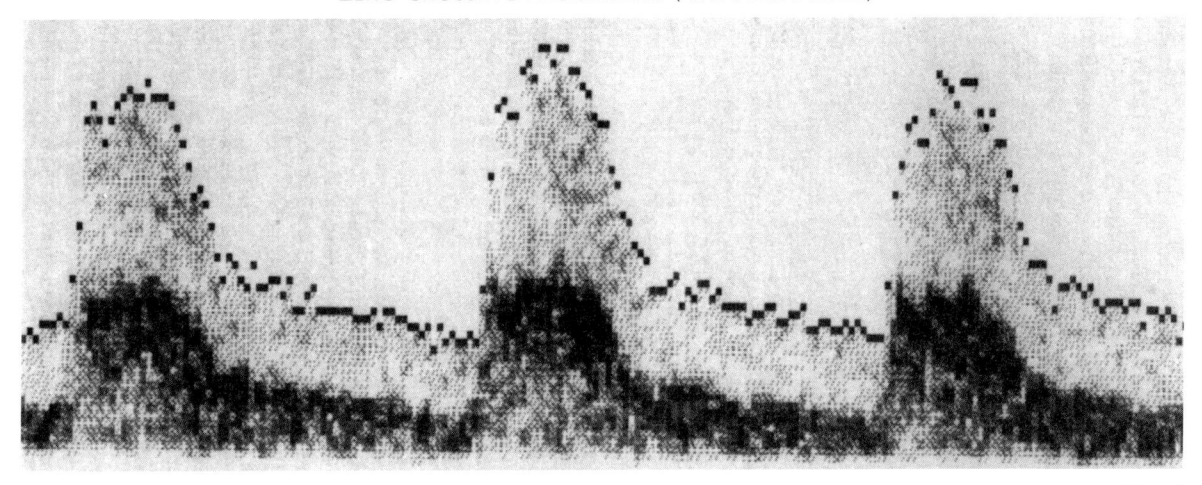

Figure 8.23 Sonogram with superimposed maximum frequency enveloped. This particular display resulted from accidental simultaneous insonation of two vessels in the neck of a normal subject. The threshold of the microcomputer has been set to pick out the vessel giving the larger Doppler shifts, but could equally well have been set to follow the maximum frequency outline of the other vessel. The maximum frequency follower is the only processor that will allow such separation and it is only possible with a system that superimposes the maximum frequency envelope directly on to the sonographic display (reproduced by permission from Evans 1985a)

noise than other followers; used in conjunction with a spectral display it can be very much better.

8.5 ZERO-CROSSING PROCESSORS (RMS FOLLOWERS)

Zero-crossing processors were the first type of envelope follower to be successfully employed with Doppler ultrasound velocimeters (Franklin et al 1961) and are still in use today in simple commercial devices. For the case of band-limited noise (such as the Doppler audio signal) the output of such a device is proportional to the root mean square (RMS) frequency of the input signal (Rice 1944). These devices have remained in use primarily because of their simplicity, but are subject to many errors (Flax et al 1970, 1973, Lunt 1975, Johnston et al 1977).

8.5.1 Statistical Noise and Output Filters

Flax et al (1970) have studied the statistical noise generated in the zero-crossing detector both theoretically and experimentally and have highlighted the conflicting conditions necessary for both a low noise level and an adequate frequency response at the output. They calculated the relative standard deviation at the output of a low pass filter receiving three types of pulse trains, and showed that for band-limited white noise the ratio of the standard deviation to the mean of the signal at the output is given by:

$$\frac{\sigma}{\mu} = \left(\frac{1 - 2\zeta\lambda}{2\lambda\kappa}\right)^{1/2} \qquad 8.58$$

where λ is the mean zero-crossing density, κ the filter time constant and ζ the time interval after the occurrence of a pulse, during which the probability of the occurrence of a second pulse is considered to be zero. They also showed that ζ could be approximated by $0.48/\lambda$ and eqn. 8.58 may therefore be rewritten:

$$\frac{\sigma}{\mu} = \frac{0.2}{(2\lambda\kappa)^{1/2}} \qquad 8.59$$

For low values of λ, that is for portions of the cardiac cycle where there is a low flow velocity, the signal-to-noise ratio at the output may be very poor unless the filter time constant is long. Unfortunately, a long time constant will smooth the output waveform and prevent the accurate observation of rapid flow changes, such as those that occur at the beginning of systole: therefore, some compromise between required frequency response and tolerable signal-to-noise ratio must be reached.

8.5.2 Relationship to Mean Frequency

The relationship between the mean frequency and RMS frequency of the Doppler power spectrum depends on the shape of the velocity profile. If the profile is flat, the two frequencies will be equal; if the profile is parabolic (and uniformly insonated), the RMS frequency will be 15% higher than the mean frequency (Woodcock et al 1972, Evans 1982).

In pulsatile flow the velocity profile changes during the cardiac cycle and therefore so does the relationship between the output of a zero-crossing detector and the mean frequency. As Johnston et al (1977) have pointed out, a number of authors have tested the linearity of the zero-crossing detector using steady flow regimens, and have erroneously concluded that its output accurately follows changes in mean velocity.

8.5.3 Effect of Non-uniform Insonation

The Doppler spectrum may be dramatically distorted by non-uniform insonation of the blood vessel (Section 7.3.2), and this affects the behaviour of the RMS follower. Evans (1982, 1985a) has discussed the changes in the output of such devices resulting from partial insonation of vessels containing parabolic velocity profiles.

8.5.4 Effect of Attenuation, Spectral Broadening and Filters

The effects of attenuation, spectral broadening and filters on the output of RMS followers are qualitatively similar to their effects on the output of mean followers. These have not been investigated in detail, presumably because of other more significant limitations of RMS followers.

8.5.5 Effect of Signal-to-noise Ratio

A major problem with zero-crossing detectors is that they require a high signal-to-noise ratio (SNR) to operate properly. The ideal zero-crossing

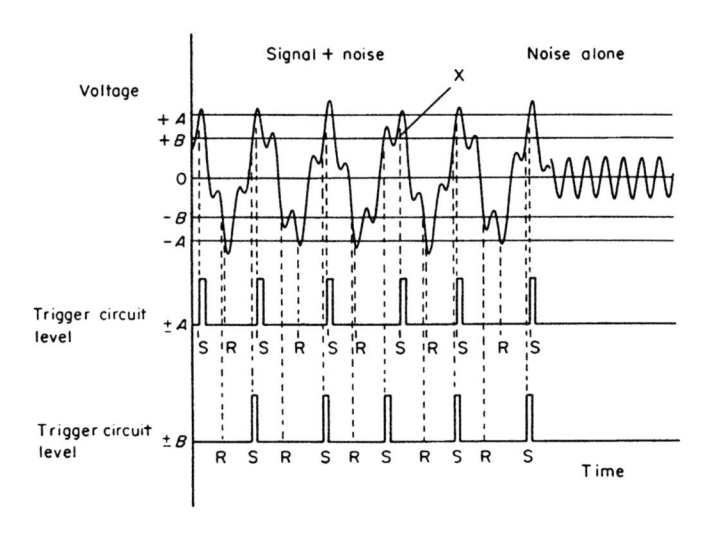

Figure 8.24 The SET–RESET zero-crossing detector with noise. The output is correct with trigger levels $\pm A$ and $\pm B$. Note that the noise alone has more 'true' zero-crossings than the signal + noise (reproduced by permission of Elsevier Science, from Lunt 1975, *Ultrasound in Medicine and Biology*, © World Federation of Ultrasound in Medicine and Biology)

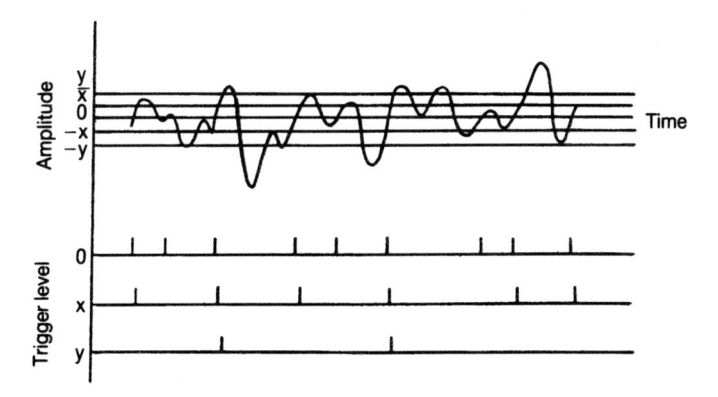

Figure 8.25 The effect of signal amplitude on the output of a zero-crossing detector. With a complex signal, the number of pulses depends on the trigger level or, if the trigger level is kept constant, on the mean amplitude of the signal (reproduced by permission of Elsevier Science, from Lunt 1975, *Ultrasound in Medicine and Biology*, © World Federation of Ultrasound in Medicine and Biology)

detector would function by generating an output pulse each time the signal crossed its own mean level. Such a detector is, however, impracticable because even in the absence of signal the electrical noise from the system would generate a large output signal. This problem is overcome in practice by using a SET–RESET system (Lunt 1975), which produces a pulse when two trigger levels placed either side of the zero line are passed in succession (Fig. 8.24). Provided the peak-to-peak amplitude of the noise is less than the difference between the two trigger levels, it cannot trigger the detector. Note that in the example shown an 'ideal' zero-crossing detector would produce more pulses from the noise alone than from the noise plus signal.

For the simple signal shown in Fig. 8.24 the zero-crossing detector is insensitive to amplitude (provided it exceeds the difference between the SET and RESET level), but for the complex noisy signals encountered from blood flow this is not so and the number of output pulses depends on the signal amplitude and threshold levels (Fig. 8.25). If the SNR at the input of the detector is high, the detector output will be independent of signal amplitude over a wide range, and will be proportional to the RMS frequency of the input; if the SNR is low, there may be no plateau region when the output is independent of input signal amplitude, and the output will be unreliable.

8.6 OTHER SIGNAL LOCATION ESTIMATORS

Mean, maximum and RMS frequency processors are at present much more common than any other type of signal location estimator, but a number of others have been suggested from time to time. Three frequency followers which may have some potential are the first moment follower, the mode frequency follower, and the median frequency follower.

8.6.1 First Moment Followers

The first moment of the Doppler power spectrum is a particularly interesting quantity because it is theoretically more closely related to instantaneous volumetric flow that is the intensity-weighted mean frequency (Saini et al 1983).

The intensity-weighted mean frequency of the Doppler power spectrum $\bar{f}(t)$ was defined in eqn. 8.31 as its first moment, M_1, divided by its zeroth moment, M_0, where:

$$M_1 = \int_f P(f) f \, df \qquad 8.60$$

and

$$M_0 = \int_f P(f) \, df \qquad 8.61$$

Note that M_0 represents the total power in the Doppler spectrum and that for conditions of uniform insonation this is proportional to the cross-sectional area of the target blood vessel.

Under ideal conditions the instantaneous mean velocity of flow is proportional to $\bar{f}(t)$ and may be calculated using the standard Doppler equation (eqn. 1.1). The instantaneous volumetric flow may then be calculated by multiplying the instantaneous mean velocity by the cross-sectional area of the target blood vessel. If the vessel cross-section is relatively constant throughout the cardiac cycle, a single area measurement will suffice; if, however, the vessel is very pulsatile, then ideally this should be measured continuously. Since M_0, the zeroth moment of the power spectrum, is proportional to vessel cross-sectional area, multiplying $\bar{f}(t)$ by M_0 (or in practice refraining from dividing M_1 by M_0) results in a quantity proportional to instantaneous flow.

The value of the first moment follower as a means of monitoring changes in mean flow has been investigated *in vitro* by Saini et al (1983), and it has also been used *in vivo* in waveform analysis applications by Maulik et al (1982) and Thompson et al (1985, 1986). It is slightly easier to implement a first moment follower than a mean frequency follower, whether using analogue or digital techniques, because less processing is required.

In principle the first moment follower seems to have very desirable characteristics for monitoring proportional changes in flow, but unfortunately the total power in the Doppler spectrum is affected by factors other than the vessel cross-section, and in particular may be reduced very significantly during some parts of the cardiac cycle by the high pass filters designed to removed the high amplitude low-frequency power from static and nearly static structures. Also, any slight change in the relationship or coupling between the Doppler probe and sample volume can lead to very dramatic changes in the received Doppler power, and so the first moment follower is only suitable for short recordings in geometrically very stable conditions.

8.6.2 Mode Frequency Followers

An off-line mode frequency follower has been described by Greene and colleagues (Greene et al 1982, Knox et al 1982). The algorithm first finds the region of the spectrum containing the signal by locating the median frequency which exceeds a fixed threshold, and then searches its immediate vicinity for the signal mode. This estimator ignores the high-amplitude narrow-bandwidth noise around the zero flow region, and is claimed to be insensitive to most cases of background noise and quadrature channel crosstalk. Such a follower may be of use because of its noise-resistant properties. Voyles et al (1985) found the mode frequency to be more accurate than either the mean or maximum frequency for estimating flow in a hydraulic model.

8.6.3 Median Frequency Followers

A median frequency follower which operates off-line on the output of a spectrum analyser has been described by Thomson (1980). During a first pass, an arithmetic logical unit functions as an accumulator and computes the total amplitude of the spectrum. This amplitude is divided by two, and then during a second pass the amplitude of each component of the spectrum is subtracted from this value until it reaches zero. The bin in which this occurs is stored as the median. Strictly speaking it would seem more correct to operate on the intensity of each bin rather than the amplitude.

8.7 ENVELOPE AVERAGING TECHNIQUES

The possibility of reducing spectral variance by averaging has been discussed in Section 8.2.2. The signal-to-noise ratio (SNR) of the Doppler envelope waveform may similarly be improved by using a coherent averaging technique (note that the 'noise' in this context may either be due to noise in the true sense, or to the stochastic properties of the Doppler signal). The principle of such a technique is that if the waveforms are properly aligned (by time locking them to a suitable feature) and summed, the signal amplitude will increase more rapidly than noise, which has a random time relationship to the signal. If there is perfect alignment and the noise is of a random Gaussian nature, the signal amplitude will rise in proportion to the number of the waveforms summed, N, whilst the noise will increase only as the square root of N, leading to an improvement in the SNR of root N (English and Woollons 1982). Thus, for example, if 64 Doppler waveforms are coherently averaged there will be an improvement in the SNR of 800%. An example of the improvement achieved by summing only 17 waveforms is shown in Fig. 8.26.

The feature used to time-lock the averager can be part of the signal itself, for example the foot of the systolic up-slope (Evans et al 1985, Evans 1988) or an associated external event, such as the R-wave of the ECG. In practice, physiological waveforms are never exactly repetitive, and the interval between the R-wave and the flow pulse is slightly variable; nevertheless, considerable improvement in SNR is possible.

Kitney and Giddens (1983) have reported the use of phase shift averaging with velocity waveforms derived from hot film anemometry, which we have found useful for some Doppler waveforms. This elegant method helps to reduce phase variations due to physiological variability and trigger errors. An initial ensemble average is calculated by using the R-wave as the fiducial point. Each individual waveform is then cross-correlated with this average, and a new ensemble average calculated by aligning the waveforms on the correlation maxima. The process may then be repeated until there is no further change in the ensemble average or until some other convergence criterion is satisfied.

8.8 COMPARISON AND CHOICE OF ENVELOPE EXTRACTION TECHNIQUES

Each of the envelope extraction techniques discussed has a number of advantages and disadvantages, and their utility must to some extent depend on their intended use.

For volumetric blood flow measurements, the mean frequency processor has traditionally been the method of choice, but both the first moment processor and the maximum processor may offer advantages. Whilst the output of the mean processor is proportional to instantaneous mean velocity, the output of the first moment processor is (ideally) proportional to instantaneous volumetric flow, and thus may be more accurate when there are significant changes in the vessel cross-section over the cardiac cycle (although absolute calibration is not easy). The maximum frequency processor has the advantage that its output is essentially independent of the shape and size of the ultrasound beam and may be of value for flow measurement when the velocity profile is flat, or where the flow is fully established and relatively non-pulsatile.

The relative merits of different processing techniques when used prior to waveform analysis have not been studied in any detail, even though the use of different methods may considerably

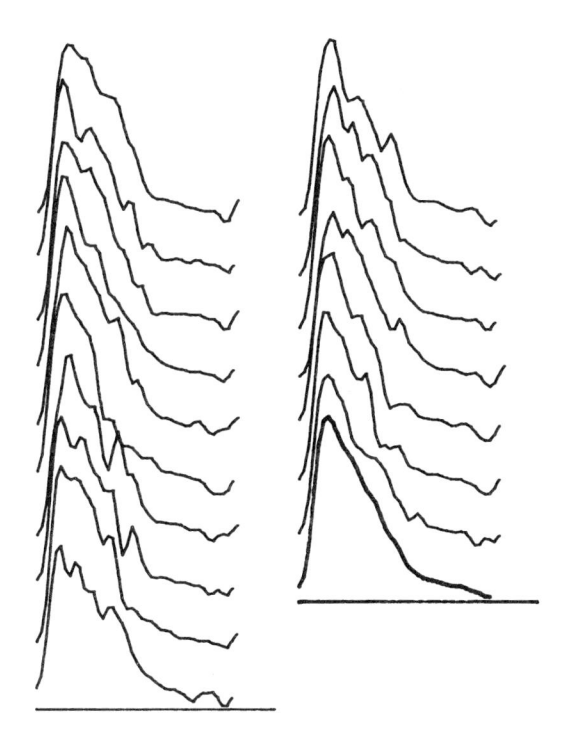

Figure 8.26 Maximum frequency envelopes derived from a consecutive series of Doppler recordings from the anterior cerebral artery of a neonate. The final waveform is an ensemble average and shows a considerable improvement in SNR over the individual components

modify the waveform shape (Evans 1985a), and there is little evidence that any particular processing technique improves the sensitivity of any of the waveform analysis techniques discussed in the next chapter. It is, however, very important that each laboratory establishes normal and abnormal ranges for each technique they intend to use, as these may be significantly affected by the choice of ultrasound transducer and envelope extraction technique. Since there is little evidence to suggest the use of any particularly processor, it seems expedient to use a method that is relatively insensitive to noise and other distorting influences. Both the maximum follower and the mode follower are valuable in this respect, and the former is particularly useful because it is easy to monitor its performance by superimposing its output on a sonogram.

8.9 SUMMARY

The Doppler shift signal contains a wealth of information about blood flow, but since this information is encoded in the frequency content of the signal, it is necessary either directly to extract a characteristic frequency envelope (such as a mean frequency envelope or maximum frequency envelope) from the signal or, more usually, to transform the signal into the frequency domain and then extract information from the resulting 'sonogram'.

A number of techniques have been used both for spectral estimation, and envelope extraction, and these have been surveyed in this chapter. By far the most widespread spectral estimation technique is that of fast Fourier transformation, and for most purposes this remains the best method of transforming signals into the frequency domain. Other methods, such as autoregressive and autoregressive moving average modelling and the various time-frequency distribution methods, may have benefits to offer when the Doppler signal is changing rapidly in time, but at present their uses are more or less confined to research applications.

There are several methods available for extracting an envelope signal from the full Doppler signal. Methods in widespread use include those that extract the instantaneous mean, maximum and RMS frequencies from the power spectrum, and other methods such as first moment and mode frequency followers have been suggested as being valuable. Each of the followers has a number of advantages and disadvantages and none is ideal for all purposes.

Under ideal conditions, the output of the mean frequency follower is proportional to instantaneous mean velocity and the output of the first moment follower to instantaneous volumetric flow, but both are susceptible to any influence that modifies the Doppler power spectrum, particularly non-uniform insonation, noise and filters.

The maximum frequency follower is probably the method of choice in waveform shape applications and under some circumstances may be used for flow measurement. Its major advantages are that it is resistant to noise and to measures designed to reduce noise, such as high pass filters, and that its output is fairly independent of the way in which the ultrasound beam samples the blood vessel. It is, however, particularly susceptible to the effects of intrinsic spectral broadening. For the best results, its output should be superimposed on a real-time sonogram display, so that an operator can ensure that it is following the desired part of the spectrum.

The zero-crossing detector (or RMS follower) is cheap and simple to construct but has little else to recommend its use.

8.10 NOTATION

a_k	Autoregressive coefficients
b_k	Moving average coefficients
B_d	Doppler bandwidth due to intrinsic spectral broadening
B_0	-6 dB bandwidth of transmitted pulse
BW	Bandwidth of transmitted pulse
c	Ultrasound velocity
$C(d)$	Local image measure along direction d
$C(t, \omega, \phi)$	Cohen class of time–frequency distributions
D	Active Doppler aperture width of transducer

NOTATION

f	Frequency	$\hat{P}_{MF}(f_d)$	Multifrequency Doppler power spectral estimate
$f_{c(atten)}$	Downshifted centre frequency of pulse due to attenuation	$\hat{P}(v)$	'Velocity matched' spectral estimate
f_c	Centre frequency of transmitted ultrasound pulse	$\hat{P}_{conv.}(v)$	Conventional spectral estimate
f'_c	Centre frequency of received ultrasound pulse	$P(\omega)$	Underlying power spectral density
		$\hat{P}(\omega)$	Estimated power spectral density
f_d	Doppler shift frequency	$P_{1D}(f_d)$	Conventional 1-D Fourier spectrum of Doppler signal
$\bar{f}_d$	Mean Doppler shift frequency		
$f_{d(max)}$	Maximum Doppler shift frequency	$P_{2D}(f_{RF}, f_d)$	2-D Fourier spectrum of received ultrasound signal
$f'_{d(max)}$	Maximum Doppler shift in presence of ISB	r	Radial co-ordinate
		$r.u.(\bullet)$	Relative uncertainty in parameter $\bullet$
f_{error}	Error in estimation of mean frequency	R	Vessel radius
		S	Signal-to-noise ratio
f_{HP}	Cut-off frequency of high pass filter	t	Time
$\hat{f}_i$	Frequency estimate derived from τ_i	T	Pulse repetition period
$\bar{f}_n$	Centroid of noise spectrum	T_x	Data length
f_{RF}	Frequency (spectrum) of transmitted ultrasound pulse	$u(n)$	Input driving sequence
		$U(f)$	Fourier transform of a pulse's complex envelope
$\bar{f}(t)$	Intensity-weighted mean frequency		
$|H(f)|$	Frequency-dependent filter function	v	Target velocity
$I(x, y)$	Input to an image filter	v_{max}	Maximum velocity
j	Square-root of -1	$\bar{v}$	True mean velocity
L_T	Threshold level	$\hat{\bar{v}}$	Estimate of true mean velocity
L'_T	Modified threshold level	$var[\bullet]$	Variance of quantity enclosed by the brackets
$m(d)$	Mean value of elements aligned along a direction d		
		$w(k)$	Window function
$m(x, y)$	Local mean of terms inside a filter window	W_n	Weighting function or window
		$W_x(t, \omega)$	Wigner distribution
M_0	Zeroth moment	x_n	values in a time series
M_1	First moment	$x(n)$	values in a time series
N	Length of transmitted pulse in cycles	$x(\mu)$	Time signal
N_0	Spectral density of white Gaussian noise	$x(t, k)$	Complex base-band signal
		Z	Tissue propagation distance
$O(x, y)$	Output of an image filter		
P_{AR}	Autoregressive power spectral density	α	Scaling factor
		α_0	Frequency-dependent attenuation coefficient
P_{ARMA}	Autoregressive moving average power spectral density		
		β	Rate of modulation (in kHz/ms) or parameter describing velocity profile
P_{MA}	Moving average power spectral density		
$\hat{P}_{AR(m)}(f)$	Autoregressive estimate of AR PSD for mth order model	$\Delta f_{c(atten)}$	Decrease in centre frequency due to attenuation
$\hat{P}_{ML}(f)$	Maximum likelihood estimator	$\Delta f_{c(scat)}$	Increase in centre frequency due to scattering
$P(f)$	Doppler power spectrum		
$\hat{P}(f)$	Estimated Doppler power spectrum	Δv	Range of velocities in sample volume
$\hat{P}_m$	Periodogram estimate of a power spectrum		
		$\Delta \tau$	Duration of signal segment
$\hat{P}'_m$	Modified periodogram estimate	λ	Mean zero-crossing density
$\hat{P}(m_1, m_2)$	2-D periodogram estimate	μ	Mean value

ξ	Frequency lag
σ	Standard deviation or parameter in Choi–Williams distribution
σ_{fm}	RMS width of broadening due to frequency variation
σ_m	RMS width of total broadening due to non-stationarity and windowing
σ_t	RMS width of Gaussian window (in ms)
σ_{wb}	RMS width of broadening due to window
σ^2	Variance
τ	Time lag
τ_i	Time delay between two adjacent zero-crossings
$\phi(f)$	Integrated Doppler power spectrum
$\phi(\xi, \tau)$	Kernel of Cohen distribution
ω	Angular frequency ($=2\pi f$)
ω_0	Fundamental angular frequency

8.11 REFERENCES

Allard L, Langlois YE, Durand L-G, Roederer GO, Cloutier G (1992) Effect of ensemble averaging on amplitude and feature variabilities of Doppler spectrograms recorded in the lower limb arteries. Med Biol Eng Comput 30:267–276

Angelsen BAJ (1980a) Spectral estimation of a narrow-band Gaussian process from the distribution of the distance between adjacent zeros. IEEE Trans Biomed Eng BME-27:108–110

Angelsen BAJ (1980b) A theoretical study of the scattering of ultrasound from blood. IEEE Trans Biomed Eng BME-27:61–67

Angelsen BAJ (1981) Instantaneous frequency, mean frequency, and variance of mean frequency estimators for ultrasonic blood velocity Doppler signals. IEEE Trans Biomed Eng BME-28:733–741

Angelsen BAJ, Kristoffersen K (1983) Zerocrossing density for ultrasonic Doppler signals obtained from computer simulations. Ultrasound Med Biol 9:661–665

Arts MGJ, Roevros JMJG (1972) On the instantaneous measurement of bloodflow by ultrasonic means. Med Biol Eng 10:23–34

Atkinson P, Woodcock JP (1982) Doppler Ultrasound and Its Use in Clinical Measurement, pp. 91–106. Academic Press, London

Baggeroer AB (1976) Confidence intervals for regression (MEM) spectral estimates. IEEE Trans Inform Theory 22:534–545

Baker DW, Rubenstein SA, Lorch GS (1977) Pulsed Doppler echocardiography: principles and applications. Am J Med 63:69–80

Bonnefous O, Pesque P (1986) Time domain formulation of pulse-Doppler ultrasound and blood velocity estimation by cross correlation. Ultrason Imaging 8:73–85

Brigham EO (1974) The Fast Fourier Transform. Prentice-Hall, Englewood Cliffs, NJ

Burckhardt CB (1981) Comparison between spectrum and time interval: histogram of ultrasound Doppler signals. Ultrasound Med Biol 7:79–82

Burg JP (1972) The relationship between maximum entropy spectra and maximum likelihood spectra. Geophysics 37:375–376

Cardoso JCS, Ruano MG, Fish PJ (1996) Nonstationarity broadening reduction in pulsed Doppler spectrum measurements using time-frequency estimators. IEEE Trans Biomed Eng 43:1176–1186

Censor D, Newhouse VL, Vontz T, Ortega HV (1988) Theory of ultrasound Doppler spectra velocimetry for arbitrary beam and flow configurations. IEEE Trans Biomed Eng 35:740–751

Chabria Y, Newhouse VL (1997) Estimation of axial blood velocity using the Doppler equation corrected for broadening. Ultrasound Med Biol 23:967–968

Choi H-I, Williams WJ (1989) Improved time-frequency representation of multicomponent signals using exponential kernels. IEEE Trans Acoust Speech, Signal Processing 37:862–871

Cicilioni MG, Cevenini G, Barbini P, Blardi P, DiPerri T (1998) Comparison of time-frequency distribution techniques applied to a simulated carotid Doppler signal. In: Jan J, Kozumplk J, Szabó Z (eds) Analysis of Biomedical Signals and Images, pp. 115–117. Proceedings of the 14th Biennial International Conference—Biosignal '98. Vutium Press, Brno

Claasen TACM, Mecklenbräuker WFG (1980a) The Wigner distribution—a tool for time-frequency signal analysis. Part 1: Continuous-time signals. Philips J Res 35:217–250

Claasen TACM, Mecklenbräuker WFG (1980b) The Wigner distribution—a tool for time-frequency signal analysis. Part II: Discrete-time signals. Philips J Res 35:276–300

Claasen TACM, Mecklenbräuker WFG (1980c) The Wigner distribution—a tool for time-frequency signal analysis. Part III: Relations with other time-frequency signal transformations. Philips J Res 35:372–389

Cloutier G, Lemire F, Durand L-G, Latour Y, Langlois YE (1990) Computer evaluation of Doppler envelope area in patients having a valvular aortic stenosis. Ultrasound Med Biol 16:247–260

Cloutier G, Allard L, Guo Z, Durand L-G (1992) The effect of averaging cardiac Doppler spectrograms on the reduction of their amplitude variability. Med Biol Eng Comput 30:177–186

Cloutier G, Shung KK, Durand L-G (1993) Experimental evaluation of intrinsic and nonstationary ultrasonic Doppler spectral broadening in steady and pulsatile flow loop models. IEEE Trans Ultrason Ferroelec Freq Contr 40:786–795

Cobbold RSC, Veltink PH, Johnston KW (1983) Influence of beam profile and degree of insonation on the CW Doppler ultrasound spectrum and mean velocity. IEEE Trans Sonics Ultrason SU-30:364–371

Cochran WT, Cooley JW, Favin DL, Helms HD, Kaenel RA, Lang WW, Maling GC, Nelson DE, Rader CM, Welch PD (1967) What is the fast fourier transform? IEEE Trans Audio Electroacoust AU-15:45–91

Cohen L (1966) Generalized phase-space distribution functions. J Math Phys 7:781–786

Cohen L (1989) Time-frequency distributions—a review. Proc IEEE 77:941–981

Cooley JW, Tukey JW (1965) An algorithm for the machine calculation of complex Fourier series. Math Comp 19:297–301

Daigle RE, Baker DW (1977) A readout for pulsed Doppler velocity meters. ISA Trans 16:41–44

Daigle RJ, Stavros AT, Lee RM (1990) Overestimation of velocity and frequency values by multielement linear array Dopplers. J Vasc Tech 14:206–213

D'Alessio T (1985) 'Objective' algorithm for maximum frequency estimation in Doppler spectral analysers. Med Biol Eng Comput 23:63–68

David J-Y, Jones SA, Giddens DP (1991) Modern spectral analysis techniques for blood flow velocity and spectral measurements with pulsed Doppler ultrasound. IEEE Trans Biomed Eng 38:589–596

de Jong DA, Megens PHA, De Vlieger M, Thon H, Holland WPJ (1975) A directional quantifying Doppler system for measurement of transport velocity of blood. Ultrasonics 13:138–141

D'Luna LJ, Newhouse VL (1981) Vortex characterization and identification by ultrasound Doppler. Ultrason Imaging 3:271–293

Eicke BM, Kremkau FW, Hinson H, Tegeler CH (1995) Peak velocity overestimation and linear-array spectral Doppler. J Neuroimag 5:115–121

Embree PM, O'Brien WD (1990) Pulsed Doppler accuracy assessment due to frequency-dependent attenuation and Rayleigh scattering error sources. IEEE Trans Biomed Eng 37:322–326

English MJ, Woollons DJ (1982) Basic methods—preprocessing and signal averaging. In: Jones NB (ed) Digital Signal Processing, pp. 76–89. Peter Peregrinus, Stevenage, Herts

Evans DH (1982) Some aspects of the relationship between instantaneous volumetric blood flow and continuous wave Doppler ultrasound recordings—I. The effect of ultrasonic beam width on the output of maximum, mean and RMS frequency processors. Ultrasound Med Biol 8:605–609

Evans DH (1985a) Doppler signal processing. In: Altobelli SA, Voyles WF, Greene ER (eds) Cardiovascular Ultrasonic Flowmetry, pp. 239–261. Elsevier Science, New York

Evans DH (1985b) On the measurement of the mean velocity of blood flow over the cardiac cycle using Doppler ultrasound. Ultrasound Med Biol 11:735–741

Evans DH (1988) A pulse-foot-seeking algorithm for Doppler ultrasound waveforms. Clin Phys Physiol Meas 9:267–271

Evans DH, Macpherson DS (1982) Some aspects of the relationship between instantaneous volumetric blood flow and continuous wave Doppler ultrasound recordings—II. Ultrasound Med Biol 8:611–615

Evans DH, Archer LNJ, Levene MI (1985) The detection of abnormal neonatal cerebral haemodynamics using principal component analysis of the Doppler ultrasound waveform. Ultrasound Med Biol 11:441–449

Evans DH, Schlindwein FS, Levene MI (1989a) An automatic system for capturing and processing ultrasonic Doppler signals and blood pressure signals. Clin Phys Physiol Meas 10:241–251

Evans DH, Schlindwein FS, Levene MI (1989b) The relationship between time averaged intensity weighted mean velocity, and time averaged maximum velocity in neonatal cerebral arteries. Ultrasound Med Biol 15:429–435

Fan L (1994) Spectral and time-frequency analysis of ultrasonic Doppler signals. PhD Dissertation, University of Leicester

Fan L, Evans DH (1994a) Differences in the power structures of Fourier transform and autoregressive spectral estimates of narrow band Doppler signals. IEEE Trans Biomed Eng 41:387–390

Fan L, Evans DH (1994b) Extracting instantaneous mean frequency information from Doppler signals using the Wigner distribution function. Ultrasound Med Biol 20:429–443

Fan L, Evans DH (1994c) A real-time and fine resolution analyser used to estimate the instantaneous energy distribution of Doppler signals. Ultrasound Med Biol 20:445–454

Ferrara KW, Algazi VR (1991a) A new wideband spread target maximum likelihood estimator for blood velocity estimation—Part I: Theory. IEEE Trans Ultrason Ferroelec Freq Contr UFFC-38:1–16

Ferrara KW, Algazi VR (1991b) A new wideband spread target maximum likelihood estimator for blood velocity estimation—Part II: Evaluation of estimators with experimental data. IEEE Trans Ultrason Ferroelec Freq Contr UFFC-38:17–26

Ferrara KW, Algazi R, Liu J (1992) The effect of frequency dependent scattering and attenuation on the estimation of blood velocity using ultrasound. IEEE Trans Ultrason Ferroelec Freq Contr 39:754–767

Fish PJ (1991) Nonstationarity broadening in pulsed Doppler spectrum measurements. Ultrasound Med Biol 17:147–155

Flax SW, Webster JG, Updike SJ (1970) Statistical evaluation of the Doppler ultrasonic blood flowmeter. Bio-Med Sci Inst 7:201–222

Flax SW, Webster JG, Updike SJ (1973) Pitfalls using Doppler ultrasound to transduce blood flow. IEEE Trans Biomed Eng BME-20:306–309

Forsberg F (1989) An assessment of artefacts in Doppler blood flow velocity measurement. PhD Dissertation, Technical University of Denmark, Lyngby

Forsberg F (1991) On the usefulness of singular value decomposition—ARMA models in Doppler ultrasound. IEEE Trans Ultrason Ferroelec Freq Contr 38:418–428

Forsberg F, Oung H (1995) Comparison of time-frequency distributions for Doppler spectral estimation. In: Levy M, Schneider SC, McAvoy BR (eds) Proceedings of the 1995 IEEE Ultrasonics, pp. 1519–1522. IEEE, Piscataway

Forsberg F, Oung H (1996) Study of AQD spectral estimator on simulated and *in vivo* Doppler data. In: Levy M, Schneider SC, McAvoy BR (eds) Proceedings of the 1996 IEEE Ultrasonics Symposium, pp. 1237–1240. IEEE, Piscataway

Forsberg F, Oung H, Needleman L (1999) Doppler spectral estimation using time–frequency distributions. IEEE Trans Ultrason Ferroelec Freq Contr 46:595–608

Fort A, Manfredi C, Rocchi S (1995) Adaptive SVD-based

AR model order determination for time-frequency analysis of Doppler ultrasound signals. Ultrasound Med Biol 21:793–805

Franklin DL, Schlegel W, Rushmer RF (1961) Blood flow measured by Doppler frequency shift of back-scattered ultrasound. Science 134:564–565

Gerzberg L, Meindl JD (1977) Mean frequency estimator with applications in ultrasonic Doppler flowmeters. In: White D, Brown RE (eds) Ultrasound in Medicine, vol 3B, pp. 1173–1180. Plenum, New York

Gerzberg L, Meindl JD (1980a) Power-spectrum centroid detection for Doppler systems applications. Ultrason Imaging 2:236–261

Gerzberg L, Meindl JD (1980b) The root f power-spectrum centroid detector: system considerations, implementation, and performance. Ultrason Imaging 2:262–289

Gibbons DT, Evans DH, Barrie WW, Cosgriff PS (1981) Real-time calculation of ultrasonic pulsatility index. Med Biol Eng Comput 19:28–34

Gill RW (1979) Performance of the mean frequency Doppler modulator. Ultrasound Med Biol 5:237–247

Giovannelli J-F, Demoment G, Herment A (1996) A Bayesian method for long AR spectral estimation: a comparative study. IEEE Trans Ultrason Ferroelec Freq Contr 43:220–233

Gosling RG, King DH, Newman DL, Woodcock JP (1969) Transcutaneous measurement of arterial blood-velocity by ultrasound. J Ultrason USI Papers:16–23

Greene FM, Beach K, Strandness DE, Fell G, Phillips DJ (1982) Computer based pattern recognition of carotid arterial disease using pulsed Doppler ultrasound. Ultrasound Med Biol 8:161–176

Guo Z, Durand LG, Allard L, Cloutier G, Lee HC, Langlois YE (1993a) Cardiac Doppler blood-flow signal analysis. Part 1. Evaluation of the normality and stationarity of the temporal signal. Med Biol Eng Comput 31:237–241

Guo Z, Durand LG, Allard L, Cloutier G, Lee HC, Langlois YE (1993b) Cardiac Doppler blood-flow signal analysis. Part 2. Time-frequency representation based on autoregressive modelling. Med Biol Eng Comput 31:242–248

Guo Z, Durand L-G, Allard L, Cloutier G, Lee HC (1994a) Classification of lower limb arterial stenoses from Doppler blood flow signal analysis with time-frequency representation techniques. Ultrasound Med Biol 20:335–346

Guo Z, Durand L-G, Lee HC (1994b) Comparison of time-frequency distribution techniques for analysis of simulated Doppler ultrasound signals of the femoral artery. IEEE Trans Biomed Eng 41:332–342

Guo Z, Durand L-G, Lee HC (1994c) The time-frequency distribution of non-stationary signals based on a Bessel kernel. IEEE Trans Sig Proc 42:1700–1707

Harris FJ (1978) On the use of windows for harmonic analysis of the discrete Fourier transform. Proc IEEE 66:51–83

Heringa A, Alsters J, Hopman J, van Dam I, Daniels O (1988) Computer processing of cardiac Doppler signals. Med Biol Eng Comput 26:147–152

Holland SK, Orphanoudakis SC, Jaffe CC (1984) Frequency-dependent attenuation effects in pulsed Doppler ultrasound: experimental results. IEEE Trans Biomed Eng BME-31:626–631

Hoskins PR, Loupas T, McDicken WN (1990) A comparison of three different filters for speckle reduction of Doppler spectra. Ultrasound Med Biol 16:375–389

Hoskins PR, Li SF, McDicken WN (1991a) Velocity estimation using duplex scanners. Ultrasound Med Biol 17:L195-L199

Hoskins PR, Loupas T, McDicken WN (1991b) An investigation of simulated umbilical artery Doppler waveforms. I. The effect of three physical parameters on the maximum frequency envelope and on pulsatility index. Ultrasound Med Biol 17:7–21

Hutchison KJ, Karpinski E (1985) *In vivo* demonstration of flow recirculation and turbulence downstream of graded stenoses in canine arteries. J Biomech 18:285–296

Hutchison KJ, Karpinski E (1988) Stability of flow patterns in the *in vivo* post-stenotic velocity field. Ultrasound Med Biol 14:269–275

Jensen JA, Buelund C, Jorgensen A, Munk P (1996) Estimation of the blood velocity spectrum using a recursive lattice filter. In: Levy M, Schneider SC, McAvoy BR (eds) Proceedings of the 1996 IEEE Ultrasonics Symposium, pp. 1221–1224. IEEE, Piscataway

Jeong J, Williams WJ (1992) Mechanism of the cross-terms in spectrograms. IEEE Trans Sig Proc 40:2608–2613

Johnston KW, Maruzzo BC, Cobbold RSC (1977) Errors and artifacts of Doppler flowmeters and their solution. Arch Surg 112:1335–1342

Jones DL, Parks TW (1992) A resolution comparison of several time-frequency representations. IEEE Trans Sig Proc 40:413–420

Jones SA (1993) Fundamental sources of error and spectral broadening in Doppler ultrasound signals. Crit Rev Biomed Eng 21:399–483

Kaluzynski K (1987) Analysis of application possibilities of autoregressive modelling to Doppler blood flow signal spectral analysis. Med Biol Eng Comput 25:373–376

Kaluzynski K (1989a) Order selection in Doppler blood flow signal spectral analysis using autoregressive modelling. Med Biol Eng Comput 27:89–92

Kaluzynski K (1989b) Selection of a spectral analysis method for the assessment of velocity distribution based on the spectral distribution of ultrasonic Doppler signals. Med Biol Eng Comput 27:463–469

Kaluzynski K, Palko T (1993) Effect of method and parameters of spectral analysis on selected indices of simulated Doppler spectra. Med Biol Eng Comput 31:249–256

Kaluzynski K, Tedgui A (1989) Asymmetry of Doppler spectrum in stenosis differentiation. Med Biol Eng Comput 27:456–462

Kasai C, Namekawa K, Koyano A, Omoto R (1985) Real-time two-dimensional blood flow imaging using an auto-correlation technique. IEEE Trans Sonics Ultrason SU-32:458–464

Kay SM, Marple SL (1981) Spectrum analysis—a modern perspective. Proc IEEE 69:1380–1419

Kikkawa S, Yamaguchi T, Tanishita K, Sugawara M (1987) Spectral broadening in ultrasonic Doppler flow-meters due to unsteady flow. IEEE Trans Biomed Eng BME-34:388–391

Kitney RI, Giddens DP (1983) Analysis of blood velocity waveforms by phase shift averaging and autoregressive spectral estimation. J Biomech Eng 105:398–401

Kitney RI, Giddens DP (1986) Linear estimation of blood flow waveforms measured by Doppler ultrasound. MEDINFO 86:672–677

Kitney RI, Talhami H (1987) The zoom Wigner transform and its application to the analysis of blood velocity waveforms. J Theor Biol 129:395–409

Knox RA, Greene FM, Beach K, Phillips DJ, Chikos PM, Strandness DE (1982) Computer based classification of carotid arterial disease: a prospective assessment. Stroke 13:589–594

Kontis S (1987) Algorithms for fast computation of the intensity weighted mean Doppler frequency. Med Biol Eng Comput 25:25–26

Lacoss RT (1971) Data adaptive spectral analysis methods. Geophysics 36:661–675

Langlois Y, Roederer GO, Chan A, Phillips DJ, Beach KW, Martin D, Chikos PM, Strandness DE (1983) Evaluating carotid artery disease—the concordance between pulsed Doppler/spectrum analysis and angiography. Ultrasound Med Biol 9:51–63

Langlois YE, Greene FM, Roederer GO, Jager KA, Phillips DJ, Beach KW, Strandness DE (1984) Computer based pattern recognition of carotid artery Doppler signals for disease classification: prospective validation. Ultrasound Med Biol 10:581–595

Lee JS (1980) Digital enhancement and noise filtering by use of local statistics. IEEE Trans Pattern Anal Machine Intell 2:165–168

Lee YH, Kassam SA (1985) Generalised median filtering and related nonlinear filtering techniques. IEEE Trans Acoust Speech, Signal Processing 33:672–682

Li S, Hoskins PR, Anderson T, McDicken WN (1993) Measurement of mean velocity during pulsatile flow using time-averaged maximum frequency of Doppler ultrasound waveforms. Ultrasound Med Biol 19:105–113

Light LH (1990) Effect of selective tissue attenuation on pulsed Doppler frequency. Ultrasound Med Biol 16:317–318

Lim JS (1990) Two-dimensional Signal and Image Processing. Prentice-Hall, London

Loughlin PJ, Pitton JW, Atlas LE (1993) Bilinear time-frequency representations: new insights and properties. IEEE Trans Sig Proc 41:750–767

Loupas T (1988) Digital image processing for noise reduction in medical ultrasonics. PhD Dissertation, University of Edinburgh

Loupas T, Gill RW (1994) Multifrequency Doppler: improving the quality of spectral estimation by making full use of the information present in the backscattered RF echoes. IEEE Trans Ultrason Ferroelec Freq Contr 41:522–531

Loupas T, Ellwood DA, Gill RW, Bruce S, Fay RA (1995) Computer analysis of the early diastolic notch in Doppler sonograms of the uterine artery. Ultrasound Med Biol 21:1001–1011

Lunt MJ (1975) Accuracy and limitations of the ultrasonic Doppler blood velocimeter and zero crossing detector. Ultrasound Med Biol 2:1–10

Marasek K, Nowicki A (1994) Comparison of the performance of three maximum Doppler frequency estimators coupled with different spectral estimation methods. Ultrasound Med Biol 20:629–638

Markou CP, Ku DN (1991) Accuracy of velocity and shear rate measurements using pulsed Doppler ultrasound: a comparison of signal analysis techniques. Ultrasound Med Biol 17:803–814

Marple SL (1987) Digital Spectral Analysis with Applications, p. 175. Prentice-Hall, Englewood Cliffs, NJ

Martin W, Flandrin P (1985) Wigner–Ville spectral analysis of nonstationary processes. IEEE Trans Acoust Speech, Signal Processing ASSP-33:1461–1470

Maulik D, Saini VD, Nanda NC, Rosenzweig MS (1982) Doppler evaluation of fetal hemodynamics. Ultrasound Med Biol 8:705–710

Mayo WT, Embree PM (1990): Two dimensional processing of pulsed Doppler signals. USA Patent No. 4930513, issued 5 Jun 1990 (to US Philips Corporation, New York)

Mo LYL, Cobbold RSC (1992) A unified approach to modeling the backscattered Doppler ultrasound from blood. IEEE Trans Biomed Eng 39:450–461

Mo LYL, Yun LCM, Cobbold RSC (1988) Comparison of four digital maximum frequency estimators for Doppler ultrasound. Ultrasound Med Biol 14:355–363

Moraes R, Aydin N, Evans DH (1995) The performance of three maximum frequency envelope detection algorithms for Doppler signals. J Vasc Invest 1:126–134

Namekawa K, Kasai C, Tsukamoto M, Koyano A (1982) Realtime bloodflow imaging system utilizing autocorrelation techniques. In: Lerski RA, Morley P (eds) Ultrasound '82, pp. 203–208. Pergamon, New York

Newhouse VL, Ehrenwald AR, Johnson GF (1977) The effect of Rayleigh scattering and frequency dependent absorption on the output spectrum of Doppler blood flowmeters. In: White D, Brown RE (eds) Ultrasound in Medicine, vol 3B, pp. 1181–1191. Plenum, New York

Nowicki A, Karlowicz P, Piechocki M, Secomski W (1985) Method for the measurement of the maximum Doppler frequency. Ultrasound Med Biol 11:479–486

Ophir J, Jaeger P (1982) Spectral shifts of ultrasonic propagation through media with nonlinear dispersive attenuation. Ultrason Imaging 4:282–289

Oppenheim AV, Schafer PW (1975) Digital Signal Processing, pp. 545–549. Prentice-Hall, Englewood Cliffs, NJ

Oung H, Reid JM (1990) The analysis of nonstationary Doppler spectrum using a modified Wigner distribution. pp. 460–461. In: Proceedings of the Annual International Conference of the IEEE EMB Society, vol 12 (No.1). IEEE, Piscataway

Phillips DJ, Greene FM, Langlois Y, Roederer GO, Strandness DE (1983) Flow velocity patterns in the carotid bifurcations of young, presumed normal subjects. Ultrasound Med Biol 9 (suppl 1):39–49

Poots JK, Johnston KW, Cobbold RSC, Kassam M (1986) Comparison of CW Doppler ultrasound spectra with the spectra derived from a flow visualisation model. Ultrasound Med Biol 12:125–133

Priestley MB (1981) Spectral Analysis and Time Series (sixth printing, 1989), p. 398. Academic Press, New York

Prytherch DR, Evans DH (1985) Versatile microcomputer based system for the capture, storage and processing of spectrum-analysed Doppler ultrasound blood flow signals. Med Biol Eng Comput 23:445–452

Rabiner LR, Gold B (1975) Theory and Application of Digital Signal Processing, pp. 356–437. Prentice-Hall, Englewood Cliffs, NJ

Reid JM, Klepper J (1984) Frequency errors in pulse Doppler systems (Abstract). Ultrason Imaging 6:212

Rice SO (1944) Mathematical analysis of random noise. Bell Syst Tech J 23:282–332

Rickey DW, Fenster A (1996) Evaluation of an automated real-time spectral analysis technique. Ultrasound Med Biol 22:61–73

Roevros JMJG (1974) Analogue processing of C.W. Doppler flowmeter signals to determine average frequency shift momentaneously without the use of a wave analyser. pp. 43–54. In: Reneman RS (ed) Cardiovascular Applications of Ultrasound. Proceedings of an International Symposium held at Janssen Pharmaceutica, Beerse, Belgium, 1973. North-Holland, Amsterdam

Round WH, Bates RHT (1987) Modification of spectra of pulses from ultrasonic transducers by scatterers in non-attenuating and in attenuating media. Ultrason Imaging 9:18–28

Routh HF, Powrie CW, Bellevue RB, Peterson R (1994): Continuous display of peak and mean blood flow velocities. USA Patent No. 5287753, issued 22 February 1994 to Advanced Technology Laboratories, Inc., Bothell, Washington, DC

Ruano MG, Nocetti DFG, Fish PJ, Fleming PJ (1993) Alternative parallel implementations of an AR-modified covariance spectral estimator for diagnostic ultrasonic blood flow studies. Parallel Comput 19:463–476

Saini VD, Maulik D, Nanda NC, Rosenzweig MS (1983) Computerized evaluation of blood flow measurement indices using Doppler ultrasound. Ultrasound Med Biol 9:657–660

Sainz A, Roberts VC, Pinardi G (1976) Phase-locked loop techniques applied to ultrasonic Doppler signal processing. Ultrasonics 19:128–132

Schlindwein FS, Evans DH (1989) A real-time autoregressive spectrum analyzer for Doppler ultrasound signals. Ultrasound Med Biol 15:263–272

Schlindwein FS, Evans DH (1990) Selection of the order of autoregressive models for spectral analysis of Doppler ultrasound signals. Ultrasound Med Biol 16:81–91

Schlindwein FS, Evans DH (1992) Autoregressive spectral analysis as an alternative to fast Fourier transform analysis of Doppler ultrasound signals. In: Labs KH, Jäger KA, Fitzgerald DE, Woodcock JP, Neuerberg-Heusler D (eds) Diagnostic Vascular Ultrasound, pp. 74–84. Edward Arnold, London

Schlindwein FS, Smith MJ, Evans DH (1988) Spectral analysis of Doppler signals and computation of the normalised first moment in real time using a digital signal processor. Med Biol Eng Comput 26:228–232

Schuster A (1898) On the investigation of hidden periodicities with application to a supposed 26 day period of meteorological phenomena. Terr Mag 3:13–41

Schuster A (1899) The periodogram of magnetic declination as obtained from the records of the Greenwich Observatory during the years 1871–1895. Trans Cambridge Phil Soc 18:107–135

Skidmore R, Follett DH (1978) Maximum frequency follower for the processing of ultrasonic Doppler shift signals. Ultrasound Med Biol 4:145–147

Smith JL, Evans DH, Fan L, Thrush AJ, Naylor AR (1994) Processing Doppler ultrasound signals from blood-borne emboli. Ultrasound Med Biol 20:455–462

Stetson PF, Jensen JA (1997) Real-time blood flow estimation using a recursive least-squares lattice filter. In: Schneider SC, Levy M, McAvoy BR (eds) Proceedings of the 1997 IEEE Ultrasonics Symposium, pp. 1259–1262. IEEE, Piscataway

Talhami HE, Kitney RI (1988) Maximum likelihood frequency tracking of the audio pulsed Doppler ultrasound signal using a Kalman filter. Ultrasound Med Biol 14:599–609

Tessler FN, Kimme-Smith C, Sutherland ML, Schiller VL, Perrella RR, Grant EG (1990) Inter- and intraobserver variability of Doppler peak velocity measurements: an *in vitro* study. Ultrasound Med Biol 16:653–657

Thomas N (1995) On the application of the Doppler effect in pulsed Doppler flowmeters and the effect of certain propagation and scattering artefacts. PhD Dissertation, London University

Thomas N, Leeman S (1995) The attenuation effect in pulsed Doppler flowmeters. Acoust Imaging 21:543–552

Thomas N, Deane I, Leeman S (1989) Artefacts in pulsed Doppler measurement. In: Ultrasonics International '89 Conference Proceedings, pp. 1179–1185.

Thompson RS, Trudinger BJ, Cook CM (1985) Doppler ultrasound waveforms in the fetal umbilical artery: quantitative analysis technique. Ultrasound Med Biol 11:707–718

Thompson RS, Trudinger BJ, Cook CM (1986) A comparison of Doppler ultrasound waveform indices in the umbilical artery. II. Indices derived from the mean velocity and the first moment waveforms. Ultrasound Med Biol 12:845–854

Thomson FJ (1980) Refreshed display of ultrasonic Doppler spectrograms and measurement of haemodynamic parameters. Med Biol Eng Comput 18:33–38

Thrush AJ, Evans DH (1995) Intrinsic spectral broadening: a potential cause of misdiagnosis of carotid artery disease. J Vasc Invest 1:187–192

Torp H, Kristoffersen K (1995) Velocity matched spectrum analysis: a new method for suppressing velocity ambiguity in pulsed-wave Doppler. Ultrasound Med Biol 21:937–944

Tortoli P, Guidi G, Newhouse VL (1995) Improved blood velocity estimation using the maximum Doppler frequency. Ultrasound Med Biol 21:527–532

Vaitkus PJ, Cobbold RSC (1988) A comparative study and assessment of Doppler ultrasound spectral estimation techniques. Part I: Estimation methods. Ultrasound Med Biol 14:661–672

Vaitkus PJ, Cobbold RSC, Johnston KW (1988) A comparative study and assessment of Doppler ultrasound spectral estimation techniques. Part II: Methods and results. Ultrasound Med Biol 14:673–688

Voyles WF, Altobelli SA, Fisher DC, Greene ER (1985) A

comparison of digital and analog methods of Doppler spectral analysis for quantifying flow. Ultrasound Med Biol 11:727–734

Wang Y, Fish PJ (1996) Comparison of Doppler signal analysis techniques for velocity waveform, turbulence and vortex measurement: a simulation study. Ultrasound Med Biol 22:635–649

Wang Y, Fish PJ (1997) Correction for nonstationarity and window broadening in Doppler spectrum estimation. IEEE Sig Process Lett 4:18–20

Welch PD (1967) The use of fast Fourier transform for the estimation of power spectra: a method based on time averaging over short, modified periodograms. IEEE Trans Audio Electroacoust AU-15:70–73

Wells PNT (1988) Instrumentation including color flow mapping. In: Taylor KJW, Burns PN, Wells PNT (eds) Clinical Applications of Doppler Ultrasound, pp. 26–45. Raven, New York

Wigner EP (1932) On the quantum correction for thermodynamic equilibrium. Phys Rev 40:749–759

Willink R (1999) Mean blood velocity measurement with a narrow ultrasound beam and an asymmetric velocity profile. IEEE Trans Biomed Eng 46:362–364

Willink R, Evans DH (1994) A mean blood velocity statistic for the Doppler signal from a narrow beam. IEEE Trans Biomed Eng 41:322–331

Willink R, Evans DH (1995a) Statistical bias and variance in blood flow estimation by spectral analysis of Doppler signals. Ultrasound Med Biol 21:919–935

Willink R, Evans DH (1995b) Volumetric blood flow calculation using a narrow ultrasound beam. Ultrasound Med Biol 21:203–216

Willink RD, Evans DH (1996) The effect of noise and high-pass filtering on the estimation of mean blood velocity using wide and narrow ultrasound beams. IEEE Trans Biomed Eng 43:229–237

Wilson LS (1991) Description of broad-band pulsed Doppler ultrasound processing using the two-dimensional Fourier transform. Ultrason Imaging 13:301–315

Winkler A, Wu J (1995) Correction of intrinsic spectral broadening errors in Doppler peak velocity measurements made with phased sector and linear array transducers. Ultrasound Med Biol 21:1029–1035

Womersley JR (1957) The mathematical analysis of the arterial circulation in a state of oscillatory motion. Wright Air Development Center, Technical Report WADC-TR56-614

Woodcock J, Gosling R, King D, Newman D (1972) Physical aspects of blood velocity measurement by Doppler-shifted ultrasound. In: Roberts VC (ed) Blood Flow Measurement, pp. 19–23. Sector, London

Yule GU (1927) On a method of investigating periodicities in disturbed series, with special reference to Wolfer's sunspot numbers. Phil Trans Roy Soc Lond 226:267–298

Zeira A, Zeira E, Holland SK (1994) Pseudo-Wigner distribution for analysis of pulsed Doppler ultrasound. IEEE Trans Ultrason Ferroelec Freq Contr 41:346–352

Zhao Y, Atlas LE, Marks RJ (1990) The use of cone-shaped kernels for generalized time-frequency representations of nonstationary signals. IEEE Trans Acoust Speech, Sig Process 38:1084–1091

9

Waveform Analysis and Pattern Recognition

9.1 INTRODUCTION

In the previous chapter we considered various ways in which the Doppler shift signal may be processed to achieve either a flow velocity waveform or a Doppler power spectrum. In this chapter we will explore methods of extracting clinically useful information from these two types of output.

It may seem strange that the effort that has been expended to date on attempts to quantify the shape of velocity waveforms has apparently exceeded that on developing methods of measuring volumetric flow, but there are two reasons why this has been so. Firstly, volumetric flow is remarkably difficult to measure with any degree of accuracy (see Chapter 12) and requires a knowledge not only of the velocity waveform (derived under certain fairly stringent conditions) but also of the vessel dimensions and orientation and thus necessitates much more sophisticated hardware. Secondly, the shape of the waveform may provide information that a simple measurement of mean flow cannot. One example of this is in the assessment of arterial disease, where it is well documented (Lee et al 1978, Farrar et al 1979) that the pulsatile components of a flow waveform are affected by lesser degrees of proximal stenosis than is mean flow. Even if mean flow is reduced in such circumstances, it is impossible to tell whether this is a result of proximal or distal disease, but the shape of the waveform may give a clue as to where the problem lies. Another example is found in the study of fetal aortic blood flow, where waveform analysis is more sensitive than volumetric flow in predicting outcome (Laurin et al 1987a).

9.2 PATTERN RECOGNITION PRINCIPLES

It is convenient to think of the interpretation of the shape of Doppler waveforms as a process of pattern recognition. The object of waveform analysis is to recognise those waveforms that are abnormal, even if the details of why a particular physiological or pathological change gives rise to a particular change in waveform shape is not fully understood.

The process of pattern recognition may be split into three stages: transduction, feature extraction and classification (see Fig. 9.1). Essentially, transduction consists of deriving some type of pattern vector (for example the velocity waveform or the Doppler power spectrum) from the blood flow in the artery; feature extraction consists of extracting and combining salient features of the pattern vector into a feature vector (for example pulsatility index or an index of spectral broadening); and classification consists of deciding whether such a

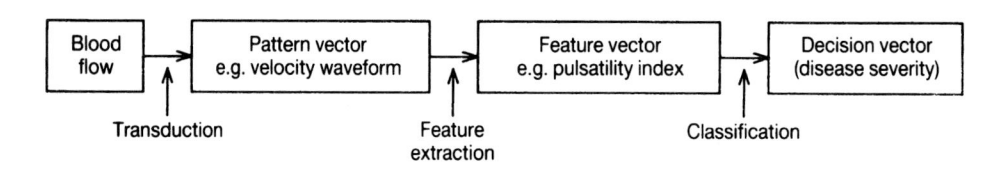

Figure 9.1 Schematic representation of the complete pattern recognition process

vector was obtained from a normal or abnormal artery.

The transduction process has largely been dealt with in previous chapters and is further considered in the next section. The rest of this chapter is devoted to feature extraction and classification. It is important to note that each stage in the process is dependent on the previous stages, and that therefore different methods of interrogating the blood flow, different signal processing techniques and different types of envelop detection will all influence the details and perhaps the performance of the rest of the recognition sequence.

9.3 PRE-PROCESSING

Feature extraction may be carried out on raw waveforms or spectra, but the quality of the input signal will affect the ultimate results (Hoskins et al 1991, Kaluzynski and Palko 1993) and various pre-processing strategies may be adopted to compress the data or improve its signal-to-noise ratio before this stage is commenced. Sophisticated examples of preprocessing can be found in the studies of Greene et al (1982), Thompson et al (1985) and Loupas et al (1995), who analysed the Doppler signals from carotid arteries, fetal umbilical arteries and uterine arteries, respectively. Greene and his colleagues (Greene et al 1982, Knox et al 1982a) collected the spectra from 20 heartbeats conforming to criteria intended to eliminate the effects of cardiac arrhythmias, and formed an ensemble average (Section 8.7) using the ECG R-wave as the fiducial point, over a period extending from −100 ms to +600 ms with respect to the R-wave. They then found the mode frequency for each 25 ms time slice using a signal location estimator, and the 3 dB and 9 dB down points both above and below the mode. All further processing was carried out using the mode and four contour frequencies. For features based on the time relationships, they derived a frequency–time waveform from an average of these five frequencies weighed according to their relative amplitudes, whilst for features based on frequency relationships they averaged each contour over a ±12.5 ms window centred about a specified point in time relative to peak systole. The peak of systole itself was found using the median smoothed first derivative of the mode waveform.

Thompson et al (1985) performed their ensemble averaging on the maximum velocity, first moment, amplitude sum and mean velocity waveforms by aligning each individual waveform with respect to a point corresponding to the beginning of each maximum velocity waveform, as identified using an autocorrelation method. They then fitted an analytical function to each ensemble average waveform using a least squares method, and carried out their further analysis on this function.

Loupas at al (1995) developed an analysis technique for automatically identifying and quantifying the 'early diastolic notch' (Section 9.4.3.7) in sonograms recorded from uterine arteries. They firstly estimated the mean level and standard deviation of the noise level present in the sonogram and used this information to derive an initial approximation of the maximum frequency envelope. They then divided the sonogram into segments, beginning at the start of the systolic upslope of each heart cycle, and calculated an ensemble average sonogram. A final maximum frequency envelope was then calculated from the averaged sonogram using the same strategy as for the raw sonogram, and used for all further processing.

In the vast majority of feature extraction methods described in this chapter, little or no preprocessing has been attempted, but clearly it could be of considerable benefit.

9.4 FEATURE EXTRACTION

Its is useful to treat the sequence of N numbers that comprise the velocity waveform or power spectrum as the components of a vector in N-dimensional 'pattern space' and the problem of separating normal and abnormal waveforms or spectra as one of separating the tips of these vectors in pattern space. In general, N is too large to tackle this problem directly and therefore it is necessary to extract a limited number of significant features from the pattern vectors and thus produce a 'feature vector' of considerably lower dimensionality. Some feature extractors achieve this by combining individual features of the waveform (for example, the maximum amplitude, minimum amplitude, mean amplitude, maximum acceleration) into one or more indices, whilst other make use of either a model or a transform to describe the overall wave shape.

Over the years many methods of feature extraction have been tried and recommended in many applications, but only a very limited number are in widespread use. The glossary of methods that follows is not exhaustive, but includes the most commonly used methods and many that are seldom used. These are included to illustrate the variety of methods of tackling the problem. No attempt to evaluate the relative efficacy of these methods is made because of the wide range of clinical problems to which they have been applied, and the wide range of opinions that exist in the literature.

9.4.1 Subjective Interpretation

The human brain is remarkably adept at pattern recognition and an experienced observer can tell a great deal about blood flow simply by listening to a Doppler shift signal, or by looking at the sonogram of the signal. It may not be possible for such an observer to explain exactly what characteristics influence his or her judgement, but there are many, some obvious and some more subtle, which may be subconsciously taken into account. Some of the more obvious features are listed in Table 9.1.

The normal sonogram does, of course, vary from site to site and not all the features listed in the table are of value at every site; furthermore, a feature considered pathological at one site (for example reverse flow in the cerebral arteries) may be a sign of normality at another site (for example in the femoral artery). Figure 9.2 shows sonograms which were taken from the common femoral arteries of a normal volunteer and three patients with peripheral vascular disease, and which exhibit a variety of abnormal features.

Subjective interpretation may be very powerful in experienced hands (Walton et al 1984), but objective methods do not rely on the expertise of the user, are not subject to observer bias, allow methods to be transported between centres, and may be able to distinguish much more subtle changes in the waveforms. At present, however, most objective methods concentrate on one particular aspect of the sonogram (for example the outline shape) and may thus sometimes ignore features that may be obvious to the human observer.

9.4.2 Maximum Frequency

When blood flows through a stenosis its velocity must increase to maintain the same flow rate as in the pre- and post-stenotic region of the vessel. The detection of the high velocities within a stenosis is the basis of a test described by Spencer and Reid (1979) and subsequently widely adopted for the detection of internal carotid artery stenosis (Zwiebel et al 1982, Johnston et al 1982a, Manga et al 1986, Bluth et al 1988). The Doppler shift is of course proportional not only to the velocity, but also to the frequency of the transmitted ultrasound and the cosine of the angle between the blood vessel and the ultrasound probe (eqn. 1.1), and these variables must be allowed for. Particular care must also be taken to compensate for the effects of spectral broadening, which can significantly affect the measurement of maximum frequency (Section 8.4.5). Furthermore, a very tight stenosis will reduce the volumetric flow through the artery, and the increase in peak velocity is therefore less than would be predicted on purely geometric grounds. A more sophisticated test based on the comparison of the velocities within and beyond the stenosis is described in Section 9.4.9.3.

9.4.3 Simple Single-site Normalised Frequency Indices

Several of the most popular feature extraction techniques are based on finding the ratio of the height of one feature of a waveform to that of

Table 9.1 Some major factors that may influence subjective interpretation of the sonogram

1. Is the signal more difficult to obtain then is usual for a patient of a particular build?
2. Is there any flow during diastole? If so, how much?
3. Is there any reverse flow?
4. Is the height of the sonogram roughly as expected for the combination of transmitted frequency, site and angle?
5. Is there a window under the sonogram? If there is spectral broadening, when does it occur and for how long?
6. Are there any vortical spikes on the sonogram?

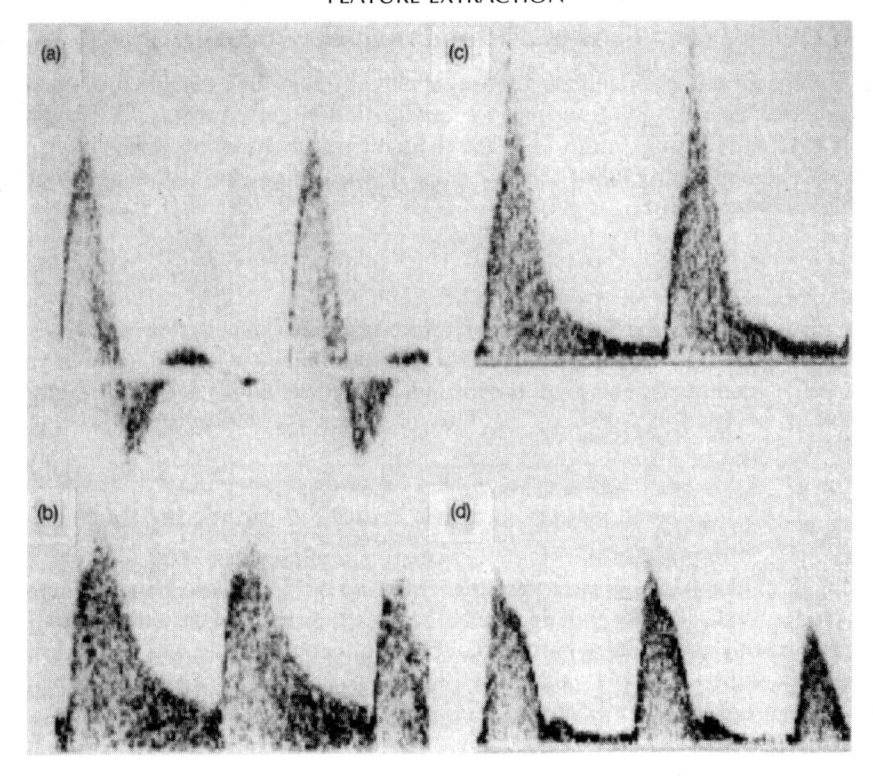

Figure 9.2 Sonograms recorded from the common femoral arteries of a normal subject and three patients with peripheral vascular disease. (a) Normal triphasic sonogram. (b) Damped sonogram resulting from an upstream stenosis. (c) Sonogram with a vortical spike originating from an upstream flow disturbance. (d) Sonogram with a shoulder on the downslope of the systolic peak associated with an occluded superficial femoral artery

another. The advantage of taking such a ratio is that both the numerator and the denominator include the cosine of the angle between the Doppler probe and the blood vessel, and the transmitted frequency, and that the index is thus independent of these variables.

9.4.3.1 Pulsatility Index

Perhaps the most widely used index of all is pulsatility index (*PI*). This index was originally defined (Gosling et al 1971) as the total oscillatory energy in the flow-velocity waveform divided by the energy in the mean flow-velocity over the cardiac cycle, i.e:

$$PI_{\text{orig}} = \frac{\sum_{n=1}^{\infty} a_n^2}{M^2} \qquad 9.1$$

where a_n is the amplitude of the nth harmonic, and M the mean value of amplitude over the cardiac cycle. The purpose of this index was to summarise the degree of pulse-wave damping at different arterial sites; the smaller the *PI*, the greater the degree of damping.

This index requires the calculation of the Fourier transform of the velocity waveform, which at that time presented some computational difficulties, and it was soon superseded (Gosling and King 1974) by a similar but simpler *PI* calculated by dividing the maximum vertical excursion of the waveform by its mean height (Fig. 9.3), i.e:

$$PI = (S - D)/M \qquad 9.2$$

where S and D are the maximum and minimum values of amplitude during the cardiac cycle (for pulsatile waveforms such as that shown in

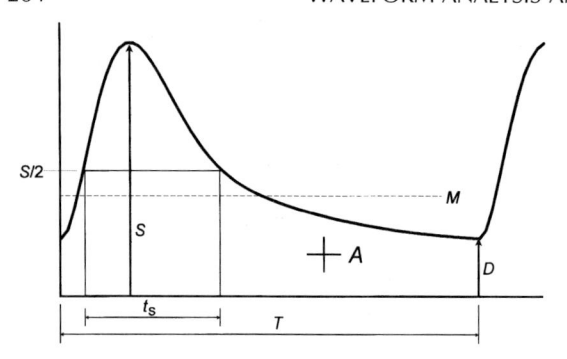

Figure 9.3 Diagram illustrating the variables involved in the definitions of pulsatility index, resistance index, S/D ratio, constant flow ratio and height–width index. S is the maximum height of the waveform, D the minimum (for PI and HWI) or end-diastolic height (for RI and S/D), M the mean height over the cardiac cycle, T the length of the cardiac cycle, A the area under the curve, and t_s the duration of the systolic peak (measured between half amplitude points)

Fig. 9.2a, D may be negative and may not occur at end diastole).

The two PI values are correlated (Gosling and King 1974, Johnston et al 1978) but are not numerically equal and therefore cannot be used interchangeably. The simplicity of PI as defined in eqn. 9.2 allows it to be calculated on-line using a simple microprocessor-based system (Gibbons et al 1981, Johnston et al 1982b), and it has been widely accepted as one of the standard ways of quantifying the 'pulsatility' of a Doppler waveform. It could be argued that of the two definitions, the earlier has more physical meaning and that, since on-line Fourier analysis is a relatively simple task with modern microcomputer-based systems, this method should be revived in the future.

PI is a useful objective description of the pulsatility of a waveform, but great care must be taken with its interpretation in the clinical setting as it (like all simple indices) can be influenced by many factors, included proximal stenosis (Johnston and Tarashuk 1976, Evans et al 1980, Johnston et al 1983), distal stenosis (Thiele et al 1983, Junger et al 1984, Macpherson et al 1984), peripheral resistance (Evans et al 1980), heart rate (van den Wiijngaard et al 1988, Hoskins et al 1989, Mari et al 1991) and in some circumstances even myocardial contractility (van Bel et al 1992). Some insight into factors that cause changes in waveform shapes can be gained from mathematical and computer models, although to date these have tended to concentrate on the fetal and maternal circulations (Mo et al 1988, Adamson et al 1989, Thompson and Stevens 1989, Thompson and Trudinger 1990, Guiot et al 1992, Todros et al 1992, Surat and Adamson 1996).

9.4.3.2 *Pourcelot's Resistance Index*

A second widely used index of pulsatility is Pourcelot's resistance index (RI), which is defined (Fig. 9.3) as:

$$RI = (S - D)/S \qquad 9.3$$

In this case D is defined as the height of the waveform at end diastole rather than its minimum value during the cardiac cycle as for PI, but these are usually identical in clinical situations where RI is used (i.e. with low impedance distal beds and continuous forward flow throughout the cardiac cycle). Resistance index was first used on waveforms from the common carotid artery (Planiol and Pourcelot 1973, Pourcelot 1976) as an indicator of the circulatory resistance beyond the measurement point, and has subsequently been widely used in obstetrics (Thompson et al 1988, Hendricks et al 1989) and for the study of neonatal cerebral haemodynamics (Perlman 1985).

Unfortunately, considerable confusion has arisen between PI as defined in eqn. 9.2 and RI as defined in eqn. 9.3, and the latter is now generally termed 'pulsatility index' in the North American neonatal literature. The term 'resistance index' is preferred here to avoid ambiguity but even this term is not ideal, since the value of RI may be influenced by many factors, of which distal resistance is only one.

To add further confusion over RI, some workers have defined D as the minimum value of the waveform, although this does not usually make any difference unless reverse flow is present. Using the 'standard' definition, RI may only vary between 0 and 1, but if D is defined as the minimum value of the waveform, then it will take on negative values in the presence of reverse flow and the value of RI will exceed unity.

RI is extremely simple to calculate and has been used to detect waveform changes in a wide variety of pathological conditions, including internal carotid stenosis (Pourcelot 1976), birth asphyxia (Bada et al 1979, Archer et al 1986), intraventricular haemorrhage (Bada et al 1979), patent ductus arteriosus (Perlman et al 1981, Lipman et al 1982), pneumothorax (Hill et al 1982) and hydrocephalus (Hill and Volpe 1982).

9.4.3.3 S/D and D/S Ratios

A slight variation of the RI has been used by a number of workers in the obstetric field to describe changes that occur in the shape of the velocity waveform in the umbilical artery with gestational age (Stuart et al 1980, Trudinger et al 1985b) and in some high-risk pregnancies (Trudinger et al 1985b). This index, which is often called the AB or A/B ratio, is simply calculated by dividing the maximum systolic height (S) by the end diastolic height (D) (Fig. 9.3). This ratio is called the S/D ratio here to distinguish it from the A/B ratio, which as been used for the evaluation of common carotid waveforms and is described in Section 9.4.3.5.

S/D is in fact a simple transformation of RI and may be rewritten;

$$S/D = 1/(1 - RI) \qquad 9.4$$

One drawback of the S/D ratio is that, as D tends to zero, S/D tends to infinity, and there is a discontinuity at $D = 0$ with S/D equal to infinity at $D = 0^+$ and to minus infinity at $D = 0^-$. By contrast, values of RI measured from the umbilical artery are distributed normally at any given gestational age (Thompson et al 1988, Hendricks et al 1989).

A further variation of the RI, the D/S ratio, has been used by Trudinger and his colleagues (1985a) to describe the velocity waveform in the uteroplacental circulation. This quantity is simply related to RI by eqn. 9.5:

$$D/S = 1 - RI \qquad 9.5$$

The index avoids the discontinuity at $D = 0$.

More recently, Arbeille and his colleagues (1995) have defined the 'high resistance index' (HRI) for vessels with high resistance to flow (and they give examples of the lower limbs and placentas with vascular diseases) as:

$$HRI = D/S \qquad 9.6$$

where S is the amplitude of the systolic peak and D the amplitude of the reverse flow peak. In the cited study they showed HRI to be highly sensitive to vascular resistance changes in an animal model. Actually, this is a rediscovery of one of the indices suggested by Rittenhouse and colleagues (1976) for quantifying changes in peripheral vascular resistance, and tested using the femoral arteries of dogs.

9.4.3.4 Constant Flow Ratio

Thompson et al (1985) have suggested the 'constant flow ratio' (CFR) as a method of describing fetal umbilical artery waveforms. They define this quantity (Fig. 9.3) as:

$$CFR = DT/A \qquad 9.7$$

where D is the minimum height of the waveform, T the duration of the waveform and A the area under the curve. The ratio A/T is the mean height of the waveform over the cardiac cycle (M), and therefore eqn. 9.7 may be written:

$$CFR = D/M \qquad 9.8$$

This index is related to PI and RI by eqn. 9.9:

$$CFR = PI/RI - PI \qquad 9.9$$

9.4.3.5 A/B Ratio

The A/B ratio was introduced as a means of characterising the shape of the signals from the common carotid and supraorbital arteries (Gosling 1976). The waveforms from both these sites exhibit two peaks (A and B) during systole (Fig. 9.4), and the ratios of their amplitudes change with both age and disease. Tests for internal carotid disease using a combination of the A/B ratios in the common carotid and supraorbital arteries (Baskett et al 1977) and in the common carotid and supratrochlear arteries (Prichard et al 1979) have been described.

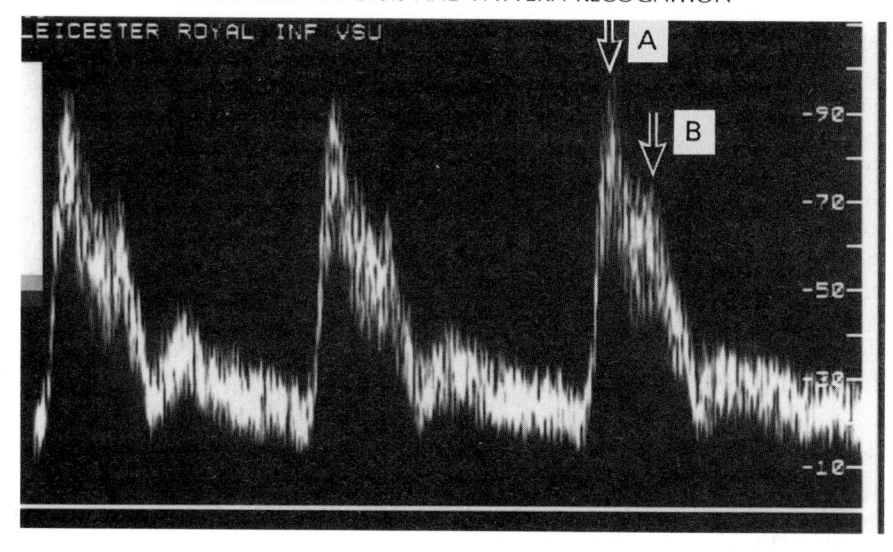

Figure 9.4 Sonogram recorded from the common carotid artery of a normal subject showing the two systolic peaks, A and B

9.4.3.6 A/C or S/N Ratio

The presence of a prominent early diastolic notch (*EDN*) in the sonogram from the uterine artery is reported to have an improved ability to predict abnormal pregnancy outcomes when compared with *RI* (Bower et al 1993, Fay et al 1994). North and colleagues (North et al 1994, Ferrier et al 1994) have attempted to quantify the *EDN* using a simple ratio of the peak systolic flow velocity to the early diastolic flow velocity. North and colleagues called these heights *A* and *C*, respectively, but the symbols *S* and *N* are preferred here for notational consistency with other methods of quantifying the notch (Fig. 9.5).

9.4.3.7 Early Diastolic Notch Pulsatility Index

A much more sophisticated approach to the analysis of the early diastolic notch of the uterine artery has been attempted by Loupas et al (1995). The pre-processing regime they applied to derive a maximum frequency envelope has already been discussed in Section 9.3. From this envelope they first identified the peak systolic velocity (*S*) and end-diastolic velocity (*D*) and then the *EDN* (*N*) (Fig. 9.5). They then calculated the '*EDN* Fourier pulsatility index' (*NFPI*), which they defined as:

$$NFPI = 100 \frac{\sum_{n=1}^{\infty} a_n^2}{(M_{N,D})^2} \qquad 9.10$$

where $M_{N,D}$ represents the mean value of the maximum frequency envelope segment between points N and D, and a_n denotes the amplitude of the nth harmonic, calculated by the Fourier transform, of the same segment after its linear

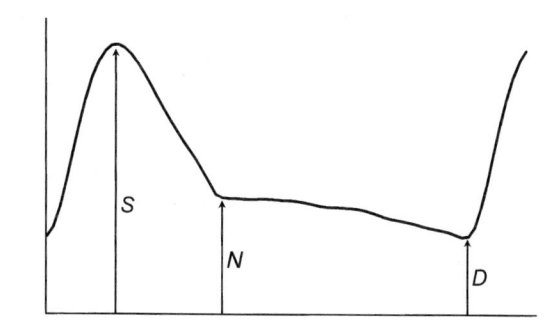

Figure 9.5 Diagram illustrating the variables involved in the definition of the *S/N* ratio and the early diastolic notch pulsatility index. *S* is the peak systolic velocity, *D* the end-diastolic velocity, and *N* the early diastolic or 'notch' velocity

trend has been removed by means of a least-squares fit line. Comparison of this equation with eqn. 9.1 shows this to be analogous to the original definition of Gosling's pulsatility index.

One further attempt to quantify the EDN has been described by Hütter and colleagues (1993), in which the flow velocity waveform was broken down into orthogonal polynomial components up to third order. This method was also claimed to provide a higher sensitivity than simple pulsatility and resistance indices.

9.4.4 Simple Single-site Indices that Include Time

Changes in the circulation affect not only the height of the recognisable features of a waveform, but also the time relationship between those features. The upslope of the waveform is particularly susceptible to changes in the cardiac impulse and the circulation proximal to the site of measurement, whilst the decay of the velocity after peak systole is more influenced by the distal circulation. The time relationships within the waveform are very easy to measure, and have been used both alone and in combination with the heights of discernible features to characterise the shape of the waveform.

9.4.4.1 Velocity Acceleration and Deceleration

Perhaps the simplest measurements that include time are those of acceleration and deceleration of the systolic peak, which were used (together with other features) by Fronek and colleagues (1976) in an attempt to assess peripheral vascular disease. They defined acceleration and deceleration as follows:

$$\text{Acceleration} = \frac{\text{Peak velocity}}{\text{Pulse rise time}} \quad 9.11$$

$$\text{Deceleration} = \frac{\text{Peak velocity}}{\text{Pulse decay time}} \quad 9.12$$

These definitions are only suitable for very pulsatile flow (such as that depicted in Fig. 9.2a) where the systolic peak rises from and falls to zero, but eqns 9.11 and 9.12 can easily be modified for use on less pulsatile flows (see, for example, Fig. 9.2b) simply by making both the velocity and time measurements over a limited region of the systolic peak.

9.4.4.2 Systolic Decay Time Index

Few indices have been reported that use purely the time relationships within a single cardiac cycle, but the systolic decay time index ($SDTI$) reported by Thompson et al (1985) is one such. It describes the ratio of the normalised rise and decay slopes close to peak systole and is defined (Fig. 9.6) as:

$$SDTI = (t_r/t_1)/(t_d/t_2) \quad 9.13$$

where t_1 is the time from the beginning of systole to peak systole, $t_2 (= T - t_1)$ the time from peak systole to the end of diastole, t_r the rise time from $0.75S$ to S, and t_d the decay time from S to $0.75S$. Thompson et al show it to be of some use in detecting abnormal fetal umbilical artery waveforms.

9.4.4.3 Height–Width Index

The height–width index (HWI) suggested by Johnston and his colleagues (Johnston et al 1984) combines information about both the pulsatility of, and the time relationships within, the arterial waveform. Waveforms recorded distal to arterial stenoses have relatively small pulsatile components and relatively wide systolic peaks, and both of these serve to decrease the value of HWI, which is defined (Fig. 9.3) as:

$$HWI = [(S - D)/M](T/t_s) \quad 9.14$$

where t_s is the duration of the systolic peak measured between the half amplitude points. The expression in the square bracket is equal to pulsatility index and so eqn. 9.14 may be rewritten:

$$HWI = PI(T/t_s) \quad 9.15$$

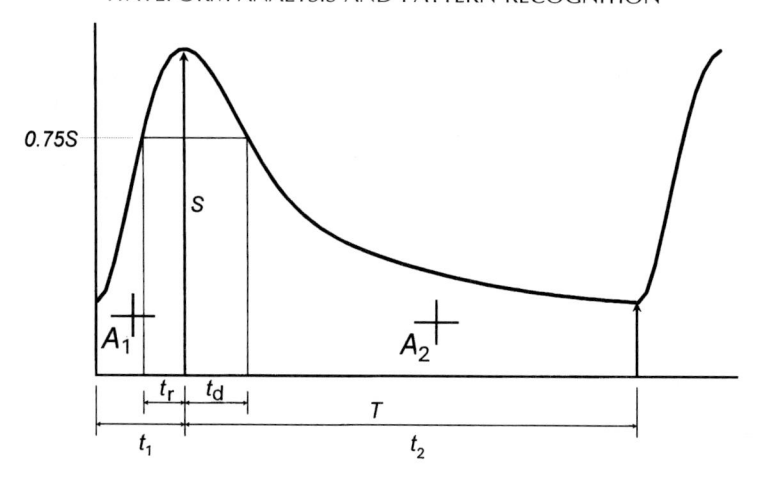

Figure 9.6 Diagram illustrating the variables involved in the definition of systolic decay time index and relative flow rate index. t_1 is the time from the beginning of systole to peak systole, t_2 the time from peak systole to the end of diastole, t_r the rise time from $0.75S$ to S, t_d the decay time from S to $0.75S$, A_1 the area under the curve before peak systole and A_2 the area under the curve after peak systole

9.4.4.4 Path Length Index

A further index of pulsatility introduced by Johnston et al (1984) is the path length index (*PLI*). This exploits the decrease in the total path length traced out over the cardiac cycle as a waveform becomes more damped. It is normalised to remove angle dependence and heart rate, and is defined (Fig. 9.7) as:

$$PLI = \sum_{i=0}^{N-1} [(f_{i+1} - f_i)^2/M^2 + (t_{i+1} - t_i)^2/T^2]^{1/2}$$

9.16

where f_i is the *i*th value of the waveform amplitude and t_i the *i*th value of time.

9.4.4.5 Relative Flow Index

The relative flow index (*RFRI*) describes the ratio of the average flow rate before the systolic peak to the average flow rate during the rest of the cardiac cycle, and was introduced by Thompson et al (1985) for use with fetal umbilical artery waveforms. It is defined (Fig. 9.6) as:

$$RFRI = (A_1/A_2)/(t_1/t_2) = M_1/M_2 \quad 9.17$$

where A_1 and A_2 are the areas under the curve before and after peak systole, t_1 is the time from the beginning of systole to peak systole, t_2 the time from peak systole to end diastole, M_1 the mean height of the curve before the systolic peak, and M_2 the mean height of the curve during the rest of the cardiac cycle.

9.4.4.6 Curve Broadening Index

A variation on the systolic time decay index, the curve broadening index (*CBI*) has been reported by Windeck and colleagues (1992) as being useful

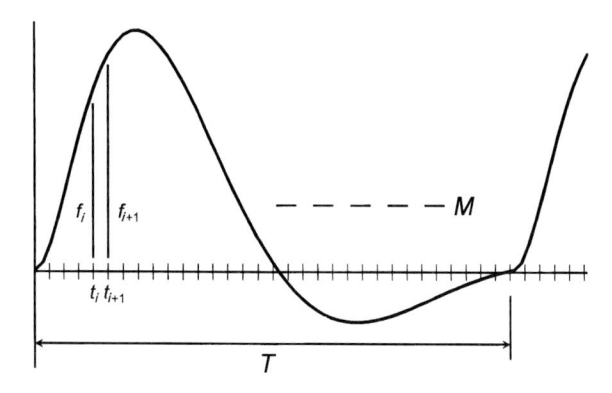

Figure 9.7 Diagram illustrating the variables involved in the definition of path length index. M is the mean height of the waveform, T the cycle length, f_i the *i*th value of the waveform amplitude and t_i the *i*th value of time

in detecting the success of percutaneous transluminal angioplasty in patients with isolated lesions in the distal superficial femoral artery, and as being superior to *PI*, *SDTI* and *HWI* for this purpose. *CBI* was defined (Fig. 9.6) as:

$$CBI = t_1/t_r \times t_2/t_d \qquad 9.18$$

9.4.4.7 Trans-systolic Time or Systolic Width

Hanlo and colleagues (1995) have introduced an index, which they have named trans-systolic time (*TST*), in an attempt to quantify the changes in transcranial Doppler waveforms resulting from changes in intracranial pressure. The index resulted from studies of a hydrodynamic model of the cerebral circulation and its electrical analogue, and Hanlo et al claim that *TST* more specifically reflects the properties of the intracranial system than either *PI* or *RI*. *TST* is defined as the duration of the systolic peak measured at a velocity level of V_{tst} given by:

$$V_{tst} = \frac{S + D}{2} \qquad 9.19$$

where S is the velocity at peak systole, and D the velocity at end diastole. In fact this index is identical to 'systolic width' used by Sillesen and Schroeder (1988) to predict pressure reduction across internal carotid artery stenoses.

9.4.5 Spectral Broadening Indices

Each of the indices mentioned so far has been concerned with the shape of the velocity waveform (whether it be maximum, mean or RMS). These indices ignore the information contained in the power spectrum concerning the distribution of velocities within the ultrasound beam, which may sometimes be of diagnostic importance. In particular it has been found that the shape of the spectrum measured from the internal carotid artery at or around peak systole is influenced by quite moderate degrees of proximal disease (Barnes et al 1976, Blackshear et al 1979, Reneman and Spencer 1979). Sonograms recorded from normal vessels exhibit a clear window under the systolic peak, resulting from a relatively flat velocity profile at this point in the cardiac cycle (Fig. 9.8a) whilst those from diseased vessels show a degree of spectral broadening (Fig. 9.8b), which is related to the degree of proximal stenosis and is believed to be the result of disturbed non-axial flow.

One of the earliest attempts to quantify spectral broadening was that of Bodily et al (1980), who defined fractional broadening to be s_f/f_{mean}, where f_{mean} is the mean frequency at peak systole and s_f the standard deviation of frequency at the same time. The mean frequency was estimated to be half-way between the minimum and maximum frequency and the standard deviation assumed to be proportional to the difference between the two frequencies. This led to the empirical relationship for a spectral broadening index (*SBI*):

$$\text{Fractional broadening } (SBI(1)) = k\frac{f_{max} - f_{min}}{f_{max} + f_{min}}$$
$$9.20$$

where f_{max} and f_{min} are the maximum and minimum frequency at peak systole, and k is an experimentally derived constant found to be equal to 0.47.

Since that time many indices have been tried both *in vitro* and *in vivo* with varying degrees of success, and a selection of these are enumerated in Table 9.2. A problem with many of these indices is the rather loose way in which the minimum, mean and maximum frequencies have been defined, which makes it difficult to repeat the original studies. There is also a tremendous variation in the time period over which the data has been collected, the way in which the data from different time slices have been combined, the amount of averaging that has been used, and the size of the sample volume. This last factor is particularly important as it may change the Doppler spectrum completely (Evans 1982, Knox et al 1982b, van Merode et al 1983, Wijn et al 1987).

9.4.6 Use of Multiple Features

The feature extraction methods discussed so far have all been based to some degree on experience

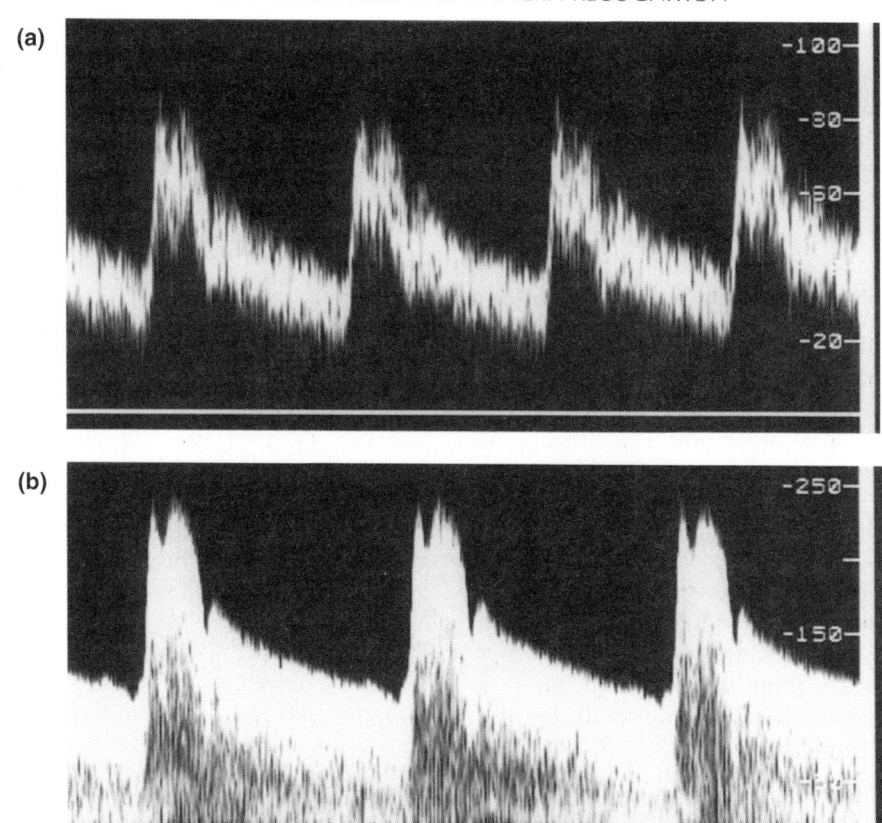

Figure 9.8 Sonograms recorded from internal carotid arteries of (a) a normal subject, exhibiting a clear window under the systolic peak, and (b) a patient with disease at the origin of the internal carotid artery, showing spectral broadening. Note that the frequency scale of (b) is more than twice that of (a)

Table 9.2 Selection of spectral broadening indices

Reference	Definition	Comments
Bodily et al 1980	$SBI(1) = 0.47(f_{max} - f_{min})/(f_{max} + f_{min})$	PW
Johnston et al 1981	$SBI(2) = (f_{max} - f_{min})/f_{max}$	CW; same information content as $SBI(1)$
Woodcock et al 1982	$SBI(3) = (f_{max} - f_{min})/f_{mean}$	PW
Brown et al 1982	$SBI(4) = (f_{max} - f_{mean})/f_{max}$	CW: f_{mean} calculated using analogue frequency follower
Woodcock et al 1983	$SBI(5) = (f_{max} - f_{min})/f_{median}$	PW
Rittgers et al 1983	$SBI(6) = f_{min}/f_{max}$	CW; same information content as $SBI(1)$; calculated from 12 dB down points from t_0 to $t_0 + 100$ ms
Sheldon et al 1983	$SBI(7) = f_{max}/f_{mean}$	PW; same information content as $SBI(4)$; averaged over 16–32 cycles
Kalman et al 1985	$SBI(8) = s_f/f_{mean}$	CW; four lines of spectral data from around peak systole, averaged
Kalman et al 1985	$SBI(9) = m3_f/(s_f)^3$	As for $SBI(8)$
Kalman et al 1985	$SBI(10) = m4_f/(s_f)^4 - 3$	As for $SBI(8)$

For notation, see Section 9.7.

or intuition, have resulted in a one-dimensional feature vector, and have concentrated on one particular attribute of the Doppler signal (either the shape of the velocity waveform or the degree of spectral broadening). Another approach is to evaluate a large number of features which may assist with the classification process, and to select a set of these on the basis of their ability to discriminate between disease states. This approach has been adopted by Rutherford et al (1977), who selected five out of nine features from the common carotid artery waveform; Greene and his colleagues (Greene et al 1982, Knox et al 1982a, Langlois et al 1984), who selected two or three out of 94 features, from recordings taken from four sites in the internal and common carotid arteries, for each of three binary decisions concerning disease severity; Van Asten et al (1991), who selected varying numbers of parameters from 20 candidate features taken from common femoral artery waveforms both at rest and during reactive hyperaemia; and Allard et al (1991), who selected various subsets of 19 raw diagnostic features from various sites in the lower limb to make two binary decisions concerning arterial diameter reduction in each arterial segment. The advantage of such strategies is that they are able to highlight features or combinations of features which are unexpectedly good discriminators. An example of this is 'feature I.3' of Greene et al (1982), which was a measure of the relative increase in spectral width between peak systole and peak systole +100 ms, and proved to be very important for separating normal and diseased carotid arteries. Feature selection cannot be entirely divorced from classification with these methods since the selection of features is based upon their ability, to discriminate between disease class. The classifier ultimately chosen may, however, be independent of the selection algorithms.

The first step in feature selection is to choose a number of candidate features that may potentially assist in the classification process. These may be simple values such as the height of the waveform at different times during the cardiac cycle, derived features such as pulsatility index or fractional broadening, or any combination of these. Simple analyses tend to concentrate on features known or thought to be most closely correlated with disease, but more complicated analyses may evaluate several dozen features. It is then necessary to eliminate a proportion of the features, either because they are poor discriminators on their own or because they share the same discriminating information as other features and are thus redundant. This may either be done in a stepwise manner or by testing a large number of possible feature subsets.

In the former case, the analysis proceeds in a step-wise manner; the feature that provides the greatest univariate discrimination is first selected and then paired with all the remaining features to ascertain the combination that produces the greatest discrimination. The procedure is then repeated to find the best triplet and so forth, until the addition of further features no longer improves the discrimination or a predetermined number of features has been selected. Although the method described produces an optimal set of discriminating features, it may not necessarily be the best combination, and various strategies that combine 'forward' selection (adding one feature at a time) and 'backward' selection (dropping one feature at a time) can be used to improve the final selection (Klecka 1980).

The study of Allard et al (1991) provides an example of the second approach, where at each decision node all possible feature subsets varying between one and $N/5$ (where N was the smallest training sample size in a class for a specific node) were used to test the discrimination power of combinations of features.

9.4.7 Laplace Transform Analysis

A method of feature extraction based on a description of the shape of the entire blood velocity waveform in terms of a Laplace transform was introduced by Skidmore and Woodcock (1978) and explored in depth in a series of papers (Skidmore and Woodcock 1980a,b, Skidmore et al 1980). The envelope signal is first transformed into the frequency domain, and then fitted to the third-order Laplace equation:

$$H(S) = 1/(S^2 + 2\delta\omega_0 S + \omega_0^2)(S + \gamma) \quad 9.21$$

where $S = j\omega$. Equation 9.21 can be represented in graphical form by plotting its poles on an Argand diagram. The three poles are found by equating the denominator to zero and are given by:

$$S_{1,2} = -\delta\omega_0 \pm j\omega_0(1 - \delta^2)^{1/2} \qquad 9.22$$

and

$$S_3 = -\gamma \qquad 9.23$$

The S-plane representation of eqn. 9.21 is shown in Fig. 9.9. The particular virtue of the Laplace transform model is that in health, at least, each of the terms ω_0, γ and δ appears to be related to a different aspect of the circulation, specifically arterial stiffness, distal impedance and proximal lumen size. This is clearly advantageous, as it allows the effects of distal and proximal disease on the Doppler waveform to be deconvolved. Several studies have evaluated δ as a measure of proximal stenosis (Baird et al 1980, Johnston et al 1984, Macpherson et al 1984, Junger et al 1984).

Although Skidmore and his colleagues calculated the coefficients ω_0, γ and δ by fitting the Fourier transform of the waveform to a third-order Laplace equation, an equally valid technique is to fit the waveform to the inverse Laplace transform of eqn. 9.21, i.e:

$$f(t) = \frac{\gamma(\alpha^2 + \beta^2)}{(\gamma - \alpha)^2 + \beta^2} \\ \times \left[e^{-\gamma t} + e^{-\alpha t} \left(\frac{\gamma - \alpha}{\beta} \sin \beta t - \cos \beta t \right) \right] \qquad 9.24$$

where $\alpha = \omega_0 \delta$ and $\beta^2 = \omega_0^2 - \alpha^2$. This was the method used by Johnston et al (1984).

9.4.8 Principal Component Analysis

Another technique used to describe the shape of the entire Doppler waveform is that of principal component analysis. This type of analysis has been applied to signals from the common carotid arteries (Martin et al 1980), the common femoral arteries (Macpherson et al 1984, Evans 1992) and the anterior cerebral arteries of newborn infants (Evans et al 1985).

Principal component analysis is analogous to Fourier analysis in that the waveform is described in terms of the coefficients of a predetermined orthogonal set of waveforms, but rather than using sines and cosines, the orthogonal set is chosen so that it describes the waveforms from the study population most efficiently, with the smallest number of terms. Because the transform is so efficient, only two or three terms are required to reconstruct the original waveform to a high degree of accuracy, and therefore the waveform can be represented by two or three coefficients and plotted in two- or three-dimensional feature space (Fig. 9.10).

The methodology consists of two distinct stages, that of defining the principal components (PCs) of the study population, and that of calculating the coefficients of the PCs for each test waveform. The required PCs can be shown to be the eigenvectors of the covariance matrix of the population, which may be estimated from a sufficiently large and representative sample of waveforms. The first step is to calculate the sample mean record (SMR),

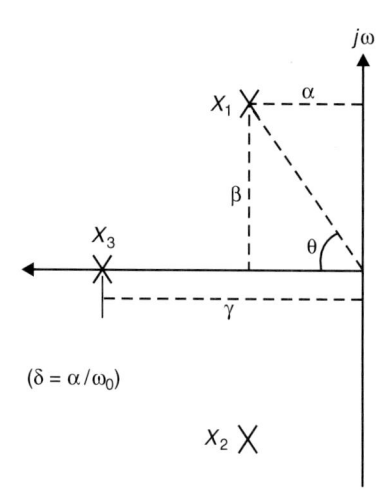

Figure 9.9 S-plane representation of eqn. 9.21, showing the position of two complex poles and one real pole

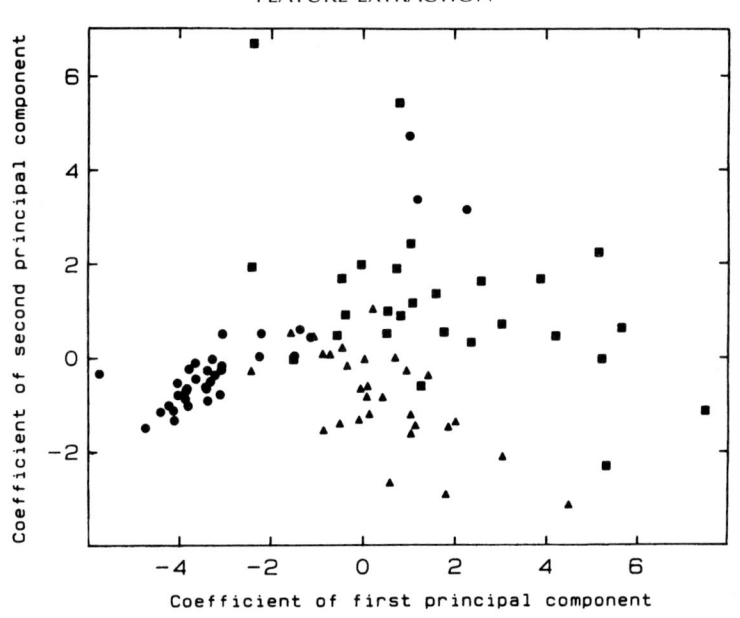

Figure 9.10 The coefficients of the first two principal components of waveforms recorded from the femoral arteries of patients with various disease patterns. The three groups were: severe aorto-iliac disease (circles), probably normal aorto-iliac segment, but blocked femoral artery (triangles), and probably normal aorto-iliac segment and patent superficial femoral artery (squares) (reproduced by permission of Elsevier Science from Evans 1984)

which is simply the ensemble average of the sample waveforms, i.e:

$$SMR_i = \sum_{n=1}^{N} f_{in}/N \qquad 9.25$$

where SMR_i is the ith element of the SMR, f_{in} is the ith element of the nth waveform, and N is the total number of waveforms in the sample. Each element of the covariance matrix can then be written:

$$C_{ij} = \sum_{n=1}^{N} (f_{in} - SMR_i)(f_{jn} - SMR_j)/(N-1) \qquad 9.26$$

The Eigenvalues and the corresponding eigenvectors of the covariance matrix can be found using a standard computer package. The first principal component is the eigenvector corresponding to the largest eigenvalue and so on, and it can be shown that, if only K PCs out of a possible P are used, the efficiency of the transform is given by:

$$E = \sum_{i=1}^{K} \lambda_i \bigg/ \sum_{i=1}^{P} \lambda_i \qquad 9.27$$

where λ_i is the ith eigenvalue.

Once the PCs have been defined, the calculation of their coefficients for each test waveform is straightforward. If b_k is the coefficient of the kth PC, then:

$$b_k = \sum_{i=1}^{I} (f_i - SMR_i)(r_{ki}) \qquad 9.28$$

where f_i is the ith element of the test waveform and r_{ki} the ith element of the kth PC. The initial calculation of the PCs requires a fair amount of computing power, but the second stage is simple and can be carried out on-line on a small microcomputer system (Prytherch et al 1982).

Although principal component analysis has been applied mainly to the waveform shape, it is possible to describe the entire sonogram in terms of a principal component series, and this approach has been explored by Martin and his colleagues (Martin et al 1981, Sherriff et al 1982).

9.4.9 Two Site Measurements

Most feature extraction techniques rely on the analysis of the Doppler signal recorded from a single point in the circulation; there are, however, a number of methods that involve the comparison of signals from two or more points. Several of these have been evaluated by Humphries et al (1980) and Baker et al (1986).

9.4.9.1 Damping Factor and Transit Time

The first reported attempts to quantify disease severity by comparing Doppler signals from two sites were those of Woodcock and his colleagues (Woodcock 1970, FitzGerald et al 1971, Woodcock et al 1972). They examined the effects of disease in the femoral arteries on both the damping of the Doppler waveform and its velocity as it propagated from the common femoral artery at the groin to the popliteal artery behind the knee. For this purpose they introduced two new quantities, the damping factor (DF) and the transit time (TT) (Fig. 9.11), which they defined as:

$$DF = PI_{\text{fem}}/PI_{\text{pop}} \qquad 9.29$$

and

$$TT = t_{\text{pop}} - t_{\text{fem}} \qquad 9.30$$

PI_{fem} and PI_{pop} are the pulsatility indices in the common femoral and popliteal arteries, and t_{fem} and t_{pop} the times of arrival of the 'feet' of the Doppler waveforms at these two sites. Increasing disease severity was shown to result in an increase in both TT and DF.

Normalised versions of TT and DF have since been described (Gosling 1976, Gosling and King 1978), both to eliminate site dependence and to allow for variations in mean blood pressure, which can significantly influence the pulse wave velocity. Pulse wave velocity is discussed further in Chapter 2.

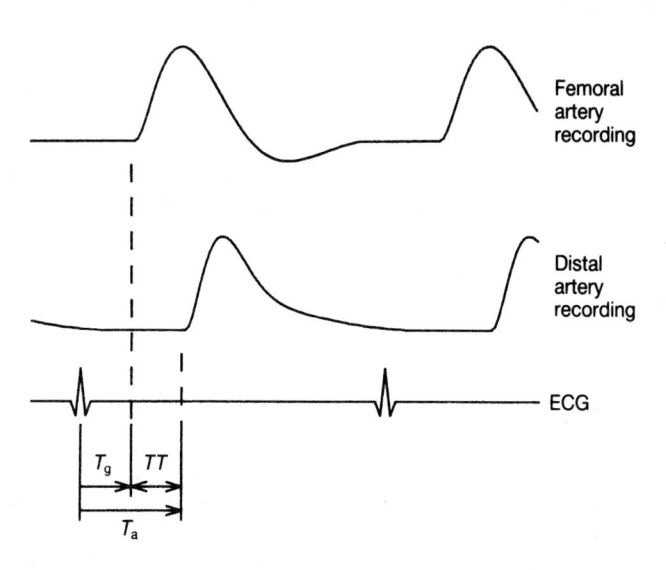

Figure 9.11 Diagram illustrating the variables involved in the definitions of transit time, damping factor and transit time ratio. T_g is the time lapse between the ECG R-wave and the arrival of the pulse at the groin, T_a the time lapse from the R-wave to the arrival of the pulse at the ankle, TT the time lapse between the arrival of the pulse at the two arterial sites. In the case of transit time ratio the arrival of the pressure pulse was defined as the mid-way point on the up-slope of the systolic peak, rather than the foot of the waveform

9.4.9.2 Transit Time Ratio

Another index based on the transit time of the arterial pulse, the transit time ratio (*TTR*) was suggested by Craxford and Chamberlain (1977). This index was defined (Fig. 9.11) as:

$$TTR = T_a/T_g \qquad 9.31$$

where T_a is the time from the ECG R-wave to the arrival of the flow waveform at the posterior tibial artery, and T_g the time from the R-wave to the arrival of flow at the common femoral artery. In both cases the reference point on the flow waveform was taken as the mid-way point on the upslope of the systolic pulse. It was suggested that this index was capable of separating normal subjects and patients with proximal, distal and mixed arterial disease.

9.4.9.3 Frequency Ratio

A further method of using information gathered from two different sites is to compare the maximum Doppler shift frequencies recorded at different locations, either in an unbranched segment of artery (Spencer and Reid 1979, Manga et al 1986, Legemate et al 1991) or in two separate arteries (Blackshear et al 1980). If the artery is unbranched, the flow at every point must be equal and therefore any decrease in the cross-sectional area of the vessel will cause an increase in velocity, and it should be possible to measure the degree of stenosis by comparing the highest Doppler shift frequency within the stenosis with that found in a normal segment of vessel. This method is better than simply measuring the maximum frequency (Section 9.4.2) because it is not upset if the stenosis is severe enough to reduce the flow through the artery. Spencer and Reid (1979) showed that the ratio f_2/f_1 (where f_1 is the Doppler shift frequency recorded from the stenosis and f_2 the downstream frequency) was more closely related to the change in radiological diameter of the vessel than the diameter squared, but this was thought to be a result of asymmetrical plaque development and the intrinsic limitations of arteriography.

9.4.10 Indices Derived from Multiple Sample Volumes

The widespread use of colour flow imaging systems means that it is now possible to gather simultaneous information about velocity from multiple sample volumes, and this provides an entirely new set of candidate features for pattern recognition processes. To date most of the sophisticated work in this area seems to have been confined to *in vitro* models such as those described by Rittgers and colleagues (Rittgers and Fei 1988, Vattyam et al 1992) and by Bolger et al (1988), but simple measurements (such as jet size) are frequently made and it seems inevitable that more refined analytical methods will be developed for future clinical use.

9.4.10.1 Quantification of Fluid Jets

Bolger et al (1988), Simpson et al (1989) and Holen et al (1990) have used *in vitro* models to study the properties of fluid jets as revealed by colour flow imaging. The first group calculated the jet area and the jet energy both from single colour frames and over the time of the injection used to produce the jet. The jet area was calculated simply as the total number of pixels in all forward velocity ranges (presumably with an implicit threshold), whilst the summed jet energy (*SJE*) was calculated as the sum of all pixel velocities squared integrated over all frames of the injection, i.e:

$$SJE = \sum_{f=0}^{F} \sum_{z=0}^{Z} \sum_{r=0}^{R} v_{fzr}^2 \qquad 9.32$$

where v_{fzr} represents the velocity calculated from each pixel, f the colour frame number, z the axial direction of the jet, and r the radial direction. The rationale behind calculating this quantity is that each pixel represents a similar volume of blood with a certain mass moving at a known velocity, and that the product of each mass with its velocity squared will produce a measure of the kinetic energy in that slice through the jet being imaged. Similar measurements, together with jet length and jet width, were reported by Simpson et al, whilst Holen et al concentrated on measurements of colour jet dimensions.

Great care must be taken in interpreting the size of jets as imaged with colour flow systems because not only are they sensitive to machine set-up (Bolger et al 1988, Utsunomiya et al 1990), but may also be more dependent on jet velocity than actual jet size (Losordo et al 1993). Furthermore jet size itself may not be a totally reliable index of disease severity (Simpson et al 1995).

9.4.10.2 Reverse Area Index

Reverse area index (RAI) was used by Vattyam et al (1992) to provide a quantitative measure of the amount of flow separation occurring downstream of a stenosis, and is a quantitative measure of the area of flow reversal in a Doppler colour image. It is simply defined as:

$$RAI = \frac{N_r}{(N_r + N_f)} \qquad 9.33$$

where N_r is the number of pixels with low (defined by Vattyam et al as less than 2 cm s^{-1}) or reverse velocity, and N_f is the number of pixels with forward velocity.

9.4.10.3 Field Profile Index

Field profile index (FPI) has been used by Rittgers and colleagues (Rittgers and Fei 1988, Vattyam et al 1992) as a measure of the coefficient of variance of the velocities occurring downstream of a stenosis, and was defined as:

$$FPI = \frac{1}{\bar{v}} \sqrt{\frac{1}{(N_p - 1)} \times \sum_{z=0}^{Z} \sum_{r=0}^{R} (v_{zr} - \bar{v})^2} \qquad 9.34$$

where v_{zr} represents the velocity calculated from each pixel, $\bar{v}$ is the mean image velocity, N_p is the total number of pixels in the image, and z and r have the same meanings as in eqn. 9.32.

9.4.10.4 Velocity Gradient Index

The most complicated of the indices described by Vattyam et al (1992) was the velocity gradient index (VGI), which was used to quantify the gradient of velocities downstream of a stenosis and was defined separately for the axial (VGI_z) and radial (VGI_r) directions, according to the following equations:

$$VGI_z = \frac{\sqrt{\sum_{z=0}^{Z} \sum_{r=0}^{R} \Delta v_z^2 / N}}{\bar{v}/R} \qquad 9.35$$

and

$$VGI_r = \frac{\sqrt{\sum_{z=0}^{Z} \sum_{r=0}^{R} \Delta v_r^2 / N}}{\bar{v}/R} \qquad 9.36$$

where $\bar{v}$ is the mean velocity over the image, R is the radius of the tube, N is the total number of gradients, and Δv_z and Δv_r are defined by:

$$\Delta v_z = (v_{i-1} - v_{i+1}) \qquad 9.37$$

and

$$\Delta v_r = (v_{k-1} - v_{k+1}) \qquad 9.38$$

where v_i is the velocity of the ith pixel in the z (axial) direction and v_k the velocity of the kth pixel in the r (radial) direction.

9.4.10.5 Contour Length and Contour/Area Ratio

Veyrat and colleagues (1994) have advocated the use of the length of the perimeter of colour flow jets, together with jet area (the contour/area ratio) as a method of studying aortic stenosis.

9.4.10.6 Miscellaneous Vascularity Measurements

Carson and colleagues (Carson et al 1993, 1998) have described a number of 'vascularity measures' which they studied in the hope of distinguishing between malignant and benign breast masses. These measures were calculated from 3-D regions of interest in a 3-D tissue volume. The measures they chose may have applications for studying tissue vascularity in other contexts, and the quantities used in their most recent study are reproduced below for reference purposes.

Power-weighted pixel density:

$$PD = \sum_{i=1}^{N_b} P_i / N_t \qquad 9.39$$

Normalised mean power in coloured pixels:

$$NMPCP = \sum_{i=1}^{N_b} \frac{P_i}{P_b} / N_b \qquad 9.40$$

Normalised power-weighted pixel density:

$$NPD = \sum_{i=1}^{N_b} \frac{P_i}{P_b} / N_t \qquad 9.41$$

Speed-weighted pixel density:

$$SWD = \sum_{i=1}^{N_b} V_i / N_t \qquad 9.42$$

Mean speed in coloured pixels:

$$\bar{v}_{cp} = \sum_{i=1}^{N_b} V_i / N_b \qquad 9.43$$

Speed and power-weighted normalised pixel density:

$$SNPD = \sum_{i=1}^{N_b} \frac{V_i P_i}{P_b} / N_t \qquad 9.44$$

Peak mean speed times peak NPD:

$$v_m \cdot NPD_m = \max(\bar{v}_{cp}) \cdot \max(NPD) \qquad 9.45$$

In each of the above equations, N_b is the number of pixels with flow in them, N_t the total number of pixels in the region of interest, P_b the average Doppler power value measured in a large vessel, P_i the Doppler power value in the ith pixel, and V_i the Doppler speed from the modulus of the mean frequency shift in the ith pixel, assuming the cosine of the Doppler angle = 0.5.

9.4.11 Miscellaneous Methods

9.4.11.1 Reactive Hyperaemia Tests

Doppler ultrasound has been used to quantify the hyperaemia that occurs after an artery has been artificially occluded for a short period of time. Both Fronek et al (1973) and Ward and Martin (1980) used this technique to detect aorto-iliac disease. They established baseline velocity readings, inflated a blood pressure cuff on the upper thigh to a suprasystolic pressure for a period of some minutes and then recorded changes in the Doppler shift frequency following deflation. Fronek et al found a smaller velocity increase and a longer $T_{1/2}$ (the time in seconds for the velocity to return half-way to the control baseline) in patients with disease; Ward and Martin did not report on the percentage velocity increase, but found a similar augmentation in $T_{1/2}$.

9.4.11.2 Back Pressure

Gosling and colleagues (1991) have suggested that a quantity they term 'back-pressure' (P_z), may be more useful than PI for characterising vascular beds where the impedance is largely resistive (for example the cerebral, renal and utero-placental circulations). The definition of P_z arose from a simple electrical model and embodies information about the blood pressure waveform shape as well as the velocity waveform shape. Back pressure is defined as:

$$P_z = \bar{P}\left(1 - \frac{PI_{pressure}}{PI}\right) \qquad 9.46$$

where $\bar{P}$ is the mean arterial pressure (calculated from diastolic pressure plus one-third of pulse pressure), PI is as defined in eqn. 9.2, and $PI_{pressure}$ is defined as pulse pressure divided by mean arterial pressure.

9.4.11.3 Frequency Contours

A method of attempting to classify lower limb arterial stenoses described by Guo et al (1994) combined a number of techniques, including frequency contour extraction, to arrive at 15 features, which were then reduced in number using selection techniques described in Section 9.4.6. The contour extraction was performed to find the extent of significant Doppler power in the sonogram or time–frequency representation (*TFR*) of the Doppler signal and this information used for further processing. The raw sonogram images were first smoothed using a 3×3 mean filter, followed by a 3×3 median filter, and a Laplacian edge detector was used to determine the boundaries of the *TFR*. A contour extraction algorithm was then used to generate a single closed contour for each cardiac cycle. Of the 15 features then used to characterise the waveform, two related to area ratios found from the contour, five to amplitude distribution features of the *TRF*, and eight to the shape of the contour. In order to define the shape of the contour it was described by a 1-D sequence of radial distances measured from the contour centroid to each point on the contour. After scaling, the shape of the resultant curve was modelled by an autoregressive (AR) process of order 7, and the resulting coefficients, together with the variance of the error, used as the eight features.

9.4.11.4 Curve Matching

Marsál and colleagues (Malcus et al 1991) have used a curve matching (CM) technique to describe the shapes of velocity waveforms from fetal aortic and umbilical arteries. This method grew out of an earlier method reported by the same group (Laurin et al 1987a,b) in which fetal aortic velocity waveforms were divided into four semi-quantitative 'blood flow classes' (BFCs) 0–III, formed by combining the *PI* with the shape of the diastolic portion of the velocity curve. BFC-0 indicated a normal *PI* and a normal blood flow velocity curve with positive flow throughout the cardiac cycle; BFC-I was characterised by a positive flow throughout the cardiac cycle, but an increased *PI* ($\geq$ mean + 2 SD); BFC-II indicated that end-diastolic flow was not detectable; whilst BFC-III indicated there was an absence of flow during the main part of diastole, or the presence of reverse flow.

The new CM technique is much more sophisticated with waveforms being compared with two sets of standard curves, one for the fetal descending

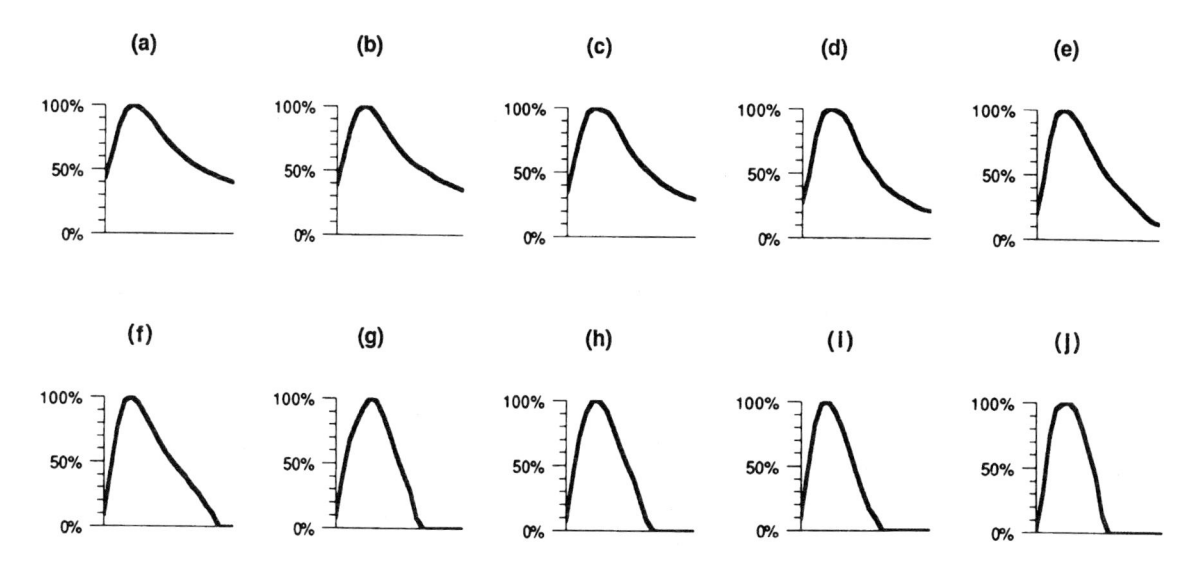

Figure 9.12 Ten standard fetal umbilical waveforms used in the waveform matching technique. In (a)–(e) there is a steadily decreasing amount of diastolic flow, in (f)–(j) there is an increasing absence of diastolic flow (reproduced by permission of Elsevier Science from Malcus et al 1991, *Ultrasound in Medicine and Biology*, © World Federation of Ultrasound in Medicine and Biology)

aorta, the other for the umbilical artery. In all there were 11 curves for the aorta (A–K) and 10 curves for the umbilical artery (a–j), which were empirically grouped with respect to diminishing diastolic flow. Figure 9.12 shows the standard set of curves for umbilical artery waveforms. In order to determine the best match for a new set of waveforms, each waveform was divided into 20 equal time intervals and normalised by its mean value. Ten cycles were then averaged together to produce a normalised average waveform that was compared to the appropriate set of standard waveforms. Normalised waveforms with absent end-diastolic flow were compared with curves F to K and f to j for fetal aortic and umbilical arteries, respectively, and a match found for the interval in which the first 'zero flow' occurred. If there was no absent diastolic flow, a best fit between the test curve and the remaining standard curves was found based on the least summed squared difference.

9.5 CLASSIFICATION

The final stage of the pattern recognition process is to assign the feature vectors obtained during feature extraction to one of a number of classes. There may be only two classes, usually normal or abnormal, or several consisting of different degrees or distributions of disease. The complexity of classification depends on the dimensionality of the feature vectors. If they are one-dimensional, as for pulsatility index and all the other indices discussed in Sections 9.4.3–9.4.5, it is only necessary to define one or more thresholds which will be used to decide the class. If the feature vectors are 2-D or more, as for example those resulting from stepwise selection algorithms (Section 9.4.6) or principal component analysis (Section 9.4.8), then both the form and the position of the separating surfaces must be decided.

9.5.1 Sensitivity, Specificity and ROC Curve Analysis

The simplest classification problem is that of separating one-dimensional feature vectors into two groups. In this situation the only choice that needs to be made is where to locate the decision threshold. If there is no overlap between the magnitudes of the vectors obtained from patients belonging to the two classes, the threshold can simply be chosen to separate the classes completely. In general, however, the results from the two classes do overlap and so, depending on where the threshold is placed, some signals from normal subjects will be adjudged abnormal and/or some signals from abnormals will be adjudged normal. The best choice of threshold will then depend on a number of factors, including the consequences of making both types of false classifications (false positive and false negative) and the prevalence of disease in the target population. Various aspects of this problem have been considered in depth by Metz (1978) and O'Donnell et al (1980).

There are two important measures of the performance of a diagnostic test; sensitivity (or true positive fraction) and specificity (or true negative fraction) which are defined as:

$$\text{Sensitivity}(TPF) = \frac{\text{Number of true positive decisions}}{\text{Number of actually positive cases}} \quad 9.47$$

and

$$\text{Specificity}(TNF) = \frac{\text{Number of true negative decisions}}{\text{Number of actually negative cases}} \quad 9.48$$

The two are not independent since they are both affected by the position of the decision threshold; and as the threshold is moved to increase sensitivity, so specificity falls. Two closely related quantities often quoted in the literature are positive and negative predictive values. The former is the probability that a positive result in a test indicates a genuine positive results, the latter that a negative result indicates a genuine negative result. These may be calculated as follows:

$$\text{Positive predictive value} = \frac{\text{Number of true positive decisions}}{\text{Total number of positive decisions}} \quad 9.49$$

and

Negative predictive value =

$$\frac{\text{Number of true negative decisions}}{\text{Total number of negative decisions}} \quad 9.50$$

One further performance indicator sometimes used is accuracy, which is defined as:

$$\text{Accuracy} = \frac{\text{Number of correct decisions}}{\text{Total number of cases}} \quad 9.51$$

and is related to sensitivity and specificity by the expression:

Accuracy = Sensitivity
  × Fraction of the study population
    that is actually positive
  + Specificity
  × Fraction of the study population
    that is actually negative   9.52

Since the accuracy is determined by the prevalence of disease in the study population, it is actually a very poor measure of performance for a diagnostic test.

The best method of assessing the value of a test and defining an appropriate decision threshold is to plot a receiver operating characteristic (ROC) curve for the test. Such a curve is derived by varying the decision threshold in small steps and determining the TPF and TNF for each new threshold value. In this way, curves of the type shown in Fig. 9.13 are built up. Conventionally, the false positive fraction (FPF) or (1-specificity) is plotted along the abscissa, and TPF or sensitivity plotted along the ordinate. A good test—curve (a)—is one for which TPF rises rapidly and FPF hardly increases at all until TPF is high; a poor test—curve (c)—is one for which TPF and FPF increases at similar rates. Plotting ROC curves for two or more tests in this way enables their relative diagnostic values to be determined. It should, however, be noted that the locus of the ROC curve for a particular tests is also affected by the 'gold standard' against which it is compared (O'Donnell et al 1980). If the gold standard is made more strict

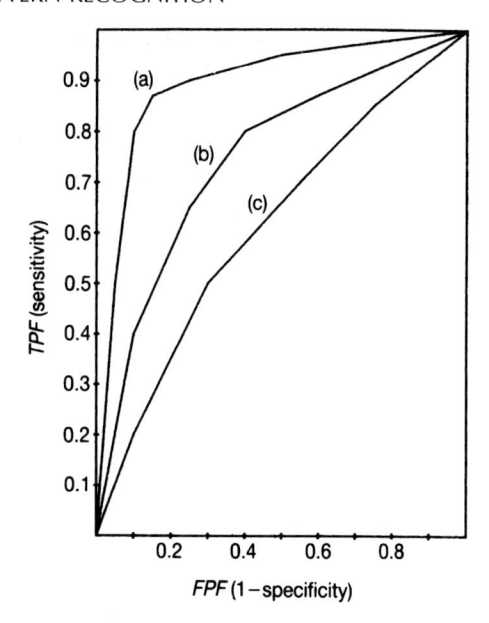

Figure 9.13 Receiver operator characteristic curves for three hypothetical tests of varying utility

(in the sense that only severely diseased arteries are placed in the abnormal group) then, for a given threshold, the apparent sensitivity of the test will increase and the apparent specificity will fall.

The 'best' decision threshold for any test depends on the shape of the ROC curve, the prevalence of disease in the population to be studied, and the costs (both financially and in terms of medical consequences) of missing disease when it is present (false negatives), and of diagnosing disease when it is absent (false positives). If the disease is rare in the study population (as for example in screening situations), the threshold should be strict, i.e. placed near the lower left portion of the ROC curve, otherwise almost all positive decisions will be false positives. If the disease is common, the threshold can be relaxed. If the cost of a false positive is much higher than that of a false negative (as for example when the treatment for disease is potentially harmful to healthy patients, and of limited benefit to diseased patients), then once again the threshold should be relatively strict, whereas if the cost of a false negative is much greater than that of a false positive the threshold should be set towards the upper right portion of the ROC curve. Methods of calculating the optimal threshold are discussed by Metz (1978).

9.5.2 Classification in Two or More Dimensions

When the feature vectors are of two or more dimensions, the first step in the classification procedure is to find a suitable shape and orientation for the separating surfaces. It may then, depending on a procedure adopted, be possible to vary a threshold to produce ROC curves and pick an optimum operating point as for the one-dimensional problem discussed in the last section. There are many approaches to the N-dimensional classification problem (Andrews 1972, Tou and Gonzalez 1974), but there have as yet been few reports of their use with Doppler signals. Nicolaides et al (1976), Rutherford et al (1977) and Keagy et al (1982) have all used discriminant analysis applied to simple features extracted from the Doppler waveform. Greene et al (1982) compared the performance of five training algorithms for classifying the results of their step-wise feature-selection algorithms (Section 9.4.6) and Evans and Caprihan (1985) compared four techniques of classifying the results of principal component analysis (Section 9.4.8). In a further paper, Evans et al (1985) applied two classification techniques to principal component analysis data collected from neonatal cerebral arteries, and compared the results using ROC curve analysis. A detailed discussion of N-dimensional classification techniques is beyond the scope of this chapter, but a brief outline of three commonly used methods follows. Each is an example of 'supervised' learning, that is to say the classification algorithms are trained on a set of data each of which is of known class. Before applying any of these techniques, the features should be transformed to have a population mean of zero and a variance of unity so as to remove their dimensions (Sebestyen 1962).

9.5.2.1 Nearest Neighbour Algorithms

The nearest neighbour (NN) method of classification requires no assumptions to be made about the statistical distribution of the data and is simple to implement. A training set of data, each of known class, is stored and serves as a reference against which new data are compared. The training set may consist of one or many vectors from each class, and the classification algorithm may assign the new vector to the class of its NN or the majority of its q NNs (where q is the integer greater than two, and usually odd for the two-class problem). In the simplest case where are there only two classes, each represented by a single prototype, the NN rule leads to a separating function which is a linear hyperplane. Multiple prototypes lead to piecewise-linear separating functions. The NN technique is particular valuable where the members of each class form distinct clusters, i.e. the intraclass distances are small compared with the interclass distances. Pattern classification by distance function is dealt with in detail by Tou and Gonzalez (1974, Chapter 3), and Evans and Caprihan (1985) have illustrated the NN method with two sets of Doppler data.

9.5.2.2 Multivariate Discriminant Analysis

Discriminant analysis is an example of a statistical classification technique. Each class is assumed to be drawn from a population with a multivariate normal distribution, and the covariance matrices of all classes are assumed to be identical, or nearly so. The classification functions that result under these circumstances are simple linear combinations of the discriminating variable. Details of discriminant analysis and its implementation can be found in Thorndike (1978) and Klecka (1980), and the method has been used to classify Doppler data by Rutherford et al (1977) and Keagy et al (1982). Statistical techniques are potentially much more powerful than distribution-free methods, such as the nearest neighbour method, because they allow probabilities to be assigned to each classification, and the decision threshold to be varied and optimised for disease prevalence and the consequences of both correct and incorrect classifications (Section 9.5.1).

9.5.2.3 The Bayes Method

A second method which uses the statistical properties of the data is the Bayes method. In this method the relative probabilities that a given feature vector belongs to each possible class are calculated from the probability density function and *a priori* probability of each class. The advantage of this method over that discussed in the previous section is that the probability density function may be estimated using either parametric or non-parametric methods, and that if a multivariate normal density function is used (as is most usual) there is no requirement that the covariance matrices of each class are identical or even similar. Because of this the Bayes method seems more appropriate for separating the feature vectors derived from Doppler signals as it is unlikely that the normal and abnormal classes will have similar multivariate normal distributions in N-dimensional feature space. Details of the Bayes method can be found in Tou and Gonzalez (1974, Chapter 4) and it has been used to classify Doppler data by Evans and Caprihan (1985) and Evans et al (1985).

9.6 SUMMARY

Waveform analysis is a powerful diagnostic tool which is complementary to, rather than a substitute for, volumetric blood flow measurement. Formally, the pattern recognition process involved in waveform analysis may be split into transduction, feature extraction and selection, and classification. A variety of approaches have been tried for each of these stages, and the best combination is influenced by the recording sites and the objectives of the analysis. The widespread availability of microcomputers should allow the introduction of more sophisticated methods of feature selection and classification, and the provision of on-line diagnostic information.

9.7 NOTATION

a_n	Amplitude of the nth Fourier harmonic
A	Area under the curve, or height of the first peak of twin peaked waveforms
A_1	Area under Doppler curve up to the systolic peak
A_2	Area under Doppler curve from peak systole onwards
b_k	Coefficient of the kth principal component
B	Height of the second peak of a twin-peaked waveform
BFC	Blood flow class
CBI	Curve broadening index
CFR	Constant flow ratio
C_{ij}	One element of a covariance matrix
D	Minimum or end-diastolic height of a Doppler waveform
DF	Damping factor
E	Efficiency of a transform
EDN	Early diastolic notch
f_i	The ith value of the waveform amplitude
f_{in}	The ith value of the nth waveform
f_{max}	Maximum frequency at peak systole
f_{mean}	Mean frequency at peak systole
f_{median}	Median frequency at peak systole
f_{min}	Minimum frequency at peak systole
f_1	Doppler frequency within a stenosis
f_2	Doppler frequency beyond a stenosis
FPF	False positive fraction
FPI	Field profile index
HRI	High resistance index
HWI	Height–width index
$H(s)$	Laplace transform
j	Square root of -1
$m3_f$	Third moment of frequency about the mean
$m4_f$	Fourth moment of frequency about the mean
M	Mean value of amplitude of a waveform over the cardiac cycle
$M_{N,D}$	Mean value of amplitude of a waveform from the early diastolic notch to end diastole
M_1	Mean height of Doppler curve up to systolic peak
M_2	Mean height of Doppler curve from peak systole onwards
N	Early diastolic or 'notch' velocity; or total number of waveforms
N_b	Number of pixels containing flow
N_f	Number of pixels in an image with forward velocities
N_p	Total number of pixels in an image
N_r	Number of pixels in an image with reverse velocities
N_t	Total number of pixels in a region of interest

$NFPI$	'Notch' Fourier pulsatility index	T	Total duration of a waveform
$NMPCP$	Normalised mean power in coloured pixels	T_a	Time from ECG R-wave to arrival of flow waveform at posterior tibial artery
NPD	Normalised power-weighted pixel density	T_g	Time from ECG R-wave to arrival of flow waveform at common femoral artery
$\bar{P}$	Mean arterial blood pressure		
P_b	Average Doppler power value measured in a large vessel	TFR	Time–frequency representation
P_i	Doppler power value in the ith pixel	TNF	True negative fraction
P_z	Back pressure	TPF	True positive fraction
PI	Pulsatility index	TST	Trans-systolic time
PI_{fem}	Pulsatility index in common femoral artery	TT	Transit time
		TTR	Transit time ratio
PI_{orig}	Original (Fourier) pulsatility index	$\bar{v}$	Mean velocity over an image
PI_{pop}	Pulsatility index in popliteal artery	$\bar{v}_{cp}$	Mean speed in coloured pixels
$PI_{pressure}$	Pulse pressure divided by mean arterial pressure	v_{fzr}	Velocity corresponding to a pixel for a given colour frame number and at given axial and radial co-ordinates
PLI	Path length index		
r	Co-ordinate in radial direction of jet	v_{tst}	Velocity mid-way between D and S
r_{ki}	The ith element of the kth principal component	v_{zr}	Velocity corresponding to a pixel at given axial and radial co-ordinates
RAI	Reverse area index	V_i	Doppler speed from the modulus of the mean frequency shift from the ith pixel, assuming the cosine of the Doppler angle to be 0.5
$RFRI$	Relative flow rate index		
RI	Pourcelot's resistance index		
s_f	Standard deviation of frequency		
S	Maximum height of a Doppler waveform (or $j\omega$)	VGI	Velocity gradient index
		z	Co-ordinate in axial direction of a jet
$SBI(\bullet)$	Spectral broadening index (numbered 1–10)	δ	Coefficient derived from Laplace transform analysis related to proximal arterial narrowing
$SDTI$	Systolic decay time index		
SJE	Summed jet energy	γ	Coefficient derived from Laplace transform analysis related to vasoconstriction/vasodilatation
SMR	Sample mean record		
$SNPD$	Speed and power-weighted normalised pixel density		
		λ_i	The ith eigenvalue of a covariance matrix
SWD	Speed-weighted pixel density		
t_0	Time of peak systole	ω_0	Coefficient derived from Laplace transform analysis related to arterial elastic modulus
t_{0+n}	n ms after peak systole		
t_1	Time from beginning of systole to peak systole		
t_2	Time from peak systole to the end of diastole		
t_i	The ith value of time		
t_d	Decay time from S to $0.75S$		
t_{fem}	Time of arrival of waveform foot at common femoral site		
t_{pop}	Time of arrival of waveform foot at popliteal site		
t_r	Rise time from $0.75S$ to S		
t_s	Duration of systolic peak (measured between the half amplitude points)		

9.8 REFERENCES

Adamson SL, Morrow RJ, Bascom PAJ, Mo LYL, Ritchie JWK (1989) Effect of placental resistance, arterial diameter, and blood pressure on the uterine arterial velocity waveform: a computer modeling approach. Ultrasound Med Biol 15:437–442

Allard L, Langlois YE, Durand L-G, Roederer GO, Beaudoin M, Cloutier G, Roy P, Robillard P (1991) Computer analysis and pattern recognition of Doppler blood flow spectra for disease classification in the lower limb arteries. Ultrasound Med Biol 17:211–223

Andrews HC (1972) Introduction to mathematical techniques in pattern recognition. Wiley-Interscience, New York

Arbeille P, Berson M, Achaibou F, Bodard S, Locatelli A (1995) Vascular resistance quantification in high flow resistance areas using the Doppler method. Ultrasound Med Biol 21:321–328

Archer LNJ, Levene MI, Evans DH (1986) Cerebral artery Doppler ultrasonography for prediction of outcome after perinatal asphyxia. Lancet ii (15 November):1116–1118

Bada HS, Hajjar W, Chua C, Sumner DS (1979) Noninvasive diagnosis of neonatal asphyxia and intraventricular hemorrhage by Doppler ultrasound. J Pediat 95:775–779

Baird RN, Bird DR, Clifford PC, Lusby RJ, Skidmore R, Woodcock JP (1980) Upstream stenosis—its diagnosis by Doppler signals from the femoral artery. Arch Surg 115:1316–1322

Baker AR, Evans DH, Prytherch DR, Bell PRF (1986) Haemodynamic assessment of the femoropopliteal segment: comparison of pressure and Doppler methods using ROC curve analysis. Br J Surg 73:559–562

Barnes RW, Bone GE, Reinertson J, Slaymaker EE, Hokanson DE, Strandness DE (1976) Noninvasive ultrasonic carotid angiography: prospective validation by contrast arteriography. Surgery 80:328–335

Baskett JJ, Beasley MG, Murphy GJ, Hyams DE, Gosling RG (1977) Screening for carotid junction disease by spectral analysis of Doppler signals. Cardiovasc Res 11:147–155

Blackshear WM, Phillips DJ, Thiele BL, Hirsch JH, Chikos PM, Marinelli MR, Ward JK, Strandness DE (1979) Detection of carotid occlusive disease by ultrasonic imaging and pulsed Doppler spectrum analysis. Surgery 86:698–706

Blackshear WM, Phillips DJ, Chikos PM, Harley JD, Thiele BL, Strandness DE (1980) Carotid artery velocity patterns in normal and stenotic vessels. Stroke 11:67–71

Bluth EI, Wetzner SM, Stavros AT, Aufrichtig D, Marich KW, Baker JD (1988) Carotid duplex sonography: a multicenter recommendation for standardized imaging and Doppler criteria. RadioGraphics 8:487–506

Bodily KC, Zierler RE, Marinelli MR, Thiele BL, Greene FM, Strandness DE (1980) Flow disturbances following carotid endarterectomy. Surg Gynecol Obstet 151:77–80

Bolger AF, Eigler ML, Pfaff JM, Resser KJ, Maurer G (1988) Computer analysis of Doppler color flow mapping images for quantitative assessment of *in vitro* fluid jets. J Am Coll Cardiol 12:450–457

Bower S, Schuchter K, Campbell S (1993) Doppler ultrasound screening as part of routine antenatal scanning: prediction of pre-eclampsia and intrauterine growth retardation. Br J Obstet Gynaecol 100:989–994

Brown PM, Johnston KW, Kassam M, Cobbold RSC (1982) A critical study of ultrasound Doppler spectral analysis for detecting carotid disease. Ultrasound Med Biol 8:515–523

Carson PL, Li X, Pallister J, Moskalik A, Rubin JM, Fowlkes JB (1993) Approximate quantification of detected fractional blood volume and perfusion from 3-D color flow and Doppler power signal imaging. In: Proceedings of the 1993 IEEE Ultrasonics Symposium, pp. 1023–1026. IEEE, Piscataway, NJ

Carson PL, Fowlkes JB, Roubidoux MA, Moskalik AP, Govil A, Normolle D, LeCarpentier G, Nattakom S, Helvie M, Rubin JM (1998) 3-D color Doppler image quantification of breast masses. Ultrasound Med Biol 24:945–952

Craxford AD, Chamberlain J (1977) Pulse wave form transit ratios in the assessment of peripheral vascular disease. Br J Surg 64:449–452

Evans DH (1982) Some aspects of the relationship between instantaneous volumetric blood flow and continuous wave Doppler ultrasound recording—III. The calculation of Doppler power spectra from mean velocity waveforms, and the results of processing these spectra with maximum, mean, and RMS frequency processors. Ultrasound Med Biol 8:617–623

Evans DH (1984) The interpretation of continuous wave ultrasonic Doppler blood velocity signals viewed as a problem in pattern recognition. J Biomed Eng 6:272–280

Evans DH (1992) Principal component analysis applied to the diagnosis of arterial disease. In: Labs KH, Jager KA, Fitzgerald DE, Woodcock JP, Neuerburg-Heusler D (eds) Diagnostic Vascular Ultrasound, pp. 85–94. Edward Arnold, London

Evans DH, Caprihan A (1985) The application of classification techniques to biomedical data, with particular reference to ultrasonic Doppler blood velocity waveforms. IEEE Trans Biomed Eng BME-32:301–311

Evans DH, Barrie WW, Asher MJ, Bentley S, Bell PRF (1980) The relationship between ultrasonic pulsatility index and proximal arterial stenosis in a canine model. Circ Res 46:470–475

Evans DH, Archer LNJ, Levene MI (1985) The detection of abnormal neonatal cerebral haemodynamics using principal component analysis of the Doppler ultrasound waveform. Ultrasound Med Biol 11:441–449

Farrar DJ, Green HD, Peterson DW (1979) Noninvasively and invasively measured pulsatile haemodynamics with graded arterial stenosis. Cardiovasc Res 13:45–57

Fay RA, Ellwood DA, Bruce S, Turner A (1994) Colour Doppler imaging of the uteroplacental circulation in the mid-trimester: features of the uterine artery flow-velocity waveform that predict abnormal pregnancy outcome. Aust NZ J Obstet Gynaecol 34:515–519

Ferrier C, North RA, Becker G, Long D, Hallo J, Kincaid-Smith P (1994) Uterine artery Doppler waveform indices in the second trimester: resistance index, notch measurement and placental position. J Obstet Gynaecol 14:237–243

FitzGerald DE, Gosling RG, Woodcock JP (1971) Grading dynamic capability of arterial collateral circulation. Lancet 1:66–67

Fronek A, Johansen KH, Dilley RB, Bernstein EF (1973) Noninvasive physiologic tests in the diagnosis and characterization of peripheral arterial occlusive disease. Am J Surg 126:205–214

Fronek A, Coel M, Bernstein EF (1976) Quantitative ultrasonographic studies of lower extremity flow velocities in health and disease. Circulation 53:957–960

Gibbons DT, Evans DH, Barrie WW, Cosgriff PS (1981) Real-time calculation of ultrasonic pulsatility index. Med Biol Eng Comput 19:28–34

Gosling RG (1976) Extraction of physiological information

from spectrum-analysed Doppler-shifted continuous wave ultrasound signals obtained non-invasively from the arterial system. In: Hill DW, Watson BW (eds) IEE Medical Monographs: Monographs 18–22, pp. 73–125. Peter Peregrinus, Stevenage

Gosling RG, King DH (1974) Continuous wave ultrasound as an alternative and complement to X-rays in vascular examinations. In: Reneman RS (ed) Cardiovascular Applications of Ultrasound, pp. 266–282. North-Holland, Amsterdam

Gosling RG, King DH (1978) Processing arterial Doppler signals for clinical data. In: deViieger et al (eds) Handbook of Clinical Ultrasound, pp. 613–646. Wiley, New York

Gosling RG, Dunbar G, King DH, Newman DL, Side CD, Woodcock JP, FitzGerald DE, Keates JS, MacMillan D (1971) The quantitative analysis of occlusive peripheral arterial disease by a non-intrusive ultrasonic technique. Angiology 22:52–55

Gosling RG, Lo PTS, Taylor MG (1991) Interpretation of pulsatility index in feeder arteries to low-impedance vascular beds. Ultrasound Obstet Gynecol 1:175–179

Greene FM, Beach K, Strandness DE, Fell G, Phillips DJ (1982) Computer based pattern recognition of carotid arterial disease using pulsed Doppler ultrasound. Ultrasound Med Biol 8:161–176

Guiot C, Pianta PG, Todros T (1992) Modelling the fetoplacental circulation: 1. A distributed network predicting umbilical haemodynamics throughout pregnancy. Ultrasound Med Biol 18:535–544

Guo Z, Durand L-G, Allard L, Cloutier G, Lee HC (1994) Classification of lower limb arterial stenoses from Doppler blood flow signal analysis with time-frequency representation techniques. Ultrasound Med Biol 20:335–346

Hanlo PW, Peters RJA, Gooskens RHJM, Heethaar RM, Keunen RWM, van Huffelen AC, Tulleken CAF, Willemse J (1995) Monitoring intracranial dynamics by transcranial Doppler—a new Doppler index: trans-systolic time. Ultrasound Med Biol 21:613–621

Hendricks SK, Sorensen TK, Wang KY, Bushnell JM, Seguin EM, Zingheim RW (1989) Doppler umbilical artery waveform indices—normal values from fourteen to forty-two weeks. Am J Obstet Gynecol 161:761–765

Hill A, Volpe JJ (1982) Decrease in pulsatile flow in the anterior cerebral arteries in infantile hydrocephalus. Pediatrics 69:4–7

Hill A, Perlman JM, Volpe JJ (1982) Relationship of pneumothorax to occurrence of intraventricular hemorrhage in the premature newborn. Pediatrics 69:144–149

Holen J, Nanna M, Lockhart J, Waag R (1990) Doppler color flow in echocardiography: Analytical and *in vitro* investigations of the quantitative relationship between orifice flow and color jet dimensions. Ultrasound Med Biol 16:543–551

Hoskins PR, Johnstone FD, Chambers SE, Haddad NG, White G, McDicken WN (1989) Heart rate variation of umbilical artery Doppler waveforms. Ultrasound Med Biol 15:101–105

Hoskins PR, Loupas T, McDicken WN (1991) An investigation of simulated umbilical artery Doppler waveforms. I. The effect of three physical parameters on the maximum frequency envelope and on pulsatility index. Ultrasound Med Biol 17:7–21

Hütter W, Grab D, Sterzik K, Terinde R, Wolf A (1993) Polynomial analysis of placental flow patterns in growth-retarded fetuses. Gynecol Obstet Invest 35:155–161

Humphries KN, Hames TK, Smith SWJ, Cannon VA (1980) Quantitative assessment of the common femoral to popliteal arterial segment using continuous wave Doppler ultrasound. Ultrasound Med Biol 6:99–105

Johnston KW, Taraschuk I (1976) Validation of the role of pulsatility index in quantitation of the severity of peripheral arterial occlusive disease. Am J Surg 131:295–297

Johnston KW, Maruzzo BC, Cobbold RSC (1978) Doppler methods for quantitative measurement and localization of peripheral arterial occlusive disease by analysis of the blood flow velocity waveform. Ultrasound Med Biol 4:209–223

Johnston KW, de Morais D, Kassam M, Brown PM (1981) Cerebrovascular assessment using a Doppler carotid scanner and real-time frequency analysis. J Clin Ultrasound 9:443–449

Johnston KW, Brown PM, Kassam M (1982a) Problems of carotid Doppler scanning which can be overcome by using frequency analysis. Stroke 13:660–666

Johnston KW, Kassam M, Cobbold RSC (1982b) Online identifying and quantifying Doppler ultrasound waveforms. Med Biol Eng Comput 20:336–342

Johnston KW, Kassam M, Cobbold RSC (1983) Relationship between Doppler pulsatility index and direct femoral pressure measurements in the diagnosis of aortoiliac occlusive disease. Ultrasound Med Biol 9:271–281

Johnston KW, Kassam M, Koers J, Cobbold RSC, MacHattie D (1984) Comparative study of four methods for quantifying Doppler ultrasound waveforms from the femoral artery. Ultrasound Med Biol 10:1–12

Junger M, Chapman BLW, Underwood CJ, Charlesworth D (1984) A comparison between two types of waveform analysis in patients with multisegmental arterial disease. Br J Surg 71:345–348

Kalman PG, Johnston KW, Zuech P, Kassam M, Poots K (1985) *In vitro* comparison of alternative methods for quantifying the severity of Doppler spectral broadening for the diagnosis of carotid arterial occlusive disease. Ultrasound Med Biol 11:435–440

Kaluzynski K, Palko T (1993) Effect of method and parameters of spectral analysis on selected indices of simulated Doppler spectra. Med Biol Eng Comput 31:249–256

Keagy BA, Pharr WF, Thomas D, Bowes DE (1982) A quantitative method for the evaluation of spectral analysis patterns in carotid artery stenosis. Ultrasound Med Biol 8:625–630

Klecka WR (1980) Discriminant analysis. Sage University Paper Series on Quantitative Applications in the Social Sciences. Sage, Beverly Hills, CA, and London

Knox RA, Greene FM, Beach K, Phillips DJ, Chikos PM, Strandness DE (1982a) Computer based classification of carotid arterial disease: a prospective assessment. Stroke 13:589–594

Knox RA, Phillips DJ, Breslau PJ, Lawrence R, Primozich J, Strandness DE (1982b) Empirical findings relating sample volume size to diagnostic accuracy in pulsed Doppler cerebrovascular studies. J Clin Ultrasound 10:227–232

Langlois YE, Greene FM, Roederer GO, Jager KA, Phillips DJ, Beach KW, Strandness DE (1984) Computer based pattern recognition of carotid artery Doppler signals for disease classification: prospective validation. Ultrasound Med Biol 10:581–595

Laurin J, Lingman G, Marsal K, Persson PH (1987a) Fetal blood flow in pregnancies complicated by intrauterine growth retardation. Obstet Gynecol 69:895–902

Laurin J, Marsal K, Persson PH, Lingman G (1987b) Ultrasound measurement of fetal blood flow in predicting fetal outcome. Br J Obstet Gynaecol 94:940–948

Lee BY, Assadi C, Madden JL, Kavner D, Trainor FS, McCann WJ (1978) Hemodynamics of arterial stenosis. World J Surg 2:621–629

Legemate DA, Teeuwen C, Hoeneveld H, Ackerstaff RGA, Eikelboom BC (1991) Spectral analysis criteria in duplex scanning of aortoiliac and femoropopliteal arterial disease. Ultrasound Med Biol 17:769–776

Lipman B, Serwer GA, Brazy JE (1982) Abnormal cerebral hemodynamics in preterm infants with patent ductus arteriosus. Pediatrics 69:778–781

Losordo DW, Pastore JO, Coletta D, Kenny D, Isner JM (1993) Limitations of color-flow Doppler imaging in the quantification of valvular regurgitation—velocity of regurgitant jet, rather than volume, determines size of color Doppler image. Am Heart J 126:168–176

Loupas T, Ellwood DA, Gill RW, Bruce S, Fay RA (1995) Computer analysis of the early diastolic notch in Doppler sonograms of the uterine artery. Ultrasound Med Biol 21:1001–1011

Macpherson DS, Evans DH, Bell PRF (1984) Common femoral artery Doppler wave-forms: a comparison of three methods of objective analysis with direct pressure measurements. Br J Surg 71:46–49

Malcus P, Andersson J, Marsal K, Olofsson PA (1991) Waveform pattern recognition—a new semiquantitative method for analysis of fetal aortic and umbilical artery blood flow velocity recorded by Doppler ultrasound. Ultrasound Med Biol 17:453–460

Manga P, Dhurandhar RW, Stockard B (1986) Doppler frequency ratio and peak frequency in the assessment of carotid artery disease: a comparative study with angiography. Ultrasound Med Biol 12:573–576

Mari G, Moise KJ, Deter RL, Carpenter RJ, Wasserstrum N (1991) Fetal heart rate influence on the pulsatility index in the middle cerebral artery. J Clin Ultrasound 19:149–153

Martin TRP, Barber DC, Sherriff SB, Prichard DR (1980) Objective feature extraction applied to the diagnosis of carotid artery disease using a Doppler ultrasound technique. Clin Phys Physiol Meas 1:71–81

Martin TRP, Sherriff SB, Barber DC, Lakeman JM (1981) Analysis of the total Doppler signal obtained from the common carotid artery. Ultrasonics 2:269–276

Metz CE (1978) Basic principles of ROC analysis. Semin Nucl Med 8:283–298

Mo LYL, Bascom PAJ, Ritchie K, McCowan LME (1988) A transmission line modelling approach to the interpretation of uterine Doppler waveforms. Ultrasound Med Biol 14:375–376

Nicolaides AN, Gordon-Smith I, Dayandas J, Eastcott HHG (1976) The value of Doppler blood velocity tracings in the detection of aortoiliac disease in patients with intermittent claudication. Surgery 80:774–778

North RA, Ferrier C, Long D, Townend K, Kincaid-Smith P (1994) Uterine artery Doppler flow velocity waveforms in the second trimester for the prediction of pre-eclampsia and fetal growth retardation. Obstet Gynecol 83:378–386

O'Donnell TF, Pauker SG, Callow AD, Kelly JJ, McBride KJ, Korwin S (1980) The relative value of carotid noninvasive testing as determined by receiver operator characteristic curves. Surgery 87:9–19

Perlman JM (1985) Neonatal cerebral blood flow velocity measurement. Clin Perinatol 12:179–193

Perlman JM, Hill A, Volpe JJ (1981) The effect of patent ductus arteriosus on flow velocity in the anterior cerebral arteries: Ductal steal in the premature newborn infant. J Pediat 99:767–771

Planiol T, Pourcelot L (1973) Doppler effect study of the carotid circulation. In: de Vleiger M, White DN, McCreedy VR (eds) Ultrasonics in Medicine, pp. 104–111. Elsevier, New York

Pourcelot L (1976) Diagnostic ultrasound for cerebral vascular diseases. In: Donald I, Levi S (eds) Present and Future of Diagnostic Ultrasound, pp. 141–147. Kooyker, Rotterdam

Prichard DR, Martin TRP, Sherriff SB (1979) Assessment of directional Doppler ultrasound techniques in the diagnosis of carotid artery diseases. J Neurol Neurosurg Psychiatry 42:563–568

Prytherch DR, Evans DH, Smith MJ, Macpherson DS (1982) On-line classification of arterial stenosis severity using principal component analysis applied to Doppler ultrasound signals. Clin Phys Physiol Meas 3:191–200

Reneman RS, Spencer MP (1979) Local Doppler audio spectra in normal and stenosed carotid arteries in man. Ultrasound Med Biol 5:1–11

Rittenhouse RE, Maixner W, Burr JW, Barnes RW (1976) Directional arterial flow velocity: A sensitive index of changes in peripheral vascular resistance. Surgery 79:350–355

Rittgers SE, Fei D-Y (1988) Flow dynamics in a stenosed carotid bifurcation model—Part II: Derived indices. Ultrasound Med Biol 14:33–42

Rittgers SE, Thornhill BM, Barnes RW (1983) Quantitative analysis of carotid artery Doppler spectral waveforms: diagnostic value of parameters. Ultrasound Med Biol 9:255–264

Rutherford RB, Hiatt WR, Kreutzer EW (1977) The use of velocity wave form analysis in the diagnosis of carotid artery occlusive disease. Surgery 82:695–702

Sebestyen G (1962) Decision Making Processes in Pattern Recognition. Macmillan, New York

Sheldon CD, Murie JA, Quin RO (1983) Ultrasonic Doppler spectral broadening in the diagnosis of internal carotid artery stenosis. Ultrasound Med Biol 9:575–580

Sherriff SB, Barber DC, Martin TRP, Lakeman JM (1982) Use of principal component factor analysis in the detection of carotid artery disease from Doppler ultrasound. Med Biol Eng Comput 20:351–356

Sillesen H, Schroeder T (1988) Changes in Doppler waveforms can predict pressure reduction across internal carotid artery stenoses. Ultrasound Med Biol 14:649–655

Simpson IA, ValdesCruz LM, Sahn DJ, Murillo A, Tamura T, Chung KJ (1989) Doppler color flow mapping of simulated in vitro regurgitant jets: evaluation of the effects of orifice size and hemodynamic variables. J Am Coll Cardiol 13:1195–1207

Simpson IA, deBelder MA, Kenny A, Martin M, Nihoyannopoulos P (1995) How to quantitate valve regurgitation by echo Doppler techniques. Br Heart J 73 (suppl 2):1–9

Skidmore R, Woodcock JP (1978) Physiological significance of arterial models derived using transcutaneous ultrasonic flowmeters. J Physiol 277:29–30

Skidmore R, Woodcock JP (1980a) Physiological interpretations of Doppler-shift waveforms—I: Theoretical considerations. Ultrasound Med Biol 6:7–10

Skidmore R, Woodcock JP (1980b) Physiological interpretations of Doppler-shift waveforms—II: Validation of the Laplace transform method for characterisation of the common femoral blood-velocity/time waveform. Ultrasound Med Biol 6:219–225

Skidmore R, Woodcock JP, Wells PNT, Bird D, Baird RN (1980) Physiological interpretation of Doppler-shift waveforms—III: Clinical results. Ultrasound Med Biol 6:227–231

Spencer MP, Reid JM (1979) Quantitation of carotid stenosis with continuous-wave (C-W) Doppler ultrasound. Stroke 10:326–330

Stuart B, Drumm J, FitzGerald DE, Duignan NM (1980) Fetal blood velocity waveforms in normal pregnancy. Br J Obstet Gynaecol 87:780–785

Surat DR, Adamson SL (1996) Downstream determinants of pulsatility of the mean velocity waveform in the umbilical artery as predicted by a computer model. Ultrasound Med Biol 22:707–717

Thiele BL, Bandyk DF, Zierler RE, Strandness DE (1983) A systematic approach to the assessment of aortoiliac disease. Arch Surg 118:477–481

Thompson RS, Stevens RJ (1989) Mathematical model for interpretation of Doppler velocity waveform indices. Med Biol Eng Comput 27:269–276

Thompson RS, Trudinger BJ (1990) Doppler waveform pulsatility index and resistance, pressure and flow in the umbilical placental circulation: an investigation using a mathematical model. Ultrasound Med Biol 16:449–458

Thompson RS, Trudinger BJ, Cook CM (1985) Doppler ultrasound waveforms in the fetal umbilical artery: quantitative analysis technique. Ultrasound Med Biol 11:707–718

Thompson RS, Trudinger BJ, Cook CM, Giles WB (1988) Umbilical artery velocity waveforms: normal reference values for A/B ratio and Pourcelot ratio. Br J Obstet Gynaecol 95:589–591

Thorndike RM (1978) Discriminant analysis. pp. 203–223. In: Thorndike RM (ed) Correlational Procedures for Research. Gardner Press, New York

Todros T, Guiot C, Pianta PG (1992) Modelling the fetoplacental circulation: 2. A continuous approach to explain normal and abnormal flow velocity waveforms in the umbilical arteries. Ultrasound Med Biol 18:545–551

Tou JT, Gonzalez RC (1974) Pattern Recognition Principles. Addison-Wesley, Reading, MA

Trudinger BJ, Giles WB, Cook CM (1985a) Uteroplacental blood flow velocity-time waveforms in normal and complicated pregnancy. Br J Obstet Gynaecol 92:39–45

Trudinger BJ, Giles WB, Cook CM, Bombardieri J, Collins L (1985b) Fetal umbilical artery flow velocity waveforms and placental resistance: clinical significance. Br J Obstet Gynaecol 92:23–30

Utsunomiya T, Ogawa T, King W, Sunada E, Moore GW, Henry WL, Gardin JM (1990) Effect of machine parameters on variance display in Doppler color flow mapping. Am Heart J 120:1395–1402

Van Asten WNJC, Beijneveld WJ, Pieters BR, Van Lier HJJ, Wijn PFF, Skotnicki SH (1991) Assessment of aortoiliac obstructive disease by Doppler spectrum analysis of blood flow velocities in the common femoral artery at rest and during hyperemia. Surgery 109:633–639

van Bel F, Steendijk P, Teitel DF, Peter de Winter J, Van Der Velde ET, Baan J (1992) Cerebral blood flow velocity: the influence of myocardial contractility on the velocity waveform of brain supplying arteries. Ultrasound Med Biol 18:441–449

van den Wijngaard JAGW, van Eyck J, Wladimiroff JW (1988) The relationship between fetal heart rate and Doppler blood flow velocity waveforms. Ultrasound Med Biol 14:593–597

van Merode T, Hick P, Hoeks APG, Reneman RS (1983) Limitations of Doppler spectral broadening in the early detection of carotid artery disease due to the size of the sample volume. Ultrasound Med Biol 9:581–586

Vattyam HM, Shu MCS, Rittgers SE (1992) Quantification of Doppler colour flow images from a stenosed carotid artery model. Ultrasound Med Biol 18:195–203

Veyrat C, Sainte Beuve D, El Yafi W, Sebaoun G, Kalmanson D (1994) A new Doppler imaging measurement in aortic stenosis: the contour length of the jet origin flow area. Relationships between both, with usual Doppler data and left ventricular hypertrophy. Ultrasound Med Biol 20:831–839

Walton L, Martin TRP (1984) Prospective assessment of the aorto-iliac segment by visual interpretation of frequency analysed Doppler waveforms—a comparison with arteriography. Ultrasound Med Biol 10:27–32

Ward AS, Martin TP (1980) Some aspects of ultrasound in the diagnosis and assessment of aortoiliac disease. Am J Surg 140:260–265

Wijn PFF, van der Sar P, Gootzen THJM, Tilmans MHJ, Skotnicki SH (1987) Value of the spectral broadening index in continuous wave Doppler measurements. Med Biol Eng Comput 25:377–385

Windeck P, Karl-Heinz L, Jaeger KA (1992) How useful are acceleration- and deceleration-based Doppler indices? A trial on patients with percutaneous transluminal angioplasty. Ultrasound Med Biol 18:525–534

Woodcock JP (1970) The transcutaneous ultrasonic flowmeter and the significance of changes in the velocity-time waveform in occlusive arterial disease of the leg. PhD Dissertation, University of London

Woodcock JP, Gosling RG, FitzGerald DE (1972) A new non-invasive technique for assessment of superficial femoral artery obstruction. Br J Surg 59:226–231

Woodcock JP, Shedden J, Skidmore R, Machleder H, Evans J, Wells PNT (1982) Doppler spectral broadening and anomalous vessel wall movement in the study of atherosclerosis of the carotid arteries. Ultrasound Med Biol 8 (suppl 1):211

Woodcock JP, Shedden J, Aldoori M, Skidmore R, Burns R, Evans J (1983) Doppler spectral broadening and anomalous vessel wall movement in the study of atherosclerosis of the carotid arteries. In: Lerski RA, Morley P (eds) Ultrasound, '82, pp. 235–237. Pergamon, Oxford

Zwiebel WJ, Zagzebski JA, Crummy AB, Hirscher M (1982) Correlation of peak Doppler frequency with lumen narrowing in carotid stenosis. Stroke 13:386–391

10

Colour Flow Imaging (CFI) Systems

10.1 INTRODUCTION

Colour flow imaging was introduced in 1982 by Aloka Co. Ltd (Japan) based on signal processing developed and later reported by Kasai et al (1985). Since then the technology has steadily developed and matured and is now widely applied clinically. Indeed, it can be said that colour Doppler imaging is the technique which has established Doppler methods in medicine and also helped in the application of the older Doppler spectral techniques (Evans 1993, Routh 1996, Ferrara and DeAngelis 1997, Hoskins and McDicken 1997).

The versatility of ultrasonic and signal processing techniques means that it is necessary to identify carefully what specific Doppler method is being used. The following list includes the more common modalities, the terminology for which has still to be standardised. For the foreseeable future terminology is more likely to be determined by common usage rather than logical definition. The term 'Doppler' is often employed by clinical users specifically to mean spectral Doppler.

1. *Colour Doppler imaging (colour flow imaging)* — in which the mean frequency ('velocity') of the Doppler signal and direction of flow is depicted in each pixel.
2. *Power Doppler imaging* — in which the power of the Doppler signal is depicted in each pixel.
3. *Directional power Doppler imaging* — in which the direction of flow information is used to colour code the power Doppler image.
4. *Harmonic colour Doppler imaging* — in which the image is created using the Doppler information in the harmonics of the reflected ultrasound rather than that in the fundamental frequency. Increased harmonics in the ultrasound signal are usually produced by introducing microbubble contrast agents into the blood. It may result from distortion of the transmitted pulse as it passes through tissue but the harmonic signal resulting from the scattering of such a distorted signal from blood is weak and difficult to exploit.
5. *Harmonic power Doppler imaging* — in which the power of the harmonic Doppler signal is depicted in the pixels.
6. *Colour Doppler M-mode* — in M-mode the ultrasound beam direction is fixed, and the display shows time along the horizontal axis, and depth (or 'fast time') along the vertical axis. In conventional M-mode the size of the returning echoes are coded using a grey-scale, but in colour Doppler M-mode the mean frequency of the Doppler signal and direction of flow is depicted in each pixel.
7. *Doppler tissue imaging* — in which the technology has been adapted to image the high-amplitude, low-frequency Doppler signals from tissue rather than the low-amplitude, high-frequency signals from blood. Harmonic Doppler tissue imaging is feasible using high amplitude transmission pulses which lead to non-linear propagation.

The signal processing techniques associated with these methods, for example to estimate velocity and discriminate blood from tissue, are described in Chapter 11. Each manufacturer usually gives a different label and associated acronym to each type of processing or image production. An example of this is the use of the terms 'Power Doppler' (Diasonics), 'Energy Doppler' (Acuson), 'Angio Doppler' (ATL), and 'CVI-Angio' (Philips), all to describe essentially the same imaging modality. The need for these different labels is presumably related to marketing and the legalities of patents and copyrights.

10.2 ULTRASONIC IMAGING

Ultrasonic imaging methods have been widely described and they continue to develop rapidly (McDicken 1991, Wells 1977, Kremkau 1998). In this section they will be briefly reviewed from the point of view of Doppler imaging. The vast majority of ultrasonic images are produced by sweeping a narrow pulsed beam through a scan plane (Fig. 10.1). Echo signals, resulting from reflection and scattering in soft tissue and blood, are detected by the transducer and processed to produce B-mode and Doppler images. Typically the transmit/receive cycle is carried out at 200 scan line positions across the field-of-view which may be around 15 cm wide. Since the velocity of ultrasound in tissue is high, 1540 m s^{-1} on average, echo information from each scan line is collected very quickly, for instance in less than 200 μs. It is this very high velocity which makes real-time imaging possible and which has been of crucial importance in the development of both B-mode and Doppler imaging. The types of transducer used and the shapes of the beams they produce are described in Chapter 5. The availability of small hand-held transducers which can generate narrow beams is an important feature of diagnostic ultrasound technology. Although mechanical transducers are used in some Doppler imaging systems, the intrinsic relative motion of the scanned tissues and the transducer means that low blood flow velocities cannot readily be separated from the velocities of the tissue. Electronic linear or phased array transducers are therefore commonly used. As noted in Section 5.2.4.1, where digital beam forming is described, both the transmission and reception of ultrasonic signals is extensively controlled by computer, as is image construction and display. Display and recording technology are presented in Appendix 3. Options for colour-coding velocity and power information are described below.

In any imaging technique, artefacts must be recognised and discounted during the interpretation of scans. A knowledge of Doppler artefacts is required for each application and machine. Doppler artefacts can be described and classified into groups relating to their origin, i.e. from ultrasound propagation in tissue, the electronics, or technique.

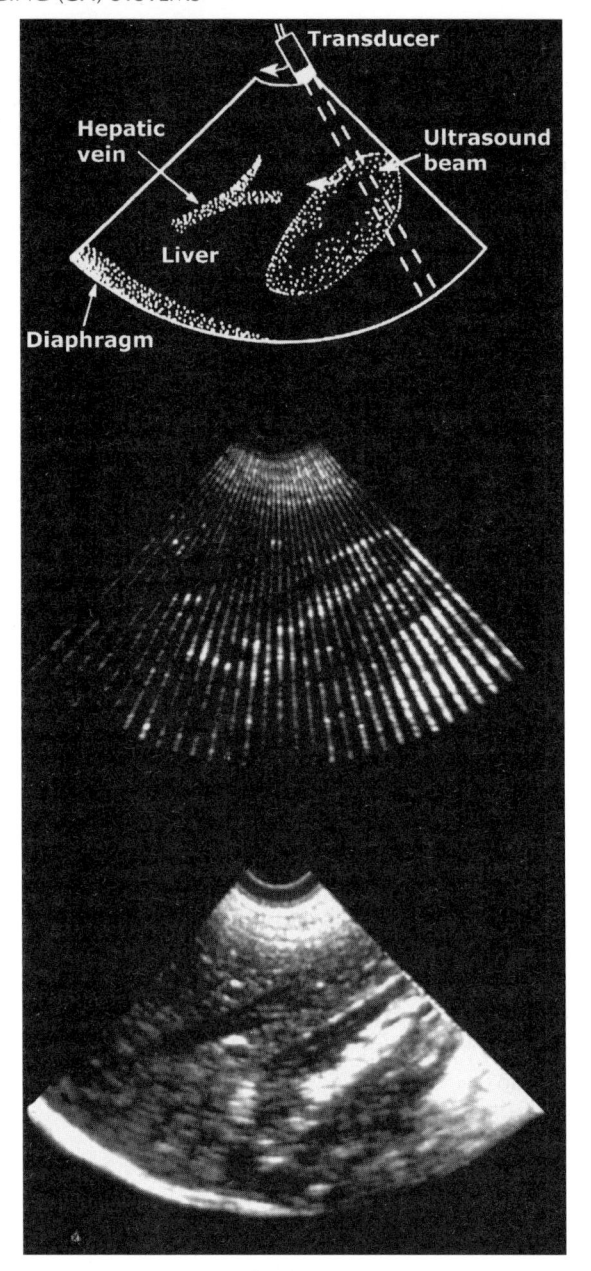

Figure 10.1 Production of an ultrasound image. A narrow beam sweeps through a scan plane and echo information is collected from each beam direction. Echo signals are processed to extract amplitude or Doppler information, which is then displayed along the beam scan line. With a sufficient number of lines a complete cross-sectional image is produced

Each new Doppler mode needs to be thoroughly assessed for artefacts. An understanding of the achievable resolution and factors which affect it are also required for the accurate interpretation of Doppler images. Resolution can be regarded as having three components namely spatial, contrast and temporal. The first two are not completely independent—a high contrast region can be accurately defined in space. The temporal resolution is determined by the image frame rate and is therefore usually quite poor. For high temporal resolution with Doppler methods, spectral Doppler or colour Doppler M-mode is used. As for B-mode imaging, the spatial resolution in the image is determined by the beam width and the pulse length. It is worth remembering that beam width depends on the combination of the shape of the transmitted field and the reception zone. Contrast resolution depends on the noise affecting the velocity or power being depicted in each pixel. In some situations the Doppler image information and the B-mode information would ideally be generated using pulse shapes and lengths optimised for each task. However, it is more common now to use the same pulse for both imaging modes, particularly since the advent of wideband transducers and digital signal processing. The question of the achievable resolution with any ultrasonic imaging method cannot be answered in general but must be related to the particular application. For instance, in carotid studies spatial resolution of less than 0.25 mm should be achieved in both B-mode and Doppler imaging with a 10 MHz state-of-the-art scanner. However, the resolution depends on the depth of the vessel and its disease state. Few test objects have been developed for the assessment of Doppler imaging methods. They are discussed in Appendix 2.

One pulse transmit/receive cycle is sufficient to produce a scan line of echo information in B-mode imaging, although several may be used if it is desired to step the transmit focus along the beam. On the other hand, Doppler imaging requires several consecutive echoes from a moving structure to allow Doppler information to be calculated. Since several transmit pulses are required along each scan line, the frame rate in Doppler imaging is slower than in B-mode. Commonly the B-mode may operate with 50 frames per second, whereas a typical Doppler rate is 15 s^{-1} and the latter is only achieved by reducing the width of the field of view. With more modern multi-beam scanning, B-mode frame rates can be in excess of 100 s^{-1} and Doppler as high as 40 s^{-1}. High frame rates are desirable in Doppler imaging, since flow events can change rapidly. Images are often reviewed using a slowed down cine-loop to assist interpretation.

10.3 DOPPLER IMAGING METHODS

10.3.1 Colour Doppler Imaging

A colour Doppler image depicts in each pixel the mean Doppler shift frequency detected in the direction along the scanning beam. Commonly flow toward the transducer is shown as red to yellow colours and away from the transducer in blue to green, although this is not always the case (Fig. 4.17: see, Plate II). Mean velocity is usually calculated using phase-shift autocorrelation signal processing (Kasai et al 1985) and occasionally time-domain cross-correlation processing (Bonnefous and Pesque 1986, Hein and O'Brien 1993), as discussed in Chapter 11. These calculations also provide the variance in the frequencies present in the Doppler signal. Regions of increased variance can be depicted in an image and are considered to show turbulence, for example associated with high velocity jets. The field of view is often restricted for the colour Doppler image, compared to the pulse-echo B-mode image, to achieve a reasonable frame rate, e.g. 15 s^{-1}, although this is becoming less necessary with the introduction of multi-beam techniques. Commercial colour Doppler imagers are available using ultrasonic frequencies of 3–10 MHz, i.e. to scan the whole range from deep to superficial vessels. Apart from the obvious resolution factors affecting image quality, the ability of an imager to discriminate between slow-moving tissue and blood is of importance in giving a good performance. Care has to be taken in the interpretation of colour images, particularly with regard to changes in colour due changes in beam/vessel (beam/flow) angle (Fig. 4.18: see Plate II). Aliasing is also a common artefact producing colour changes. A colour Doppler facility can be added to all types of ultrasonic

B-mode scanner, provided that the beam is not permanently at 90° to the direction of flow. The latter situation is rare and usually only occurs in certain types of catheter scanner.

10.3.2 Power Doppler Imaging

In power Doppler imaging, it is the power of the Doppler signal at each pixel in the scan plane that is presented in the image. As noted in the next chapter, in recent years this technique has been refined by improved filtering and signal averaging, which have made it possible to image slow flow in small vessels (Rubin et al 1994). Figures 4.23–4.25 (see Plate III) show examples of Power Doppler images. Since the backscatter from blood is fairly constant except where there is turbulent flow, the vessels are presented as a uniform colour. Near the vessel wall the power of the signal appears to be lower but this is a partial volume artefact where the beam is partly in the vessel and partly in the neighbouring tissue. The blood also moves more slowly next to the wall and the Doppler frequency drops below the wall thump filter level. This imaging mode is less sensitive to beam/vessel angle than Doppler velocity techniques and does not suffer from the aliasing artefact. Although the latter is regarded as an advantage, it only arises since power Doppler does not attempt to measure velocity in the first place. A more recent development has been to make use of the direction of flow information to colour-code the power Doppler image (Fig. 10.2: see Plate IV). Power Doppler imaging has proved popular in clinical use since it is sensitive and gives a more complete image of the vasculature, providing an ultrasound image analogous to an X-ray angiogram. We will see later that the completeness of this type of image is helpful in 3-D imaging.

10.3.3 Harmonic Colour Doppler and Harmonic Power Doppler Imaging

There are two ways in which additional harmonic frequencies can be generated in echo signals. The first to be discovered is that which occurs when ultrasound is scattered from contrast agent microbubbles. In addition to oscillating at the frequency of the incident ultrasound, the bubbles also oscillate at higher harmonic frequencies. This adds harmonic frequencies to the ultrasound scattered from the bubbles, which can therefore be used to identify them (Schrope et al 1992, Schrope and Newhouse 1993, de Jong et al 1994a, 1994b, Burns et al 1994, Powers et al 1997). Harmonic contrast techniques are discussed in a later section of this chapter (10.4.3) and in Section 13.2.3. The second way in which harmonics are generated is during the propagation of high-amplitude ultrasound through tissue. Non-linear propagation of the high-amplitude ultrasound results in distortion of the normally smooth sinusoidal type waveform and hence additional harmonic frequency components in the ultrasound pulse (Section 3.2.5). This phenomenon is of interest since it occurs most strongly in the high-amplitude central beam and not in the weaker side-lobes and spurious multiple reflection echoes. When the harmonics from tissue echoes are filtered for further processing, it is primarily the central beam echoes that contribute to the image, which therefore contain fewer extraneous echo signals. The extracted harmonic echo components can be used to produce B-mode images and also Doppler images of moving tissue. Harmonic imaging of pure blood is difficult due to the low level of the scattered echo signal. Different combinations of techniques lead to a variety of possible imaging methods, some of which are just now being developed and assessed.

1. B-mode grey-shade imaging
2. B-mode grey-shade harmonic imaging
3. B-mode grey-shade contrast imaging
4. B-mode grey-shade harmonic contrast imaging
5. Doppler velocity imaging
6. Doppler velocity harmonic imaging
7. Doppler velocity contrast imaging
8. Doppler velocity harmonic contrast imaging
9. Doppler power imaging
10. Doppler power harmonic imaging
11. Doppler power contrast imaging
12. Doppler power harmonic contrast imaging

Stimulated emission imaging, described in Section 10.4.3, relies on a different process for signal production. Pulse inversion Doppler imaging with contrast agents is discussed in Section 13.2.3.

Some of these techniques are at an early stage of development and may not prove to be sufficiently sensitive or of value. However, the list does illustrate the need to specify imaging techniques precisely. Most recent activity is in the evaluation of B-mode grey-shade harmonic imaging and Doppler harmonic contrast imaging. Note that in this text 'B-mode' is used to denote a non-Doppler technique.

10.3.4 Doppler Tissue Imaging

Doppler methods can also be employed to study tissue motion. The most obvious field of application is examination of contractility of heart muscle where colour Doppler imaging and colour Doppler M-mode have demonstrated velocities and velocity gradients within the tissue (McDicken et al 1992, Fleming et al 1994, Miyatake et al 1995) (Fig. 10.3: see Plate IV). The signal processing techniques used for Doppler blood flow have been shown to be applicable to Doppler tissue motion. The technique can also be applied to other moving tissues, such as artery walls. Forced motion of tissue by an external applicator is being studied to characterise tissue by measuring its elasticity, a technique sometimes referred as 'sonoelasticity' (Lerner et al 1990, Gao et al 1996, deKorte et al 1997). Doppler tissue imaging is appearing on many commercial scanners under various labels and acronyms. It is discussed further in Section 13.10.

10.3.5 Duplex and Triplex Imaging

Duplex imaging presents B-mode and spectral Doppler simultaneously on the screen, the Doppler sample volume being located at a site of interest in the B-mode image. Triplex imaging presents B-mode, a colour Doppler mode and spectral Doppler on the screen (Fig. 10.4: see Plate V). There is usually some time sharing in the collection of echo data for the imaging methods and the spectral mode, so they are not truly simultaneous, but this is of little practical significance. Several other combinations of images are possible, utility will no doubt determine what is used in the future. The combination of imaging and spectral Doppler illustrates the value of the different type of information supplied by the latter.

10.3.6 Colour Doppler M-mode

Just as a pulse-echo M-mode can be generated by fixing the ultrasound beam along a direction of interest and sweeping the line of echoes across the display to trace the motion of echo-producing structures, a colour Doppler beam can be fixed and the line of colour Doppler signals swept across the screen to show the change with time of velocity components at sites along the beam (Fig. 4.16: see Plate II). Although this technique is not widely used, it is the colour Doppler method with the best temporal resolution. It has found application in the measurement of velocity gradients in heart muscle (Fleming et al 1994) (see Fig. 13.29: Plate XI).

10.4 SPECIALISED DOPPLER IMAGING METHODS

10.4.1 Three-dimensional Doppler Imaging

The images we have considered so far have been two-dimensional, in which the plane of scan slices through a three-dimensional structure. It seems logical to try to collect ultrasonic echoes from the whole structure and display them as a 3-D B-mode or Doppler image. 3-D imaging is at an early stage of development and its value in most fields of clinical application has still to be proven. 3-D scanning is normally performed by collecting a series of adjacent 2-D scans in a computer memory and then presenting the information by one of several 3-D display techniques. 3-D display can be quite problematic, since external structures tend to obscure internal ones. Considerable research effort is expended in trying to find the optimum data processing and display methods for each application. With modern computer power the display aspects of 3-D imaging can be performed in real time. However a more fundamental barrier to true real-time 3-D imaging is the finite time required to collect the echo signals. With one scanning beam it typically takes 10 s to scan a 15 cm × 15 cm × 15 cm cube of tissue with an adequate scan line density for pulse-echo imaging. For Doppler imaging, where perhaps 10 pulses are transmitted along each scan line, 100 s are required. It is evident that true real-time 3-D Doppler imaging will be difficult to achieve. The problem is being tackled by generating several simultaneous beams and by wideband signal processing. For accurate 3-D echo collection the positions of the scan planes in space should be known to within 1 mm or less. Electromagnetic position measuring devices are available which are adequate for this task. In some early 3-D systems reliance was placed on the operator to produce a steady sweep of the scan plane through the 3-D structure. The electronics of the display then assumed that the scan planes were evenly separated in space and a 3-D display was generated. Appreciation of different locations in space of the scan planes was assisted by changing the colouring of the 2-D scans as the sweep was carried out. At present virtually all companies are developing 3-D scanners and the availability of easy-to-use devices will permit an assessment of the clinical value of this modality. It may well be that one of the most relevant applications of 3-D imaging is to help resolve the shape and flow patterns of complex vascular beds. The total vascularity of tumours may also be of value. 3-D Doppler imaging is further discussed in Section 13.9, and some images produced using 3-D techniques are shown in Fig. 13.22 (see Plate X), and Figs 13.24–13.26 (see Plates X–XI) and Fig. 13.27 (page 343).

10.4.2 Catheter Doppler Imaging

To date Doppler methods with catheters are of the Doppler wire type, as described in Section 5.2.6, and do not produce images. However, in theory it is possible to design a catheter scanner that could produce colour Doppler images in a field of view ahead of the catheter tip. Commercially available intravascular ultrasound scanners (IVUS) have scan planes perpendicular to the catheter axis, i.e. they are side viewing (Fig. 5.12). Images of flowing blood can be obtained with these devices, although strictly speaking they do not use a Doppler effect. Instead, decorrelation techniques are employed to

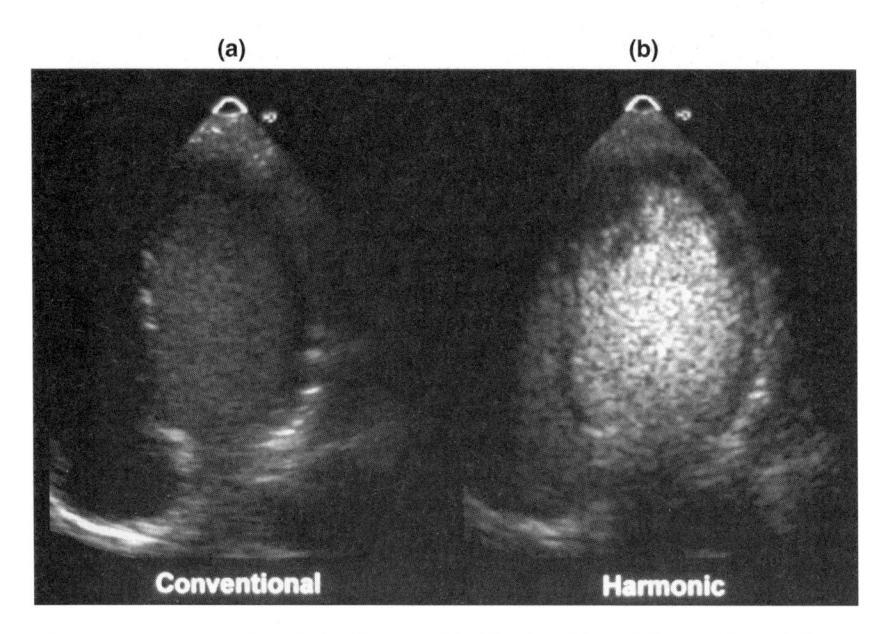

Figure 10.5 B-mode contrast images showing (a) visualisation of the blood pool in the left ventricle, and (b) the increased sensitivity of harmonic B-mode imaging (courtesy of ATL)

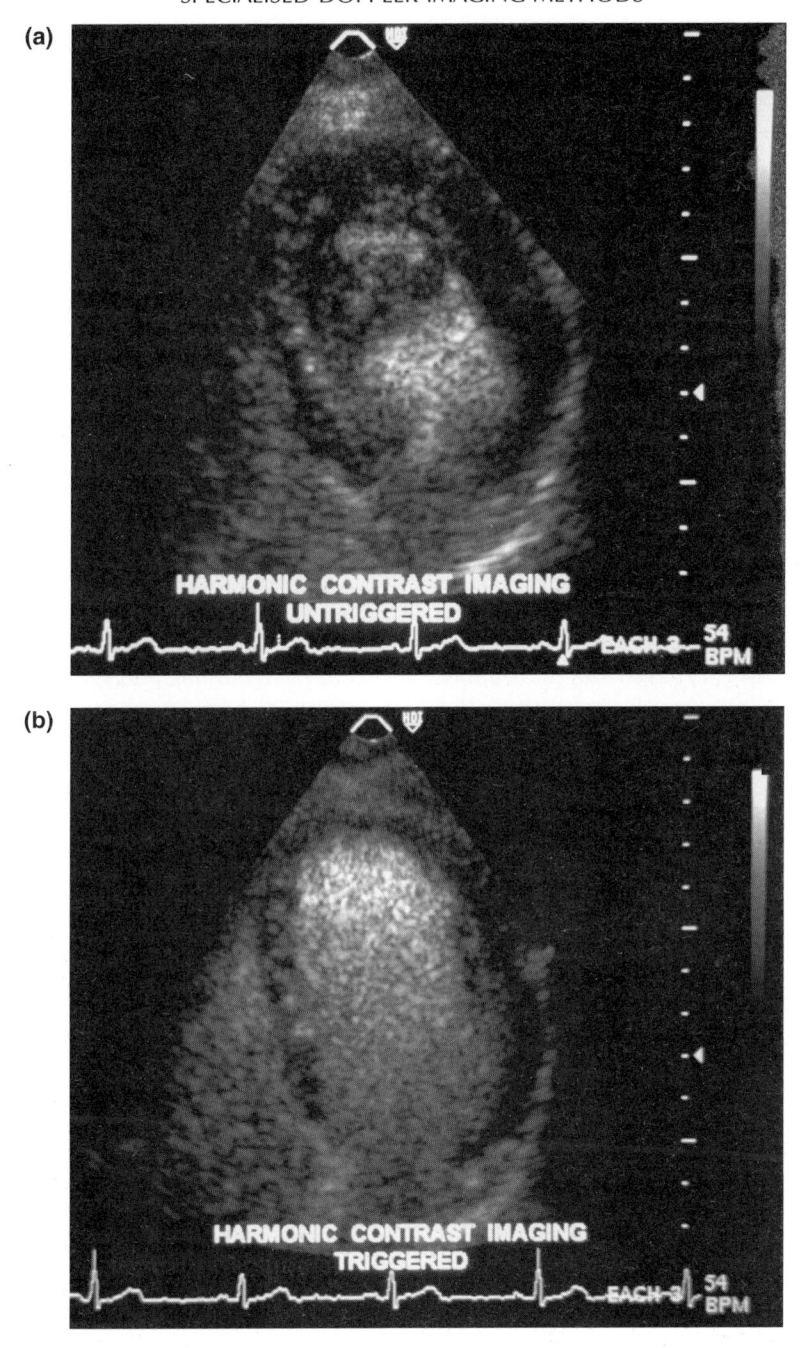

Figure 10.7 Transient harmonic B-mode imaging of the left ventricle: (a) untriggered microbubble contrast agent image; (b) triggered image (courtesy of ATL)

detect changes in the scattering speckle pattern of blood, which is related to blood velocity (Li et al 1998) (Fig. 13.20: see Plate X). The catheter B-mode image obtained simultaneously with the flow image enables the cross-sectional area of the vessel to be measured and hence quantitative flow to be calculated. This technique is further discussed in Section 13.8.

10.4.3 Contrast Doppler Imaging

The exploitation of Doppler methods and contrast agents is producing several imaging techniques which are advancing rapidly at present. They are discussed more fully in Section 13.2, where it is noted that they are based on the strong scattering of encapsulated microbubbles which can pass through the lung after intravenous injection. The increase in the echo signal when such an agent is injected into blood depends also on the bubble concentration and the type of agent. For the moment we will note the new types of images which are in theory possible with velocity and power Doppler signal processing of the scattered echoes from microbubble agents. Currently some of them are more commonly applied to pulse-echo B-mode techniques, hence a mixture of B-mode and Doppler examples are used in the following figures. For further discussions of these techniques, see Chapter 13.

1. *Low-amplitude linear scattered echo images.* Figure 10.5a illustrates how contrast agents can increase the level of backscatter from blood in the left ventricle to the extent that it can be imaged using a conventional B-scan imaging system. Figure 13.1 (see Plate VII) shows how this simple method may be used to enhance power Doppler images of the Circle of Willis in the brain.
2. *High-amplitude non-linear scattered echo images (harmonic images).* Figure 10.5b illustrates the increased sensitivity attained using

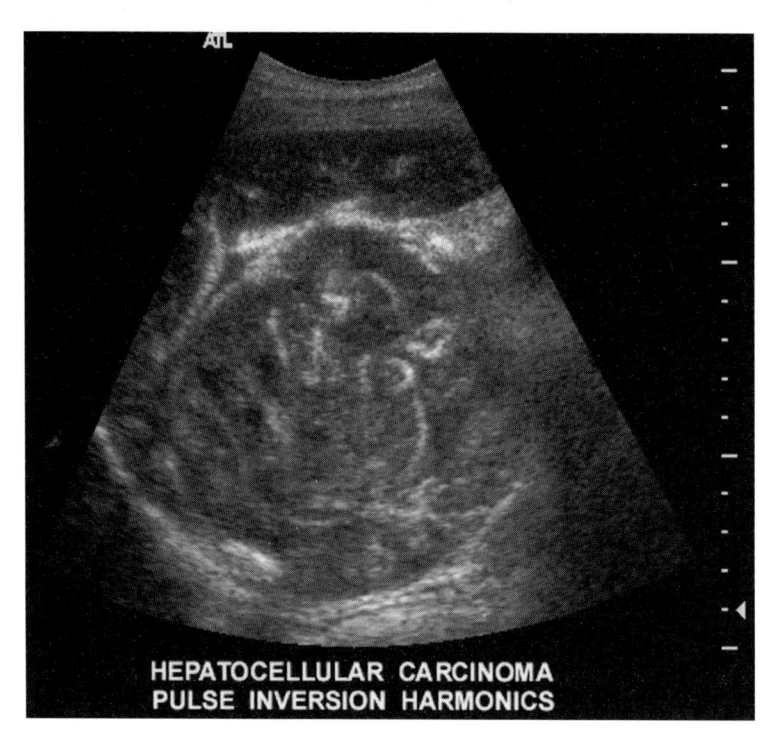

Figure 10.8 Pulse inversion contrast harmonic B-mode image showing vascularity of hepatocellular carcinoma. The vascularity is enhanced compared to other techniques (courtesy of S Wilson, P Burns and DH Simpson)

B-mode harmonic imaging in comparison to standard contrast imaging; Fig. 10.6 (see Plate V) is an example of harmonic power Doppler imaging.
3. *Transient (intermittent) B-mode imaging.* Figure 10.7 illustrates how transient (triggered) imaging can further enhance the backscatter from an ultrasonic contrast agent. The technique can also be used in Doppler imaging.
4. *Pulse inversion Doppler imaging.* Figure 10.8 is a pulse inversion B-mode pulse-echo image. At the time of writing, pulse inversion Doppler imaging systems are not commercially available.
5. *Stimulated acoustic emission imaging.* Figure 10.9 (see Plate V) shows the use of this technique to image liver metastases.

Some of these imaging approaches can be combined to give an overall improvement in sensitivity.

10.4.4 Extended Field of View Imaging

The field of view of real-time B-mode and Doppler imaging can be relatively small compared to the size of the structures to be examined. If the real-time aspect is sacrificed, it is possible to extend the field of view by moving the transducer in a direction parallel to the scan plane. The initial image information is retained as new echoes are gathered and displayed adjacent to it during the motion of the transducer. During the sweep there is continuous re-registration of the consecutive images to maintain the resolution of the extended image. The 'SieScape^{TM}' imaging mode developed by Siemens GMbH can be applied in both B-mode and Doppler imaging. It is of value in depicting images of large vascular structures (Fig. 10.10: see Plate VI).

10.5 FEATURES OF DOPPLER IMAGERS

Doppler imaging is relatively new and the features will only become standardised as the technology matures. The terminology used to describe Doppler facilities is far from standardised. Nevertheless some common features can be identified.

10.5.1 Colour-coding

The production of quality colour images by the mixing of primary colour images is an extensive topic and needs to be studied in considerable detail to make use of the large dynamic range of colour displays (see Rogowitz and Treinish 1998 for an introduction to this topic). However, the needs of colour Doppler are modest and therefore such detail is of some interest to machine designers, but not to the routine user. Machine specifications will often quote colour pallets with exceedingly large ranges of colours which are obviously just what is theoretically possible. Users of machines normally select one or two simple colour-coding schemes, since there is little value in having changes in colour which do not correspond to useful changes in information. Very commonly, dark to light red covers the range of velocities toward the transducer and dark to light blue velocities away from it. Other colours can be introduced to indicate regions of turbulence, making use of the velocity variance information, or to tag particular velocities. Shades of orange are commonly used in Power Doppler images, unless they are also colour-coded to show direction (Section 10.3.2).

10.5.2 Sensitivity Control

The clarity of a colour Doppler image can be highly dependent on the gain setting, which affects the Doppler signal. This is to be expected; however, the effect of gain on colour images is more dramatic than for B-mode images. This may be due to the fact that the addition or subtraction of colour to an image is more striking than alteration of the shades of grey. Particular attention therefore needs to be paid to the setting of the sensitivity control to optimise the image detail.

10.5.3 Scanning Strategy

The pulse sequences for Doppler imaging are quite demanding of time, for example, 10 pulses along a scan line for the measurement of typical velocities and pulsing over a longer duration for low velocities. This consumption of time leads to a restricted frame rate or field-of-view size unless ways can be found round it. Normally the size of the field of view is sacrificed for the sake of a suitable frame rate and the colour Doppler image is presented within a box in the B-mode image. As the dimensions of this box are altered, the machine maximises frame rate.

The effect of the need for several pulses for each scan line can be reduced significantly by having a number of simultaneous beams, a more satisfactory but more demanding solution than field-of-view size manipulation (see Fig. 5.15, page 86).

When slow pulsing rates are adequate to measure low velocities, several scan lines can be examined simultaneously by spreading a high pulsing rate between these scan lines so that a number of colour vectors are built up simultaneously using a technique of 'interleaved signal processing'.

Image persistence is generally regarded as undesirable, since blood flow imaging requires good temporal resolution. It is, however, a common feature of Power Doppler imaging where the emphasis is on high sensitivity and noise reduction.

10.5.4 Cine-loop

The rapid rate of change of blood flow patterns cannot always be interpreted in a colour flow image. Cine-loop storage, of say 512 frame capacity, enables the image frames to be replayed at a slower rate and is regularly used to assist interpretation. Modern scanners may offer several cine-loops.

10.5.5 Display and Recording Techniques

Doppler techniques are not specially demanding of display and recording techniques apart from the large amount of data to be stored, indeed we can often benefit from advances in the domestic market. They are further discussed in Appendix 3.

10.5.6 Real-time Zoom

The flow pattern of diagnostic interest can often occur in a limited region of the field of view. The real-time zoom facility enlarges this region and maintains good image quality by allocating an increased number of pixels to the zoomed area.

10.5.7 Slow Flow Detection

Considerable effort has gone into expanding the capabilities of Doppler equipment to create images of slow flow so that small vessels can be examined. The option to go into a 'slow flow mode' may exist on a scanner. In this mode the PRF is reduced and samples are taken over a longer period to enable low Doppler shift frequencies to be catered for. As noted above in the section on imaging strategies, the scanning beam may be moved to other locations between sampling for slow flow along a particular beam direction.

10.5.8 Priority Encoding

The success of colour Doppler imaging depends to a large extent in the instrument's ability to discriminate between soft tissue and blood. This discrimination usually depends on making use of the different velocities and scattering powers of tissue and blood (see Section 11.2.5). Some machines have a control which allows the operator to alter the discrimination level to suit each application.

10.5.9 Frame Averaging and Interpolation

The number of true lines of velocity information in a Doppler image may be rather low so that the line pattern is prominent. Artificial lines may be created and interpolated between true lines by deriving information from adjacent true lines, resulting in a smoother and more pleasing picture. Smoothing may also be achieved by averaging information from consecutive frames. Image persistence also results from this averaging, however some control is usually permitted by altering the amount of averaging.

10.5.10 Mobile Units

To date, most Colour Doppler machine have been of the large state-of-the-art variety. Now, as the technology is being better understood and as increased computing power is available cheaply, smaller mobile units are appearing on the market.

10.6 PERFORMANCE AND USE OF COLOUR DOPPLER IMAGING

10.6.1 Scanning Technique

The designer of ultrasonic equipment needs to bear in mind that all successful ultrasonic techniques are real-time in nature allowing the operator to function in an interactive way and results to be obtained while the patient is being examined. Disease in arteries occurs at well-known sites, often associated with complex flow patterns, for example at bifurcations. It is also worth remembering that disease at a particular site may not be clinically significant if there has been time for a collateral blood supply to have been established. In any blood flow study it is also essential to know the physiological status of the patient with regard to heart rate, exercise, temperature, anxiety, posture, food, smoking and other drugs. Considerable expertise is required for high quality ultrasonic vascular examinations. Apart from the knowledge of normal anatomy and variants of it, each type of examination is approached in a particular way and 'tricks of the trade' are employed as required. Examples of the latter are squeezing tissues to augment venous flow, avoiding pressure on superficial vessels, making use of anatomical landmarks such as the pairing of arteries and veins, correlating longitudinal and transverse images, and using Power Doppler and contrast agents to increase the sensitivity of the method. Finally, the experienced operator has a knowledge of artefacts, the more significant of which are described below (Section 10.7).

10.6.2 Multi-mode Scanning

Multi-mode scanning is the rule rather than the exception. The simplest example of this is the powerful combination of B-mode to show structure and Doppler to show related flow patterns. It is not always easy to relate the information in these two modes, particularly where plaque casts shadows or the vessels are too small to be imaged. The combination of imaging and spectral Doppler is surprisingly popular, not just to give detailed velocity at specific sites, but also to confirm the interpretation of Doppler images. Doppler imaging and Doppler M-mode applied to tissue motion has been shown to be of value in displaying changing velocity patterns across layers of tissue and for noting changes in velocity gradients. Doppler M-mode is not extensively used in blood flow studies, where spectral Doppler is favoured. It may be that the good temporal resolution of Doppler M-mode is not fully appreciated. The advent of 3-D Doppler will increase further the use of multi-mode imaging.

10.6.3 Spatial Resolution

Spatial resolution in Doppler images is the smallest spatial separation of two moving targets for which they can be displayed separately in the image. Since a significant part of Doppler technology and its application is to do with imaging, the resolution in the images is important. Accurate interpretation of images and measurements from them require knowledge of the resolution achievable, otherwise errors can arise from a false sense of high resolution. Determination of the spatial resolution of a system is not a simple matter since it depends on three components, the sample volume length (axial resolution), the in-plane beam width (lateral resolution) and the out-of-plane beam width (elevation resolution). The effective values of these components depend on the sensitivity setting of the system. Ideally a spatial resolution test object would consist of two small moving targets with the same scattering properties as blood, which could be moved anywhere in the field of view and whose separation could be varied. The targets would also be immersed in a liquid whose acoustic propagation properties would be the same as the average for soft tissue. Some methods that have been proposed for measuring spatial resolution are described in

Sections A2.3.3 and A2.5.6. Early colour Doppler systems had spatial resolution inferior to that of pulse-echo B-mode but with the advent of wideband signal processing the difference is being reduced and can now be around 1 mm at 3 MHz.

10.6.4 Velocity and Power Resolution (Contrast Resolution)

Velocity and power resolution are the smallest difference in these parameters that can be detected in the appropriate type of image, i.e. the smallest change in contrast that can be detected in the image. They have not been studied to any significant extent and no proven test objects have been reported at the time of writing. This situation reflects the fact that it is only recently that colour Doppler images have acquired a quality which shows more than just forward and reverse flow within the confines of a vessel. At present velocity resolution is usually worse than 10%.

10.6.5 Temporal Resolution

Temporal resolution is the minimum separation in time of two events which can be detected in an image. It is rarely measured in any ultrasonic system but is more deduced from the frame rate of the scanner. With colour Doppler frame rates of over 40 s^{-1} now available, it is possible to achieve temporal resolution as high as approximately 25 ms.

10.6.6 Calibration

The need for calibration of Colour Doppler instruments to date has not been high. However, as the image quality continues to improve and more complex diagnostic tasks are taken on, confidence in the accuracy of the equipment will be important. Test devices for Doppler systems are described in detail in Appendix 2. Spatial calibration in terms of image scale and accuracy of measurement callipers is normally carried out for the B-mode image by scanning a test object of known dimensions (Clark 1988). If the colour Doppler image is consistently located precisely within the walls of blood vessels, the spatial calibration of the Doppler mode is assumed to be satisfactory. Velocity calibration for a colour Doppler image or a spectrum is performed by injecting signals of well-defined frequency into the device, either electronically or acoustically (Appendix 2). Since power Doppler is used for non-quantitative imaging, means to calibrate the power level of Doppler signals have not been developed. Since many factors affect the signal whose power is displayed in an image, the development of power calibration techniques will not be easy.

10.6.7 Measurement from Images and Spectra

Once images have been created it is often useful to make measurements on them, since the exact site of the source of data is known. In B-mode imaging there is considerable interest in measuring the dimensions of structures and their growth; however, in Doppler imaging there is less interest in measurement directly from the image. Doppler images are used instead to identify the relevant sites, and then PW Doppler is utilised to obtain velocity spectra from these sites. A few quantities are measured from images, for example a particular value of velocity can be observed throughout the image by tagging it with a selected colour. The extent of turbulence beyond a stenosis is measured by incorporating the velocity variance into the image with appropriate colour-coding. In cardiology, the extent of regurgitant jets associated with diseased valves can be used as a guide to valvular incompetence. In research into plaque formation at arterial walls there is a desire to measure velocity gradients adjacent to them, although it is not clear whether such gradients can be quantified with sufficient accuracy. When Doppler imaging is further refined, we may see more quantification from images.

10.6.8 Quantitative Flow Measurement

The advent of high resolution PW Doppler and colour Doppler imaging has resulted in mean velocity information being available in pixels across blood vessels. If a 1-D velocity profile is measured and circular symmetry is assumed, a 2-D profile is derived which, combined with measurement of the cross-sectional area, gives quantitative flow (Picot and Embree 1994) (Section 12.3). It is known that this approach can easily produce errors in excess of 20% and even as high as 100% in small vessels. An advance on this approach exploits the true 2-D velocity profile using the mean velocities in each pixel of an image, and again flow is calculated (Picot et al 1995) (Section 12.4). With this method errors of less than 5% are achievable. For further discussion of quantitative flow measurement, see Chapter 12.

10.7 ARTEFACTS IN DOPPLER TECHNIQUES

Artefacts in Doppler detection and imaging techniques have been described elsewhere (McDicken 1991, Pozniak et al 1992). Many arise from commonly known factors such as attenuation, refraction, reflection and beam width, which are related to the basic processes involved in propagation through tissue and beam generation. In this section only those artefacts which have a more electronic origin and which may be removed by improved instrument design are highlighted.

10.7.1 Noise

When the gain of the scanner is increased and the noise signal becomes large enough to be recorded in a colour velocity image, the mean frequency of the noise is immediately displayed randomly as red or blue pixels over wide regions of the image (Fig. 10.11A: see Plate VI). The pattern of small coloured dots degrades the image and the Doppler gain cannot be increased further. With Power Doppler imaging, the power of the signal is measured and increases more gradually with gain, hence degrading the image more slowly (Fig. 10.11B: see Plate VI). This is one reason for the greater sensitivity of power Doppler imaging.

10.7.2 Spectral Broadening

Intrinsic spectral broadening is due to the fact that the finite aperture of the active region of the transducer can be quite large, particularly with linear arrays, and hence ultrasound interrogates the sample volume from a range of angles (Fig. 10.12). This had not been fully appreciated until quite recently, when it was shown that errors as large as 50% could arise in the measurement of maximum velocity, with very significant consequences for patient management (Thrush and Evans 1995, Hoskins 1996). This error can be avoided by taking the beam angle to be that of the ultrasound from the side of the aperture that makes the most acute angle with the direction of flow. To date this has only been implemented in one machine manufactured by Diasonics Ultrasound. Care obviously has to be taken to distinguish intrinsic spectral broadening from true spectral broadening

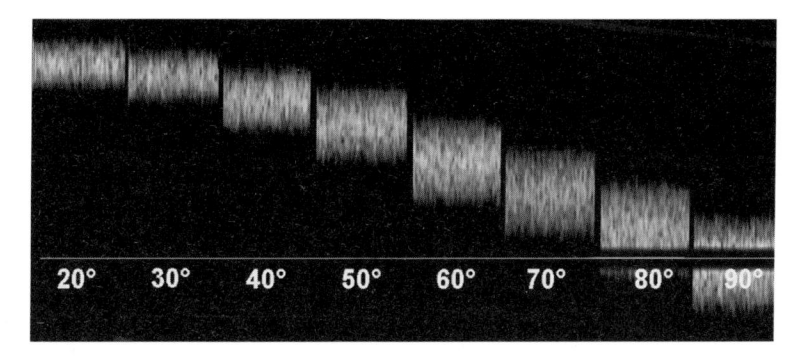

Figure 10.12 Doppler sonograms recorded from a moving string phantom at various Doppler angles. Since all the 'targets' are moving with the same velocity, spectra would be relatively much narrower in the vertical direction in the absence of intrinsic spectral broadening. Note how the broadening increases as the Doppler angle tends to 90°. (courtesy of PR Hoskins)

when the latter arises from haemodynamic effects and is used to try to assess conditions such as turbulence. Intrinsic spectral broadening is discussed in detail in Sections 7.3.4 and 7.4.3.

10.7.3 Aliasing

As noted in Section 4.2.2, since PW and colour Doppler imaging are sampling processes, they can suffer from aliasing if the Doppler signal is undersampled. Artefactual reverse flow presentation in PW and colour Doppler is commonly encountered where there is high velocity flow (Fig. 4.18: see Plate II). There is a small black gap in an image between regions of true reverse and forward flow due to filtering out of very low velocities. With aliasing there is continuity in the image as the high colour-coded velocities wrap round. Aliasing is used to identify the presence of high velocity signals in spectra and images, but this is essentially a way of trying to make best use of an unfortunate artefact. A high PRF mode seems to have been the most successful technique for the reduction of aliasing with non-imaging Doppler systems. With this mode, since echoes from one transmission have not died out before the next is made, range ambiguity arises as to the origin of the echoes being processed. However, in situations where there is known to be one dominant source of Doppler signal, this is acceptable. Employing low frequency transmitted ultrasound and fairly large beam/vessel angles are also helpful in reducing aliasing. A variety of methods that have been proposed for overcoming aliasing are described in Section 13.11.

10.7.4 Speckle

The power of the signal in a Doppler sonogram fluctuates, producing a noise-like appearance. This is due to fluctuations in the train of signals from the random distribution of blood cell scatterers (Section 7.2.5). We cannot, therefore, interpret the individual pixel values in the sonogram as being an accurate measure of the number of cells moving with the particular velocity corresponding to the pixel. Image processing such as smoothing or averaging can improve the appearance of the sonogram with little degradation of the temporal resolution but this has not been taken up to any extent (Fig. 8.12) (Hoskins et al 1990).

10.7.5 Flow Direction Sensing

It cannot always be assumed that flow direction sensing circuitry is working correctly. Small design or setting-up errors can result in erroneous indication of direction of flow. A check with a test-rig, a flow pattern at a well-understood vascular site, or comparison with another instrument is advisable.

10.7.6 Filtering

Filtering is designed to separate the signals from slow moving strongly reflecting tissue structures and those from faster moving weakly reflecting blood. For most spectral Doppler studies this is satisfactory, since very sharp frequency cut-offs can be implemented. In colour Doppler imaging, however, the situation is not so straightforward since the small number of pulse returns available makes the implementation of sharp filters difficult, and additional information, such as the size of the reflected echoes, can be misleading because some tissue boundaries are quite weakly reflecting (for example the intima/blood interface in arteries and the muscle/blood interface in cardiac chambers). The decision as to which pixels to code for tissue and which to code for blood is difficult. The success of a colour Doppler imaging machine can depend to a significant extent on this filtering. Tissue/blood priority encoding is further discussed in Section 11.2.5.

10.7.7 Large Signal Distortion

It is quite common in some spectral Doppler devices to see the presence of high frequency harmonics due to the distortion of large signals in the electronics. Reduction in gain can usually remedy this artefact. Artefactual harmonic generation is not evident in Doppler imaging, presumably because a shift in mean frequency is more difficult to observe or since expensive state-of-the-art machines have greater dynamic range.

10.7.8 Beam/Vessel (Flow) Angle

The commonest artefact in a Doppler colour image arises from variation of the angle between the direction of blood flow and the direction of the beam as it scans through the field of view. As Fig. 10.13 (see Plate VI) illustrates, the beam/flow angle can vary substantially in different parts of the image even although the direction and magnitude of the flow is constant across the scan plane. It always has to be borne in mind that the colour Doppler images show in each pixel the Doppler frequency related to the velocity component along the beam axis. It should also be remembered that the direction of flow in a vessel is not always parallel to the vessel wall, indeed full specification of velocity needs three components measured along orthogonal axes. Images have to be interpreted with care, particularly localised areas of apparent flow reversal, which may actually be due to the beam/flow angle changing from less than 90° to greater than 90° in neighbouring regions.

The quality of Doppler signals depends on the beam/flow angle and deteriorates above 70°. In theory, at 90° no signal is obtained but in practice some is registered due to spectral broadening and disturbed flow. Power Doppler images appear more complete since the technique is less sensitive to angle effects. Vector Doppler imaging is being developed to help overcome some of the problems associated with beam/flow angle (Section 13.6).

10.7.9 Flash Artefact

When the transducer is moved fairly quickly over the patient, or major tissues structures move, Doppler shifts can be induced in echo signals from large areas of the field of view, resulting in corresponding colour-coding in the image. This flash artefact is obviously very distracting. Manufacturers incorporate image processing to detect this widespread colour-coding and fairly effectively reduce the artefact. Power Doppler imaging, which is more sensitive than colour Doppler imaging, is more prone to the flash artefact.

10.7.10 Side-lobes and Grating-lobes

Although side-lobes and grating-lobes are of lower intensity than the main central beam, they can produce Doppler signals, especially if they interrogate large moving structures. These Doppler signals are registered as though they have been detected in the main beam.

10.7.11 Machine Defects

Colour flow imaging technology is developed and marketed rapidly. Occasionally this results in design defects in machines reaching the user. One example of this was the variation in the transducer aperture size at different positions along a linear array which gave variable intrinsic broadening to the Doppler signals. The blood/tissue discrimination algorithms vary from machine to machine and at high Doppler gain some may cause anechoic regions, such as the gall bladder, to be colour-coded (Mitchell et al 1990). Spectral Doppler can be used to confirm that a colour-coded region has in fact flow within it. Such defects can produce problems in clinical studies. These types of problem demonstrate the value of studying the performance of complex scanners with a simple well-defined test-device such as a string or flow phantom.

10.8 SUMMARY

Since Doppler methods are now extensively used to produce several types of image, the quality of which is steadily improving, the basic concepts of ultrasound imaging were considered. These concepts included components of resolution, beam widths, pulse length, field of view, frame rate, line density and transducers. Nine types of image were briefly studied. The technology was considered from the points of view of instrument features, performance and use. As with all imaging methods, artefacts are present and are commonly dealt with by the operator building up a knowledge of them. The need for care in the interpretation of Doppler images was emphasised.

10.9 REFERENCES

Bonnefous O, Pesque P (1986) Time domain formulation of pulse-Doppler ultrasound and blood velocity estimation by cross correlation. Ultrason Imaging 8:73–85

Burns PN, Powers JE, Simpson DH, Kolin A, Chin CT, Uhlendorf V, Fritzch T (1994) Harmonic power mode Doppler using microbubble contrast agents: an improved method for small vessel flow imaging. In: Proc. 1994 IEEE Ultrasonics Symposium, pp. 1547–1550. IEEE, Piscataway, NJ

Clark PD (1988) Performance checks. In: Lerski RA (ed) Practical Ultrasound, pp. 53–83. IRL Press, Oxford

de Jong N, Cornet R, Lancée CT (1994a) Higher harmonics of vibrating gas-filled microspheres. Part One: Simulations. Ultrasonics 32:447–453

de Jong N, Cornet R, Lancée CT (1994b) Higher harmonics of vibrating gas-filled microspheres. Part Two: Measurements. Ultrasonics 32:455–459

de Korte CL, van der Steen AFW, Dijkman BHJ (1997) Performance of time delay estimation methods for small time shifts ultrasonic signals. Ultrasonics 35:263–274

Evans DH (1993) Techniques for color-flow imaging. In: Wells PNT (ed) Advances in Ultrasound Techniques and Instrumentation—Clinics in Diagnostic Ultrasound, vol 28, pp. 87–107. Churchill Livingstone, New York

Ferrara KW, DeAngelis G (1997) Color flow mapping. Ultrasound Med Biol 23:321–345

Fleming AD, Xia X, McDicken WN, Sutherland GR, Fenn L (1994) Myocardial velocity gradients detected by Doppler imaging. Br J Radiol 67:679–688

Gao L, Parker KJ, Lerner RM, Levinson SF (1996) Imaging of elastic properties of tissue—a review. Ultrasound Med Biol 22:959–977

Hein IA, O'Brien WD (1993) Current time-domain methods for assessing tissue motion by analysis from reflected ultrasound echoes—a review. IEEE Trans Ultrason Ferroelec Freq Contr 40:84–102

Hoskins PR (1996) Accuracy of maximum velocity estimates made using Doppler ultrasound systems. Br J Radiol 69:172–177

Hoskins PR, McDicken WN (1997) Colour ultrasound imaging of blood flow and tissue motion. Br J Radiol 70:878–890

Hoskins PR, Loupas T, McDicken WN (1990) A comparison of three different filters for speckle reduction of Doppler spectra. Ultrasound Med Biol 16:375–389

Kasai C, Namekawa K, Koyano A, Omoto R (1985) Real-time two-dimensional blood flow imaging using an autocorrelation technique. IEEE Trans Sonics Ultrason SU-32:458–464

Kremkau FW (1998) Diagnostic Ultrasound: Principles and Instrumentation, 5th edn. Saunders, Philadelphia, PA

Lerner RM, Huang SR, Parker KJ (1990) Sonoelasticity images derived from ultrasound signals in mechanically vibrated tissues. Ultrasound Med Biol 16:231–239

Li W, van der Steen AFW, Lancée CT, Céspedes I, Bom N (1998) Blood flow imaging and volume flow quantitation with intravascular ultrasound. Ultrasound Med Biol 24:203–214

McDicken WN (1991) Diagnostic Ultrasonics: Principles and Use of Instruments, 3rd edn. Churchill Livingstone, London

McDicken WN, Sutherland GR, Moran CM, Gordon LN (1992) Colour Doppler velocity imaging of the myocardium. Ultrasound Med Biol 18:651–654

Mitchell DG, Burns P, Needleman L (1990) Color Doppler artifacts in anechoic regions. J Ultrasound Med 9:255–260

Miyatake K, Yamagishi M, Tanaka N, Uematsu M, Yamazaki N, Mine Y, Sano A, Hirama M (1995) New method for evaluating left-ventricular wall-motion by color-coded tissue Doppler imaging. J Am Coll Cardiol 25:717–724

Picot PA, Embree PM (1994) Quantitative volume flow estimation using velocity profiles. IEEE Trans Ultrason Ferroelec Freq Contr 41:340–345

Picot PA, Fruitman M, Rankin RN, Fenster A (1995) Rapid volume flow rate estimation using transverse colour Doppler imaging. Ultrasound Med Biol 21:1199–1209

Powers JE, Burns PN, Souquet J (1997) Imaging instrumentation for ultrasound contrast agents. In: Nanda NC, Schlief R, Goldberg BB (eds) Advances in Echo Imaging Using Contrast Enhancement, pp. 139–170. Kluwer Academic, Dordrecht

Pozniak MA, Zagzebski JA, Scanlan KA (1992) Spectral and color Doppler artifacts. RadioGraphics 12:35–44

Rogowitz BE, Treinish LA (1998) Data visualization: the end of the rainbow. IEEE Spectrum 35 (12):52–59

Routh HF (1996) Doppler ultrasound. IEEE Eng Med Biol 15 (6):31–40

Rubin JM, Bude RO, Carson PL, Bree RL, Adler RS (1994) Power Doppler US: a potentially useful alternative to mean frequency-based color Doppler US. Radiology 190:853–856

Schrope BA, Newhouse VL (1993) Second harmonic ultrasonic blood perfusion measurement. Ultrasound Med Biol 19:567–579

Schrope B, Newhouse VL, Uhlendorf V (1992) Simulated capillary blood flow measurement using a nonlinear ultrasonic contrast agent. Ultrason Imaging 14:134–158

Thrush AJ, Evans DH (1995) Intrinsic spectral broadening: a potential cause of misdiagnosis of carotid artery disease. J Vasc Invest 1:187–192

Wells PNT (1977) Biomedical Ultrasonics. Academic Press, London

11

Signal Processing for Colour Flow Imaging

11.1 INTRODUCTION

The basic principles of colour flow imaging (CFI) techniques have been introduced in Chapter 4 (4.5.3), whilst CFI systems have been described in some detail in Chapter 10. In this chapter signal processing techniques for CFI will be discussed. This will include a discussion of each of the major building blocks required to produce a colour flow image, i.e. the clutter rejection filters, the Doppler shift frequency estimation algorithms, priority encoding and artefact suppression algorithms, and post-processing algorithms. Section 11.2 deals with instruments based on the autocorrelation technique of frequency estimation, which was the first method to be used in commercial colour flow systems and still the most widely used today. Section 11.3 deals with a selection of other phase shift estimation algorithms, some of which are inferior to the autocorrelation algorithm, although others may have potential advantages. Section 11.4 discusses wide-band systems, which are based on measuring the time delay rather than the phase shift between successively returning ultrasound pulses, Section 11.5 introduces 'direction of arrival' methods, whilst Section 11.6 deals briefly with future developments.

It should be noted that whilst the separation of velocity estimation algorithms into phase shift and time shift, or narrow-band and wide-band estimators is appropriate for simple algorithms, this is less so for more complex algorithms, where the distinctions become much more blurred. For example, phase shift estimators may be used in conjunction with wide-band pulses if a separate estimate is also made of the returned RF centre frequency (by autocorrelation), and some time shift estimation algorithms based on modelling the cross-correlation function only work satisfactorily with narrow-band signals. In fact it turns out that algorithms derived from the standpoint of phase shift estimation and those derived from the standpoint of time domain estimation ultimately converge.

11.2 CURRENT NARROW-BAND COLOUR FLOW IMAGING SYSTEMS

As stated in the Introduction, most contemporary colour flow imaging systems are based on the autocorrelation technique for estimating blood flow velocity, and the important building blocks for such a system are described in this section.

11.2.1 Current CFI Systems—General Layout

Figure 11.1 illustrates the general layout of a CFI system, although even at this superficial level the details will vary from machine to machine. The RF signals from the beam-former are quadrature demodulated by mixing them with quadrature signals derived from the system master oscillator, and stored in memory. In early systems the information required to calculate each colour vector was acquired in turn, but with modern systems that use interleaved acquisition of the data to improve velocity resolution without sacrificing frame rate, the build-up of the data for each flow vector is no longer necessarily sequential.

Once the data from each sample volume are complete, they are read, filtered to reject clutter components (unless this is purely carried out as part of the velocity estimation algorithm), and processed to form estimates of the power, mean frequency and bandwidth of the Doppler signal from that sample volume. The results from each sample volume are stored in a colour

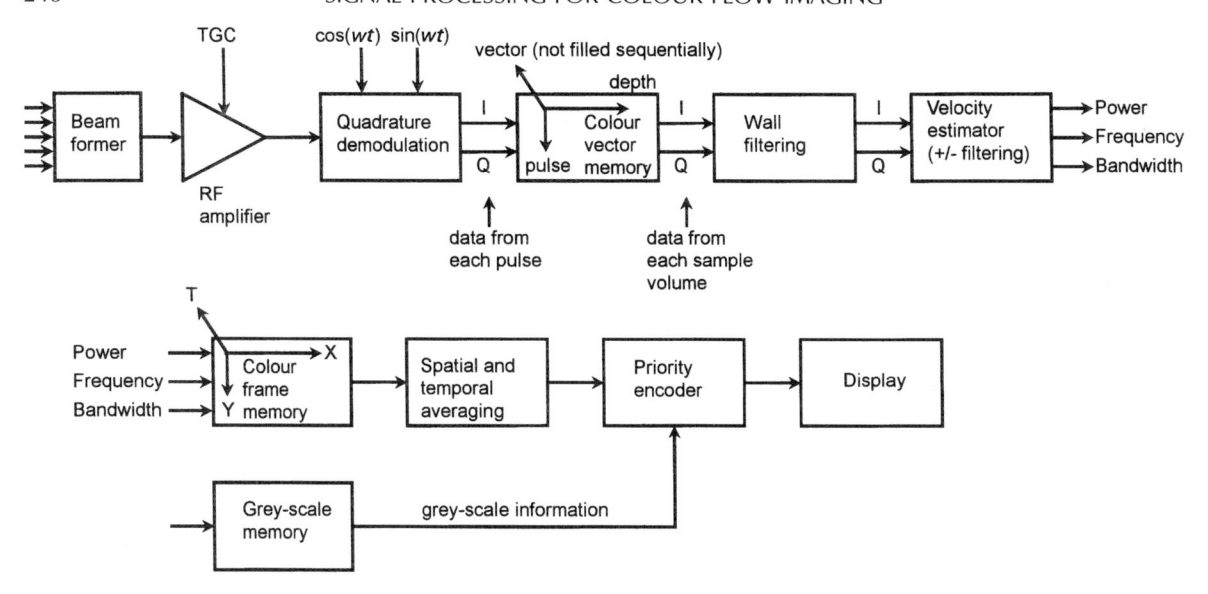

Figure 11.1 General layout of the Doppler path for a colour flow imaging system

frame memory, which allows both spatial and temporal averaging of these data before they are combined with grey-scale information from the same sample volume to determine the probability that flow is present. Depending on the result of this determination, either an appropriately coded colour or a grey pixel is written to the display memory.

11.2.2 Filters and Clutter Rejection

The purpose of the wall movement or clutter filters is exactly the same as the wall movement filters in conventional CW or PW Doppler systems, that is to reject the very high-amplitude but low-frequency echoes from stationary or near-stationary targets. The clutter signals, typically with frequencies below 1 kHz, may be 40–60 dB higher than the scattered signals from blood (Jensen 1993c, Thomas and Hall 1994, Brands et al 1995, Heimdal and Torp 1997) and must be removed if an accurate estimate of the flow velocity is to be obtained. Because of the very few samples of signal available, due to the time constraints of CFI, this poses a difficulty for system designers and is usually regarded as one of the most difficult technical challenges to be overcome.

A number of different approaches to clutter rejection have been adopted and these are described in the next few subsections.

11.2.2.1 Finite Impulse Response Filters (Echo Cancellers)

The simplest type of clutter rejection filter is the simple single echo canceller, the structure of which is shown in Fig. 11.2a. The output is derived simply by subtracting each successive echo from its predecessor. Signals from stationary objects are unchanged and thus cancelled, whereas those from moving objects change, and thus are preserved to a greater or lesser extent. This filter is non-recursive in the sense that it computes the 'current' output only from the current and past inputs, and not past outputs (Williams 1986). Non-recursive filters are also known as finite impulse response (FIR) filters, since the length of the non-zero portion of the response to an impulse is dictated by the number of coefficients or delays in the filter.

The response of the single echo canceller is shown in Fig. 11.3a, and in practice is inadequate for most CFI applications because of its poor roll-off (6 dB/octave) and wide transition band (note the -3 dB frequency is at 50% of the Nyquist frequency). The transition band (the

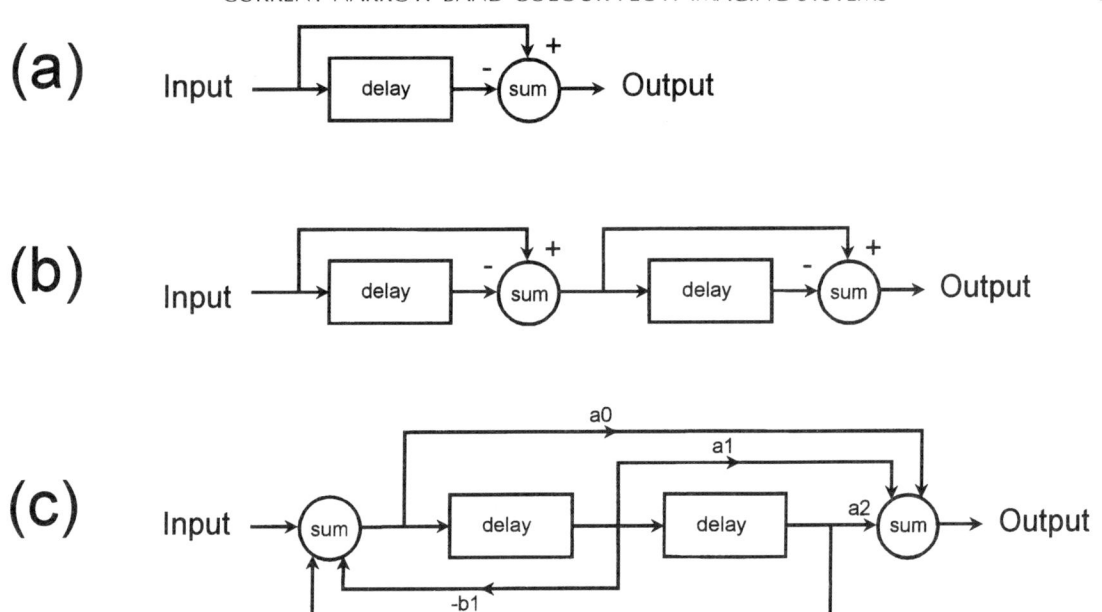

Figure 11.2 Clutter rejection filters: (a) simple single echo canceller; (b) simple double echo canceller; (c) second order infinite impulse response filter. All delays are equal to one pulse repetition period

region between the filter pass-band and stop-band) must be relatively narrow, otherwise blood flow signals over a considerable portion of the lower part of the available frequency band (0–0.5 PRF) will be significantly attenuated and leave only Doppler signals corresponding to high flow rates.

The roll-off of the simple echo canceller can be increased by cascading two such filters (Fig. 11.2b), but this only exacerbates the problem of the wide transition band (Fig. 11.3b). More complex FIR filters can be constructed to produce adequately narrow transition bands, but this necessitates much higher order filters (longer filter kernels) which require a large number of data samples (in fact the longer the impulse response of a filter, and therefore the number of stages used with an FIR, the more faithfully a desired frequency response can be constructed). Unfortunately large numbers of data samples are a luxury which cannot be afforded in CFI systems, where in order to maintain acceptable frame rates it is necessary to calculate velocities from approximately 4–16 pulses. Even a second-order filter reduces the number of samples available for the velocity calculation by two, and increases the variance of the velocity estimate, particularly for the smaller sample lengths.

The use of FIR filters in conjunction with the classic autocorrelation algorithm has been discussed by Willemetz et al (1989) and Nowicki et al (1990), both of whom showed that these filters introduce significant bias into the estimate of mean frequency under poor signal-to-noise ratio (SNR) conditions. Rajaonah et al (1994) have explained this in terms of the significant colouring of the noise spectrum produced by FIR filters with their wide transition zones. This is discussed further in Section 11.2.3.

11.2.2.2 Infinite Impulse Response Filters

The frequency characteristics of digital filters may be dramatically improved by the use of recursive techniques, that is to say filters that derive their inputs not only from the present and past inputs, but also from the present and past outputs. Such filters are also known as infinite impulse response (IIR) filters because of their theoretically infinite settling times. One possible implementation of a second-order IIR filter is illustrated in Fig. 11.2c,

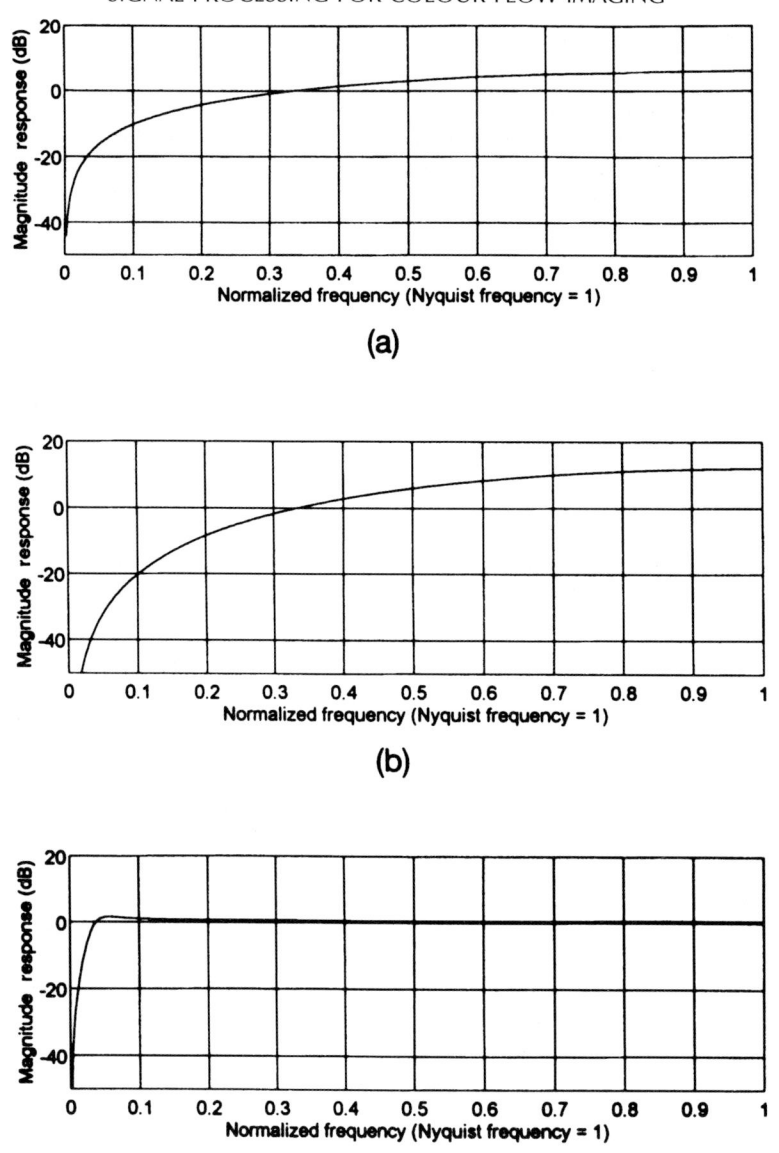

Figure 11.3 Frequency responses of clutter rejection filters: (a) simple single echo canceller; (b) simple double echo canceller; (c) second order infinite impulse response filter. (reproduced by permission of Elsevier Science, from Tysoe and Evans 1995, *Ultrasound in Medicine and Biology*, © World Federation of Ultrasound in Medicine and Biology)

whilst the frequency response of the second-order IIR filter described by Shariati et al (1993) is shown in Fig. 11.3c. Clearly the frequency characteristics of the second-order IIR filter are vastly superior to those of the second-order FIR filter shown in Fig. 11.3b. The problem with IIR filters, however, is their long transient responses or settling times, which means that unless appropriate steps are taken to initialise them appropriately, once a new signal is applied to the filter a large number of output values must be discarded before valid data becomes available for frequency estimation. This is of course not possible with CFI systems, which have few data samples from each sample volume, and therefore initialisation becomes very important (this problem is of course not unique to Doppler ultrasound; phased array radars in particular have long had this difficulty).

A number of techniques have been suggested as useful for the initialisation of IIR filters, and these have been reviewed by Peterson and his colleagues (Peterson 1993, Peterson et al 1994). The simplest of these, step initialisation, simply sets the values of the filter 'state vector' by assuming the first complex sample value had existed since time sample $n = -\infty$ (Fletcher and Burlage 1972). This technique eliminates the transient response due to zero-frequency energy but does not eliminate the response due to other frequencies, and step initialising IIR filters has been shown to introduce significant bias into the output of frequency estimation algorithms for a significant time period (Kadi and Loupas 1995, Tysoe and Evans 1995). A more sophisticated approach to initialisation was suggested by Chornoboy (1992), which he named 'projection initialisation'. The result of this method for a second-order filter is to load the initial filter state vector as if the samples with negative time indices had formed a straight line passing through the first two samples of the sequence (Peterson et al, 1994). In general this method, whilst more complicated, appears to be superior to step initialisation under most circumstances (Peterson 1993, Peterson et al 1994). Peterson and his colleagues suggested a further method which they termed 'exponential initialisation'. With this method a model of the input data is created using the data contained in the observation window, and the filter is then initialised using a frequency and initial phase extracted from the model. It is claimed that this method should yield an immediate steady state response when operating on complex sinusoids (Peterson 1993, Peterson et al 1994) and appears to be a good alternative to other IIR filter initialisation methods when applied to CFI wall filtering.

In summary, IIR filters can be designed to produce more nearly ideal frequency characteristics than FIR filters using a relatively small number of delays. Thus, IIR filters have the advantages that they are able to preserve echoes from flows with lower velocities whilst rejecting the clutter components, and that under poor SNR conditions white noise is not unduly coloured and therefore does not bias the velocity estimate unduly. On the other hand, IIR filters have very poor transient responses, and this in itself can lead to significant bias in mean frequency calculations, and whilst appropriate initialisation can reduce this problem, in practice it cannot be eliminated.

11.2.2.3 Regression Filters

Hoeks et al (1991) introduced an entirely new approach to the removal of the low frequency Doppler signals (i.e. the clutter components) in CFI systems, based on the assumption that the slowly varying clutter components of the Doppler signal can be approximated by a polynomial of low order which can be determined by performing least-squares regression analysis. Once the slowly varying signal component has been approximated, it can then be subtracted from the raw Doppler signal to leave the component originating from blood flow. It should be noted that one significant advantage of this technique is that it does not reduce the number of output samples available for frequency estimation. Hoeks et al (1991) examined the properties of the first order (linear) regression filter and showed that it exhibited a rapid transition from pass-band to rejection-band, with a roll-off of 12 dB/octave, and suggested that a sharper roll-off could be obtained by replacing the regression line by a curve of higher order at the expense of complexity and processing time. Kadi and Loupas (1995) compared the performance of regression filters with orders of up to four with step-initialised IIR clutter filters. They reported that the ability of both filters to suppress clutter varied considerably, depending on factors such as the clutter/flow signal ratio and ensemble length, but that the former were found to offer significantly better performance than the latter under heavy clutter conditions. They did, however, add the caution that improved initialisation techniques for the IIR filter might influence this result.

Bjaerum and Torp (1997) have described an adaptive regression filter, where the regression is performed with a clutter signal that is estimated from the signal statistics. They found that the adaptive filter performed significantly better than a standard regression filter when there were 'large probe movements' during data acquisition.

Torp (1997) has published a description of a theoretical approach to linear clutter filters applied to discrete Doppler base-band signals, in which the filter is represented as a complex valued matrix, and the frequency response for general linear filters defined. Both time-invariant FIR and IIR filters, and time-variant regression filters are discussed within this framework.

11.2.2.4 Other Approaches

Thomas and Hall (1994) have described an approach to clutter rejection which they call 'DC removal', which depends on the assumption that the clutter signal is always the dominant component of the total Doppler signal. The mean frequency of the total Doppler signal (which should therefore be a very good approximation to the mean frequency of the clutter signal) is estimated, and this frequency is then used to demodulate the Doppler signal itself. The zero component of the demodulated signal is then rejected by subtracting its (complex) average from the total signal, and the blood flow velocity estimated using the residual signal.

A similar technique has been suggested by Brands et al (1995) for use in the RF domain, although having shifted the power spectrum to a lower frequency, they use a conventional high-pass filter, with a lower cut-off frequency than would otherwise be possible if the shift had not taken place.

Ledoux et al (1997) have used a filtering technique based on singular value decomposition, which, unlike correlation filters, makes use of information from both the temporal and spatial directions. Simulation results with this filter are reported to be good in comparison to linear regression filters, but at present, because of their considerable complexity, they are not a practical option for use with colour flow systems.

11.2.3 Autocorrelation Algorithm

Once the Doppler signal has been filtered to remove low-frequency high-amplitude components due to tissue motion, it is sent to the velocity estimator, which has the task of calculating its mean frequency, bandwidth and power. Despite the fact that many methods of frequency estimation have been described in the literature, the vast majority of colour flow systems still use the autocorrelation algorithm used in the very first such systems (Namekawa et al 1982, Kasai et al 1985). This algorithm is also variously referred to as the 'first order autoregressive estimator', the 'covariance estimator', the 'correlation phase estimator', or the 'correlation angle estimator', and has been explored in some detail in the general signal processing literature (Miller and Rochwarger 1972, Sirmans and Bumgarner 1975, Zrnic 1977, 1979), where it has been shown to have a number of favourable properties—the mean estimate is unbiased even at low SNRs, its uncertainty is relatively low even for poor SNRs, and the influence of noise on the standard deviation of the estimate is negligible for SNRs of greater than 15 dB (Sirmans and Bumgarner 1975). Barber et al (1985) showed experimentally that the technique performs well with Doppler ultrasound signals in a wide range of SNR environments. For favourable SNRs its performance was comparable to that of the Fourier transform approach, but for low SNRs (0-8 dB) it was superior.

There are a number of ways of deriving the autocorrelation estimator, one of which is given in Section 4.5.3.1. The following more formal derivation is taken from Kasai et al (1985), which appears in turn to be based on that of Miller and Rochwarger (1972), who derived and explicitly solved the maximum-likelihood equations for the mean frequency and spectral width of samples of a signal plus noise process (both assumed to be mean-zero stationary complex Gaussian processes) in pairs separated by a time T. The general derivation is applicable to non-contiguous pulse pairs, but generally colour flow systems operate by stepping along contiguous pulse pairs (i.e. the information derived from pulse 1 is compared to that derived from pulse 2, then the information derived from pulse 2 is compared to that derived from pulse 3, etc.).

The mean angular frequency, $\bar{\omega}$, of a Doppler power spectrum, $P(\omega)$, may be defined as:

$$\bar{\omega} = \frac{\int_{-\infty}^{\infty} \omega P(\omega)\, d\omega}{\int_{-\infty}^{\infty} P(\omega)\, d\omega} \qquad 11.1$$

whilst its variance, σ^2, is given by:

$$\sigma^2 = \frac{\int_{-\infty}^{\infty} (\omega - \bar{\omega})^2 P(\omega)\, d\omega}{\int_{-\infty}^{\infty} P(\omega)\, d\omega} \qquad 11.2$$

$$= \overline{\omega^2} - (\bar{\omega})^2 \qquad 11.3$$

Furthermore, the autocorrelation function $R(\tau)$ is related to $P(\omega)$ by the Wiener–Khinchin theorem, i.e:

$$R(\tau) = \int_{-\infty}^{\infty} P(\omega) e^{j\omega\tau}\, d\omega \qquad 11.4$$

Differentiating eqn. 11.4 with respect to τ leads to two further relationships, namely:

$$\dot{R}(\tau) = j \int_{-\infty}^{\infty} \omega P(\omega) e^{j\omega\tau}\, d\omega \qquad 11.5$$

and

$$\ddot{R}(\tau) = -\int_{-\infty}^{\infty} \omega^2 P(\omega) e^{j\omega\tau}\, d\omega \qquad 11.6$$

Equations 11.1 and 11.3 may now be written in terms of the autocorrelation functions for zero lag to give:

$$\bar{\omega} = -j\frac{\dot{R}(0)}{R(0)} \qquad 11.7$$

and

$$\sigma^2 = \left(\frac{\dot{R}(0)}{R(0)}\right)^2 - \frac{\ddot{R}(0)}{R(0)} \qquad 11.8$$

Note also that the average power of the signal, $\bar{W}$, is given by the autocorrelation function at $\tau = 0$, i.e:

$$\bar{W} = R(0) \qquad 11.9$$

It is possible to evaluate eqns 11.7 and 11.8 directly but, as Kasai et al (1985) pointed out, this is rather time-consuming, and they showed that if the autocorrelation function is treated as:

$$R(\tau) = |R(\tau)| e^{j\phi(\tau)} = A(\tau) e^{j\phi(\tau)} \qquad 11.10$$

where $A(\tau)$ is a real even function of τ, and $\phi(\tau)$ a real odd function of τ, then the following approximations are valid:

$$\bar{\omega} = \dot{\phi}(0) \approx \phi(T)/T$$

$$= \frac{1}{T} \arctan \frac{\operatorname{Im} R(T)}{\operatorname{Re} R(T)} \equiv \frac{1}{T} \arg R(T) \qquad 11.11$$

and

$$\sigma^2 \approx \frac{2}{T^2}\left(1 - \frac{|R(T)|}{R(0)}\right) \qquad 11.12$$

where T is the time between subsequent ultrasonic pulses. The mean angular frequency and variance of the Doppler signal may thus be calculated from its autocorrelation magnitudes and phases at lags of $\tau = 0$ and $\tau = T$.

Miller and Rochwarger (1972) derived expressions for the maximum-likelihood estimates of $R(T)$ and $R(0)$ when the observations of a complex process Z consist of pairs of samples each spaced T seconds apart (their equations numbered 11). Colour flow mapping systems, however, generally use a number (generally known as the ensemble or package length, PL) of equally spaced lines to generate their frequency estimates, so that all but the first and last samples are involved in two paired comparisons. Rewriting Miller and Rochwarger's equations for the case in which the observations of Z consist of $PL-1$ pairs of samples $Z(i-1)$, $Z(i)$, spaced T seconds apart, leads to:

$$\hat{R}(T) = \frac{1}{PL-1} \sum_{i=1}^{PL-1} Z(i-1)Z^*(i) \quad 11.13$$

and

$$\hat{R}(0) = \frac{1}{2(PL-1)} \sum_{i=1}^{PL-1} [|Z(i-1)|^2 + |Z(i)|^2] \quad 11.14$$

where the asterisk denotes complex conjugation.

Rewriting $Z(i)$ as $(I(i) + jQ(i))$, the in-phase and quadrature magnitudes of the Doppler signal, replacing the values of $R(T)$ and $R(0)$ in equations 11.11 and 11.12 by their estimates $\hat{R}(T)$ and $\hat{R}(0)$ from equations 11.13 and 11.14, and writing $N = PL - 1$ for convenience, leads to:

$$\hat{\bar{\omega}} = \frac{1}{T} \tan^{-1} \left\{ \frac{\sum_{i=1}^{N} Q(i)I(i-1) - I(i)Q(i-1)}{\sum_{i=1}^{N} I(i)I(i-1) + Q(i)Q(i-1)} \right\} \quad 11.15$$

which is the same as equation 4.12, and

$$\hat{\sigma}^2 = \frac{2}{T^2} \times \left(1 - \frac{\left[\left(\frac{1}{N} \sum_{i=1}^{N} Q(i)I(i-1) - I(i)Q(i-1) \right)^2 + \left(\frac{1}{N} \sum_{i=1}^{N} I(i)I(i-1) + Q(i)Q(i-1) \right)^2 \right]^{1/2}}{\frac{1}{N} \sum_{i=0}^{N} Q^2(i) + I^2(i)} \right) \quad 11.16$$

The estimate of average power is given by:

$$\hat{\bar{W}} = \hat{R}(0) = \frac{1}{N} \sum_{i=0}^{N} Q^2(i) + I^2(i) \quad 11.17$$

The algorithm for determining $\hat{\bar{\omega}}$ from eqn. 11.15 is represented schematically in Fig. 11.4.

The output of the autocorrelation estimator ideally is unbiased (Miller and Rochwarger 1972, Barber et al 1985, van Leeuwen et al 1986, Nowicki et al 1990, Jensen 1996) because adding white noise to the signal does not affect the autocorrelation at finite lags, and this has been

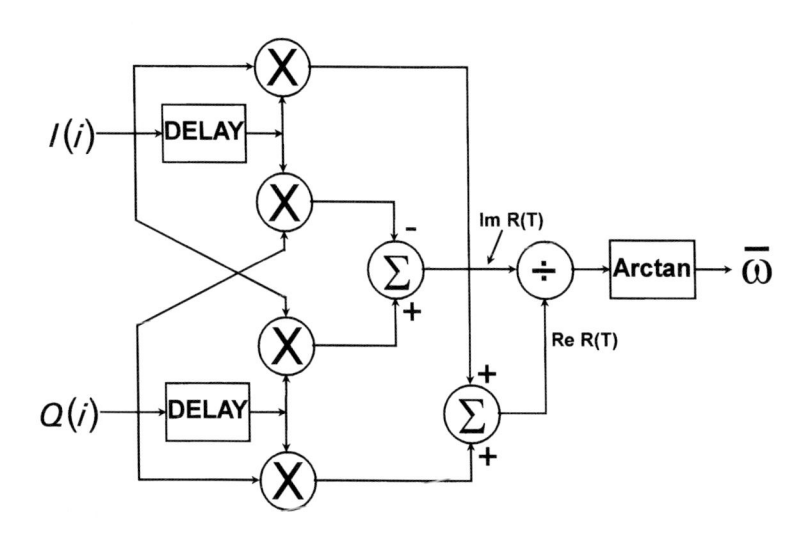

Figure 11.4 Schematic representation of the signal processing path necessary to implement the standard autocorrelation algorithm

confirmed in various practical simulations (Sirmans and Bumgarner 1975, van Leeuwen et al 1986, Jensen 1996), although a decreasing SNR leads to an increase in the variance of the estimate. There appears, however, to be no general treatment of the statistical properties of the autocorrelation estimator, although Kristoffersen (1988) has dealt with the case where strong filtering is used, a condition which is rarely met in real-time colour flow imaging.

Although the autocorrelator has a high immunity to uncorrelated noise, it is affected by correlated noise, and unfortunately wall motion filters inevitably introduce a correlated component into the noise (Willemetz et al 1989, Nowicki et al 1990). To some extent the resulting bias can be eliminated by estimating the noise from a region of tissue without flow and using this to compensate the mean frequency estimate, but the use of IIR filters as opposed to FIR filters reduces the bias, and indeed, with second order IIR filters the bias can be reduced to close to zero (Willemetz et al 1989).

Rajaonah et al (1994) have explained this phenomenon in terms of the effect of the wall filters on the shape of the otherwise (assumed) white noise spectrum. If no wall filter is used and the noise is white, the autocorrelation function of the noise at lag one is zero and the estimate unbiased. In the presence of a filter, however, the noise will no longer be white, the autocorrelation function no longer zero at lag one, and the estimated mean frequency biased. Alternatively, the required conditions for an unbiased estimate is that the total spectrum (consisting of the signal and the noise) should be symmetrical around the signal mean frequency when considering a Nyquist interval centred at that mean frequency. White noise does not bias the estimate if no wall filter is used, but the use of a filter breaks the symmetry and introduces bias. These explanations suggest that the reason why IIR filters cause less bias than FIR filters is their superior roll-off characteristics.

One major advantage of the autocorrelator over a number of other narrow-band (phase shift) estimation methods is that it gives an unambiguous output for narrow bandwidth signals over the range $\pm \pi$ radians (as would be predicted from eqn. 11.15) and also is well-behaved with wide bandwidth signals with mean frequencies of up to close to $\pm \pi$ radians because the autocorrelator has the important property of correctly accounting for partial aliasing of a continuous spectrum (van Leeuwen et al 1986). In simulations, van Leeuwen et al showed that for signals with bandwidths of up to 0.3 PRF (the pulse repetition frequency), the autocorrelator output could correctly track the mean frequency level beyond 0.4 PRF.

Clearly the larger the number of pulse pairs (i.e. the packet length) over which the summations in equations 11.15–11.17 can be calculated, the better will be their statistics, but in practice the upper limit for this is constrained by the necessity to maintain adequate colour flow frame rates (see Section 10.5.3). Torp and Bjaerum (1996) have published the results of a study in which they assessed the effect of packet size on the bias and standard deviation of mean velocity, bandwidth and power estimates, calculated using the standard autocorrelation algorithm on data taken from the left ventricle. Further studies of the trade-off between packet size and the statistics of the autocorrelation estimate are needed, although in practice the best compromise in any clinical situation is usually left in the hands of the scanner operator, who can directly control packet size (albeit called something less technical).

Many other frequency estimation techniques have been suggested for use with Doppler ultrasound signals, some of which can be regarded as extensions of the simple autocorrelation technique. Kim (1989) and Liu et al (1991) have described spatial vector averaging techniques. Herment et al (1991, 1993, 1996) have suggested adaptive estimators which are based on the weighted summation of all available correlation lags, rather than the first lag alone. Torp et al (1994) and Loupas et al (1995a,b) have studied the complex 2-D autocorrelation function (autocorrelation as a function of both temporal and radial lags). Hoeks and colleagues (Hoeks et al 1994b, Collaris 1996) have explored the use of 'sub-sample volume Doppler processing', in which the Doppler signal is split into sub-samples, the autocorrelation coefficient for each sub-sample estimated, and the estimates then combined. A detailed survey of techniques based on phase shift estimation can be found in Section 11.3.

11.2.4 Post-processing

Once estimates of power, mean frequency and bandwidth have been calculated for each sample volume, this information, together with grey-scale information from the same region, is used to determine whether valid flow is present in the sample volume and, if so, how it should be colour-encoded. Doppler signals are, however, stochastic in nature, which causes the estimated parameters to fluctuate in a random manner, which can lead to a display with a very mottled appearance that changes rapidly with time, and in particular may lead to regions of colour drop-out in flow-containing areas and flashes of colour in flow-free areas. This problem can be particularly severe close to vessel walls, where parameter values tend to be close to decision thresholds, and leads to the borders of vessels having a 'frayed' appearance that changes from frame to frame (Collaris and Hoeks, 1994). As also pointed out by Collaris and Hoeks, this can be particularly distracting because the human visual system is very sensitive to rapid changes. Because of these problems, CFI systems employ various post-processing techniques which may include linear and/or non-linear spatial and temporal filtering of the Doppler parameter data to improve the acceptability of colour images. These may be applied before or after priority encoding or both.

Whilst there appears to be no systematic discussion of post-processing algorithms in the literature, a number of authors have made reference to suitable techniques. Bohs et al (1993a,b) used a spatial 2-D smoothing filter which replaced each velocity with the median of itself and its four nearest neighbours (in conjunction with a speckle tracking system), whilst Torp and Bjaerum (1996) used (presumably) linear spatial averaging over three sample volumes in both the radial and lateral directions to derive 'reference' estimates for their study of the effect of quality versus frame rate changes in CFI. Forestieri (1995) describes median temporal filtering prior to linear temporal filtering, primarily for use in power Doppler applications. Collaris and Hoeks (1994) have described a sophisticated scheme for non-linear temporal filtering, which they classify as an attack-sustain filter. If the velocity or the frequency of the Doppler signal increases from frame to frame, or there is sudden reversal in velocity direction, the original signal is preserved. If, on the other hand, the velocity decreases from frame to frame, then a low pass filter is applied. The simplest embodiment of this filter may be written:

If $\quad (|V_{in}(t)| > |V_{out}(t-1)|)$

or $\quad (\text{sgn}(V_{in}(t)) = -\text{sgn}(V_{out}(t-1)))$

then $\quad V_{out}(t) = V_{in}(t)$

else $\quad V_{out}(t) = V_{out}(t-1)$
$\quad\quad\quad + C.(V_{in}(t) - V_{out}(t-1))\quad$ 11.18

where $V_{in}(t)$ and $V_{out}(t)$ are the algorithm's input and output at time t, and sgn( ) is the sign operator which returns 1 if $V > 0$, -1 if $V < 0$, and 0 if $V = 0$. Parameter C ($0 < C < 1$) controls the time constant of the sustain filter.

Collaris and Hoeks claim that the attack-sustain filter 'combines a presentation of unfiltered peak velocity values with the filling of sudden holes and notches in the velocity profiles'. In the same paper they describe two modifications to the above scheme which are said to improve the utility of the filter. In the first, the parameter C is made a function of the modulus of the latest output of the algorithm, and the second is designed to make the filter exhibit an even faster response to the majority of velocities whilst retaining the ability to fill holes and suppress border jitter. In their experimental implementation, Collaris and Hoeks preceded their attack-sustain filter with a threshold or priority encoding algorithm (see next section) and a 3×3 spatial median filter. The threshold algorithm consisted of applying a 3×3 median filter to the Doppler bandwidth data, and then thresholding the result to produce a binary acceptance/rejection mask (Collaris, 1996).

Before leaving this section, it should be noted that in CFI systems the spectral Doppler data may also need some post-processing to fill in the missing Doppler signal segment from the period when image acquisition is occurring. Kristoffersen and Angelsen (1988) have published the description of one approach to this problem, which consists of generating an FIR filter based on the Doppler signal immediately prior to the imaging interrupt, and then using this filter to generate a synthetic Doppler signal by driving it with white noise. Such a signal will, of course, have spectral properties (and thus also audible sound) similar to the real Doppler segment on which it is based.

11.2.5 Priority Encoding/Artefact Suppression Algorithms

The autocorrelator estimator will calculate values for the Doppler signal power, frequency and bandwidth in each sample volume, irrespective of whether or not there is any flow or even movement present, and it is therefore necessary to determine whether the outputs correspond to a signal from blood flow or are purely artefactual in nature. The priority encoder determines the probability that the grey-scale signal originated from solid tissue or blood according to its amplitude, and the probability that the colour signal originated from real movement or simply noise according to the various parameters calculated by the autocorrelator, and hence the overall probability that there is real flow present in any sample volume. If no valid flow data is present, then only grey-scale information will be displayed; if valid flow data is present, then either only velocity data will be displayed (exclusive display) or both grey-scale and colour coded velocity data will be displayed (additive display).

The details of priority encoding will vary from machine to machine, but a number of common themes are likely to be present in most:

- *Doppler signal magnitude threshold.* If the (post-clutter removal) estimate of the average Doppler signal power, $\hat{\overline{W}}$ (eqn. 11.17), is less than a minimum threshold level, then only grey-scale information will be displayed. This is because reliable velocity estimates cannot be calculated from small signals when both the numerator and denominator in eqn. 11.15 approach zero. The level of this threshold is set by the operator using the colour gain control.
- *Velocity threshold.* If the velocity calculated by the autocorrelator is very low, it is likely that the Doppler spectrum is dominated by clutter signals that have leaked through the wall filters and should be ignored. Therefore a threshold is applied to the velocity estimate, and if it falls below this level only grey-scale information will be displayed.
- *Maximum echo intensity threshold.* If the amplitude of the grey-scale echoes is very large, then the probability is that the signal is from a region containing solid tissue, rather than blood, and only grey-scale information should be displayed. Once again the threshold for this, the 'colour write enable', can usually be set by the operator.
- *Doppler bandwidth threshold.* The bandwidth of the Doppler signal (calculated from eqn. 11.16) from blood targets is relatively narrow. On the other hand, the bandwidth of the background noise in the absence of a true Doppler signal is relatively wide. Therefore, a threshold can be introduced such that if the estimated variance of the Doppler power spectrum exceeds a certain value, the velocity signal is assumed to be invalid.

Few publications appear to have given details of priority encoding procedures, but Lipshutz (1988) described an algorithm which firstly applies a Doppler signal magnitude threshold and, provided this is exceeded, makes a weighted comparison between the grey-scale echo intensity (I_{gs}) and the magnitude of the calculated velocity (V_{mag}). The principle of this comparison is that as I_{gs} becomes larger, it becomes less and less likely that a given pixel should be displayed in colour; on the other hand, as V_{mag} increases it becomes more and more likely that true flow is present. Lipshutz expressed this decision as:

If $\quad X \times I_{gs} > Y \times V_{mag}$

then   display grey-scale information

else   display colour information.   11.19

The values of X and Y that are used in this comparison depend on experience, and the application.

11.2.6 Power Doppler

The general principles and the advantages and disadvantages of power Doppler (PD) have been discussed in Sections 4.5.3.3 and 10.3.2. Basically, in this technique the CFI system displays the estimate of the average Doppler signal power, $\overline{\hat{W}}$ (eqn. 11.17), rather than the estimate of the signal mean velocity, $\hat{\overline{\omega}}$ (eqn. 11.15), so that a display of the amount of flowing blood in each part of the tissue, rather than the speed with which it is flowing, is created (Rubin et al 1994).

The detailed requirements of such a system differ slightly from those for conventional CFI, and therefore PD systems are optimised in a slightly different way. Since it is the absence or presence of blood flow that is of interest, temporal resolution is not so important, and PD systems are able to use ensemble lengths that are at the higher end of the range normally used in conventional applications, together with much more frame-to-frame temporal averaging, which serves both to eliminate pulsatile information and to increase the SNR. Since it is power rather than velocity that is displayed, it is no longer necessary to decide whether or not to display the colour signal depending on whether the Doppler power exceeds a threshold, and therefore the colour gain control as such is redundant. Also, because it is power that is displayed, flash artefacts due to the movement of solid tissue can become particularly troublesome, and specific steps must be taken to reduce this problem, for example by the use of temporal median filtering (Forestieri, 1995). Aliasing is not a problem with PD, because whilst aliasing of the signal still occurs, this has no effect on total measured signal power, and therefore the pulse repetition frequency can be set relatively low, which improves the performance of clutter removal filters and hence facilitates the detection of flow with low velocities.

11.3 OTHER NARROW-BAND (PHASE SHIFT ESTIMATION) METHODS

Although the autocorrelator algorithm was the first, and still is the most widely used, method for estimating mean frequency in colour flow imaging systems, several other methods have been suggested as being suitable for use in multigate and colour flow imaging systems, and many of them appear under different names in different places, leading to some confusion. A number of such algorithms are described below.

11.3.1 Zero-crossing Detector

Zero-crossing detectors are frequently used in simple CW and PW Doppler systems because of their low cost and ease of construction, even though they are known to have a number of significant limitations (Lunt, 1975). They have also been used in multi-gate systems (Hoeks et al 1981,1984), apparently with considerable success. They function by detecting each occasion on which either the 'direct' or 'quadrature' signal passes through zero, and from this information estimate the root mean square frequency of the signal. This process is depicted in Fig. 11.5. The signal vector illustrated is rotating in a counter-clockwise direction, and the quadrature signal changes its sign from negative to positive between $t = t_2$ and $t = t_3$, the direct signal changes from positive to negative between $t = t_4$ and $t = t_5$, and the quadrature signal goes negative again between $t = t_6$ and $t = t_7$. From these three zero-crossings, the only information available is that the phase of the signal vector has changed in a positive direction, by somewhere between 180° and 360° in six sample periods, and thus the mean frequency is between $1/(12T)$ and $1/(6T)$, where T is the sampling period. It can be seen that with a large number of zero-crossings a reasonable estimate of frequency should be possible, but because of the timing considerations associated

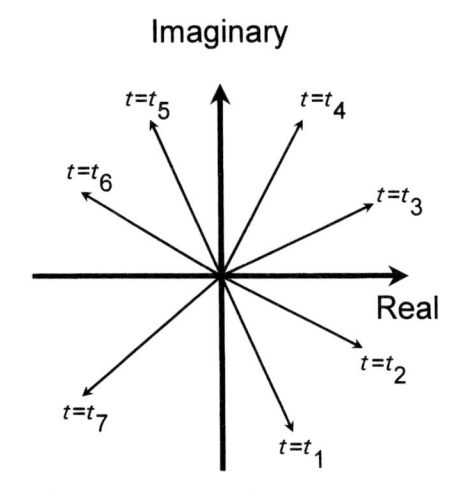

Figure 11.5 Successive positions of a signal vector relative to real and imaginary axes as it rotates in a counter-clockwise direction

with CFI systems there are seldom more than 16 samples available for analysis, and usually far fewer, and it is quite clear that under these circumstances the zero-crossing detector is not a viable system. At low velocities, and particularly when very few data samples are available, there may actually be no zero-crossings at all. From this example it can be seen that CFI systems operating in the phase domain must effectively measure the angles between successive signal vectors, so that small phase changes can be detected, if they are to be successful. Both the autocorrelation technique already described, and the rest of the methods described in this section work on this principle.

11.3.2 Phase Detector

Estimators based on the 'phase detector' algorithm or part of it have appeared under various names, including 'phase detector' (Grandchamp, 1975; Brandestini, 1975), 'infinite gate' (Nowicki and Reid, 1981), 'I/Q algorithm' (Barber et al, 1985), and 'double correlation centroid detector' (Gerzberg and Meindl, 1980). The basis of the technique can be simply explained with reference to Fig. 11.6, which shows the positions of a signal vector during two successive samples, $(i-1)$ and (i). The angular frequency, ω, of the vector is defined as its rate of change of phase, or

$$\omega = \frac{d\phi}{dt} \qquad 11.20$$

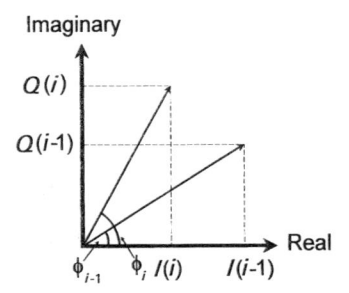

Figure 11.6 Position of a rotating signal vector during two successive samples $(i-1)$ and (i), showing in-phase components $I(i-1)$ and $I(i)$, and quadrature components $Q(i-1)$ and $Q(i)$

The phase of the signal for sample i, ϕ_i, is simply the arctangent of $[Q(i)/I(i)]$, and substituting this into 11.20 leads to the relationship:

$$\omega = \frac{I(i)\dot{Q}(i) - Q(i)\dot{I}(i)}{I^2(i) + Q^2(i)} \qquad 11.21$$

where a dot indicates a derivative with respect to time. Now if the derivatives are replaced by the finite backward differences between the samples (i) and $(i-1)$ (Nowicki et al, 1990), and the result averaged over a number of samples, then equation 11.21 may be rewritten:

$$\bar{\omega} = \frac{1}{T} \frac{\sum_{i=1}^{N} I(i)Q(i-1) - Q(i)I(i-1)}{\sum_{i=1}^{N} I^2(i) + Q^2(i)} \qquad 11.22$$

where T is the pulse repetition interval.

The performance of the phase detector has been discussed by a number of authors (van Leeuwen et al, 1986; Nowicki et al, 1990; Barber et al, 1985). Its most obvious drawbacks are that its frequency output can only be unambiguous over the range $\pm \pi/2$ radians (i.e. $\pm PRF/4$ or half the Nyquist frequency), and that because of the squared terms in the denominator the estimation will be biased by any non-Doppler-shifted components applied to the input. In theory, the majority of non-shifted signals will have been removed by the DLCs, but inevitably some will leak through, as will uncorrelated electronic noise. Both of these will lead to an increase in the size of the denominator, and a reduction in sensitivity to flow of the estimator. Simulation studies to evaluate the stationary and dynamic performance of the phase detection algorithm have also shown that its performance is rather poor (van Leeuwen et al, 1986). Only with relatively narrow-band signals (bandwidth ≤0.1 PRF), and SNRs of 30 dB or greater, were acceptable values for the relative error obtained for centre frequencies of up to 0.2 PRF. Either increasing the signal bandwidth or decreasing the SNR led to a rapid deterioration in the performance of the estimator.

11.3.3 Instantaneous Frequency Detector

The instantaneous frequency detector (IFD) was apparently first used in a multigate system (Brandestini 1978, Brandestini and Forster 1978). Its basic operation can once again be explained by making reference to Fig. 11.6. The angular frequency can, as before, be written as the rate of change of phase (equation 11.20), and this may be approximated by:

$$\omega \approx \frac{\phi_i - \phi_{i-1}}{T} \qquad 11.23$$

Both ϕ_i and ϕ_{i-1} may be written in terms of arctangents of the in-phase and quadrature signals, which after averaging over a number of samples N, leads to the expression:

$$\bar{\omega} = \frac{1}{NT} \sum_{i=1}^{N} \left[\arctan\frac{Q(i)}{I(i)} - \arctan\frac{Q(i-1)}{I(i-1)} \right] \qquad 11.24$$

Simulations have been performed on this detector by Hoeks and colleagues (van Leeuwen et al, 1986; Hoeks et al, 1984), and they have shown its performance to be considerably superior to that of the phase detector. For narrow bandwidth signals with a high SNR, its output is unambiguous over a range of $\pm\pi$ radians (i.e. $\pm PRF/2$). As with the phase detector, the output of the IFD falls away when either the SNR decreases or the bandwidth increases. This is mainly caused by mapping of instantaneous frequencies outside the ± 0.5 PRF range back into the -0.5 to $+0.5$ interval. Because of this, its performance (and that of other phase domain detectors) also depends on the shape of the spectrum being processed. For example, if a frequency spectrum is skewed so that the mean frequency is close to the maximum frequency, then the performance of the IFD will be better than it would be for a symmetrical spectrum with the same mean frequency and bandwidth. This is because in the skewed case the maximum frequency is further from the Nyquist frequency, and therefore fewer mapping errors occur.

The IFD also tends to overestimate velocity for very low Doppler frequencies (below approximately PRF/10–PRF/20) (van Leeuwen et al, 1986), in exactly the same way as for conventional CW and PW Doppler units (Gill, 1979); the reason for this is that the high-pass filters (implemented by DLCs in CFI systems) remove not only the unwanted low-frequency clutter signals but also the low-frequency blood flow signals.

11.3.4 Autocorrelation Estimator with Sub-sampling

Hoeks et al (1994b) have suggested a modification of the autocorrelation method of estimating mean frequency, which involves splitting segments of the Doppler signal into sub-samples prior to applying the autocorrelation algorithm. Their rationale for this approach is that, contrary to the usual implicit assumption, the phase of the received high frequency ultrasound signal does not change linearly with depth because of the random spatial distribution of the scattering centres within the ultrasound beam, and therefore averaging the demodulated signal over an observation window covering a number of periods of the received signal does not improve the estimate of the instantaneous quadrature components of the Doppler signal originating from a given depth (Hoeks et al 1993). Thus, the accuracy of the Doppler velocity estimate is independent of the length of the observation window employed. The argument then proceeds that, if the observation window is split into sub-sample volumes, each with a length of one period at the transmitted frequency, and estimates of the autocorrelation at one lag found for each sub-sample, then the combination of these will lead to a reduction in the variance of the velocity estimate.

Specifically, if each sample is split into NP (numbers of periods) sample volumes, $m = 0, ..., NP - 1$, then equation 11.13 can be rewritten:

$$\hat{R}(T) = \frac{1}{NP(PL-1)} \sum_{m=0}^{NP-1} \sum_{i=1}^{PL-1} Z(m, i-1) Z^*(m, i) \qquad 11.25$$

Rewriting $Z(m, i)$ as $(I(m, i) + jQ(m, i))$, the in-phase and quadrature magnitudes of the Doppler signal, and replacing $R(T)$ in eqn. 11.11 with this estimate, leads to an explicit expression for the estimated Doppler shift frequency:

$$\hat{\bar{\omega}} = \frac{1}{T} \tan^{-1} \left\{ \frac{\sum_{m=0}^{NP-1} \sum_{i=1}^{PL-1} I(m,i)Q(m,i-1) - I(m,i-1)Q(m,i)}{\sum_{m=0}^{NP-1} \sum_{i=1}^{PL-1} I(m,i-1)I(m,i) + Q(m,i-1)Q(m,i)} \right\}$$

11.26

Similarly modified expressions can easily be written for both the spectral variance and average power.

Hoeks et al (1994b) compared the performance of this new method (which they named 'S-Dopp') with that of a cross-correlation interpolation method (see Section 11.4.2.1) which they had previously described (de Jong et al 1990, Hoeks et al 1993) over a wide range of signal processing conditions. They found the two methods to be comparable in terms of the bias and standard deviation of the mean frequency estimates, and the S-Dopp method to be computationally more attractive. One potential disadvantage of the S-Dopp method (like any phase shift-based method) was the potential deviation of the carrier frequency from the assumed emission frequency due to frequency-dependent attenuation by the tissue (see Section 8.3.5).

Another way of viewing this approach is to regard it as the estimation of the complex 2-D autocorrelation $R(r, \tau)$ at $r = 0$, $\tau = T$ (where r is the radial lag and τ the temporal lag) by averaging over both the radial and temporal directions, and this is exactly the approach taken (apparently completely independently of Hoeks et al) by Torp et al (1994, 1995). This approach results in an identical equation to that given in 11.25 (except that $\hat{R}(T)$ is now replaced by $\hat{R}(0, T)$), if the averaging filters in the radial and temporal directions are rectangular in shape. It should be noted that this is different from the 'full' 2-D autocorrelation approach proposed by Loupas et al (1995a,b) and introduced in the next section, although the latter reduces to the former if certain simplifying substitutions are made, and the discussion of the latter will also provide further insight into the approach adopted by Torp and colleagues.

In their 1994 paper, Torp et al discussed the statistical properties of the autocorrelation estimator in some detail, and presented results from a model that showed that, at least for the model parameters chosen, the radial averaging reduced the variance much more rapidly than temporal averaging. They also presented a series of images of Doppler data gathered from the heart, which showed very dramatically the potential for radial averaging to improve image quality more efficiently than temporal averaging.

11.3.5 Two-dimensional Autocorrelation Estimator

The 2-D autocorrelation estimator was introduced as a method of processing ultrasound signals for colour flow imaging by Loupas et al (1995a,b), and differs from all previously described colour flow estimators, in that the axial velocity is calculated from the Doppler equation using explicit estimates of both the mean Doppler and the mean RF frequency at each range gate location. This has considerable advantages over the conventional 1-D autocorrelation method, which assumes that the mean RF frequency is constant and equal to the centre frequency of the transmitted pulse, since in reality, even for constant flow, the RF frequency varies considerably, due to both the stochastic nature of the backscattered signal and the effects of frequency-dependent attenuation and scattering (Section 8.3.5). The fluctuations in RF frequency lead to perturbations in the Doppler frequency, which may be quite large but which tend to track the RF fluctuations (Bonnefous, 1992). Thus, if velocity is calculated from the ratio of the measured Doppler shift frequency to the measured RF frequency, rather than to the assumed RF frequency, a much more stable estimate is obtained.

The concept of treating the Doppler signal as a 2-D signal was introduced in Section 8.2.3, the two dimensions being depth (also called 'fast time' or 'the radial direction') and line number (also called 'slow time' or 'the temporal direction') (see Fig. 8.4). As explained in Section 8.2.3, the 2-D Fourier transform of such a 2-D signal has axes corresponding to 'fast frequency' (the received pulse RF frequency) and 'slow frequency' (the Doppler shift frequency), and constant velocity Doppler shifts map onto this plane as ellipses (see Fig. 8.5). The major axes of such ellipses pass through the origin, and the tangent of the angle between these axes and the RF axis are proportional to scatterer velocity. The 2-D autocorrelation method proposed by Loupas et al estimates mean axial velocity directly in the time domain from the slope of the line that passes through the origin of the frequency plane and the centre of mass of the 2-D spectrum in either the positive or negative frequency semiplane.

Adapting slightly the notation used by Loupas et al, the mean velocity estimate using the 2-D autocorrelation method may be written:

$$\hat{v}_{2D} = \frac{c}{2} \frac{\frac{1}{2\pi T}\tan^{-1}\left\{\frac{\text{Im}[\hat{R}_a(0,T)]}{\text{Re}[\hat{R}_a(0,T)]}\right\}}{\frac{1}{2\pi t_s}\tan^{-1}\left\{\frac{\text{Im}[\hat{R}_a(t_s,0)]}{\text{Re}[\hat{R}_a(t_s,0)]}\right\}} = \frac{c}{2}\frac{\hat{f}_d}{\hat{f}_{RF}}$$

11.27

where $\hat{R}_a(r, \tau)$ is the estimate of the complex 2-D autocorrelation function of the analytic signal version of the backscattered RF echoes at radial lag r and temporal lag τ. T is the pulse repetition period, t_s the sampling interval along depth, and c the velocity of ultrasound in soft tissue. Note that the analytic version of RF echoes is used because the 2-D Fourier transform for real signals is symmetric with respect to the origin of the frequency plane, and unless this symmetry is removed, the centre of mass of the 2-D spectrum will be at the origin.

Loupas et al (1995b) went on to show that the 2-D autocorrelation technique could also be applied to the complex demodulated (quadrature) signals, and for this case derived the following expression for mean velocity:

$$\hat{v}_{2D} = \frac{c}{2}$$

$$\times \frac{\frac{1}{2\pi T}\tan^{-1}\left\{\frac{\text{Im}[\hat{R}_{\text{dem}}(0,T)]}{\text{Re}[\hat{R}_{\text{dem}}(0,T)]}\right\}}{\frac{1}{2\pi t_s}\left(2\pi f_{\text{dem}} + \tan^{-1}\left\{\frac{\text{Im}[\hat{R}_{\text{dem}}(t_s,0)]}{\text{Re}[\hat{R}_{\text{dem}}(t_s,0)]}\right\}\right)}$$

11.28

where $\hat{R}_{\text{dem}}(r, \tau)$ is the estimate of the complex 2-D autocorrelation function of the complex demodulated signal, and f_{dem} is the frequency of the demodulation sinusoid normalised by multiplying by t_s, the sampling interval along depth.

If the complex demodulated signal is written in terms of its in-phase (I) and quadrature (Q) components which are sampled from PL lines at ND depths, then eqn. 11.28 may be re-written:

$$\hat{v}_{2D} \cong \frac{c}{2}$$

$$\times \frac{\frac{1}{T}\tan^{-1}\left\{\frac{\sum_{m=0}^{ND-1}\sum_{i=1}^{PL-1} I(m,i)Q(m,i-1) - I(m,i-1)Q(m,i)}{\sum_{m=0}^{ND-1}\sum_{i=1}^{PL-1} I(m,i-1)I(m,i) + Q(m,i-1)Q(m,i)}\right\}}{\frac{1}{t_s}\left(2\pi f_{\text{dem}} + \tan^{-1}\left\{\frac{\sum_{m=0}^{ND-2}\sum_{i=1}^{PL} I(m+1,i-1)Q(m,i-1) - I(m,i-1)Q(m+1,i-1)}{\sum_{m=0}^{ND-2}\sum_{i=1}^{PL} I(m,i-1)I(m+1,i-1) + Q(m,i-1)Q(m+1,i-1)}\right\}\right)}$$

11.29

Note that the top line of this equation is identical to eqn. 11.26, the equation derived by Hoeks et al (1993) for their 'S-Dopp' method (except that NP, the number of periods, is replaced by ND, the number of depths). Thus, the estimation of the Doppler frequency in the 2-D autocorrelation method is identical to the method used by Hoeks et al (1993) and Torp et al (1994), but the 2-D autocorrelation method differs significantly in the explicit estimation of the RF frequency by which the Doppler frequency is divided to estimate velocity. If the fluctuations in RF frequency are ignored in equation 11.29, then the velocity may be written:

$$\hat{v}_{\text{2D-DOP}} \cong \frac{c}{2} \times \frac{\frac{1}{T}\tan^{-1}\left\{\frac{\sum_{m=0}^{ND-1}\sum_{i=1}^{PL-1} I(m,i)Q(m,i-1) - I(m,i-1)Q(m,i)}{\sum_{m=0}^{ND-1}\sum_{i=1}^{PL-1} I(m,i-1)I(m,i) + Q(m,i-1)Q(m,i)}\right\}}{\frac{2\pi f_{\text{dem}}}{t_s}}$$

11.30

Furthermore, the top lines of both eqn. 11.29 and eqn. 11.30 reduce to eqn. 11.15, the conventional 1-D autocorrelation algorithm, if samples from only a single depth in tissue are used.

The 2-D autocorrelation approach is thus a generalisation of (and improvement on) the 1-D autocorrelation approach in two distinct ways. Firstly the estimate of the Doppler frequency is formed by processing samples from a number of axial depths (which has already been shown to be advantageous by Hoeks et al 1993 and Torp et al 1994), and secondly, the local RF frequency is explicitly estimated, which allows a degree of compensation for errors caused by the variations in this frequency due to the stochastic nature of the signal and frequency-dependent scattering and attenuation in the tissues.

The performance of the 2-D autocorrelator was tested by means of extensive simulations in Loupas et al (1995b) and with *in vitro* flow data in Loupas et al (1995a). Loupas et al (1995b) compared the 1-D autocorrelation algorithm (eqn. 11.15), the 'simplified 2-D autocorrelation algorithm' (eqn. 11.30), the full 2-D autocorrelation algorithm (eqn. 11.29) and the cross-correlation algorithm (Section 11.4.2) under different operating parameters, including range gate length, ensemble length, SNR and pulse bandwidth. Of the four methods, the 1-D autocorrelator (i.e. the method currently in use in most commercial machines) was found to be consistently worse than any of the other estimators. In general, the simplified 2-D algorithm performed third best, with the other two algorithms performing best and in a similar manner to each other, except under heavy noise conditions, where the 2-D autocorrelator showed noticeably better robustness. Loupas et al (1995a) used *in vitro* flow data and compared the effects of range gate and ensemble length, noise level and angle of insonation on the estimates of both mean velocity and power found using the 1-D autocorrelator and the full 2-D autocorrelator. Once again, the 2-D method was found to be greatly superior to the 1-D method in terms of both velocity and power estimation.

The apparent significant advantages of the 2-D autocorrelation estimator over its 1-D counterpart would suggest that it may well find widespread applications in colour flow mapping systems, and at least one manufacturer appears to have already incorporated it into a commercial system.

11.3.6 Extended Autocorrelation Estimator

The extended autocorrelation method (EAM) introduced by Lai et al (1997) is an extension of the 2-D autocorrelation estimator described by Loupas et al (1995a,b). Essentially, the velocity estimator used is the same as that given by Loupas et al (performed on base-band complex signals), except that use is made of both the phase and magnitude of the correlation function. The phase information is used, as in all conventional autocorrelation estimators, to measure velocity; the additional magnitude information is used to resolve any possible ambiguities in terms of aliasing.

Using the same notation as before, the maximum amplitude of the correlation function $R(r, T)$ occurs where $r = \tau_v$, where τ_v is the delay between echoes from pulses T seconds apart (see Section 11.4.1 for a more detailed discussion of the cross-correlation function). Assuming that the mean phase change, $\bar{\phi}$, in this interval can be written $-\pi < \bar{\phi} < \pi$, i.e. the sampling frequency exceeds that required by the Nyquist criterion, then τ_v may be written:

$$\tau_v = \frac{\bar{\phi}}{2\pi f_c} \qquad 11.31$$

where f_c is the centre frequency of the received signal. More generally, if no assumption is made about the sampling frequency, then:

$$\tau_n = \frac{\bar{\phi}}{2\pi f_c} + \frac{n}{f_c} \qquad n = 0, \pm 1, \pm 2 \ldots \qquad 11.32$$

Therefore, there are actually a number of candidate delays, τ_n, corresponding to a given mean phase change; if the velocity is below the Nyquist limit, then $n = 0$; if, however, the velocity is above the Nyquist limit ($|n| > 0$), aliasing will occur when standard phase shift estimation techniques are employed. The addition of the magnitude of the correlation function allows these ambiguities to be resolved. Because the global peak of the correlation function (ideally) corresponds to the true velocity, the true time delay candidate can be found from the position of the maximum amplitude of the correlation envelope $\hat{R}(\tau_n, T)$, which of course need only be evaluated for a small number of candidate values of n. In practice, for sampled signals, the candidate values of τ_n will not correspond to an integral number of sample periods, and therefore it is necessary to interpolate between the available values of $\hat{R}(r, T)$. Lai et al (1997) found simple parabolic interpolation to be satisfactory in this respect, even with low (depth) sampling rates.

Lai et al (1997) applied both the EAM and the standard cross-correlation estimator to simulated signals in which the flow velocities were up to four times the Nyquist velocity, and to experimental RF data from the subclavian artery. They concluded that the EAM has a similar performance to that of the cross-correlation estimator, but that the former is more computationally efficient. The cross-correlation approach itself is discussed further in Section 11.4.

11.3.7 Low Order Autoregressive Estimators

The use of the autoregressive (AR) spectral estimator as a means of analysing Doppler ultrasound signals in spectral Doppler applications was introduced in Section 8.2.4, where it was seen to have certain advantages over the conventional Fourier approach. The feasibility of using low order (for computational efficiency) AR models in colour flow applications has been explored by Loupas and McDicken (1990) and Ahn and Park (1991).

Loupas and McDicken examined low order AR models as methods of estimating both mean and maximum frequency from data segments consisting of a small number of samples. They examined various mean frequency estimators based on either the peak or the integral of the AR spectrum, but the one they found to have the best performance was the weighted sum of the local maxima of the AR spectrum (model order p), i.e.

$$\hat{f}_{\text{mean}}(p) = \frac{\sum_{n=1}^{N} f_i P_{\text{AR}(p)}(f_i)}{\sum_{n=1}^{N} P_{\text{AR}(p)}(f_i)} \qquad 11.33$$

where $P_{\text{AR}(p)}(f_i)$ are the local maxima at frequency f_i and N is the total number of poles of the spectrum ($N \leqslant p$).

It turns out that $\hat{f}_{\text{mean}}(1)$ is in fact identical to the standard 1-D autocorrelator (Section 11.2.3) and therefore the 1-D autocorrelator can be thought of as a first order AR estimator. In addition, Loupas and McDicken showed that $\hat{f}_{\text{mean}}(1)$ performs better than $\hat{f}_{\text{mean}}(2)$ as well as higher order estimators of the same type, and after extensive experimentation found only one estimator, known as phase-only poly-pulse-pair processing (Mahapatra and Zrnic, 1983), that performed even marginally better than the $\hat{f}_{\text{mean}}(1)$ or autocorrelation estimator, and even then only for SNRs of less than 0 dB, and at the expense of a considerable increase in computational complexity. They also drew attention to the possibility of using the second order AR estimator on the raw Doppler signal as a method of separating the Doppler and clutter signals, a method which Ahn and Park (1991) independently explored the following year in some detail.

Loupas and McDicken estimated maximum frequency from the integral of the AR spectrum by finding the frequency at which the integrated power exceeded a given percentage of the total integrated power (Section 8.4.1.2). Unlike the case for mean frequency estimation, $\hat{f}_{max}(2)$ (the maximum frequency derived using the 2nd order AR model) was found to perform better than $\hat{f}_{max}(1)$ in all cases as far as bias was concerned, and for high SNRs as far as variance was concerned. However, because $\hat{f}_{max}(2)$ deteriorates significantly for low SNRs, and is considerably more complex than $\hat{f}_{max}(1)$, Loupas and McDicken suggested that the latter holds more promise as a method of estimating maximum frequency from a data sequence containing a small number of samples.

The approach adopted by Ahn and Park (1991) differed from that explored by Loupas and McDicken in that a second order AR model was applied to the total Doppler signal, i.e. the signal without the application of clutter filters. The essential idea of this technique is that under these circumstances the required Doppler signal and the unwanted clutter signal will be modelled by the two poles of the AR model, and that the mean frequency and the variance of the Doppler signal can be estimated, respectively, from the phase and magnitude of the pole, with the larger phase among the two poles. It was argued that this would have advantages over the autocorrelation algorithm when only few samples are available for analysis, since as the clutter filters reduce the number of samples available for frequency estimation even further, a more stable mean frequency estimate can be obtained if the filtering process can be avoided. However, Ahn and Park concluded that the performance of the AR estimator could be degraded by the presence of large clutter signals, and that the relative merits of the two methods required further exploration.

Brands and Hoeks (1992) compared the second order AR method proposed by Ahn and Park with both the conventional first order AR estimator (i.e. the autocorrelator) and the complex linear regression estimator introduced in the next section, and concluded that its performance is poor in comparison to either of the latter two.

11.3.8 Complex Linear Regression Estimator

Brands and Hoeks (1992) suggested the use of a further mean frequency estimation algorithm based on an AR model, which differs from those described in the previous section in that it is based on modelling the Doppler signal as a single pole on the unit circle in the z-domain, rather than one or more poles within the unit circle. This leads to a mean frequency estimator which is equivalent to the slope of the regression fit through the unwrapped phase of the signal, and hence the name 'complex linear regression estimator'. From their simulations, Brands and Hoeks concluded that the complex linear regression estimator is superior, in terms of low variance and bias, to the conventional autocorrelation estimator for short estimation windows (between nine and 17 sample points), but that the latter is superior for longer sample windows.

11.4 WIDE BAND (TIME SHIFT ESTIMATION) METHODS

The principles of the time domain approach to the estimation of blood flow velocities have been introduced in Section 4.5.3.2. Further details are given in this section.

11.4.1 Cross-correlation Approach

The first publications reporting the use of the cross-correlation technique for measuring the velocity of blood flow appear to be those of Dotti and colleagues (Dotti et al 1976, Bassini et al 1979, 1982) in which they described firstly *in vitro* measurements of the velocity profile of steady blood flow in a plexiglas tube, and then later, *in vivo* measurements of pulsatile flow in carotid and other human arteries (albeit requiring many cardiac cycles to construct a single profile). A number of publications developing the basic theory of the cross-correlation method and describing experimental systems followed during the second half of the 1980s (Foster 1985, Bonnefous and Pesque 1986, Bonnefous et al

1986, Embree 1986, Bonnefous 1989), and over the last decade there has been a steady stream of publications on various time domain methods. However, as will emerge, these methods have so far failed to be widely introduced into commercial systems, presumably because of the large amount of computing power required for real-time implementation.

The basis of the cross-correlation method is illustrated in Fig. 11.7. At time T_1 an isolated group of scatterers is located at position P_1. If an ultrasonic pulse is transmitted at T_1, then it will take a time t_1 to travel to P_1 and return to the transducer. If a second pulse is transmitted T seconds later at $T_2 = T_1 + T$, then the group of scatterers will have moved to point P_2 and the transit time will be t_2. The axial distance, x, the scatterers have moved in the direction of the ultrasonic transducer can be calculated from the difference in the transit times of the two ultrasound pulses, $\tau = t_1 - t_2$, i.e:

$$x = \frac{\tau}{2} c \qquad 11.34$$

where c is the velocity of ultrasound in the tissue. If the direction of motion is at an angle θ to the ultrasound beam, then the component of the velocity, v, towards the transducer may be written:

$$v \cos \theta = \frac{x}{T} = \frac{\tau c}{2T} \qquad 11.35$$

or

$$v = \frac{\tau c}{2T \cos \theta} \qquad 11.36$$

Equation 11.36 is similar to the standard Doppler equation (e.g. eqn. 1.2), except that it has the time shift τ in the numerator rather than the change of frequency f_d, and the pulse repetition period, T, in the denominator rather than the transmitted frequency f_t. In practice, of course, the scatterers are not isolated, but any group of scatterers gives rise to a 'scattering signature' due to their unique distribution, which may be tracked along the vessel using a cross-correlation technique. To implement this, short segments of echo signal from each pulse are compared with segments of the same length, and at approximately the same temporal position on the echoes from subsequent pulses, to find the best match and hence to estimate the displacement of the group of scatterers in each index segment. This apparently simple process is complicated by a number of other factors, the most important of which are related to the component of motion perpendicular to the ultrasound beam and to the velocity dispersion within the range cell, both of which mean that no two scattering signatures can be identical. This means that the value of the normalised cross-correlation function never reaches a value of unity; nevertheless, the mean displacement of a group of cells can be tracked by searching for the maximum value of the cross-correlation.

Generally speaking, the cross-correlation method is applied directly to the real RF signal; there is, however, no reason why it should not be used on envelope detected data (Dickinson and Hill 1982, Trahey et al 1986), although this significantly increases axial jitter (Walker and Trahey 1994) or complex RF signals (Torp et al 1994, Brands et al 1997, Lai et al 1997). Ledoux et al (1998) have explored the correlation behaviour of analytic RF signals and point out that the method has advantages in that, like the envelope technique, it removes the frequency-dependent oscillations of the spatial correlation function but also produces smaller secondary maxima, which are less likely to be confused with the main peak.

An excellent review of correlation techniques has been given by Hein and O'Brien (1993a).

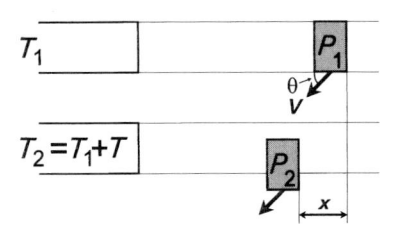

Figure 11.7 Illustration of the motion of an isolated group of scatterers between two sequential ultrasonic pulses. The scatterers have moved from P_1 to P_2 in the time T. The axial distance moved can be calculated from difference between the round trip times for the two pulses, τ. Note that whilst all the scatterers in P_1 fall within the ultrasound beam, some scatterers have moved out of the beam in P_2

11.4.2 Cross-correlation Based Algorithms

Adopting a similar approach to Jensen (1993a) and dividing each A-line into segments, each with a length of N_s samples (Fig. 11.8), the normalised correlation coefficient for a displacement of n samples between the ith segment of echo 1 (y_1), and echo 2 (y_2) may be written:

$$\hat{R}_{12}(n, i_{\text{seg}}) = \frac{\sum_{k=0}^{N_s-1} y_1[k + i_{\text{seg}} N_s] y_2[k + i_{\text{seg}} N_s + n]}{\sqrt{\sum_{k=0}^{N_s-1} (y_1[k + i_{\text{seg}} N_s])^2} \cdot \sqrt{\sum_{k=0}^{N_s-1} (y_2[k + i_{\text{seg}} N_s + n])^2}} \quad 11.37$$

Because the echoes are sampled at discrete times, the correlation coefficient can only be calculated for discrete lags, which are unlikely to coincide with the true position of the maximum correlation between the waveform segments. The accuracy of the estimate can be improved by increasing the sampling frequency, but generally this is kept as low as reasonably possible to reduce computational complexity. An alternative solution is to find the location and value of the maximum correlation coefficient, together with the values of its neighbours, and then to perform an interpolation to find a better estimate of the true correlation maximum. Foster (1985) and Foster et al (1990) have suggested fitting a parabola through the maximum correlation and its immediate neighbours, whilst Lai and Torp (1996) have tested a number of interpolation methods based on parabola fitting in conjunction with measures designed to reduce bias, as well as a matched filter approach, which proved to be particularly effective when the SNR was low. de Jong et al (1991) have reported the use of a cosine function for interpolation in conjunction with their 'correlation interpolation' method (see Section 11.4.2.1).

The echoes used to calculate the position of the maximum correlation coefficient may be from any two transmission cycles (i.e. they do not need to be adjacent), provided that a reasonable proportion of the scatterers interrogated during the first cycle, and thus contributing to the scattering signature, remain within the beam during the second cycle. Ideally, if the blood flow is slow and the *prf* high, then it may well be preferable to evaluate equation 11.37 for non-adjacent echoes, perhaps separated by several transmission cycles, otherwise the precision of measuring the small displacement of the scatterers may be quite poor (see Section 11.4.3 for a quantitative discussion of this issue). On the other hand, if the flow is very rapid there will be no problem in measuring the relatively large displacements occurring between adjacent pulses; indeed, if the *prf* is low it will be necessary to do so, because the reference group of scatterers may move completely out of the ultrasound beam before the next pulse (the likelihood of this depends on factors such as the direction of flow as well as the speed of flow). For steady flow it would be possible to optimise the *prf*, to achieve the maximum precision for any given velocity, but this is not possible with pulsatile flow. In an attempt to

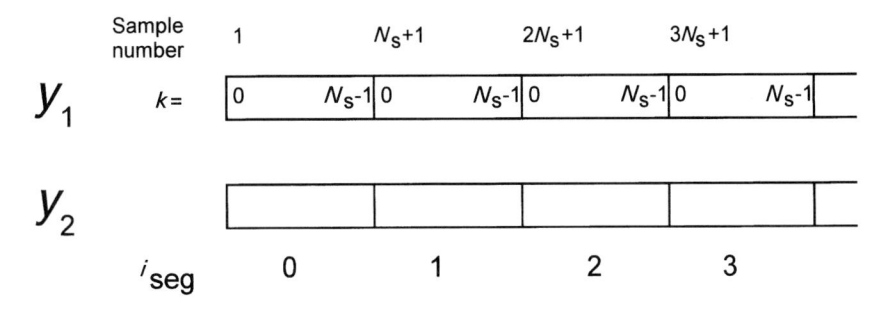

Figure 11.8 Segmentation of A-lines for the calculation of the cross-correlation function

overcome this, Embree and O'Brien (1990) have described a technique which seeks to adapt the effective *prf* of the ultrasound scanner by taking a weighted average of velocity estimates determined from echo pairs with different temporal spacing, the weighting factors being determined by the variance of the individual velocity estimates.

Rewriting eqn. 11.36 in terms of any two pulses which are q transmission cycles apart, individual estimates of velocity using the cross-correlation method are given by:

$$v_q = \frac{\tau_q c}{2qT \cos \theta} \qquad 11.38$$

where τ_q is the change in the transit time of two ultrasound pulses q cycles apart. For a total of N_e echoes there will be $N_e - q$ velocity estimates for each q. These can be averaged to provide a single velocity estimate $\bar{v}_q$ for each q. Embree and O'Brien (1990) then define an optimum average velocity estimate as:

$$\bar{v}_{\text{opt}} = \sum_{q=1}^{N_e - 1} W_q \bar{v}_q \qquad 11.39$$

where W_q is a weighting function determined by the variance of the individual $\bar{v}_q$ estimates. Estimates with a low variance are weighted more highly than those with a high variance, thus theoretically producing a better estimate than if all the estimates were simply averaged. Hein and O'Brien (1993b) have used this technique to measure pulsatile flow in arteries, but have found that the weighted average technique does not appear to improve the precision or accuracy of the estimate when compared with simple averaging, and have suggested that a better approach may be simply to use the average estimate from the pair of echoes which results in the lowest variance.

Calculation of blood-flow maps in real time using the cross-correlation method is computationally very intensive, requiring of the order of 10^9 multiplications and additions per second (Jensen 1993a). For this reason, in practice the cross-correlation is often calculated using only simple one-bit signals, as suggested by Foster (1985) and Bonnefous et al (1986). In this case the normalised cross-correlation function may simply be written as:

$$\hat{R}_{12(\text{sgn})}(n, i_{\text{seg}})$$
$$= \frac{1}{N_s} \sum_{k=0}^{N_s - 1} \text{sgn}(y_1[k + i_{\text{seg}} N]) \text{sgn}(y_2[k + i_{\text{seg}} N + n]) \qquad 11.40$$

where

$$\text{sgn}(\bullet) = \begin{cases} +1 & \text{for } (\bullet) \geqslant 0 \\ -1 & \text{for } (\bullet) < 0 \end{cases}$$

Jensen (1994) has shown that if y_1 and y_2 follow a Gaussian distribution with zero mean, then for infinitely many data:

$$\hat{R}_{12}(*) = \sin\left(\frac{\pi}{2} \hat{R}_{12(\text{sgn})}(*)\right) \qquad 11.41$$

For real data the relationship deviates from that shown in eqn. 11.41, but apparently not sufficiently to invalidate the method. Figure 11.9 is reproduced from Jensen's (1994) article and shows that for the conditions considered (32 sample segment and a *SNR* of 10) the deviations are not large, and that the maxima of the two estimates coincide. Note that as it is only necessary to find the position of the peak of the correlation coefficient it is not necessary to perform the sine operation in equation 11.41.

More recently Wang et al (1996) have proposed a correlation method that converts each sample into a two-bit representation, which includes the sign of the sample and an adaptively selected threshold. They show that the new algorithm is better than the one-bit correlation method in respect of susceptibility to noise.

A rather different approach has been proposed by Xu et al (1995), who used a wavelet transform-based cross-correlation (WTCC) technique for time delay estimation. The method comprises three steps: (a) computing the wavelet transforms of the received echoes, (b) computing the cross-correlations in the wavelet domain, and (c) estimating the time delays by maximising the

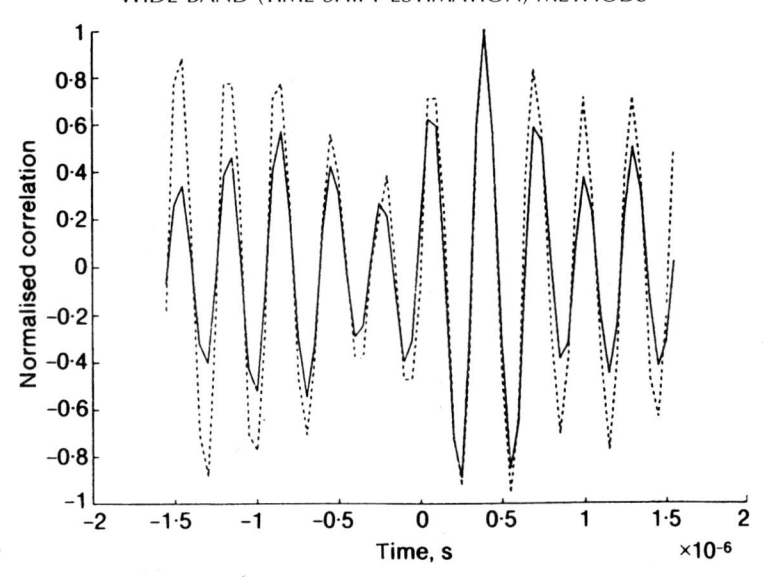

Figure 11.9 Estimation of cross-correlation using full data (solid line) and using only the sign data (dashed line) for a 32 sample segment and a *SNR* of 10 (reproduced by permission from Jensen 1994, © IFMBE)

cross-correlations. Xu et al reported that their computer simulations showed that the WTCC method provides a better estimate of time delay (lower failure rate and lower estimate error) and that its performance is more robust under various conditions including different window sizes, than the standard cross-correlation technique.

11.4.2.1 Correlation Interpolation Algorithm

The correlation interpolation or cross-correlation model (CCM) technique was introduced by the Maastricht group (de Jong et al 1990, 1991, Hoeks et al 1993) as a method for studying the motion of solid tissue, and particularly the left ventricular wall. It has the advantage of improving the speed of processing and lowering the hardware requirements for finding velocity from the maximum of the cross-correlation function. The technique works by calculating the 2-D cross-correlation at five points in the vicinity of the correlation maximum, and using an interpolation algorithm to estimate the true maximum. In the case of a narrow-band signal, the 2-D cross-correlation $R(r, \tau)$, where r is depth (corresponding to 'fast time') and τ is time (corresponding to 'slow time'), can be modelled as (de Jong et al, 1990):

$$R(r, \tau) = 2\beta S_0 \cos(2\pi f_t [r - v\tau]) \quad 11.42$$

where β is bandwidth, S_0 power density, f_t centre frequency, and v velocity. $\hat{R}(r, \tau)$ is evaluated at the points $\hat{R}(0, 0)$, $\hat{R}(0, T)$, $\hat{R}(t_s, 0)$, $\hat{R}(t_s, T)$ and $\hat{R}(-t_s, T)$ (where t_s and T are the sampling intervals in depth and time respectively), and the results used to determine the model parameters and hence velocity. The estimate for mean velocity can then be found from (Hoeks et al, 1993):

$$\hat{\bar{v}} = \frac{c \cdot prf}{2f_s} \times \frac{\arctan 2\left[\hat{R}', \hat{R}(0, T)\sin\left\{\arccos\left(\frac{\hat{R}(t_s, 0)}{\hat{R}(0, 0)}\right)\right\}\right]}{\arccos\left(\frac{\hat{R}(t_s, 0)}{\hat{R}(0, 0)}\right)}$$

$$11.43$$

where

$$\hat{R}' = \frac{\hat{R}(t_s, T) - \hat{R}(-t_s, T)}{2} \quad 11.44$$

de Jong et al (1991) tested the method with a rotating agar dish used as a phantom of the heart, and concluded that the accuracy of the displacement measured was sufficient to determine wall thickening in the heart. The method does appear to have disadvantages when compared with conventional cross-correlation methods, in that it works best with narrow-band signals (which of course implies poorer axial resolution) and is subject to aliasing in the same way as the standard Doppler phase domain method. Nevertheless, it is reported to perform better than the standard autocorrelation technique over a wide range of conditions (Hoeks et al 1993). Hoeks et al (1994a) have discussed the effects of using a reduced number of sample points over depth on the bias and standard deviation of the CCM method.

Brands et al (1997) examined the CCM technique further, and introduced what they termed the 'RF complex cross-correlation model' (C3M), which is based on cross-correlation of the complex rather than the real signal. The new estimate of the mean velocity they derived may be written:

$$\hat{v} = \frac{c \cdot prf}{2f_s} \frac{\arg(\hat{R}(0, T))}{\arg(\hat{R}(t_s, 0))} \quad 11.45$$

$$= \frac{c}{2} \frac{\frac{1}{2\pi T} \arctan\left\{\frac{\text{Im}[\hat{R}(0, T)]}{\text{Re}[\hat{R}(0, T)]}\right\}}{\frac{1}{2\pi t_s} \arctan\left\{\frac{\text{Im}[\hat{R}(t_s, 0)]}{\text{Re}[\hat{R}(t_s, 0)]}\right\}} \quad 11.46$$

which is of course identical to eqn. 11.27, the equation derived by Loupas et al (1995b) which they called the 2-D autocorrelation estimator. Brands et al (1997) acknowledge that the results are identical to those of Loupas et al, but assert that their method of derivation has the advantage of also providing expressions for both the signal-to-noise ratio and the spectral bandwidth of the received RF signals (which they show can be written in terms of $\hat{R}(0,0)$ $\hat{R}(0,T)$ and $\hat{R}(t_s, 0)$).

Brands et al (1997) compared the performance of the CCM and the C3M methods, in terms of the effects of SNR, bandwidth and sample frequency, by means of numerical simulations. Under the conditions tested, the C3M estimator was shown to offer the better performance in terms of exhibiting no bias and a low standard deviation of the estimate.

11.4.2.2 Maximum Likelihood Estimators

Each of the velocity estimation algorithms described so far has been based on either the measurement of the rate of change of phase of the signals originating from a given sample gate, or the change of the round trip time to a particular group of scatterers in the inter-pulse interval. The maximum likelihood estimators (MLEs), introduced by Ferrara and Algazi (1990, 1991a,b, 1992, Ferrara et al 1996) combine these two separate approaches by correlating the complex envelope of the received signal with a model of the complex envelope of the received signal that describes the changes that occur in the signal due to the motion of the scatterers. Rather than searching over all delays, as is the case for cross-correlation, a search is conducted over all velocities. The received signals are converted to baseband and then, for each candidate velocity, a correlation is performed over a series of short windows along a trajectory determined by the candidate velocity. For each window the phase of the signal is adjusted to allow for the phase change predicted by the model (Fig. 11.10).

In general, it is possible to write the likelihood function of velocity as:

$$\hat{l}(v) = \int_{t_1}^{t_2} \int_{t_1}^{t_2} r'^*(t) h'(t, u; v) r'(u) \, dt \, du \quad 11.47$$

where t_1 and t_2 are the initial and final times over which the estimate is computed, $r'( )$ is the complex envelope of the received signal, $h'(t, u; v)$ is a linear filter dependent on the two

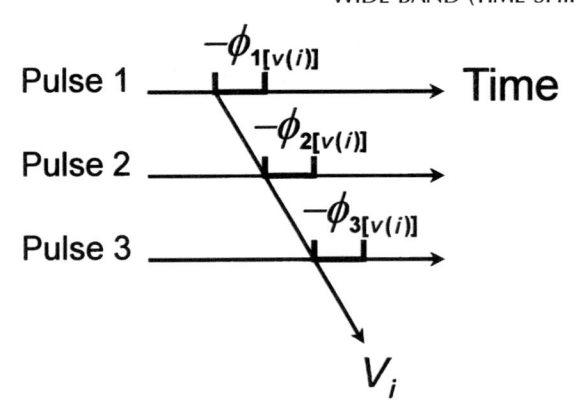

Figure 11.10 Illustration of the wideband maximum likelihood method. For each candidate velocity, v_i, a correlation is performed over a series of short windows along a trajectory determined by the candidate velocity. In each case the phase of the signal in each window is adjusted in accordance with the phase change predicted by the model

time variables, t and u, and * denotes complex conjugation. The filter $h'(t, u; v)$ could in principle take many forms, and Ferrara and Algazi (1991a) describe two suitable models. The first, 'the wideband range spread MLE', describes the returned signal from wide-band illumination of a range and velocity spread target ('a doubly spread target'). The second simpler estimator, 'the wide-band point MLE', is based on a 'slowly fluctuating point target'. Ferrara and Algazi also make the point that if an estimate of tissue type and depth can be made, it would be possible to include the effects of frequency-dependent attenuation and scattering in the filters.

For the wide-band point MLE, the likelihood of a velocity, v_i, (calculated over k pulses), is given by:

$$\hat{l}(v_i) = \left| \sum_k \int_{t_1}^{t_2} r'(t) s'^* \right.$$
$$\left. \times (t - d - kT[1 + 2v_i/c]) \exp[-j\eta v_i t] \, dt \right|^2$$

11.48

where $r'(t)$ is the complex envelope of the received signal, $s'(\cdot)$ is the model of the complex envelope of the received baseband signal, d is the transit time between the transducer and the index sample, T is the pulse repetition period, c is the ultrasound velocity, and $\eta = 2\pi f_c / c$ where f_c is the expected frequency of the returned signal.

The maximum likelihood axial velocity of the scatterers, $\hat{v}_{max}$, is then given by the velocity corresponding to the maximum of $\hat{l}(v)$. The mean velocity, $\hat{v}_{mean}$, can be calculated from the ratio of the first moment of the likelihood function to the zeroth moment, i.e:

$$\hat{v}_{mean} = \sum_i v_i \hat{l}(v_i) \bigg/ \sum_i \hat{l}(v_i) \qquad 11.49$$

Ferrara and colleagues have studied the properties and performance of various maximum likelihood estimators under various conditions using experimental data (Ferrara and Algazi 1990, 1991b, Ferrara et al 1996) and have concluded that the use of wide-band MLEs improve velocity estimation using a limited number of transmitted pulses, when compared with narrow-band estimators. Ferrara and DeAngelis (1997) suggest that the MLE technique has advantages over the cross-correlation technique in terms of a reduction in computing load. This presumably refers to the wide-band point MLE, as the wide-band range-spread MLE is considerably more complex.

One particular attraction of the MLE methods is the form of the output, i.e. likelihood versus velocity. The shape of the likelihood peak will depend on the velocity distribution within the sample volume. If all the scatterers within the measurement sample are similar, then the likelihood peak will be narrow and tall, whilst if there is significant velocity dispersion, the peaks will be wider and smaller. Figure 11.11 (taken from Ferrara and Algazi 1991b) shows a display of the wide-band point MLE calculated from relatively slow steady flow of Sephadex particles in a 7 mm diameter straight plastic tube. Potentially this type of display can highlight areas of disturbed flow. Ferrara et al (1995) have produced 2-D colour maps of the normalised likelihood function, with the goal of using the magnitude of the signal correlation to detect changes in velocity and shear rate which occurs in regions of disturbed flow.

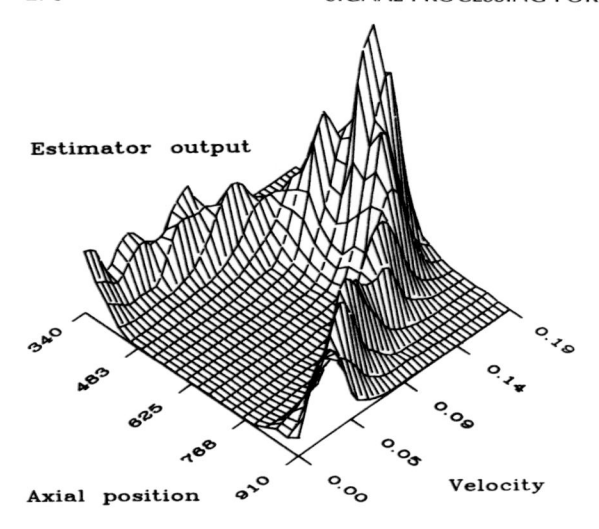

Figure 11.11 Three-dimensional display of wideband point MLE calculated from the steady flow of Sephadex particles in a 7 mm diameter straight plastic tube (reproduced by permission from Ferrara and Algazi 1991b, © IEEE, 1991).

11.4.2.3 Butterfly Search Method

Alam and Parker (1995, 1996) have described an estimator which, like the MLE estimators, is based on tracking echoes from a group of moving scatterers in two dimensions (slow time and fast time—denoted by n and t), by performing searches along trajectories that describe lines of constant velocity. The time domain version of this method is illustrated in Fig. 11.12a, which shows a series of echo envelopes from a single moving reflector. As the reflector moves, the echoes are time-shifted. If a series of searches are carried out along the butterfly lines (each of which corresponds to a particular velocity), then on the line corresponding to the correct velocity, all the data samples will have the same value and their variance will be zero. In practice, there will be noise present and the variance will not be zero, and therefore to find the correct butterfly line it is necessary to search for minimum variance. A similar technique can also be applied to the RF signal, but for demodulated quadrature components a slightly different approach is needed, which combines aspects of time domain and frequency domain analyses. The concept as enunciated by Alam and Parker is that for a single target moving at a constant velocity, the quadrature components sampled along the correct butterfly line will be a single-frequency sinusoid, and the frequency of that sinusoid will be determined by the target velocity (Fig. 11.12b). Therefore, when the search is made along each butterfly line, the result is checked for the unique frequency that the sampled complex envelope would have if the target had a velocity corresponding to that butterfly line. With noise present, the appropriate frequency would still be expected to have the maximum energy.

Alam and Parker (1995) show that the most likely velocity, $\hat{v}$, and the mean velocity, $\hat{v}_{\mathrm{mean}}$, can be written:

$$\hat{v} = \max_{v} \{\hat{L}(v)\} \qquad 11.50$$

and

$$\hat{v}_{\mathrm{mean}} = \sum_{i} v_{i} \hat{L}(v_{i}) \bigg/ \sum_{i} \hat{L}(v_{i}) \qquad 11.51$$

respectively, where $\hat{L}(v)$ is given by:

$$\hat{L}(v) = \frac{\left| \sum_{n} \tilde{r}_{Bv}[n] \exp[j4\pi f_c n(v/c)T] \right|^2}{\sum_{n} |\tilde{r}_{Bv}[n]|^2},$$

11.52

and f_c is the ultrasound centre frequency, n the slow time index, v the candidate velocity, c the velocity of ultrasound, T the pulse repetition period, and $\tilde{r}_{Bv}[n]$ denotes the re-sampling of the complex envelope along the butterfly line for velocity v, to estimate the velocity at depth z, which may be written:

$$\tilde{r}_{Bv}[n] = \tilde{r}(n, t)\delta\left(t - 2\frac{z}{c} + 2n\frac{v_i}{c}T\right) \qquad 11.53$$

where $\tilde{r}(n, t)$ is the complex envelope of the demodulated signal, which is a function of slow time and fast time. Interestingly, it can be shown

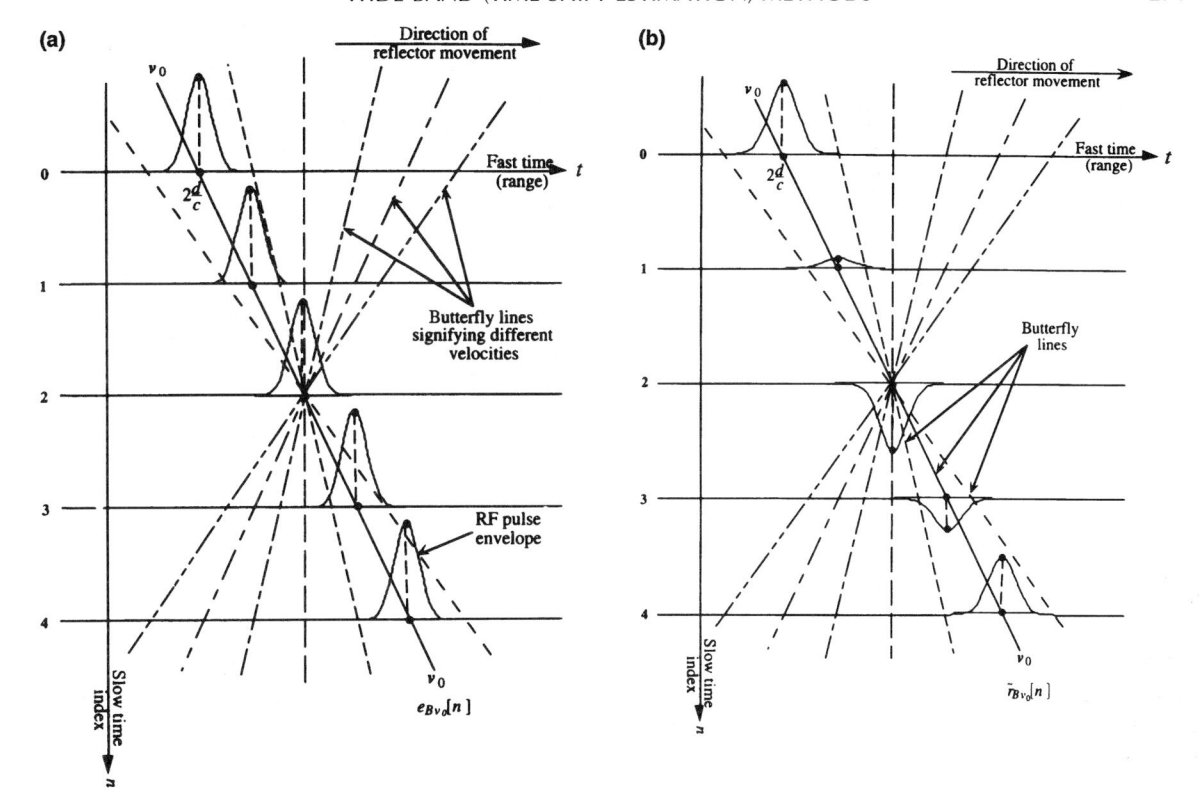

Figure 11.12 (a) Illustration of butterfly search on a series of echo envelopes from a single moving reflector. Only on the solid line will all the samples have the same value. (b) Illustration of butterfly technique applied on quadrature components. Only the solid line would sample a constant amplitude sinusoidal function $\tilde{r}_{Bv_0}[n]$ (reproduced by permission of Elsevier Science, from Alam and Parker 1995, *Ultrasound in Medicine and Biology*, © World Federation of Ultrasound in Medicine and Biology)

that the wide-band point MLE derived by Ferrara and Algazi (1991a) and given by eqn. 11.48 reduces to eqn. 11.52 if a number of simplifying assumptions and modifications are made (Alam and Parker, 1995).

Alam and Parker (1995) stated that the butterfly search on quadrature components shows good noise and aliasing immunity, with relatively few (down to three) successive sample lines, and was found to outperform both the standard autocorrelation and cross-correlation algorithms in simulations and phantom experiments with strong noise. In their 1996 paper, Alam and Parker described modifications (whereby a pre-multiplication operator or a Hilbert transform operator is used) that significantly reduce the computational complexity of butterfly searches on quadrature components, and thus reduce the hardware requirements and processing time necessary to implement the technique.

11.4.3 Cross-correlation—Analysis of Performance

There have been several theoretical studies of the properties of the cross-correlation method as a method measuring blood flow velocity, but unfortunately the individual results of these studies do not seem to be entirely compatible. Indeed, the substitution of numerical values into the various derived equations leads to a wide range of solutions for quantities such as measurement precision. This is presumably as a result of the validity of the simplifying assumptions made in each derivation. Some of the theoretical expressions that have been derived are presented here, without comment on their merits, for the interested reader to pursue in more depth.

Bonnefous (1989) derived an expression for the maximum value of the normalised cross-correlation function, $R_{12(\max)}$, which occurs at delay $\tau = \tau_m$ and may be written:

$$R_{12(\max)} = \left(1 - \frac{\overline{D}}{BW}\right)(1 - 2\pi^2 f_t^2 \sigma_{[\tau_m]}^2) \quad 11.54$$

where $\overline{D}$ is the mean transverse displacement of the index group of scatterers between ultrasound pulses, BW the ultrasound beam width, f_t the transmitted ultrasound frequency, and $\sigma_{[\tau_m]}^2$ the variance of τ_m caused by the velocity dispersion within the measurement sample volume. As Bonnefous pointed out, both the transverse displacement and the dispersion of the velocity distribution cause a decrease in the correlation peak, and unfortunately there is no way to distinguish between these two effects. Nevertheless, the size of the correlation peak can be used as a measure of disturbed flow, as demonstrated by Bonnefous using recordings from healthy and diseased carotid arteries.

Ferrara and Algazi (1994a, 1994b) have published two interesting studies which demonstrate the effects of the transverse velocity and velocity dispersion on the normalised correlation coefficient. In the first of these studies, they derived an expression for the autocorrelation of the returned RF signal from laminar blood flow in terms of the ultrasound beam shape, blood velocity distribution, signal properties and number of pulse repetition periods between the correlations, and evaluated it numerically under a number of conditions. Figure 11.13(a)–(d) are all taken from Ferrara and Algazi (1994a) and show a number of important effects (see figure caption for parameter details). Figures (a)–(c) were all derived using flows with a single constant velocity (0.45, 4.0, and 21.3 cm s^{-1}, respectively) and it can be seen that, as the velocity increases, two separate effects occur: firstly, of course, the correlation peak moves more rapidly in each interpulse interval; and secondly, because the angle between the ultrasound beam and the blood vessel is non-zero (45°), blood from the original sample volume moves out of the ultrasound beam more rapidly, and the maximum correlation as a function of time decays more rapidly. Note that for the slow flow rate, the signal correlation remains very high for the 200 pulse intervals shown, and Ferrara and Algazi quote a value of 0.93 after a lag of 199 pulses. Comparable figures for 4.0 and 21.3 cm s^{-1} are quoted as 0.63 after 49 pulses and 0.23 after 23 pulses, respectively. Figure (d) is derived using a velocity profile containing a range of velocities from 1 to 9 cm s^{-1} and should be compared with (b), which is calculated using the same mean velocity. In this case the normalised correlation function decreases much more rapidly because of the velocity dispersion. Ferrara and Algazi concluded that when a significant range of velocity components are present in a sample volume, it is that range that is the limiting factor in the length of the correlated signal interval, and therefore the use of wide-band signals, which reduce the sample volume, produces a signal that may be correlated over a longer time period. In a separate study, Ferrara (1995) discussed the effect of the beam-vessel angle on the received acoustic signal from blood in the context of exploring the feasibility of algorithms to determine the three-dimensional velocity magnitude from the received ultrasonic blood echoes from a single line of sight. As for previous studies, it was shown that the two major effects limiting the correlated signal interval are the spread of axial velocities within the sample volume and the transit time across the lateral beam width.

In their second study, Ferrara and Algazi (1994b) present a theoretical and experimental analysis of the received signal from disturbed blood flow in stenotic flow phantoms. As would be predicted, during systole there was a dramatic decrease in the correlation function in comparison with that found in laminar flow, although in some flow regions during diastole the correlation actually increased, due to a decrease in both mean velocity and velocity spread.

One of the earliest, and possibly the most widely cited, detailed analyses of the time domain correlation flowmeter appears to be that of Embree (Embree 1986, Foster et al 1990). This analysis is based on classical radar theory, in which

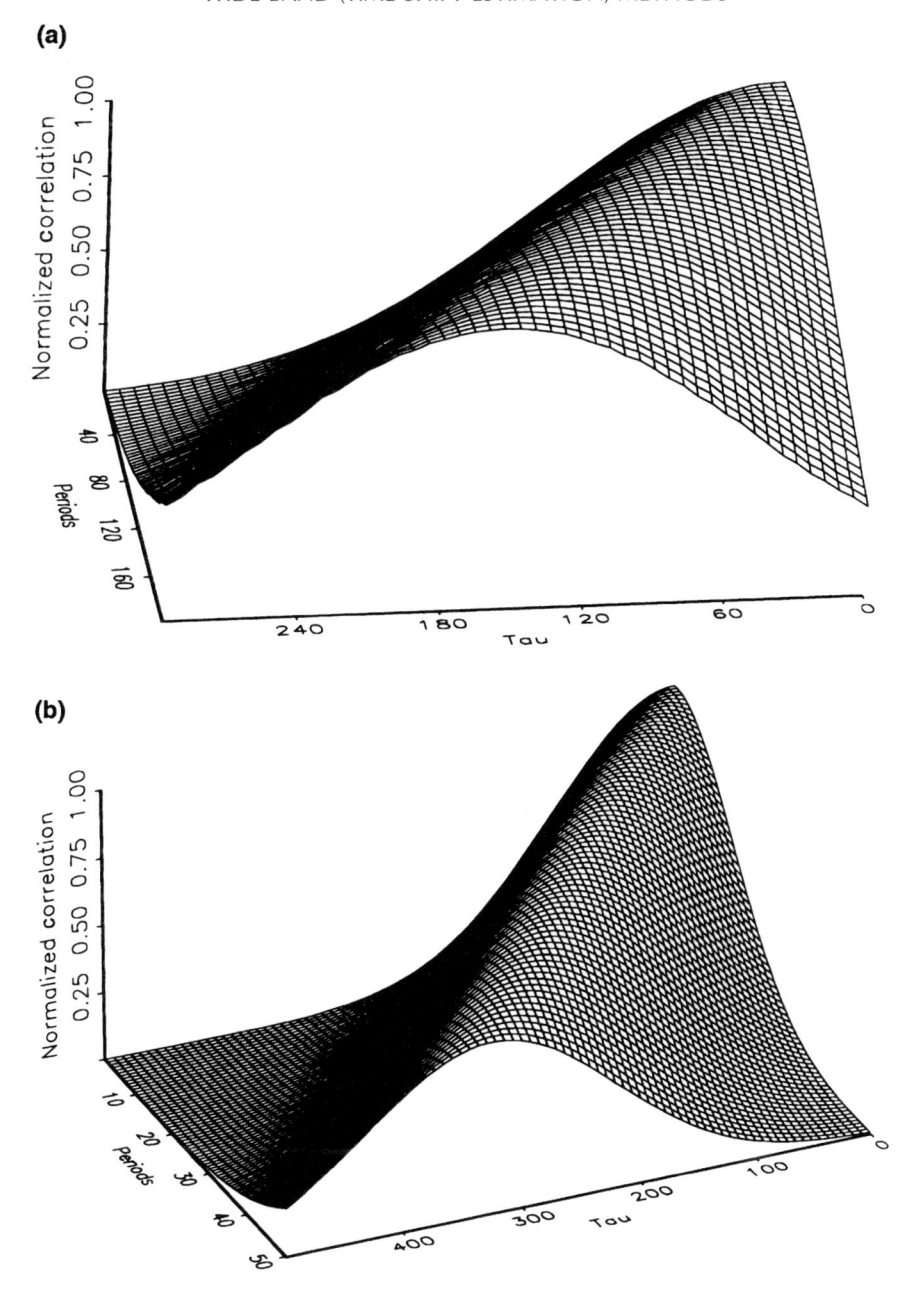

Figure 11.13 Normalised correlation functions plotted against time delay τ (ms) and inter-pulse period for laminar flow under different flow conditions. In each case the centre frequency used was 7.5 MHz, the pulse repetition frequency 8013 Hz, the envelope of the transmitted signal was Gaussian with a standard deviation of 75 ns, the acoustic velocity was 1545 ms^{-1}, the lateral beam modulation function was Gaussian with a standard deviation of 0.3 mm, the beam to vessel angle was 45°, and the correlation window was 300 ns. (a) Constant velocity of 0.45 cm s^{-1}. (b) Constant velocity of 4 cm s^{-1}. (c) Constant velocity of 21.3 cm s^{-1}. (d) Range of velocities from 1 to 9 cm s^{-1} within the sample volume (reproduced by permission from Ferrara and Algazi 1994a, © IEEE, 1994)

it is assumed that the signal level is large compared with the noise level, and that the window length is large compared with the wavelength. It is also assumed that the additive white noise present in the two signals causes a small deviation (much less than a wavelength) in the time delay estimate. Based on these assumptions, Foster et al wrote the precision of the time domain correlation velocity

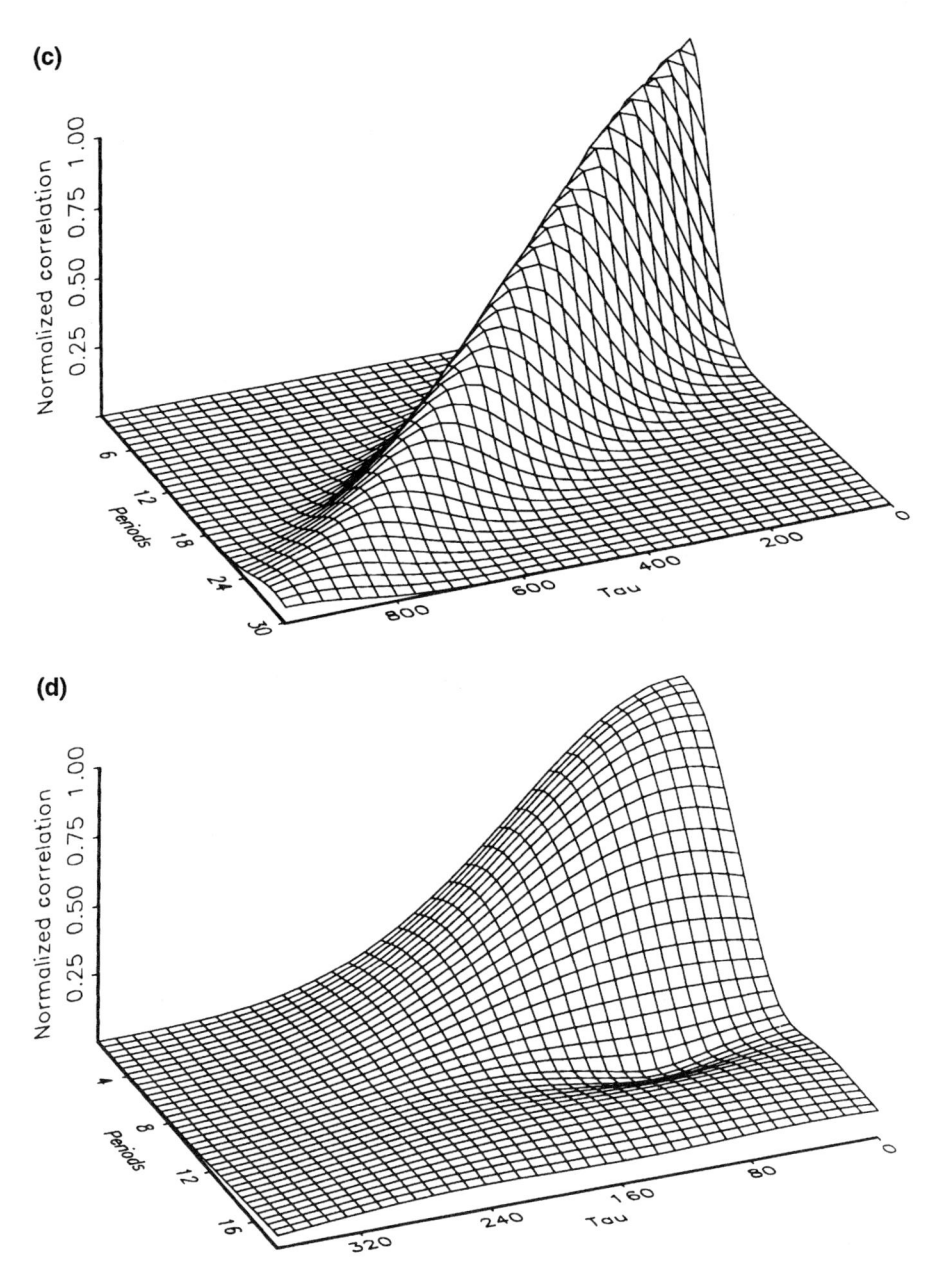

Figure 11.13 *Continued*

estimator as:

$$\text{Precision}[\hat{\tau}_m] = \frac{\sigma_{\hat{\tau}_m}}{\tau_m} = \frac{\sqrt{2}}{2\pi \beta_{rms} R_{12(max)} \tau_m SNR} \quad 11.55$$

where τ_m is the true time delay, $\sigma_{\hat{\tau}_m}$ the variance of the estimate, β_{rms} the RMS bandwidth of the received echoes, $R_{12(max)}$ the maximum correlation coefficient of the echoes without noise, and SNR is the signal to noise ratio. From eqn. 11.55, the precision of the estimate is seen to be inversely proportional to the product of the SNR and bandwidth. It is also dependent on the time delay, but not in an entirely simple way, because $R_{12(max)}$ is also dependent on τ_m. This is because, as the time that elapses between successive interrogating pulses becomes greater (and thus the greater the value of τ_m for a given velocity), the greater the number of scatterers that enter and leave the sample volume and, therefore, the smaller the value of the maximum correlation. Thus, if the PRF is too low, the precision is poor because $R_{12(max)}$ is low; if the PRF is too high, the precision is also poor because τ_m is too small.

Foster went on to write an equation for $R_{12(max)}$, based on theoretical results taken from Foster (1985) and from experimental studies, which may be written:

$$R_{12(max)} = \left(1 - 1.2 \frac{D}{BW_{3dB}}\right)$$

$$\text{for} \left(0 < \frac{D}{BW_{3dB}} < 0.6\right) \quad 11.56$$

where D is the transverse displacement of the index group of scatterers between ultrasound pulses, and BW_{3dB} the 3 dB beamwidth (cf. eqn. 11.54). Noting that D may be written $vT\sin\theta$ where v is the (uniform) velocity of the targets towards the transducer, and θ is the usual Doppler angle, and substituting for vT from eqn. 11.36, leads to:

$$R_{12(max)} = \left(1 - 0.6\left\{\frac{\tau_m c \tan\theta}{BW_{3dB}}\right\}\right) \quad 11.57$$

Substituting eqn. 11.57 into eqn. 11.55 leads to an expression for precision in terms of time delay τ_m, i.e:

$$\frac{\sigma_{\hat{\tau}_m}}{\tau_m} = $$

$$\times \left[\sqrt{2}\,\pi\beta_{rms}\tau_m SNR\left(1 - 0.6\left\{\frac{\tau_m c \tan\theta}{BW_{3dB}}\right\}\right)\right]^{-1} \quad 11.58$$

Figure 11.14 illustrates this relationship for the system evaluated by Foster et al (1990) and Hein and O'Brien (1993a), for which the SNR was 20 dB, the RMS bandwidth 2.5 MHz and the beamwidth 0.6 mm. The precision remains good for much higher values of τ_m when the angle is relatively low because the transverse motion between pulses is less. Experimental results obtained by Embree and O'Brien (1990) are in general agreement with these theoretical results.

Foster et al (1990) also considered errors associated with windowing, those associated with scatterers moving at different velocities within the range cell, those associated with the finite duration of the system impulse response, and those associated with the intensity profile across the ultrasound beam. In the case of the errors associated with windowing, Foster et al reported the results of a computer simulation that was used to determine the bias and variance of the estimate $\hat{\tau}_m$ for two identical echoes with varying window lengths. The estimate was unbiased for all window lengths, and the variance decreased with window length—in practice, of course, two returning echoes are not identical and will become less so with increasing window length. To study errors associated with scatterers moving with different velocities, Foster et al used a computer simulation and presented the results graphically, and made the point that the worst case errors occur near the vessel wall, where the shear rate is greatest, and smallest errors occur in the centre of the vessel, where the opposite is true.

Foster at al (1990) concluded that the bias of time domain correlation measurements is directly related to the size of the range cell, and that the precision is improved by use of a narrower

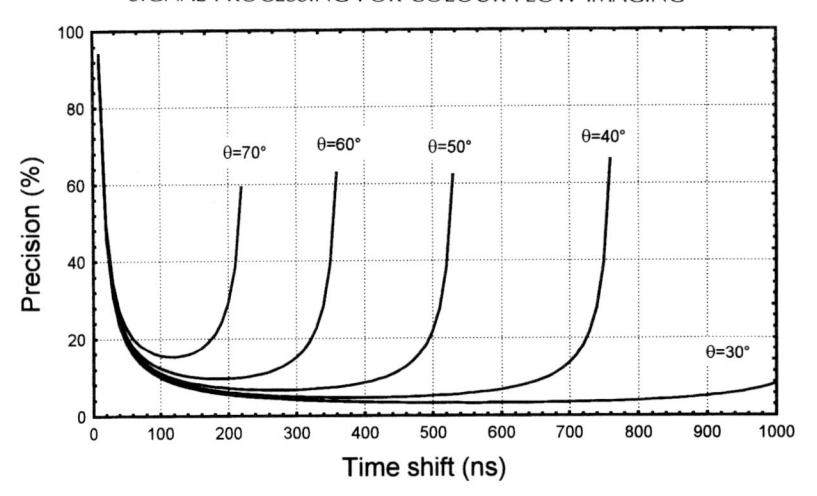

Figure 11.14 The theoretical precision versus time shift τ_m of the time-domain correlation system, as described by Foster et al (1990). SNR is 20 dB, RMS bandwidth 2.5 MHz and beamwidth 0.6 mm

beamwidth, a longer window length and a smaller velocity gradient across the range cell, but that clearly these three contributions are not independent of each other.

Another analysis of the variance of the peak location of the cross-correlation function was given by Jensen (1994) based on Bendat and Piersol (1986). This analysis differed from that of Foster et al (1990) in that it did not make the assumption of more or less negligible noise. Jensen's result may be written:

$$\sigma_{\hat{\tau}_m} \approx \frac{0.93}{\pi\beta} \{\varepsilon[\hat{R}_{12(max)}]\}^{1/2} \quad 11.59$$

where β is bandwidth, and $\varepsilon[\bullet]$ is the error in $\hat{R}_{12(max)}$, which can be written:

$$\varepsilon[\hat{R}_{12(max)}] = \frac{1}{\sqrt{\beta T_W}} \sqrt{1 + \frac{1}{SNR^2} + \frac{1}{2SNR^4}} \quad 11.60$$

T_W is the window length of the two samples, and SNR the signal-to-noise ratio. Substitution of practical values into eqns 11.59 and 11.60 yields rather high values for $\sigma_{\hat{\tau}_m}$ which would suggest that there is little point in interpolating to find the delay corresponding to the maximum correlation.

In the example quoted by Jensen, a bandwidth of 5 MHz, a SNR of 10, and a window length of 1.6 μs lead to a value for $\sigma_{\hat{\tau}_m}$ of 35 ns. Jensen, however, went on to report that simulations have in fact revealed that eqn. 11.59 yields a rather pessimistic value for the variance, and that in practise the variance is sufficiently small to make interpolation worthwhile.

Walker and Trahey (1995) have taken a slightly different approach by deriving the minimum error achievable by any unbiased delay estimation algorithm including, but not limited to, cross-correlation. This is achieved by the application of the Cramér–Rao Lower Bound to derive an analytical expression which predicts the magnitude of jitter errors incurred when estimating delays using radio frequency data from speckle targets. Their expression may be written:

$$\sigma_{\hat{\tau}_m} \geq \sqrt{\frac{3}{2f_t^3\pi^2 T_W(\beta_{fract}^3 + 12\beta_{fract})} \times \left(\frac{1}{R_{12(max)}^2}\left(1 + \frac{1}{SNR^2}\right)^2 - 1\right)} \quad 11.61$$

where f_t is the centre frequency, T_W the window length, and β_{fract} the fractional bandwidth ($=\beta/f_t$).

This expression predicts the jitter magnitude incurred when aligning broad band signals with flat power spectra that have been corrupted by electronic noise and decorrelated by physical processes. It assumes that signal decorrelation occurs uniformly over the signal bandwidth. Walker and Trahey performed a large number of numerical simulations which confirmed the validity of eqn. 11.61 under a wide range of conditions. They also quote an example of the lower bound of the standard deviation of jitter for a typical blood flow measurement scenario. Values of $f_t = 5$ MHz, $\beta_{fract} = 50\%$, $SNR = 0$ dB, $R_{12(max)} = 0.98$ and $T_W = 0.65$ µs leads to a standard deviation of 31.1 ns.

Céspedes et al (1997) discussed the combined effect of signal decorrelation and random noise on the variance of the time delay estimate, and derived an expression for the Cramér–Rao Lower Bound in terms of SNR and an 'equivalent SNR', SNR_{equiv}, due to the decorrelation of the signal, which may be written:

$$\sigma_{\hat{\tau}_m} \geq \sqrt{\frac{3}{f_t{}^3 \pi^2 T_W(\beta_{fract}{}^3 + 12\beta_{fract})} \times \left(\frac{1}{SNR_{equiv}} + \frac{1}{SNR}\right)}$$

11.62

They also showed that SNR_{equiv} can be written:

$$SNR_{equiv} = \frac{R_{12(max)}}{1 - R_{12(max)}} \qquad 11.63$$

which, if rearranged and substituted into eqn. 11.61, leads directly to eqn. 11.62 (assuming both SNR and SNR_{equiv} are much greater than unity).

Friemel et al (1998) have explored 'speckle decorrelation' due to 2-D flow gradients, which they point out have significant implications for all Doppler imaging devices, whatever their mode of operation. They show that speckle decorrelation due to flow gradients reaches a maximum at a Doppler angle of $0°$, the best angle for conventional Doppler estimation. Decorrelation due to flow gradients reaches a minimum at $90°$, where conventional Doppler devices are ineffective. They conclude that this suggests there is an optimum angle for Doppler imaging at which the contribution to the variance of the Doppler estimate due to flow gradients and non-zero beam-vessel angle reaches a minimum.

11.4.4 Implementation of Cross-correlation-based Systems

Although many correlation-based algorithms have been described in the literature, few colour flow imaging systems based on correlation principles appear to have been built in practice. The general layout of a cross-correlation system is shown in Fig. 4.22, and has been briefly discussed in Section 4.5.3.2. For most of the algorithms discussed in 11.4.2, it is necessary to digitise the raw RF data, which requires a much higher rate of digitisation than that for the demodulated signals used in the classical autocorrelation method. Whilst this was at one time extremely demanding, analogue-to-digital converters have steadily increased in speed over the years, and many high-end ultrasound machines now digitise the RF signal prior to beam forming, which means that the appropriate digitised signal should already be available. Unfortunately, however, correlation methods are extremely calculation-intensive, both because of the increased number of samples that must be handled (of the order of at least 10–20 times more than for the autocorrelation technique) and also because of the more complicated and time-consuming estimator structures.

Jensen (1996) has estimated the approximate number of operations (additions and multiplications) required to be performed by scanners based on the autocorrelation and on the cross-correlation approaches to frequency estimation. In the case of the autocorrelator, for a system using 10 lines per estimate, an echo cancelling filter with four coefficients and an eight period 3 MHz pulse, he calculated approximately 2×10^6 operations per second. In the case of the cross-correlator, for a system using 1 mm range gates, eight lines per estimate, a fourth order echo cancelling filter and a sampling frequency of 20 MHz, he calculated approximately 600×10^6 operations per second.

Whilst the former is easily achievable using current signal processors, the latter is not, and system designers are forced to adopt alternative strategies, such as the one-bit correlator suggested by Foster (1985) and Bonnefous et al (1986), or the correlation interpolation approach introduced by de Jong et al (1990, 1991), if real-time operation is required.

As for phase-shift estimation techniques, it is necessary to filter out the clutter due to large stationary and near-stationary targets, and this is achieved in the same way for cross-correlation techniques. It should be noted, however, that filtering must be carried out on every sample in the range gate, which may be of the order of 20 as opposed to only two (direct and quadrature) samples for phase-shift estimators. This in itself imposes a high computational load, and explains the suggestion of the single fixed echo canceller by Bonnefous and Pesque (1986).

The specific use of clutter filters in conjunction with cross-correlation algorithms has been discussed by Jensen (1993c) and Brands et al (1995). Practical aspects of the implementation of time domain cross-correlation algorithms have been discussed by Jensen (1993a, 1996) and Hein et al 1993). Li (1995) has described a frequency domain autocorrelator for evaluating the time shift of range-gated RF signals over subsequent lines.

11.4.5 Theoretical Advantages and Disadvantages of Wide-band/Time Domain Techniques

As mentioned in the introduction to this chapter, it is no longer possible to classify all ultrasonic velocity estimation techniques as either wide-band time-delay estimation-based algorithms or narrow-band phase-delay estimation-based algorithms. By considering the advantages and disadvantages of the simple cross-correlation estimator (CCE) as compared to the simple autocorrelation estimator (ACE), however, it is possible to understand the incentive there has been to derive new algorithms that combine the advantages of both approaches. As has already been explained, the essential difference between the two techniques is that in the former case, wide-band (short duration) pulses are transmitted into the tissue, and the mean displacement of an index group of scatterers is estimated in terms of the change of the round trip time of the ultrasound pulse from the transducer; whilst in the latter case, narrow-band (relatively long) pulses are transmitted, and the mean displacement of the scatterers is estimated in terms of the change in phase at the carrier frequency (assumed to be the same as the transmitted frequency). The difference in performance of the two approaches is a result of both the bandwidth of the transmitted signals and the method of estimating the target displacement.

11.4.5.1 Variance of Velocity Estimates

As pointed out by Bonnefous et al (1986), an inherent problem of the phase domain approach to velocity estimation is that random fluctuations in the spacing of the scatterers in the sample volume leads to random RF frequency fluctuations in the returned ultrasound signal, which in turn cause similar fluctuations in the Doppler signal. This problem should not affect the time domain approach because it is changes in delay rather than phase that are measured. As a result of this, Bonnefous and colleagues (Bonnefous et al 1986, Bonnefous 1992) claim that good velocity estimates can be obtained from fewer transmission cycles using a CCE approach when compared to an ACE approach.

This concept is to some extent challenged by Torp et al (1993) and Hoeks et al (1994b), both of whom (independently) argue that the variance of the ACE is large in comparison to that of the CCE because insufficient radial averaging is performed in the former case. Torp et al carried out radial averaging on experimental data and found that, whilst the CCE method was significantly better than the ACE method for high bandwidths, the difference for low bandwidths was minor. They argued that under poor SNR conditions, it is necessary to reduce the bandwidth anyway in order to increase sensitivity, in which case the two methods have comparable performances, provided that equal amounts of radial averaging are performed. Hoeks et al, using a computer simulation, demonstrated that their subsample volume processing method, which involves splitting the observation window into sub-samples with a length of one period of the transmitted frequency and averaging the separate autocorrelation estimates, reduced the variance of the velocity estimate to that expected of the CCE method. Both Torp's and Hoeks' approaches are discussed in more detail in Section 11.3.4.

The 2-D autocorrelation approach introduced by Loupas et al (1995b)—(see Section 11.3.5), combines the advantages of radial averaging, with compensation for any random RF frequency fluctuations that might in turn give rise to random Doppler frequency fluctuations, by performing an explicit estimation of the mean RF frequency of the data within each range gate. Loupas et al directly compared the performance of the 2-D autocorrelator with the CCE method, and reported that 'it was repeatedly observed that they were almost identical under low-noise conditions, but that the 2-D autocorrelator was shown to offer noticeably better robustness in the presence of heavy noise'.

11.4.5.2 Bias of Velocity Estimates Due to Frequency-dependent Mechanisms

The effect of frequency-dependent attenuation and scattering on mean velocity estimates have been discussed in Section 8.3.5. Essentially, because higher frequencies are attenuated more rapidly than lower frequencies, the centre frequency of an ultrasound pulse reduces as it propagates through tissue, whilst because high frequencies are scattered more strongly than low frequencies, the centre frequency of a pulse increases when it is scattered by blood (Newhouse et al 1977, Ophir and Jaeger 1982, Ferrara et al 1992). As a result of these carrier frequency modifications, mean velocity estimates derived from phase delay measurements may be biased either up or down depending on whether scattering or attenuation is the dominant mechanism in a particular set of circumstances. Furthermore, the shorter the transmitted pulse, and therefore the wider the bandwidth, the greater will be the bias, and this is an important reason why the standard ACE method cannot be used with very short pulses. Cross-correlation techniques, on the other hand, because they measure time delay rather than phase change, are not influenced by the frequency-dependent mechanisms discussed above (Bonnefous et al 1986). Ferrara et al (1992) have explicitly shown that neither the CCE nor their own wide-band maximum likelihood estimator are biased by the effects of frequency dependent scattering and attenuation, whilst the ACE is.

The 2-D autocorrelation approach described by Loupas et al (1995a,b) should, in theory, overcome this problem because it explicitly estimates the mean received RF frequency rather than using the transmitted frequency.

11.4.5.3 Aliasing

A further theoretical advantage of the CCE over the ACE is that it should be immune to aliasing, since it is the time shift rather than the phase shift that is estimated (Bonnefous et al 1986). This may be the case when there is a strong signal and little noise, because the cross-correlation will have an unambiguous global maximum. Because of the limited bandwidth of the signal and its consequent periodic nature, however, under low SNR conditions and when successive A-lines are derived from slightly different scatterer paths, there may be ambiguity about which correlation maximum to use and thus the appearance of aliasing. When such aliasing does occur, it occurs at an identical maximum velocity in relation to the carrier frequency and the pulse repetition frequency as for the ACE method. Jensen (1993b) has discussed aliasing in the context of time delay-estimation techniques, and points out that one possible method for increasing the probability of correct detection is to lower the side lobe peaks in the cross-correlation function by using shorter pulses (which yields a narrower autocorrelation). However, as Jensen further points out, frequency-dependent attenuation will gradually lengthen the pulse as it propagates through tissue, and therefore even the shortest pulses will still give rise to false detections. Jensen then goes on to discuss possible strategies for avoiding false detections.

Of the more sophisticated algorithms described in this chapter, it should be noted that the extended autocorrelation estimator (Section 11.3.6) behaves more like the CCE than the 'ACE family' from which it is derived, in terms of aliasing, because it uses the magnitude of the correlation function to determine the correct candidate delay, whilst the correlation interpolation algorithm (Section

11.4.5.4 Range Resolution and Sensitivity

As with standard ultrasonic pulse-echo imaging techniques, the achievable range resolution and sensitivity of colour flow systems are intimately connected. In principle it is possible to obtain better spatial resolution using wide-band systems than narrow-band systems, because the former can be made to transmit very short pulses. On the other hand, wide-band systems are subject to considerably more receiver noise and, under poor SNR conditions which frequently occur in blood flow measurement situations, the transmitted pulse bandwidth must be decreased in order to be able to detect the blood signal (Torp et al 1993).

11.4.5.5 Implementation

The implementation of CCE systems has already been discussed in Section 11.4.4, where it was seen to be much more demanding in terms of hardware and computing power than the implementation of ACE systems. This has encouraged the development of a number of simpler methods of estimating the cross-correlation function, and several of these have been discussed in Section 11.4.2. It should, however, be noted that colour flow systems use the same transducers for both pulse-echo imaging and velocity imaging, and that pulse echo transducers are of necessity better suited to wide-band than narrow-band operation.

11.4.5.6 Safety

If wide-band and narrow-band pulses contain an equal amount of energy, then the peak negative pressure of the wide-band pulse will be greater than that of the narrow-band pulse. Whilst this may not usually represent a hazard, this difference should at least be considered when moving to wider bandwidths.

11.5 DIRECTION OF ARRIVAL METHODS

A new type of approach to velocity estimation for colour flow imaging, based on 'direction-of-arrival' (DOA) estimation techniques, as used in sonar and radar array processing, has been suggested by Allam and colleagues (Allam and Greenleaf 1996, Allam et al 1996) and Vaitkus and colleagues (Vaitkus and Cobbold 1998, Vaitkus et al 1998). The object of DOA processing is to determine the angle at which incoming narrow-band plane waves strike a (usually) linear array of equally spaced passive sensors. The problem is essentially one of determining the progressive shift in time of arrival of the wave-fronts at each array element, and is the same as determining the progressive increase or decrease in the round trip time for individual ultrasonic pulses as they are scattered by a given group of targets in medical applications. The equivalence of PW Doppler ultrasound velocity estimation and radar and sonar DOA estimation is illustrated in Fig. 11.15, whilst a list of analogous parameters for the two techniques is given in Table 11.1.

Allam and Greenleaf (1996) discussed a number of alternative DOA estimation techniques, including the minimum variance (MV) estimate (Capon 1969) and the multiple signal classification (MUSIC) algorithm (Schmidt 1986, Schmidt and Franks 1986, Porat and Friedlander 1988), which is based on the eigendecomposition of an estimate of the covariance matrix of the received signal. The principle behind the latter powerful method is that the covariance matrix of the observed signal is decomposed into two orthogonal subspaces on the

Table 11.1 List of analogous parameters in DOA estimation and Doppler ultrasound (modified from Allam and Greenleaf 1996)

Direction of arrival	Doppler ultrasound
Speed of propagation of signals	Speed of propagation of signals
Distance between array elements (D)	Pulse repetition period ($1/prf$)
Direction of arrival (unknown)	Velocity of scatterers (unknown)
Nyquist limit given by $D < \lambda_{(min)}/2$	Nyquist limit given by $prf > 2f_{d(max)}$

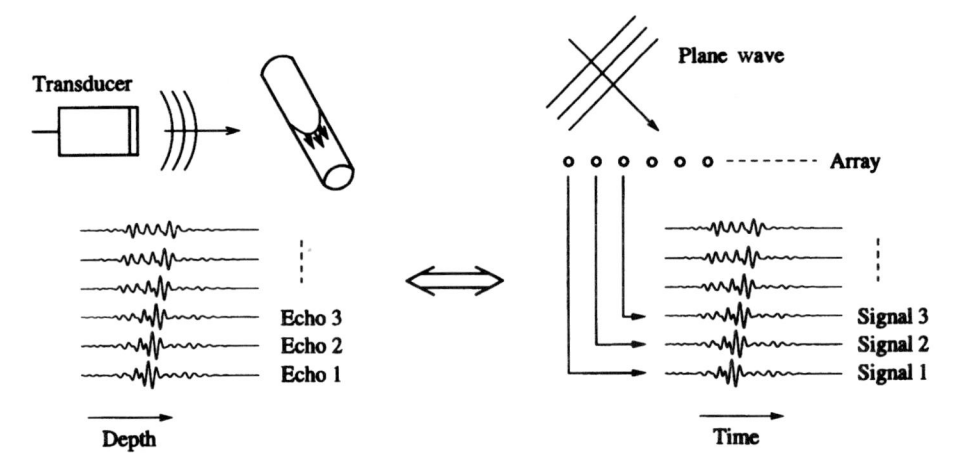

Figure 11.15 The equivalence of pulsed wave Doppler ultrasound velocity estimates and direction-of-arrival estimates (reproduced by permission from Allam and Greenleaf 1996, © IEEE, 1996)

basis of the relative sizes of its eigenvalues. The subspace spanned by the eigenvectors corresponding to the smallest eigenvalues is called the 'noise subspace', whilst its orthogonal compliment, spanned by the remaining eigenvectors, is called the 'signal (or more correctly the signal + noise) subspace'. The MUSIC algorithm computes a spatial spectrum (velocity spectrum in the Doppler case) from the noise subspace, and the DOAs (velocities) are determined from the dominant peaks of the spectrum.

Allam and Greenleaf point out that such methods cannot be applied directly to ultrasound signals where wide-band pulses are used, and consequently describe a number of techniques for wide-band to narrow-band conversion. Of these methods, the best appears to be that based on a radial projection of the signal in 2-D Fourier space onto a pre-selected projection frequency (see Section 8.2.3 for a discussion of 2-D Fourier transform methods).

One of the particular advantages of DOA-based techniques is their high spectral resolution, which means that they can be used to identify multiple velocities within the sample volume (subject to an adequate number of received pulses). An important further implication of this is that it is possible to use the technique as an alternative approach to clutter rejection, and Allam and Greenleaf (1996) point out that the eigenvector(s) corresponding to the clutter signal(s) can be identified (by virtue of the large size of the corresponding eigenvalue(s)) and included as part of the noise space rather than the signal subspace.

In a companion paper, Allam et al (1996) present experimental results using a combination of their 2-D projection method and both the MV estimator and the MUSIC estimator. They report experiments with moving string phantoms and flow rigs, which suggest that DOA techniques require fewer pulses than more conventional approaches to yield satisfactory results, and may therefore hold some promise in Doppler applications. On the negative side, DOA estimators are quite computationally demanding; the MV estimator requires inversion of the covariance matrix, whilst the MUSIC estimator requires eigendecomposition of the covariance matrix.

Vaitkus and colleagues (Vaitkus and Cobbold 1998, Vaitkus et al 1998) adopted a similar approach to that of Allam and colleagues, but used the root-MUSIC algorithm (Barabell 1983), which is similar to the (spectral) MUSIC algorithm, except that the DOAs are determined from the roots of a polynomial formed from the noise subspace. Unlike the MUSIC algorithm, which can be used with arbitrary sensor arrays, the root-MUSIC algorithm is only suitable for linear equally spaced arrays, which in the pulsed Doppler situation corresponds to a constant pulse repetition frequency, as is the normal case in colour flow mapping. Under these conditions, the root-MUSIC algorithm has been shown to be superior to the spectral-MUSIC algorithm on theoretical grounds (Rao and Hari 1989). In experiments with flow phantoms, Vaitkus and colleagues show the root-MUSIC algorithm to have a similar performance to the classical autocorrelation technique (Section 11.2.3), the time domain correlation technique (Section 11.4.1) and the wide-band maximum likelihood estimator (Section 11.4.2.2) at high (20 dB) SNRs, but to be superior at poor (0 dB and −3 dB) SNRs. Vaitkus and colleagues also draw attention to the advantage of using the root-MUSIC algorithm technique for clutter rejection, in that it eliminates the need not only for the filters themselves, but also the consequent need for filter initialisation techniques to reduce the biasing effects of the filter transients (Section 11.2.2.2).

11.6 FUTURE DIRECTIONS

There have been tremendous advances in colour flow imaging systems over the past decade, and there is no reason not to expect at least as dramatic improvements over the next decade.

There have been very significant advances in transducer technology in terms of matching, bandwidth, and efficiency, and it seems likely $1\frac{1}{2}$-D and even 2-D arrays will become commonplace in the next few years. Leading machines already digitise the RF signal from each active transducer element prior to beam-forming, and with the ever-increasing speed of analogue-to-digital converters, and signal processing chips, it seems inevitable that most or all colour flow systems will migrate in this direction. Filtering algorithms have shown dramatic improvements, but with increases in computing power it is likely they will become much better still.

It is amazing that most colour flow systems still estimate the mean velocity of flow using what is basically the algorithm described by Namekawa et al in 1982, and this will undoubtedly be an area of rapid progress in the near future. Many of the algorithms described in Sections 11.3 and 11.4 are very significantly better than the simple autocorrelation algorithm, and are capable of implementation using hardware already available.

Perhaps more fundamentally, 2-D and even 3-D vector Doppler and speckle tracking techniques, together with 3-D imaging, will become widespread. It also seems likely that solid tissue imaging, tissue perfusion measurements, and harmonic imaging and ultrasonic contrast agents, both together and separately, will become important in colour flow imaging. These developments, together with other emerging Doppler techniques, are considered further in Chapter 13.

11.7 SUMMARY

Signal processing for colour flow imaging has advanced dramatically over the past decade, but to some extent the pace of progress on commercial machines appears to have been restricted more by the availability of adequate computing power than by any lack of new ideas from workers in this field. Researchers have devised and implemented off-line many exciting algorithms, both for clutter rejection and velocity estimation, and it is likely that very significant improvements in both these areas will ensue as computing power increases.

The purpose of this chapter has been both to discuss currently used algorithms in some detail, and to survey other algorithms that have been described in the literature but not yet implemented on colour flow machines. By gathering these algorithms together, it is hoped to give the reader a taste of what is to come but, more importantly, to provide a common framework so that the various algorithms and their relationships with each other are more easily understood by the non-specialist in the area of colour flow signal processing.

11.8 NOTATION

BW	Beam width
BW_{3dB}	3 dB beam width
c	Velocity of ultrasound
C	Filter parameter
d	Transit time between transducer and index sample volume
$\overline{D}$	Mean transverse displacement of scatterers between ultrasound pulses

NOTATION

f_c	Centre frequency of received signal		samples between the ith segment of echo 1 and echo 2
f_d	Doppler shift frequency	$R_{12(\text{sgn})}(n, i_{\text{seg}})$	Normalised cross-correlation function for a displacement of n samples between the ith segment of echo 1 and echo 2, calculated using only sign information
f_{dem}	Frequency of demodulation sinusoid		
$f_{\max}(p)$	Maximum frequency derived from AR estimator of order p		
$f_{\text{mean}}(p)$	Mean frequency derived from AR estimator of order p	$s'(\ )$	Model of the complex envelope of a received baseband signal
f_{RF}	Received pulse RF frequency	S_0	Power density
f_s	Sampling frequency	SNR	Signal-to-noise ratio
f_t	Transmitted ultrasound frequency	SNR_{equiv}	Equivalent SNR due to signal decorrelation
$h'(t, u; v)$	Linear filter dependent on two time variables, t and u	t	Time
		t_s	Sampling interval in radial or depth direction
$I(i)$	Real or in-phase part of $Z(i)$		
$I(m, i)$	Real or in-phase part of $Z(m, i)$	T	Pulse repetition period
I_{gs}	Grey scale echo intensity	T_W	Window length
$l(v)$	Likelihood function of velocity	v	Velocity
N	Summation limit	$v_{\max}$	Maximum velocity
ND	Number of depths; number of sample volumes	v_{mean}	Mean velocity
		$\bar{v}_{\text{opt}}$	Optimal mean velocity
NP	Number of periods; number of sample volumes	v_q	Calculated velocity from two-pulses q cycles apart
N_s	Number of samples in a segment of an A-line	$V_{\text{in}}(t)$	Input to filter algorithm
		V_{mag}	Magnitude of calculated velocity
prf	Pulse repetition frequency	$V_{\text{out}}(t)$	Output from filter algorithm
$P(\omega)$	Power spectrum	$\overline{W}$	Average power
PL	Package length or ensemble length	W_q	Weighting function
		x	Axial distance moved by scatterers
$Q(i)$	Imaginary or quadrature part of $Z(i)$	$y_n[m]$	mth sample of echo number n
		z	depth
$Q(m, i)$	Imaginary or quadrature part of $Z(m, i)$	$Z(i)$	Observations of a complex process (i denotes pulse number)
r	Radial lag	$Z(m, i)$	Observations of a complex process (m denotes sample volume; i denotes pulse number)
$r'(\ )$	Complex envelope of received signal		
$\tilde{r}(n, t)$	Complex envelope of demodulated signal as a function of slow time and fast time	β	Bandwidth
		β_{fract}	Fractional bandwidth ($=\beta/f_t$)
$R(\tau)$	One-dimensional autocorrelation function	β_{rms}	RMS bandwidth
		η	$2\pi f_c/c$
$R(r, \tau)$	Two-dimensional autocorrelation function	θ	The Doppler angle
		σ^2	Variance
$R_{\text{dem}}(r, \tau)$	Two-dimensional autocorrelation function of demodulated signal	τ	Temporal lag or time shift
		τ_q	Change in transit time of two ultrasound pulses q cycles apart
$R_{12(\max)}$	Maximum value of normalised cross-correlation function	ϕ	Phase
$R_{12}(n, i_{\text{seg}})$	Normalised cross-correlation function for a displacement of n	ω	Angular frequency
		$\bar{\omega}$	Mean angular frequency

11.9 REFERENCES

Ahn YB, Park SB (1991) Estimation of mean frequency and variance of ultrasonic Doppler signal by using second-order autoregressive model. IEEE Trans Ultrason Ferroelec Freq Contr UFFC-38:172–182

Alam SK, Parker KJ (1995) The butterfly search technique for estimation of blood velocity. Ultrasound Med Biol 21:657–670

Alam SK, Parker KJ (1996) Reduction of computational complexity in the butterfly search technique. IEEE Trans Biomed Eng 43:723–733

Allam ME, Greenleaf JF (1996) Isomorphism between pulsed-wave Doppler ultrasound and direction-of-arrival estimation—Part I: Basic principles. IEEE Trans Ultrason Ferroelec Freq Contr 43:911–922

Allam ME, Kinnick RR, Greenleaf JF (1996) Isomorphism between pulsed-wave Doppler ultrasound and direction-of-arrival estimation—Part II: Experimental results. IEEE Trans Ultrason Ferroelec Freq Contr 43:923–935

Barabell AJ (1983) Improving the resolution performance of eigenstructure-based direction-finding algorithms. In: International Conference on Acoustics, Speech and Signal Processing, '83 Proceedings, Boston, MA, pp. 336–339. IEEE, Piscataway, NJ

Barber WD, Eberhard JW, Karr SG (1985) A new time domain technique for velocity measurements using Doppler ultrasound. IEEE Trans Biomed Eng BME-32:213–229

Bassini M, Dotti D, Gatti E, Pizzolati P, Svelto V (1979) An ultrasonic non-invasive blood flowmeter based on cross-correlation techniques. Proceedings of Ultrason International, 79, pp. 273–278

Bassini M, Gatti E, Longo T, Martinis G, Pignoli P, Pizzolati PL (1982) In vivo recordings of blood velocity profiles and studies in vitro of profile alterations induced by known stenoses. Texas Heart Inst J 9:185–194

Bendat JS, Piersol AG (1986) Random Data: Analysis and Measurement Procedures, 2nd edn, pp. 276–277. Wiley, Chichester

Bjærum S, Torp H (1997) Optimal adaptive clutter filtering in color flow imaging. In: Schneider SC, Levy M, McAvoy BR (eds) Proceedings of the 1997 IEEE Ultrasonics Symposium, pp. 1223–1226. IEEE, Piscataway, NJ

Bohs LN, Friemel BH, McDermott BA, Trahey GE (1993a) Real-time system for angle-independent US of blood flow in two dimensions: initial results. Radiology 186:259–261

Bohs LN, Friemel BH, McDermott BA, Trahey GE (1993b) A real time system for quantifying and displaying two-dimensional velocities using ultrasound. Ultrasound Med Biol 19:751–761

Bonnefous O (1989) Statistical analysis and time correlation processes applied to velocity measurement. In: Proceedings of the 1989 IEEE Ultrasonics Symposium, pp. 887–892. IEEE, Piscataway, NJ

Bonnefous O (1992) Time domain colour flow imaging: methods and benefits compared to Doppler. Acoustic Imaging 19:301–309

Bonnefous O, Pesque P (1986) Time domain formulation of pulse-Doppler ultrasound and blood velocity estimation by cross correlation. Ultrason Imaging 8:73–85

Bonnefous O, Pesque P, Bernard X (1986) A new velocity estimator for colour flow mapping. In: Proceedings of the 1986 Ultrasonics Symposium, pp. 855–860. IEEE, Piscataway, NJ

Brandestini M (1975) Application of the phase detection principle in a transcutaneous velocity profile meter. In: Kazner E, de Vlieger M, Muller HR, McCready VR (eds) Ultrasonics in Medicine. Proceedings of the 2nd European Congress on Ultrasonics in Medicine, Munich 12–16 May, 1975, pp. 144–152. Excerpta Medica, Amsterdam

Brandestini M (1978) Topoflow—a digital full range Doppler velocity meter. IEEE Trans Sonics Ultrason SU-25:287–293

Brandestini MA, Forster FK (1978) Blood flow imaging using a discrete-time frequency meter. In: Proceedings of the 1978 IEEE Ultrasonics Symposium, pp. 348–352. IEEE, Piscataway, NJ

Brands PJ, Hoeks APG (1992) A comparison method for mean frequency estimators for Doppler ultrasound. Ultrason Imaging 14:367–386

Brands PJ, Hoeks APG, Reneman RS (1995) The effect of echo suppression on the mean velocity estimation range of the RF cross-correlation model estimator. Ultrasound Med Biol 21:945–959

Brands PJ, Hoeks APG, Ledoux LAF, Reneman RS (1997) A radio frequency domain complex cross-correlation model to estimate blood flow velocity and tissue motion by means of ultrasound. Ultrasound Med Biol 23:911–920

Capon J (1969) High resolution frequency-wavenumber spectrum analysis. Proc IEEE 38:1408–1418

Céspedes I, Ophir J, Alam SK (1997) The combined effect of signal decorrelation and random noise on the variance of time delay estimation. IEEE Trans Ultrason Ferroelec Freq Contr 44:220–225

Chornoboy ES (1992) Initialization for improved IIR filter performance. IEEE Trans Sig Proc 40:543–550

Collaris RJ (1996) New concepts in ultrasound image processing and their potential for improving data presentation. PhD Dissertation, University of Limburg, Maastricht

Collaris RJ, Hoeks APG (1994) Postprocessing of velocity distributions in real-time ultrasonic color velocity imaging. Ultrason Imaging 16:249–264

de Jong PGM, Arts T, Hoeks APG, Reneman RS (1990) Determination of tissue motion velocity by correlation interpolation of pulsed ultrasonic echo signals. Ultrason Imaging 12:84–98

de Jong PGM, Arts T, Hoeks APG, Reneman RS (1991) Experimental evaluation of the correlation interpolation technique to measure regional tissue velocity. Ultrason Imaging 13:145–161

Dickinson RJ, Hill CR (1982) Measurement of soft tissue motion using correlation between A-scans. Ultrasound Med Biol 8:263–271

Dotti D, Gatti E, Svelto V, Ugge A, Vidali P (1976) Blood flow measurements by ultrasound correlation techniques. Energia Nucleare 23:571–575

Embree PM (1986) The accurate measurement of volume flow of blood by time domain correlation. PhD Dissertation, University of Illinois, Urbana-Champaign, IL

Embree PM, O'Brien WD (1990) Volumetric blood flow via time-domain correlation: experimental verification. IEEE Trans Ultrason Ferroelec Freq Contr UFFC-37:176–189

Ferrara KW (1995) Effect of the beam-vessel angle on the received acoustic signal from blood. IEEE Trans Ultrason Ferroelec Freq Contr 42:416–428

Ferrara KW, Algazi VR (1990) Improved color flow mapping using the wideband maximum likelihood estimator. In: Proc. 1990 IEEE Ultrasonics Symposium, pp. 1517–1521. IEEE, Piscataway, NJ

Ferrara KW, Algazi VR (1991a) A new wideband spread target maximum likelihood estimator for blood velocity estimation—Part I: Theory. IEEE Trans Ultrason Ferroelec Freq Contr UFFC-38:1 16

Ferrara KW, Algazi VR (1991b) A new wideband spread target maximum likelihood estimator for blood velocity estimation—Part II: Evaluation of estimators with experimental data. IEEE Trans Ultrason Ferroelec Freq Contr UFFC-38:17–26

Ferrara K, Algazi VR (1992) Comparison of estimation strategies for color flow mapping. Acoust Imaging 19:317–322

Ferrara KW, Algazi VR (1994a) A statistical analysis of the received signal from blood during laminar flow. IEEE Trans Ultrason Ferroelec Freq Contr 41:185–198

Ferrara KW, Algazi VR (1994b) A theoretical and experimental analysis of the received signal from disturbed blood flow. IEEE Trans Ultrason Ferroelec Freq Contr 41:172–184

Ferrara KW, DeAngelis G (1997) Color flow mapping. Ultrasound Med Biol 23:321–345

Ferrara KW, Algazi R, Liu J (1992) The effect of frequency dependent scattering and attenuation on the estimation of blood velocity using ultrasound. IEEE Trans Ultrason Ferroelec Freq Contr 39:754–767

Ferrara KW, Ostromogilsky M, Rosenberg S, Sokil-Melgar J (1995) Parameter mapping for the detection of disturbed blood flow. Ultrasound Med Biol 21:517–525

Ferrara KW, Zagar BG, Sokil-Melgar JB, Silverman RH, Aslanidis IM (1996) Estimation of blood velocity with high frequency ultrasound. IEEE Trans Ultrason Ferroelec Freq Contr 43:149–157

Fletcher RH, Burlage DW (1972) An initialization technique for improved MTI performance in phased array radars. Proc IEEE 60:1551–1552

Forestieri SF (1995) Median temporal filtering of ultrasonic data. USA Patent No. 5413105, issued 9 May 1995 to Diasonics Ultrasound, Inc., Milpitas, CA

Foster SG (1985) A pulsed ultrasonic flowmeter employing time domain methods. PhD Dissertation, University of Illinois, Urbana-Champaign, IL

Foster SG, Embree PM, O'Brien WD (1990) Flow velocity profile via time-domain correlation: error analysis and computer simulation. IEEE Trans Ultrason Ferroelec Freq Contr UFFC-37:162–175

Friemel BH, Bohs LN, Nightingale KR, Trahey GE (1998) Speckle decorrelation due to two-dimensional flow gradients. IEEE Trans Ultrason Ferroelec Freq Contr 45:317–327

Gerzberg L, Meindl JD (1980) Power-spectrum centroid detection for Doppler systems applications. Ultrason Imaging 2:236–261

Gill RW (1979) Performance of the mean frequency Doppler modulator. Ultrasound Med Biol 5:237–247

Grandchamp PA (1975) A novel pulsed directional Doppler velocimeter: the phase detection profilometer. In: Kazner E, de Vlieger M, Muller HR, McCready VR (eds) Ultrasonics in Medicine, pp. 137–143. Proceedings of the 2nd European Congress on Ultrasonics in Medicine, 12–16 May 1975, Excerpta Medica, Amsterdam

Heimdal A, Torp H (1997) Ultrasound Doppler measurements of low velocity blood flow: limitations due to clutter signals from vibrating muscles. IEEE Trans Ultrason Ferroelec Freq Contr 44:873–881

Hein IA, O'Brien WD (1993a) Current time-domain methods for assessing tissue motion by analysis from reflected ultrasound echoes—a review. IEEE Trans Ultrason Ferroelec Freq Contr 40:84–102

Hein IA, O'Brien WD (1993b) A real-time ultrasound time-domain correlation blood flowmeter: Part II—performance and experimental verification. IEEE Trans Ultrason Ferroelec Freq Contr 40:776–785

Hein IA, Chen JT, Jenkins WK, O'Brien WD (1993) A real-time ultrasound time-domain correlation blood flowmeter: Part I—theory and design. IEEE Trans Ultrason Ferroelec Freq Contr 40:768–775

Herment A, Demoment G, Guglielmi JP, Dumee P, Pellot C (1991) Adaptive estimation of the mean frequency of a Doppler signal from short data windows. Ultrasound Med Biol 17:901–919

Herment A, Demoment G, Dumée P, Guglielmi J-P, Delouche A (1993) A new adaptive mean frequency estimator: application to constant variance color flow mapping. IEEE Trans Ultrason Ferroelec Freq Contr 40:796–804

Herment A, Demoment G, Dumee P (1996) Improved estimation of low velocities in color Doppler imaging by adapting the mean frequency estimator to the clutter rejection filter. IEEE Trans Biomed Eng 43:919–927

Hoeks APG, Reneman RS, Peronneau PA (1981) A multigate pulsed Doppler system with serial data processing. IEEE Trans Sonics Ultrason SU-28:242–247

Hoeks APG, Peeters HHPM, Ruissen CJ, Reneman RS (1984) A novel frequency estimator for sampled Doppler signals. IEEE Trans Biomed Eng BME-31:212–220

Hoeks APG, van de Vorst JJW, Dabekaussen A, Brands PJ, Reneman RS (1991) An efficient algorithm to remove low frequency Doppler signals in digital Doppler systems. Ultrason Imaging 13:135–144

Hoeks APG, Arts TGJ, Brands PJ, Reneman RS (1993) Comparison of the performance of the RF cross correlation and Doppler autocorrelation technique to estimate the mean velocity of simulated ultrasound signals. Ultrasound Med Biol 19:727–740

Hoeks APG, Arts TGJ, Brands PJ, Reneman RS (1994a) Processing scheme for velocity estimation using ultrasound RF cross correlation techniques. Eur J Ultrasound 1:171–182

Hoeks APG, Brands PJ, Arts TGJ, Reneman RS (1994b) Subsample volume processing of Doppler ultrasound signals. Ultrasound Med Biol 20:953–965

Jensen JA (1993a) Implementation of ultrasound time-domain cross-correlation blood velocity estimators. IEEE Trans Biomed Eng 40:468–474

Jensen JA (1993b) Range/velocity limitations for time-domain blood velocity estimation. Ultrasound Med Biol 19:741–749

Jensen JA (1993c) Stationary echo canceling in velocity estimation by time-domain cross-correlation. IEEE Trans Med Imaging 12:471–477

Jensen JA (1994) Artifacts in blood velocity estimation using ultrasound and cross-correlation. Med Biol Eng Comput 32:S165–S170

Jensen JA (1996) Estimation of Blood Velocities Using Ultrasound. Cambridge University Press, Cambridge

Kadi AP, Loupas T (1995) On the performance of regression and step-initialised IIR clutter filters for colour Doppler systems in diagnostic medical ultrasound. IEEE Trans Ultrason Ferroelec Freq Contr 42:927–937

Kasai C, Namekawa K, Koyano A, Omoto R (1985) Realtime two-dimensional blood flow imaging using an autocorrelation technique. IEEE Trans Sonics Ultrason SU-32:458–464

Kim JH (1989) Doppler velocity processing method and apparatus. USA Patent No. 4800891, issued 31 Jan 1989 to Siemens Medical Laboratories Inc

Kristoffersen K (1988) Time-domain estimation of the center frequency and spread of Doppler spectra in diagnostic ultrasound. IEEE Trans Ultrason Ferroelec Freq Contr 35:484–497

Kristoffersen K, Angelsen BAJ (1988) A time-shared ultrasound Doppler measurement and 2-D imaging system. IEEE Trans Biomed Eng BME-35:285–295

Lai X, Torp H (1996) Interpolation method for time delay estimation in the RF-signal crosscorrelation technique for blood velocity measurement. In: Levy M, Schneider SC, McAvoy BR (eds) Proceedings of the 1996 IEEE Ultrasonics Symposium, pp. 1211–1216. IEEE, Piscataway, NJ

Lai X, Torp H, Kristoffersen K (1997) An extended autocorrelation method for estimation of blood velocity. IEEE Trans Ultrason Ferroelec Freq Contr 44:1332–1342

Ledoux LAF, Brands PJ, Hoeks APG (1997) Reduction of the clutter component in Doppler ultrasound signals based on singular value decomposition: a simulation study. Ultrason Imaging 19:1–18

Ledoux LAF, Willigers JM, Brands PJ, Hoeks APG (1998) Experimental verification of the correlation behavior of analytic ultrasound radiofrequency signals received from moving structures. Ultrasound Med Biol 24:1383–1396

Li T (1995) Frequency–domain autocorrelator for estimation of blood flow velocity. Jap J Appl Phys 34:2817–2821

Lipschutz D (1988) Combined color flow map and monochrome image. USA Patent No. 4761740, issued 2 Aug 1988 to Hewlett-Packard Company, Palo Alto, CA

Liu DC, Kim J, Schardt M (1991) Modified autocorrelation method compared with maximum entropy method and RF cross-correlation method as mean frequency estimator for Doppler ultrasound. In: Proceedings of the 1991 IEEE Ultrasonics Symposium, pp. 1285–1290. IEEE, Piscataway, NJ

Loupas T, McDicken WN (1990) Low-order complex AR models for mean and maximum frequency estimation in the context of Doppler color flow mapping. IEEE Trans Ultrason Ferroelec Freq Contr UFFC-37:590–601

Loupas T, Peterson RB, Gill RW (1995a) Experimental evaluation of velocity and power estimation for ultrasound blood flow imaging, by means of a two-dimensional autocorrelation approach. IEEE Trans Ultrason Ferroelec Freq Contr 42:689–699

Loupas T, Powers JT, Gill RW (1995b) An axial velocity estimator for ultrasound blood flow imaging, based on a full evaluation of the Doppler equation by means of a two-dimensional autocorrelation approach. IEEE Trans Ultrason Ferroelec Freq Contr 42:672–688

Lunt MJ (1975) Accuracy and limitations of the ultrasonic Doppler blood velocimeter and zero crossing detector. Ultrasound Med Biol 2:1–10

Mahapatra PR, Zrnic DS (1983) Practical algorithms for mean velocity estimation in pulsed Doppler weather radars using a small number of samples. IEEE Trans Geosc Rem Sens GE-21:491–501

Miller KS, Rochwarger MM (1972) A covariance approach to spectral moment estimation. IEEE Trans Inform Theory IT-18:588–596

Namekawa K, Kasai C, Tsukamoto M, Koyano A (1982) Realtime bloodflow imaging system utilizing autocorrelation techniques. In: Lerski RA, Morley P (eds) Ultrasound '82, pp. 203–208. Pergamon, New York

Newhouse VL, Ehrenwald AR, Johnson GF (1977) The effect of Rayleigh scattering and frequency dependent absorption on the output spectrum of Doppler blood flowmeters. In: White D, Brown RE (eds) Ultrasound in Medicine, vol 3B, pp. 1181–1191. Plenum, New York

Nowicki A, Reid JM (1981) An infinite gate pulse Doppler. Ultrasound Med Biol 7:41–50

Nowicki A, Reid J, Pedersen PC, Schmidt AW, Oung H (1990) On the behavior of instantaneous frequency estimators implemented on Doppler flow imagers. Ultrasound Med Biol 16:511–518

Ophir J, Jaeger P (1982) Spectral shifts of ultrasonic propagation through media with nonlinear dispersive attenuation. Ultrason Imaging 4:282–289

Peterson RB (1993) A comparison of IIR initialization techniques for improved color Doppler wall filter performance. MSEE Dissertation, University of Washington

Peterson RB, Atlas LE, Beach KW (1994) A comparison of IIR initialization techniques for improved color Doppler wall filter performance. 1994 IEEE Ultrasonics Symposium:1705–1708

Porat B, Friedlander B (1988) Analysis of the asymptotic relative efficiency of the MUSIC algorithm. IEEE Trans Acoust Speech, Signal Processing 36:532–544

Rajaonah J-C, Dousse B, Meister J-J (1994) Compensation of the bias caused by the wall filter on the mean Doppler frequency. IEEE Trans Ultrason Ferroelec Freq Contr 41:812–819

Rao BD, Hari KVS (1989) Performance analysis of root-MUSIC. IEEE Trans Acoust Speech, Signal Processing 37:1939–1949

Rubin JM, Bude RO, Carson PL, Bree RL, Adler RS (1994) Power Doppler US: a potentially useful alternative to mean frequency-based color Doppler US. Radiology 190:853–856

Schmidt RO (1986) Multiple emitter location and signal parameter estimation. IEEE Trans Antennas Propagat AP-34:276–280

REFERENCES

Schmidt RO, Franks RE (1986) Multiple source DF signal processing: an experimental system. IEEE Trans Antennas Propagat AP-34:281–290

Shariati MA, Dripps JH, McDicken WN (1993) Deadbeat IIR based MTI filtering for color flow imaging systems. In: Proceedings of the 1993 IEEE Ultrasonics Symposium, vol 2, pp. 1059–1063. IEEE, Piscataway, NJ

Sirmans D, Bumgarner B (1975) Numerical comparison of five mean frequency estimators. J Appl Meteorology 14:991–1003

Thomas L, Hall A (1994) An improved wall filter for flow imaging of low velocity flow. 1994 IEEE Ultrasonics Symposium:1701–1704

Torp H (1997) Clutter rejection filters in color flow imaging: a theoretical approach. IEEE Trans Ultrason Ferroelec Freq Contr 44:417–424

Torp H, Bjaerum S (1996) Quality versus frame rate in color flow imaging: an experimental study based on off-line processing of RF-signals recorded from patients. In: Levy M, Schneider SC, McAvoy BR (eds) Proceedings of the 1996 IEEE Ultrasonics Symposium, pp. 1229–1232. IEEE, Piscataway, NJ

Torp H, Lai XM, Kristoffersen K (1993) Comparison between cross-correlation and auto-correlation technique in color flow imaging. In: Proceedings of the 1993 IEEE Ultrasonics Symposium, pp. 1039–1042. IEEE, Piscataway, NJ

Torp H, Kristoffersen K, Angelsen BAJ (1994) Autocorrelation techniques in color flow imaging: signal model and statistical properties of the autocorrelation estimates. IEEE Trans Ultrason Ferroelec Freq Contr 41:604–612

Torp H, Kristoffersen K, Angelsen BAJ (1995) On the joint probability density function for the autocorrelation estimates in ultrasound color flow imaging. IEEE Trans Ultrason Ferroelec Freq Contr 42:899–906

Trahey GE, Smith SW, von Ramm OT (1986) Speckle pattern correlation with lateral aperture translation: experimental results and implications for spatial compounding. IEEE Trans Ultrason Ferroelec Freq Contr UFFC-33:257–264

Tysoe C, Evans DH (1995) Bias in mean frequency estimation of Doppler signals due to wall clutter filters. Ultrasound Med Biol 21:671–677

Vaitkus PJ, Cobbold RSC (1998) A new time-domain narrowband velocity estimation technique for Doppler ultrasound flow imaging. Part I: Theory. IEEE Trans Ultrason Ferroelec Freq Contr 45:939–954

Vaitkus PJ, Cobbold RSC, Johnston KW (1998) A new time-domain narrowband velocity estimation technique for Doppler ultrasound flow imaging. Part II: Comparative performance assessment. IEEE Trans Ultrason Ferroelec Freq Contr 45:955–971

van Leeuwen GH, Hoeks APG, Reneman RS (1986) Simulation of real-time frequency estimators for pulsed Doppler systems. Ultrason Imaging 8:252–271

Walker WF, Trahey GE (1994) A fundamental limit on the performance of correlation based phase correction and flow estimation techniques. IEEE Trans Ultrason Ferroelec Freq Contr 41:644–654

Walker WF, Trahey GE (1995) A fundamental limit on delay estimation using partially correlated speckle signals. IEEE Trans Ultrason Ferroelec Freq Contr 42:301–308

Wang L-M, Shung KK, Camps OI (1996) Two bit correlation—an adaptive time delay estimation. IEEE Trans Ultrason Ferroelec Freq Contr 43:473–481

Willemetz JC, Nowicki A, Meister JJ, De Palma F, Pante G (1989) Bias and variance in the estimate of the Doppler frequency induced by a wall motion filter. Ultrason Imaging 11:215–225

Williams CS (1986) Designing Digital Filters, pp. 18–23. Prentice-Hall, Englewood Cliffs, NJ

Xu XL, Tewfik AH, Greenleaf JF (1995) Time delay estimation using wavelet transform for pulsed-wave ultrasound. Ann Biomed Eng 23:612–621

Zrnic DS (1977) Spectral moment estimates from correlated pulse pairs. IEEE Trans Aerosp Electron Syst 13:344–354

Zrnic DS (1979) Estimation of spectral moments for weather echoes. IEEE Trans Geosci Electron 4:113–128

12

Volumetric Blood Flow Measurement

12.1 INTRODUCTION

One of the most exciting applications of Doppler ultrasound is the measurement of volumetric blood flow, and there are now available a wide variety of commercial machines with facilities designed to aid such measurements. At present the majority of such machines use a 'duplex' mode for this purpose, i.e. they combine CW or single-gate PW Doppler and real-time ultrasonic B-scanning, but some systems use multigate methods and may take the form either of stand-alone instruments or, more normally, are part of colour flow scanning systems. In addition, a number of alternative techniques have been suggested or used to determine flow or one or more of its component parts.

Perhaps the greatest advantage of Doppler flow measurements is that they can be made without in the least interfering with flow. They can be made on conscious patients totally non-invasively and may therefore be repeated at will so as to study normal and abnormal physiology, and to monitor changes such as the progress of a disease process or the effect of a therapy.

In this chapter we first examine the most frequently used methods of flow measurement and some of the errors these may involve, and then describe some of the alternative approaches to the flow measurement problem.

12.2 FLOW MEASUREMENT WITH DUPLEX SCANNERS

Most ultrasonic duplex scanners now have facilities for calculating and displaying volumetric blood flow. The pulse-echo system is used to image the blood vessel of interest, and allows the operator to place the Doppler sample volume accurately to encompass the vessel totally whilst avoiding signals from nearby structures, to measure the angle between the ultrasound beam and the axis of the blood vessel (θ), and to measure the diameter of the blood vessel (Fig. 12.1). The Doppler system is used to estimate the mean velocity of flow in the direction of the ultrasound beam (using a uniform insonation technique) which may then, with a knowledge of θ, be converted to the mean velocity parallel to the vessel axis. This in turn is multiplied by the vessel cross-sectional area to yield mean flow. The method, which has been explored by many workers (e.g. Gill 1979, 1982, Avasthi et al 1984, Eik-Nes et al 1984, Gill et al 1984, Gill 1985, Griffin et al 1985, Qamar et al 1985, Evans 1986, Lewis et al 1986), has considerable potential but it is important that

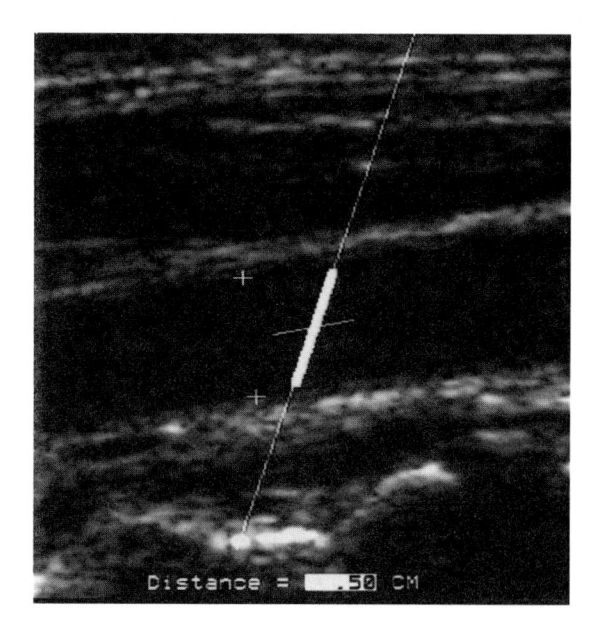

Figure 12.1 Duplex image of a common carotid artery showing the placement of the sample gate (thick white line in arterial lumen), the angle correct cursor (thin white line crossing the sample gate cursor) and the diameter measurement cursors (two small crosses on either edge of arterial lumen)

users appreciate the large errors that may occur under unfavourable circumstances.

12.2.1 Theoretical Aspects

The time-average volumetric flow through a vessel, $\bar{Q}$, is given by the time-averaged product of the cross-sectional area of the vessel $A(t)$ and the spatial mean velocity of flow within the vessel $\bar{v}(t)$, and may be written:

$$\bar{Q} = 1/T \int_{t=0}^{T} A(t)\bar{v}(t)\,\mathrm{d}t \qquad 12.1$$

If the vessel is uniformly insonated, the mean velocity may be calculated from the mean Doppler shift frequency using the standard Doppler equation, i.e:

$$\bar{v}(t) = \bar{f}_\mathrm{d}(t)c/2f_\mathrm{t}\cos\theta \qquad 12.2$$

where $\bar{f}_\mathrm{d}(t)$ is the instantaneous mean Doppler shift, f_t the transmitted zero-crossing frequency, θ the angle between the ultrasound beam and the blood vessel axis, and c the velocity of ultrasound in soft tissue (strictly speaking, c should be the velocity of ultrasound in blood, but it will be seen in Section 12.2.4 that using the value appropriate to soft tissue in eqn. 12.2 corrects for refraction occurring at the soft-tissue/blood interface, which would otherwise cause an overestimate in calculated velocity). Substituting eqn. 12.2 into eqn. 12.1 leads to:

$$\bar{Q} = (c/2f_\mathrm{t}\cos\theta) \int_{t=0}^{T} (A(t)\bar{f}_\mathrm{d}(t)/T)\,\mathrm{d}t \qquad 12.3$$

Ordinary commercial duplex machines have no method of following changes in cross-sectional area over the cardiac cycle and therefore it is implicitly assumed that such changes are insignificant or at least cancel out, and the equation that is actually evaluated is a simplified form of eqn. 12.3, namely:

$$\bar{Q} = (cA/2f_\mathrm{t}\cos\theta) \int_{t=0}^{T} (\bar{f}_\mathrm{d}(t)/T)\,\mathrm{d}t \qquad 12.4$$

where A is the 'effective vessel area'. The zero-crossing frequency of the transmitted signal is easily measured and the velocity of the sound in soft tissue is more or less constant, but each of the other terms in eqn. 12.4 ($\bar{f}_\mathrm{d}$, A and θ) must be measured.

12.2.2 Measurement of Mean Doppler Shift

The determination of mean Doppler shift frequency, and hence mean blood velocity, has already been dealt with at some length in Chapter 8. There it was shown that non-uniform insonation of the blood vessel (Section 8.3.1), differential attenuation between soft tissue and blood (Section 8.3.3), frequency-dependent attenuation and scattering (Section 8.3.5), high pass filters designed to reject high-amplitude low-frequency Doppler shifts (Section 8.3.6) and a poor signal-to-noise ratio (Section 8.3.7) could all affect the measured mean frequency.

The most important of these errors is probably that due to non-uniform vessel insonation (Evans 1985), and caution must be exercised when interpreting the mean velocity derived by duplex scanners if the ultrasound beam is significantly narrower than the vessel diameter. Since on average the velocity at the centre of the vessel is higher than at the periphery, a narrow ultrasound beam passing through the centre of the vessel leads to an overestimation of the mean velocity. The magnitude of this error has been considered in Section 8.3.1. and is summarised for a simple beam shape (rectangular) and velocity profile (parabolic) in Fig. 12.2. Willink and Evans (1994, 1995a) have suggested a method of reducing this source of error by the use of an estimator designed for small beam width/vessel diameter ratios, and this has been discussed in Section 8.3.1. Fei (1995) has discussed a compensation method for the mismatch, based on profile measurements made with a multigate system.

12.2.3 Measurement of Vessel Area

Most duplex scanners offer a choice of two methods for determining vessel cross-section. The

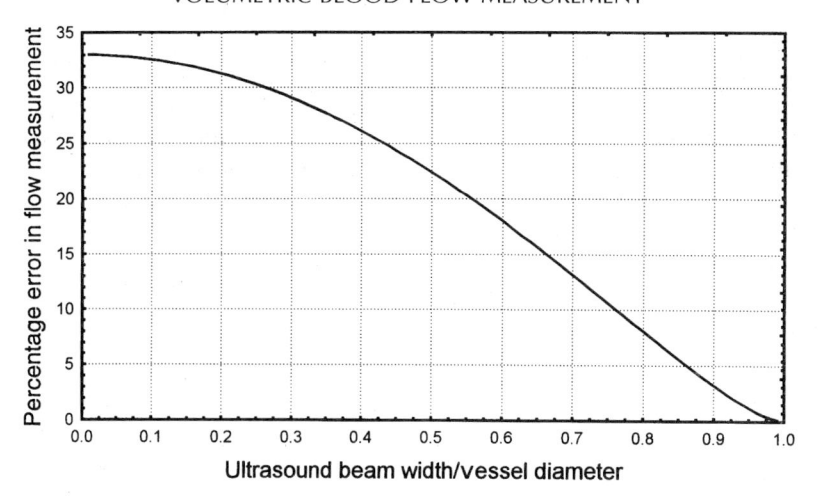

Figure 12.2 Percentage error in flow measurement due to partial insonation, by a centrally placed rectangular beam of ultrasound, of a blood vessel containing flow whose time-averaged velocity profile is parabolic

simpler and most usual method is to make a diameter measurement of the vessel using the same image used to place the Doppler sample volume (see Fig. 12.1), and then to calculate the area by assuming the vessel to be circular in cross-section. Alternatively the scan plane of the imaging device may be rotated around the vertical axis to produce a cross-sectional view of the vessel, whose area can then be measured directly. In either case, the operation of most scanners encourages the user to make only a single measurement of a quantity which is known to vary over the cardiac cycle, and this can cause quite significant errors in addition to those intrinsic to the measurement of a single unchanging diameter.

Short of evaluating eqn. 12.3 directly by constantly monitoring both vessel diameter and mean Doppler shift frequency (Hartley et al 1978, Struyk et al 1985, Powalowski 1988, Willink and Evans 1995b, Brodszki et al 1998), the error due to changing vessel size cannot be completely eliminated, but it may be reduced by measuring the vessel diameter a number of times and using the mean or median value in the flow calculations, and all users should be encouraged to do this if a vessel is at all pulsatile. In an investigation of the blood flow in the descending aortas of fetal lambs, Struyk et al (1985) found that errors of between +9% and −19% were possible if the aortic diameter happened to be measured at its maximum or minimum value, but that the use of a time-averaged diameter led to a systematic error of −5%. In a study of pulsatile arterial diameter variations in young healthy volunteers, Eriksen (1992) found median peak-to-peak diameter variations of 2.8% in their femoral arteries and 6.7% in their common carotid arteries, whilst the errors in flow measurement due to using average arterial diameter rather than true instantaneous diameter were +2.2% (range 1.5–3.8%) and +1.3% (0.4–3.6%), respectively.

Other important sources of error in determining the vessel area include incorrect assumptions about the vessel shape (i.e. the cross-section may not be circular, possibly due to disease), the limited axial resolution of the pulse-echo system, and incorrect calliper velocity settings. Axial resolution is important because echoes from the inner and outer surfaces of the blood vessel (and possibly from different layers within the wall) tend to merge together and therefore only the first echo from each wall is reliable. Because of this, workers in the field of fetal blood flow often measure from the outer aspect of the proximal wall to the inner aspect of the distal wall (Eik-Nes et al 1982, Teague et al 1985). This produces an overestimate of the vessel diameter, but this known systematic error is less serious than a similarly sized unknown random error. It is doubtful if a similar approach would be valid in adult arteries, the walls of which may have been affected by pathological changes.

Li et al (1993b) have pointed out that when vessel diameter measurements are made by setting callipers on the inner diameter of a vessel displayed on a B-scan image, the lumen size is underestimated by half the ultrasonic pulse-length (assuming the beam strikes the vessel at approximately 90°) and that this error can be quite easily corrected for. They examined four commonly available transducer/machine combinations and measured pulse lengths of between 454 ns and 962 ns, corresponding to errors of 0.35 mm to 0.74 mm, which would lead to quite significant underestimates of flow in small vessels if not corrected for. Although correction for pulse length is not currently carried out, users should be encouraged to introduce the practice.

A further minor source of error is a result of the finite width of the ultrasound beam. Even if the beam is centred on a true diameter of the vessel, echoes will be received from parts of the beam traversing chords parallel to this diameter, which must of course be shorter than the diameter. This can cause a slight underestimation of vessel diameter (Nielsen et al 1990).

Even if the axial resolution is very good, the best achievable accuracy will be of the order of a wavelength of the imaging ultrasound which may be quite significant when compared with the diameter of small vessels. Figure 12.3 shows the percentage error in measured flow plotted against vessel diameter for various errors in diameter measurement. Given that the wavelengths of ultrasound with frequencies of 3, 5 and 10 MHz are 0.5, 0.3 and 0.15 mm, respectively, it can be seen that potential errors in measurements on vessels of less than 4 mm diameter may be very large indeed.

Calliper setting errors may arise because the velocity of ultrasound in blood ($\sim$1580 ms^{-1}) is significantly greater than that in soft tissue ($\sim$1540 ms^{-1}), and if this is overlooked the flow may be underestimated by about 5%.

12.2.4 Measurement of Angle

The angle between the ultrasound beam and the blood vessel axis is usually found by rotating a dedicated cursor on the B-scan image to align it with the axis of the vessel (Fig. 12.1). Since it is the cosine of θ that determines the component of velocity measured by the Doppler probe, the accuracy of this measurement becomes much more critical at angles approaching 90°, where the cosine function varies very rapidly. The percentage error in the flow measurement due to a given error in the measurement of the angle θ is plotted as a function of the true value of θ in Fig. 12.4. With care, on a straight vessel, it is

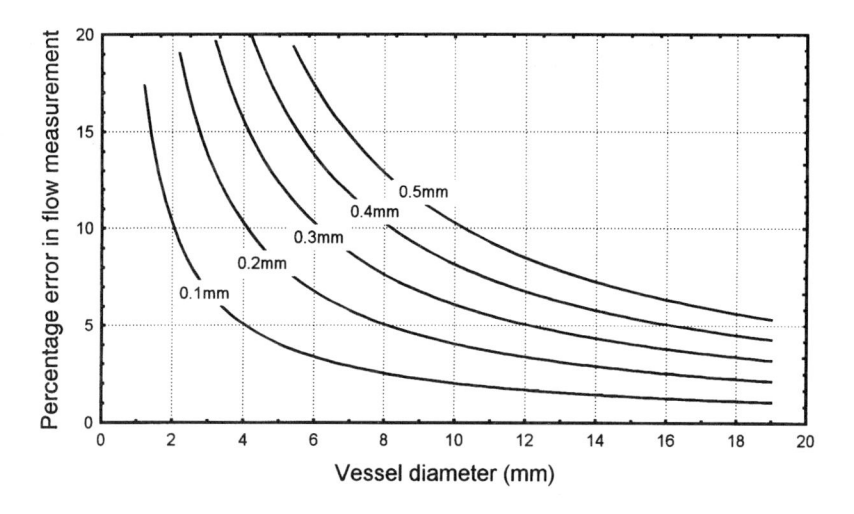

Figure 12.3 Percentage error in flow measurement due to the uncertainty in the diameter measurement ($\pm$0.1 mm to $\pm$0.5 mm) plotted as a function of diameter

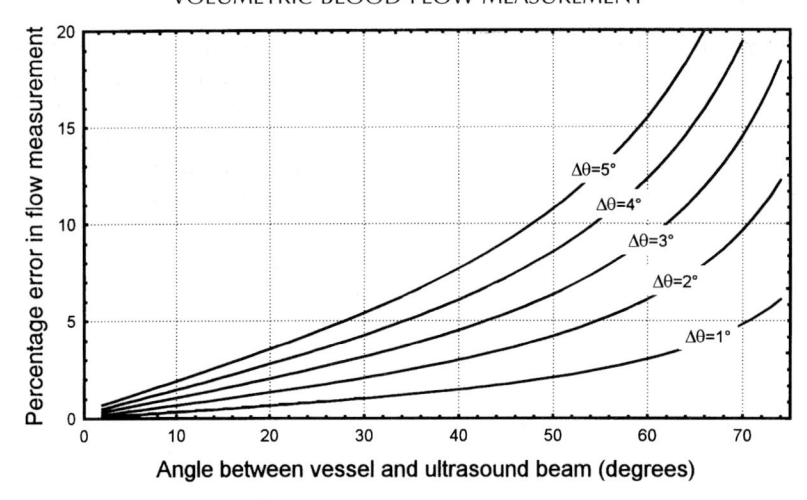

Figure 12.4 Percentage error in flow measurement due to the uncertainty in the measurement of the angle (±1° to ±5°) between the ultrasound beam and the axis of the blood vessel, plotted as a function of angle

usually possible to measure θ to ±2° and thus, provided that θ itself is kept below 60° and preferably below 45°, errors from this source should not be excessive. It should be noted, however, that there is an implicit assumption that the flow vectors are parallel to the vessel wall; this may be a reasonable assumption in long straight sections of vessels away from bifurcations, branches, curves and particularly stenoses, but whenever there are geometry changes the validity of this assumption needs to be questioned carefully.

It has been assumed in the derivation of Fig. 12.4 that the plane of scan is coincident with the axis of the vessel; deviations from this will cause the flow to be underestimated, and therefore efforts should be made to ensure that a reasonable length of the vessel is in the plane of the scan wherever possible. Fortunately, for angles of less than 15° the error due to misalignment of the scan plane and vessel axis is less than 3%, but at 20°, 25° and 30° the errors are 6%, 9% and 13% respectively.

The different velocities of ultrasound in blood and soft tissues (approximately 1580 ms^{-1} as opposed to 1540 ms^{-1}) has caused various authors to raise the role of refraction at the vessel wall/blood interface as a source of error (Kremkau 1990, Li et al 1993c, Christopher et al 1995). However, it appears that the error from this source is cancelled by another potential error, that due to using the velocity of ultrasound in soft tissue in the Doppler equation rather than the velocity in blood (Oates 1989). The proof of this, first given by Li et al (1993c), is presented in a modified form below.

Referring to Fig. 12.5, the Doppler equation may be written:

$$v = \frac{f_d}{2f_t} \cdot \frac{c_b}{\cos \theta_b} \qquad 12.5$$

where v, f_d, and f_t have their usual meanings (see eqn. 12.2), c_b is the velocity of ultrasound in blood, and θ_b is the 'true' Doppler angle in blood, which differs from the measured Doppler angle θ_t in soft tissue. It is possible to relate c_b, θ_b, θ_t and c_t

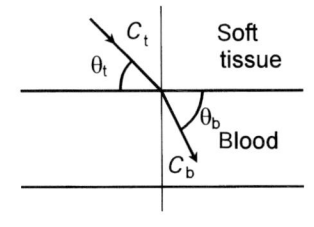

Figure 12.5 Refraction of ultrasound beam as it passes from soft tissue into blood in which ultrasound has a higher velocity

(the velocity of ultrasound in soft tissue) by Snell's law, which gives:

$$\frac{c_t}{\sin(\pi/2 - \theta_t)} = \frac{c_b}{\sin(\pi/2 - \theta_b)} \qquad 12.6$$

or

$$\frac{c_t}{\cos\theta_t} = \frac{c_b}{\cos\theta_b} \qquad 12.7$$

Substituting for $\cos\theta_b$ in eqn. 12.5 leads to:

$$v = \frac{f_d}{2f_t} \cdot \frac{c_t}{\cos\theta_t} \qquad 12.8$$

so that provided the velocity of ultrasound in soft tissue is used in the Doppler equation, refraction at the vessel boundary can be ignored. Note also that no matter how many parallel layers of medium separate tissue and blood, the Doppler equation can be rewritten to be independent of the speed of sound in blood, but dependent on the speed of sound in the first tissue layer. Christopher et al (1995) have extended this discussion to consider the effects of non-parallel tissue layers.

12.2.5 Practical Limitations

There are numerous sources of error in volumetric flow measurements made using duplex scanners, but many of them can be practically eliminated by careful attention to technique. The method is as its best when applied to medium-sized vessels of between about 4 mm and 8 mm diameter. For smaller vessels it is very difficult to make accurate determinations of the vessel size, whereas for larger vessels it becomes progressively more difficult to ensure uniform insonation.

12.3 FLOW MEASUREMENT WITH MULTIGATE SYSTEMS—1-D PROFILES

A second type of flow measuring system that is available is the multigate flowmeter which, rather than relying on uniform insonation to sample all parts of the vessel equally, actually measures the shape of the instantaneous velocity profiles using a series of small sample volumes. In order to do this efficiently it is important that the ultrasound beam is considerably narrower than the vessel and that the sample length is such that several may be placed across the vessel lumen. These constraints limit the use of the method to fairly large superficial vessels.

The multigate technique may be applied using a single gate, which is stepped across the vessel over a number of cardiac cycles (Peronneau et al 1972, Histand et al 1973), but is much better carried out with a true multigate system, which may consist either of a large number of finite gates operating in parallel (Baker 1970, Keller et al 1976) or a single serial digital signal processor capable of behaving as a multigate system (Brandestini 1978, Hoeks et al 1981). Systems designed primarily to produce colour flow images (by whatever method) sample larger blood vessels at several positions across their lumen, and can also be used to make multigate flow measurements if their sample volumes can be made sufficiently small and their frame rates sufficiently high, and a number of authors have recently reported such measurements (Bohs et al 1995, Forsberg et al 1995, Picot and Embree 1994, Deane and Markus 1997).

As with Duplex scanning, an independent measure of the angle θ between the ultrasound beam and the direction of blood flow is required, and if the multigate system does not incorporate an imaging mode, another approach to measuring the Doppler angle is needed.

12.3.1 Theoretical Aspects

If the cross-section of a blood vessel is split into a large number of small elements, ΔA_i, the instantaneous total flow through the vessel, $Q(t)$, is given by the sum of the flows through each of these elements, i.e:

$$Q(t) = \sum_i \Delta A_i v_i(t) \qquad 12.9$$

where $v_i(t)$ is the velocity of the blood travelling through ΔA_i. If the velocity, $v_n(t)$, is uniform within a semi-annulus (see Fig. 12.6), then eqn. 12.9 may be rewritten:

$$Q(t) = \pi \sum_n r_n \Delta r_n v_n(t) \quad 12.10$$

where Δr_n is the thickness of the semi-annulus at distance r_n from the vessel centre. $v_n(t)$ may be found from the instantaneous mean Doppler shift frequency, f_{dn}, measured from the appropriate range gate of a multigate system, i.e:

$$v_n(t) = f_{dn}(t) c / 2 f_t \cos \theta \quad 12.11$$

Substituting eqn. 12.11 into eqn. 12.10, and remembering that in practice the Doppler gates are uniformly spaced and that the suffix n may therefore be dropped from Δr_n, we obtain:

$$Q(t) = (c \pi \Delta r / 2 f_t \cos \theta) \sum_n r_n f_{dn}(t) \quad 12.12$$

Finally the mean flow may be determined by integrating eqn. 12.12 over the cardiac cycle, i.e.:

$$\bar{Q} = (c \pi \Delta r / 2 f_t \cos \theta T) \int_{t=0}^{T} \sum_n r_n f_{dn}(t) \, dt \quad 12.13$$

The velocity of ultrasound, c, is a known constant, and the transmitted zero-crossing frequency, f_t, for a given system is constant, and therefore in order to determine volumetric flow only Δr, r_n, θ and $f_{dn}(t)$ need to be found.

12.3.2 Measurement of Δr

The separation of the semi-annuli may be calculated in one of two ways, both of which are subject to errors. The first relies on the fact that the separation of the gates, s, for a particular system is known, and that the angle θ must be measured to calculate the blood flow velocity parallel to the axis of the vessel; thus, Δr may be simply calculated by multiplying s by $\sin \theta$ (Fig. 12.7). The major problem with this method is that it is critically dependent on the accurate measurement of θ, particularly when θ is small (see 12.3.4).

The second method is to measure the diameter of the vessel directly using ultrasound imaging techniques. In this case Δr may be calculated by dividing the diameter of the vessel, D, by the number of Doppler gates that detect a blood flow signal. There are two sources of inaccuracy in this determination, those related to measuring the diameter accurately and the difficulty of defining precisely the edge of the vessel lumen.

12.3.3 Measurement of Radius

The measurement of r_n is very closely related to the measurement of Δr, since r_n is an integral multiple of Δr. Therefore, the same considerations apply to the accuracy of measuring r_n as to Δr and since the two quantities are multiplied together the error in flow that results from an error in the estimate of Δr is doubled in percentage terms. This is exactly analogous to the duplex case, where an error in measuring the vessel diameter leads to twice the percentage error in the flow measurement.

12.3.4 Measurement of Angle

Non-colour flow imaging multigate systems do not usually incorporate a B-scan imager, but the angle θ may still be found in any of a number of ways (see Section 12.9.2). The MAVIS systems (GEC Medical Ltd) uses a pair of cross-sectional

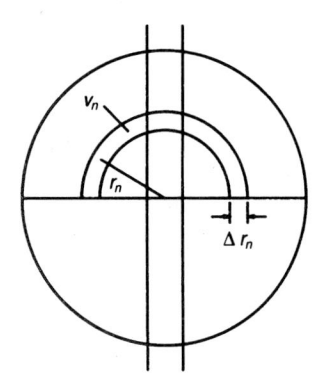

Figure 12.6 Schematic diagram of a narrow ultrasound beam passing through the centre of a blood vessel containing flow that is uniform within any small semi-annulus

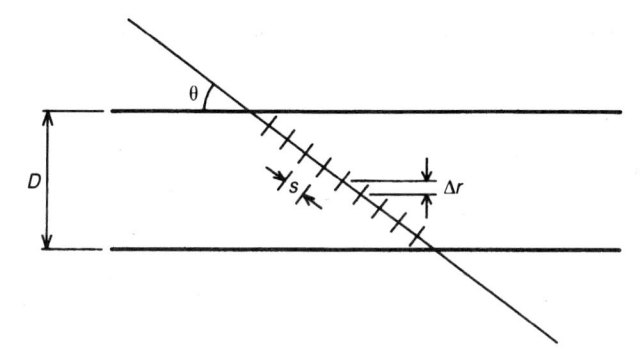

Figure 12.7 Schematic diagram of a multigate Doppler beam intersecting a blood vessel of diameter D at an angle θ

Doppler images from either side of the flow measurement point to establish the axis of the vessel (Section 12.9.2.1), whilst the EBF system (Novamed) uses a second transducer to find the direction perpendicular to the flow axis, both by measuring the size of the echoes from the vessel walls and by minimising the Doppler shift frequency (Section 12.9.2.2).

The results of the multigate method are critically dependent on accurate measurement of the Doppler angle θ. This can be demonstrated by rewriting eqn. 12.13 to show the angle dependence of the flow measurement. Replacing Δr by $s \sin \theta$, and r_n by $d_n \sin \theta$, where d_n is the distance of the semi-annulus from the vessel centre along the beam (i.e. $n.s$), leads to:

$$\overline{Q} = (\sin^2 \theta / \cos \theta)(c\pi s / 2 f_t T) \int_{t=0}^{T} \sum_n d_n f_{dn}(t) \, dt$$

12.14

Thus, the measured flow is proportional to $(\sin^2 \theta / \cos \theta)$ or $(\sin \theta . \tan \theta)$. The percentage error in eqn. 12.14 due to a given error in the measurement of the angle θ is plotted as a function of the true value of θ in Fig. 12.8. It can be seen that even small errors in θ lead to substantial errors in measured flow. Even at ideal angles of between 45° and 65°, the error in flow measurement due to a small error in angle measurement is approximately 5–6% per degree. For values of θ of less than 45°, errors in calculated diameter increase rapidly, whilst for values of θ of greater than 65°, errors in calculated velocity increase rapidly; in either case the overall error from these two sources added together rises considerably.

12.3.5 Measurement of Velocity

In theory each sample volume of a multigate system should be so small that only a very narrow band of Doppler shift frequencies is present. This being the case, f_{dn} could be determined by any of a number of methods, but the ones normally chosen (for non-colour flow systems) are to use a mean frequency follower or zero-crossing detector. In this particular application, uniform insonation is not an issue and so the major sources of error in the measurement of f_{dn} are the limited signal-to-noise ratio, the effects of the high pass filters and, because of the short pulse durations necessary for short gate lengths, frequency-dependent attenuation and scattering. These have been discussed in Chapter 8 (Sections 8.3.6, 8.3.7, 8.3.5.)

12.3.6 Practical Limitations

There are, in addition to the errors already discussed, two major limitations of the multigate technique. These are that the method assumes that there is partial symmetry in the velocity profile, and that in practice the sample volumes are of a finite size in both length and width. The former means that measurements should only be made remote from branches, bifurcations, curves,

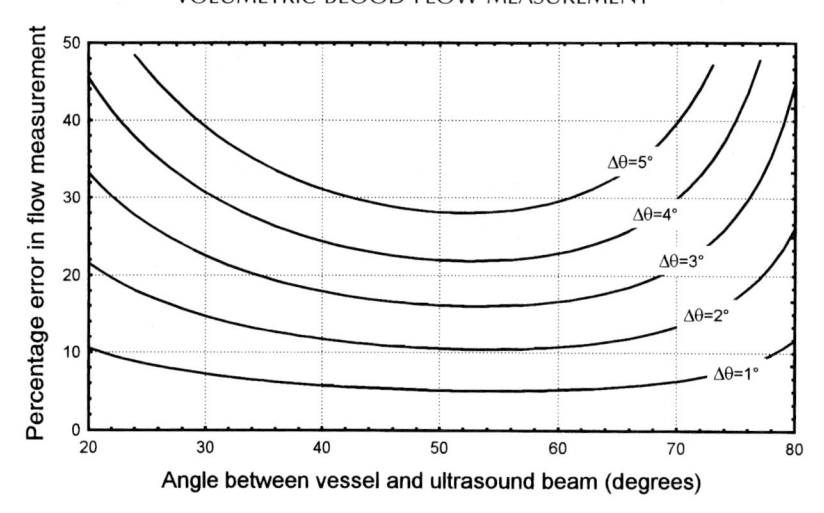

Figure 12.8 Percentage error in flow measurement using a 1-D multigate system due to the uncertainty in the measurement of the angle ($\pm 1°$ to $\pm 5°$) between the ultrasound beam and the axis of the blood vessel, plotted as a function of angle

disease sites and any other geometry changes that might disturb the symmetry of the flow profile (see Chapter 2), whilst the latter limits the technique to use on large vessels where the sample volumes will be small compared with the vessel diameter. The effect of the finite sample volume is to distort the shape of the measured velocity profile, and hence the measured value of flow. This has been discussed in some depth by several authors including Jorgensen et al (1973), Jorgensen and Garbini (1974), Baker et al (1978), Picot and Embree (1994), and Flaud et al (1997). Picot and Embree (1994) have calculated the effect of different beamwidths in conjunction with different vessel diameters on measured flow, whilst the other four publications discuss the use of deconvolution techniques to estimate the true velocity profile, and hence to obtain more accurate flow measurements. A deconvolution process has been used in the MAVIS device to improve its accuracy (Fish 1981), but still the method may not be used with small blood vessels.

A further practical problem of the multigate method is ensuring that the narrow beam of ultrasound passes through the centre of the vessel. If this is not achieved precisely, the flow may be seriously underestimated, since both arterial diameter and mean velocity are underestimated. Picot and Embree (1994) have studied these errors numerically and experimentally.

12.4 FLOW MEASUREMENT WITH MULTIGATE SYSTEMS—2-D PROFILES

A logical development of the estimation of flow from 1-D velocity profiles is to extend the method to the full 2-D profile and therefore avoid having to make the assumption of axial flow symmetry, and this approach has now been explored by a number of authors.

O'Brien and colleagues (Suorsa and O'Brien 1988, Embree and O'Brien 1990) attempted to measure volumetric flow (under continuous flow conditions) by using a time domain correlation technique to estimate the velocity at many points across the lumen of a circular tube (Fig. 12.9b). They then fitted a series of constant flow ellipses to these data (Fig. 12.9a). The diameter of the tube was estimated from the minor axis of the ellipse, where the velocity fell to zero, and the 'Doppler' angle from the eccentricity of the several constant velocity ellipses. Finally, the total flow was estimated by summing the flow contributions from each of the eccentric annular flow regions. Hein and O'Brien (1989) reported that the same method worked well with pulsatile flow, but it should be noted that their data acquisition was not in real time, and the results apparently took several minutes to calculate, even for a single profile.

Kitabatake et al (1990) plotted 2-D distributions of the flow velocity in the ascending aortas of

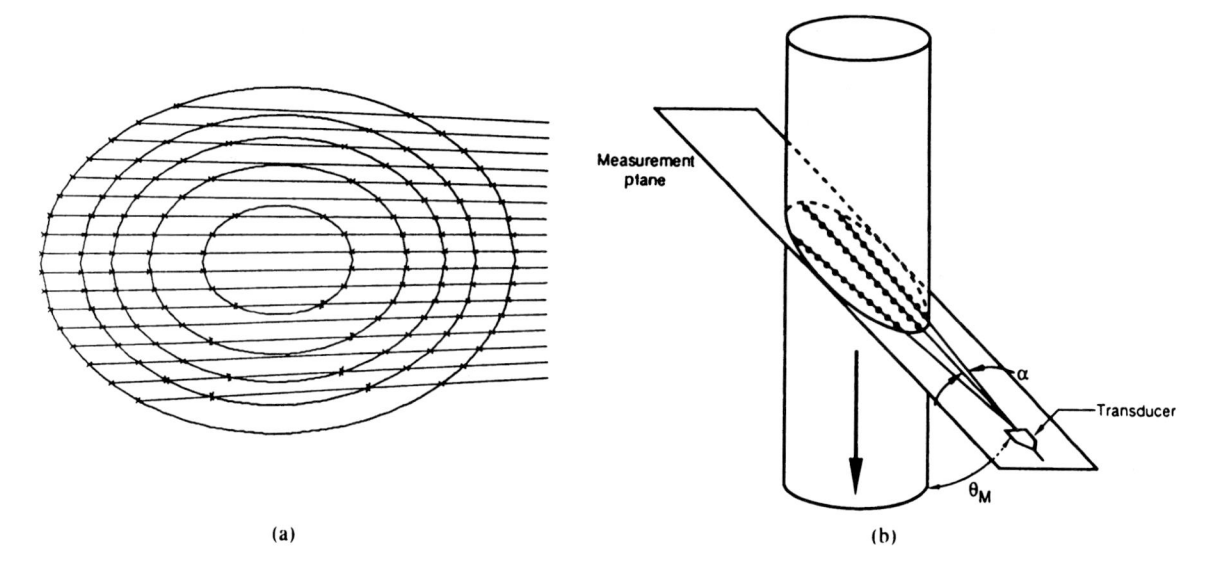

Figure 12.9 (a) Constant velocity ellipses in a circular tube at a 45° measurement angle. Straight lines indicate the beam paths, and Xs indicate constant flow velocity points. (b) Schematic diagram of method used by Embree and O'Brien to measure 2-D velocity profiles (reproduced by permission from Embree and O'Brien 1990, © 1990 IEEE)

patients with normal and diseased aortic valves, and used this information to calculate the total flow rate through these vessels. The Doppler angle was calculated from the eccentricity of the colour area on the colour flow image.

More recently, Picot et al (1995) have published a detailed description and evaluation of a system using a similar methodology, although in this case the Doppler angle was estimated by making (non-angle corrected) Doppler measurements of temporal average flow rates at the same site from two different viewing angles, and then solving the two resulting equations simultaneously for θ (this is similar to one of the techniques described in Section 12.9.2.3).

Li and colleagues (Li et al 1996, 1998, van der Steen et al 1997, Céspedes et al 1998) have made volumetric flow measurements with intravascular catheters by summing velocity components derived using their decorrelation technique over the entire vessel area (see Section 13.8).

12.4.1 Theoretical Aspects

Following the derivation of Picot et al, the true axial blood velocity in the ith pixel of the image at time t is given by:

$$v_{i(\text{axial})}(t) = v_i(t)/\cos\theta \qquad 12.15$$

where $v_i(t)$ is the component of velocity towards the transducer, and θ is the Doppler angle. The pixel area projected onto the vessel cross-section is given by:

$$\Delta A_{(\text{true})} = \Delta A \sin\theta \qquad 12.16$$

where ΔA is the measured pixel area. Therefore, the flow rate $\Delta Q_i(t)$ through each pixel at time t is given by:

$$\Delta Q_i(t) = \frac{v_i(t)}{\cos\theta} \Delta A \sin\theta \qquad 12.17$$

and the total volumetric blood flow is obtained by summing the contributions from all pixels with a non-zero value for velocity, i.e:

$$Q(t) = \Delta A \tan\theta \sum_{i}^{npix} v_i(t) \qquad 12.18$$

where *npix* is the number of non-zero pixels.

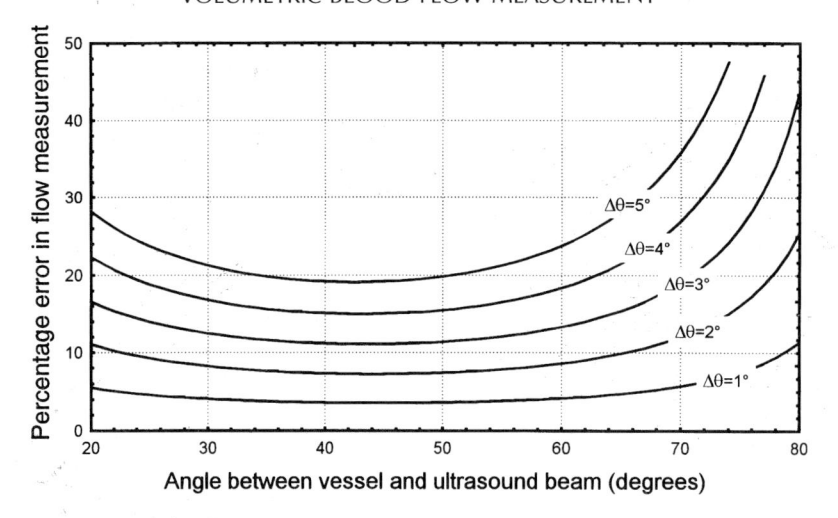

Figure 12.10 Percentage error in flow measurement using a 2-D multigate system due to the uncertainty in the measurement of the angle (±1° to ±5°) between the ultrasound beam and the axis of the blood vessel, plotted as a function of angle

12.4.2 Practical Limitations

The 2-D velocity profile technique has clear advantages over the 1-D technique in that it is not sensitive to translational positioning of the ultrasound transducer and removes the need for the assumptions of both vessel circularity and flow profile symmetry. It does, however, make the assumption that the ultrasound sample volumes are small compared with the vessel, so that the vessel is sampled at several locations across its diameter, and (like most other techniques) that all the flow streamlines in the vessel at the measurement site are parallel. These last two assumptions mean that the technique is likely to be best suited to flow measurements in fairly large non-diseased vessels.

As for the 1-D velocity profile method, the accuracy of the 2-D method is critically dependent on the accurate measurement of the Doppler angle θ. The percentage error in eqn. 12.18 due to a given error in the measurement of θ is plotted as a function of the true value of θ in Fig. 12.10, where it can be seen that for angles of between 45° and 60° the error in flow measurement due to a small error in angle measurement is approximately 4% per degree.

Further studies are needed of the overall accuracy and limitations of this technique.

12.5 FLOW MEASUREMENT—ATTENUATION-COMPENSATED METHOD

The attenuation-compensated (AC) volume flowmeter is an ingenious variation of the uniform insonation technique first described by Hottinger and Meindl (1979). It differs from all flow techniques so far described in that it makes use not only of the Doppler shift frequencies, but also of the total power of the Doppler signal.

A Doppler transducer is arranged to transmit two concentric pulsed beams. The sample volume of one totally encompasses the vessel (Fig. 12.11a) and is used to make both Doppler shift and power measurements. The sample volume of the other lies totally within the vessel and is used to make power measurements for the purpose of calibrating the power measurements from the first (Fig. 12.11b).

12.5.1 Theoretical Aspects

Provided the vessel is uniformly insonated, the total Doppler power is proportional to the vessel cross-section and inversely proportional to the cosine of θ, and therefore we may write the area

Therefore, the volumetric flow can be calculated from the integrated product of power and Doppler shift frequency, provided that the constant $M(z)$ can be evaluated.

Using the notation of Hottinger and Meindl (1979), $M(z)$ may be shown to be given by:

$$M(z) = T(z)\eta I_1(z)\Delta z \qquad 12.21$$

where $T(z)$ is the round trip transmission efficiency (representing the effects of attenuation caused by the tissue between the transducer and vessel at range z), η is the volumetric scattering coefficient representing the scattering of the blood, $I_1(z)$ is a parameter indicating the transducer sensitivity, and Δz is the Doppler sample volume length. Both $T(z)$ and η are dependent on the measurement configuration and must therefore be evaluated for each flow determination, and it is the method of so doing that is the crux of the AC method. A second smaller ultrasound beam, concentric with the first, is used to make power measurements from entirely within the lumen (Fig. 12.11b). This power, P_2, may be written:

$$P_2 = T(z)\eta I_2(z)A_2(z)\Delta z \qquad 12.22$$

where $I_2(z)$ indicates the sensitivity of transducer 2, and $A_2(z)$ is the cross-sectional area of the sample volume. Substituting eqn. 12.22 into eqn. 12.21 leads to:

$$M(z) = P_2[I_1(z)/A_2(z)I_2(z)] \qquad 12.23$$

The expression in the square brackets is, for a particular transducer design, dependent only upon the range z, and can be either calculated or, better, measured experimentally, and therefore $M(z)$ is easily calculated from P_2, given z.

12.5.2 Practical Limitations

The greatest practical problem to be overcome with the AC method is that of producing two ultrasound beams, one of which is uniform over the vessel lumen, and the other which is smaller than the lumen. The former is technically the more difficult to overcome, and any non-uniformity will affect both the mean Doppler shift (in the same

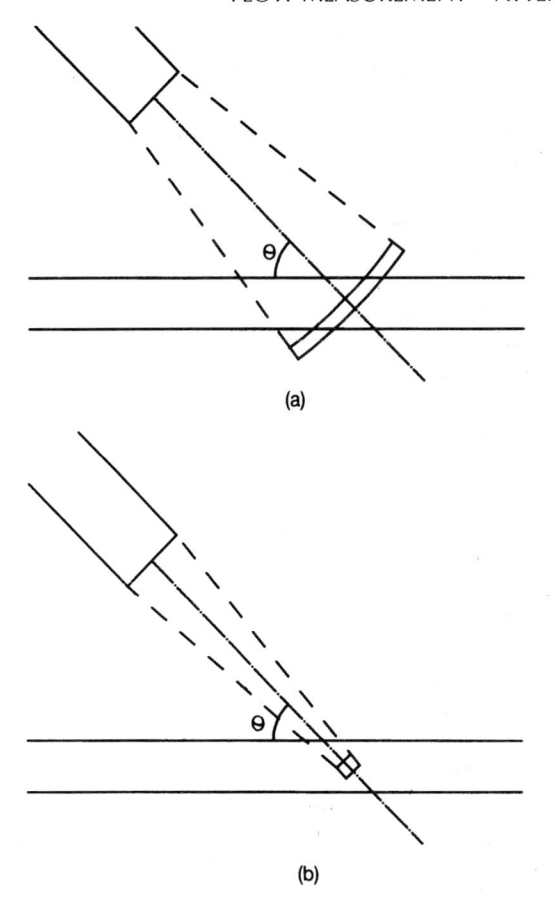

Figure 12.11 The sample volumes interrogated by the two ultrasound beams necessary for the attenuation compensated flowmeter. (a) The entire cross-section of the vessel is insonated. (b) The sample volume lies entirely within the vessel lumen (redrawn by permission of the Institute of Physics and Engineering in Medicine from Evans et al 1986

as:

$$A(t) = P(t)\cos\theta/M(z) \qquad 12.19$$

where $P(t)$ is the total power in the Doppler signal and $M(z)$ is a depth-dependent constant. If eqns 12.2 and 12.19 are now substituted into eqn. 12.1, the cosine term vanishes and we are left with:

$$\overline{Q} = (c/2M(z)f_t)\int_{t=0}^{T}(P(t)\overline{f}_d(t)/T)\,dt \qquad 12.20$$

way as for the duplex scanner) and the total power received. At present, the most promising solution to producing a large uniform beam seems to be to use an annular array transducer (Fu and Gerzberg 1983, Evans et al 1986, Evans et al 1989b).

The production of a small sample volume is not very demanding and may be achieved with the same transducer used to produce the large beam, but the AC method may not be used on small vessels, where difficulty arises in keeping the sample volume entirely within the blood vessel lumen.

A major advantage of the AC method is that both the cross-sectional area of the vessel and the angle θ are automatically taken into account, but it is susceptible to the same problems with high pass filters and limited signal-to-noise ratios that all other Doppler flowmeters have.

Evans et al (1989b) have described a practical implementation of the AC technique for non-invasively measuring cardiac output, and the same authors (Evans et al 1989a) reported an excellent correlation with results obtained using a thermo-dilution method in 54 patients. More recently, Gibson et al (1994) have described, in principle, an intravascular volumetric flowmetry technique based on the AC method, and carried out some *in vitro* measurements using scaled-up models.

12.6 FLOW MEASUREMENT—ASSUMED VELOCITY PROFILE METHOD

The assumed velocity profile method is a variation of the Duplex method. The cross-sectional area of the vessel and the angle θ are determined as for that method, but the mean velocity, $\bar{v}$, is determined from the time-averaged maximum Doppler shift. The equation evaluated is a slight modification of eqn. 12.4, i.e:

$$\bar{Q} = (KcA/2f_t \cos\theta) \int_{t=0}^{T} (\hat{f}_d(t)/T) \, dt \quad 12.24$$

where $\hat{f}_d(t)$ is the instantaneous maximum Doppler shift, and K a constant which depends on the time-averaged velocity profile. The instantaneous maximum Doppler shift frequency may be found using a maximum frequency follower and is generally more reliable than the estimated mean frequency because of its immunity to the effects of attenuation, noise and high pass filters (see Section 8.4). It should, however, be recalled that the maximum frequency follower is very susceptible to the effects of intrinsic spectral broadening (Section 8.4.5).

The assumed velocity profile technique is of value in two distinct circumstances, when the time-averaged velocity profile is flat and when it is parabolic. The former is at least approximately true of the flow in the ascending aorta and the arch of the aorta, and a number of workers (e.g. Light and Cross 1972, Mackay 1972, Light 1974, Huntsman et al 1983) have used the assumed velocity profile method to measure and to monitor changes in cardiac output. Since the velocity profile is almost flat, the maximum velocity and the mean velocity are very similar and therefore the constant K in eqn. 12.24 is approximately equal to unity. In practice, skewing of the velocity profile will cause a slight spread in the velocities across the vessel lumen and the mean velocity is therefore slightly overestimated.

In vessels in which the flow is fully established (Sections 2.3.2 and 2.4.4) the time-averaged velocity profile is parabolic, and in this case the time-averaged maximum velocity is twice the mean velocity, provided that the lamina with the maximum velocity is always in the centre of the vessel (Evans 1985). Under these circumstances, K in eqn. 12.24 has a value of 0.5. It is as yet uncertain which arterial sites meet the appropriate criteria sufficiently closely for this method to be of value, but the common carotid artery is one likely candidate, and the method has been shown to be of value for measuring mean velocities in the neonatal cerebral arteries (Evans et al 1989c) (unfortunately, flow is unobtainable because the vessel diameters are too small to be measured). Li et al (1993a) studied the relationship between mean velocity and time averaged maximum velocity in a flow rig and confirmed that, in a long straight tube, maximum velocity could be used to derive a good estimate of mean velocity.

12.7 FLOW MEASUREMENT—C-MODE DOPPLER TECHNIQUES

Doppler ultrasound techniques for blood flow measurement are usually aimed at measuring the volume of blood flowing across a plane that is perpendicular to the vessel axis. Thus, velocity measurements are 'angle-corrected' so that the velocity components parallel to the vessel axis are estimated, and area measurements are made in the plane perpendicular to the vessel axis. In fact, it is quite possible to measure total blood flow by measuring the flow across any arbitrarily shaped surface across the vessel, provided that the velocity component normal to each point on that surface is considered.

More formally, volume flow is defined as the surface integral:

$$Q = \iint_{\text{surface}} \vec{v} \cdot d\vec{A} \qquad 12.25$$

where $\vec{v} \cdot d\vec{A}$ is the scalar product of the velocity at each point on the surface and the corresponding element of the surface.

In general, this formulation is unhelpful with regard to Doppler systems, except in the special circumstance in which the arbitrary surface is chosen to be perpendicular to the ultrasound beam, in which case the component of the velocity required is exactly that component of the velocity measured by the Doppler system. Furthermore, under such circumstances, each elemental area is determined by the effective beam cross-section at a given depth (except when the beam is only partially in the blood vessel).

The use of C-mode Doppler to measure blood flow in this way appears to have been first suggested by Hottinger and Meindl (1975) and has been developed by Moser and colleagues (Moser et al 1992a,b, 1995, Schumacher et al 1995). Figure 12.12 illustrates the basic idea of the method. A 2-D ultrasound transducer is used to define a measurement slice through the vessel perpendicular to the direction of the ultrasound beams, and the flow through that slice calculated by summing the elemental flow components parallel to the beams through that slice, i.e:

$$Q = \sum_{n=1}^{N} \sum_{m=1}^{M} \bar{v}_{m,n} \delta a \qquad 12.26$$

where $\bar{v}_{m,n}$ is the component of the mean velocity parallel to the ultrasound beam in each element m,

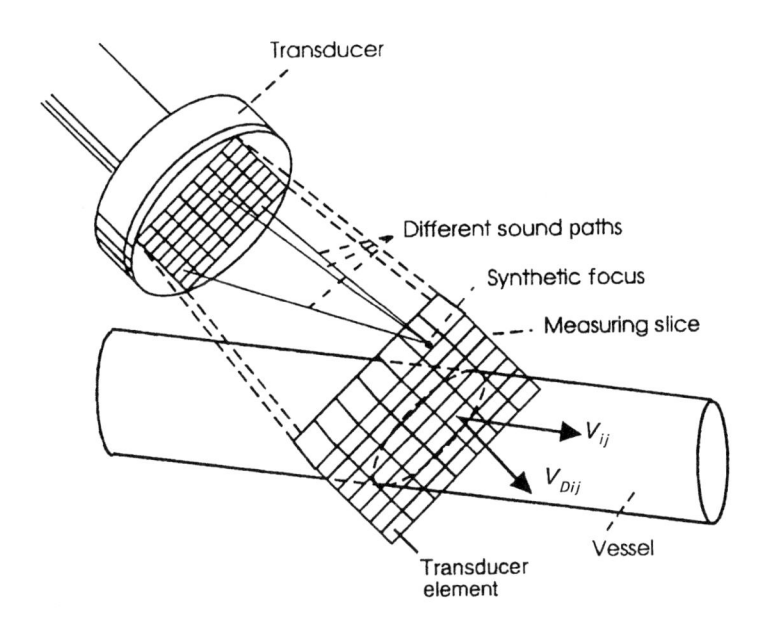

Figure 12.12 Illustration of the C-Mode Doppler method for measuring volumetric blood flow. A 2-D array transducer allows a group of voxels lying in a single plane intersecting the blood vessel to be interrogated. Only the velocity components perpendicular to the plane need to be calculated (reproduced by permission from Moser et al 1995, © Plenum Publishing Corp)

n, and δa is the effective cross-section of the element. This may of course easily be rewritten in terms of Doppler shift frequency, i.e:

$$Q = \frac{c}{2f_t} \sum_{n=1}^{N} \sum_{m=1}^{M} \bar{f}_{d[m,n]} \delta a \qquad 12.27$$

where c is the velocity of ultrasound, f_t the transmitted ultrasound frequency, and $\bar{f}_{d[m,n]}$ the mean Doppler shift frequency from each element.

Some of the advantages and limitations of the C-mode method have been discussed by Moser et al (1995). Perhaps its major advantage is that theoretically it is totally independent of flow direction, and therefore it is neither necessary to estimate the angle between the ultrasound beam and the vessel nor to assume that the flow is parallel to the vessel axis. A further advantage is that the method is insensitive to small ultrasonic field variations and, indeed, to spatial and/or temporal changes in the reflection coefficient of blood, because it uses only local estimates of mean frequency and not the power of the Doppler signal. The main factors which limit its accuracy are its limited spatial resolution and the partial volume effects which occur near the edge of the vessel. Equations 12.26 and 12.27 are written on the basis that the effective cross-section of each element, δa, is identical, but near the vessel edge this is not the case, as some elements will only be partially within the lumen, and unless this is compensated for, an overestimation of the flow will occur.

Two slight variations on the 'standard' C-mode approach have recently been suggested. The first, from Liu and Burns (1997), is designed to compensate for the partial volume error by introducing a correction factor, calculated from the ratio of the power measured from each sample volume, divided by a reference power measured from a sample volume which is known to be entirely within the vessel lumen. Equation 12.27 may be rewritten to include this correction, i.e:

$$Q = \frac{c}{2f_t} \sum_{n=1}^{N} \sum_{m=1}^{M} \bar{f}_{d[m,n]} (P_{meas[m,n]}/P_{ref}) \delta a \qquad 12.28$$

where P_{ref} is the reference power corresponding to a point where the ultrasound beam is completely within the vessel, and $P_{meas[m,n]}$ is the power measured from each sample volume. In practice, the correction term is combined with the calculation of $\bar{f}_{d[m,n]}$ which contains $P_{meas[m,n]}$ in the denominator (see eqn. 8.31). Liu and Burns also found it necessary, when using this technique, to make corrections for the effects of the wall filters in order to obtain satisfactory measurements from their *in vitro* test rig. It should be noted that this partial volume compensation procedure negates one of the advantages of the C-mode method, that is its immunity to variations in the ultrasound field and changes in the reflection coefficient of blood. It is also not clear what effects the differing thicknesses of attenuating tissue overlying different parts of the vessel would have on the method *in vivo*.

The second variation on the standard method is that suggested by Poulsen and Kim (1996), in which a multiplanar transducer is used to scan at points on the surface of a sphere intersecting the blood vessel (Fig. 12.13). In principle this is identical to the standard C-mode method, in that the surface integral as given in eqn. 12.25 is

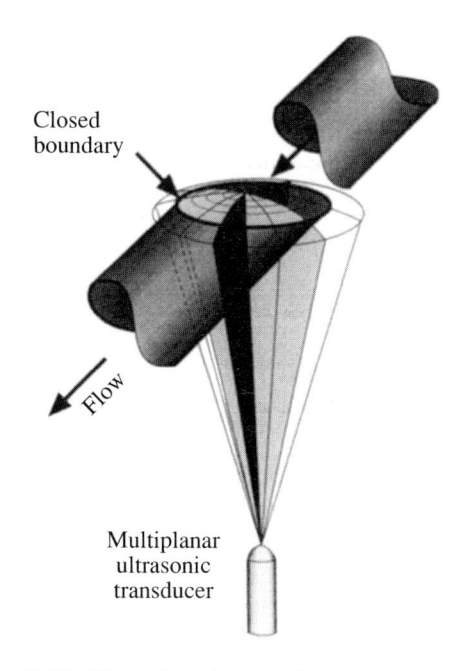

Figure 12.13 Illustration of the multiplanar C-mode Doppler method. In this case the closed boundary is part of the surface of a sphere (reproduced by permission from Poulsen and Kim 1996, © 1996 IEEE)

evaluated, and only velocity estimates along the ultrasound beam are required, but the arbitrary surface in now non-planar. Poulsen (1997) has discussed both the standard C-mode approach and the spherical surface approach, together with the possibility of using cylindrical and fan shaped surfaces.

12.8 FLOW MEASUREMENTS USING CONTRAST AGENTS

The advent of ultrasonic contrast agents has raised the possibility of measuring blood flow using indicator-dilution washout curves. Schwarz et al (1993) have compared three ultrasonic methods, including back-scattered Doppler power, for generating echo contrast time intensity curves *in vitro*, and shown the washout rate after bolus injection into a system of tubes and chambers to be linearly related to volumetric flow rate, but also to be dependent on the volume of mixing. Relative changes in flow rate could be determined directly from changes in the washout rate, independent of the mixing volume. One suggested application for the method was for making coronary artery blood flow measurements in the catheter laboratory. The use of contrast agents to quantify flow perfusion is further discussed in Section 13.3.1.

Shung and colleagues (Shung and Flenniken 1995, Wang and Shung 1995) have suggested that contrast agents may assist ultrasonic echo-tracking methods of blood flow measurement by enhancing the signal-to-noise ratio from the otherwise weak scattering centres in blood.

12.9 MISCELLANEOUS TECHNIQUES

12.9.1 First Moment Followers

The first moment follower is not a method of measuring flow, but is mentioned here because its output is, at least theoretically, more closely related to flow in a highly pulsatile vessel than the output of a mean frequency follower. Essentially it works in the same way as the attenuation-compensated flowmeter (Section 12.5), except that no absolute calibration is performed, and therefore its output is only proportional to flow, and the constant of proportionality changes with the measurement configuration. The reader is referred to Section 8.6.1 for more details.

12.9.2 Angle Estimation Techniques

A wide variety of methods have been used to measure the angle, θ, between the direction of the ultrasound beam and the flow axis. Some of them rely on imaging the vessel, and some on combining or comparing the output from two or more Doppler transducers.

12.9.2.1 Imaging Techniques

The most widely used method of measuring θ is the imaging one described in Section 12.2.4 and illustrated in Fig. 12.1. In addition it is possible to define the axis of the vessel, and hence θ, by making cross-sectional images of the vessel some distance apart but close to the Doppler measurement site, using either a Doppler imaging technique (Fish 1978), or a B-scan technique (Gill 1979). The orientation of the line joining the centre of these two images is an estimate of the vessel axis, and θ can thus be calculated by simple 3-D geometry.

12.9.2.2 Definition of Normal

Another approach to measuring θ is to start by defining a normal to the vessel, either by minimising the Doppler shift frequency from a Doppler transducer (Histand et al 1973) or by maximising the size of the vessel wall echoes using an A-mode transducer (Doriot et al 1975, Uematsu 1981) or a combination of both (Marquis et al 1983).

There are two variations of this technique. In one, the ultrasound probe is constructed from two transducers held at a fixed angle, ϕ, to each other so that when the angle-finding transducer is perpendicular to the vessel, the angle of the other with respect to the vessel axis is known to be $\pi/2 - \phi$ (Fig. 12.14a). In the second variation, the same Doppler transducer is rotated through a known angle from the known perpendicular, and

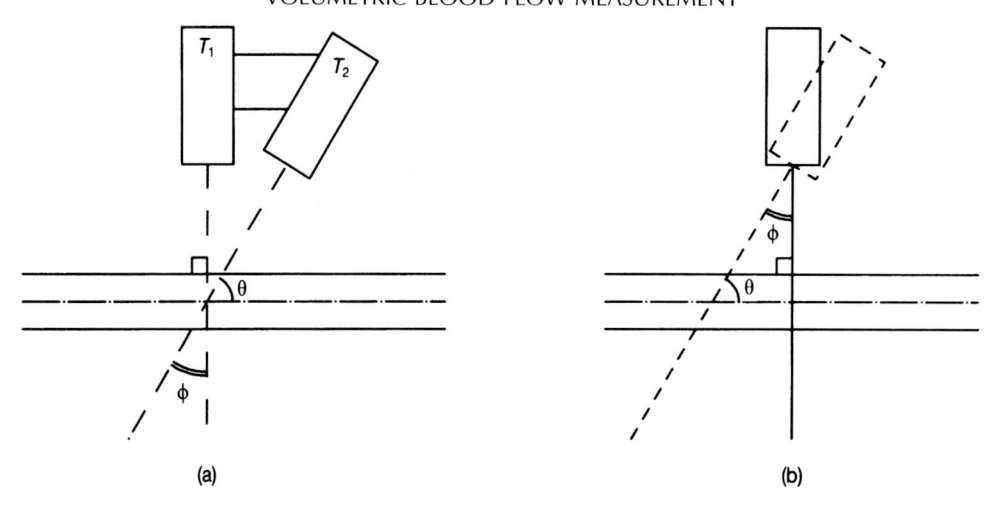

Figure 12.14 Angle estimation by the prior definition of the normal (a) using a fixed pair of transducers, and (b) by rotating a single transducer through a known angle. Note that the two ultrasound beams in (a) may not necessarily intersect with the same part of the vessel

the angle of insonation derived in the same way (Fig. 12.14b).

12.9.2.3 Dual and Triple Transducer Systems

There are a number of methods of measuring θ by comparing the signals from two transducers set at a known angle to each other (see Section 13.6). The simplest, which is only applicable to vessels that are parallel or nearly parallel to the skin, is to arrange two crystals symmetrically about a perpendicular so that they intersect within or near the vessel lumen (Fig. 12.15a) The orientation of the transducer pair is then adjusted until the Doppler shifts detected by each crystal are equal and opposite, in which case θ is equal to $\alpha/2$, where α is the angle between the two crystals (Safar et al 1981, Levenson et al 1981, Wang and Yao 1982).

A more sophisticated method is to use a similar transducer arrangement set at an arbitrary angle to the vessel (Fig. 12.15b) and to use the relationship between the two Doppler shift frequencies to determine the angles (Peronneau et al 1972, 1977, Wang and Shao 1986, Vilkomerson et al 1994, 1997).

Referring to Fig. 12.15b, the Doppler shift measured by transducer 1 (f_{d1}) may be written $Kv \cos \theta_1$, where K is a constant equal to $2f_t/c$. Similarly, the shift measured by transducer 2 (f_{d2}) will be $-Kv \cos \theta_2$. Combining these two expressions to eliminate Kv gives:

$$f_{d2} \cos \theta_1 = -f_{d1} \cos \theta_2 \qquad 12.29$$

Writing θ_2 in terms of θ_1 and γ, the angle between the two ultrasound beams, gives:

$$f_{d2} \cos \theta_1 = f_{d1} \cos(\theta_1 + \gamma) \qquad 12.30$$

Finally, expanding the term in the bracket and rearranging the equation leads to:

$$\theta_1 = \arctan \frac{f_{d1} \cos \gamma - f_{d2}}{f_{d1} \sin \gamma} \qquad 12.31$$

A similar approach is adopted in the system described by Uematsu (1981), except that a single transmitting crystal is flanked by two receiving crystals (Fig. 12.15c).

A variation of the two-transducer approach, described by Borodzinski et al (1976), is to measure the apparent diameters of the vessel as viewed by each transducer in a pulsed Doppler mode (Fig. 12.15d); θ is then calculated from the two diameters and the angle between the two beams.

Referring to Fig. 12.15d, the distance measured by transducer 1 is aa' and may be written $D/\sin \theta_1$, where D is the true diameter of the blood vessel. Similarly, distance bb' measured by transducer 2,

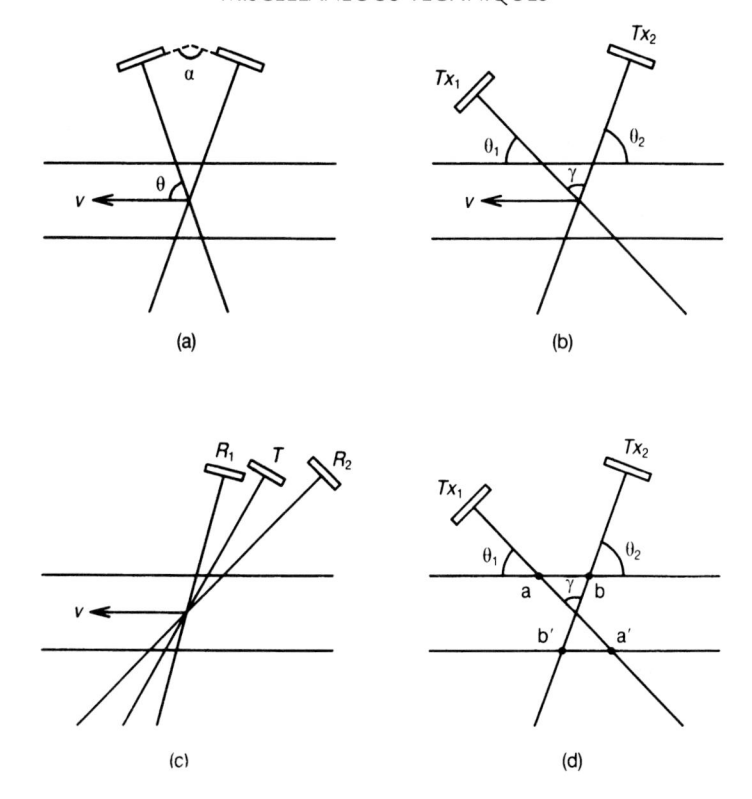

Figure 12.15 Angle estimation by (a) equalising forward and reverse Doppler shifts, (b,c) comparing the Doppler shift frequencies from two transducers, and (d) measuring two apparent diameters with a pulsed Doppler system

may be written $D/\sin\theta_2$. Combining these expressions and eliminating D gives:

$$aa' \sin\theta_1 = bb' \sin\theta_2 \quad 12.32$$

Writing θ_2 in terms of θ_1 and γ gives:

$$aa' \sin\theta_1 = bb' \sin(\theta_1 + \gamma) \quad 12.33$$

Finally, expanding the terms in the bracket and rearranging the equation leads to:

$$\theta_1 = \arctan\frac{bb' \sin\gamma}{aa' - bb' \cos\gamma} \quad 12.34$$

An even more elaborate design of transducer was used by Daigle et al (1975) to make aortic blood velocity measurements with an oesophageal probe, and is illustrated in Fig. 12.16. The centre transducer is mounted perpendicular to the axis of the probe and used for wall motion measurements. The three outer crystals, which are used for the velocity measurements, are spaced at the vertices of an equilateral triangle, and allow triangulation of the blood velocity vector at the ultrasound beam intersection point to determine the orientation of the velocity vector with respect to the probe body. The Doppler angle and velocity magnitude can be found from a simple combination of the velocities measured by the three probes; the derivation of the relevant equations can be found in Daigle (1974).

Vector Doppler techniques and, in particular, angle independent colour flow imaging are further discussed in Section 13.6.

12.9.3 Diameter Estimation Techniques

The majority of diameter and area measuring techniques have already been described earlier in

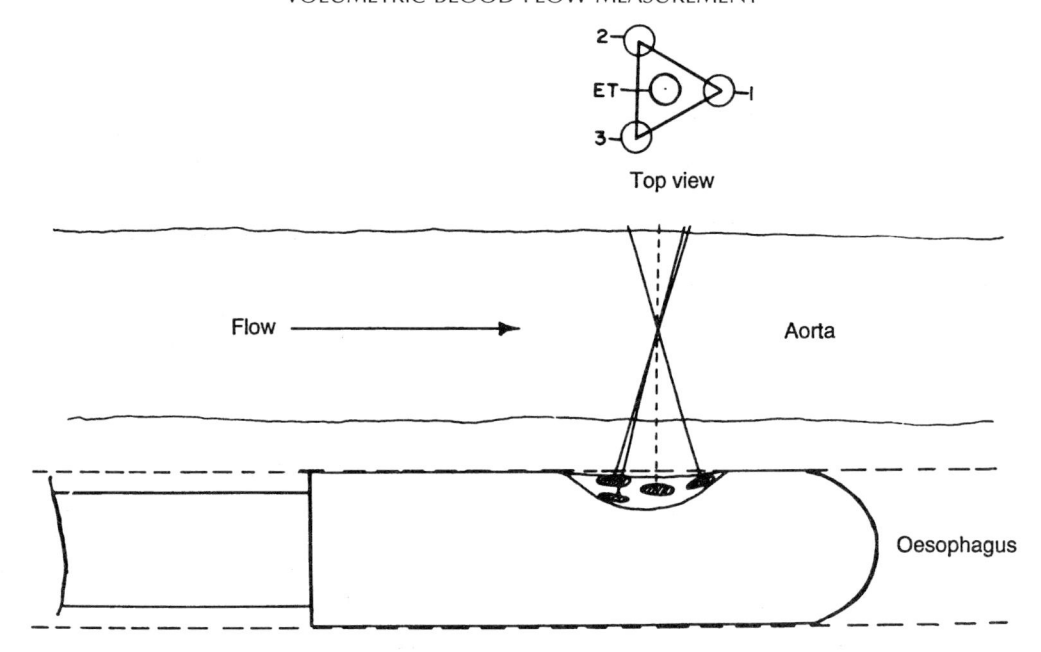

Figure 12.16 Schematic diagram of oesophageal probe used by Daigle et al. Transducers 1, 2, and 3 were used for the Doppler blood velocity measurements, whilst the centre transducer was used to track the motion of the arterial walls (reproduced by permission from Daigle et al 1975, © The American Physiological Society

this chapter. These comprise direct measurement from a B-scan image (Section 12.2.3), measurement using an A-scan transducer orientated at 90° to the vessel axis (Section 12.3.2) and the method of measuring the ratio of two powers used in the attenuation-compensated method (Section 12.5).

One additional method is a pulsed Doppler one, where a range gate is moved across the vessel to establish the points at which the Doppler signal first appears and then eventually disappears. The angle θ must be other than 90°, since a finite Doppler shift is required, and therefore θ must be found at the same time in order to calculate the true diameter. This method tends to overestimate the size of the vessel because of the finite sample volume size (Jorgensen et al 1973, Borodzinski et al 1976, Levenson et al 1981, Safar et al 1981).

A number of methods have also been described for tracking the motion of the vessel wall throughout the cardiac cycle (Hokanson et al 1970, 1972, Wildi et al 1980, Groves et al 1982, Hoeks et al 1985, Struyk et al 1985, Eriksen 1987, Hoeks et al 1990, Wilson et al 1990, Kool et al 1994, Stadler et al 1996) and there have been attempts to combine instantaneous vessel diameter and mean velocity measurements (see Section 12.2.3).

12.9.4 Doppler Vector Tomography

In an entirely different approach to Doppler blood flow measurement, Jansson et al (1997) have described a 'Doppler vector tomography' system. Such a system is only suitable for anatomical sites (such as the breast) that can be completely encircled by the transducer-scanning trajectory, and currently requires extremely long processing times (of the order of hours), but could potentially have application if the latter difficulty can be overcome.

12.9.5 Detected Fractional Moving Blood Volume

Carson and colleagues (Carson et al 1993, Rubin et al 1995) have described a technique for estimating the fraction of moving blood in a region of interest in the tissues, by comparing the integrated power spectrum from that region with that from a reference region, consisting entirely of flowing blood (i.e. entirely within a blood vessel) at the same depth, and with the same overlying tissues. This does not, of course, give a direct measure of either volumetric flow or perfusion, but may have some clinical value in identifying organs with increased vascularity. The method also has a number of limitations, the most serious of which appears to be that flows that are too slow to produce adequate Doppler shifts will not contribute to the estimate. Unfortunately, this includes flow in the capillaries and venules. Furthermore, the method is critically dependent on the adequate rejection of clutter signals from slowly moving solid tissue. Other methods which are intended to quantify perfusion are discussed in Chapter 13 (13.3).

12.10 FLOW MEASUREMENT ERRORS

Each flow measurement method is subject to a number of errors, which limits its use under some conditions. The exact types of error are dependent on the method in use, but all ultimately stem from the same sources.

It is important to distinguish between systematic and random errors, since the former may be of little import if they affect all measurements to a similar degree, so that comparisons of flow between patients and within the same patient are valid. Random errors, on the other hand, may invalidate such comparisons, although many of them may be reduced by repeated measurement. It is worth noting that one of the largest contributions to the overall error of measuring volumetric flow is usually that of vessel size measurement, and therefore mean velocity may be evaluated much more accurately then volumetric flow, and may sometimes be a useful alternative measurement.

12.11 SUMMARY

The problem of measuring volumetric flow may be split into three parts: the measurement of the component of velocity parallel to the axis of the Doppler transducer, the measurement of the angle between the flow axis and the transducer axis, and the measurement of the cross-sectional area of the blood vessel. There are a variety of approaches to each of these measurements. Velocity may be measured using a uniform insonation method, or by measuring or assuming the shape of the velocity profile. The angle θ may be found using imaging techniques, by first finding a normal to the blood vessel using a Doppler or A-scan technique, or by comparing the Doppler shift frequencies or vessel diameter measured by different transducers. The cross-sectional area of the blood vessel may be measured from a cross-sectional image, or calculated from a diameter found either by imaging, by an A-scan technique, or using pulsed Doppler. One method, the attenuation compensated method, makes use of power measurements to eliminate θ from the Doppler equation and to measure the vessel cross-section.

Each method of measurement is subject to errors and it is important that they are appreciated by the user if the best possible accuracy is to be achieved, and if realistic confidence limits are to be placed on the measurement; with care, reasonably accurate measurements of blood flow can be made in a variety of vessels.

As with other ultrasound techniques, flow measurement by Doppler ultrasound has the great virtue of not interfering with the parameters being measured, and flows can therefore be measured under normal physiological conditions, and this benefit may far outweigh the cost of any inaccuracies when compared with other methods.

12.12 REFERENCES

Avasthi PS, Greene ER, Voyles WF, Eldridge MW (1984) A comparison of echo-Doppler and electromagnetic renal blood flow measurements. J Ultrasound Med 3:213–218

Baker DW (1970) Pulsed ultrasonic Doppler blood-flow sensing. IEEE Trans Sonics Ultrason SU-17:170–185

Baker DW, Forster FK, Daigle RE (1978) Doppler principles and techniques. In: Fry FJ (ed) Ultrasound: Its Applications in Medicine and Biology, pp. 161–287. Vol 3 of Methods and Phenomena: Their Applications in Science and Technology (Part 1). Elsevier Scientific, Amsterdam

Bohs LN, Friemel BH, Trahey GE (1995) Experimental velocity profiles and volumetric flow via two-dimensional speckle tracking. Ultrasound Med Biol 21:885–898

Borodzinski K, Filipczynski I, Nowicki A, Powalowski T

(1976) Quantitative transcutaneous measurements of blood flow in carotid artery by means of pulse and continuous wave Doppler methods. Ultrasound Med Biol 2:189–193

Brandestini M (1978) Topoflow—a digital full range Doppler velocity meter. IEEE Trans Sonics Ultrason SU-25:287–293

Brodszki J, Gardiner HM, Eriksson A, Stale H, Marsál K (1998) Reproducibility of ultrasonic fetal volume blood flow measurements. Clin Physiol 18:479–485

Carson PL, Li X, Pallister J, Moskalik A, Rubin JM, Fowlkes JB (1993) Approximate quantification of detected fractional blood volume and perfusion from 3-D color flow and Doppler power signal imaging. In: Proceedings of the 1993 IEEE Ultrasonics Symposium, pp. 1023–1026. IEEE, Piscataway, NJ

Céspedes EI, Carlier S, Li W, Mastik F, van der Steen AFW, Bom N, Verdouw P, Serruys PW (1998) Blood flow assessment using a mechanical intravascular ultrasound catheter: initial evaluation *in vivo*. J Vasc Invest 4:39–44

Christopher DA, Burns PN, Hunt JW, Foster FS (1995) The effect of refraction and assumed speeds of sound in tissue and blood on Doppler blood velocity measurements. Ultrasound Med Biol 21:187–201

Daigle RE (1974) Aortic flow sensing using an ultrasonic esophageal probe. PhD Dissertation, Colorado State University, Fort Collins, CO

Daigle RE, Miller CW, Histand MB, McLeod FD, Hokanson DE (1975) Nontraumatic aortic blood flow sensing by use of an ultrasonic esophageal probe. J Appl Physiol 38:1153–1160

Deane CR, Markus HS (1997) Colour velocity flow measurement: *in vitro* validation and application to human carotid arteries. Ultrasound Med Biol 23:447–452

Doriot PA, Casty M, Milakara B, Anliker M, Bollinger A, Siegenthaler W (1975) Quantitative analysis of flow conditions in simulated vessels and large human arteries and veins by means of ultrasound. In: Kazner E, de Vlieger M, Muller HR, McCready VR (eds) Ultrasonics in Medicine, pp. 160–168. Proceedings 2nd European Congress on Ultrasonics in Medicine, Munich, 12–16 May 1975, Excerpta Medica, Amsterdam

Eik-Nes SH, Marsal K, Brubakk AO, Kristofferson K, Ulstein M (1982) Ultrasonic measurement of human fetal blood flow. J Biomed Eng 4:28–36

Eik-Nes S, Marsal K, Kristoffersen K (1984) Methodology and basic problems related to blood flow studies in the human fetus. Ultrasound Med Biol 10:329–337

Embree PM, O'Brien WD (1990) Volumetric blood flow via time-domain correlation: experimental verification. IEEE Trans Ultrason Ferroelec Freq Contr UFFC-37:176–189

Eriksen M (1987) Noninvasive measurement of arterial diameters in humans using ultrasound echoes with prefiltered waveforms. Med Biol Eng Comput 25:189–194

Eriksen M (1992) Effect of pulsatile arterial diameter variations on blood flow estimated by Doppler ultrasound. Med Biol Eng Comput 30:46–50

Evans DH (1985) On the measurement of the mean velocity of blood flow over the cardiac cycle using Doppler ultrasound. Ultrasound Med Biol 11:735–741

Evans DH (1986) Can ultrasonic duplex scanners really measure volumetric flow? In: Evans JA (ed) Physics in Medical Ultrasound, pp. 145–154. IPSM Report No. 47, IPSM, London

Evans DH, Schlindwein FS, Levene MI (1989c) The relationship between time averaged intensity weighted mean velocity, and time averaged maximum velocity in neonatal cerebral arteries. Ultrasound Med Biol 15:429–435

Evans JM, Skidmore R, Wells PNT (1986) A new technique to measure blood flow using Doppler ultrasound. In: Evans JA (ed) Physics in Medical Ultrasound, pp. 141–144. IPSM, London

Evans JM, Skidmore R, Baker JD, Wells PNT (1989a) A new approach to the noninvasive measurement of cardiac output using an annular array Doppler technique—II. Practical implementation and results. Ultrasound Med Biol 15:179–187

Evans JM, Skidmore R, Luckman NP, Wells PNT (1989b) A new approach to the noninvasive measurement of cardiac output using an annular array Doppler technique—I. Theoretical considerations and ultrasonic fields. Ultrasound Med Biol 15:169–178

Fei DY (1995) A theory to correct the systematic error caused by the imperfectly matched beam width to vessel diameter ratio on volumetric flow measurements using ultrasound techniques. Ultrasound Med Biol 21:1047–1057

Fish PJ (1978) Doppler vessel imaging and its aid to flow measurement. In: Woodcock JP, Sequera RF (eds) Doppler Ultrasound in the Study of the Central and Peripheral Circulation, pp. 50–54. Bristol University Press, Bristol

Fish PJ (1981) A method of transcutaneous blood flow measurement—accuracy considerations. Reprint from: Recent Advances in Ultrasound Diagnosis 3, Proceedings of the 4th European Congress on Ultrasonics in Medicine, Dubrovnik, May 1981

Flaud P, Bensalah A, Peronneau P (1997) Deconvolution process in measurement of arterial velocity profiles via an ultrasonic pulsed Doppler velocimeter for evaluation of wall shear rate. Ultrasound Med Biol 23:425–436

Forsberg F, Liu JB, Russell KM, Guthrie SL, Goldberg BB (1995) Volume flow estimation using time domain correlation and ultrasonic flowmetry. Ultrasound Med Biol 21:1037–1045

Fu CC, Gerzberg L (1983) Annular arrays for quantitative pulsed Doppler ultrasonic flowmeters. Ultrason Imaging 5:1–16

Gibson WGR, Cobbold RSC, Johnston KW (1994) Principles and design feasibility of a Doppler ultrasound intravascular volumetric flowmeter. IEEE Trans Biomed Eng 41:898–908

Gill RW (1979) Pulsed Doppler with B-mode imaging for quantitative blood flow measurement. Ultrasound Med Biol 5:223–235

Gill RW (1982) Accuracy calculations for ultrasonic pulsed Doppler blood flow measurements. Australas Phys Eng Sci Med 5:51–57

Gill RW (1985) Measurement of blood flow by ultrasound: accuracy and sources of error. Ultrasound Med Biol 11:625–641

Gill RW, Kossoff G, Warren PS, Garrett WJ (1984) Umbilical

venous flow in normal and complicated pregnancy. Ultrasound Med Biol 10:349–363

Griffin DR, Teague MJ, Tallet P, Willson K, Bilardo C, Massini L, Campbell S (1985) A combined ultrasonic linear array scanner and pulsed Doppler velocimeter for the estimation of blood flow in the fetus and adult abdomen. II. Clinical evaluation. Ultrasound Med Biol 11:37–41

Groves DH, Powalowski T, White DN (1982) A digital technique for tracking moving interfaces. Ultrasound Med Biol 8:185–190

Hartley CJ, Hanley HG, Lewis RM, Cole JS (1978) Synchronized pulsed Doppler blood flow and ultrasonic dimension measurement in conscious dogs. Ultrasound Med Biol 4:99–110

Hein IA, O'Brien WD (1989) Volumetric measurement of pulsatile flow via ultrasound time-domain correlation. J Cardiovasc Technol 8:339–348

Histand MB, Miller CW, McLeod FD (1973) Transcutaneous measurement of blood velocity profiles and flow. Cardiovasc Res 7:703–712

Hoeks APG, Reneman RS, Peronneau PA (1981) A multigate pulsed Doppler system with serial data processing. IEEE Trans Sonics Ultrason SU-28:242–247

Hoeks APG, Ruissen CJ, Hick P, Reneman RS (1985) Transcutaneous detection of relative changes in artery diameter. Ultrasound Med Biol 11:51–59

Hoeks APG, Brands PJ, Smeets FAM, Reneman RS (1990) Assessment of the distensibility of superficial arteries. Ultrasound Med Biol 16:121–128

Hokanson DE, Strandness DE, Miller CW (1970) An echo-tracking system for recording arterial-wall motion. IEEE Trans Sonics Ultrason SU-17:130–132

Hokanson DE, Mozersky DJ, Sumner DS, Strandness DE (1972) A phase-locked echo tracking system for recording arterial diameter changes *in vivo*. J Appl Physiol 32:728–733

Hottinger CF, Meindl JD (1975) Unambiguous measurement of volume flow using ultrasound. Proc IEEE:984–985

Hottinger CF, Meindl JD (1979) Blood flow measurement using the attenuation-compensated volume flowmeter. Ultrason Imaging 1:1–15

Huntsman LL, Stewart DK, Barnes SR, Franklin SB, Colocousis JS, Hessel EA (1983) Noninvasive Doppler determination of cardiac output in man. Circulation 67:593–602

Jansson T, Almqvist M, Stråhlén K, Eriksson R, Sparr G, Persson HW, Lindström K (1997) Ultrasound Doppler vector tomography measurements of directional blood flow. Ultrasound Med Biol 23:47–57

Jorgensen JE, Garbini JL (1974) An analytical procedure of calibration for the pulsed ultrasonic Doppler flow meter. J Fluids Eng:158–167

Jorgensen JE, Campau DN, Baker DW (1973) Physical characteristics and mathematical modelling of the pulsed ultrasonic flowmeter. Med Biol Eng 11:404–421

Keller HM, Meier WE, Anliker M, Kumpe DA (1976) Non-invasive measurement of velocity profiles and blood flow in the common carotid artery by pulsed Doppler ultrasound. Stroke 7:370–377

Kitabatake A, Tanouchi J, Yoshida Y, Masuyama T, Uematsu M, Kamada T (1990) Quantitative color flow imaging to measure the two-dimensional distribution of blood flow velocity and the flow rate. Jap Circ J 54:304–309

Kool MJF, van Merode T, Reneman RS, Hoeks APG, Struyker Boudier HAJ, Van Bortel LMAB (1994) Evaluation of a reproducibility of a vessel wall movement detector system for assessment of large artery properties. Cardiovasc Res 28:610–614

Kremkau F (1990) Doppler angle error due to refraction. Ultrasound Med Biol 16:L523–L524

Levenson JA, Peronneau PA, Simon A, Safar ME (1981) Pulsed Doppler: determination of diameter, blood flow velocity, and volumetric flow of brachial artery in man. Cardiovasc Res 15:164–170

Lewis P, Psaila JV, Davies WT, McCarty K, Woodcock JP (1986) Measurement of volume flow in the human common femoral artery using a Duplex ultrasound system. Ultrasound Med Biol 12:777–784

Li S, Hoskins PR, Anderson T, McDicken WN (1993a) Measurement of mean velocity during pulsatile flow using time-averaged maximum frequency of Doppler ultrasound waveforms. Ultrasound Med Biol 19:105–113

Li S, McDicken WN, Hoskins PR (1993b) Blood vessel diameter measurement by ultrasound. Physiol Meas 14:291–297

Li S, McDicken WN, Hoskins PR (1993c) Refraction in Doppler ultrasound. Ultrasound Med Biol 19:593–594

Li W, Lancée CT, van der Steen AFW, Gussenhoven EJ, Bom N (1996) Blood velocity estimation with high frequency intravascular ultrasound. In: Levy M, Schneider SC, McAvoy BR (eds) Proceedings of the 1996 IEEE Ultrasonics Symposium, pp. 1485–1488. IEEE, Piscataway, NJ

Li W, van der Steen AFW, Lancée CT, Céspedes I, Bom N (1998) Blood flow imaging and volume flow quantitation with intravascular ultrasound. Ultrasound Med Biol 24:203–214

Light H, Cross G (1972) Cardiovascular data by transcutaneous aortovelography. In: Roberts C (ed) Blood Flow Measurement, pp. 60–63. Sector, London

Light LH (1974) Initial evaluation of transcutaneous aorto-velography—a new non-invasive technique for haemodynamic measurements in the major thoracic vessels. In: Reneman RS (ed) Cardiovascular Applications of Ultrasound, pp. 325–360. North-Holland, Amsterdam

Liu GY, Burns PN (1997) The attenuation compensated C-mode flowmeter: a new Doppler method for blood volume flow measurement. In: Schneider SC, Levy M, McAvoy BR (eds) Proceedings of the 1997 IEEE Ultrasonics Symposium, pp. 1285–1289. IEEE, Piscataway, NJ

Mackay RS (1972) Non-invasive cardiac output measurement. Microvasc Res 4:438–452

Marquis C, Meister JJ, Mirkovitch V, Depeusinge C, Mooser E, Mosimann R (1983) Femoral blood flow determination with a mutichannel digital pulsed Doppler: an experimental study on anesthetized dogs. Vasc Surg 17:95–103

Moser U, Anliker M, Schumacher P, Vieli A, Pinter P (1992a) C-Mode Doppler: a quantitative ultrasonic real-time procedure for the measurement of 2-D velocity fields and volume flow in large vessels. Proc Eurodop '92:63–64

Moser U, Vieli A, Schumacher P, Pinter P, Basler S, Anliker M

(1992b) Ein Doppler-Ultraschall-Gerät zur Bestimmung des Blut-Volumenflusses. Ultraschall in Med 13:77–79

Moser U, Schumacher PM, Anliker M (1995) Benefits and limitations of the C-mode Doppler procedure. Acoust Imaging 21:509–514

Nielsen TH, Iversen HK, Tfelt-Hansen P (1990) Determination of the luminal diameter of the radial artery in man by high frequency ultrasound: a methodological study. Ultrasound Med Biol 16:787–791

Oates CP (1989) The Doppler shift and speed of sound in blood. Ultrasound Med Biol 15:75

Peronneau P, Xhaard M, Nowicki A, Pellet M, Delouche P, Hinglais J (1972) Pulsed Doppler ultrasonic flowmeter and flow pattern analysis. In: Roberts VC (ed) Blood Flow Measurement, pp. 24–28. Sector, London

Peronneau P, Sandman W, Xhaard M (1977) Blood flow patterns in large arteries. In: White D, Brown RE (eds) Ultrasound in Medicine, vol 3B, pp. 1193–1208. Plenum, New York

Picot PA, Embree PM (1994) Quantitative volume flow estimation using velocity profiles. IEEE Trans Ultrason Ferroelec Freq Contr 41:340–345

Picot PA, Fruitman M, Rankin RN, Fenster A (1995) Rapid volume flow rate estimation using transverse colour Doppler imaging. Ultrasound Med Biol 21:1199–1209

Poulsen JK (1997) Four ultrasonic methods for the measurement of volumetric flow with no angle correction. Acoust Imaging 23:197–202

Poulsen JK, Kim WY (1996) Measurement of volumetric flow with no angle correction using multiplanar pulsed Doppler ultrasound. IEEE Trans Biomed Eng 43:589–599

Powalowski T (1988) Ultrasonic system for noninvasive measurement of hemodynamic parameters of human arterial-vascular system. Arch Acoustics 13:89–108

Qamar MI, Read AE, Skidmore R, Evans JM, Williamson RCN (1985) Transcutaneous Doppler ultrasound measurement of coeliac axis blood flow in man. Br J Surg 72:391–393

Rubin JM, Adler RS, Fowlkes JB, Spratt S, Pallister JE, Chen J-F, Carson PL (1995) Fractional moving blood volume: estimation with Power Doppler US. Radiology 197:183–190

Safar ME, Peronneau PA, Levenson JA, Toto-Moukouo JA, Simon AC (1981) Pulsed Doppler: diameter, blood flow velocity and volumic flow of the brachial artery in sustained essential hypertension. Circulation 63:393–400

Schumacher PM, Moser U, Anliker M (1995) *In vitro* measurements of the true volume flow with C-mode Doppler. Acoustical Imaging 21:515–522

Schwarz KQ, Bezante GP, Chen X, Mottley JG, Schlief R (1993) Volumetric arterial flow quantification using echo contrast. An *in vitro* comparison of three ultrasonic intensity methods: radio frequency, video and Doppler. Ultrasound Med Biol 19:447–460

Shung KK, Flenniken RR (1995) Time-domain ultrasonic contrast blood flowmetry. Ultrasound Med Biol 21:71–78

Stadler RW, Karl WC, Lees RS (1996) New methods for arterial diameter measurement from B-mode images. Ultrasound Med Biol 22:25–34

Struyk PC, Pijpers L, Wladimiroff JW, Lotgering FK, Tonge M, Bom N (1985) The time-distance recorder as a means of improving the accuracy of fetal blood flow measurements. Ultrasound Med Biol 11:71–77

Suorsa VT, O'Brien WD (1988) Effects of ultrasonic attenuation on the accuracy of the blood flow measurement technique utilizing time domain correlation. In: Proceedings of the 1988 IEEE Ultrasonics Symposium, pp. 989–994. IEEE, Piscataway, NJ

Teague MJ, Willson K, Battye CK, Taylor MG, Griffin DR, Campbell S, Roberts VC (1985) A combined ultrasonic linear array scanner and pulsed Doppler velocimeter for the estimation of blood flow in the foetus and adult abdomen—1: Technical aspects. Ultrasound Med Biol 11:27–36

Uematsu S (1981) Determination of volume of arterial blood flow by an ultrasonic device. J Clin Ultrasound 9:209–216

van der Steen AFW, Li W, Céspedes EI, Carlier S, Eberle M, Verdouw PD, Serruys PW, Bom N (1997) *In vivo* validation of blood flow estimation using the decorrelation of radio frequency intravascular echo signals. In: Schneider SC, Levy M, McAvoy BR (eds) Proc. 1997 IEEE Ultrasonics Symposium, pp. 1247–1250. IEEE, Piscataway, NJ

Vilkomerson D, Lyons D, Chilipka T (1994) Diffractive transducers for angle-independent velocity measurements. In: Proceedings of the 1994 IEEE Ultrasonics Symposium, pp. 1677–1682. IEEE, Piscataway, NJ

Vilkomerson D, Lyons D, Chilipka T, Lopath P, Shung KK (1997) Diffraction-grating transducers. In: Schneider SC, Levy M, McAvoy BR (eds) Proceedings of the 1997 IEEE Ultrasonics Symposium, pp. 1691–1696. IEEE, Piscataway, NJ

Wang LM, Shung KK (1995) Contrast medium assisted fluid flow measurements. IEEE Trans Ultrason Ferroelec Freq Contr 42:309–315

Wang WQ, Shao QM (1986) Reduced error in double beam Doppler ultrasound flow velocity measurement. Ultrasound Med Biol 12:L413–L417

Wang W-Q, Yao L-X (1982) A double beam Doppler ultrasound method for quantitative blood flow velocity measurement. Ultrasound Med Biol 8:421–425

Wildi E, Knutti JW, Allen HV, Meindl JD (1980) Dynamics and limitations of blood/muscle interface detection using Doppler power returns. IEEE Trans Biomed Eng BME-27:565–573

Willink R, Evans DH (1994) A mean blood velocity statistic for the Doppler signal from a narrow beam. IEEE Trans Biomed Eng 41:322–331

Willink R, Evans DH (1995a) Volumetric blood flow calculation using a narrow ultrasound beam. Ultrasound Med Biol 21:203–216

Willink R, Evans DH (1995b) Volumetric blood flow measurement by simultaneous Doppler signal and B-mode image processing: a feasibility study. Ultrasound Med Biol 21:481–492

Wilson LS, Dadd MJ, Gill RW (1990) Automatic vessel tracking and measurement for Doppler studies. Ultrasound Med Biol 16:645–652

13

Miscellaneous Doppler Techniques

13.1 INTRODUCTION

The purpose of this chapter is to introduce the reader to a range of Doppler techniques which have not been discussed in detail in previous chapters. Some of these techniques (for example transverse Doppler) have been available for many years, but have not yet found significant clinical applications and appear to have an uncertain future. Others (such as ultrasonic contrast agents) have become well established over the last few years. Yet other techniques (such as vector Doppler) have not yet found their way into clinical practice, but promise to be of great importance in years to come.

As with the other chapters in this book, the emphasis will be on the physics of the techniques rather than the many clinical applications these methods have spawned.

13.2 CONTRAST AGENTS

Blood is a relatively weak scatterer of ultrasound in the low MHz region, and it is possible to increase very significantly the intensity of the signal it scatters by injecting 'ultrasonic contrast agents' into an artery or vein (de Jong et al 1991, Campani et al 1998, Calliada et al 1998). Contrast is used in B-mode and also in Doppler imaging techniques when the agent moves with the blood (Fig. 13.1: see Plate VII). The power of spectral Doppler signals can also be dramatically increased with contrast agent (Fig. 13.2: see Plate VII) (Seidel et al 1996, Forsberg et al 1997). In this discussion we will concentrate on techniques which combine Doppler methods and contrast agents. The combination of Doppler and contrast is popular since both improve the sensitivity of blood detection. Doppler velocity and power contrast imaging are of interest for examinations of small vessels, deep vessels, and for depicting large vessels more completely. In about 80% of reported studies, contrast agents are employed just to enhance the signal from the blood pool; in the remaining 20%, some attempt at quantification is made. Quantification is sometimes labelled 'perfusion'; however, in Section 13.3 it will be seen that this term is often used rather loosely. Contrast agent has also been used to enhance the blood signal to make speckle tracking more successful (Shung and Flenniken 1995)

13.2.1 Types of Contrast Agent

Several types of material have been proposed as being suitable as ultrasound contrast agents (Ophir and Parker 1989, Hindle and Perkins 1995). These include free gas bubbles (Gramiak and Shah 1968, Gramiak et al 1969, Kremkau et al 1970, Ziskin et al 1972, Meltzer et al 1980a, 1980b, 1981, 1985), encapsulated gas bubbles (Carroll et al 1980, Barnhart et al 1990, Fritzsch et al 1988, 1990, Simon et al 1990, Unger et al 1993), colloidal suspensions (Ophir et al 1980, 1985, Mattrey et al 1982, 1983, 1987, Mattrey 1989, Parker et al 1987, 1990, Tarter 1991) and aqueous solutions (McWhirt 1979, Ophir et al 1979, Tyler et al 1981). By far the most successful of these at present are those based on encapsulated microbubbles. In fact, free gas bubbles make excellent scatterers of ultrasound and their disadvantages are related mainly to their inability to survive intact in the circulation for an adequate time period; encapsulation is used as a method of stabilising them.

The scattering cross-section, σ, of a particle of radius a, suspended in a medium with different acoustic properties may in the long wavelength situation (i.e. $\lambda \gg a$) be written (Morse and Ingard, 1968):

$$\sigma = \frac{4\pi k^4 a^6}{9}\left[\left(\frac{\beta_p - \beta_s}{\beta_s}\right)^2 + \frac{1}{3}\left(\frac{3\rho_p - 3\rho_s}{2\rho_p + \rho_s}\right)^2\right] \quad 13.1$$

where k is the wave-number ($=2\pi/\lambda$), β_p and β_s are the adiabatic compressibility of the particle and suspending medium, respectively, and ρ_p and ρ_s are the corresponding densities (note this is slightly different in form from eqns 7.3 and 7.4 because it refers to scattering cross-section rather than differential cross-section at a particular angle; see Shung and Thieme, 1993). Substitution of realistic values (see Table 13.1) into eqn. 13.1 for a 1 μm diameter rigid sphere of iron (with high density and low compressibility) and a 1μm diameter sphere of air (low density and high compressibility) immersed in water shows that the scattering cross-section for the gas is 2×10^8 times greater than that for the iron. Obviously this ratio will depend on ultrasound frequency, the size of the scatterer and the materials chosen, but this example does serve to demonstrate that scattering from gaseous targets far exceeds that from solid (or, of course, liquid) targets immersed in a liquid.

There is, however, another extremely important effect to consider when discussing the scattering behaviour of gas bubbles, that of resonance. The fundamental resonant frequency, f_0, of a free gas bubble may be written:

$$f_0 \approx \frac{1}{2\pi a}\sqrt{\frac{3\gamma p_0}{\rho_0}} \quad 13.2$$

Table 13.1 Comparison of scattering cross-sections at 3 MHz for air bubbles and iron particles of 1 μm diameter suspended in water. Values are from de Jong et al (1991)

Material	Density (kg m^{-3})	Compressibility (m^{2}N^{-1})	σ (in water) (m^2)
Water	10^3	450×10^{-12}	–
Air	1.2	7.65×10^{-6}	10^{-11}
Iron	7.8×10^3	5.5×10^{-12}	5×10^{-20}

where a is the bubble radius, γ the adiabatic ideal gas constant, p_0 the ambient fluid pressure, and ρ_0 the ambient density of the surrounding medium (Kinsler et al 1982).

Assuming a value for γ of 1.4 (the value for diatomic gases), an ambient pressure of 1 atmosphere (10^5 N m^{-2}) and a medium density equal to that for water (10^3 kg m^{-3}), eqn. 13.2 may be simplified to give:

$$f_0 \approx \frac{3.3}{a} \quad 13.3$$

where a is in metres and f_0 in Hz, or a is in μm and f_0 in MHz. Thus, a bubble of approximately the size of an erythrocyte has a resonant frequency in the low megahertz region. Below resonance, the scattering cross-section, σ, increases with the fourth power of frequency, as predicted by eqn 13.1, but around the resonant peak (which is relatively narrow) σ increases dramatically and can reach a value 10^3 times greater than would otherwise be the case. Above the resonance, the bubble has a scattering cross-section which is equal to the physical cross-section (i.e. is frequency-independent) and acts as a normal reflector (de Jong et al 1991). Figure 13.3 illustrates the resonant behaviour of free gas bubbles of various sizes.

The fact that micron-sized bubbles resonate at low MHz frequencies is an important coincidence which means that bubbles with a similar size to red blood cells (and which can therefore pass through the microcirculation if suitably protected) scatter sufficient ultrasound power to be easily detected at fairly low concentrations.

As has already been mentioned, whilst free gas bubbles have extremely good scattering properties, bubbles of the most suitable sizes do not survive for an adequate time in the circulation (see Table 13.2), and in particular are absorbed by the capillaries of the lungs, which means that for their successful use intra-arterial injection would be required either to study the systemic arterial circulation or to elicit contrast in the soft tissues.

In order to stabilise gas bubbles, Carroll et al (1980) manufactured an agent based on nitrogen-filled gelatin capsules. These were used to image tumours in rabbits, but their large size (80 μm)

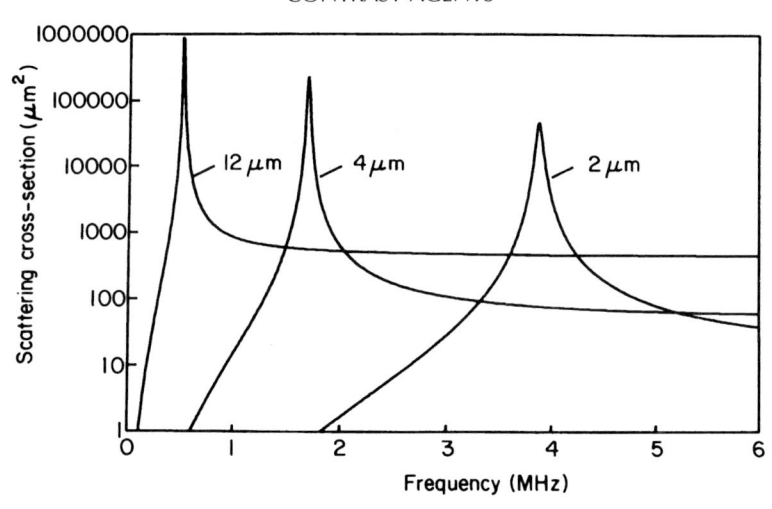

Figure 13.3 The frequency dependence of the scattering cross-section of ideal gas bubbles of three different sizes (diameters 2 μm, 4 μm, 12 μm). It is assumed that the bubbles resonate in a loss-less medium (reproduced by permission of Elsevier Science, from de Jong et al (1991), © 1991

Table 13.2 Dissolution time (in seconds) for air bubbles in water (surface tension 70×10^{-3} N m^{-1}, temperature = 300 K). C/C_s is the ratio of dissolved gas concentration to the saturation concentration. Values are from de Jong (1996)

Sphere radius (μm)	Dissolution time (s) ($C/C_s = 0$)	Dissolution time (s) ($C/C_s = 1$)
1000	1×10^4	6×10^6
100	100	5×10^3
10	1	6
1	≪ 1	≪ 1

precluded intravenous administration. The real breakthrough came in the mid-1980s, when air-filled sorbitol (Feinstein et al 1984a,b) and later albumin (Keller et al 1987) microbubbles were produced using a sonication technique, which by virtue of their small size and encapsulation could survive the passage through the lungs. Since that time there has been tremendous activity in this area, not least because of the perceived commercial value of developing successful ultrasonic contrast agents. Microbubbles of mean diameter similar to that of red blood cells, i.e. ranging in size from 1 to 10 μm are now created by a number of processes. Their encapsulating shells are typically 10 nm in thickness, since thicker shells reduce their oscillatory response to ultrasound waves. Microbubbles of diameter close to 5 μm appear to have similar rheological properties to those of blood cells (Keller et al 1989, Jayaweera et al 1994). Table 13.3 illustrates the characteristics of some microbubbles which are currently being evaluated in clinical practice. At present only three or four bubble agents are commercially available; however, it has been estimated that at least 30 are on trial.

Encapsulated bubbles have similar resonant behaviour to free gas bubbles, but some modification of the theory given above is necessary to account for the stiffness of the shell which both increases the resonant frequency and broadens the resonant peak (de Jong et al 1992). For more details of the theory of both free and encapsulated bubbles, the reader is referred to Medwin (1977), Anderson and Hampton (1980) and de Jong et al (1992).

13.2.2 Interaction of Ultrasound with Microbubble Contrast Agents

The interaction of an ultrasound wave and a microbubble at its resonance frequency can result in a number of different effects. These effects are still being clarified and obviously depend on the

Table 13.3 Properties of some common intravenous, lung-crossing contrast agents (courtesy of C Moran)

Left heart agent	Manufactured by	Type of agent	Capsule	Gas	Bubble size	Dose (concentration)
LEVOVIST	Schering	Lipid stabilised bubble	Palmitic acid	Air	3–5 μm	0.8–3.2 g
SHU 563A (Sonovist)	Schering	Solid microspheres	Cyano-acrylate	Air	Mean 2 μm	0.1–1 μl/kg (200 × 10^8 μ bubbles/ml)
DEFINITY (DMP-115)	ImaRx/Du Pont	Encapsulated bubble	Lipid	Perfluoropropane	Mean 2.5 μm	3–5μl/kg (10 × 10^8 μ bubbles/ml)
QUANTISON^{TM}	Quadrant Healthcare	Rigid microsphere	Albumin	Air	Mean 3.2 μm Range (<2%, >6 μm)	Infused at 1 ml/min (5 × 10^8 μ bubbles/ml)
OPTISON	Mallinckrodt	Encapsulated microsphere	Albumin	Octafluoropropane	Mean 3.7 μm	1 ml @ fundamental 0.5 ml @ second harmonic (5–8 × 10^8 μ bubbles/ml)
BR1 (Sonovue^{TM})	Bracco	Stabilised bubble	Phospholipids	SF$_6$	2–3 μm (90% < 8 μm)	5–40 μl/kg (1–5 × 10^8 μ bubbles/ml)
ALBUNEX (Infoson in Europe)	Nycomed (Europe) Mallinckrodt (USA)	Encapsulated bubble	Albumin	Air	Mean 4 μm Range 2–10 μm	0.025–1.0 ml/kg (3–5 × 10^8 μ bubbles/ml)

structure of the bubble. A new effect can often be exploited to extend existing contrast imaging methods or to provide a new modality.

At low wave pressure, e.g. 0.05 MPa, the bubble resonates in a linear fashion. As the pressure is increased, say into the range 0.05–0.2 MPa the oscillations become non-linear and the bubble scatters ultrasound at both the incident frequency and at higher harmonics. Around these pressures or at higher pressures, the bubble may leak and a short-lived unencapsulated bubble forms next to the original bubble capsule. At even higher pressures, typically around 1.0 MPa, some bubbles break suddenly or even explode in a cavitation process. The interaction of ultrasound and microbubbles is complex and is still being investigated by both acoustic methods and high-speed photography (Leighton 1994, Dayton et al 1997, 1999).

13.2.3 Imaging Techniques with Contrast Agents

Imaging methods, both B-mode and Doppler, are being developed based on the above interaction effects. At the low pressure amplitude, the echo signals are detected and contrast images are produced. The sensitivity is sufficient to allow moderate to large blood vessels and heart chambers to be clearly seen. Such enhancement of Doppler images was one of the first successful applications of contrast agents (Schlief et al 1993).

Harmonic contrast imaging, in which the second harmonic produced by the non-linear oscillations of the bubbles of contrast agent are used instead of the fundamental, gives an additional boost in sensitivity of around 30 dB. This is because although the second harmonic response of contrast agents is considerably reduced relative to the fundamental (typically measured to be between −15 dB and −30 dB; de Jong at al 1994b, Burns 1996, Chang et al 1996), the clutter signal at the second harmonic due to non-linear propagation of ultrasound in tissue, is proportionally much smaller (de Jong et al 1994b, Schrope et al 1992). An additional advantage of the harmonic technique is that it is less prone to the shadowing artefact than basic fundamental frequency imaging because attenuation due to the contrast medium

peaks around the resonant frequency (Forsberg et al 1997). Burns (1996) has published a useful tutorial article on harmonic imaging with contrast agents.

The use of the harmonic signals generated by ultrasound contrast agents was first explored in the early 1990s. Schrope et al (1992) and Schrope and Newhouse (1993) attempted to make perfusion measurements both *in vitro* on excised sheep kidneys and *in vivo* in rabbits. de Jong et al (1994a,b) compared simulated results for harmonic signals with *in vitro* experimental data, using Albunex. Burns et al (1994) used contrast harmonic imaging to study blood flow in the abdominal aortas and kidneys of rabbits, and demonstrated a gain of more than 30 dB in signal-to-clutter ratio and superior imaging of the microvasculature. Since then the method has been widely explored with considerable success (Chang et al 1996, Forsberg et al 1996, Mulvagh et al 1996, Becher et al 1997, Dipak et al 1997, Krishna and Newhouse 1997). Figure 10.6 shows a harmonic contrast image of the heart. Harmonic contrast imaging should be clearly distinguished from harmonic tissue imaging (Section 3.2.5).

When the bubbles are moving, a Doppler effect can be detected in the harmonic frequencies and used to generate velocity and power Doppler images. Chang et al (1995) show that in the case of harmonic Doppler, the standard Doppler equation (eqn 1.1) can be written in a slightly more general form. The Doppler shift, f_{dn}, for the nth harmonic of the transmitted frequency, f_t, is given by:

$$f_{dn} = nf_t - f_r = (2nf_t v \cos\theta)/c \qquad 13.4$$

where $n = 1/2, 1, 2, 3, \ldots$ (sub-harmonic, first harmonic, second harmonic, third harmonic, ...), v is the target velocity, θ is the standard Doppler angle, and c is the velocity of ultrasound. Verbeek et al (1998) derive an equivalent equation for pulsed wave Doppler written in terms of the phase difference between two subsequently received echoes.

A third ultrasound contrast agent technique depends on the phenomenon whereby high amplitude ultrasound causes encapsulated bubbles to leak and to form adjacent free air bubbles, which are then dissolved quickly. The free unencapsulated bubbles give a short-lived but large increase in sensitivity, which has been found to be very useful in tissue blood flow imaging. Since the bubbles are destroyed in a few scanning sweeps of the ultrasound beam (Moran et al 1998), time must be allowed for the flow of blood to replenish the agent in the scan plane before further imaging is attempted (for example, scanning may be stopped for one heart cycle). This type of imaging is called 'intermittent' or 'transient response' imaging (Porter et al 1995). The signals generated by this method are rich in second harmonics and therefore a combination of intermittent, harmonic and power Doppler imaging (IHPD) is a very sensitive approach (Broillet et al 1998).

A related but separate technique for contrast imaging uses bubbles which break suddenly under the influence of the ultrasound beam and scatter the echoes with random phases. The Doppler imaging circuitry detects these phase changes and displays a colour image in the regions of agent uptake. This technique has been called 'loss of correlation' (LOC) imaging, 'stimulated acoustic emission' (SAE) imaging, 'sono-scintigraphy', and 'flash echo' imaging (Blomley et al 1998). Its use in parenchymal imaging is being investigated at present (Hauff et al 1997, Blomley et al 1998). The interaction between ultrasound and encapsulated microbubbles is complex, and the details of the physical mechanisms involved in both this and the previous technique still require elucidation. An image formed using the LOC technique can be found in Fig. 10.9 (see Plate V).

Finally, Simpson and Burns (1997) have recently introduced a technique for the detection of contrast agents called 'pulse inversion doppler'. In this method every second pulse in the pulse Doppler sequence of transmitted pulses is inverted. Then, using subtraction and addition signal processing, the linearly and non-linearly scattered components of the echoes, predominantly from tissue and contrast agent, respectively, can be separated (this is possible because in a non-linear process, negative and positive portions of the pressure wave behave differently). These components can be used to produce grey-shade B-mode images; however, conventional and harmonic Doppler processing may also be applied. The Doppler-shifted signals from linear and non-linear scatterers behave in different ways, and the portion of the Doppler signal appearing between $-prf/4$ and $+prf/4$ contains only Doppler signals arising from non-linear scattering which can therefore be analysed separately (the rest of the spectrum, from $-prf/2$ to $-prf/4$ and $+prf/4$ to $+prf/2$, contains signals from both linear and non-linear scattering). The Nyquist limit to maintain this separation is half that for normal pulsed Doppler techniques, but this is not a serious problem, since the increased ability to detect contrast agent is primarily required in small vessels where there is slow flow. The advantage over normal harmonic imaging is that wideband signals can be used, since separation of the components does not rely on filtering and hence better axial resolution is obtained. Pulse inversion Doppler also appears to give up to 10 dB more agent-to-tissue contrast than conventional harmonic Doppler and up to 16 dB more than conventional broadband Doppler. Since only software modifications are required, pulse inversion Doppler can readily be incorporated into PW, colour velocity and power Doppler machines.

13.2.4 Applications of Contrast Agents

Contrast agent is used, often very successfully, as a means of increasing the echo signal magnitude and hence the clinical sensitivity; indeed, contrast agents are sometimes referred to as 'image enhancers'. In addition to enhancing the signal from small and deep vessels, the introduction of agent may increase the number of small vessels observed. Contrast agents may also be exploited to make 3-D Doppler images more complete (Delker and Turowski 1997). Some quantification of the echo signals is attempted as described below in the discussion on perfusion. The main clinical applications are in assessing the vascularity of tumours to discriminate between malignant and benign (Cosgrove 1994, Bell et al 1995, Fein et al 1995), in depicting the uptake of agent in the myocardium (Villanueva and Kaul 1995, Keller et al 1990, Mor-Avi et al 1993, Rovai et al 1993) and in improving the signal quality in transcranial blood flow studies (Ries et al 1993, Bogdahn et al 1993, Kaps et al 1997, Murphy et al 1997, Baumgartner et al 1997, Postert et al 1998). A small number of more specialised studies have been reported such as distinguishing fetal and maternal blood flow in the placenta, flow through implanted shunts, and delineating cardiac chambers during stress tests (Schmiedl et al 1998, Furst et al 1998, Simpson et al 1998, Bazzocchi et al 1998, Leischik et al 1997, Distante and Rovai 1997). In the future, agents may be organ-specific or be used in drug delivery where they could be destroyed at the site of delivery. In cardiac applications to observe blood flow in the myocardium, an infusion rather than bolus approach is advocated at present to reduce the effect of attenuation by the agent in the heart chambers, although at the expense of a slow build-up of background signal as a result of recirculation (Albrecht et al 1998).

Contrast agents are fragile, and for reproducible results considerable care needs to be taken in defining the study protocol and handling procedures. The safety of contrast applications is discussed in Section 14.6.

13.3 PERFUSION TECHNIQUES

Perfusion of tissue may be defined as the blood flow through the very fine capillaries that control the transport of nutrients and cell products to and from the tissue (Eriksson et al 1991), and is generally measured in units of volume of blood per unit mass of tissue per unit time (e.g. ml per 100 g of tissue per minute).

Perfusion as defined in these terms would clearly require detection of slow blood flow in capillaries. The term 'perfusion' is, however, often used fairly loosely in the ultrasound literature and usually refers to flow in small vessels (based implicitly or explicitly on the assumption that the blood flow in the sample volume is proportional to capillary flow) (Dymling et al 1991, Eriksson et al 1991, 1995, Schrope and Newhouse 1993) and not to true perfusion. The use of the terms 'tissue vascularization' (Fein et al 1995) or 'tissue vascularity' (Hirsch et al 1995) when more appropriate would help to reduce the confusion in this area. It has not been demonstrated that even the most sensitive of power Doppler devices can detect perfusion, although there is some evidence of flow in very small vessels in the kidney being depicted (Durick et al 1995, Dacher et al 1996). Attempts continue to quantify power Doppler images by assuming that the power of the signal is related to the amount of moving blood in the tissue (Rubin et al 1995). Since blood in tissue may be moving very slowly and at times almost stopped, it is questionable whether Doppler techniques are appropriate for true perfusion studies. Since no Doppler signal is detected from very slow blood movement, measurement cannot be made of the whole blood pool. It may be that in time a clinically useful Doppler tool will be established even with this problem, but care will obviously be required in building up experience of interpreting its data.

Contrast agents, as described in Section 13.2, have been employed with Doppler methods to provide a larger backscatter signal and hence greater sensitivity in perfusion-type studies (Greenberg et al 1996, Taylor et al 1996, 1998a,b, Huber et al 1998, Hartley et al 1993, Schwarz et al 1996). The lack of Doppler signal from very slow microbubbles is a problem, as for red cells. A new type of contrast agent, which disintegrates instantly under the influence of the ultrasonic pulse and generates phase-shifted echoes which are detected by Doppler circuitry, may prove useful for perfusion studies, since the echo signal depends on the presence of the agent rather than its motion (Hauff et al 1997, Blomley et al 1998) (see Section 13.2.3). Perfusion techniques with real-time B-mode imaging obviously can obtain echo signals from slow or static microbubbles; however, they are still being developed to improve their sensitivity.

13.3.1 Quantification of Passage of Contrast Agent

The uptake of contrast agent to assess blood flow in heart muscle is one of the main fields for perfusion studies and has been undertaken with both grey shade B-mode imaging and power Doppler methods. In this application the use of the agent permits qualitative assessment of blood supply to the myocardium, but not an accurate measure of perfusion (Bos 1997, Villanueva and Kaul 1995). There is also interest in measuring perfusion in the parenchyma of organs in general.

Time–intensity curves are often plotted for the uptake and clearance of contrast agent, and parameters are extracted from them in attempts to assess the perfusion (Fig. 13.4) (Bos 1997). Typical parameters are:

1. Time from injection to appearance of contrast at site of interest.
2. Time from appearance to peak contrast.
3. Half-life of contrast at site of interest.
4. Time from appearance of contrast to half peak during clearance.
5. Peak contrast.
6. Area under the curve.

Analogies are drawn with approaches which utilise radiopharmaceuticals and fluorescent microspheres (and thorough studies often use these techniques as gold standards for ultrasonic contrast developments). Unfortunately, ultrasound perfusion techniques have several hurdles to overcome before they can match these analogous methods. It is worth listing some of these hurdles:

1. There is a range of bubble sizes in each injection and a corresponding range in acoustic responses, e.g. resonance frequency and backscatter cross-section.
2. The bubbles have a finite life-time, which is affected by physiological pressure, acoustic beam pressure and their chemical environment.
3. Bubbles may stick in parts of the circulation.
4. Variation of Doppler power signal with velocity due to electronic filter settings, particularly at low velocities.
5. Difficulty in standardising the imaging protocol.
6. With venous agents, the flow input function to the site of interest is strongly affected by the transit through the lungs.
7. The background signal may vary as the agent passes, making it difficult to subtract.
8. Shadowing due to attenuation by the agent can be a serious problem, in terms of both its

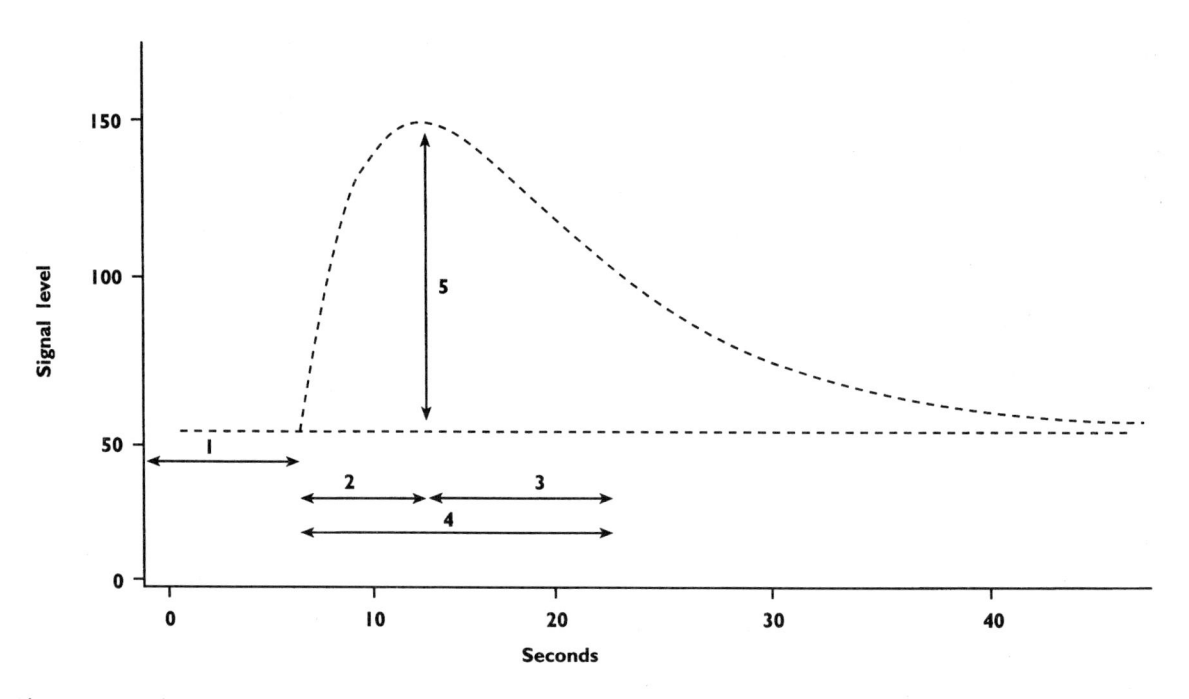

Figure 13.4 Schematic time–intensity curve, showing uptake and clearance of contrast agent. Some typical parameters are indicated which may be of value in attempts to make the technique more quantitative. 1, Time from injection to appearance of contrast; 2, time from appearance to peak contrast; 3, half-life of contrast at site of interest; 4, time from appearance to half peak during clearance; 5, peak contrast

magnitude and its variation with the passage of the agent.

9. Video signals are compressed to suit displays and the human eye, and are therefore not accurately related to agent concentration. Digitisation of the basic RF echo signal can avoid this problem.

Classical indicator dilution modelling is not adequate in this situation. The use of high amplitude ultrasound fields, which propagate in a non-linear fashion, and of harmonic Doppler techniques increase the difficulties of quantification. A more achievable goal may be reproducible qualitative imaging rather than perfusion.

13.4 EMBOLI DETECTION

It has been known for many years that ultrasound can be used to detect bubbles or particulate matter in blood both *in vitro* (Austen and Howry 1965, Patterson and Kessler 1969) and *in vivo* (Gillis et al 1968, Herndon et al 1975), but it was the recognition that Doppler ultrasound provides a sensitive method for detection of emboli entering the cerebral circulation in many clinical situations (Padayachee et al 1986, Pugsley 1986, Spencer et al 1990) that gave a real impetus to the development of Doppler embolic detection systems.

The basic principle of emboli detection is simply that, as an embolus passes through the Doppler sample volume, provided that its scattering cross-section is sufficiently large, it will give rise to an additional component of the Doppler signal which can be heard (embolic signals are variously described as sounding like a 'snap' a 'chirp' or a 'moan'; Ackerstaff et al 1995) or seen on a Doppler display. Figure 13.5 (see Plate VII) shows embolic events as they might appear on both a standard sonogram display and an oscilloscope screen. Such events can be characterised in terms of their frequency, intensity and duration. The size and characteristics of the embolic Doppler signal greatly influences both the success of embolic detection and the way in which signals from emboli are best processed, and the next two sections deal with these two aspects of embolic signals.

13.4.1 Embolic Signal Power

Because the Doppler power from an artery fluctuates very significantly in the short term due to the random nature of the signal scattered from blood (Angelsen 1980, Mo and Cobbold 1986), increases in returned Doppler power can only be reliably detected if the power generated by the embolus is significantly greater than that generated by the surrounding blood. Furthermore, even if the power scattered by the embolus is large, because of the unknown attenuation by the tissues between the ultrasonic transducer and the embolus, it is impossible to make direct measurements of this power, and therefore the signal from the surrounding blood is used as a 'standard scatterer' with which the embolic signal can be compared. Thus 'embolic power' is usually expressed in decibels relative to the signal from blood. This ratio will be referred to as the measured embolic power ratio ($MEPR$) and can be written as:

$$MEPR = 10 \log_{10} \frac{\text{Doppler power measured when embolus is present}}{\text{Doppler power measured when embolus is absent}} \quad 13.5$$

There are many factors which combine to determine $MEPR$, and which ultimately set a limit on the accuracy with which information can be derived from Doppler signals.

A useful starting point in the discussion of $MEPR$ is the closely related concept of 'embolus to blood ratio' (EBR) introduced by Moehring and Klepper (1994), and described as the ratio of the acoustic power backscattered from the embolus to that from the moving blood surrounding the embolus. Although EBR is similar to $MEPR$, the latter is determined by EBR together with a number of additional factors (such as the trajectory the embolus follows through the Doppler sample volume and the type of signal processing employed by the Doppler instrumentation). Mathematically, Moehring and Klepper defined EBR as:

$$EBR = \frac{\sigma_B + \sigma_E}{\sigma_B} = 1 + \frac{\sigma_E}{V\alpha} \quad 13.6$$

where σ_E and σ_B are the backscatter cross-sections of the embolus and flowing blood within the sample volume, respectively, V is the volume of flowing blood within the Doppler sample volume, and α is the volume backscatter coefficient of blood.

The backscatter cross-sections of spherical particles and bubbles suspended in a surrounding homogeneous medium has been explored by Lubbers and van den Berg (1976) who calculated the scattering from gaseous microemboli, red blood cell aggregates and fat emboli in flowing blood by extending Rschevkin's (1963) solution for the scattering of low amplitude plane waves by a sphere. The resulting equations are complex, but show that the backscatter cross-section is a function of embolus size, ultrasound frequency and the composition of both the embolus and its surrounding medium (as characterised by their density and the velocity of propagation of ultrasound). Figure 13.6 shows the solution to Lubbers and van den Berg's equations for the backscatter cross-section of air bubbles and red cell aggregates as a function of embolus diameter, for three transmitted ultrasound frequencies. These graphs reveal a number of important points. Firstly, it is clear that the power of the returned signals from gaseous emboli are much higher than those from similarly sized solid emboli (by two to three orders of magnitude); secondly, there is an overlap between the backscatter cross-section of large solid emboli; and small gaseous emboli; and thirdly, although the backscatter cross-section of gas increases monotonically with size, the same is not true for solid particles (at least in this idealised model).

As noted in eqn 13.6, the backscatter cross-section of the blood volume is determined by the product of the volume of blood within the effective sample volume and the volume backscatter coefficient of blood. The backscatter coefficient of blood for ultrasound has been explored by many authors (Section 7.2), but may be thought of as a function of the scattering cross-section of red cells in plasma, the distribution of aggregate sizes and the packing factor (Mo and Cobbold 1986). Fortunately, these parameters do not change greatly under normal flow conditions in large arteries, and so this quantity can be calculated to a sufficient degree of accuracy to predict the approximate magnitude of the volume backscatter coefficient, and hence theoretical values of EBR for different situations.

Figure 13.7a shows the theoretical EBR (in dB) for two sizes of gaseous emboli at three different ultrasound frequencies, plotted as a function of the effective sample volume. A number of points emerge from these graphs. The first is that the

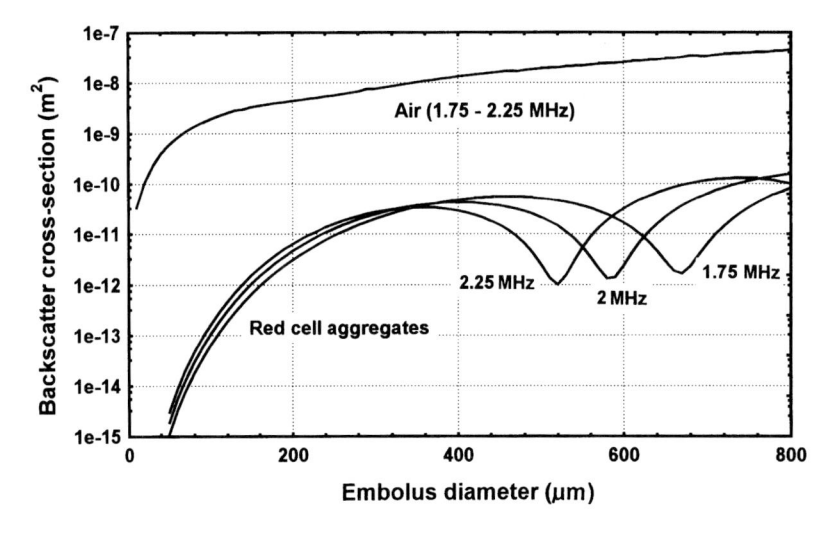

Figure 13.6 Theoretical values of back-scattering cross-section for gaseous and particulate (red cell aggregate) emboli plotted as a function of embolus diameter for three different ultrasound frequencies. The three sets of results for the gaseous emboli almost coincide and therefore have been plotted as a single line

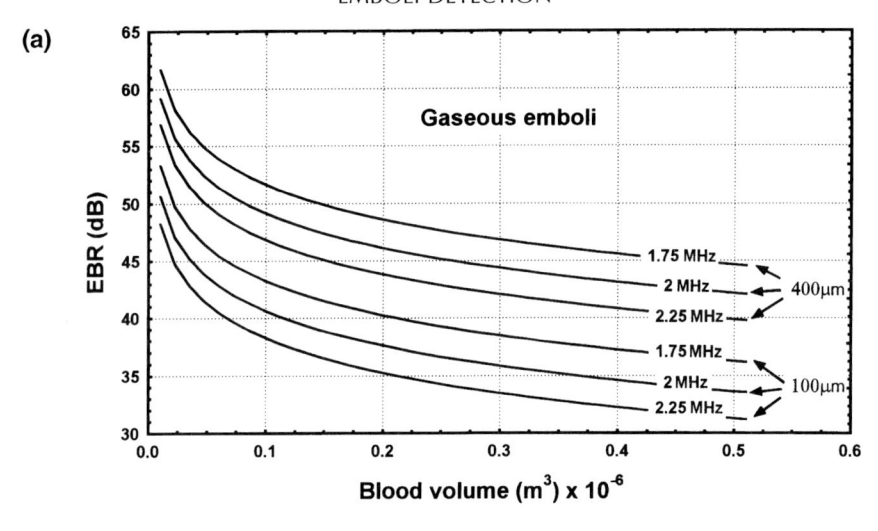

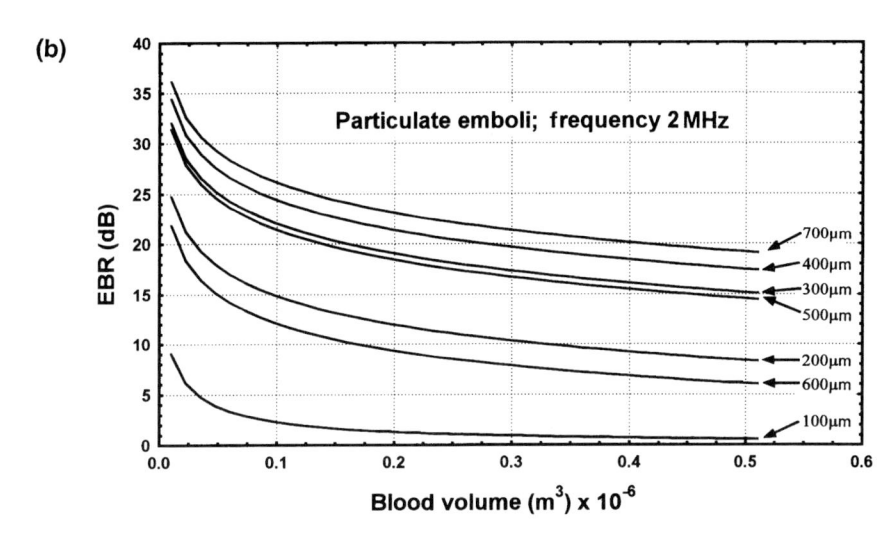

Figure 13.7 (a) Embolus-to-blood power ratio (*EBR*) for gaseous emboli of two different diameters (100 μm and 400 μm) plotted against the sample volume for blood at three different ultrasound frequencies. (b) EBR for particulate emboli (red cell aggregates) of several different diameters (100–700 μm) plotted against the sample volume for blood. The transmitted ultrasound frequency is 2 MHz

smaller the sample volume, the higher the *EBR*, and hence theoretically the easier it is to detect an embolus (because the background signal from blood is less likely to swamp it). One of the practical consequences of this is that it is easier to detect emboli in small rather than large arteries. A second point to arise from Fig. 13.7a is that the lower the ultrasound frequency, the higher the *EBR*. The reason for this is that in the low MHz frequency range, blood acts as a Rayleigh scatterer and, under these conditions, the backscattered power is proportional to the fourth power of frequency (Shung et al 1976), whilst the emboli of the size shown are outside the Rayleigh range and their backscatter power increases more slowly with frequency, giving rise to an overall decrease in *EBR* with frequency. This is one reason why although emboli are easily detected in the middle cerebral artery using trans-cranial Doppler (TCD) systems which operate at 2 MHz, they are not normally detected in the carotid arteries, where ultrasound frequencies are usually at least double those used for TCD.

Figure 13.7b shows the theoretical *EBR* for several sizes of particulate emboli at 2 MHz. As for Fig. 13.7a *EBR* increases with reducing sample volume, but in this case *EBR* does not increase monotonically with embolic size. The reason for this can be understood by reference to Fig. 13.6, which shows that the backscattering cross-section of solid emboli do not, at least in this theoretical model, rise monotonically with embolic diameter. In normal monitoring situations, however, it seems probable that most microemboli fall below the size range where the departure from monotonicity would become significant.

Although *EBR* has an important influence in determining *MEPR*, there are many other factors which are significant in this respect, the most important of which are the shape of the ultrasound beam, the embolic trajectory along the artery, and the interaction of these two parameters. Other factors that influence the *MEPR* are the way in which the Doppler power from the blood is measured and the technique used to measure the returned power from the embolus itself.

13.4.2 Embolic Signal Characteristics

From first principles the Doppler signal from a single 'point' target travelling at a more or less constant speed through a classical 'tear-shaped' sample volume (Section 5.4.2) would take the form of an amplitude modulated sine wave, such as that shown in Fig. 13.5B (see Plate VII). If the embolus were to gradually change velocity (i.e. speed or direction) as it travelled through the sample volume, then in addition some degree of frequency modulation would also be present, as shown in Fig. 13.5C.

The likely duration of such signals can be calculated from the velocity of blood flow in the artery and the Doppler sample volume length. It should be noted, however, that emboli have different physical properties to the surrounding blood, and may not move in the same way as an equivalent sized blood element. Furthermore, it should be remembered that the sample volume of a Doppler instrument is not a fixed quantity, but rather depends on the power returned by the target, and so is larger for heavily scattering targets such as gas bubbles. Allowing these cautions, and taking the middle cerebral artery as an example, blood flow velocity is of the order of $0.2-0.5$ ms^{-1} and sample volume lengths generally of the order of 5 mm, which gives an approximate duration for embolic signals of the order of 10–50 ms (bearing in mind that not all the emboli will travel in the centre stream). These values are compatible with experimental measurements of embolic signal duration made during carotid endarterectomy (Evans et al 1997).

The other key attribute of embolic signals is their magnitude, or more specifically their *MEPR*, and these can be estimated from the *EBR* values given in Figs 13.7a and 13.7b, which suggest that *MEPR* values for particulate emboli might vary between 0 dB and 25 dB, and those for gaseous emboli between about 30 dB and 50 dB. Once again these figures are compatible with experimental measurements made during carotid endarterectomy (Evans et al 1997).

Thus, Doppler signals from emboli can be thought of as amplitude-modulated sine waves, which may also contain frequency modulations. Most embolic signals have durations of between 2 ms and 100 ms, whilst the majority of signals from solid emboli are probably at the lower end of this range. The power received from an embolus may be as much as 65 dB above the signal from blood, but once again the majority of signals from solid emboli are probably at the lower end of this range. These singular characteristics influence the way in which Doppler signals from emboli should be processed.

13.4.3 Processing Embolic Signals

A common way to process ultrasound Doppler signals is to subject them to Fourier transformation, and to display the results in the form of a sonogram. It is questionable, however, whether this is the best way to process and display embolic signals, and certainly it must be modified to some extent if it is to be used for this purpose. The reasons for this are related to the characteristics of embolic signals discussed in the last section, i.e. their extremely large dynamic range and their sometimes extremely short durations.

The dynamic range of interest for signals from blood flow rarely exceeds 20 or even 15 dB, and since most stand-alone pulsed Doppler units were originally designed to process blood flow signals, their dynamic range is generally no more than 25–30 dB. As seen in the previous section, however, embolic signals can be as much as 55 dB greater than the signal from blood, and therefore if the gain of a conventional Doppler unit is set sufficiently high to detect blood flow signals, it will become overloaded by signals from emboli with large scattering cross-sections. This leads to waveform clipping in the time-domain display and to noise spikes in the frequency domain display, which occupy most or all of the spectral display, in both the positive and negative frequency directions (Fig. 13.8: see Plate VIII), which renders any meaningful measurements on the embolic velocity or *MEPR* impossible.

The extremely short durations of some embolic signals also means that Fourier transform techniques which have a spectral resolution that is limited to the reciprocal of the temporal resolution (Section 8.2.2) are unsuitable for detailed analysis, and that alternative spectral estimation approaches, such as Wigner analysis (Section 8.2.5), are necessary if detailed measurements of both the frequency and duration of the embolic signal are required (Smith et al 1994).

Although modern spectral analysis techniques may have considerable advantages over the Fourier method, it is possible that just as much information could be gathered from measurements made directly on the time domain signal. In the case of blood flow, it is necessary to transform the Doppler signal into the frequency domain, as there is a wide distribution of velocities present. In the case of embolic signals, however, when only one target is present in the ultrasound sample volume, most if not all the information required from the signal can be readily estimated from the time domain representation.

For a more detailed discussion of technical aspects of detection and processing of Doppler signals from emboli the reader is referred to Evans (1999).

13.5 TRANSVERSE DOPPLER

In general it is considered desirable to make Doppler ultrasound measurements on blood vessels using a relatively small Doppler angle (usually ≤60°). One important reason for this is that this configuration maximises the Doppler shift frequency from blood flow, whilst at the same time minimising the Doppler shift frequency from radial vessel motion, thereby facilitating the separation and rejection of the wall clutter components. A second reason is that, as the standard Doppler equation (eqn. 1.1) shows, the Doppler shift frequency is proportional to the cosine of the Doppler angle, θ, which of course varies more slowly as a function of θ when θ is small. Therefore, errors introduced into estimates of velocity by small errors in the measurement of the angle θ, or by non-axial flow, have a less serious impact on the accuracy of velocity estimates when a small Doppler angle is used (see Section 12.2.4).

An interesting variation on the standard Doppler technique of estimating the velocity axial to the ultrasound beam by measuring the Doppler shift frequency, is to estimate the velocity perpendicular to the ultrasound beam by measuring the Doppler bandwidth—the so-called transverse Doppler method. This technique, which appears to have originally been suggested by Newhouse et al (1987) and which has been explored in a number of subsequent publications (Newhouse and Reid 1990, Tortoli et al 1992, 1993a,b, Cloutier et al 1993, Tortoli et al 1994, Newhouse et al 1994, McArdle et al 1995, Lu 1996, Yeung 1998), can of course only work under specific circumstances where the Doppler bandwidth is determined largely or solely by intrinsic spectral broadening (broadening due to the properties of the measurement system—see Section 7.3.4) rather than other broadening mechanisms such as non-stationarity.

Perhaps the most obvious application of the transverse Doppler technique is for estimating blood velocities in vessels at beam-to-flow angles at or near 90°, when for one reason or another it is not possible to achieve a more conventional

Doppler angle of 60° or less. Other possible applications include 'blind' Doppler measurements, where the ultrasound transducer is adjusted so that the Doppler spectrum is symmetrical about the zero frequency line, a condition that only exists at 90° (Tortoli et al 1993b), and the combination of the conventional and transverse Doppler methods to allow the vectorial evaluation of blood velocity (Newhouse et al 1994, McArdle et al 1995).

13.5.1 Transverse Doppler Theory

It has been shown in Section 7.3.4.3 (eqn. 7.22), and much more rigorously by Censor et al (1988), that the bandwidth of the Doppler signal from flow with uniform velocity, v, through the waist of a focused ultrasound beam may be written:

$$B_d \approx \frac{2f_t}{c} \frac{D}{F} v \sin \theta \qquad 13.7$$

where f_t is the transmitted ultrasound frequency, c the velocity of ultrasound, D is the transducer diameter, and F its focal length. Thus, under the special conditions described above, the measured bandwidth of the Doppler signal could be used to determine v. Fortunately, and perhaps unexpectedly, as has been shown theoretically by Newhouse and Reid (1990) and experimentally by Tortoli et al (1992), this same result holds true for straight line flow with the same orientation anywhere in the beam.

Equation 13.7 is written in terms of a single velocity, v, and the question naturally arises as to the effect on bandwidth, if the Doppler sample volume is sufficiently large to encompass a range of velocities, and this issue has been addressed by Tortoli et al (1993a). The required bandwidth may simply be written as the difference between the maximum Doppler shift frequency recorded, $f'_{d(max)}$, and the corresponding minimum frequency $f'_{d(min)}$, i.e:

$$B_d = f'_{d(max)} - f'_{d(min)} \qquad 13.8$$

The maximum Doppler shift frequency (including the effects of spectral broadening) may be written:

$$f'_{d(max)} = f_{d(max)} + B_{d(max)}/2 \qquad (13.9)$$

where $f_{d(max)}$ is the Doppler frequency due to the maximum velocity (without the effects of broadening) and $B_{d(max)}$ is the broadening of that component. This expression may easily be rewritten:

$$f'_{d(max)} = \frac{2v_{max}f_t}{c} \left\{ \cos \theta + \frac{D}{2F} \sin \theta \right\} \qquad 13.10$$

(see Section 8.4.5). Calculation of the minimum frequency is not quite so straightforward, because since spectral broadening is proportional to centre frequency, depending on the geometry, the minimum frequency component may either arise from broadening of the minimum component or from broadening of the maximum component. These two cases may be written as:

$$f'_{d(min)} = f_{d(min)} - B_{d(min)}/2 \qquad (13.11)$$

and

$$f'_{d(min)} = f_{d(max)} - B_{d(max)}/2 \qquad (13.12)$$

In fact it is easily shown that the value of $f'_{d(min)}$ given by eqn. 13.11 is always less than that given by 13.12, provided that the fractional broadening, B_d/f_d, is less than 2, i.e. $f'_{d(min)} > 0$. The critical value of the Doppler angle, θ, at which 13.12 becomes less than 13.11 is when one extreme ray of the ultrasound beam becomes perpendicular to the flow axis, and may be written:

$$\theta_{crit} = \arctan \left(\frac{2F}{D} \right). \qquad 13.13$$

For values of θ less than θ_{crit}, $f'_{d(min)}$ will be given by:

$$f'_{d(min)} = \frac{2v_{min}f_t}{c} \left\{ \cos \theta - \frac{D}{2F} \sin \theta \right\} \qquad 3.14$$

and therefore the bandwidth may be written:

$$B_d = \frac{2f_t}{c}\left\{ v_{max}\left(\cos\theta + \frac{D}{2F}\sin\theta\right) - v_{min}\left(\cos\theta - \frac{D}{2F}\sin\theta\right)\right\} \quad 13.15$$

whilst for $\theta \geq \theta_{crit}$,

$$f'_{d(min)} = \frac{2v_{max}f_t}{c}\left\{\cos\theta - \frac{D}{2F}\sin\theta\right\} \quad 13.16$$

and

$$B_d = \frac{2v_{max}f_t}{c}\left\{\frac{D}{F}\sin\theta\right\} \quad 13.17$$

Note also that if the sample volume is sufficiently large to encompass an entire vessel which contains uni-directional flow, then v_{min} will be zero, and the bandwidth expression for $\theta < \theta_{crit}$ given in eqn 13.15 reduces to the expression for the maximum frequency given in eqn 13.10, i.e:

$$B_d = \frac{2v_{max}f_t}{c}\left\{\cos\theta + \frac{D}{2F}\sin\theta\right\} \quad 13.18$$

To summarise these relationships (Tortoli et al 1993a):

1. For range cells which only include the axial flow region, the dependence of bandwidth on beam to flow angle is given by eqn 13.17.
2. For range cells which include the whole vessel diameter, if the Doppler angle is greater than the critical angle, θ_{crit}, then the bandwidth is also given by eqn 13.17.
3. For range cells which include the whole vessel diameter, if the Doppler angle is less than the critical angle, θ_{crit}, then the bandwidth is given by eqn 13.18.
4. The critical angle, θ_{crit}, is that at which one of the beam edges is normal to the vessel axis, and is given by eqn 13.13.

As has already been mentioned, the above relationships only hold where the Doppler bandwidth is determined solely or largely by intrinsic spectral broadening, rather than non-stationary broadening (see Section 8.2.2.1). Cloutier and colleagues (1993) have made experimental measurements (using porcine blood) of the total spectral broadening in steady and pulsatile flow loop models and have shown that for pulsatile laminar flow, the relative Doppler bandwidth invariance theorem (Newhouse and Reid 1990, Tortoli et al 1992) does not hold during flow acceleration and deceleration due both to non-stationary broadening and to spectral skewness. The spectral analysis technique used for this study was Fourier analysis, and Cloutier et al make the point that the non-stationarity broadening could probably be reduced by using more modern spectral analysis techniques, such as parametric modelling (Section 8.2.4) or time–frequency distributions (Section 8.2.5). In the same publication, Cloutier et al also reported results from pulsatile turbulent flow and found, not surprisingly, additional spectral broadening due to the wider velocity distribution.

In a recent development, Lu (1996) has suggested a method of improving the accuracy of transverse Doppler measurements by using a limited diffraction beam designed to produce peaks at the boundaries of the Doppler spectrum.

13.6 VECTOR DOPPLER TECHNIQUES

A major limitation of all conventional Doppler techniques is that, however good they may be, they only provide the component of the velocity vector along the ultrasound beam. In order to interpret the results obtained from Doppler measurement of blood flow, it is therefore usual to image the vessel wall and to make the assumption that the flow direction is parallel to the walls of the vessel, and to adjust the measured velocity appropriately (hence the cosine θ term in the standard Doppler equation). The assumption that the flow is parallel to the vessel wall may be reasonable in long straight undiseased vessels, but any geometrical changes, such as curves, branches, bifurcations or narrowings may invalidate this (see Chapter 2), and therefore there has for many years been interest in measuring more than one component of the velocity vector in order to obtain true angle and magnitude information.

In general, if a sample volume can be interrogated from two non-collinear directions, then it is possible to calculate the component of the velocity vector in the plane defined by the two directions of interrogation (although no information is available about the component perpendicular to that plane). Furthermore, if a sample volume can be interrogated from three different non-coplanar directions, then it is possible to calculate the true magnitude and direction of the vector in 3-D space. Many schemes for measuring the velocity vectors in both 2-D and 3-D have been published over the years, and many of these are reviewed below. For the sake of simplicity, these have been divided into techniques that are primarily designed to derive vectorial information from a single sample volume, and those designed to derive vectorial information from an extended plane, although some of those in the former category could at least in theory be adapted for many sample volumes.

13.6.1 Vector Doppler Techniques: Individual Sample Volumes

One of the earliest descriptions of the use of more than one Doppler transducer for determining the magnitude and direction of velocity vectors in 2-D is that given by Peronneau et al (1972), who demonstrated the principle for steady and oscillatory flow in straight and curved pipes. Their system consisted of two separate pulsed Doppler transducers, one pointing upstream and one pointing downstream, inclined at a fixed angle to each other, such that their beams crossed within the flow channel. The system was able to construct velocity profiles by taking samples from a number of positions across the pipe, using serial measurements. Because of the fixed geometry of the system, the sample volumes defined by the two transducers only coincided at one radial position in the tube, therefore for other positions, pairs of results from the same radial displacement but different axial positions were combined. The 'quadrature diffractive transducer' recently described by Vilkomerson et al (1994) measures the 2-D velocity vector in essentially the same way as described by Peronneau et al, and is a simple and convenient way of implementing this approach if the transducer can be placed close to the velocity to be measured.

Apparently the first *in vivo* use of 3-D 'vector Doppler' was by Daigle et al (1975) who constructed and tested a four-transducer oesophageal probe for measuring aortic blood flow in the dog (Fig. 12.16). A central pulse-echo transducer, which was mounted perpendicular to the axis of the probe and used for wall motion measurements, was surrounded by three PW Doppler transducers mounted at 60° to the plane of the central crystal and spaced at the vertices of an equilateral triangle with 7 mm sides. The beams from the three Doppler transducers converged at a point 7 mm away from the face of the probe, which was roughly in the centre of the aorta of an average-sized dog. Each Doppler transducer was used independently as both a transmitter and receiver to obtain three separate projections of the true velocity vector. Curiously enough, in view of what has been said above, this information was used to determine the axis of the aorta by assuming that the velocity vector must be parallel to this axis during high forward flow. Once the data for determining the orientation of the aorta had been acquired, one of the PW transducers (the 'tip' transducer) was used to scan across the aorta and make velocity measurements at 0.6 mm increments so as to construct a velocity profile. In addition, the time-varying diameter of the aorta was measured with the central pulse-echo transducer. Daigle and colleagues carried out a series of *in vivo* comparisons between measurements made using this device, and measurements made using implanted electromagnetic and PW Doppler flow transducers, and after correcting for refraction caused by the different acoustic properties of the coupling gel and the oesophagus, which increased the nominal Doppler angle from 60° to 64°, obtained good agreement. Daigle et al also reported 'significant radial components of the blood velocity vector during flow reversal and the transition between forward and reverse flow'. Although this early system worked well, a clear problem (and one that applies to the majority of methods described in this section) is that because of the fixed geometry, 3-D velocity measurements could only be made in one point in space relative to the transducer.

Peronneau et al (1977) published details of measurements made with their two-transducer system (described above) on large vessels in the thorax of the dog. By placing the transducer in various positions on the aortic arch and brachio-cephalic trunks, they were able to reconstruct 2-D velocity profiles along various lines through these vessels, at several phases of the cardiac cycle. Figure 13.9 reproduces two such sets of results.

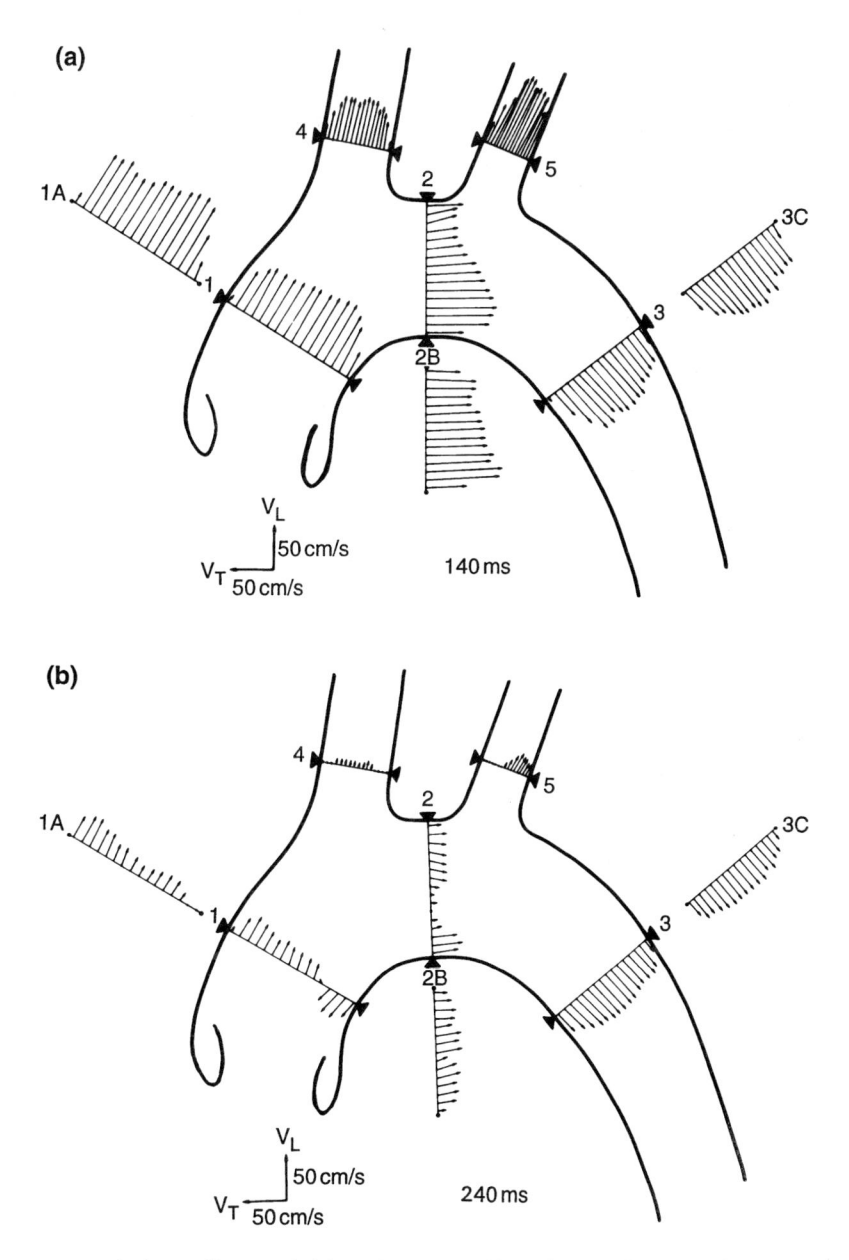

Figure 13.9 Instantaneous velocity profiles recorded from the aorta and brachio-cephalic trunks of a dog at two different phases of the cardiac cycle recorded by Peronneau and colleagues: (a) $t = 140$ ms; (b) $t = 240$ ms. The total cardiac cycle length was 400 ms (reproduced by permission from Diebold and Peronneau 1985)

The following year Fox (1978) proposed a multiple crossed-beam ultrasound Doppler velocimeter that was intended to determine, independently, three orthogonal velocity components from a sample volume within the body, using three separate crossed ultrasound beam systems. Two such systems determined the velocity in two orthogonal directions parallel to the face of the transducer, by each simultaneously transmitting along two crossed beams, and detecting the resulting scattered ultrasound signal from the region where the four beams crossed with a separate receiving transducer. The third crossed-beam system, intended to determine the velocity component axial to the transducer face, acted in a conventional transmit–receive mode. Experimental measurements made using a revolving turntable produced encouraging results, but the complex arrangement of transducers necessary appeared to limit its potential for clinical measurements.

Sato and Sasaki (1979) suggested a new approach to determining the 3-D velocity vector, which consisted of transmitting pulses along a single direction through the tissue, and receiving the resulting scattered signals from a single sample volume with three separate transducers pointing in non-coplanar directions. The particularly novel feature of this approach was that the three signals were then heterodyned and their three cross-bispectra calculated, and these results were interpreted to yield the magnitude and direction of the velocity vector. Sato and Sasaki claimed that this arrangement is less susceptible to the effects of noise than a conventional system using separate determinations of different components of the velocity vector, because the bispectrum of any Gaussian random signal is zero. They were able to demonstrate the basic operation of the system using a phantom consisting of a rotating acrylic disc of random thickness.

Uematsu (1981) published details of a system for measuring volumetric flow which was developed by Yoshimura and colleagues (Furuhata et al 1978, 1979). The ultrasound probe, illustrated in Fig. 13.10, consisted of a pulse-echo transducer (A) for measuring the diameter of the blood vessel, together with three coplanar Doppler transducers for measuring blood flow velocity. The central transducer (T) was used as a transmitter, whilst the two outside Doppler transducers (R_a and R_b) were used as receivers. Measurements of volumetric flow using this configuration were compared with measurements made with an electromagnetic flowmeter, both for steady flow *in vitro* and for pulsatile flow in an animal model, with encouraging results.

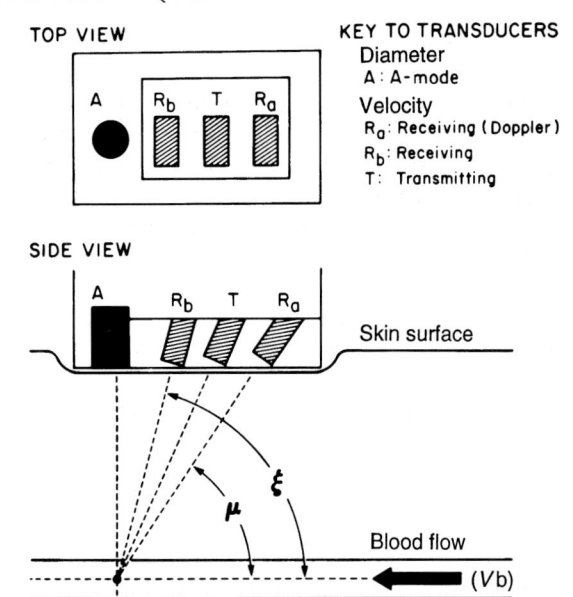

Figure 13.10 Probe arrangement developed by Yoshimura and colleagues for measuring volumetric flow. Transducer A is an A-mode transducer used to measure vessel diameter; the other three transducers are for Doppler measurements. Transducer T is used purely for transmission, whilst R_a and R_b are used simultaneously for reception (reproduced by permission of Churchill Livingstone, from Uematsu 1981)

Fox and Gardiner (1988) used three non-coplanar transducers, each of which was used in a conventional transmit–receive mode. They developed a closed-form solution of the Doppler equation for the magnitude and direction of the 3-D velocity vector, and used both turntable and flow phantom experiments to validate their approach. In the same year, Ashrafzadeh et al (1988) published an analysis of velocity estimation errors for single-transducer, dual-transducer and triple-transducer systems designed, respectively, for 1-D, 2-D and 3-D determinations of the velocity vector, and concluded, not surprisingly, that a triple transducer design is the most appropriate for studying flow in situations, such as turbulence, where the velocity vector may be pointing in any direction.

Since the 1980s there have been many publications relating to the use of vector Doppler techniques. Schrank et al (1990) described a method for measuring the 2-D flow vector using a conventional duplex scanner in conjunction with a 'position location system' which allowed the scan head to be repositioned on the surface of the skin to insonate a given sample volume from two different directions. The system was used to obtain flow velocity data from two locations over the human femoral artery and to reconstruct and plot 2-D velocity vectors. The Seattle group have also published details of both 2-D (Overbeck et al 1992) and 3-D (Dunmire et al 1995, Beach et al 1996) vector Doppler systems, comprising a central transmitting transducer with a very narrow focused beam, surrounded by either two (for 2-D), or four (for 3-D) receiving transducers with wide unfocused sensitivity zones (Fig. 13.11). These authors argue that this configuration is better than one which uses two or three non-coplanar Doppler transceivers, in that it is less susceptible to the effects of refraction. Overbeck et al (1992) show that for simple models consisting of parallel layers of tissues with differing ultrasound velocity values, the sample volume misregistration due to refraction in the case of the one-transmitter/two-receiver configuration is less than that in the case of the two-transceiver configuration. Overbeck et al also describe validation experiments with a moving string phantom, and results from the carotid artery of a healthy volunteer. Dunmire et al (1995) describe experiments with steady flow in various stenosis models, and conclude that their 3-D system is 'capable of detecting parabolic flow, the region of high speed para-axial flow corresponding to jet flow, the recirculation zone between the jet and the tube wall, and the regions of jet break-up'. Figure 13.12 (see Plate VIII) illustrates the results obtained by Dunmire and colleagues for a steady flow rate of approximately 2.46 l min^{-1} through a 40% stenosis in a 15 mm diameter tube.

An entirely different approach to determining the 3-D flow vector, which only uses two separate Doppler transducers, has been suggested by Newhouse et al (1991, 1994). The method, which consists of measuring not only the mean frequency shifts detected by the two transducers but also their bandwidths, is a combination of a conventional 2-D vector method and a transverse Doppler method (Section 13.5). The basic configuration of the

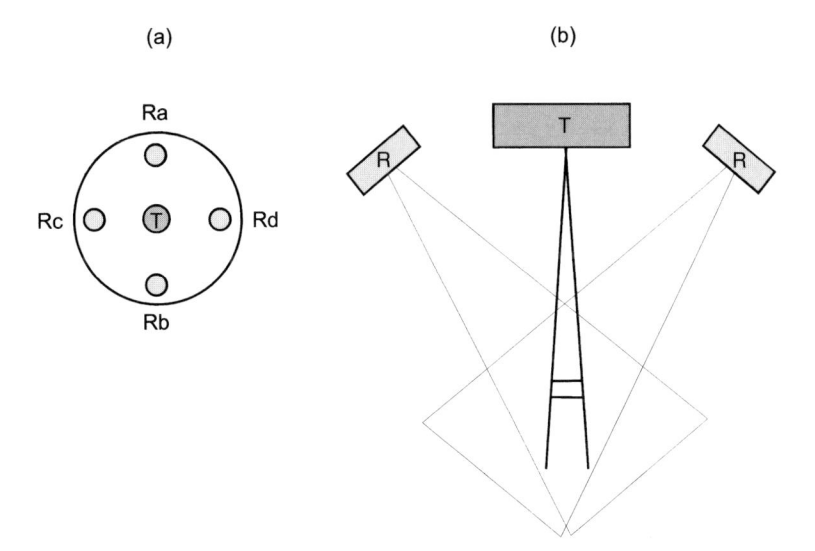

Figure 13.11 Transducer arrangement used by the Seattle group for vector Doppler measurements. (a) Plan view of transducer showing transmitting crystal surrounded by four receiving crystals. (b) Section through transducer showing narrow transmitted beam and wide reception zones (redrawn by permission from Dunmire et al 1995, © 1995 IEEE)

system described by Newhouse et al (1994) is shown in Fig. 13.13a. A pair of circular aperture transducers, focused so that their sample volumes coincide, transmit and receive in tandem. This results in a triple-peaked Doppler spectrum, as shown in Fig. 13.13b. Each side peak is due to transmission by one transducer, and reception by the same transducer. The middle peak is due to transmission by one transducer and reception by the other and, since this represents the combined cross-reception of the two transducers, is twice as high as the outer peaks. Newhouse et al (1994) show that the separation of the outer peaks, S_p, is proportional to the velocity component in the plane of the two transducers and normal to the line bisecting the angle formed by their beams, i.e. V_x. The shift of the central peak from zero, C_p, is proportional to the radial velocity parallel to the line bisecting the two beams, V_z. Finally, the spectral width of the two outer peaks can be written in terms of V_x, V_y and V_z, and therefore since V_x and V_z are already known, V_y can be found. McArdle et al (1995) compared this method of determining the 3-D flow vector in a flow phantom with the 2-D and 3-D methods described by the Seattle group (Overbeck et al 1992, Dunmire et al 1995). They found that the two-transducer method of Newhouse followed the results given by the five-transducer 3-D method, and reduced the errors given by the three-transducer 2-D method.

Two further recently described techniques (Anderson 1997, 1998, Jensen and Munk 1998) impose a lateral spatial modulation on the ultrasound field, and estimate the lateral velocity components of targets moving through the field from the resulting modulation of the received signals.

In addition to the Doppler-based approaches to velocity vector measurements, some authors have described correlation-based methods. Routh et al (1990) described a dual-beam 2-D cross-correlation method. An aluminium lens was used to generate two focal spots a fixed distance apart in the flow channel, and the flow velocity estimated from the time taken for an ensemble of scatterers to pass from one focus to the other (determined by cross-correlating the two received ultrasound signals). Routh et al also suggested that the method could be extended to measure 3-D velocities. Dotti et al (1992) proposed a 3-D method using a two-sector transducer probe, which was used 'to create three partially superimposed sensitivity areas between which it is possible to measure correlation functions in such a way as to reveal possible displacements of the scattering centres moving within these areas'. Dotti and colleagues (Cantarelli et al 1994, Dotti and Lombardi 1996) have also described a correlation-based technique which requires only a single ultrasound transducer, and which is capable of determining both blood flow velocity and the angle between the ultrasound beam and the direction of flow (see Section 13.8 for more details). Hein (1995, 1997) has described an

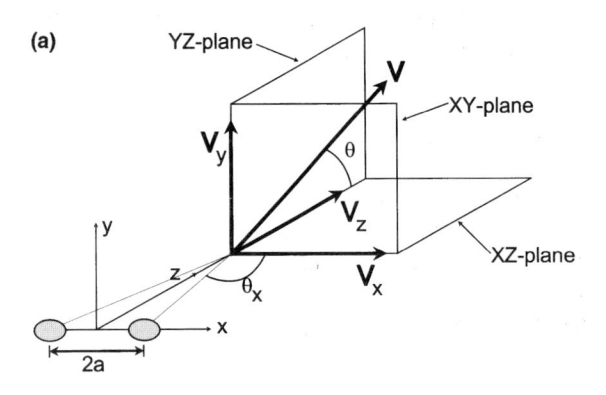

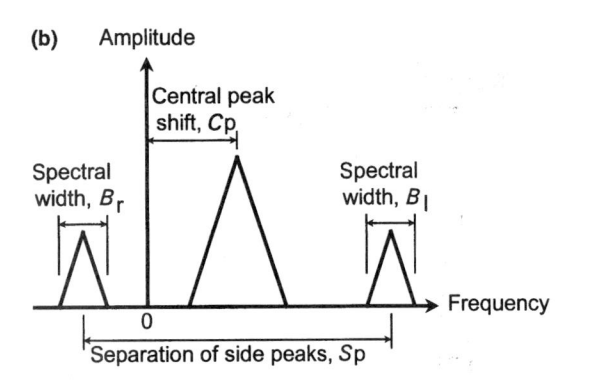

Figure 13.13 Two transducer system described by Newhouse and colleagues for determining the 3-D flow vector. (a) The 3-D vector is decomposed into three components V_X, V_Y, and V_Z. The two circular apertures are centred on the X-axis and are focused to the same position on the Z-axis. (b) The triple-peaked Doppler spectrum resulting from the arrangement shown in (a) (redrawn by permission from Newhouse et al 1994, © 1994 IEEE)

extension of the method used by Routh et al, which employs a triple-beam lens transducer (Fig. 13.14) to generate three parallel closely spaced ultrasound beams. Using this transducer it is possible, using RF correlation techniques, to track scattering centres within blood, both as they move along the beams and as they move from beam to beam, thus permitting the determination of the 3-D velocity vector. Hein (1997) presents results from continuous, fully developed flow in a flow phantom under a wide range of flow rates and flow direction, and concludes that the accurate measurement of the 3-D flow velocity vector is possible using the method described. Several workers have also explored 2-D speckle tracking techniques, which seek to provide a map of 2-D vectors in the imaging plane by tracking speckle patterns in the B-mode image from frame to frame. These techniques are discussed in Section 13.7.

13.6.2 Vector Doppler Techniques: Angle Independent Colour Flow Imaging

Although it is valuable to be able to determine the 3-D velocity vector for individual sample volumes, in a clinical situation it is much more valuable to be able to map the 2-D components of the velocity vectors over an extended region of the scan plane, since this provides a much better appreciation of the overall flow pattern, and several authors have constructed systems designed to do this. With such a system, some information about the components of the flow vectors perpendicular to a particular scan plane can of course also be gained by simply rotating the transducer to interrogate another plane. A map of the 3-D components would, in theory, be even better but at present is impracticable, and would pose considerable challenges in terms not only of data acquisition but also of display. Indeed, even the visualisation of 2-D vector maps has required new display techniques to be developed. Methods which have been used include: the superposition of vector arrows on colour flow images or diagrams (Tamura et al 1990, Xu et al 1991, Maniatis et al 1994b, Giarrè et al 1996); the use of single colour images with hue representing angle, and intensity and saturation representing velocity magnitude (Schmolke and Ermert 1988, Bohs and Trahey 1991, Vera et al 1992); and the use of separate colour-coded angle and magnitude images (Fei et al 1994, Hoskins 1997).

In order to construct 2-D maps of the 2-D components of the velocity vectors in an imaging plane, it is necessary to interrogate a tissue slice from two or more directions, and to combine the velocity information thus acquired from each sample volume. The best way of achieving this in real time is to arrange for colour flow imaging of the tissue slice from two or more directions, using either a single transducer whose steering angle can be altered in real time, or alternatively using more than one transducer (or portions of the same transducer which can be steered separately). One of the first successful attempts to implement this technique was reported by Schmolke and Ermert (1988), who used two separate linear array transducers set at 90° to each other, in a tomographic configuration, to compute a 2-D vector map. The system was reportedly used to

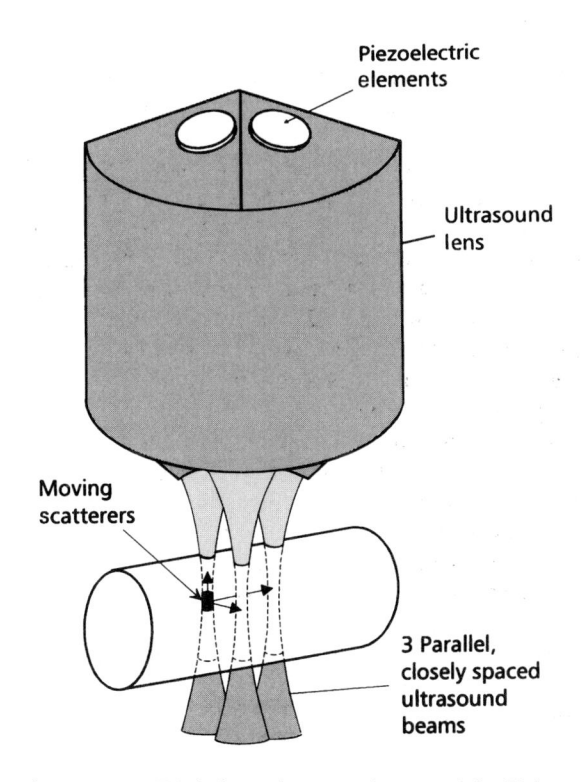

Figure 13.14 Triple beam lens transducer used by Hein to determine 3-D velocity vectors using RF correlation techniques (reproduced by permission from Hein 1997, © 1997 IEEE)

image simple test objects *in vitro*, but unfortunately, because colour reproduction was not available no experimental images are shown.

Since this early work, nearly all reports of angle-independent colour flow imaging (AICFI) have been based on the use of a single linear array transducer, the beam of which is electronically steered 'left', 'right' or 'vertical' (Tamura et al 1990, 1991, Fei et al 1993, 1994, 1997, Hoskins et al 1994, Maniatis 1994a,b, Hoskins 1997). The geometry of this method is illustrated in Fig. 13.15. One colour image is acquired from each direction of interest, and the information from these images combined to derive the angle and magnitude (parallel to the scan plane) of the velocity vectors at each point where sufficient information is present. This process is illustrated in Fig. 13.16A,B (see Plate IX). Figure 13.16A shows a series of colour flow images obtained from a flow rig containing a 73% stenosis for various beam-to-vessel angles when the beam from the linear array is steered 'left' and 'right'. Figure 13.16B shows corresponding AICFI velocity magnitude and angle plots.

Tamura et al (1990, 1991) appear to have been the first authors to report on the use of the steered linear array method. Xu et al (1991) discussed the general theory of velocity vector computations using both two-colour flow images and more than two-colour flow images (where the equations for the velocity vector are therefore over-defined), and in particular compared two possible methods of dealing with the latter situation. In the first method they derived velocity vector images from all possible pairs of raw colour flow images, and then calculated a weighted average, where the weighting factors were determined from the angle between each pair of sound beam axes. In the second method they used a least squares technique, designed to minimise the statistical noise, to compute a single vector image from all the raw colour flow images. Experimental results obtained both *in vitro* and *in vivo* suggested that the use of images from more than two directions improved the vector image quality substantially when compared to vector images derived from only two raw images.

Maniatis et al (1994b) compared the bias and variance of four reconstruction methods, essentially those suggested by Xu et al (1991), using a model of simulated laminar flow in a tube. For their 'two-component method', only two raw images were used to derive the vector results; for their 'simple average method', the vector values derived from three pairs of directions were assigned the same weights; for their 'weighted average method', the weights for each of the three pairs were assigned according to the expected variances (calculated from each vector magnitude and direction). Their 'least mean squares method' was as described by Xu et al. Maniatis et al concluded from these studies that the dominant factor affecting reconstruction accuracy is the angle between the observation directions, and that contrary to the results reported by Xu et al (1991), the simple two-component method generally has as good a performance as the more complex alternatives.

Fei et al (1993, 1994) studied the accuracy of measurements made with an AICFI system *in vitro* by comparing calculated velocities with those derived from measured volumetric flow rates, and by comparing measured angles with those calculated from the co-ordinates of the tube image. Three velocity components were calculated for each pixel, corresponding to the three possible steering angles of the linear array, and the two largest components selected for calculating velocity vector magnitude

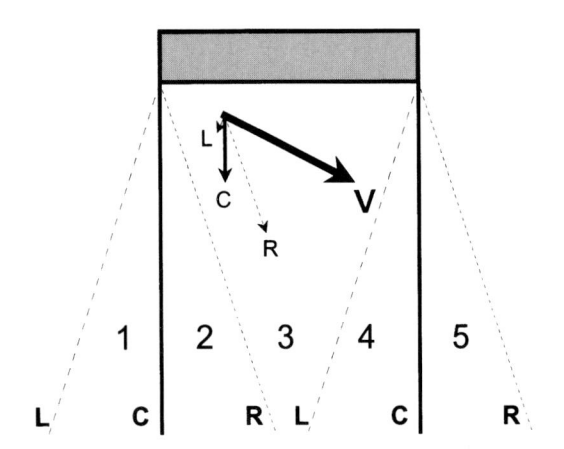

Figure 13.15 Steering of linear array transducer so as to interrogate sample Doppler sample volumes from two or three independent co-planar directions. The beam may be steered left (L), centre (C) or right (R), to find the corresponding component of the vector **V**. Regions 1 and 5 may only be interrogated from one direction, regions 2 and 4 from two directions, and region 3 from all three directions

and direction. Fei et al concluded that the AICFI method is capable of providing relatively accurate true velocity information for a 2-D flow field.

Giarrè et al (1996) described a method of spatially filtering the raw colour flow image, designed to improve the accuracy of vector reconstruction. The new method was tested on data obtained from a single transducer that was translated between measurements so as to produce a set of results from two overlapping sectors. The flow vector reconstructed without filtering had an error of more than 20% associated with it, as opposed to a 10% error observed if the prefiltering technique was used.

One notable exception to the use of a steered linear array for real-time measurements is the method used by Phillips et al (1995), in which a linear array transducer is electronically split into two sub-apertures, one of which transmits but both of which are simultaneously used for reception. This method is illustrated in Fig. 13.17. Note that in this arrangement the receiver for sub-aperture A must track echoes along the direction of the acoustic pulse transmitted from sub-aperture B, at the same time that echoes are received by sub-aperture B. Velocity vector reconstructions are carried out using the same principles as for the steered single-aperture linear array method, although the two interrogation angles are constantly changing. There are, however, a number of significant differences between the two methods. Firstly, the sub-aperture technique has a wider field of view than the steered single-aperture technique, as both apertures can be simultaneously aimed at a point well outside the normal region of overlap for the latter technique. Secondly, because of the range of angles available in the sub-aperture technique, the difference in the two insonation angles, particularly near the probe, can be greater, which is known to improve the accuracy of vector reconstruction (Maniatis et al 1994b). Thirdly, the information from a similar field of view can be obtained more rapidly, since each transmitted burst contributes to results from two lines of sight. Whilst the sub-aperture technique appears to have advantages when compared with the single-aperture technique, it is more difficult for non-commercial research groups to implement, and this may explain why the latter method is at present much more popular. As vector techniques become incorporated in commercial machines, it will be interesting to see whether the new technique overtakes the older technique, or whether it has hitherto unrecognised disadvantages.

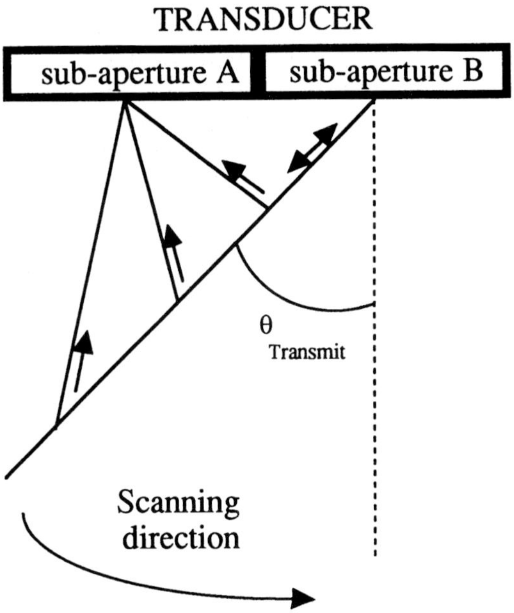

Figure 13.17 Scanning configuration suggested by Phillips et al in which the linear array is electronically split into two sub-apertures. Sub-aperture B scans in a conventional backscattered phased array format, while sub-aperture A receives angular-scattered echoes by tracking the transmitted beam formed by sub-aperture B (reproduced by permission of Elsevier Science, from Phillips et al 1995, *Ultrasound in Medicine and Biology*, © World Federation of Ultrasound in Medicine and Biology)

A number of interesting results obtained using the AICFI technique have now begun to appear in the literature. These include studies of the flow in an end-to-side anastomosis model (Maniatis et al 1994a), studies of spiral laminar flow *in vivo* (Hoskins et al 1994, Stonebridge et al 1996), peak velocity estimates in arterial stenosis models (Hoskins 1997), and *in vivo* studies of flow in human carotid arteries (Fei et al 1997).

13.7 SPECKLE TRACKING TECHNIQUES

Vector Doppler techniques in general work by interrogating each sample volume of interest from two or more directions, and estimating the velocity components along each of these directions in order to calculate either a 2-D or 3-D velocity vector. Another approach to determining motion in two dimensions is by frame-to-frame analysis of the movement of the speckle patterns in B-mode ultrasound images. Such techniques are clearly not Doppler techniques, but are introduced here because they are closely related. In fact the (non-Doppler) ultrasound time shift estimation methods described in Section 11.4 can be thought of 1-D speckle tracking techniques, and it will emerge that some 2-D speckle tracking methods are merely generalisations of these methods.

Hein and O'Brien (1993) have provided an excellent review of both 1-D and 2-D time domain methods for assessing tissue motion from reflected ultrasound echoes. They point out that as a volume of tissue moves, it can either retain its shape and simply be translated or rotated, or it can be deformed and/or compressed. In their review, Hein and O'Brien describe both 2-D block matching techniques, which are most appropriate where the motion is mostly by translation, as in the case of blood flow, and optical flow techniques, which are suitable for more complicated motions, which include rotation, deformation and compression (such as might occur when a muscle contracts). Only block-matching techniques are described here.

The basic idea of block-matching techniques is illustrated in Fig. 13.18. A regular grid of kernel regions of $k \times l$ pixels are identified in the first image X (only one kernel is shown in the figure). Each such kernel region is then compared with all possible $k \times l$ regions within a $2M \times 2N$ search region in a successive image Y, and the best match used to define the displacement vector, $\vec{D}$. The displacement vector can then be used to calculate the velocity of the kernel region (both magnitude and direction), which may be superimposed on the grey-scale image in any of a number of ways, including by the use of colour coding.

It is important to note that whilst the above technique has the important advantage over standard 1-D methods of providing 2-D directional information, the axial and lateral resolutions of the method are very different; indeed, the lateral resolution will be several times coarser than the axial resolution (Bohs and Trahey 1991, Hein and O'Brien 1993). Furthermore, if the ultrasound system produces a sector image, the

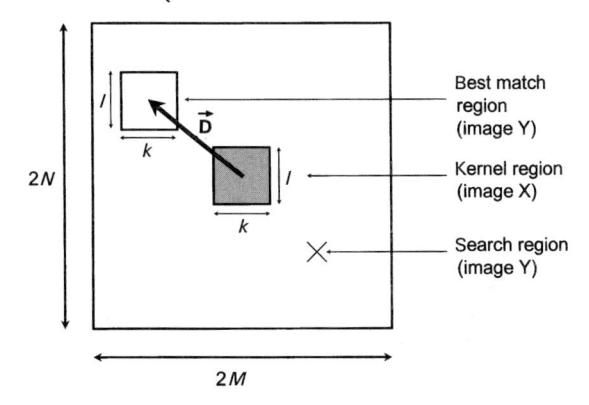

Figure 13.18 Geometry for 2-D speckle tracking. A kernel region (shaded) of $(k \times l)$ pixels is identified in the first image, X, and compared with all possible matching locations in a $(2M \times 2N)$ search region in a successive image, Y. The best match is used to define the displacement vector $\vec{D}$

lateral resolution becomes poorer at greater depths, since the pixel spacing becomes wider (Hein and O'Brien 1993). In this respect the vector Doppler techniques described in the last section, which in general depend on the measurement of axial displacements along two or more ultrasound beams, appear to have significant advantages. Note also that the performance of block-matching techniques is critically dependent on the size of the kernel region selected. In general, larger window sizes produce better results since they have a 'more unique' speckle pattern, but larger windows also lead to decreased spatial resolution and increased computation times (Ramamurthy and Trahey 1991).

Yeung et al (1998) have described a multi-level block-matching algorithm that attempts to overcome the inherent trade-off between resolution and noise immunity found in conventional block-matching techniques. The method is based around the use of matching blocks and search windows of variable size. Initially a large block size is used to provide a course resolution estimate of overall motion. Each subsequent level uses a smaller block size and search window so as to increase spatial resolution, without sacrificing the noise immunity of previous searches.

Various approaches to the problem of determining the best match between regions in successive frames are possible, and some of these are described below.

13.7.1 2-D Speckle Tracking: 2-D Correlation Method

The 2-D correlation method of tracking blood flow introduced by Trahey et al (1987) is simply a generalisation of the 1-D correlation method described in Section 11.4.2. The basic methodology is as described above, with the best match being defined as the position where the normalised correlation, $\rho_{m,n}$, is a maximum. The normalised correlation function is calculated using the following equation:

$$\rho_{m,n} = \frac{\sum_{i=1}^{k}\sum_{j=1}^{l}(X_{i,j} - \overline{X})(Y_{i+m,j+n} - \overline{Y})}{\sqrt{\sum_{i=1}^{k}\sum_{j=1}^{l}(X_{i,j} - \overline{X})^2 \sum_{i=1}^{k}\sum_{j=1}^{l}(Y_{i+m,j+n} - \overline{Y})^2}}$$

13.19

where $X_{i,j}$ are the values of the pixels in the kernel region of image X, and $Y_{i+m,j+n}$ are the values of the pixels in the search region of image Y corresponding to a displacement m, n. $\overline{X}$ and $\overline{Y}$ are the mean pixel values in the corresponding windows.

Trahey et al (1988) have expanded on the description of the correlation method introduced in their 1987 paper, and have reported the results of test tank experiments which define the performance of the method under a variety of conditions. Ramamurthy and Trahey (1991) have provided a detailed analysis of the potential and limitations of 2-D correlation methods, and assert that the same fundamental relationships apply to all 2-D motion detection algorithms. Ultimately the performance of such methods is limited by the second-order statistical characteristics of B-mode images; the auto-covariance function and the cross-covariance function. The performance of motion-detection algorithms depends in particular on the direction of the analysis with respect to the ultrasound beam (i.e. lateral or axial) and the type of echo data used (i.e. radio-frequency or envelope data). Processing the raw RF is better than processing the envelope data, particularly in the axial direction, because the autocorrelation length of the speckle is much smaller in the former case.

13.7.2 2-D Speckle Tracking: Sum-absolute-difference Method

A major problem with the 2-D correlation method for speckle tracking is that computationally it is extremely demanding and is not suitable for real-time implementation. A far simpler method for determining the similarity between image regions uses a 'sum-absolute-difference' (SAD) algorithm (Barnea and Silverman, 1972), which was first suggested as a suitable method for imaging blood flow and tissue motion by Bohs and Trahey (1991). The SAD method is identical to the correlation method described above, except that the best match is now defined as the point where the SAD, $\varepsilon_{m,n}$, defined as follows, is a minimum.

$$\varepsilon_{m,n} = \sum_{i=1}^{k}\sum_{j=1}^{l} | X_{i,j} - Y_{i+m,j+n} | \qquad 13.20$$

$X_{i,j}$ and $Y_{i+m,j+n}$ are as defined above.

This method requires only one absolute difference operation per pixel (compared to eight more complex operations for the normalised cross-correlation method), making it particularly suitable for real-time implementation. Bohs and Trahey (1991) reported quantitative studies using speckle-generating targets translated by fixed amounts, both axially and laterally, which indicated that the SAD method tracks moving speckle as accurately as the correlation method.

In subsequent publications, Bohs et al (1993a, 1993b) have described a real-time system for the implementation of the SAD method. The parallel architecture of the system they used allowed the calculation of up to 2600 motion vectors in each successive image pair at a rate of 30 frames per second. A programmable graphics processor encodes individual velocity vectors with colour and displays them superimposed on the B-mode image in real time. Figure 13.19 (see Plate IX), reproduced from Bohs et al (1993b), shows some of the initial results obtained using the system to image flow in a test phantom. *In vitro* tests indicated that the system could track velocities well over the Doppler aliasing limit in any direction in the scan plane with greater than 94% accuracy. Bohs et al (1995a) have also described the use of the SAD method for deriving velocity profiles and volumetric flow *in vitro* (see Section 12.3). Friemel et al (1995) compared the relative performance of the normalised correlation, the non-normalised correlation, and the SAD algorithms by computer simulation, and concluded that the difference in performance between the SAD and normalised correlation algorithm was not statistically significant for any kernel sizes or noise levels tested. The performance of the non-normalised correlation algorithm was not so good, particularly for small kernel regions.

13.7.3 2-D Speckle Tracking: Miscellaneous Approaches

In addition to the two methods described above, a number of other speckle tracking methods have appeared in the literature, but apparently not been followed up at the time of writing. Gardiner and Fox (1989) described a method of detecting motion in B-scan images simply by comparing successive frames using various Boolean operations. Whilst this method can highlight regions of flow, it cannot be used for flow quantification.

Wilson and Gill (1993) have suggested an ultrasonic speckle projection technique for the measurement of 2-D blood velocity vectors, which involves the calculation of the 2-D Fourier transforms of projections of the 3-D data set (x, y, t) onto its orthogonal boundaries (x, t) and (y, t).

Bohs and colleagues (Bohs et al 1995b, Geiman et al 1996, Bohs et al 1998) have suggested an 'ensemble tracking method' which is essentially a development of the SAD method. An ensemble of 2-D echo patterns are acquired in temporal sequence along a single line of sight using parallel receive processing (Shattuck et al 1984), which allows the simultaneous formation of multiple receive beams from each transmitted burst. Since the time between subsequent acquisitions of 2-D data sets is very short (i.e. a single pulse repetition interval), the movement of the speckle pattern between subsequent frames is very small, which means, firstly, that there is very little decorrelation due to flow gradients, and secondly, that the search region can be kept very small. On the other hand, because the motion is so small (typically less than the spatial sampling interval) it is necessary to utilise interpolation techniques to estimate the movement to the required fractional pixel accuracy. This is achieved by firstly calculating the SAD values at integral pixel displacements, and then interpolating between these values to find the 'true' minimum of the SAD function. Additionally, multi-lag processing may be employed to generate multiple velocity estimates at each lag (for example, single-lag estimates between the first and second frame, and the second and third frames, and so forth, and then double-lag estimates between the first frame and the third frame, and the second and fourth frames, and so on). Interpolation methods that have been employed include cubic spline interpolation (Bohs et al 1995b) and 'grid slopes' interpolation (Geiman et al 1996, 1997).

Bashford and von Ramm (1996) have described a method they claim to be suitable for 3-D blood flow velocity measurement, which works by tracking features of the ultrasonic speckle pattern in 3-D. A 3-D volume of data is acquired, and 3-D maxima in the speckle pattern identified, and then tracked through subsequent 3-D data sets. Bashford and von Ramm report off-line experiments using a tissue phantom and a real-time volumetric ultrasound imaging system, and show that the local maximum detected values of the speckle signal may be identified and tracked for measuring velocities typical of human blood flow. They also point out that the 3-D method overcomes the problem common to 1-D and 2-D tracking systems, of speckle decorrelation due to the movement of scatterers in and out of the sample volume.

Wang and Shung (1996) have compared three bit-pattern correlation algorithms designed for block matching (the so-called 'sum-of-approximate-match', the 'coincident bit count' and the 'video two-bit correlation') with the standard normalised cross-correlation algorithm in terms of both accuracy and computation time. RF and video signals were collected from a flow phantom containing fresh porcine blood, and flow velocity profiles determined using the four methods. The results showed good agreement between the velocity profiles estimated by the three-bit pattern correlation algorithms and the full cross-correlation function, together with very significant savings in computation times.

13.8 DECORRELATION-BASED TECHNIQUES

A number of methods of estimating blood flow based on correlation methods have been described in Section 11.4. In general such techniques work by estimating the difference in the round-trip time of two or more ultrasonic pulses to specific groups of scatterers, by finding the time lag which maximises the cross-correlation function between pairs of echoes. A variation on this approach, which has been found to be of value in intravascular applications, is to determine the rate of decorrelation between pairs of echoes (Li et al 1996, Crowe et al 1996). The particular circumstances which make this attractive are that intravascular ultrasound (IVUS) catheters designed for imaging vessel walls ideally transmit and receive ultrasound at right angles to the vessel wall, which makes conventional Doppler techniques inappropriate, and that intravascular systems operate at relatively high ultrasound frequencies (usually at least 20 MHz), at which frequency blood becomes a strong scatterer of ultrasound (blood acts as a Rayleigh scatterer to at least 15 MHz, and therefore scattering increases with the fourth power of frequency—see Section 7.2).

The decorrelation method depends for its success on there being a linear relationship, over a significant range, between lateral motion of scatterers through the ultrasound beam and the decorrelation of the RF signals returning from each sample volume of interest. Li et al (1997) have shown this to be a reasonable assumption in the near field of unfocused IVUS catheters. It should be noted, however, that the slope of the relationship will be dependent on depth and must be determined by experiment for each type of catheter.

Assuming that the model of linear decorrelation is appropriate, and adopting the notation of Li et al (1998), the displacement correlation function $\rho(z, \Delta x)$ may be written:

$$\rho(z, \Delta x) = 1 - \alpha_d(z)\Delta x \qquad 13.21$$

where α_d is the displacement decorrelation slope, z is depth, and Δx is lateral displacement (relative to the axis of the ultrasound beam).

Similarly, it is possible to write an expression for the time correlation function, i.e:

$$\rho'(z, \Delta t) = 1 - \alpha_t(z)\Delta t \qquad 13.22$$

where α_t is the time decorrelation slope and Δt is time delay. Assuming now that in a normal measurement situation it is only the lateral displacement that causes decorrelation over the measurement time (Li et al 1997), then we can equate $\rho(z, \Delta x)$ in eqn. 13.21 with $\rho'(z, \Delta t)$ in eqn. 13.22 to derive an expression for the component of velocity across the beam at depth z, i.e:

$$V_x(z) = \frac{\Delta x}{\Delta t} = \frac{\alpha_t(z)}{\alpha_d(z)} \qquad 13.23$$

Thus, the transverse velocity component can be estimated from the ratio between the calibrated decorrelation slope, α_d (estimated by experiment or theory), and the measured decorrelation slope, α_t.

Li et al (1996, 1998) have described their method of determining velocities and ultimately volumetric flow in some detail. The decorrelation function is evaluated using a sequence of N RF traces $(S_1, S_2, \ldots S_N)$ at a fixed pulse interval, from the same angular direction. The traces are first aligned using a phase-matching procedure to remove any axial velocity components due to the ultrasound beam not being absolutely perpendicular to the direction of flow, and then the correlation coefficient between trace S_n ($n = 1, 2, \ldots N - 1$) and S_m ($m = 2, \ldots N$) calculated for all values of n and m:

$$\rho_{(n,m)} = \frac{\sum_{k=1}^{K} S_n(k) S_m(k)}{\sqrt{\sum_{k=1}^{K} [S_n(k)]^2 \sum_{k=1}^{K} [S_m(k)]^2}} \qquad 13.24$$

where K is the number of samples in each range gate. Finally, a linear model is fitted to the values of $\rho_{n,m}$ (only values of $\rho_{n,m} > 0.5$ are used) in order to estimate the time decorrelation slope $\alpha_t(z)$ at each depth. Note also that the mean time shift necessary to align the RF traces is available, which allows an estimate of the axial component of velocity.

The estimated velocity components from each range may be superimposed as a colour-coded image on a standard grey-scale IVUS image (Fig. 13.20: see Plate X), and may be used for further processing. Li and colleagues (Li et al 1996, 1998) have used the method to calculate volumetric blood flow by summing the velocity components over the vessel cross-section; i.e:

$$\hat{Q} = \sum_A V_x(r, \theta) \Delta A(r, \theta) \qquad 13.25$$

(where r and θ are polar coordinates, $\Delta A(r, \theta) = r\Delta r\Delta\theta$, and Δr and $\Delta\theta$ are the scanning step sizes in the radial and angular directions, respectively), and obtained relatively good agreement of such measurements with electromagnetic flowmeter measurements (van der Steen et al 1997, Céspedes et al 1998).

The decorrelation method shows considerable promise in the context of intravascular blood flow measurements, but is at an early stage of development and it remains to be seen whether it will become clinically useful. Decorrelation measurements, even with the strong echoes returned by blood at high frequencies, are of course sensitive to errors caused by decorrelation unrelated to flow velocity, such as noise and velocity gradients within the sample volume (see Section 11.4.3). They also tend to have a very high variance associated with them because, in order to obtain good axial resolution and reduce the error due to velocity gradients within the range cell, it is necessary to use relatively short-range windows. This means that considerable averaging is necessary to obtain useful results. Another drawback with such methods is that currently they require off-line processing and so the results are not available in real time.

In an entirely separate application, the rate of decorrelation of ultrasound echoes has been used as a method of estimating the angle, θ, between the ultrasound beam and the direction of flow (Cantarelli et al 1994, Dotti and Lombardi 1996). The first step in the procedure is to derive a correlation map. The correlation function $R(\Delta x, \Delta t)$ is measured in a 2-D domain, where Δx is the distance along the beam axis and Δt is the time between pairs of pulses. If $s(x, t)$ is a sample from the echo signals at time t, originating from scattering particles at a distance x from the ultrasonic transducer, then the correlation function may be written:

$$R(\Delta x, \Delta t) = \langle s(x, t)s(x + \Delta x, t + \Delta t)\rangle \qquad 3.26$$

An example of such a correlation map is shown in Fig. 13.21. A series of contour lines join points in the 2-D correlation space which have the same correlation value. Negative correlation values are shown in grey. It can be seen that such a map consists of a series of ridges and troughs running at an angle to the x and t axes, the heights and depths of which decrease as they are followed (in this case) from bottom left to top right. The main lobe which passes through $(0, 0)$ is identified, and values for both the axial velocity and the angle between the ultrasound beam and flow direction, θ, derived from the characteristics of this ridge. The axial velocity is found from the angle the ridge subtends to the axes (the direction of the ridge corresponds to the case where the time shifts best track the change in position of the scatterers). The angle θ is found from the decay rate of the ridge. The more acute the angle θ, the more rapidly decorrelation will occur, because a higher percentage of scatterers leave and enter the sample volume. Indeed, hypothetically, if $\theta = 0$ and there were no other decorrelation mechanisms present, then the lobe would not decay at all.

In order to determine the angle in practice, Dotti and colleagues define a line along the centre of the main lobe which begins at the origin and ends where the height of the ridge falls to 50% of its value at the origin. The projection of this line on the Δx axis is then proportional to the tangent of θ. The constant of proportionality depends on the probe focusing and on the distance between the probe and the measuring location, but can be determined off-line by careful measurement.

Dotti and colleagues make the point that although individual estimates of angle determined in this way will be subject to large degrees of uncertainty, the angle θ is, in many measurement situations, independent of both time and measuring location, and therefore good estimates of θ can be achieved by averaging estimates over time and space. Attempts to implement this method *in vitro* have given some promising results, but have

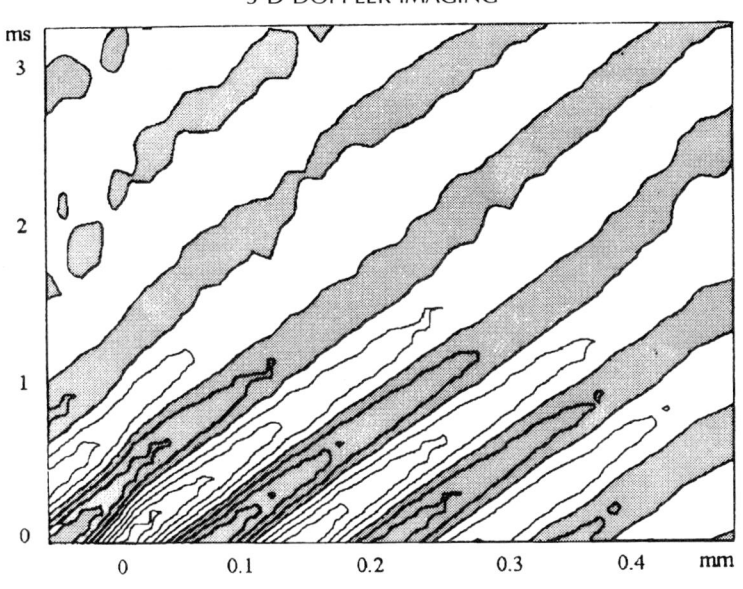

Figure 13.21 Example of a correlation map. The abscissa represents the distance along the beam axis, the ordinate the time between pairs of pulses. Contours join points in correlation space which have the same value. Areas of negative correlation are shown in grey (reproduced by permission from Dotti and Lombardi 1996, © 1996 IEEE)

shown that the measurement points must be carefully chosen to be close to the focal point of the transducer and inside an area of nearly uniform blood velocity, and that an extended averaging time is necessary to achieve reliable results (Cantarelli et al 1994, Dotti and Lombardi 1996).

13.9 3-D DOPPLER IMAGING

Just as an ultrasonic beam can be swept through a 2-D scan plane to give a 2-D image, so it can be swept through a 3-D volume to provide a 3-D image. Often the 3-D image is stored mentally as the operator examines neighbouring scan planes. However, systems are now available commercially which allow 3-D data sets to be stored in computer memory, processed and presented as a 3-D image (Fenster and Downey 1996). The display of 3-D images is problematic, since external structures obscure internal ones. Image manipulation techniques are therefore incorporated in the display system, for example to peel off the external layers or for the whole image to be rotated for viewing from different directions. A fundamental problem of both 3-D B-mode and Doppler mode is the length of time it takes to collect echo data from a 3-D volume. It will be recalled that 2-D B-mode imaging typically uses all the available echo collection time to produce around 30 frames per second. In Doppler imaging the problem is even more severe, since several pulses are transmitted along each scan line. Therefore, although computer processing is becoming more and more able to supply fast data processing for real-time images, the time required for echo collection is a problem for real-time 3-D imaging in both pulse-echo and Doppler modes. Techniques employing several simultaneous beams are under development to try to overcome this limitation (Section 13.9.1.3). Blood flow imaging may be one of the areas where 3-D imaging has a clinical pay-off, for example in the display of complex flow patterns in the diseased carotid artery (Picot et al 1993). The vascularity of tumours may also be beneficially depicted by power Doppler and contrast Doppler 3-D imaging (Ferrara et al 1996). The high sensitivity and lack of beam/flow angle dependence of power Doppler appears to make it the most suitable Doppler method for 3-D imaging (Ritchie et al 1996, Downey and Fenster 1995a, 1995b, Guo and Fenster 1996).

Fenster and Downey (1996) have published an excellent review of the various approaches to 3-D ultrasound imaging, and the following sections briefly summarise some of the points contained therein. Another excellent review is that of Nelson and Pretorius (1998). In what follows more emphasis is given to image acquisition than to reconstruction and display techniques as the latter are common to many areas of 3-D medical imaging.

13.9.1 Image Acquisition Techniques

There are three basic approaches to the acquisition of a 3-D volume of (Doppler) ultrasound data— free-hand scanning, mechanical scanning and electronic scanning.

13.9.1.1 Free-hand Scanning

With the free-hand method the operators hold the ultrasound transducer in much the same way as they would for conventional 2-D scanning, manipulating the probe to acquire a series of 2-D slices that have no fixed relationship to each other. The obvious advantage of this is that operators can acquire the slices however they wish so as to cope with the surface anatomy of the patient and to obtain views from angles that provide most information. Perhaps this last point is more important for pulse-echo imaging, where specular surfaces are best imaged at close to 90°, but it also has advantages for Doppler imaging, where angles of close to 90° to the flow direction need to be avoided (except possibly when using power mode Doppler). The disadvantage of free-hand scanning is that unless steps are taken to avoid it, it is quite possible to leave gaps in the 3-D volume collected. Furthermore, since the same element of tissue may be insonated from a number of angles and at a number of different times, the question of image registration becomes particularly important. Even if the means of locating the position and angulation of the transducer were perfect, both patient movement and the small local variations in the speed of sound in tissue would degrade the registration.

The ability to locate the position and angulation of the ultrasound beam in 3-D space is, of course, an integral part of all 3-D imaging methods but is particularly challenging with the free-hand acquisition method, and at least four approaches have been used. The simplest approach is to use a mechanical system similar to that used in some early B-scan systems, where the transducer is connected to a fixed reference point by a mechanical arm system with multiple movable joints whose positions are monitored by potentiometers (Geiser et al 1982, Sawada et al 1983). Another approach to position sensing is based on acoustic ranging (which somehow seems apposite for an ultrasound system!). In this method the location and orientation of the probe is obtained by mounting three sound-emitting transducers on the ultrasound probe, and mounting an array of microphones in a fixed position above the patient. The system then automatically determines the required positional information from the times of flight from the sound-emitting transducers to the various microphones. For this system to work correctly there must be unobstructed lines of sight from the emitters to the microphones. This system has been employed by many workers (Brinkley et al 1982, Moritz et al 1983, Levine et al 1989, King et al 1990). The most common method of position sensing at present is based on magnetic position sensing (Raab et al 1979, Detmer et al 1994, Hodges et al 1994, Barry et al 1997). In this method a transmitter that produces a spatially varying magnetic field is placed close to the patient, whilst a receiver containing three orthogonal coils, which are used to measure field strength, is attached to the ultrasound probe. The main limitation of this approach is that ferrous or highly conductive metals distort the magnetic fields and must therefore be excluded from its immediate vicinity. Finally, systems incorporating several small video cameras are being developed for 3-D ultrasound beam registration.

A completely new approach to the registration of 2-D images has recently been introduced by Siemens, which is based on their SieScape™ imaging technology (see Section 10.4.4). The basis of the technique is that because of the high degree of frame-to-frame similarity between ultrasound images, it is possible to estimate transducer motion over the body by comparing sequences of images, one with another. With the new technique (named 3-Scape™) the ultrasound transducer is moved at right angles to the scan plane so as to image a volume of tissue, rather than in the same plane so as to increase the extent of the plane. It would appear to be easier to align images accurately when the motion of the transducer takes place in the same plane as the scan plane, and the accuracy of the techniques still requires evaluation. It does, however, appear to be capable of producing presentable 3-D images fairly easily, and one such image is shown in Fig. 13.22 (see Plate X).

One particular difficulty with most free-hand scanning techniques is calibration, i.e. determining the position and orientation of the 2-D slices with respect to the probe-mounted sensor. Prager et al (1998) have recently described a new technique for this purpose which they claim to be both rapid and efficacious.

13.9.1.2 Mechanical Scanning

Although free-hand scanning provides flexibility, it does have problems in relation to the acquisition of a complete data set and to precise image registration, particularly when the tissue volume to be scanned is small, and therefore some workers have used a mechanical system to move a 2-D probe so as to sweep out a well defined 3-D volume in the tissue, with 2-D images being captured at regular spatial or angular intervals. This ensures that no data are missed, precisely defines the spatial relationship between the 2-D slices and (in general) avoids the problems associated with insonating the same volume more than once. There have been three basic approaches to mechanical scanning adopted: linear or parallel scanning, fan scanning and rotational scanning (Fig. 13.23).

In linear scanning (Fig. 13.23a), the transducer is simply translated at a fixed angle to the scanning plane and parallel to the patient's skin to produce a series of parallel scans with a pre-defined spatial separation. Advantages of this method are that the data are equally spaced throughout the image which as well as avoiding changes of resolution with depth or distance from the axis (other than those due to the beam size) leads to efficient reconstruction algorithms. This method has been used both with standard colour flow Doppler imaging (Picot et al 1993, Fenster et al 1995, Guo et al 1995, Moskalik et al 1995) and power Doppler imaging (Downey and Fenster 1995a, 1995b).

In fan scanning (Fig. 13.23c) the transducer is rotated about an axis at (and parallel to) the transducer face so as to produce a fan of scans at a pre-defined angular separation (Fenster and Downy 1996, Tong et al 1996, Gilja et al 1994). The advantage of this technique is that it leads to a

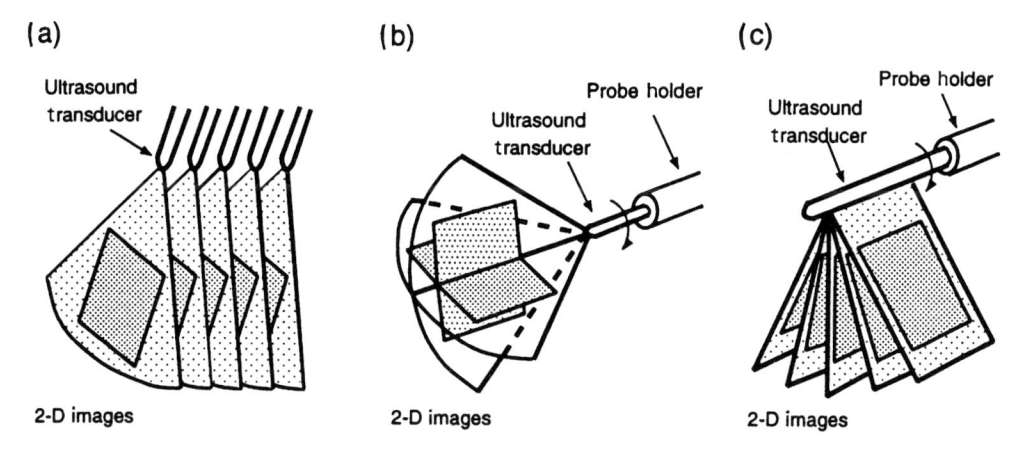

Figure 13.23 Schematic diagrams of the three methods of mechanical scanning for acquiring the 2-D images necessary for 3-D reconstructions: (a) linear scanning, where the transducer is simply translated; (b) rotational scanning, where the transducer is rotated about the central axis of the probe; (c) fan scanning, where the transducer is rotated about an axis at and parallel to the transducer face. Although (a) and (b) are illustrated for sector scans, they could equally well be rectilinear scans, and the converse applies to (c) (reproduced by permission from Fenster et al 1995, © Society of Photo-optical Instrumentation Engineers)

much more compact transducer assembly than the linear scanning approach. Its disadvantage is that since the angular separation between the scan planes is fixed, they become more widely separated the greater the distance from the transducer. Nevertheless, if the volume to be scanned does not extend over too great a depth, the angular separation can be chosen to provide an appropriate separation at the depth of interest. Whilst this method may be suitable for use with power Doppler imaging, it is unlikely to be so with standard colour flow imaging because of the changes of angle.

In rotational scanning (Fig. 13.23b) the transducer is rotated about the central axis of the probe so as to produce a series of scans at a known angular separation which intersect along the central axis of the transducer (Ghosh et al 1982, Roelandt et al 1994, Fenster and Downey 1996). As for fan scanning this produces a sampling distance that increases as a function of the distance from the axis of rotation. Also, the tissue falling within the axis of rotation will be interrogated during each scan and unless the axis remains identical will lead to artefacts. The same limitation with regard to colour flow imaging and changes of angle apply to this technique as to the previous technique.

13.9.1.3 Electronic scanning

2-D transducer arrays are becoming available (Turnbull and Foster 1991, Smith et al 1992), and a natural way in which to acquire 3-D volumes of data is to electronically steer the ultrasound beam in a pyramidal scan so as to cover the entire volume of interest. An additional, and important, advantage of this is that it provides rapid and even real-time scanning (at least for pulse-echo imaging) if parallel processing is used to generate a number of image vectors from each transmitted pulse (Shattuck et al 1984, Smith et al 1991, von Ramm et al 1991).

13.9.2 Image Reconstruction Techniques

The process of 3-D reconstruction is the generation of a 3-D representation of the ultrasound interrogated structure from the acquired set of 2-D images. Two distinct approaches have been used in the case of ultrasound images. In the first, the series of 2-D images are segmented so as to extract the required features before the 3-D reconstruction is performed. The advantages of this approach is that it greatly reduces the amount of data that needs to be handled in the 3-D reconstruction, and can be quite useful in situations where there are 'hard boundaries' between structures. It is not, however, so useful where image contrast is low, and the segmentation process may also be time-consuming and inaccurate. The second approach is to use the acquired series of 2-D images to built up a complete 3-D data set for subsequent manipulation and display. In this case the values assigned to any voxel not sampled by the 2-D images must be interpolated. The advantage of this approach is that no information is lost, and it allows a variety of rendering techniques (see Section 13.9.3). The disadvantage is that it can involve very large data sets, which are presently relatively time-consuming to manipulate.

13.9.3 Image Display Techniques

The display of 3-D image data is a problem common to all imaging fields since unfortunately there is no completely satisfactory solution to presenting 3-D data unless the image contains a trivial amount of information. It is, of course, possible to display surface views or cuts through the image in any arbitrary plane (subject to appropriate pre-processing), but this is not the same as displaying the entire 3-D data set.

The many ways of displaying 3-D data sets may be divided into three classes, i.e. surface-based, multi-planar and volume-based techniques. In surface-based techniques, the 3-D information is first segmented or classified so that the structure to which each voxel belongs can be ascertained. Once this has occurred, the surface of interest is either displayed using a wire-frame approach, where the boundaries are represented by a grid of lines that can be viewed in 3-D perspective, or using a surface-rendering approach where the surface representations are shaded and illuminated (see Fenster and Downey 1996).

Two methods of multiplane viewing have been used. The first, which is common to many medical imaging techniques, is simply to present various 2-D cuts through the 3-D volume as though the original scans were done through those planes. Often two or three perpendicular planes through a structure of interest are displayed simultaneously together with information about their relative orientations and intersection (Moskalik et al 1995) (see Fig. 13.24: Plate X). The second method is based on multiplane visualisation with texture-mapping (Picot et al 1993, Guo et al 1995, Fenster and Downey 1996, Tong et al 1996), where the image is presented as a polyhedron representing the boundaries of the reconstructed volume (Fig. 13.25: Plate XI). The polyhedron may be rotated to any required orientation, and the boundaries of the polyhedron moved in or out to view different cuts in different orientations.

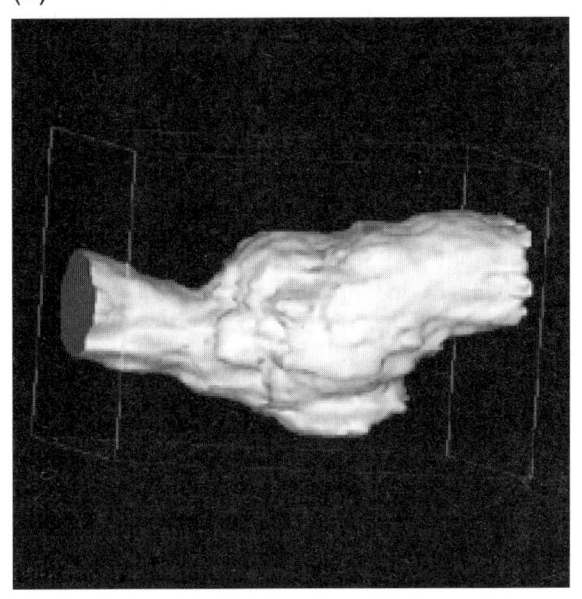

Volume rendering techniques produce displays of the entire 3-D image projected onto one or more 2-D planes, as shown in Figs 13.26 (see Plate XI) and Fig. 13.27. There are a number of alternative approaches to mapping the 3-D information onto the 2-D planes (Tuy and Tuy 1984, Levoy 1990, Fenster and Downey 1996) and the reader is referred to the relevant literature for further details. Further examples of 3-D scans (including dynamic illustrations) can be found on major manufacturers' web sites.

13.10 TISSUE MOTION IMAGING

Tissue motion can be measured or detected using modified velocity and power Doppler imaging techniques. Soft tissue motion has been studied less than blood flow but there are areas where clearly it is important, for example in cardiac and vascular disease. In other applications tissue motions are used more to identify structures or to assist procedures rather than to provide diagnosis. By relating the tissue motion to the driving force, usually either blood pressure or an externally applied force, the elasticity of the tissue

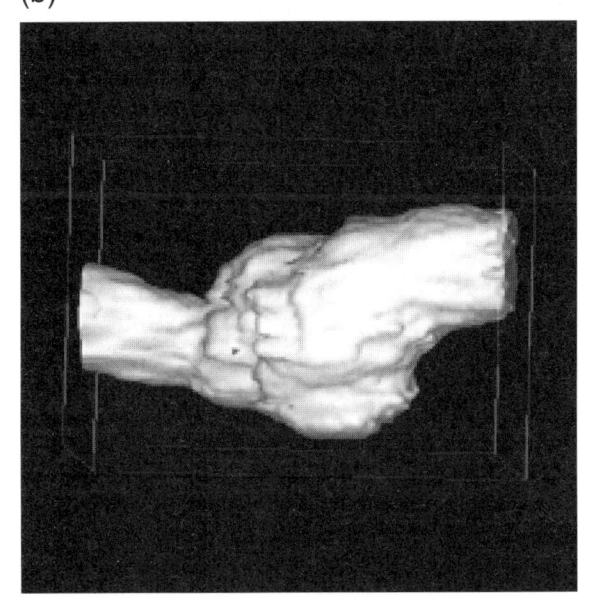

Figure 13.27 Volume-rendered images of the junction between a reverse saphenous vein jump graft and an existing femoral peroneal bypass graft. Thirty transverse power-mode images across a 2 cm length of the vein graft were acquired. The spatial positions of the images were registered with a magnetic position sensor attached to the ultrasound scanhead. These images were then 3-D scan converted into cubic voxels in 3-D space; (a) and (b) are volume renderings of this voxel data in two different orientations. Because of the high contrast between the blood and the background in the power mode, the boundary can be easily extracted by voxel intensity thresholding (courtesy of J-M Jong, University of Washington, WA)

can be calculated. This field is commonly called the 'sonoelasticity approach' to tissue characterisation. Tracking methods using changes in tissue speckle or organ wall echo signals are normally employed in these applications (Section 13.7).

It is worth remembering that B-mode and M-mode techniques, as discussed in Chapter 10, have been employed successfully to study tissue motion since the earliest days of diagnostic ultrasound. Even the A-mode, which displays echo signals from a fixed-beam direction on an oscilloscope screen, has been developed as a means of detecting very small brain tissue motions via corresponding small fluctuations in the signal amplitudes (ter Braak and de Vlieger 1965). Also, many specialised systems based on various ultrasonic techniques have been developed to track the motion of vessel walls throughout the cardiac cycle (see Section 12.9.3). New Doppler techniques need to be evaluated to see if they can provide more information on tissue motion than that already available from well-established M- and B-modes. They also need to be assessed relative to speckle tracking methods (Sections 13.7 and 13.10.3).

13.10.1 Real-time B-mode and M-mode for Tissue Motion

The basic M-mode has not developed significantly over recent years, but the B-mode, which is highly dependent on the quality and rate of production of images, has improved steadily with advances in computer and transducer technology. The ability to produce several simultaneous beams from one transducer has made possible frame rates as high as 120 s^{-1}. Such frame rates give temporal resolution of around 10 ms which accommodates well most tissue motion within the body. For higher temporal resolution, as may be required for heart valve studies, the M-mode can provide temporal resolution of less than 1 ms when pulses are transmitted along a selected beam direction with a repetition frequency of several thousand per second. M-mode scanning has the best temporal resolution. The spatial resolution of both real-time B-mode and M-mode scanning of moving tissue with around 3 MHz ultrasound is similar to that for static tissue i.e. around 1 mm.

Motion of boundaries may be more readily appreciated with real-time computerised edge-detection. Detection of cardiac boundaries is not simple due to the irregular structure of the endocardium and the low level signals from heart muscle. Abnormal wall motion is very diagnostic in cardiology.

With the tissue motion detection methods described so far, only the component of displacement and velocity along the axis of the scanning beam is readily measured, which can be misleading where there is complex tissue motion. However, M-mode and real-time B-mode are good examples of techniques which supply limited information yet are clinically very valuable. To measure two or three velocity components and hence establish more accurate measures of the true velocity vector, the moving tissue must be interrogated with beams from two or three directions as noted in Section 13.6 for blood flow. A difficulty with 2-D and 3-D vector techniques is that the sources of the beams need to be separated by a significant distance, resulting in large transducer assemblies. This gives significant problems for heart examinations, where access is usually limited to a few rib spaces. Speckle tracking does not suffer from this problem.

13.10.2 CW and PW Doppler for Tissue Motion

CW and PW Doppler instruments can record tissue motion well if suitable sensitivity and frequency filtering are employed. Fetal heart detection and monitoring were early applications of CW Doppler in which the echoes are from a mixture of blood and tissue.

Duplex Doppler (Section 4.3) has been used in adult heart wall investigations (Garcia et al 1996, Kostis et al 1972, Isaaz et al 1989) and in observing tissue motion as a result of fetal breathing actions (McDicken et al 1979). Basic duplex systems have now been replaced by Doppler tissue imagers (DTIs) for tissue motion investigations (see next section). A spectrogram from a PW Doppler examination of a specific tissue site should give more velocity information than the mean velocity in a DTI image. However, further investigation of the PW technique is required to explain the spectrograms and to identify artefacts (Fig. 13.28). Geometrical spectral broadening would not seem to be large enough to account for the wide range of velocities obtained from localised sample volumes in which it would be expected that the tissue is all moving with the same velocity.

13.10.3 Doppler Tissue Imaging (DTI)

Modifications can be made to Doppler blood flow imagers to enable images of moving tissue to be produced (McDicken et al 1992, Yamazaki et al 1994) (Fig. 10.3: see Plate IV). The Doppler M-mode has proved useful for the investigation of tissue motion (Fig. 13.29: see Plate XI). Changes in the velocity pattern, with time, of pixels along the selected beam direction are presented with very good temporal resolution, e.g. 1 ms. The changes in velocity gradient across the myocardium have been of particular interest in the study of contractility. With echo levels from soft tissue typically 30 dB above that from blood, tissue Doppler techniques have been easier to implement than those for blood flow. However, care is still required in setting up the scanner to discriminate between blood and muscle. The signal-to-noise ratio is better with Doppler tissue techniques than with the corresponding blood flow techniques, hence fewer echoes are required for the estimation of mean velocity at each pixel location. Validation of mean velocity estimation by autocorrelation signal processing has been carried out both with moving tissue phantoms and by comparison of colour Doppler M-mode and pulse-echo M-mode recordings of moving heart wall (Fleming et al 1994, 1996) (Fig. 13.30, page 346, and Fig. 13.31: see Plate XII). Typically, velocity can be measured to an accuracy of 5% using Doppler tissue imaging.

Power Doppler tissue imaging has been of some value in extracting signals from moving structures in clutter echoes generated by multiple reflections and beam sidelobes (Lange et al 1997). Not surprisingly, DTI is primarily used in clinical

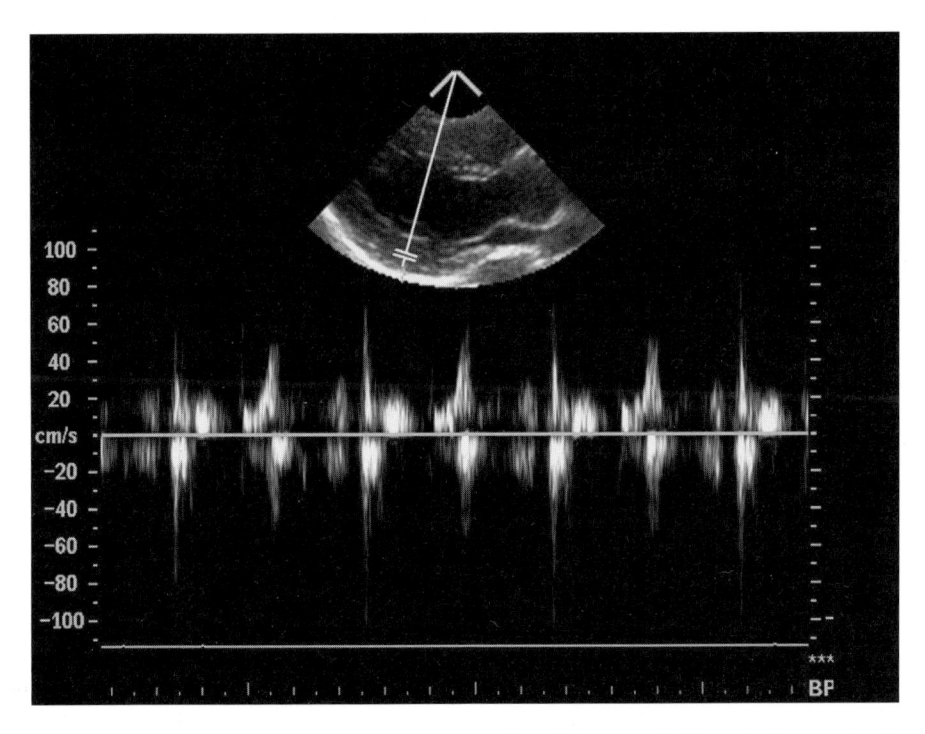

Figure 13.28 Pulsed wave Doppler used to produce a spectrogram of moving myocardial tissue in the sample volume indicated by the associated B-Mode image

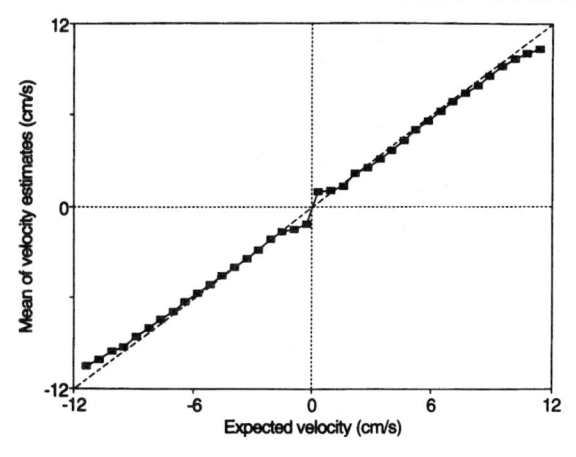

Figure 13.30 Validation of mean velocity estimation by autocorrelation signal processing of tissue-type echo signals. Doppler estimated velocities across a rotating disc of tissue equivalent material agree with known velocities to within 5%

research into heart disease, e.g. to study myocardial infarction and hypertrophy of the heart muscle (Sutherland et al 1994, Palka et al 1997). Its value in routine clinical application has still to be ascertained. Contractility of heart muscle is related closely to disease state and DTI can provide information directly on this topic. There has been some application of DTI to other muscles in the body (Grubb et al 1995) (Fig. 13.32: see Plate XII) and this may expand with the current rapid growth of musclo-skeletal ultrasound. High frequency DTI is being explored in relation to artery wall disease (Schmidt-Trucksass et al 1998). DTI is now a facility available on machines from several manufacturers.

Speckle tracking techniques (described above in Section 13.7) for blood flow can be applied to tissue motion. Some speckle tracking methods, particularly where the speckle pattern is tracked at very short time intervals along selected beam directions, closely resemble Doppler methods. Speckle tracking and Doppler methods are compared in the next section.

13.10.4 Comparison of Tissue Motion Imaging Techniques

The pulse-echo methods, i.e. A-mode, M-mode and real-time B-mode, detect the motion of tissue boundaries, which is simply observed on a display screen. Doppler methods and speckle tracking techniques can detect and measure motion of the weaker and more complex echoes from parenchymal tissue within organs as well as those from boundaries. M-mode and B-mode scans show the motion of the walls of an organ resulting from the internal motions. Although the external wall and internal tissue motions are related, they are not the same; a particular external wall motion could be the result of different internal motions. The accuracy of velocity measurement with all of these techniques can be high, better than 5%, but errors can be very large if each system is not carefully calibrated.

Standard CW Doppler instruments are very sensitive and have good temporal resolution, as low as 5 ms. Their lateral resolution is reasonable, being related to beam width; however, unless the transmission field and reception zone are made to intersect at an obtuse angle, they have very poor axial resolution. Their application in tissue motion studies is therefore of limited interest.

Pulsed wave Doppler units can have axial and lateral resolution similar to that of B-mode imagers. In practice, the axial resolution is poorer than pulse-echo B-mode, since slightly longer pulses are used to improve the signal-to-noise ratio. Typically, the axial resolution at 3 MHz is over 1 mm, whereas that of the B-mode is less than 1 mm. The lateral resolution is the same as in B-mode scanning. The temporal resolution is similar to that of CW devices, i.e. as low as 5 ms.

A strength of DTI is that at each scan line position velocity estimates are produced from echoes generated by a rapid succession of transmitted pulses in that direction. Tissue displacement is small during this process; for a tissue velocity of 10 cm s^{-1} it might move 0.1 mm. A second attraction is that DTI can readily be incorporated into Doppler blood flow imaging equipment. A disadvantage of DTI is the need for several transducer positions and beam directions to implement vector Doppler.

A desirable feature of speckle tracking is that it can make measurements of velocity components in the two dimensions of the image plane. There is also ongoing research into the possibility of 3-D speckle tracking. Speckle tracking techniques fit well with the general field of image processing, but there is difficulty in relating the speckle patterns in consecutive image frames. The time interval between frames is typically 10 ms or more, during which the tissue might move 1 mm and decorrelation of the signal pattern may occur. If the speckle pattern is the result of interference of echoes from a uniform tissue made up of many small scattering centres, the motion of the speckle may not correspond to the motion of the tissue. This artefact was reported several years ago (Morrison et al 1983) and studied in detail more recently with a computer model (Kallel et 1994). The modelling demonstrated that the artefact is due to the curved nature of the point response function of the imager. When the tissue contains stronger discrete scattering centres as well as those of the small scatterers, interpretation may be more reliable. Real-time implementation has been achieved for the simpler speckle tracking algorithms but could be a problem for more complex ones (Bohs and Trahey 1991). Further problems with speckle tracking are that the accuracy of velocity measurement is direction-dependent, due to the different image resolution in axial and lateral beam directions, and depth-dependent in sector scans due to the increasing line separation with depth.

The spatial resolution of DTI and speckle tracking are similar, both being slightly inferior to B-mode imaging, although it appears that this difference is decreasing with the use of wide bandwidth imaging methods. The temporal resolution of DTI is usually superior to that of speckle tracking; however, that of the latter is very dependent on the algorithm employed.

The best and most convenient ways of measuring tissue motion have still to be determined. At present it has been easier to incorporate Doppler methods into scanning machines.

13.11 ANTI-ALIASING TECHNIQUES

The problem of aliasing in pulsed-wave Doppler systems has been discussed in Section 4.2.2. In essence, the problem arises because the Doppler shift signal is reconstructed by determining the phase changes that occur between the returns from each successive transmitted pulse. Because phase is cyclic, if the Doppler shift frequency is not sampled sufficiently frequently, the phase change between samples may exceed $\pm \pi$ radians but, in the absence of additional information, will still be interpreted as lying within this range.

The obvious solution to overcome aliasing is to sample the signal more frequently, but in pulsed Doppler applications there is an upper limit to this because of the finite propagation time of ultrasound, and the fact that range ambiguity will arise if an ultrasound pulse is transmitted before the previous pulse has had time to return from targets at the maximum range of interest. Whilst such depth ambiguity may, under some circumstances, be acceptable in conventional pulsed Doppler systems, it is clearly not so in colour flow imaging systems. Combining the requirements that the sampling frequency must be at least twice the maximum Doppler shift frequency, and that each pulse must have time to return from the target before another is transmitted, means that the maximum velocity, $V_{\max}$, that can be unambiguously measured at any depth Z is given by:

$$V_{\max} = c^2/8Zf_t \cos\theta \qquad 13.27$$

where c is the velocity of ultrasound, f_t is the transmitted ultrasound frequency and θ is the usual Doppler angle (see Section 4.2.2.4 for a derivation of this expression). Substitution of realistic values into this equation shows that aliasing can easily be a problem in clinical practice (for example, if $Z = 10$ cm, $f_t = 5$ MHz and $\theta = 30°$, then $V_{\max}$ is only 0.68 ms^{-1}), particularly when high velocity flows need to be interrogated from deep within the body (for example, when measuring flow through a stenosed heart valve). There has therefore been considerable interest in developing techniques for overcoming aliasing, and several approaches designed to do just this have been published in the literature.

The earliest attempts to overcome aliasing in Doppler systems relied on the transmission of random and pseudo-random noise rather than tone bursts (Section 4.2.3.1). Because such systems use random signals, they should not in principle exhibit range ambiguity and therefore should not be subject to aliasing. Unfortunately, in practice, they do not work in the presence of the very large clutter signals found in clinical environments. A system described by Bendick and Newhouse (1974), which used continuous transmission of a random signal, worked with simple targets but had difficulty coping with the complex target situation found in the body. Cathignol et al (1980) improved on the technique of Bendick and Newhouse by using bursts of pseudo-random noise, and claimed to be able to reduce range ambiguity by a factor of two; however, Cathignol later showed (1988) that such systems can only function where the signal-to-clutter ratio is relatively high (greater than -23 dB in the particular example Cathignol cited).

Newhouse et al (1980) suggested an alternative method of alias rejection based on the use of non-equally spaced pulse transmission sequences. The idea behind this method is that in a pulsed Doppler system the Doppler spectra consists of the 'true' Doppler spectrum centred at f_d together with a series of replicas of itself, centred at frequencies $nf_r \pm f_d$, where f_r is the pulse repetition frequency and $n = 0, 1, 2, \ldots$. If, then, the system is operated alternatively with two repetition frequencies, all the replica Doppler spectra will displace with their corresponding repetition frequencies, but the true Doppler spectrum will not, thus allowing its identification. Newhouse et al carried out *in vitro* simulations and proved the technique for narrow spectra, but pointed out that the method can only work in situations where the Doppler spectrum of the observed flow has a fractional bandwidth of much less than unity. Nishiyama et al (1988) and Katakura et al (1994) have also suggested the use of non-equally spaced transmission pulses for the resolution of frequency aliases, but it is not clear how their method differs from that of Newhouse et al.

Several attempts to overcome the effects of aliasing have been based on the use of *a priori* information about the behaviour of blood flow to assign velocity values outside the normal Nyquist range. Hartley (1981) discussed the resolution of frequency aliases by making use of physiological information (the continuity of flow and some knowledge of the direction and/or timing of the blood flow being interrogated), and introduced the idea of frequency tracking beyond the Nyquist limit. He pointed out that although it is usual to interpret all Doppler frequencies as lying between $-PRF/2$ and $+PRF/2$, if the band-width of the signal is strictly limited to less than PRF, they can be interpreted as lying anywhere between f_x and $f_x + PRF$, where f_x is a frequency determined using physiological information. Thus, if it is known that flow in a certain vessel never reverses, f_x could be set as zero, and frequencies anywhere between 0 and PRF be satisfactorily dealt with. In CFI terms this is the equivalent of 'rotating the colour map', so that instead of interpreting phase changes of between 0 radians and $-\pi$ radians as reverse flow, they are interpreted as rapid forward flow. Frequency tracking takes this concept one step further, so that instead of fixing the value of f_x, it is allowed to vary dynamically and can be reset following each estimate of frequency f_e, as $f_e - PRF/2$. In this way it is possible to successfully track the signal from estimate to estimate, provided that the changes between estimates are not themselves greater than $\pm PRF/2$. Clearly this method can only work if the true velocity is known at some point so that it can be initialised appropriately, the most obvious being during diastole when the velocity should be relatively low and the Doppler shift in the range between $-PRF/2$ and $+PRF/2$.

Lucas et al (1983) described a method of frequency tracking in which three different first moment estimates of Doppler shift frequency were calculated simultaneously, each based on the assumption of a different phase-change to frequency mapping. The particular estimate which was then selected at any moment of time as the 'correct' estimate for output depended on a set of heuristic rules which determined whether the current Doppler signals corresponded to a 'systolic state', a 'diastolic state' or a 'base-line state'. The method described by Iinuma (1984) was even simpler; the output from the Doppler instrument being displayed as a sonogram with a frequency axis corresponding to $-PRF$ to $+PRF$, making no assumptions about the correct phase-change to frequency mapping. Thus, a phase change of $\pi/4$ would be displayed both as a frequency of $PRF/4$ and $-3PRF/4$. The correct visual interpretation of the sonogram is then fairly simple if one or more cardiac cycles are represented, particularly if the signal bandwidth is relatively low. The obvious disadvantages of this approach are that it is of no use when automatic measurements are required, and is of course of no value in colour flow systems. Frequency tracking beyond the Nyquist limit has also been described by Tortoli and colleagues (Tortoli 1989, Tortoli et al 1989), who have implemented a system which allows a continuous adaptive variation of the phase-change to frequency mapping based on the instantaneous mean frequency of the signal (in theory allowing for its known under-sampling if necessary).

A variation on the theme of frequency tracking was suggested by Baek et al (1989), who described tracking along the spatial axis rather than the temporal axis, based on the assumption of continuity of the velocity profile along the beam direction. This technique appeared to work quite well on the bench and has advantages over the temporal axis technique in that it requires less storage, is easy to initialise (in that flow near to a vessel wall is usually relatively slow), each time frame is independent, which means that it does not 'miss' information whilst the ultrasound beam is interrogating other vectors, and that errors are not automatically propagated into subsequent calculations.

Another method based on the use of *a-priori* information is that proposed by Stewart (1998), in which a fluid dynamic model was fitted to the colour flow map. To suppress the aliasing errors, jet velocities were fitted iteratively to a model constrained to match the orifice velocity measured without aliasing, using CW Doppler. At each iteration the model was used to detect aliased velocities, which were excluded during the next iteration. Stewart obtained a good correlation between measured and calculated flow rates in experimental and computational studies.

A method of resolving aliases, based on the simultaneous transmission of two carrier frequencies has been suggested by Fehr (Fehr 1991, Fehr et al 1991). Since the points at which any aliasing occurs depends on the carrier frequency (eqn 13.27), the interpretation of velocities which give rise to frequencies outside the $\pm PRF/2$ ranges will be different, depending on which transmitted frequency is used (Fig. 13.33). By combining the information from the two carrier frequencies, it is possible to determine the value of the true velocity. The new Nyquist limit for such a combined system is determined by the difference between the two carrier frequencies, so that if 4 MHz and 5 MHz carriers are used, the system will have the aliasing characteristics of a system operating at 1 MHz. Early testing of a commercial system based on the dual frequency technique showed promising results (Fan et al 1990).

A limitation of the method described by Fehr and colleagues is that it is only suitable for determining mean velocity, which is fine for colour flow systems but not for pulsed Doppler systems, where often the entire Doppler spectrum is required for further analysis. To overcome this limitation, Nitzpon et al (1995) suggested a two-stage approach whereby the non-aliased mean Doppler frequency is first determined by a method such as that of Fehr, and knowledge of the correct mean frequency is then used as a basis for interpolating new values of phase between the measured values, to reconstruct an appropriately sampled Doppler signal (i.e. one in which the

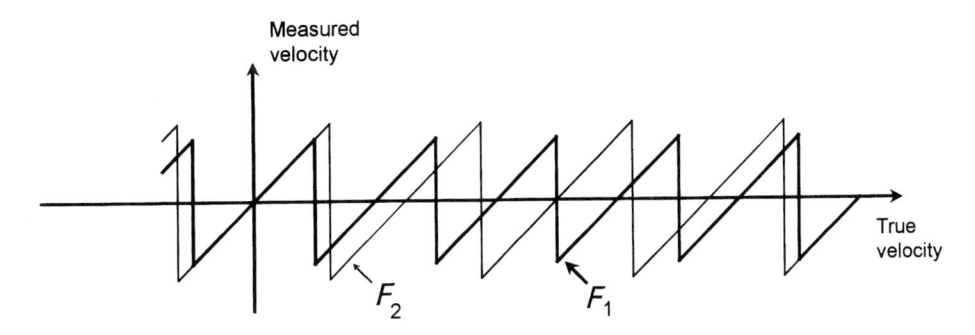

Figure 13.33 Aliasing characteristics of two pulsed Doppler systems using slightly different carrier frequencies. F_1 is higher than F_2 and therefore produces higher Doppler shift frequencies for a given velocity. If both systems use the same pulse repetition frequency, the system with the higher carrier frequency aliases more frequently than that with the lower frequency. By combining information about the velocities measured with each of the systems, it is possible to calculate the true velocity even after a number of aliases have occurred

phase rotation between individual samples is always less than $\pm \pi$ radians). The reconstructed signal can then be subjected to Fourier analysis in the normal way to derive the full Doppler spectrum. Results published by Nitzpon et al demonstrated the effective reconstruction of highly aliased Doppler spectra recorded both *in vitro* and *in vivo*.

Other recent approaches to alias rejection include 'velocity-matched' spectrum analysis (Torp and Kristoffersen 1995), which has been described in Section 8.2.3.2, and the 'extended autocorrelator' method (Lai et al 1997), which has been described in Section 11.3.6.

Finally, one of the theoretical advantages of time-shift estimation techniques (such as cross-correlation-based algorithms) over phase-shift estimation techniques (such as autocorrelation-based techniques) is that since they measure time delays rather than phase changes, they should apparently be immune to aliasing. This, however, is not always the case in practice, for reasons discussed in Section 11.4.5.3.

13.12 B-FLOW

A technique named 'B-flow' which allows the real-time visualisation of arterial blood flow has recently been introduced by GE Ultrasound. This technique, which is not a Doppler technique, allows the operator to gain an impression of arterial blood flow speed and direction by displaying in real-time the changing speckle pattern from the blood. This has been possible for many years in large veins where the low shear rates lead to erythrocyte aggregation and an increase in the backscatter of ultrasound from blood (Section 7.2.4.1), but not in arteries where the backscatter has been too small to allow direct imaging of blood flow. The advance that has made this possible is the introduction of pulse coding techniques. Coded signal methods have already been considered in the context of Doppler ultrasound in Section 4.2.3. Essentially these techniques provide an increase in signal to noise ratio (SNR) without increasing peak power by spreading the pulse power over a longer duration (without necessarily decreasing bandwidth and therefore spatial resolution). This additional SNR can either be used to reduce peak power, or to improve depth penetration, and to detect weaker signals such as those from flowing blood. Coded excitation for pulse echo systems has been described by O'Donnell (1992), Li et al (1992), Rao (1994) and Welch and Fox (1998).

B-flow produces dramatic images and has a number of attractive features, but also appears to have a number of limitations. Its main role is likely to be similar to that of power Doppler, i.e. for detecting the presence or absence of flow rather than for quantification. Because it is a pulse-echo technique it does not require a large number of pulses to form an estimate, and therefore it has rather better temporal resolution than Doppler techniques, which allows higher frame rates over a full-field of view. It also has excellent spatial resolution, and does not appear to suffer from the problems of overwriting at the vessel wall. On the negative side, the improved SNR is achieved by increasing the average power of the transmitted pulse, which is undesirable from a bioeffects point of view. Also there is currently no way of quantifying the velocity, and it is important to remember that the motion of the speckle pattern may not correspond to the motion of the blood (Section 13.10.4).

13.13 SUMMARY

Doppler ultrasound is still in a phase of rapid growth, and many exciting Doppler and 'Doppler-related' techniques are continuing to emerge. A range of such techniques has been described in this chapter.

13.14 REFERENCES

Ackerstaff RGA, Babikian VL, Georgiadis D, Russell D, Siebler M, Spencer MP, Stump D (1995) Basic identification criteria of Doppler microemboli signals: consensus committee of the 9th International Cerebral Hemodynamics Symposium. Stroke 26:1123

Albrecht T, Urbank A, Mahler M, Bauer A, Dore CJ, Blomley MJK, Cosgrove DO, Schlief R (1998) Prolongation and optimization of Doppler enhancement with microbubble US contrast agent by using continuous infusion: preliminary experience. Radiology 207:339–347

REFERENCES

Anderson AL, Hampton LD (1980) Acoustics of gas-bearing sediments 1. Background. J Acoust Soc Am 67:1865–1889

Anderson ME (1997) Spatial quadrature: A novel technique for multi-dimensional velocity estimation. In: Schneider SC, Levy M, McAvoy BR (eds) Procedings of the 1997 IEEE Ultrasonics Symposium, pp. 1233–1238. IEEE, Piscataway, NJ

Anderson ME (1998) Multi-dimensional velocity estimation with ultrasound using spatial quadrature. IEEE Trans Ultrason Ferroelec Freq Contr 45:852–861

Angelsen BAJ (1980) A theoretical study of the scattering of ultrasound from blood. IEEE Trans Biomed Eng BME-27:61–67

Ashrafzadeh A, Cheung JY, Dormer KJ (1988) Analysis of velocity estimation error for a multidimensional Doppler ultrasound system. IEEE Trans Ultrason Ferroelec Freq Contr 35:536–544

Austen WG, Howry DH (1965) Ultrasound as a method to detect bubbles or particulate matter in the arterial line during cardiopulmonary bypass. A preliminary report. J Surg Res 6:283–284

Baek KR, Bae MH, Park SB (1989) A new aliasing extension method for ultrasonic 2-dimensional pulsed Doppler systems. Ultrason Imaging 11:233–244

Barnea DI, Silverman HF (1972) A class of algorithms for fast digital image registration. IEEE Trans Comput C-21:179–186

Barnhart J, Levene H, Villapando E, Maniquis J, Fernandez J, Rice S, Jablonski E, Gjoen T, Tolleshaug H (1990) Characteristics of Albunex: air-filled albumin microspheres for echocardiography contrast enhancement. Invest Radiol 25:S162–S164

Barry CD, Allott CP, John NW, Mellor PM, Arundel PA, Thomson DS, Waterton JC (1997) Three-dimensional freehand ultrasound: image reconstruction and volume analysis. Ultrasound Med Biol 23:1209–1224

Bashford GR, von Ramm OT (1996) Ultrasound three-dimensional velocity measurement by feature tracking. IEEE Trans Ultrason Ferroelec Freq Contr 43:376–384

Baumgartner RW, Gonner F, Arnold M, Muri RM (1997) Transtemporal power- and frequency-based color-coded duplex sonography of cerebral veins and sinuses. Am J. Neuroradiol 18:1771–1781

Bazzochi M, Quaia E, Zuiani C, Moroldo ML (1998) Transcranial Doppler: state of the art. Eur J Radiol 27:S141–S148

Beach KW, Dunmire B, Overbeck JR, Waters D, Billeter M, Labs K-H, Strandness DE (1996) Vector Doppler systems for arterial studies. Part I: Theory. J Vas Invest 2:155–165

Becher H, Tiemann K, Schlief R, Luderitz B, Nanda NC (1997) Harmonic power Doppler contrast echocardiography: preliminary clinical results. Echocardiography 14:637–642

Bell DS, Bamber JC, Eckersley RJ (1995) Segmentation and analysis of colour Doppler images of tumour vasculature. Ultrasound Med Biol 21:635–647

Bendick PJ, Newhouse VL (1974) Ultrasonic random-signal flow measurement system. J Acoust Soc Am 56:860–865

Blomley M, Albrecht T, Cosgrove D, Jayaram V, Patel N, Butler-Barnes J, Eckersley R, Bauer A, Schlief R (1998) Stimulated acoustic emission imaging ('sono-scintigraphy') with the ultrasound contrast agent Levovist. Academic Radiol 5 (suppl 1):S236-S239

Bogdahn U, Becker G, Schlief R, Reddig J, Hassel W (1993) Contrast-enhanced transcranial color-coded real-time sonography. Results of a phase-two study. Stroke 24:676–684

Bohs LN, Trahey GE (1991) A novel method for angle independent ultrasonic imaging of blood flow and tissue motion. IEEE Trans Biomed Eng BME-38:280–286

Bohs LN, Friemel BH, McDermott BA, Trahey GE (1993a) Real-time system for angle-independent US of blood flow in two dimensions: initial results. Radiology 186:259–261

Bohs LN, Friemel BH, McDermott BA, Trahey GE (1993b) A real time system for quantifying and displaying two-dimensional velocities using ultrasound. Ultrasound Med Biol 19:751–761

Bohs LN, Friemel BH, Trahey GE (1995a) Experimental velocity profiles and volumetric flow via two-dimensional speckle tracking. Ultrasound Med Biol 21:885–898

Bohs LN, Geiman BJ, Nightingale KR, Choi CD, Friemel BH, Trahey GE (1995b) Ensemble tracking: a new method for 2D vector velocity measurement. In: Levy M, Schneider SC, McAvoy BR (eds) Proceedings of the 1995 IEEE Ultrasonics Symposium, pp. 1485–1488. IEEE, Piscataway, NJ

Bohs LN, Geiman BJ, Anderson ME, Breit SM, Trahey GE (1998) Ensemble tracking for 2D vector velocity measurement: experimental and initial clinical results. IEEE Trans Ultrason Ferroelec Freq Contr 45:912–924

Bos LJ (1997) The application of contrast echocardiography for the assessment of myocardial perfusion. PhD Dissertation, University of Amsterdam

Brinkley JF, Muramatsu SK, McCallum WD, Popp RL (1982) *In vitro* evaluation of an ultrasonic three-dimensional imaging and volume system. Ultrason Imaging 4:126–139

Broillet A, Puginier J, Ventrone R, Schneider M (1998) Assessment of myocardial perfusion by intermittent harmonic power Doppler using 'Sono Vue', a new ultrasound contrast agent. Invest Radiol 33:209–215

Burns PN (1996) Harmonic imaging with ultrasound contrast agents. Clin Radiol 51 (suppl 1):50–55

Burns PN, Powers JE, Simpson DH, Kolin A, Chin CT, Uhlendorf V, Fritzch T (1994) Harmonic power mode Doppler using microbubble contrast agents: an improved method for small vessel flow imaging. In: Proceedings of the 1994 IEEE Ultrasonics Symposium, pp. 1547–1550. IEEE, Piscataway, NJ

Calliada F, Campani R, Bottinelli O, Bozzini A, Sommargua MG (1998) Ultrasound contrast agents—basic principles. Eur J Radiol 27:S157–S160

Campani R, Calliada F, Bottinelli O, Bozzini A, Sommaruga MG, Draghi F, Anguissoia R (1998) Contrast enhancing agents in ultrasonography: clinical applications. Eur J Radiol 27:S161–S170

Cantarelli M, Dotti D, Lombardi R (1994) Blood velocity direction estimated by correlation. In: Proceedings of the 1994 IEEE Ultrasonics Symposium, pp. 1687–1690. IEEE, Piscataway, NJ

Carroll BA, Turner RJ, Tickner EG, Boyle DB, Young SW (1980) Gelatin encapsulated nitrogen microbubbles as ultrasonic contrast agents. Invest Radiol 15:260–266

Cathignol D (1988) Pseudo-random correlation flow measurement. In: Newhouse VL (ed) Progress in Medical Imaging, pp. 247–279. Springer-Verlag, New York

Cathignol DJ, Fourcade C, Chapelon J-Y (1980) Transcutaneous blood flow measurements using pseudorandom noise Doppler system. IEEE Trans Biomed Eng BME-27:30–36

Censor D, Newhouse VL, Vontz T, Ortega HV (1988) Theory of ultrasound Doppler-spectra velocimetry for arbitrary beam and flow configurations. IEEE Trans Biomed Eng 35:740–751

Céspedes EI, Carlier S, Li W, Mastik F, van der Steen AFW, Bom N, Verdouw P, Serruys PW (1998) Blood flow assessment using a mechanical intravascular ultrasound catheter: initial evaluation in vivo. J Vasc Invest 4:39–44

Chang PH, Shung KK, Wu S-j, Levene HB (1995) Second harmonic imaging and harmonic Doppler measurements with Albunex. IEEE Trans Ultrason Ferroelec Freq Contr 42:1020–1027

Chang PH, Shung KK, Levene HB (1996) Quantitative measurements of second harmonic Doppler using ultrasound contrast agents. Ultrasound Med Biol 22:1205–1214

Cloutier G, Shung KK, Durand L-G (1993) Experimental evaluation of intrinsic and nonstationary ultrasonic Doppler spectral broadening in steady and pulsatile flow loop models. IEEE Trans Ultrason Ferroelec Freq Contr 40:786–795

Cosgrove DO (1994) Doppler ultrasound of the breast. Imaging 6:185–196

Crowe JR, Shapo BM, Stephens DN, Bleam D, Eberle MJ, Wu CC, Muller DWM, Kovatch JA, Lederman RJ, O'Donnell M (1996) Coronary artery flow imaging with an intraluminal array. In: Levy M, Schneider SC, McAvoy BR (eds) Proceedings of the 1996 IEEE Ultrasonics Symposium, pp. 1481–1484. IEEE, Piscataway, NJ

Dacher JN, Pfister C, Monroc M, Eurin D, Ledosseur P (1996) Power Doppler sonographic pattern of acute pyelonephritis in children—comparison with CT. Am J Roentgenol 166:1451–1455

Daigle RE, Miller CW, Histand MB, McLeod FD, Hokanson DE (1975) Nontraumatic aortic blood flow sensing by use of an ultrasonic esophageal probe. J Appl Physiol 38:1153–1160

Dayton PA, Morgan KE, Klibanov AL, Brandenburger G, Nightingale KR, Ferrara KW (1997) A preliminary evaluation of the effects of primary and secondary radiation forces on acoustic contrast agents. IEEE Trans Ultrason Ferroelec Freq Contr 44:1264–1277

Dayton PA, Morgan KE, Klibanov AL, Brandenburger GH, Ferrara KW (1999) Optical and acoustical observations of the effects of ultrasound on contrast agents. IEEE Trans Ultrason Ferroelec Freq Contr 46:220–232

de Jong N (1996) Improvements in ultrasound contrast agents. IEEE Eng Med Biol 15 (6):72–82

de Jong N, Ten Cate FJ, Lancée CT, Roelandt JRTC, Bom N (1991) Principles and recent developments in ultrasound contrast agents. Ultrasonics 29:324–330

de Jong N, Hoff L, Skotland T, Bom N (1992) Absorption and scatter of encapsulated gas filled microspheres: theoretical considerations and some measurements. Ultrasonics 30:95–103

de Jong N, Cornet R, Lancée CT (1994a) Higher harmonics of vibrating gas-filled microspheres. Part One: Simulations. Ultrasonics 32:447–453

de Jong N, Cornet R, Lancée CT (1994b) Higher harmonics of vibrating gas-filled microspheres. Part Two: Measurements. Ultrasonics 32:455–459

Delker A, Turowski B (1997) Diagnostic value of three-dimensional transcranial contrast duplex sonography. J Neuroimag 7:139–144

Detmer PR, Bashein G, Hodges T, Beach KW, Filer EP, Burns DH, Strandness DE Jr (1994) 3D ultrasonic image feature localization based on magnetic scanhead tracking: in vitro calibration and validation. Ultrasound Med Biol 20:923–936

Diebold B, Peronneau PA (1985) Hemodynamics of the aorta: the normal and pathological. In: Altobelli SA, Voyles WF, Greene ER (eds) Cardiovascular Ultrasonic Flowmetry, pp. 101–124. Elsevier, New York

Dipak IA, Malhotra S, Nanda NC, Truffa CC, Agrawal G, Thakur AC, Jamil F, Taylor GW, Becher H (1997) Harmonic power Doppler contrast echocardiography: preliminary experimental results. Echocardiography 14:631–635

Distante A, Rovai D (1997) Stress echocardiography and myocardial contrast echocardiography in viability assessment. Eur Heart J 18:714–715

Dotti D, Lombardi R (1996) Estimation of the angle between ultrasound beam and blood velocity through correlation functions. IEEE Trans Ultrason Ferroelec Freq Contr 43:864–869

Dotti D, Lombardi R, Piazzi P (1992) Vectorial measurement of blood velocity by means of ultrasound. Med Biol Eng Comput 30:219–225

Downey DB, Fenster A (1995a) Three-dimensional power Doppler detection of prostatic cancer. Am J Roentgenol 165:741

Downey DB, Fenster A (1995b) Vascular imaging with a three-dimensional power Doppler system. Am J Roentgenol 165:665–668

Dunmire BL, Beach KW, Labs KH, Detmer PR, Strandness DE (1995) A vector Doppler ultrasound instrument. In: Levy M, Schneider SC, McAvoy BR (eds) Proceedings of the 1995 IEEE Ultrasonics Symposium, pp. 1477–1480. IEEE, Piscataway, NJ

Durick JE, Winter TC, Schmeidl UP, Cyr DR, Starr FL, Mack LA (1995) Renal perfusion pharmacologic changes depicted with power Doppler ultrasound in an animal model. Radiology 197:615–617

Dymling SO, Persson HW, Hertz CH (1991) Measurement of blood perfusion in tissue using Doppler ultrasound. Ultrasound Med Biol 17:433–444

Eriksson R, Persson HW, Dymling SO, Lindström K (1991) Evaluation of Doppler ultrasound for blood perfusion measurements. Ultrasound Med Biol 17:445–452

Eriksson R, Persson HW, Dymling SO, Lindström K (1995) Blood perfusion measurement with multifrequency Doppler ultrasound. Ultrasound Med Biol 21:49–57

Evans DH (1999) Detection of microemboli. In: Babikian VL, Wechsler LR (eds) Transcranial Doppler Ultrasonography, 2nd edn. pp 141–155. Butterworth-Heinemann, Boston

Evans DH, Smith JL, Naylor AR (1997) Characteristics of

Doppler ultrasound signals recorded from cerebral emboli. Ultrasound Med Biol 23 (suppl 1):S140

Fan P, Nanda NC, Cooper JW, Cape E, Yoganathan A (1990) Color Doppler assessment of high flow velocities using a new technology: *in-vitro* and clinical studies. Echocardiography 7:763–769

Fehr R (1991) Doppler flow velocity meter. USA Patent No. 5046500, issued 10 Sep 1991 to Kontron Instruments Holdings

Fehr R, Dousse B, Grossniklaus B (1991) New advances in colour flow mapping: quantitative velocity measurement beyond the Nyquist limit. Br J Radiol 64:651

Fei D-Y, Fu C-T, Brewer WH, Kraft KA (1993) The accuracy of angle independent Doppler color imaging in velocity and volumetric flow measurements. Proceedings of the 15th Annual International Conference IEEE/EMBS:214–215

Fei D-Y, Fu C-T, Brewer WH, Kraft KA (1994) Angle independent Doppler color imaging: determination of accuracy and a method of display. Ultrasound Med Biol 20:147–155

Fei D-Y, Liu DD, Fu C-T, Makhoul RG, Fisher MR (1997) Feasibility of angle independent Doppler color imaging for *in vivo* application: preliminary study on carotid arteries. Ultrasound Med Biol 23:59–67

Fein M, Delorme S, Weisser G, Zuna I, Van Kaick G (1995) Quantification of color Doppler for the evaluation of tissue vascularization. Ultrasound Med Biol 21:1013–1019

Feinstein SB, Shah PM, Bing RJ, Meerbaum S, Corday E, Chang B-L, Santillan G, Fujibayashi Y (1984a) Microbubble dynamics visualized in the intact capillary circulation. J Am Coll Cardiol 4:595–600

Feinstein SB, ten Cate FJ, Zwehl W, Ong K, Maurer G, Tei C, Shah PM, Meerbaum S, Corday E (1984b) Two-dimensional contrast echocardiography. *In vitro* development and quantitative analysis of echo-contrast agents. J Am Coll Cardiol 3:14–20

Fenster A, Downey DB (1996) 3-D Ultrasound imaging: a review. IEEE Eng Med Biol 15 (6):41–51

Fenster A, Tong S, Sherebrin S, Downey DB, Rankin RN (1995) Three-dimensional ultrasound imaging. Soc Photooptical Instrumentation Engineers Proc 2432:176–184

Ferrara KW, Zagar B, Sokil-Melgar J, Algazi VR (1996) High resolution 3D color flow mapping: applied to the assess-ment of breast vasculature. Ultrasound Med Biol 22:293–304

Fleming AD, McDicken WN, Sutherland GR, Hoskins PR (1994) Assessment of colour Doppler tissue imaging using test-phantoms. Ultrasound Med Biol 20:937–951

Fleming AD, Palka P, McDicken WN, Fenn LN, Sutherland GR (1996) Verification of cardiac Doppler tissue images using grey-scale M-mode images. Ultrasound Med Biol 22:573–581

Forsberg F, Goldberg BB, Liu JB, Merton DA, Rawool NM (1996) On the feasibility of real-time, *in vivo* harmonic imaging with proteinaceous microspheres. J Ultrasound Med 15:853–860

Forsberg F, Wu Y, Makin IRS, Wang W, Wheatley MA (1997) Quantitative acoustic characterisation of a new surfactant-based ultrasound contrast agent. Ultrasound Med Biol 23:1201–1208

Fox MD (1978) Multiple crossed-beam ultrasound Doppler velocimetry. IEEE Trans Sonics Ultrason SU-25:281–286

Fox MD, Gardiner WM (1988) Three-dimensional Doppler velocimetry of flow jets. IEEE Trans Biomed Eng 35:834–841

Friemel BH, Bohs LN, Trahey GE (1995) Relative performance of two-dimensional speckle-tracking techniques: normalized correlation, non-normalized correlation and sum-absolute-difference. In: Levy M, Schneider SC, McAvoy BR (eds) Proceedings of the 1995 IEEE Ultrasonics Symposium, pp. 1481–1484. IEEE, Piscataway, NJ

Fritzsch T, Schartl M, Siegert J (1988) Preclinical and clinical results with an ultrasonic contrast agent. Invest Radiol 23:S302-S305

Fritzsch T, Hilmann J, Kampfe M, Muller N, Schobel C, Siegert J (1990) SHU-508, a transpulmonary echo-contrast agent: initial experience. Invest Radiol 25:S160-S161

Furst G, Malms J, Heyer T, Saleh A, Cohnen M, Frieling T, Weule J, Hofer M (1998) Transjugular intrahepatic portosystemic shunts: improved evaluation with echo-enhanced color Doppler sonography, power Doppler sonography, and spectral duplex sonography. Am J Roentgenol 170:1047–1053

Furuhata H, Kanno R, Kodaira K et al (1978) Development of ultrasonic volume flow meter. Proc 17th Jap Soc Med Eng Biomed Eng 2:33, July 1978 (cited by Uematsu 1981)

Furuhata H, Kanno R, Kodaira K, Fujishiro K, Hayashi J, Matsumoto M, Yoshimura S (1979) An ultrasonic quantitative blood flow measuring system to measure the absolute volume flow rate. pp. 9–11. In: Proc 12th Intl Conf Med Biol Engineering, (cited by Tamura et al 1990)

Garcia MJ, Rodriguez L, Ares M, Griffin BP, Klein AL, Stewart WJ, Thomas JD (1996) Myocardial wall velocity assessment by pulsed Doppler tissue imaging: characteristic findings in normal subjects. Am Heart J 132:648–656

Gardiner WM, Fox MD (1989) Color-flow US imaging through the analysis of speckle motion. Radiology 172:866–868

Geiman B, Bohs L, Czenszak S, Anderson M, Trahey G (1996) Initial experimental results using ensemble tracking for 2D vector velocity measurement. In: Levy M, Schneider SC, McAvoy BR (eds) Proceedings of the 1996 IEEE Ultrasonics Symposium, pp. 1241–1244. IEEE, Piscataway, NJ

Geiman B, Bohs L, Anderson M, Breit S, Trahey G (1997) A comparison of algorithms for tracking sub-pixel speckle motion. In: Schneider SC, Levy M, McAvoy BR (eds) Proceedings of the 1997 IEEE Ultrasonics Symposium, pp. 1239–1242. IEEE, Piscataway, NJ

Geiser EA, Christie LG, Conetta DA, Conti CR, Gossman GS (1982) A mechanical arm for spatial registration of two-dimensional echocardiographic sections. Catheterization Cardiovasc Diagn 8:89–101

Ghosh A, Nanda NC, Maurer G (1982) Three-dimensional reconstruction of echo-cardiographic images using the rotation method. Ultrasound Med Biol 8:655–661

Giarré M, Dousse B, Meister JJ (1996) Velocity vector reconstruction for color flow Doppler: experimental evaluation for a new geometrical method. Ultrasound Med Biol 22:75–88

Gilja OH, Thune N, Matre K, Hausken T, Odegaard S, Berstad A (1994) In vitro evaluation of three-dimensional ultrasonography in volume estimation of abdominal organs. Ultrasound Med Biol 20:157–165

Gillis MF, Peterson PL, Karagianes MT (1968) In vivo detection of circulating gas emboli associated with decompression sickness using the Doppler flowmeter. Nature 217:965–967

Gramiak R, Shah PM (1968) Echocardiography of the aortic root. Invest Radiol 3:356–366

Gramiak R, Shah PM, Kramer DH (1969) Ultrasonic cardiography: contrast studies in anatomy and function. Radiology 92:939–948

Greenberg RS, Taylor GA, Stapleton JC, Hillsley CA, Spinak D (1996) Analysis of regional cerebral blood-flow in dogs, with an experimental microbubble-based US contrast agent. Radiology 201:119–123

Grubb NR, Fleming A, Sutherland GR, Fox KAA (1995) Skeletal muscle contraction in healthy volunteers: assessment with Doppler tissue imaging. Radiology 194:837–842

Guo Z, Fenster A (1996) Three-dimensional power Doppler imaging: a phantom study to quantify vessel stenosis. Ultrasound Med Biol 22:1059–1069

Guo Z, Moreau M, Rickey DW, Picot PA, Fenster A (1995) Quantitative investigation of in vitro flow using three-dimensional colour Doppler ultrasound. Ultrasound Med Biol 21:807–816

Hartley CJ (1981) Resolution of frequency aliases in ultrasonic pulsed Doppler velocimeters. IEEE Trans Sonics Ultrason SU-28:69–75

Hartley CJ, Cheirif J, Collier KR, Bravenec JS, Mickelson JK (1993) Doppler quantification of echo-contrast injections in vivo. Ultrasound Med Biol 19:269–278

Hauff P, Fritzsch T, Reinhardt M, Weitschies W, Luders F, Uhlendorf V, Heldmann D (1997) Delineation of experimental tumors in rabbits by a new ultrasound contrast agent and stimulated acoustic emission. Invest Radiol 32:94–99

Hein IA (1995) Triple-beam lens transducers for three-dimensional ultrasonic fluid flow estimation. IEEE Trans Ultrason Ferroelec Freq Contr 42:854–869

Hein IA (1997) 3-D Flow velocity vector estimation with a triple-beam lens transducer—experimental results. IEEE Trans Ultrason Ferroelec Freq Contr 44:85–95

Hein IA, O'Brien WD (1993) Current time-domain methods for assessing tissue motion by analysis from reflected ultrasound echoes—a review. IEEE Trans Ultrason Ferroelec Freq Contr 40:84–102

Herndon JH, Bechtol CO, Crickenberger DP (1975) Use of ultrasound to detect fat emboli during total hip replacement. Acta Orthop Scand 46:108–118

Hindle AJ, Perkins AC (1995) History and basic principles of echo-contrast media. Br Med Ultrasound Soc Bull 3 (1 February):17–23

Hirsch W, Bell DS, Crawford DC, Kale SG, McCready VR, Bamber JC (1995) Colour Doppler image analysis for tissue vascularity and perfusion: a preliminary clinical evaluation. Ultrasound Med Biol 21:1107–1117

Hodges TC, Detmer PR, Burns DH, Beach KW, Strandness DE (1994) Ultrasonic three-dimensional reconstruction: in vitro and in vivo volume and area measurement. Ultrasound Med Biol 20:719–729

Hoskins PR (1997) Peak velocity estimation in arterial stenosis models using colour vector Doppler. Ultrasound Med Biol 23:889–897

Hoskins PR, Fleming A, Stonebridge P, Allan PL, Cameron D (1994) Scan-plane vector maps and secondary flow motions in arteries. Eur J Ultrasound 1:159–169

Huber S, Helbich T, Kettenbach J, Dock W, Zuna I, Delorme S (1998) Effects of a microbubble contrast agent on breast tumors: computer-assisted quantitative assessment with color Doppler US—early experience. Radiology 208:485–489

Iinuma K (1984) Doppler effect blood flow sensing device lying within a band width related to sampling frequency. USA Patent No. 4485821, issued 4 Dec 1984 to Tokyo Shibaura Denki Kabushiki Kaisha, Kawaski, Japan

Isaaz K, Thompson A, Ethevenot K, Cloez JL, Brembilla B, Pernot C (1989) Doppler echocardiographic measurement of low velocity motion of the left ventricular posterior wall. Am J Cardiol 64:66–75

Jayaweera AR, Edwards N, Glasheen WP, Villanueva FS, Abbott RD, Kaul S (1994) In vivo myocardial kinetics of air-filled microbubbles during myocardial contrast echocardiography. Circ Res 74:1157–1165

Jensen JA, Munk P (1998) A new method for estimation of velocity vectors. IEEE Trans Ultrason Ferroelec Freq Contr 45:837–851

Kallel F, Bertrand M, Meunier J (1994) Speckle motion artifact under rotation. IEEE Trans Ultrason Ferroelec Freq Contr 41:105–122

Kaps M, Schaffer P, Beller KD, Seidel G, Bliesath H, Diletti E (1997) Characteristics of transcranial Doppler signal enhancement using a phospholipid-containing echocontrast agent. Stroke 28:1006–1008

Katakura K, Nishiyama H, Ogawa T (1994) Non-equally-spaced pulse transmission for non-aliasing ultrasonic pulsed Doppler measurement. Ultrasound Med Biol 20 (suppl 1):S69

Keller MW, Feinstein SB, Watson DD (1987) Successful left ventricular opacification following peripheral venous injection of sonicated contrast agent: an experimental evaluation. Am Heart J 114:570–575

Keller MW, Segal SS, Kaul S (1989) The behaviour of sonicated albumin microbubbles within the microcirculation: a basis for their use during myocardial contrast echocardiography. Circ Res 65:458–467

Keller MW, Spotniz WD, Matthew TL, Glasheen WP, Watson DD, Kaul S (1990) Intraoperative assessment of regional myocardial perfusion using quantitative myocardial contrast echocardiography: an experimental evaluation. J Am Coll Cardiol 16:1267–1279

King DL, King DLJ, Shao Y-C (1990) Three-dimensional spatial registration and interactive display of position and orientation of real-time ultrasound images. J Ultrasound Med 9:525–532

Kinsler LE, Frey AR, Coppens AB, Sanders JV (1982) Fundamentals of Acoustics, 3rd edn, p. 228. Wiley, New York

Kostis JB, Mavrogeorgis EM, Slater A, Bellet E (1972) Use of a range-gated, pulsed ultrasonic Doppler technique for continuous measurement of velocity of the posterior heart wall. Chest 62:597–604

Kremkau FW, Gramiak R, Carstensen EL (1970) Ultrasonic detection of cavitation at catheter tips. Am J Roentgenol 110:177–183

Krishna PD, Newhouse VL (1997) Second harmonic characteristics of the ultrasound contrast agents albunex and FSO69. Ultrasound Med Biol 23:453–459

Lai X, Torp H, Kristoffersen K (1997) An extended autocorrelation method for estimation of blood velocity. IEEE Trans Ultrason Ferroelec Freq Contr 44:1332–1342

Lange A, Palka P, Caso P, Fenn LN, Olszewski R, Ramo MP, Shaw TRD, Nowicki A, Fox KAA, Sutherland GR (1997) Doppler myocardial imaging vs. B-mode grey-scale imaging: a comparative *in vitro* and *in vivo* study into their relative efficacy in endocardial boundary detection. Ultrasound Med Biol 23:69–75

Leighton TG (1994) The Acoustic Bubble. Academic Press, London

Leischik R, Kuhlmann C, Bruch C, Jeremias A, Buck T, Erbel R (1997) Reproducibility of stress echocardiography using intravenous injection of ultrasound contrast agent (BY 963). Int J Cardiac Imag 13:387–394

Levine RA, Handschumacher MD, Sanfilippo AJ, Hagege AA, Harrigan P, Marshall JE, Weyman AE (1989) Three-dimensional echocardiographic reconstruction of the mitral valve, with implications for the diagnosis of mitral valve prolapse. Circulation 80:589–598

Levoy M (1990) A hybrid ray tracer for rendering polygon and volume data. IEEE Computer Graphics Appl 10:33–40

Li P-C, Ebbini E, O'Donnell M (1992) A new filter design technique for coded excitation systems. IEEE Trans Ultrason Ferroelec Freq Contr 39:693–699

Li W, Lancée CT, van der Steen AFW, Gussenhoven EJ, Bom N (1996) Blood velocity estimation with high frequency intravascular ultrasound. In: Levy M, Schneider SC, McAvoy BR (eds) Proceedings of the 1996 IEEE Ultrasonics Symposium, pp. 1485–1488. IEEE, Piscataway, NJ

Li W, Lancée CT, Céspedes EI, van der Steen AFW, Bom N (1997) Decorrelation of intravascular echo signals: potentials for blood velocity estimation. J Acoust Soc Am 102:3785–3794

Li W, van der Steen AFW, Lancée CT, Céspedes I, Bom N (1998) Blood flow imaging and volume flow quantitation with intravascular ultrasound. Ultrasound Med Biol 24:203–214

Lu J-Y (1996) Improving accuracy of transverse velocity measurement with a new limited diffraction beam. In: Levy M, Schneider SC, McAvoy BR (eds) Proceedings of the 1996 IEEE Ultrasonics Symposium, pp. 1255–1260. IEEE, Piscataway, NJ

Lubbers J, van den Berg JW (1976) An ultrasonic detector for microgas emboli in a bloodflow line. Ultrasound Med Biol 2:301–310

Lucas CL, Keagy BA, Hsiao HS, Wilcox BR (1983) Software analysis of 20 MHz pulsed Doppler quadrature data. Ultrasound Med Biol 9:641–655

Maniatis TA, Cobbold RSC, Johnston KW (1994a) Flow imaging in an end-to-side anastomosis model using two-dimensional velocity vectors. Ultrasound Med Biol 20:559–569

Maniatis TA, Cobbold RSC, Johnston KW (1994b) Two-dimensional velocity reconstruction strategies for color flow Doppler ultrasound images. Ultrasound Med Biol 20:137–145

Mattrey RF (1989) Perfluorooctylbromide: a new contrast agent for CT, sonography and MR imaging. Am J Roentgenol 152:247–252

Mattrey RF, Scheible FW, Gosink BB, Leopold GR, Lond DM, Higgins CB (1982) Perfluorooctylbromide: a liver/spleen-specific and tumour-imaging ultrasound contrast agent. Radiology 145:759–762

Mattrey RF, Leopold GR, von Sonnenberg E, Gosink BB, Scheible FW, Long D (1983) Perfluorochemicals as liver and spleen-seeking ultrasound contrast agents. J Ultrasound Med 2:173–176

Mattrey RF, Strich G, Shelton RE, Gosink BB, Leopold GR, Lee T, Forsythe J (1987) Perfluorochemicals as US contrast agents for tumor imaging and hepatosplenography: preliminary clinical results. Radiology 163:339–343

McArdle A, Newhouse VL, Beach KW (1995) Demonstration of three-dimensional vector flow estimation using bandwidth and two transducers on a flow phantom. Ultrasound Med Biol 21:679–692

McDicken WN, Anderson T, McHugh R, Bow CR, Boddy K, Cole R (1979) An ultrasonic real-time scanner with pulsed Doppler and T-M facilities for foetal breathing and other obstetrical studies. Ultrasound Med Biol 5:333–339

McDicken WN, Sutherland GR, Moran CM, Gordon LN (1992) Colour Doppler velocity imaging of the myocardium. Ultrasound Med Biol 18:651–654

McWhirt RE (1979) Speed of sound measurements in potential contrast agents for use in diagnostic ultrasound. Master's Dissertation, University of Kansas

Medwin H (1977) Counting bubbles acoustically: a review. Ultrasonics 15:7-13

Meltzer RS, Serruys PW, McGhie J, Verbaan N, Roelandt J (1980a) Pulmonary wedge injections yielding left sided echocardiographic contrast. Br Heart J 44:390–394

Meltzer RS, Tickner EG, Popp RL (1980b) Why do the lungs clear ultrasonic contrast. Ultrasound Med Biol 6:263–269

Meltzer RS, Serruys PW, Hugenholtz PG, Roelandt J (1981) Intravenous carbon dioxide as an echocardiographic contrast agent. J Clin Ultrasound 9:127–131

Meltzer RS, Klig V, Teichholz LE (1985) Generating precision microbubbles for use as an echocardiographic contrast agent. J Am Coll Cardiol 5:978–982

Mo LYL, Cobbold RSC (1986) A stochastic model of the backscattered Doppler ultrasound from blood. IEEE Trans Biomed Eng BME-33:20–27

Moehring MA, Klepper JR (1994) Pulse Doppler ultrasound detection, characterization and size estimation of emboli in flowing blood. IEEE Trans Biomed Eng 41:35–44

Moran CM, Anderson T, Sboros V, Sutherland GR, Wright R, McDicken WN (1998) Quantification of the enhanced backscatter phenomenon from an intravenous and an intra-arterial contrast agent. Ultrasound Med Biol 24:871–880

Mor-Avi V, David D, Askelrob S, Bitton Y, Choshniak I (1993) Myocardial regional blood flow: quantitative measurement by computer analysis of contrast enhanced echocardiographic images. Ultrasound Med Biol 19:619–633

Moritz WE, Pearlman AS, McCabe DH, Medema DK, Ainsworth ME, Boles MS (1983) An ultrasonic technique for imaging the ventricle in three-dimensions and calculating its volume. IEEE Trans Biomed Eng BME-30:482–492

Morrison DC, McDicken WN, Smith DSA (1983) A motion artifact in real-time ultrasound scanners. Ultrasound Med Biol 9:201–203

Morse PM, Ingard KU (1968) Theoretical Acoustics, pp. 319–333. McGraw-Hill, New York

Moskalik A, Carson PL, Meyer CR, Fowlkes JB, Rubin JM, Roubidoux MA (1995) Registration of three-dimensional compound ultrasound scans of the breast for refraction and motion correction. Ultrasound Med Biol 21:769–778

Mulvagh SL, Foley DA, Aeschbacher BC, Klarich KK, Seward JB (1996) 2nd harmonic imaging of an intravenously administered echocardiographic contrast agent— visualization of coronary arteries and measure ment of coronary blood flow. J Am Coll Cardiol 27:1519–1525

Murphy KJ, Bude RO, Dickinson LD, Rubin JM (1997) Use of intravenous contrast material in transcranial sonography. Acad Radiol 4:577–582

Nelson TR, Pretorius DH (1998) Three-dimensional ultrasound imaging. Ultrasound Med Biol 24:1243–1270

Newhouse VL, Reid JM (1990) Invariance of Doppler bandwidth with flow axis displacement. In: McAvoy BR (ed) IEEE Ultrasonics Symposia, vol 3, pp. 1533–1536. IEEE, Piscataway, NJ

Newhouse VL, LeCong P, Furgason ES, Ho CT (1980) On increasing the range of pulsed Doppler systems for blood flow measurement. Ultrasound Med Biol 6:233–237

Newhouse VL, Censor D, Vontz T, Cisneros JA, Goldberg BB (1987) Ultrasound Doppler probing of flows transverse with respect to beam axis. IEEE Trans Biomed Eng 34:779–789

Newhouse VL, Cathignol D, Chapelon J-Y (1991) Study of vector flow estimation with transverse Doppler. In: Proceedings of the 1991 IEEE Ultrasonics Symposium, pp. 1259–1263. IEEE, Piscataway, NJ

Newhouse VL, Dickerson KS, Cathignol D, Chapelon JY (1994) Three-dimensional vector flow estimation using two transducers and spectral width. IEEE Trans Ultrason Ferroelec Freq Contr 41:90–95

Nishiyama H, Katakura K, Toshio O (1988) Nonaliasing ultrasonic pulsed Doppler measurement. Ultrason Imaging 10(abstr):70

Nitzpon HJ, Rajaonah JC, Burchkardt CB, Dousse B, Meister JJ (1995) A new pulsed wave Doppler ultrasound system to measure blood velocities beyond the Nyquist limit. IEEE Trans Ultrason Ferroelec Freq Contr 42:265–279

O'Donnell M (1992) Coded excitation system for improving the penetration of real-time phased-array imaging systems. IEEE Trans Ultrason Ferroelec Freq Contr 39:341–351

Ophir J, Parker KJ (1989) Contrast agents in diagnostic ultrasound. Ultrasound Med Biol 15:319–333

Ophir J, McWhirt RE, Maklad NF (1979) Aqueous solutions as potential ultrasonic contrast agents. Ultrason Imaging 1:265–279

Ophir J, Gobuty A, McWhirt RE, Maklad NF (1980) Ultrasonic backscatter from contrast producing collagen microspheres. Ultrason Imaging 2:67–77

Ophir J, Gobuty A, Maklad NF, Tyler T, Jaeger P, McWhirt RE (1985) Quantitative assessment of in vivo backscatter enhancement from gelatin microspheres. Ultrason Imaging 7:293–299

Overbeck JR, Beach KW, Strandness DE (1992) Vector Doppler: Accurate measurement of blood velocity in two dimensions. Ultrasound Med Biol 18:19–31

Padayachee TS, Gosling RG, Bishop CC, Burnand K, Browse NL (1986) Monitoring middle cerebral artery blood velocity during carotid endarterectomy. Br J Surg 73:98–100

Palka P, Lange A, Wright RA, Starkey IR, Fleming AD, Bouki KP, Sutherland GR, Shaw TRD, Fox KAA (1997) Myocardial velocity gradient measured throughout the cardiac cycle in dilated cardiomyopathy hearts—a potential new parameter of systolic and diastolic myocardial function by Doppler myocardial imaging. Eur J Ultrasound 5:141–154

Parker KJ, Tuthill TA, Lerner RM, Violante MR (1987) A particulate contrast agent with potential for ultrasound imaging of liver. Ultrasound Med Biol 13:555–566

Parker KJ, Baggs RB, Lerner RM, Tuthill TA, Violante MR (1990) Ultrasonic contrast for hepatic tumors using IDE particles. Invest Radiol 25:1135–1139

Patterson RH, Kessler J (1969) Microemboli during cardiopulmonary bypass detected by ultrasound. Surg Gynecol Obstet 129:505–510

Peronneau P, Xhaard M, Nowicki A, Pellet M, Delouche P, Hinglais J (1972) Pulsed Doppler ultrasonic flowmeter and flow pattern analysis. In: Roberts VC (ed) Blood Flow Measurement, pp. 24–28. Sector, London

Peronneau P, Sandman W, Xhaard M (1977) Blood flow patterns in large arteries. In: White D, Brown RE (eds) Ultrasound in Medicine, vol 3B, pp. 1193–1208. Plenum, New York

Phillips PJ, Kadi AP, von Ramm OT (1995) Feasibility study for a two-dimensional diagnostic ultrasound velocity mapping system. Ultrasound Med Biol 21:217–229

Picot PA, Rickey DW, Mitchell R, Rankin RN, Fenster A (1993) Three-dimensional colour Doppler imaging. Ultrasound Med Biol 19:95–104

Porter TR, Xie F (1995) Transient myocardial contrast after initial exposure to diagnostic ultrasound pressures with minute doses of intravenously injected microbubbles: demonstration and potential mechanisms. Circulation 92:2391–2395

Postert T, Braun B, Federlein J, Przuntek H, Köster O, Büttner T (1998) Diagnosis and monitoring of middle cerebral artery occlusion with contrast-enhanced transcranial color-coded real-time sonography in patients with inadequate acoustic bone windows. Ultrasound Med Biol 24:333–340

Prager RW, Rohling RN, Gee AH, Berman L (1998) Rapid calibration for 3-D freehand ultrasound. Ultrasound Med Biol 24:855–869

Pugsley W (1986) The use of Doppler ultrasound in the assessment of microemboli during cardiac surgery. Perfusion 4:115–122

Raab FH, Blood EB, Steiner TO, Jones HR (1979) Magnetic position and orientation tracking system. IEEE Trans Aerosp Electron Syst AES-15:709–717

Ramamurthy BS, Trahey GE (1991) Potential and limitations of angle-independent flow detection algorithms using radiofrequency and detected echo signals. Ultrason Imaging 13:252–268

Rankin RN, Fenster A, Downey DB, Munk PL, Levin MF, Vellet AD (1993) Three-dimensional sonographic reconstruction: techniques and diagnostic applications. Am J Roentgenol 161:695–702

Rao NAHK (1994) Investigation of a pulse compression technique for medical ultrasound: a simulation study. Med Biol Eng Comput 32:181–188

Ries F, Honisch C, Lambertz M, Schlief R (1993) A transpulmonary contrast medium enhances the transcranial Doppler signal in humans. Stroke 24:1903–1909

Ritchie CJ, Edwards WS, Mack LA, Cyr DR, Kim Y (1996) Three-dimensional ultrasonic angiography using power-mode Doppler. Ultrasound Med Biol 22:277–286

Roelandt JRTC, ten Cate FJ, Vletter WB, Taams MA (1994) Ultrasonic dynamic three-dimensional visualization of the heart with a multiplane transesophageal imaging transducer. J Am Soc Echo 7:217–219

Routh HF, Pusateri TL, Waters DD (1990) Preliminary studies into high velocity transverse blood flow measurement. In: Proceedings of the 1990 IEEE Ultrasonics Symposium, pp. 1523–1526. IEEE, Piscataway, NJ

Rovai D, Ghelardini G, Lombardi M, Trivella MG, Nevola E (1993) Myocardial washout of sonicated iopamidol does not reflect the transmural distribution of coronary blood flow. Eur Heart J 14:1072–1078

Rschevkin SN (1963) A Course of Lectures on the Theory of Sound. Pergamon, Oxford

Rubin JM, Adler RS, Fowlkes JB, Spratt S, Pallister JE, Chen J-F, Carson PL (1995) Fractional moving blood volume: estimation with power Doppler US. Radiology 197:183–190

Sato T, Sasaki O (1979) Ultrasonic Doppler velocimeter using cross-bispectral analysis. Ultrason Imaging 1:144–153

Sawada H, Fujii J, Kato K, Onoe M, Kuno Y (1983) Three dimensional reconstruction of the left ventricle from multiple cross sectional echocardiograms. Value for measuring left ventricular volume. Br Heart J 50:438–442

Schlief R, Schurmann R, Balzer T, Petrick J, Urbank A, Zomack M, Niendorf H-P (1993) Diagnostic value of contrast enhancement in vascular Doppler ultrasound. In: Nanda N, Schlief R (eds) Advances in Echo Imaging Using Contrast Enhancement, pp. 309–323. Kluwer Academic, Dordrecht

Schmidt-Trucksass A, Grathwohl D, Schmid A, Boragk R, Upmeier C, Keul J, Huonker M (1998) Assessment of carotid wall motion and stiffness with tissue Doppler imaging. Ultrasound Med Biol 24:639–646

Schmiedl UP, Kormarniski K, Winter TC, Luna JA, Cyr DR, Ruppenthal G, Schlief R (1998) Assessment of fetal and placental blood flow in primates using contrast enhanced ultrasonography. J Ultrasound Med 17:75–80

Schmolke JK, Ermert H (1988) Ultrasound pulse Doppler tomography. In: Proceedings of the 1988 IEEE Ultrasonics Symposium, pp. 785–788. IEEE, Piscataway, NJ

Schrank E, Phillips DJ, Moritz WE, Strandness DE (1990) A triangulation method for the quantitative measurement of arterial blood velocity magnitude and direction in humans. Ultrasound Med Biol 16:499–509

Schrope BA, Newhouse VL (1993) Second harmonic ultrasonic blood perfusion measurement. Ultrasound Med Biol 19:567–579

Schrope B, Newhouse VL, Uhlendorf V (1992) Simulated capillary blood flow measurement using a nonlinear ultrasonic contrast agent. Ultrason Imaging 14:134–158

Schwarz KQ, Chen X, Bezante GP, Phillips D, Schlief R (1996) The Doppler kinetics of microbubble echo contrast. Ultrasound Med Biol 22:453–462

Seidel G, Beller K-D, Kaps M (1996) Pharmacokinetic studies of different echo-contrast agents in the cerebral circulation of dogs. Ultrasound Med Biol 22:1037–1042

Shattuck DP, Weinshenker MD, Smith SW, von Ramm OT (1984) Explososcan: a parallel processing technique for high speed ultrasound imaging with linear phased arrays. J Acoust Soc Am 75:1273–1282

Shung KK, Flenniken RR (1995) Time-domain ultrasonic contrast blood flowmetry. Ultrasound Med Biol 21:71–78

Shung KK, Thieme GA (eds) (1993) Ultrasonic Scattering in Biological Tissues, pp. 9–12. CRC Press, London

Shung KK, Sigelmann RA, Reid JM (1976) Scattering of ultrasound by blood. IEEE Trans Biomed Eng 23:460–467

Simon RH, Ho SY, DiArrigo J, Wakefield A, Hamilton SG (1990) Lipid-coated ultrastable microbubbles as a contrast agent in neurosonography. Invest Radiol 25:1300–1304

Simpson DH, Burns PN (1997) Pulse inversion Doppler: a new method for detecting nonlinear echoes from microbubble contrast agents. In: Schneider SC, Levy M, McAvoy BR (eds) Proceedings of the 1997 IEEE Ultrasonics Symposium, pp. 1597–1600. IEEE, Piscataway, NJ

Simpson NAB, Nimrod C, deVermette R, Leblanc C, Fournier J (1998) Sonographic evaluation of intervillous flow in early pregnancy: use of echo-enhancement agents. Ultrasound Obstet Gynecol 11:204–208

Smith JL, Evans DH, Fan L, Thrush AJ, Naylor AR (1994) Processing Doppler ultrasound signals from blood-borne emboli. Ultrasound Med Biol 20:455–462

Smith SW, Pavy HG, von Ramm OT (1991) High-speed ultrasound volumetric imaging system—Part I: Transducer design and beam steering. IEEE Trans Ultrason Ferroelec Freq Contr 38:100–108

Smith SW, Trahey GE, von Ramm OT (1992) Two-dimensional arrays for medical ultrasound. Ultrason Imaging 14:213–233

Spencer MP, Thomas GI, Nicholls SC, Sauvage LR (1990) Detection of middle cerebral artery emboli during carotid endarterectomy using transcranial Doppler ultrasonography. Stroke 21:415–423

Stewart SFC (1998) Aliasing-tolerant color Doppler quantification of regurgitant jets. Ultrasound Med Biol 24:881–898

Stonebridge PA, Hoskins PR, Allan PL, Belch JFF (1996) Spiral laminar flow *in vivo*. Clin Sci 91:17–21

Sutherland GR, Stewart MJ, Groundstroem KWE, Moran CM, Fleming AD, Guell-Peris FJ, Riemersma RA, Fenn LN, Fox KAA, McDicken WN (1994) Colour Doppler myocardial imaging—a new technique for the assessment of myocardial function. J Am Soc Echo 7:441–458

Tamura T, Cobbold RSC, Johnston KW (1990) Determination of 2-D velocity vectors using color Doppler ultrasound. In: Proceedings of the 1990 IEEE Ultrasonics Symposium, pp. 1537–1540. IEEE, Piscataway, NJ

Tamura T, Cobbold RSC, Johnston KW (1991) Determination of two-dimensional flow velocity vector fields using color Doppler ultrasound. In: Proceedings of the American Society of Mechanical Engineers (ASME) Columbus, OH. 1991 Biomechanics Symposium, pp. 17–20. AMD-Vol 120. ASME, New York

Tarter VM, Satterfield R, Schumacher DJ, Tran P, Mattrey RF (1991) Comparison of Imagent(R) BP (PFOB) and Fluosol-DA 20% as ultrasound contrast agents. Ultrasound Med Biol 10:S17-S18

Taylor GA, Ecklund K, Dunning PS (1996) Renal cortical perfusion in rabbits—visualization with color amplitude imaging and experimental microbubble-based US contrast agent. Radiology 201:125–129

Taylor GA, Barnewolt CE, Adler BH, Dunning PS (1998a) Renal cortical ischemia in rabbits revealed by contrast-enhanced power Doppler sonography. Am J Roentgenol 170:417–422

Taylor GA, Barnewolt CE, Dunning PS (1998b) Excitotoxin-induced cerebral hyperemia in newborn piglets: regional cerebral blood flow mapping with contrast-enhanced power Doppler US. Radiology 208:73–79

ter Braak JWG, de Vlieger M (1965) Cerebral pulsations in echoencephalography. Acta Neurochir 12:678–694

Tong S, Downey DB, Cardinal HN, Fenster A (1996) A three-dimensional ultrasound prostate imaging system. Ultrasound Med Biol 22:735–746

Torp H, Kristoffersen K (1995) Velocity matched spectrum analysis: a new method for suppressing velocity ambiguity in pulsed-wave Doppler. Ultrasound Med Biol 21:937–944

Tortoli P (1989) A tracking FFT processor for pulsed Doppler analysis beyond the Nyquist limit. IEEE Trans Biomed Eng 36:232–237

Tortoli P, Valgimigli F, Guidi G (1989) Clinical evaluation of a new anti-aliasing technique for ultrasound pulsed Doppler analysis. Ultrasound Med Biol 15:749–756

Tortoli P, Guidi G, Mariotti V, Newhouse VL (1992) Experimental proof of Doppler bandwidth invariance. IEEE Trans Ultrason Ferroelec Freq Contr 39:196–203

Tortoli P, Guidi G, Newhouse VL (1993a) Invariance of the Doppler bandwidth with range cell size above a critical beam-to-flow angle. IEEE Trans Ultrason Ferroelec Freq Contr 40:381–386

Tortoli P, Guidi G, Pignoli P (1993b) Transverse Doppler spectral analysis for a correct interpretation of flow sonograms. Ultrasound Med Biol 19:115–121

Tortoli P, Guidi G, Guidi F, Atzeni C (1994) A review of experimental transverse Doppler studies. IEEE Trans Ultrason Ferroelec Freq Contr 41:84–89

Trahey GE, Allison JW, von Ramm OT (1987) Angle independent ultrasonic detection of blood flow. IEEE Trans Biomed Eng 34:965–967

Trahey GE, Hubbard SM, von Ramm OT (1988) Angle independent ultrasonic blood flow detection by frame-to-frame correlation of B-mode images. Ultrasonics 26:271–276

Turnbull DH, Foster FS (1991) Beam steering with pulsed two-dimensional transducer arrays. IEEE Trans Ultrason Ferroelec Freq Contr 38:320–333

Tuy HK, Tuy LT (1984) Direct 2-D display of 3-D objects. IEEE Computer Graphics Appl 4:29–33

Tyler TD, Ophir J, Maklad NF (1981) *In vivo* enhancement of ultrasonic image luminance by aqueous solutions with high speed of sound. Ultrason Imaging 3:323–329

Uematsu S (1981) Determination of volume of arterial blood flow by an ultrasonic device. J Clin Ultrasound 9:209–216

Unger E, Shen DK, Fritz T, Lund P, Wu GL, Kulik B, DeYoung D, Standed J, Ovitt T, Matsunuga T (1993) Gas-filled liposomes as echocardiographic contrast agents in rabbits with myocardial infarcts. Invest Radiol 28:1155–1159

van der Steen AFW, Li W, Céspedes EI, Carlier S, Eberle M, Verdouw PD, Serruys PW, Bom N (1997) *In vivo* validation of blood flow estimation using the decorrelation of radio-frequency intravascular echo signals. In: Schneider SC, Levy M, McAvoy BR (eds) Proceedings of the 1997 IEEE Ultrasonics Symposium, pp. 1247–1250. IEEE, Piscataway, NJ

Vera N, Steinman DA, Ethier CR, Johnston KW, Cobbold RSC (1992) Visualization of complex flow fields, with application to the interpretation of colour flow Doppler images. Ultrasound Med Biol 18:1-9

Verbeek XAAM, Ledoux LAF, Brands PJ, Hoeks APG (1998) Baseband velocity estimation for second-harmonic signals exploiting the invariance of the Doppler equation. IEEE Trans Biomed Eng 45:1217–1226

Vilkomerson D, Lyons D, Chilipka T (1994) Diffractive transducers for angle-independent velocity measurements. In: Proceedings of the 1994 IEEE Ultrasonics Symposium, pp. 1677–1682. IEEE, Piscataway, NJ

Villanueva FS, Kaul S (1995) Assessment of myocardial perfusion in coronary artery disease using myocardial contrast echocardiography. Coronary Artery Dis 6:18–28

von Ramm OT, Smith SW, Pavy HG (1991) High-speed ultrasound volumetric imaging system—Part II: Parallel processing and image display. IEEE Trans Ultrason Ferroelec Freq Contr 38:109–115

Wang L-M, Shung KK (1996) Adaptive pattern correlation for two-dimensional blood flow measurements. IEEE Trans Ultrason Ferroelec Freq Contr 43:881–887

Welch LR, Fox MD (1998) Practical spread spectrum pulse compression for ultrasonic tissue imaging. IEEE Trans Ultrason Ferroelec Freq Contr 45:349–355

Wilson LS, Gill RW (1993) Measurement of two-dimensional blood velocity vectors by the ultrasonic speckle projection technique. Ultrason Imaging 15:286–303

Xu S, Ermert H, Hammentgen R (1991) Phased array pulse Doppler tomography. In: Proceedings of the 1991 IEEE Ultrasonics Symposium, pp. 1273–1276. IEEE, Piscataway, NJ

Yamazaki N, Mine Y, Sano A, Hirama M, Miyatake K, Yamagishi M, Tanaka N (1994) Analysis of ventricular wall-motion using color-coded tissue Doppler imaging-system. Jap J Appl Phys 33:3141–3146

Yeung F, Levinson SF, Parker KJ (1998) Multilevel and motion model-based ultrasonic speckle tracking algorithms. Ultrasound Med Biol 24:427–441

Yeung K-WW (1998) Angle-insensitive flow measurement using Doppler bandwidth. IEEE Trans Ultrason Ferroelec Freq Contr 45:574–580

Ziskin MC, Bonakdapour A, Weinstein DP, Lynch PR (1972) Contrast agents for diagnostic ultrasound. Invest Radiol 7:500–505

14

Safety Considerations in Doppler Ultrasound

14.1 INTRODUCTION

Ultrasound is a form of energy and, as with any form of energy, if its level is increased it will eventually produce effects in tissue, some deleterious. In the 1990s the prime concern has been with tissue heating and this is particularly relevant to Doppler techniques [National Council on Radiation Protection and Measurements (NCRP) 1992]. When developing both equipment and clinical techniques, the aim should be to keep the exposure of the patient to the minimum commensurate with obtaining a diagnosis. In the event of there being an identifiable risk from a procedure, the user then has to attempt the difficult task of weighing up benefit versus risk. For several decades diagnostic ultrasound has been considered safe if the time-averaged intensity in the region of the maximum in the beam, often the focus, is less than 100 mW cm^{-2}. Although this is a desirable upper limit for intensity, it is often too limiting and represents a rather oversimplified view of the question of potential hazard. The whole subject is discussed in further detail below. It is worth noting that the outputs of many pulsed Doppler units far exceed the 100 mW cm^{-2} level. If high intensity levels are employed, the user should certainly be aware of it and be able to justify the levels used. New indices (Thermal and Mechanical Indices) are being established to be displayed on the instrument screen to provide the user with an awareness of the ultrasound field's potential to heat and disrupt tissue (AIUM/NEMA 1992).

The aim of this chapter is to provide a summary of the material relevant to the bioeffects of ultrasound and to give a list of references for further study. There is considerable activity and revision of ideas in this field at present, so the user of ultrasonic equipment is obliged to keep up to date with current opinion and to study reports which are put out by a number of organisations.

14.2 ULTRASONIC FIELD MEASUREMENT

A more detailed discussion of ultrasonic field measurement is given in Chapter 5. From the point of view of safety, there is particular interest in intensity (I_{spta}, I_{sppa}), power (W) and rarefaction pressure (p_0^-).

Virtually all measurements of the parameters of ultrasonic fields have been performed with water as the propagating medium. If it is desired to know the value of a parameter such as intensity or pressure amplitude at a certain depth in tissue, the corresponding value in water is used, with allowance being made for attenuation in tissue. This provides an 'in situ', or 'derated' value, as described later in this chapter. The accuracy of this calculation is questionable since it does not take into account the often very variable degradation of an ultrasound beam by different tissue structures. In addition, dispersive attenuation and non-linear propagation are not included in the calculation. Nevertheless, a simple extrapolation from water to tissue is a practical way of estimating field parameters in tissue.

Extensive work has been undertaken to develop instruments capable of accurate measurement of ultrasonic fields. If care is taken it is now possible to measure power, intensity and pressure amplitude (Brendel 1985, Preston et al 1983, Preston 1991, Stewart 1982, Perkins 1989). Such instruments are commercially available.

14.2.1 Power

The power (W) of an ultrasonic beam is the rate of flow of energy through the cross-sectional area of the beam. Power at diagnostic levels is most often measured using a force (radiation pressure) balance. With a carefully designed balance, power can be measured down to levels around 1 mW with an accuracy of 0.1 mW in ideal conditions. The force balance works on the principle that when an ultrasonic beam is completely absorbed by a target, it exerts a force of W/c on the target, where W is the power of the beam and c is the velocity of sound in the propagating medium (Westervelt 1951). If the target completely reflects the ultrasound, the force on it is $2W/c$ (Section 3.2.2). A number of force balances have been described over the years (Freeman 1963, Kossoff 1965, Hill 1970, Perkins 1989). They have now been reduced in size to make them portable and therefore much more convenient for calibrating medical instruments (Farmery and Whittingham 1978, Duck et al 1985). However, the most sensitive power balances are delicate instruments and it may be better to take the ultrasound unit to the calibration laboratory. Figure 14.1 shows a sensitive power balance. A number of ultrasonic power balances are commercially available.

14.2.2 Pressure Amplitude

The pressure amplitude of a CW or PW beam is of considerable interest in hazard studies, since it can be measured throughout the beam and is of direct relevance in the explanation of non-thermal biological effects. The negative pressure amplitude (p_0^-) is of greatest interest since it is related to cavitation. However, the positive amplitude (p_0^+) is also of some interest, since it can take on a short spiked shape due to non-linear propagation (Section 3.2.5), the significance of which has not been fully assessed with regard to biological effects.

When the pressure amplitude of a CW ultrasound beam is to be measured, the position in the beam should be specified. An ultrasonic beam can be represented schematically, as shown in Fig. 14.2. The pressure amplitude is often mea-

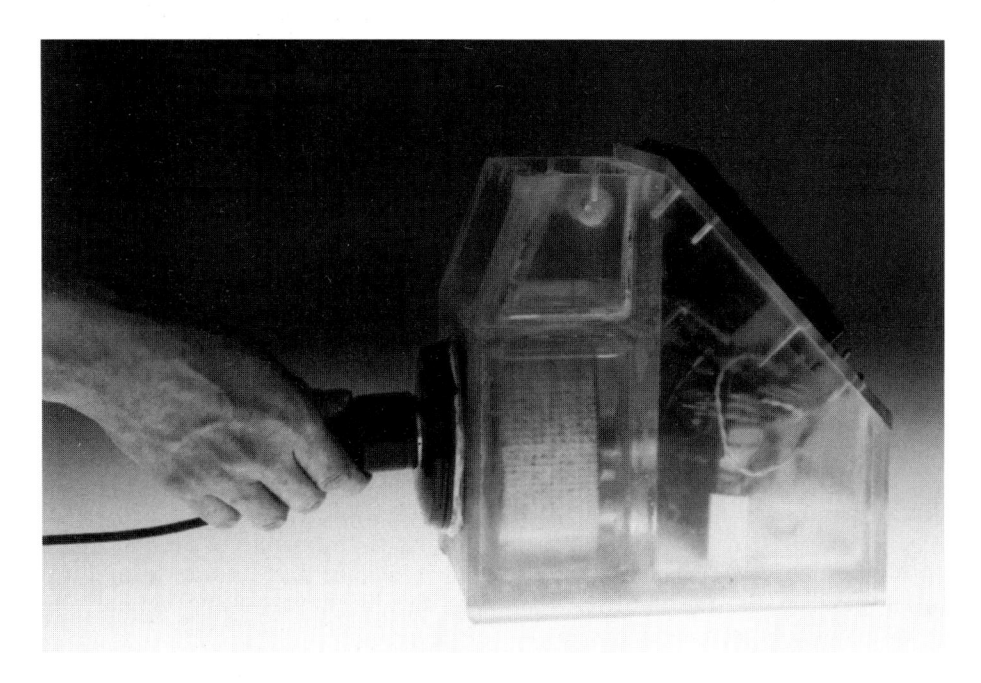

Figure 14.1 A sensitive power balance. The beam is directed horizontally toward a reflecting cone (courtesy of MA Perkins)

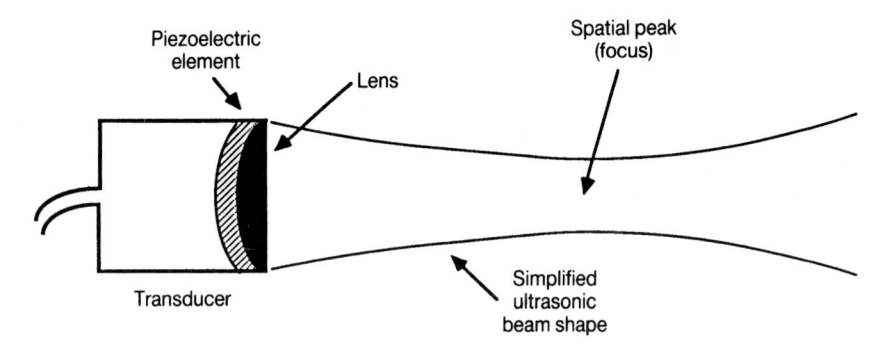

Figure 14.2 Schematic representation of an ultrasonic beam. The spatial peak intensity does not necessarily coincide with the focus

sured near the transducer or at the spatial peak (p_{sp}), which is commonly the focus, although it may be the largest axial maximum of an unfocused beam. The pressure may also be spatially averaged across the beam at the specified range (p_{sa}).

Prior to measuring the pressure amplitude of a PW ultrasound beam, it is necessary to define the exact quantity to be measured (Harris 1985). The pressure amplitude may be measured at the spatial peak and at the temporal peak of the pulse, p_{sptp}, or it may be averaged over the cross-section of the beam and over the duration of the whole exposure, p_{sata}. Another common measurement is the pressure amplitude at the spatial peak averaged over the pulse length, p_{sppa}. Other combinations of peak and average quantities are possible as shown in Table 14.1. The pressure parameters of greatest interest from the point of view of bioeffects are p_{sp}, p_{sptp}, and p_{sppa}. When the ultrasonic waveform is not symmetrical about the ambient pressure level, the above quantities may be measured for both positive and negative pressure amplitudes. The negative (rarefaction) pressure amplitude is of interest, since it provides local fluctuating tension in the propagating medium, which can give rise to small bubbles which grow and collapse (cavitation), possibly resulting in tissue damage.

14.2.3 Intensity

The intensity of a continuous or pulsed wave beam relates to thermal biological effects. It is normally calculated by inserting the square of the pressure amplitude into a simple formula:

$$I = p^2/2\rho_0 c \qquad 14.1$$

where ρ_0 is density (see equation 3.3). This formula is valid only at points in the beam where pressure and particle velocity are in phase, i.e. at the focus and in the far field but not in the near field of the transducer (Preston 1985).

Just as for pressure, the particular intensity to be measured must be precisely specified. Derived from the pressure amplitudes defined above are corresponding intensities (Table 14.1). Of most interest in the field of bioeffects are I_{sp}, I_{sata}, I_{spta} and I_{sppa}.

Table 14.1 Some parameters used to specify an ultrasonic field. The pressure parameters may be quoted for both positive and negative amplitudes

Pressure (spatial peak)	p_{sp}
Pressure (spatial average)	p_{sa}
Pressure (spatial peak, temporal peak)	p_{sptp}
Pressure (spatial average, temporal average)	p_{sata}
Pressure (spatial peak, temporal average)	p_{spta}
Pressure (spatial average, temporal peak)	p_{satp}
Pressure (spatial peak, pulse average)	p_{sppa}
Pressure (spatial average, pulse average)	p_{sapa}
Intensity (spatial peak)	I_{sp}
Intensity (spatial average)	I_{sa}
Intensity (spatial peak, temporal peak)	I_{sptp}
Intensity (spatial average, temporal average)	I_{sata}
Intensity (spatial peak, temporal average)	I_{spta}
Intensity (spatial average, temporal peak)	I_{satp}
Intensity (spatial peak, pulse average)	I_{sppa}
Intensity (spatial average, pulse average)	I_{sapa}
Intensity averaged over largest half cycle in pulse	I_m

14.3 ULTRASONIC OUTPUT FROM DOPPLER UNITS

The ultrasonic outputs from machines have increased quite substantially since 1991. For example, there has been approximately a fivefold increase in mean I_{spta} for B-mode scanners and a two-fold increase in the maximum measured value of I_{spta} and power for pulsed Doppler units (Duck and Martin 1991, Whittingham 1994, Henderson et al 1995). Although no clinically significant deleterious effects have been observed, the upward trend is giving cause for concern. The outputs are very dependent on the mode of operation and the machine settings.

14.3.1 CW Doppler

Table 14.2 presents typical pressure amplitude, power and intensity values for CW instruments. It should not be assumed that a Doppler unit will have output values close to the centre of these ranges. Intensity values have been reported over a wide range (Duck et al 1987). The lower intensities correspond to fetal heart detectors, the higher to vascular units.

14.3.2 PW Doppler

Tables 14.3 and 14.4 show pressure amplitude, power and intensity values for pulsed wave instruments. Again, the values of intensity vary over a wide range. Transcranial Doppler units are one of the few PW devices which are still used as stand-alone devices independent of an imaging mode. To compensate for attenuation in skull bone, typically they generate low frequency pulses (2 MHz) of I_{spta} around 250 mW cm^{-2}.

Table 14.2 Typical values for pressure amplitude, intensity and power for CW units. Values outside the quoted range can also be encountered (reproduced by permission from Duck and Martin 1986)

CW Doppler	
Spatial peak pressure (P_{sp})	0.01–0.12 MPa
Spatial peak intensity (I_{sp})	6–455 mW cm^{-2}
Power (P)	1–84 mW

Table 14.3 Typical values for pressure amplitude, intensity and power for PW units. PD-pulsed Doppler, CD-colour Doppler (reproduced by permission from Henderson et al 1995)

Peak negative pressure values

	Range (MPa)	Mean (MPa)	Median (MPa)
B/M mode	0.45–5.54	2.42	2.4
PD mode	0.67–5.32	2.18	2.10
CD mode	0.46–4.25	2.41	2.38

Maximum I_{spta} values

	Range (mW cm^{-2})	Mean (mW cm^{-2})	Median (mW cm^{-2})
M mode	11.2–430	121	106
B mode	0.3–991	105.9	34.0
PD mode	173–9080	1659	1180
CD mode	21–2050	344	290

Total acoustic power values

	Range (mW)	Mean (mW)	Median (mW)
M mode	1–68	20.6	9
B mode	0.3–285	77.9	75
PD mode	10–440	123.8	100
CD mode	15–440	119	90

Table 14.4 Typical values for pressure amplitude and intensity for intracavity probes. PD-pulsed Doppler, CD-colour Doppler (reproduced by permission from Henderson et al 1995)

Peak negative pressure values for intracavity probes

	Range (MPa)	Mean (MPa)	Median (MPa)
B/M mode	0.66–3.50	2.32	2.29
PD mode	0.97–3.53	2.26	2.04
CD mode	1.14–3.04	2.47	2.86

Maximum I_{spta} values for intracavity probes

	Range (mW cm^{-2})	Mean (mW cm^{-2})	Median (mW cm^{-2})
M mode	2.0–210	62.7	18
B mode	0.8–284	64.6	29
PD mode	97.1–1440	747	11
CD mode	4.1–465	148.2	8

14.3.3 Doppler Imaging (Flow Mapping)

Doppler imaging instruments are widely used in clinical practice. In real-time Doppler flow mapping (velocity or power), where the beam moves quickly, the number of pulses per second transmitted along a specified beam direction is typically one-twentieth of the number transmitted in a static pulsed Doppler beam. The time-averaged intensity in any one beam direction is therefore also reduced to one-twentieth of that of a static beam. A contribution from neighbouring beams should, however, be included when calculating the total intensity, which results in the lower value being increased by a factor in the range 2–10. The intensity (I_{spta}) from a Doppler flow mapping unit is therefore typically one-tenth to one-half of that from a static pulsed Doppler device. For a sector scanner, the contribution from neighbouring beams is obviously dependent on the depth of the point considered. The pressure amplitude of CW and PW imaging devices is not influenced by the fact that the beam is moving and has similar values to the static beam values (Table 14.3). The power of a real-time imaging beam is reduced by one-tenth to one-half compared to a static beam, just as intensity is. The total power coming out of an imaging transducer is similar to that from a static pulsed Doppler one (Table 14.3).

14.3.4 Doppler Tissue Imaging

The Doppler mode designed to image the motion of tissue such as heart muscle rather than blood flow typically uses three or four pulses along each beam direction, rather than say 12 as in a blood imaging mode, since a stronger backscatter signal is obtained from tissue (McDicken et al 1992, Fleming et al 1994). Intensity and power outputs are therefore also lower by around a factor of 4 in the tissue imaging mode. The pressure amplitude of the transmitted ultrasonic pulse for the tissue imaging mode is the same or less than that in the blood flow imaging mode.

14.3.5 Duplex Systems

When a Doppler instrument is part of a duplex system, the patient is obviously receiving ultrasound in two ways. The interaction of the pulse from the real-time B-scanner is of most relevance to bioeffects in terms of its pulse pressure amplitude but heating may also be a problem for higher output devices. On the other hand, when the system is switched to the CW or PW Doppler mode, thermal effects are of central interest. The outputs of Doppler instruments in duplex systems are similar to those of static beam devices (Table 14.3).

14.3.6 Fetal Monitoring

Fetal monitoring instruments are designed with divergent beams and also low power and intensity outputs, since the duration of their use may be several hours. The power output from a monitoring machine should be less than 10 mW.

14.3.7 Cuff Pressure Measuring Instruments

The outputs from Doppler devices in cuff pressure measuring instruments are as for other CW Doppler blood flow units (Table 14.2).

14.4 PHYSICAL EFFECTS OF ULTRASOUND

14.4.1 Mechanisms

Ultrasound interacts with tissue via thermal and non-thermal mechanisms. The non-thermal mechanisms can be further subdivided into cavitation, streaming and other 'direct' effects that need further research (Nyborg 1985, Williams 1983, Barnett et al 1994, Miller et al 1996).

Heating is considered to be an important mechanism and results when the orderly vibrational energy of the wave is converted into the random vibrational motions of heat by absorption processes described in Section 3.3.3. Table 14.5 gives a feeling for temperature rises for different intensities when perfusion and conductivity are low, as is the case for bone and the lens of the eye (Whittingham 1994). Theoretical results such as

Table 14.5 Time to increase temperature by 2°C when perfusion and conductivity are low (reproduced by permission from Whittingham 1994)

Tissue	Frequency (MHz)	Intensity (mW cm^{-2})	Time for 2°C rise
Liver	3.4	100	3.7 min
		1000	22 s
Bone	3	100	2.9 s
		1000	0.3 s
Eye lens	10	100	33 s
		1000	3.3 s

these often prove useful for making predictions of temperature rise. This heating could be a hazard for adjacent tissues. It can be seen that the time for a 2°C rise may only be a few seconds. No abnormalities have been observed in fetuses provided that the tissue temperature is below 41°C (Miller and Ziskin 1989, Edwards 1986). Chapter 1 of the report on the 1996 WFUMB symposium on safety of ultrasound in medicine (WFUMB, 1998) provides a recent update on thermal bioeffects issues.

Streaming occurs as a result of the radiation force that is generated when the wave energy is absorbed by a liquid (Section 3.2.2). Microstreaming also takes place in the vicinity of small oscillating bubbles, as discussed below in relation to cavitation (see WFUMB 1998, Chapter 3).

Cavitation takes two forms; non-inertial (stable) and inertial (transient). Non-inertial cavitation is a phenomenon whereby existing small bubbles in a liquid grow under the influence of the wave until they reach a particular size, at which resonant oscillations occur. The resonant frequency f of a bubble of radius a is given by:

$$f \approx 3.3/a \qquad 14.2$$

where a is in μm and f is in MHz (see Section 13.2.1).

Inertial cavitation takes place at higher ultrasonic pressure amplitudes and is due to the rapid growth of a bubble during the low-pressure half of the wave cycle, followed by a violent collapse in the high-pressure half. For tissue devoid of cavities, cavitation and related phenomena have not been reported for diagnostic instruments.

In both types of cavitation, non-inertial and inertial, the gas-bubble undergoes cyclic changes of volume in response to the applied acoustic field. In non-inertial cavitation, the changes are periodic and regular but cause microstreaming of the surrounding liquid, creating shear stresses in the locality of the bubbles and hence producing disruption at the cellular level (Dyson et al 1974). In inertial cavitation, the changes in volume of the bubble occur rapidly and the bubble becomes unstable and implodes violently, generating extremely high temperatures and pressure shock waves, resulting in the generation of free radicals (Carmichael et al 1986, WFUMB 1998, Chapter 4), sonoluminescence (Fowlkes and Crum 1988) and disruption of surrounding tissues. All of these effects have been seen *in vitro* but it is the presence of cavitation *in vivo* in mammalian tissue which is of direct relevance to clinical studies. There is some data suggesting that cavitation does occur *in vivo* in the hind legs of guinea-pigs using 0.75 MHz therapeutic ultrasound (ter Haar and Daniels 1981, Sommer and Pounds 1982, ter Haar et al 1982, 1986). Cavitation occurs at a specific acoustic pressure output, this threshold being determined as that pressure necessary to produce cavitational activity determined by the presence of cell lysis (Clarke and Hill 1970). Several parameters influence this cavitation threshold, e.g. the number of nuclei present in the solution, ambient pressure and viscosity (Williams 1983). Chapter 2 of WFUMB (1998) provides a useful overview of acoustic cavitation.

There is some evidence that other mechanisms may be involved in bioeffects, although they have not been precisely described and are grouped together under the collective heading of 'direct' mechanisms. If an effect is seen under conditions where heating, streaming or cavitation is not considered to be significant, the effect is attributed to a direct mechanism which is not always fully understood (ter Haar 1986, Inoue et al 1991). Examples of direct mechanisms are muscle stimulation, reported by Forester et al (1982), functional changes in the cochlea, described by Barnett (1980), and tissue repair, described by Dyson and Suckling (1978).

14.4.2 Physical Effects at Therapeutic Levels

The effects of heating, streaming and cavitation have been extensively studied at therapeutic intensity and the results can be reproduced (NCRP 1983, 1992, Wells 1987, Miller 1987). Therapeutic levels can be as much as 3 W cm^{-2} (I_{sata}) of 3 MHz ultrasound applied for 5 minutes. This could result in a temperature rise of several degrees Celsius in tissue (Lehmann and Guy 1972). In practice, the rise is highly dependent on the vascularity of the tissue. Haemolysis *in vivo* due to heating has been demonstrated above 3 W cm^{-2} at 3.4 MHz (Williams et al 1986). Evidence for the formation of cavitation bubbles *in vivo* with therapy ultrasound has been reported (ter Haar and Daniels 1981). Cavitation can affect biological material by temperature elevation, mechanical stress and free radical generation. Streaming of liquid in the vicinity of oscillating bubbles or needles has been shown to disrupt cells (Gershoy and Nyborg 1973). Studies of bioeffects at therapy levels emphasise that outputs of Doppler instruments should be kept below 1 W cm^{-2} (AIUM 1988). The IEC Report (1996) for therapy devices states that I_{sata} should not exceed 3 W cm^{-2}.

14.4.3 Physical Effects at Diagnostic Levels

At the levels of ultrasound normally encountered in diagnostic echo imaging, heating is not usually thought to be of concern in mammalian tissue. Temperature rises of up to 1 °C can be accommodated by the body, since they are experienced in normal daily variations. However, as noted in Table 14.3, some echo imaging machines have been shown to generate average intensities well in excess of 100 mW cm^{-2} (I_{spta}) (Duck et al 1985, Whittingham 1994, Henderson et al 1995), which has generally been regarded as a desirable upper limit (AIUM 1984). Also, the output intensities (I_{spta}) of several pulsed Doppler units cover a very wide range, up to 9080 mW cm^{-2} (I_{spta}). These levels are similar to those of therapy machines which can produce temperature rises of a few degrees Celsius in 2–3 minutes, as has been noted experimentally and theoretically (Table 14.5).

Streaming is not thought to be a likely occurrence in tissue with diagnostic ultrasound except where there is a significant volume of liquid, e.g. in the pregnant uterus, the bladder, the eye and cysts.

The awareness of non-linear propagation of diagnostic pulses and the improved accuracy of hydrophones caused the values of the pulse pressure amplitudes from imaging equipment to be revised upwards (Bacon 1984, Duck and Starritt 1984, Muir and Cartensen 1980). Cavitation may therefore occur for imaging with Doppler beams in tissue in the diagnostic situation (Holland et al 1996). One theoretical study has concluded that small bubble nuclei suspended in liquid could be made to generate cavitation when irradiated by short ultrasonic pulses, as utilised in diagnosis (Flynn 1982). This has been confirmed experimentally in water (Crum and Foulkes 1986, Carmichael et al 1986, Madsen et al 1991, Fowlkes and Crum 1988). The physical properties of bubbles in acoustic fields have been extensively described (Leighton 1996, Young 1989). The most successful types of ultrasound contrast agent take the form of thin-walled microbubbles which could act as cavitation centres. Contrast agents are of increasing importance and are discussed in Section 14.6.

14.5 BIOEFFECTS OF ULTRASOUND

A large number of publications have been produced on the biological effects of ultrasound on molecules, subcellular structures, cells, individual tissues, plants, insects and whole animals. The subject has been treated from the point of view of almost all biological subjects, i.e. biochemistry, embryology, physiology, genetics and epidemiology. Several textbooks, reports and reviews have appeared on the subject (Nyborg and Ziskin 1985, Williams 1983, NRCP 1983, Brent et al 1991, Barnett et al 1994, 1997, Barnett and Kossoff 1998). Most of the confirmed results are related to therapy levels of ultrasound and there is a great lack of reliable data for the diagnostic situation. It is not possible to extrapolate from the therapy results to lower intensity levels, since many ultrasonic processes are likely to have thresholds. The best-known example is the transient cavitation threshold, which is highly dependent on the conditions in the propagating medium but is well defined. The possibility of heat producing significant temperature rises is of most concern particularly in sensitive tissues such as the embryo, the fetal brain and spine, the neonatal brain and the eye. References to selected papers are listed at the end of the chapter. There has been a continual output of publications on bioeffects over several decades; the subjects that have produced most reaction and concern among users of diagnostic equipment are briefly outlined below. When studying the bioeffects literature, it is worth bearing in mind that the *in vitro* environment is very different from the *in vivo* one. Important differences are an increased likelihood of bubbles in an *in vitro* preparation, altered heating patterns, and release of contaminants from the container wall.

14.5.1 Chromosome Damage

In the early 1970s the main topic of study was the possibility of chromosome damage. A few positive results were reported but these could not be confirmed by other investigators (Macintosh and Davey 1970, Boyd et al 1971, Hill et al 1972, Brock et al 1973). In retrospect, it is felt too much attention was paid to this work, since the high level of damage reported would have led to cell death and not developmental defects (Thacker 1973).

14.5.2 Sister Chromatid Exchanges

In the early 1980s, sister chromatid exchanges (SCEs) were used to look for an effect of ultrasound at diagnostic levels. An SCE is the interchange of equivalent parts of the chromosome (Fig. 14.3). It can be detected using fluorescent labelling techniques. The process can occur spontaneously without the influence of any agent and its significance is not fully understood. Certain chemicals are well-established agents of this effect and are used as controls in experimental practice, e.g. mytomycin. The SCE test is reported to be about 100 times more sensitive than chromosome damage but this depends on the agent causing the exchange; surprisingly, ionising radiation is not particularly effective.

Both continuous and pulsed beams have been employed in the investigations of SCEs. The

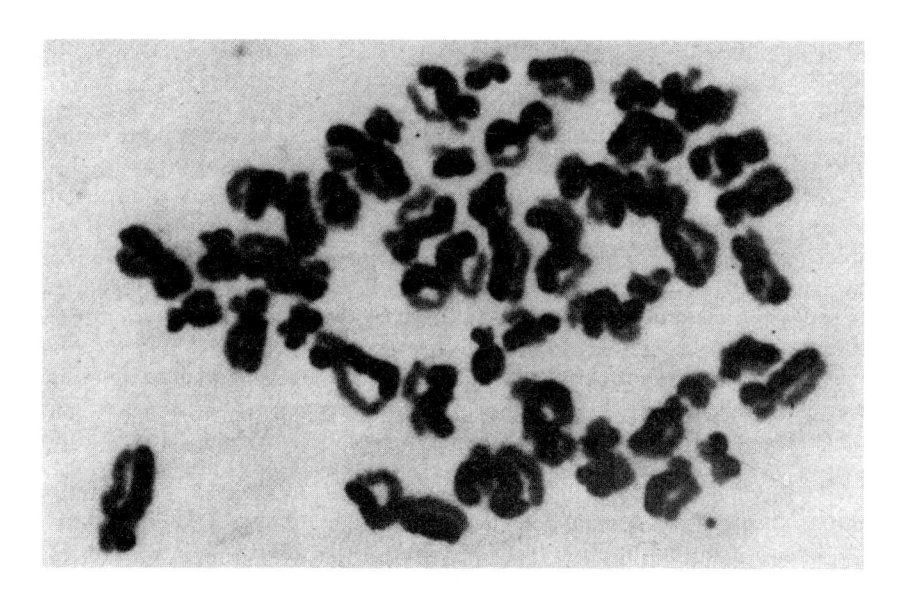

Figure 14.3 Chromatid staining and demonstration of sister chromatid exchanges (reproduced by permission of Churchill Livingstone from Jacobson-Kram 1984)

dosimetry and biology have been of variable quality. Reviews show that about one in five studies have reported positive results (Miller 1985, Goss 1984). SCEs apparently do not change the information content of DNA and therefore may not be of clinical importance.

14.5.3 Growth Retardation

Growth retardation has been reported in several studies for fetal mice and rats irradiated by therapeutic and diagnostic levels of ultrasound, e.g. with a CW therapy instrument at 1 MHz, I_{sa} 0.5–5.5 W cm^{-2} and exposure time 10–300 s (O'Brien 1983, 1984). A radiation regimen such as this can produce quite large temperature rises in the mother and fetus, e.g. 5°C in a few minutes. A dependence of fetal weight on exposure was identified in this work. Further consideration of this field led to the conclusion that growth retardation should not be of concern at diagnostic levels (Barnett and Williams 1990). Recent work using multiple exposures of ultrasound in monkeys reports a possible reduced birth weight, so the question is still being explored (Tarantal et al 1995).

14.5.4 Damage to Blood Vessels

Dyson et al (1974) have demonstrated damage to the inside wall of blood vessels by transmitting therapeutic levels of CW and PW ultrasound (1–5 MHz, up to 12 W cm^{-2} peak intensity, 15 minutes irradiation time) in an arrangement that created standing waves (Section 3.2.4). These were obviously conditions different from those found in clinical examinations but they do serve to remind designers and users of an upper limit to acceptable ultrasonic field strengths.

14.5.5 Cavitation in Moving Blood

There has been a suggestion that moving blood was a medium in which micrometre-sized bubbles could be spontaneously produced and that these bubbles might cavitate under the influence of an ultrasonic wave (Williams 1985). However, in subsequent work it has been proved that microbubbles do not exist in blood *in vivo* and that even artificially injected bubbles do not generate stable cavitation when irradiated by CW and PW ultrasound of spatial peak intensities up to 16 W cm^{-2} and frequencies in the range 0.5–1.6 MHz (Williams et al 1985).

14.5.6 Epidemiological Surveys

The most convincing evidence for the safety of ultrasound would be from large-scale population surveys. There are many obstacles to this approach, e.g. the difficulty of obtaining a control population and in guaranteeing accurate dosimetry. The latter is virtually impossible in retrospective studies. Rapidly changing techniques present another major hurdle.

A number of epidemiological studies have been carried out, with particular attention being paid to possible effects of ultrasound on childhood malignancies, neurological development, speech development and birthweight. Effects are rarely found and those that are can often be attributed to statistical chance or poor study design. There are many pitfalls in such surveys (Williams 1991, NCRP 1983, Ziskin and Petitti 1988). A few possibilities, such as low birthweight, non-right-handedness and speech delay, require further study. The situation is well summarised in a tutorial paper by the Radiation Safety Committee of the European Federation of Societies in Medicine and Biology (EFSUMB 1996). Although this type of study is difficult to perform, their importance is emphasised when it is remembered that virtually all fetuses in the developed world are now scanned by ultrasound.

14.6 CONTRAST AGENTS

The safety of contrast agents requires special consideration, since they are increasingly being used with Doppler techniques and both new agents and equipment are being actively developed. The topic can be divided into two main areas: safety with respect to possible intrinsic toxicology of the agents when injected into the human body, and the safety issues surrounding ultrasound insonation of gas-filled bubbles. A review of each of these topics can be found in Nanda (1993).

14.6.1 Toxicology

Ultrasonic contrast agents are composed of gaseous microbubbles which, as a consequence of the difference in compressibility and density between the gas within the microbubble and the surrounding liquid carrier, result in high backscatter at the interface of the microbubbles and carrier liquid. However, the injection of a gaseous medium into the bloodstream is, in a clinical context, considered extremely hazardous because of the risk of gas embolisation. As a result of this concern, the American Society of Echocardiography published a report on the safety of ultrasonic contrast agents (Bommer et al 1984). The report concluded that when using contrast agents, precautions similar to those employed during catheterisation should be used, especially in patients with right-to-left cardiac shunts. However, more recently it has been suggested (Cosgrove 1995) that injection of gaseous contrast microbubbles into the blood stream is significantly different from injection of an air-bubble directly into the blood stream in three respects. Firstly, the size of contrast microbubbles is small, they are sufficiently small to pass through the capillaries and are thus unlikely to block the capillaries. Secondly, the total volume of gas injected is small; for a 70 kg person with, say, a dose of 10 μl of agent/kg (typical of several contrast agents) and with 10^9 particles/ml, the volume of gas injected is still only in the region of 0.12 ml. Thirdly, most of the intravenous agents have a short life-time so that the risk is greatly reduced with increasing time after injection.

However, it is important to note that it is not only the toxicology of the contrast microcapsules which must be considered but also the carrier solution. Three properties of the contrast carrier solution of importance to consider are the osmolarity, viscosity and surfactant properties. A review of the importance of these properties has been published (Nanda 1993). The majority of side-effects reported in the literature have been attributed to the carrier solution. In particular, for a hyperosmolar carrier solution such as that associated with Levovist and Echovist (manufactured by Schering), a small percentage of patients experienced vasodilation, injection site pain, paraesthesia, pain and taste perversion (Bauer 1997). For contrast agents which are highly viscous in nature, adverse effects such as transient changes in left ventricular haemodynamics may be experienced (Koenig and Meltzer 1986). However, the percentage of patients experiencing these side-effects for both viscous and hyperosmolar carrier solutions are small in number and the effects have been shown to be transitory in nature.

14.6.2 Ultrasound Insonation of Agents

As noted above, cavitation occurs as the result of the interaction between ultrasound and gaseous inclusions (micronuclei) within a medium. Injecting the contrast agent increases the number of cavitation nuclei and is thought to lower the threshold for cavitational activity. Indeed, in studies by Holland and Apfel (1990) it was found that Albunex microspheres, upon injection into an *in vitro* model, appeared to be very efficient cavitation nuclei at an insonating frequency of 757 kHz. Furthermore, Miller and Thomas (1995) found that the addition of small amounts of Levovist (2, 0.2, 0.02 mg/ml) or Albunex (10^{-2}, 10^{-3}, 10^{-4} dilutions) into a non-cavitating system initiated hydrogen peroxide production, suggesting that indeed the presence of the contrast agents provided nuclei for inertial cavitation. Extension of these studies into *in vitro* studies using human blood also verified that it was possible to induce cavitation, indicated by haemolysis (Brayman and Miller 1994, Brayman et al 1996). The results from these studies corroborated work from an earlier *in vitro* study by Williams et al (1991), in which it was shown that the addition of Echovist enhanced haemolysis induced by cavitation. In more recent years, two studies using standard diagnostic scanners have shown that cavitation is possible. The first study performed *in vivo* on rat lungs showed that, using diagnostic ultrasound, it was possible to induce cavitation (Holland et al 1996). The second study indicated that haemolysis could be induced *in vivo* in murine hearts at diagnostic output pressures after an injection of Albunex (Dalecki et al 1997). However, even at insonating pressures of 10 MPa, only limited haemolysis (4.8%) occurred. Recently Doppler imaging techniques have been developed which depend on the destruction of microbubbles to produce phase changes in the echo signals (Nanda et al 1997).

At present, there is little conclusive evidence to suggest that any significant risk factors are associated with diagnostic injections of contrast.

Table 14.6 Intensity levels specified by the FDA for 'fast-track' acceptance of equipment for particular clinical applications

Use	I_{spta} in situ (mW cm^{-2})	I_{spta} in water (mW cm^{-2})
Cardiac	430	730
Peripheral vessel	720	1500
Ophthalmic	17	68
Fetal imaging and other*	94	180

* Abdominal, intraoperative, paediatric, small organ (breast, thyroid, testes), neonatal cephalic, adult cephalic.

14.7 SAFETY STANDARDS

14.7.1 Standards

Several organisations have set up working parties to search and evaluate the world literature on bioeffects and safety. These organisations include the World Health Organisation, The American Institute of Ultrasound in Medicine, the European Federation of Societies for Ultrasound in Medicine and Biology, and the Japanese Society of Ultrasonics. The available reports are referenced at the end of the chapter. These organisations have all come to the conclusion that there have been no confirmed deleterious effects of diagnostic ultrasound. Most surveys also highlight the severe lack of reliable and confirmed results on the effects of low-intensity ultrasound.

Up until around 1992 the approach to safety was to inform users of the outputs of machines, i.e. intensity (I_{spta}, I_{sppa}), acoustic pressure (p) and power (W), and to attempt to specify acceptable levels for certain applications (Table 14.6). With this approach allowance is made for attenuation in tissue. The outputs are then called 'derated' ('*in situ*') intensity and pressure amplitude. The output power (W) is considered to be a property of the instrument and no derating calculation is performed. The concept of derated output intensity and acoustic pressure is considered further below (Section 14.7.2).

Whilst the above approach is still an option, the favoured approach is now to have an on-screen display of a thermal index and a mechanical index for each machine setting (Section 14.7.4). These indices are related to heating and cavitation in tissue.

The trend in regulations now is toward the AIUM/NEMA Output Display Standard using thermal and mechanical Indices. However the Food and Drug Administration in the USA permits both of the above approaches at present, the 'fast track' using derated outputs for specific applications and the 'slow track' with thermal and mechanical indices. The International Electrotechnical Commission (IEC) is drawing up regulations on a world-wide basis, but its proposals are not in a finalised form.

14.7.2 Derated (*In Situ*) Values

The *in situ* value of intensity or pressure is the value calculated within tissue, allowing for the attenuation of the transmitted ultrasound by the tissue. Derated values are rough approximations due to the wide variations in attenuation in different clinical situations.

The derated intensity and rarefaction pressure may be calculated for any point in the acoustic beam with the following equations:

Derated intensity = $\varphi \times$ intensity in water   14.3

Derated rarefaction pressure

= $\varphi^{0.5} \times$ rarefaction pressure in water 14.4

where φ is called the derating factor. To establish the expression for φ, a particular tissue model is assumed which is appropriate to the application. For example the Homogeneous Model, which might be considered appropriate where soft tissues are the dominant structures (CDRH 1985), gives:

$$\varphi = \exp(-0.23\phi f_c z) \qquad 14.5$$

where 0.23 converts decibels to nepers, ϕ is the assumed attenuation coefficient, f_c is the centre frequency in MHz, and z is the path length to the site of interest.

Commonly the attenuation coefficient is taken to be 0.3 dB cm^{-1} MHz^{-1} and then:

$$\varphi = \exp(-0.069 f_c z). \qquad 14.6$$

The NCRP (1983) provides another expression for φ based on an attenuation of 0.435 dB cm^{-1} MHz^{-1}:

$$\varphi = \exp(-0.1 f_c z). \qquad 14.7$$

Where there is a large amount of fluid present, e.g. in obstetrics, the Fixed Attenuation or Layered Model is used. Taking a worst case scenario for fetal scanning (NCRP 1983, Carson 1988), the Layered Model comprises a 1.7 cm tissue layer and a layer of liquid of negligible attenuation between the transducer and the fetus, then:

$$\varphi = \exp(-0.17 f_c). \qquad 14.8$$

The attenuation coefficient has been taken to be 0.435 dB cm^{-1} MHz^{-1}. Note that in this scenario, φ is not dependent on depth.

By way of a numerical example for the Homogeneous Model, using the CDRH expression, if the intensity in water = 120 mW cm^{-2}, $f_c = 3$ MHz, attenuation coefficient $\phi = 0.3$ dB cm^{-1} MHz^{-1}, and $z = 5$ cm, the derating factor will be $\varphi = 0.355$ and the derated intensity will be 43 mW cm^{-2}. For the Fixed Attenuation Model and other things being equal, $\varphi = 0.60$, which gives a derated intensity of 72 mW cm^{-2}.

Calculated values in a particular application can be compared to the permitted derated maximum values of Table 14.6. Derated values are usually calculated for I_{spta}, I_{sppa} and p. Measurement and calculation procedures for the determination of derated values are given in the AIUM/NEMA 1992 publication.

14.7.3 Pre-enactment Standard

Some instrumentation requirements are such that imaging units and pulsed Doppler devices function satisfactorily with output intensities (I_{spta}) below 100 mW cm^{-2}. However manufacturers are unhappy with this restriction. In an attempt to cater for the need for higher outputs, the Food and Drug Administration of the USA has, since the mid-1980s, permitted the supply in the USA of instruments that comply with regulations which allow higher outputs. In this 'fast track' approach, the regulations specify the derated intensity limit a machine may generate in each clinical application. A consideration which determined the allowable derated intensity levels was that no new instruments should generate higher intensities than were in use prior to 28 May 1976, the 'pre-enactment date'. Table 14.6 lists the derated intensities allowable for different applications and the corresponding levels in water, where attenuation is not significant. As shown above, when an output intensity is measured in water, the derated value may be calculated from it to see if it is below the permitted level.

14.7.4 Thermal Index (*TI*) and Mechanical Index (*MI*)

From around 1993, there have been moves to have indices related to the possibility of biological hazard from ultrasound rather than try to relate hazard to intensity levels, as quoted above. Two types of index, often called biophysical indicators, have been introduced, namely the thermal index (*TI*) and the mechanical index (*MI*). In fact three thermal indices are used, the soft tissue index (*TIS*), the bone index (*TIB*) and the cranial index (*TIC*). The appropriate index is selected for the application being undertaken. It is expected that the scanner will display this index on its screen when settings for the application are chosen, but the user should also have the option of specifying the displayed index. This approach to monitoring hazard is being established at present and is described in the AIUM/NEMA Output Display Standard (1992). It permits higher outputs than were allowed under the previous regulations, I_{spta} up to 720 mW cm^{-2} and *MI* up to 1.9, but requires the indices to be displayed on the machine so that the operator is continually aware of a measure of output. There is still ongoing discussion about the accuracy of the thermal indices, so up-to-date literature and comment has to be monitored.

The *MI* is an attempt to indicate the probability that mechanical damage by non-thermal processes, in particular cavitation, might occur within tissue. The *TI* is an estimate of the rise in tissue temperature in °C under worst-case conditions. When these indices have a value of 1 or more, the possibility of hazard should be considered and justification made for the use of such output levels. Where the range of machine settings for a selected clinical application could give rise to an *MI* or *TI* greater than 1, the manufacturer is obliged to display all values greater than 0.4 as the machine controls are altered. The manufacturer decides whether it is appropriate to display an *MI* or *TI* in each situation. Although under the Standard the manufacturer is not obliged to display an index all of the time, it is hoped that they will do so.

The thermal index *TIS* is used when soft tissue is being scanned, *TIB* when bone is present, and *TIC* in transcranial studies where there is bone near the transducer. Both *TI* and *MI* are calculated using models as shown below. The calculation of the equations for the thermal index is the more complex and is based on tissue heating modelling, which allows for factors such as perfusion and bone heating. For each machine setting (i.e. transmission field, transducer and scanning mode) the maximum values for *TI* and *MI* in the insonated tissue are calculated using the appropriate model. Since they are based on a simple tissue model of the anatomy, these indices are not intended to be used blindly and the user should try to take account of any special circumstances. Responsibility still rests with the user to minimise the exposure of the patient, in other words to use the ALARA principle (As Low As Reasonably Achievable). This principle has been adopted from the field of ionising radiations and recognises that there is a risk–benefit assessment to be carried out for medical procedures. The AIUM/NEMA Output Display Standard (ODS) gives formulae for calculating *TIs* and *MI*. Regulations for the sale of equipment in the USA now refer to this standard.

The thermal index is defined as the ratio of the acoustic power emitted to the power required to heat tissue by 1°C at the point in the field of maximum temperature rise, allowing time for thermal equilibrium to be achieved. To calculate the soft tissue thermal index (*TIS*), a homogeneous attenuating medium with an attenuation coefficient of 0.3 dB cm^{-1} MHz^{-1} is employed. The models for calculating *TIB* and *TIC* have bone in them, at the focus and surface respectively.

The soft tissue thermal index, *TIS*, is calculated in two situations, where soft tissues are being scanned and also where there is bone deep within the tissue (Fig. 14.4a,b) (O'Brien 1994, Duck 1997). In both these cases the maximum temperature is seen to be close to the surface. There is some concern at the use of *TIS* in the latter situation, particularly for fetal scanning. Equation 14.9 presents the formula for *TIS* corresponding to these figures:

$$TIS(\text{scanned}) = \frac{W_1 f_c}{210} \qquad 14.9$$

where W_1 is the time-averaged acoustic power at the source emitted from the central 1 cm of the active aperture (in mW), and f_c is the centre frequency in MHz. The transcranial situations for a scanned and an unscanned beam are in shown in Figs 14.5a and 14.5b, and the *TIC* formula, which is the same for both cases, is given in eqn. 14.10. For each case the maximum temperature rise is at the surface.

$$TIC(\text{scanned and unscanned}) = 0.025 W_0 \sqrt{\pi/4A} \qquad 14.10$$

where W_0 is the time average acoustic power at the source (in mW), and A is the active aperture area (in cm^2). Figure 14.6 depicts the case of an unscanned beam insonating bone at depth in soft tissue, in which case the maximum temperature is found to be at the depth of the bone. Equation 14.11 provides the formula for *TIB*:

TIB(unscanned)

$$= \text{the lower of } \frac{\sqrt{W_{0.3} I_{0.3}}}{50} \text{ and } \frac{W_{0.3}}{4.4} \qquad 14.11$$

where $W_{0.3}$ is the time-averaged acoustic power (in mW) derated (at 0.3 dB cm^{-1} MHz^{-1}) to the

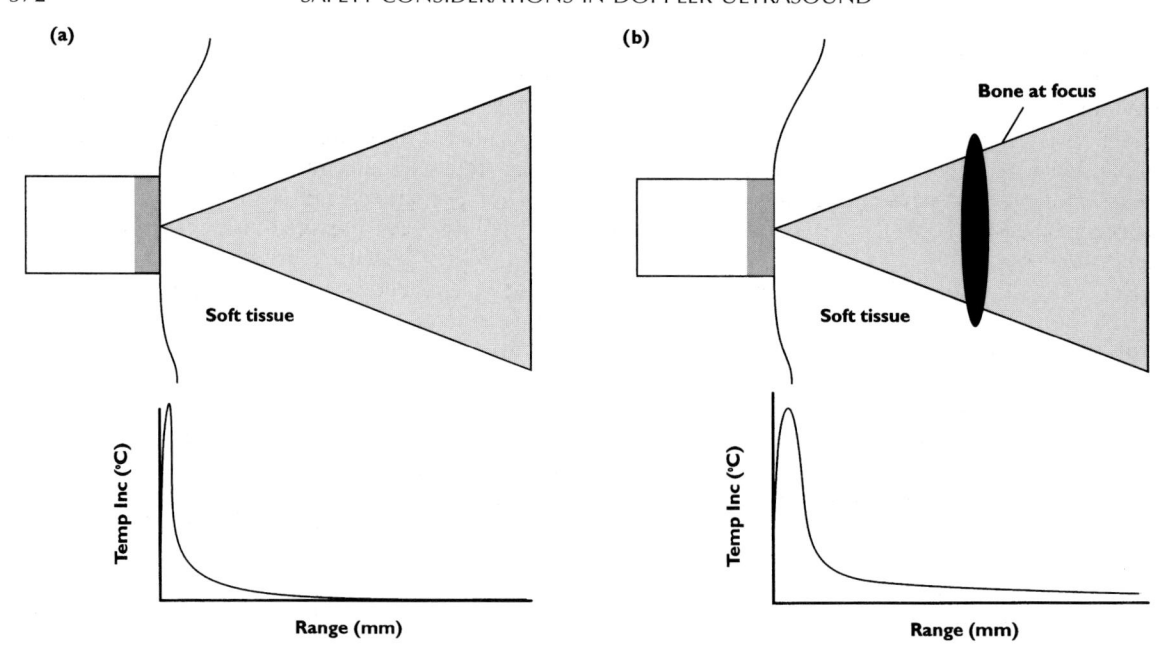

Figure 14.4 Axial temperature increase (Temp Inc) profiles for (a) a scanned ultrasound beam in soft tissue, and (b) a scanned ultrasound beam with bone at the focus (courtesy of W.D. O'Brien Jr)

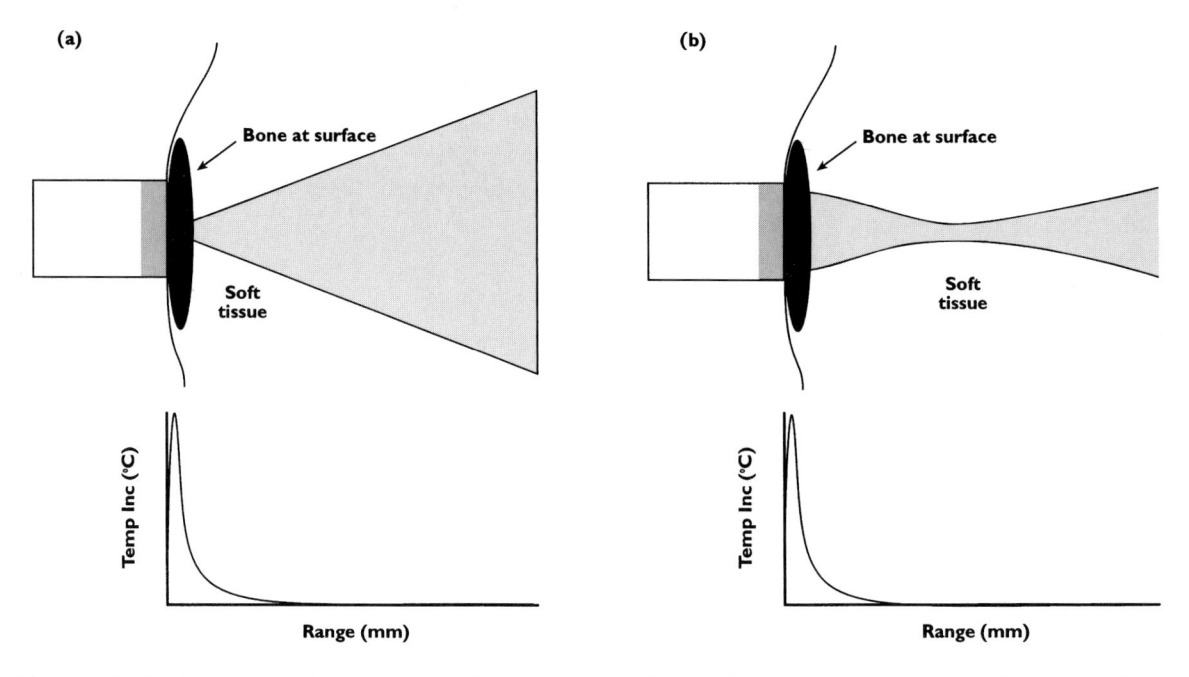

Figure 14.5 Axial temperature increase profiles for (a) a scanned ultrasound beam with bone at the surface and (b) for an unscanned beam with bone at the surface (courtesy of W.D. O'Brien Jr)

depth of the maximum temperature rise, and $I_{0.3}$ is the derated temporal average intensity (in mW cm^{-2}).

For the examination of soft tissue with an unscanned beam, the formula for TIS depends on the size of the transducer aperture, as this will influence the shape of the temperature increase profile (see Fig. 14.7).

For $A > 1$ cm^2, the TIS is given by:

TIS(unscanned)

$$= \text{the lower of } \frac{W_{0.3}f_c}{210} \text{ and } \frac{I_{0.3}f_c}{210} \quad 14.12$$

For $A < 1$ cm^2, the TIS is given by:

$$TIS(\text{scanned}) = \frac{W_0 f_c}{210} \quad 14.13$$

For further details of how TIs are calculated the reader is referred to the AIUM/NEMA Output Display Standard. In addition to the tissue model selected, the formulae allow for factors such as scanning mode and transducer aperture size.

The mechanical index is equal to the peak derated negative pressure in the ultrasound field, p^- in MPa, divided by $\sqrt{f}$, where f is the frequency in MHz and allows for the frequency dependence of cavitation, i.e:

$$MI = p^- / \sqrt{f} \quad 14.14$$

To calculate the derated p^- from the value in water, an attenuation coefficient of 0.3 dB cm^{-1} MHz^{-1} is used. MI gives the user an indication of the amplitude of the ultrasonic pulses being employed at any time.

Some problems still remain to be resolved with regard to this new labelling system (Duck 1997).

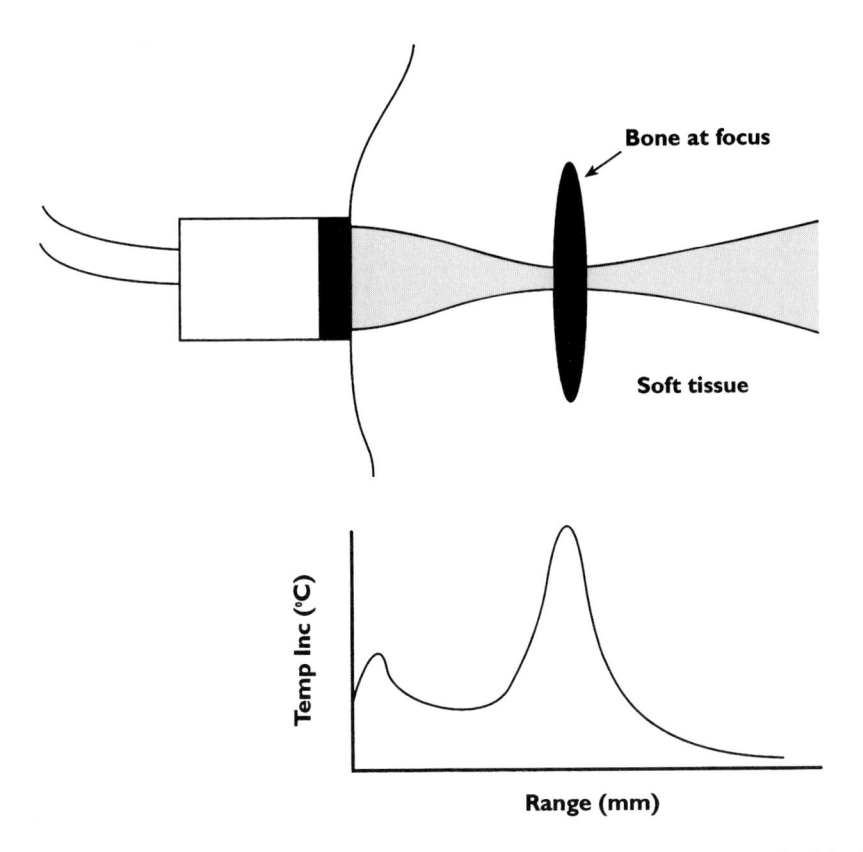

Figure 14.6 Axial temperature increase profiles for unscanned ultrasound beam insonating bone at a depth in tissue (courtesy of W.D. O'Brien Jr)

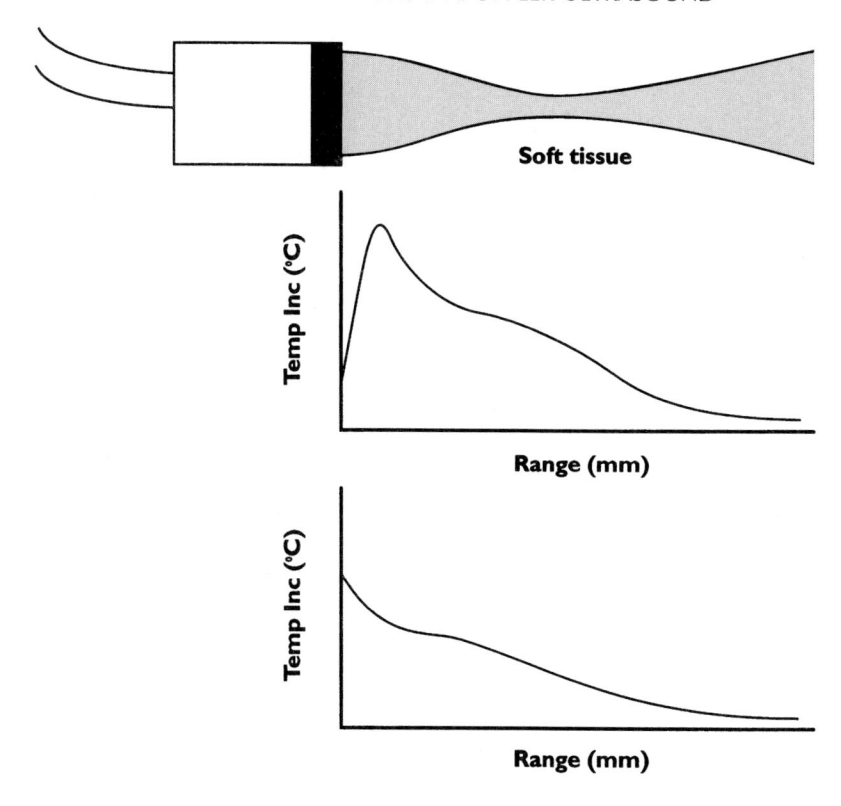

Figure 14.7 Axial temperature increase profiles for examination of soft tissue with an unscanned ultrasound beam. Upper profile, large aperture transducer; lower profile, small aperture transducer (courtesy of W.D. O'Brien Jr)

With multimode scanning, the *TI* values are added rather than the heating due to each mode and it is difficult to add the effects of scanned and unscanned modes. There is concern about the use of the soft tissue *TI* in late pregnancy, where bone has developed. Non-linear propagation in water can cause the derating calculation to be inaccurate and to underestimate the *in situ* exposures, giving low values for *MI* and *TI*, perhaps by as much as 50%. Transducer self-heating is ignored but it could be significant in internal transducers.

14.8 SAFETY STATEMENTS

Several national and international bodies issue safety statements and recommendations on a regular basis. Some of the more widely quoted statements are reproduced below.

14.8.1 WFUMB Statement on Thermal Effects in Clinical Applications. Endorsed March 1997 (WFUMB 1997, 1998)

The following safety statements were endorsed as policy of the World Federation for Ultrasound in Medicine and Biology following the recommendations from the 1991 WFUMB Symposium on Safety of Ultrasound in Medicine; Thermal Issues. The conclusions from the 1996 WFUMB Symposium on Safety of Ultrasound in Medicine are that there is no scientific evidence to alter existing safety statements on thermal issues. Hence the WFUMB Safety Statements for Thermal Bioeffects are reiterated to complete the current safety guidelines.

B-Mode Imaging
Known diagnostic ultrasound equipment as used today for simple B-mode imaging operates at acoustic outputs that are not capable of producing harmful temperature rises. Its use in medicine is

therefore not contraindicated on thermal grounds. This includes endoscopic, transvaginal, and transcutaneous applications.

Doppler

It has been demonstrated in experiments with unperfused tissue that some Doppler diagnostic equipment has the potential to produce biologically significant temperature rises, specifically at bone/soft tissue interfaces. The effects of elevated temperatures may be minimised by keeping the time for which the beam passes through any one point in tissue as short as possible. Where the output power can be controlled, the lowest available power level consistent with obtaining the desired diagnostic information should be used. Although the data on humans are sparse, it is clear from animal studies that exposures resulting in temperatures less than 38.5°C can be used without reservation on thermal grounds. This includes obstetric applications.

Transducer Heating

A substantial source of heating may be the transducer itself. Tissue heating from this source is localised to the volume in contact with the transducer.

The 1996 Symposium went on to make a number of recommendations with regard to thermal and non-thermal effects which are contained in WFUMB(1997 and 1998).

14.8.2 AIUM Statement on Non-human Mammalian *In Vivo* Biological Effects. Approved October 1992

The American Institute of Ultrasound in Medicine issues statements on many issues related to the use of ultrasound in medicine, which may be found on their web site at www.aium.org. The following statement was approved in October 1992.

Information from experiments utilising laboratory mammals has contributed significantly to our understanding of ultrasonically induced biological effects and the mechanisms that are most likely responsible. The following statement summarises observations relative to specific ultrasound parameters and indices. The history and rationale for this statement are provided in Bio-effects and Safety of Diagnostic Ultrasound (AIUM, 1993).

In the low megahertz frequency range there have been no independently confirmed adverse biological effects in mammalian tissues exposed *in vivo* under experimental ultrasound conditions, as follows:

(a) When a thermal mechanism is involved, these conditions are unfocused-beam intensities(*) below 100 mW/cm^2, focused(**)-beam intensities below 1 W/cm^2, or thermal index values less than 2. Furthermore, such effects have not been reported for higher values of thermal index when it is less than $6 - (\log_{10} t)/0.6$, where t is exposure time ranging from 1 to 250 minutes, including off-time for pulsed exposure.

(b) When a non-thermal mechanism is involved,*** in tissues that contain well-defined gas bodies these conditions are *in situ* peak rarefactional pressures below approximately 0.3 MPa or mechanical index values less than approximately 0.3. Furthermore, for other tissues no such effects have been reported.

(*) Free-field SPTA for continuous wave and pulsed exposures.
(**) Quarter-power (−6 dB) beam width smaller than four wavelengths or 4 mm, whichever is less at the exposure frequency.
(***) For diagnostically relevant ultrasound exposures.

Higher intensity is quoted for focused fields since heat is lost more readily from small focal zones. The majority of Doppler fields fall into the unfocused category.

It should be remembered that the statement is based on limited information (Hill and ter Haar 1982). For example, results are scarce on short pulse exposures and also those of long duration at low intensity. Most of the information is related to small mammals rather than to man. It was not intended that the 100 mW cm^{-2} level should be regarded as a threshold level for bioeffects below which safety is guaranteed for every application. It is more a guideline, which may be altered as new knowledge is accumulated.

14.8.3 EFSUMB Clinical Safety Statement for Diagnostic Ultrasound (Tours, March 1998)

The European Federation of Ultrasound Societies in Medicine and Biology (EFSUMB) issues safety statements annually. The latest version of their statement can be found on their web site at www.efsumb.org.

Diagnostic ultrasound has been widely used in clinical medicine for many years with no proven deleterious effects. However, as the use of ultrasound increases, with the introduction of new techniques, with a broadening of the medical indications for ultrasound examinations, and with increased exposure, continued vigilance is essential to ensure its continued safe use.

A broad range of ultrasound exposure is used in the different diagnostic modalities currently available. Doppler imaging and measurement techniques may require higher exposure than those used in B- and M-modes, with pulsed Doppler techniques having the potential for the highest levels.

Modern equipment is subject to output regulation. The recommendations contained in this statement assume that the ultrasound equipment being used is designed to international or national safety requirements and that it is used by competent and trained personnel.

B- and M-modes
Based on the scientific evidence of ultrasonically induced biological effects to date, there is no reason to withhold B- or M-mode scanning for any clinical application, including the routine clinical scanning of every woman during pregnancy.

Doppler for fetal heart monitoring (CTG)
The power levels used for fetal heart monitoring are sufficiently low that the use of this modality is not contra-indicated, on safety grounds, even when it is to be used for extended periods.

Doppler mode (colour flow imaging, power Doppler and pulsed Doppler)
Exposures used in Doppler modes are higher than for B- and M-modes. There is considerable overlap between the ranges of exposure which may be used for colour flow imaging and power Doppler, and for pulsed Doppler techniques. The clinical user should be aware that pulsed Doppler at maximum machine outputs and colour flow imaging with small colour boxes have the greatest potential for biological effects.

In general, the informed use of Doppler ultrasound is not contra-indicated. However, at maximum machine output settings, significant thermal effects at bone surfaces cannot be excluded. The user is advised to make use of any exposure information provided by the manufacturer (for example in the form of displayed safety indices) to gain awareness of the highest output conditions, and to act prudently to limit exposure of critical structures, including bone, and regions including gas. Where on-line display is not available, particular care should be taken to minimise exposure times.

Ultrasound exposure during pregnancy
The embryonic period is known to be particularly sensitive to any external influences. Until further scientific information is available, investigations using pulsed or colour Doppler ultrasound should be carried out with careful control of output levels and exposure times.

With increasing mineralisation of the fetal bone as the fetus develops, the possibility of heating fetal bone increases. The user should prudently limit exposure of critical structures such as the fetal skull or spine during Doppler studies.

14.9 MINIMISING PATIENT EXPOSURE

In hazard assessment, two quantities are of primary importance, the magnitude of the quantity describing the radiation, e.g. intensity or pressure amplitude, and the duration of the exposure. A number of investigators have drawn up diagrams of intensity versus exposure time indicating combinations of the two quantities for which bioeffects have been reported. Figure 14.8 has been derived from the work of these investigators. The boundary between the bioeffect zone (hazard zone) and the no-effect zone (safe zone) is shown as a shaded region to emphasise that for most effects thresholds have not been established. The diagnostic ultrasound zone into which most CW and PW Doppler examinations fall is indicated by the enclosed area.

A similar diagram should be drawn for wave pressure amplitude versus number of transmitted pulses but the required information is not available at present to do this.

Since the factors relevant to biological effects in patients have not been identified for low-intensity ultrasound, it is not possible to calculate the dose to the patient as might be done for ionising radiation. However, the prudent use of ultrasound dictates that the exposure of the patient should be minimised. A number of simple points add up to good practice:

1. Use the lowest transmitted power that will give a result.

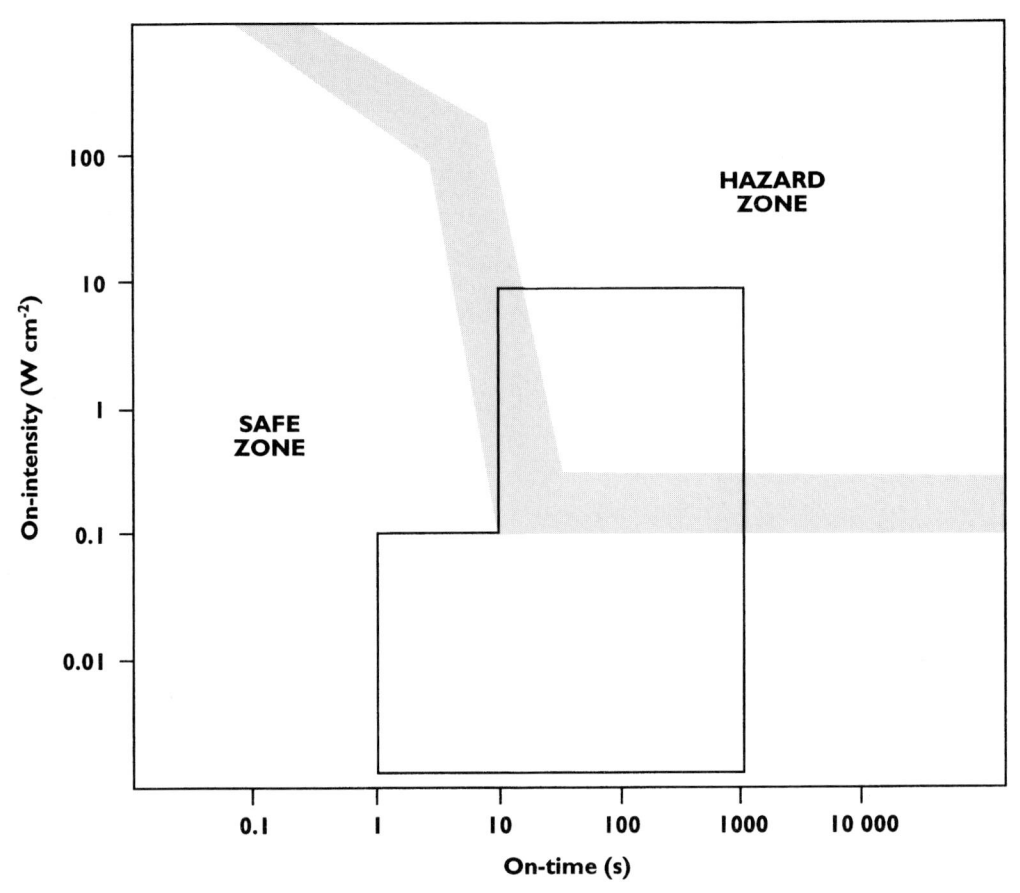

Figure 14.8 Intensity versus exposure time diagram. The enclosed area shows the region in which diagnostic exposures now fall

2. Use the minimum duration of ultrasonic scanning.
3. Use the lowest PRF of the pulsed Doppler unit that will allow the highest velocity present to be measured.
4. Use CW rather than PW Doppler if it will give a result.
5. Do not leave the Doppler beam irradiating a particular region for longer than is necessary. With a duplex system, switch back to the imaging mode as soon as the Doppler recording is completed.

These measures can easily reduce the total amount of ultrasonic energy delivered to the patient by as much as a factor of 1000.

14.10 SUMMARY

In this chapter the output parameters that specify the transmitted ultrasonic field were defined. Values of these parameters as found in practice were listed. To acquaint the user of Doppler techniques with biological effects, a brief survey of interaction mechanisms and fields of study relevant to safety was given. The concepts of 'derated' ('*in situ*') intensity and pressure amplitude and mechanical and thermal index were then discussed. Finally, current opinion and statements from national and international organisations on the safety of diagnostic ultrasound were summarised and procedures for minimising the dose to the patient were presented. The reference list provides further in-depth reading material for this extensive field.

14.11 REFERENCES

AIUM (1984) Safety Considerations for Diagnostic Ultrasound. American Institute of Ultrasound in Medicine, Bethesda, MD

AIUM (1988) Bioeffects considerations for the safety of diagnostic ultrasound. J Ultrasound Med 7:S1–S38

AIUM (1993) Bioeffects and Safety of Diagnostic Ultrasound. American Institute of Ultrasound in Medicine, Bethesda, MD

AIUM/NEMA (1992) Standard for Real-time Display of Thermal and Mechanical Output Indices on Diagnostic Ultrasound Equipment. American Institute for Ultrasound in Medicine, Rockville, MD

Bacon DR (1984) Finite amplitude distortion of the pulsed fields used in diagnostic ultrasound. Ultrasound Med Biol 10:189–195

Barnett SB (1980) Structural and functional changes in the cochlea following ultrasonic irradiation. Ultrasound Med Biol 6:25–32

Barnett SB, Kossoff G (eds) (1998) Safety of Diagnostic Ultrasound. Parthenon, London

Barnett SB, Williams AR (1990) Identification of mechanisms responsible for induced fetal weight reduction in mice. Ultrasonics 28:159–165

Barnett SB, ter Haar GR, Ziskin MC, Nyborg WL, Maeda K, Bang J (1994) Current status of research on biophysical effects of ultrasound. Ultrasound Med Biol 20:205–218

Barnett SB, Rott H-D, ter Haar GR, Ziskin MC, Maeda K (1997) The sensitivity of biological tissue to ultrasound. Ultrasound Med Biol 23:805–812

Bauer A (1997) 'Levovist': a clinical review. Adv Cardiac Echo-Contrast 5:2–34

Bommer WJ, Shah PM, Allen H, Meltzer R, Kisslo J (1984) The safety of contrast echocardiography. Report of the Committee on Contrast Echocardiography for the American Society of Echocardiography. J Am Coll Cardiol 3:6–13

Boyd E, Abdulla U, Donald I, Fleming JEE, Hall A, Ferguson-Smith MA (1971) Chromosome breakage and ultrasound. Br Med J 2:501–502

Brayman AA, Miller MW (1994) Ultrasound cell lysis *in vitro* upon fractional discontinuous exposure vessel rotation. J Acoust Soc Am 95:3666–3668

Brayman AA, Azadniv M, Cox C, Miller MW (1996) Hemolysis of Albunex-supplemented, 40% hematocrit human erythrocytes *in vitro* by 1-MHz pulsed ultrasound: acoustic pressure and pulse length dependence. Ultrasound Med Biol 22:927–938

Brendel K (1985) Ultrasonic power measurement in liquids in the frequency range 0.5 MHz to 25 MHz. International Electrotechnical Commission, Draft TC29 29D WG8

Brent RL, Jensh RP, Beckman DA (1991) Medical sonography: reproductive effects and risks. Teratology 44:123–146

Brock RD, Peacock WJ, Geard CR, Kossoff G, Robinson DE (1973) Ultrasound and chromosome aberrations. Med J Austral 2:533–536

Carmichael AJ, Mossoba MM, Riesz P, Christman CL (1986) Free radical production in aqueous solutions exposed to simulated ultrasonic diagnostic conditions. IEEE Trans Ultrason Ferroelec Freq Contr 33:148–155

Carson PL (1988) Medical ultrasound fields and exposure measurements. In: Proceedings of the 22nd Annual Meeting of the National Council on Radiation Protection, pp. 308–328. NCRP Publications, Bethesda, MD

CDRH (1985) Centre for Devices and Radiological Health: 510(k) guide for measuring and reporting acoustic output of diagnostic ultrasound devices. Food and Drug Administration, Rockville, MD

Clarke PR, Hill CR (1970) Physical and chemical aspects of ultrasonic disruption of cells. J Acoust Soc Am 47:649–653

Cosgrove D (1995) Echo enhancers—'Contrast' agents for ultrasound. Br Med Ultrasound Soc Bull 3 (1 February):34–38

Crum LA, Foulkes JB (1986) Acoustic cavitation generated by microsecond pulses of ultrasound. Nature 319:52–54

Dalecki D, Raeman CH, Child SZ, Cox C, Francis CW, Meltzer RS, Carstensten EL (1997) Hemolysis *in vivo* from exposure to pulsed ultrasound. Ultrasound Med Biol 23:307–313

Duck FA (1997) The meaning of thermal index (TI) and mechanical index (MI) values. Br Med Ultrasound Soc Bull 5 (4, November):36–40

Duck F, Martin K (1986) Acoustic output from commercial diagnostic ultrasound equipment. Br Med Ultrasound Soc Bull (40, February):12–14

Duck FA, Martin K (1991) Trends in diagnostic ultrasound exposure. Phys Med Biol 36:1423–1432

Duck FA, Starritt HC (1984) Acoustic shock generation by ultrasonic imaging equipment. Br J Radiol 57:231–240

Duck FA, Starritt HC, Aindow JD, Perkins MA, Hawkins AJ (1985) The output of pulse-echo ultrasound equipment: a survey of powers, pressures and intensities. Br J Radiol 58:989–1001

Duck FA, Starritt HC, Anderson SP (1987) A survey of the acoustic output of ultrasonic Doppler equipment. Clin Phys Physiol Meas 8:39–49

Dyson M, Suckling J (1978) Stimulation of tissue repair by ultrasound: a survey of mechanisms involved. Physiotherapy 64:105–108

Dyson M, Pond JB, Woodward B, Broadbent J (1974) The production of blood cell stasis and endothelial damage in blood vessels of chick embryos treated with ultrasound in a stationary wave. Ultrasound Med Biol 1:133–148

Edwards MJ (1986) Hyperthermia as a teratogen: a review of experimental studies and their clinical significance. Teratogen Carcinogen Mutagen 6:563–582

EFSUMB (1996) Epidemiology of diagnostic ultrasound exposure during human pregnancy. Eur J Ultrasound 4:69–71

Farmery MJ, Whittingham TA (1978) A portable radiation-force balance for use with diagnostic ultrasonic equipment. Ultrasound Med Biol 3:373–379

Fleming AD, McDicken WN, Sutherland GR, Hoskins PR

(1994) Assessment of colour Doppler tissue imaging using test-phantoms. Ultrasound Med Biol 20:937–951

Flynn HG (1982) Generation of transient cavities in liquids by microsecond pulses of ultrasound. J Acoust Soc Am 72:1926–1932

Forester GV, Roy OZ, Mortimer AJ (1982) Enhancement of contractility in rat isolated papillary muscle with therapeutic ultrasound. J Mol Cell Cardiol 14:475–477

Fowlkes JB, Crum LA (1988) Cavitation threshold measurements for microsecond length pulses of ultrasound. J Acoust Soc Am 83:2190–2201

Freeman FE (1963) Measurement of liquid-borne ultrasonic power. Ultrasonics 1:27–34

Gershoy A, Nyborg WL (1973) Perturbation of plant-cell contents by ultrasonic irradiation. J Acoust Soc Am 54:1356–1367

Goss SA (1984) Sister chromatid exchange and ultrasound. J Ultrasound Med 3:463–470

Harris GR (1985) A discussion of procedures for ultrasonic intensity and power calculations from miniature hydrophone measurements. Ultrasound Med Biol 11:803–815

Henderson J, Willson K, Jago JR, Whittingham TA (1995) A survey of the acoustic outputs of diagnostic ultrasound equipment in current clinical use. Ultrasound Med Biol 21:699–705

Hill CR (1970) Calibration of ultrasonic beams for biomedical applications. Phys Med Biol 15:241–248

Hill CR, ter Haar G (1982) Ultrasound. In: Suess MJ (ed) Non-ionising Radiation Protection, pp. 199–228. WHO Regional Publication, European Series No 10. World Health Organization, Copenhagen

Hill CR, Joshi GP, Revell SH (1972) A search for chromosome damage following exposure of Chinese hamster cells to high intensity, pulsed ultrasound. Br J Radiol 45:333–334

Holland CK, Apfel RE (1990) Thresholds for cavitation produced by pulsed ultrasound in a controlled nuclei environment. J Acoust Soc Am 88:2059–2069

Holland CK, Deng CX, Apfel RE, Alderman JL, Fernandez LA, Taylor KJW (1996) Direct evidence of cavitation *in vivo* from diagnostic ultrasound. Ultrasound Med Biol 22:917–925

IEC (1996) IEC 1685: Ultrasonics—Physiotherapy Systems—Performance Requirements and Methods of Measurement in the Frequency Range 0.5 to 5 MHz. International Electrotechnical Commission, Geneva

Inoue M, Miller MW, Church CC (1991) An alternative explanation for a non-thermal, non-cavitational ultrasound mechanisms of action on *in vitro* cells at hyperthermic temperature. Ultrasonics 28:185–189

Jacobson-Kram D (1984) The effects of diagnostic ultrasound on sister chromatid exchange frequencies: a review of the recent literature. J Clin Ultrasound 12:5–10

Koenig K, Meltzer RS (1986) Effects of viscosity on the size of microbubbles generated for use as echocardiographic contrast agents. J Cardiovasc Ultrason 5:3–4

Kossoff G (1965) Balance technique for the measurement of very low ultrasonic power outputs. J Acoust Soc Am 38:880–881

Lehmann JF, Guy AW (1972) Ultrasound therapy: interaction of ultrasound and biological tissue. In: Reid JM, Sikov MR (eds) Department of Health, Education and Welfare (DHEW) Publication, pp. 141–151. FDA, US Government Printing Office, Washington DC

Leighton TG (1996) The Acoustic Bubble. Academic Press, London

Macintosh IJC, Davey DA (1970) Chromosome aberrations induced by an ultrasonic foetal pulse detector. Br J Radiol 48:230–232

Madsen EL, Frank GR, MacDonald MC, Martin AO, Bouck NP, Iannaccone PM (1991) Method of cavitation suppressed exposure of cells and explant mouse embryos to clinical real-time and pulsed Doppler ultrasound. J Ultrasound Med 10:629–636

McDicken WN, Sutherland GR, Moran CM, Gordon LN (1992) Colour Doppler velocity imaging of the myocardium. Ultrasound Med Biol 18:651–654

Miller DL (1987) A review of the ultrasonic bioeffects of microsonation, gas body activation, and related cavitation-like phenomena. Ultrasound Med Biol 13:443–470

Miller DL, Thomas RM (1995) Ultrasound contrast agents nucleate inertial cavitation *in vitro*. Ultrasound Med Biol 21:1059–1065

Miller DL, Ziskin MC (1989) Biological consequences of hyperthermia. Ultrasound Med Biol 15:707–722

Miller MW (1985) In vitro studies: single cells and multicell spheroids. In: Nyborg WL, Ziskin MC (eds) Biological Effects of Ultrasounds, pp. 35–48. Churchill Livingstone, New York

Miller MW, Miller DL, Brayman AA (1996) A review of *in vitro* bioeffects of inertial ultrasonic cavitation from a mechanistic perspective. Ultrasound Med Biol 22:1131–1154

Muir TG, Cartensen EL (1980) Prediction of nonlinear acoustic effects at biomedical frequencies and intensities. Ultrasound Med Biol 6:345–357

Nanda NC (1993) Echocontrast enhancers—how safe are they? In: Nanda NC, Schlief R (eds) Advances in Echo Imaging Using Contrast Enhancement. Kluwer Academic, Dordecht

Nanda NC, Schlief R, Goldberg BB (eds) (1997) Advances in Echo Imaging Using Contrast Enhancement, 2nd edn. Kluwer Academic, Dordrecht

NCRP (1983) National Council on Radiation Protection and Measurements Report No. 74, Biological Effects of Ultrasound: Mechanisms and Clinical Implications. NCRP, Bethesda, MD

NCRP (1992) National Council on Radiation Protection and Measurements Report No. 113, Exposure criteria for medical diagnostic ultrasound: 1 Criteria based on thermal mechanisms. NCRP, Bethesda, MD

Nyborg WL (1985) Biophysical mechanisms of ultrasound. In: Repacholi MH, Benwell DA (eds) Essentials of Medical Ultrasound, pp. 35–72. Churchill Livingstone, New York

Nyborg WL, Ziskin MC (eds) (1985) Biological Effects of Ultrasound. Churchill Livingstone, New York

O'Brien WD (1983) Dose-dependent effect of ultrasound on fetal weight in mice. J Ultrasound Med 2:1–8

O'Brien WD (1984) Ultrasonic bioeffects: a view of experimental studies. Birth 11:149–157

O'Brien WD (1994) Output display standard: a new equipment feature. pp. 317–326. In: Proceedings of the Society of Diagnostic Medical Sonographers Conference, Chicago, 1994

Perkins MA (1989) A versatile force balance for ultrasound power measurement. Phys Med Biol 34:1645–1652

Preston RC (1985) International Electrotechnical Commission Draft TC29 29D WG8: measurement and characterisation of ultrasonic fields using hydrophones in the frequency range 0.5 MHz to 15 MHz

Preston RC (1991) Output Measurements for Medical Ultrasound. Springer-Verlag, Berlin

Preston RC, Bacon DR, Livett AJ, Rajendran K (1983) PVDF membrane hydrophone performance properties and their relevance to the measurement of the acoustic output of medical ultrasonic equipment. J Phys E: Sci Instrum 16:786–796

Sommer FG, Pounds D (1982) Transient cavitation in tissues during ultrasonically induced hyperthermia. Phys Med Biol 9:1–3

Stewart HF (1982) Ultrasonic measurement techniques and equipment output levels. In: Repacholi MH, Benwell DA (eds) Essentials of Medical Ultrasound, pp. 77–116. Humana, Clifton, NJ

Tarantal AF, Gargosky SE, Ellis DS, O'Brien WD, Hendrickx AG (1995) Hematologic and growth-related effects of frequent prenatal ultrasound exposure in the long-tailed macaque (*Macaca fascicularis*). Ultrasound Med Biol 21:1073–1081

ter Haar GR (1986) Ultrasonic biophysics. In: Hill CR (ed) Physical Principles of Medical Ultrasonics. Ellis Horwood, Chichester

ter Haar G, Daniels S (1981) Evidence for ultrasonically induced cavitation *in vivo*. Phys Med Biol 26:1145–1149

ter Haar G, Daniels S, Eastaugh KC, Hill CR (1982) Ultrasonically induced cavitation *in vivo*. Br J Cancer 45 (suppl V):151–155

ter Haar GR, Daniels S, Morton K (1986) Evidence for acoustic cavitation *in vivo*: thresholds for bubble formation with 0.75 MHz continuous-wave and pulse beams. IEEE Trans Ultrason Ferroelec Freq Contr 33:162–164

Thacker J (1973) The possibility of genetic hazard from ultrasonic radiation. Curr Topics Radiation Res Q8:235–258

Wells PNT (ed) (1987) Br J Radiol Suppl 20—The Safety of Diagnostic Ultrasound. The British Institute of Radiology, London

Westervelt P (1951) The theory of steady forces caused by sound waves. J Acoust Soc Am 23:312–315

WFUMB (1997) Conclusions and recommendations on thermal and non-thermal mechanisms for biological effects of ultrasound. WFUMB News 4 (2 July):2–4

WFUMB (1998) World Federation for Ultrasound in Medicine and Biology symposium on safety of ultrasound in medicine—conclusions and recommendations on thermal and nonthermal mechanisms for biological effects of ultrasound. Ultrasound Med Biol 24 (suppl 1)

Whittingham TA (1994) The safety of ultrasound. Imaging 6:33–51

Williams AR (1983) Ultrasound: Biological Effects and Potential Hazards. Academic Press, London

Williams AR (1985) Effect of ultrasound on blood and the circulation. In: Nyborg WL, Ziskin MC (eds) Biological Effects of Ultrasound, pp. 49–65. Churchill Livingstone, New York

Williams AR (1991) A critical evaluation of bioeffect reports and epidemiological surveys. pp. 30–33. In: Docker MF, Duck FA (eds) British Medical Ultrasound Society Report, 'The Safe Use of Diagnostic Ultrasound'. British Institute of Radiology, London

Williams AR, Gross DR, Miller DL (1985) Cavitation in mammalian blood: an *in vivo* search. In: Gill RW, Dadd MJ (eds) Proceedings of the 4th Meeting of World Federation for Ultrasound in Medicine and Biology, p. 484. Pergamon, Sydney

Williams AR, Miller DL, Gross DR (1986) Haemolysis *in vivo* by therapeutic intensities of ultrasound. Ultrasound Med Biol 12:501–509

Williams AR, Kubowicz G, Cramer E, Schlief R (1991) The effects of the microbubble suspension SHU 454 (Echovist) on ultrasonically-induced cell lysis in a rotating tube exposure system. Echocardiography 8:423–433

Young FR (1989) Cavitation. McGraw-Hill, London

Ziskin MC, Petitti DB (1988) Epidemiology of human exposure to ultrasound: a critical review. Ultrasound Med Biol 14:91–96

Appendix 1

Special Functions Arising from Womersley's Theory

M'_{10} and ε'_{10}, which appear in eqn. 2.23, are complex functions of the non-dimensional parameter α (defined in eqn. 2.24) and may be written:

$$M'_{10}(\alpha) = (1 + h_{10}^2 - 2h_{10}\cos\delta_{10})^{1/2} \quad A1.1$$

and

$$\varepsilon'_{10}(\alpha) = \tan^{-1}\left(\frac{h_{10}\sin\delta_{10}}{1 - h_{10}\cos\delta_{10}}\right) \quad A1.2$$

where

$$h_{10} = \frac{2}{\alpha}\left[\frac{(\text{Ber1})^2(\alpha) + (\text{Bei1})^2(\alpha)}{\text{Ber}^2(\alpha) + \text{Bei}^2(\alpha)}\right]^{1/2} \quad A1.3$$

and

$$\delta_{10} = \frac{3\pi}{4} - \tan^{-1}\left[\frac{(\text{Bei1})(\alpha)}{(\text{Ber1})(\alpha)}\right] + \tan^{-1}\left[\frac{\text{Bei}(\alpha)}{\text{Ber}(\alpha)}\right] \quad A1.4$$

The Kelvin functions Ber and Bei are the real and imaginary parts of a complex Bessel function of order zero, while (Ber1) and (Bei1) are the real and imaginary parts of a complex Bessel function of order one. Summations for the calculation of each of these functions have been given by Abramowitz and Stegun (1965) Handbook of Mathematical Functions, Dover Publications, New York.

The function ψ, which appears in eqn. 2.25, is also a function of α and may be written:

$$\psi = \left(\frac{\tau J_0(\tau) - \tau J_0(y\tau)}{\tau J_0(\tau) - 2J_1(\tau)}\right) \quad A1.5$$

where J_0 and J_1 are Bessel functions of the first kind and orders 0 and 1, respectively, and $\tau = \alpha i^{3/2}$.

Appendix 2

Doppler Test Devices

Contributed by PR Hoskins and KV Ramnarine
Medical Physics Department, Royal Infirmary, Edinburgh EH3 9YW, UK.

A2.1 INTRODUCTION

Doppler ultrasound test devices are of varying degrees of complexity, but their common aim is to provide reproducible Doppler ultrasound signals which are then displayed as either a spectral trace or a colour image. This provides the means by which some aspect of a Doppler ultrasound system may be investigated. Their most common use at the time of writing is as test tools in research, rather than as quality assurance devices in routine manufacture or clinical use. This chapter will describe the various types of test devices that have been described in the literature.

Test devices in which there is a moving target may be categorised according to the number of spatial dimensions of the moving target. A vibrating ball-bearing is a 'zero dimensional' target; a string phantom a '1-D' target; the vibrating plate is '2-D'; and belt and flow phantoms are '3-D'. When there is no moving target, the device is referred to as an 'injection' device, as the Doppler ultrasound signal must be synthesised electronically and injected at some point into the Doppler ultrasound system. These may be further categorised depending on the site of injection, whether it is an echo itself which is synthesised, or whether there is direct injection into the instrument electronics in the form of an RF or audio signal.

The degree of tissue equivalence of the test object must be considered. Intuitively, the flow phantom is the most tissue equivalent test device; however, a better approach to designing and choosing phantoms is to consider that the degree of tissue equivalence should be sufficient to enable the measurement of interest to be performed. It will be seen that measurement of velocity may be performed using the string phantom, provided that the backscatter is not strongly dependent on the beam-filament angle. If, however, the requirement is extended to the measurement of volume flow, then this requires flow in a vessel, hence only the flow phantom will suffice. The degree of tissue equivalence required then depends on whether the flow is pulsatile or steady; for steady flow the acoustic properties of the tissue mimic, the vessel and the blood mimic are relevant. If the flow is pulsatile, then a number of other factors must be specified; the change in velocity profile with time is dependent to some degree on the viscosity, which must therefore be matched to that of blood; the flow waveform shape should be comparable to those in the body; arteries in the body change dimension by 10% or so during the cardiac cycle, so ideally this should be modelled. These examples demonstrate that the concept of tissue equivalence is not clear-cut.

A2.2 FLOW PHANTOMS

Production of reproducible flow waveforms in tubes is of interest in the study of arterial blood flow dynamics, and a number of dedicated systems have been described for the production of steady and pulsatile flow at physiological flow rates (Issartier et al 1978, Wemeck et al 1984, Peterson 1984, Janssens et al 1989). Flow phantoms specifically designed for Doppler ultrasound systems have been described, for which Fig. A2.1 shows the main components. The characteristic of these systems is that they are concerned in varying degrees with the tissue equivalence of the materials of the flow phantom which are in the path of the ultrasound beam. The design of a tissue equivalent flow phantom is considered in more detail in the following sections.

FLOW PHANTOMS

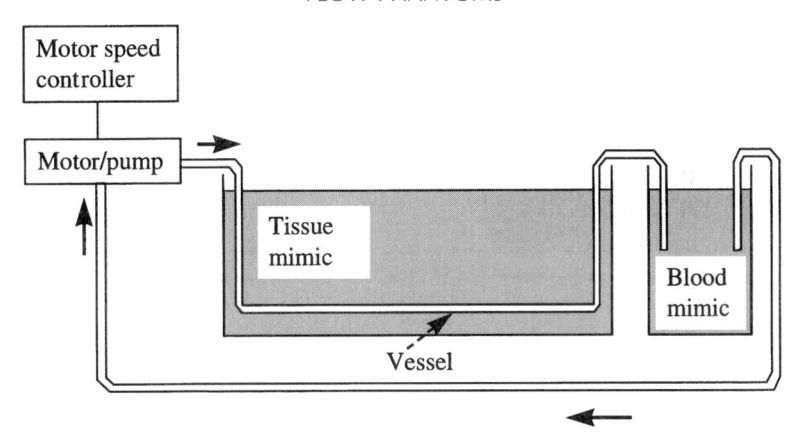

Figure A2.1 Components of a flow phantom

A2.2.1 Overall Design

The production of steady flow requires the pump output to be steady with time. The production of pulsatile flow is more demanding on the design of the flow phantom than the production of steady flow. There are two approaches to this; in the first the flow waveform shape is produced by adjustment of the pump flow rate through the flow cycle. The production of very realistic flow waveforms is possible (Fig. A2.2). The presence of reflected pressure waves from distal changes in impedance is a confounding factor which causes a lack of correspondence between the drive waveform and the Doppler waveform (Fig. A2.3). In the second approach (Shortland and Cochrane 1989), the effect of vessel compliance and the distribution of vascular resistance is modelled. The waveform shape is then mediated by the interaction between forward-going and reverse going pressure waves, as occurs in the human body. Provided that the vessel compliance and vessel resistance have been adequately chosen, it is possible to investigate the effect of distal resistance on the Doppler waveform using this latter approach.

The simplest design is a straight tube under flow conditions that are stable with distance. The inlet lengths required for this criterion are shown in Table A2.1, and it is noted that the lengths may be greater than 1 m for the larger diameters and velocities that may be of interest. This emphasises that flow phantoms are generally large, laboratory based devices. Commercial flow phantoms which are portable, are by necessity small and the flow is not within the stable flow region, unless the vessel diameters and velocities are small.

Modelling of more realistic vessel geometries such as occur in the human body requires manufacture using suitable materials. Materials which are easy to use for manufacture, such as plexiglas (Ku et al 1985, Fei et al 1988), acrylic (Fei et al 1988) or thin polyester (Frayne et al 1993), do not have appropriate acoustic properties; the acoustic velocity in plexiglas is 4000–5000 ms^{-1}, in acrylic 2756 ms^{-1}, and in polyester 2460 ms^{-1} (Selfridge 1985). To date, no suitable method has been found for the manufacture of anatomically correct models suitable for ultrasound studies.

A2.2.2 Pumps

The physiological flow range in the human is from 30 $l\ min^{-1}$ in the largest vessel, the aorta,

Table A2.1 Inlet lengths (cm) for a variety of maximum velocities and vessel diameters (reproduced by permission from Hoskins et al 1994b)

Diameter (cm)	$V_{max} =$ 1 cm s^{-1}	$V_{max} =$ 100 cm s^{-1}	$V_{max} =$ 500 cm s^{-1}
0.1	0.0055	0.55	2.75
1	0.55	55	275
2	2.2	220	1100

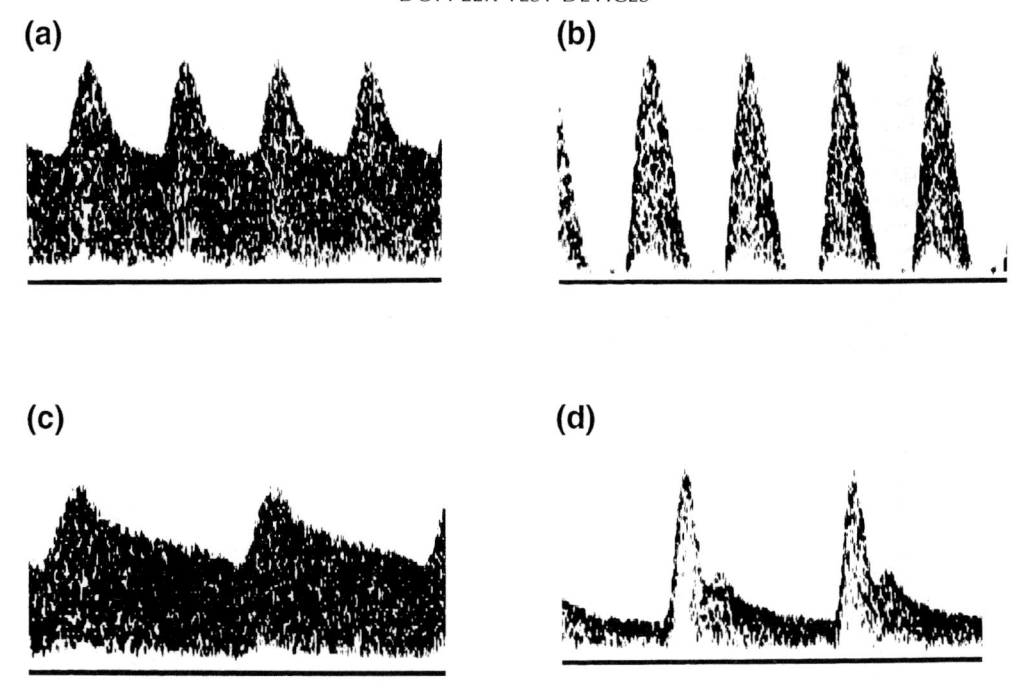

Figure A2.2 Waveforms produced from the flow phantom described by Hoskins et al (1989); (a) and (b) simulated normal and abnormal umbilical artery waveforms, (c) and (d) simulated normal and abnormal uterine artery waveforms. The tubing used was heatshrink, and the blood mimic was Sephadex in a glycerol solution (reproduced by permission from Hoskins et al 1989, © IOP Publishing Limited)

during strenuous exercise, to 10 nl min^{-1} in a single capillary. This is an enormous range, and the lower end is not achievable using current pump technology. Details of commonly used pumps are shown in Table A2.2. The production of both steady flow and pulsatile flow is of interest. Most pumps produce some slight variation in flow during steady pumping. For the gear pump this is not measurable except at very low pumping rates; however, there is a very large effect for other pumps, especially the roller pump. The use of a damping reservoir connected to the outlet of the pump is then needed to reduce flow oscillation. The use of a dampener with a fluid/air interface must be performed with caution, as disturbance of the fluid surface can produce air bubbles. A working dampener can be made using an in-line container with a thin elastic membrane. Some pumps are more suited than others to the production of pulsatile flow due to the temporal response of the pump. If a dampener is used, then this rules out the production of pulsatile flow. There is an open channel through the peristaltic and worm pump, and the momentum of the fluid will produce long response times. The gear, piston and roller pumps have short response times, so in this respect they are well-suited for the production of pulsatile flow. Systems for which the waveform shape may be altered with ease have been described by McCarty and Locke (1986) based on a combination of a centrifugal and modified roller pump, Law et al (1987) based on a modified roller pump, McDicken (1986) and Hoskins et al (1989) based on a gear pump, Holdsworth et al (1991) based on a piston pump, and Hein and O'Brien (1992) based on a combination of a roller and a piston pump. One of the earliest published systems (Michie and Fried 1973) used a hospital syringe pump to produce a flow rate of 26 ml min^{-1}, although the duration of continuous flow was limited by the size of the syringe.

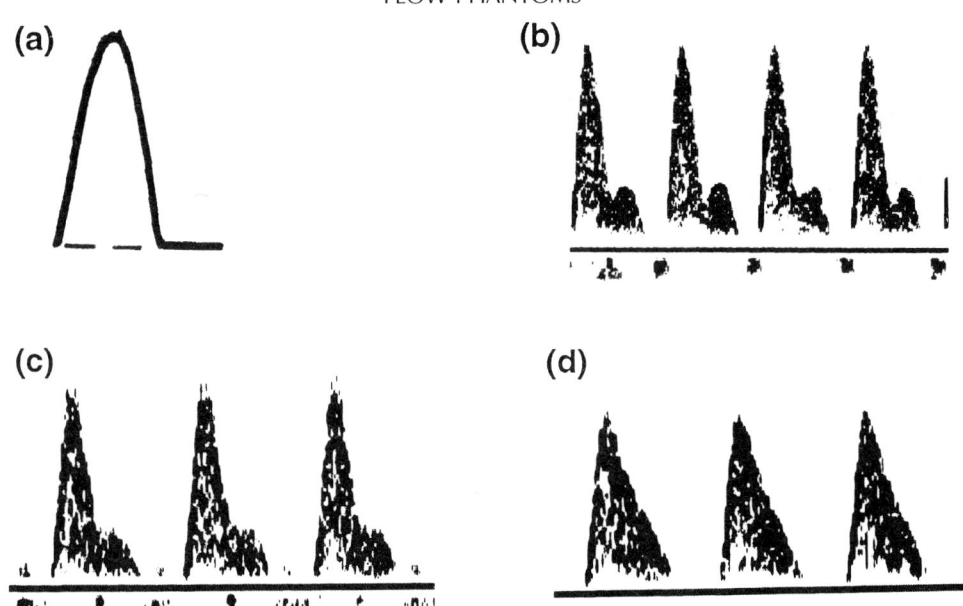

Figure A2.3 Effect of tube geometry and elasticity on the Doppler waveform. In each case the total tube length was 100 cm, and the insonation site was 50 cm from the tube end. (a) Pump drive waveform. (b) Waveforms from flow in an 8 mm diameter silicone rubber tubing. This material is very elastic. The waveform demonstrates a second peak due to the effect of reflected pressure waves from the tube end. (c) Waveforms from flow in an 8 mm diameter moulded heatshrink tube. This material is much stiffer than silicone rubber. The waveform demonstrates a second peak, (d), as for (c) except the diameter of the final 25 cm of the tube end was 4 mm. The second peak is not present, and the waveform is of a similar shape to the drive waveform in (a) (reproduced by permission from Hoskins at al 1989, © IOP Publishing Limited)

Table A2.2 Details of pumps commonly used in Doppler flow phantoms

Pump	Design	Action	Flow waveform during steady pumping	Temporal response
Gear	Interlocking gears (two or three)	Blood mimic forced between teeth of gears	Excellent: small ripple ($< 10\%$) at very low pump speeds	Very short
Piston	Tightly fitting plunger in containing outer case	Motion of plunger	Very good: transient reduction in flow when piston changes direction	Short
Centrifugal	Central inlet, with outlet at periphery of rotating veins	Centrifugal force generated by rotating veins	Excellent	Long
Peristaltic or roller	Rotating set of wheels compressing tube at outer margin	Motion of compression point along the tube	Poor: large ripple (up to 100%) using unmodified system. Less for modified system (Law et al 1987)	Short
Progressing cavity or worm	Helical shaped unit ('worm') capable of rotation about axis	Force generated by rotation of 'worm'	Very good	Long

A2.2.3 Blood Mimicking Fluids

The Doppler signal from blood arises from the red cells, as discussed earlier in this book. The blood mimic should closely match both the acoustic and physical properties of the blood. This is a tough requirement and perfect agreement has not been achieved. Table A2.3 shows values which IEC 1685 (draft standard) defines as being required of a blood-mimicking fluid.

- *Density*. Neutral buoyancy is required or the particles sink or rise giving a non-uniform particle distribution.
- *Particle size*. The red cell has an effective diameter of 5 µm. For diameters much less than the ultrasound wavelength, Rayleigh scattering occurs, in which the backscattered intensity is proportional to (diameter)^6 and (frequency)^4.
- *Particle concentration*. If the number of particles within the sample volume is sufficiently large, the Doppler signal may be described as a Gaussian random process.
- *Viscosity*. Blood is a non-Newtonian suspension; however, it is common to assume that blood is a Newtonian fluid (Pedley 1980) for vessels whose diameter is greater than about 0.1 mm. The high shear rate viscosity of human blood is 3.5–4.5 mPa s (Lowe 1988, Duck 1990).
- *Acoustic properties*. These are the backscatter power, acoustic velocity and attenuation.

Both Boote and Zagzebski (1988) and Ramnarine et al (1998) have described blood mimics for which the density, acoustic velocity and backscatter are reported to be blood-equivalent (Table A2.3). The blood mimic of Ramnarine et al (1998) is a suspension of 5 µm diameter nylon scatterers in a water base. Detergent is used to wet the particles, glycerol to obtain the correct acoustic velocity and density, and dextran to obtain the correct viscosity. The use of nylon particles in a blood mimic was first described by Oates (1991). He used a large volume concentration of nylon of 43% as opposed to the 1.8% volume concentration used by Ramnarine et al. At high volume concentrations the blood mimic is non-Newtonian.

A commonly used particle for blood mimic is Sephadex G25 (Pharmacia, Upsala, Sweden). This consists of beads of cross-linked dextran which swell on immersion in water giving a range of diameters from 20 to 70 µm. This is easily prepared, but it is difficult to formulate a stable tissue-equivalent material as the particles are dense, hence they sink at low velocities, and they are fragile, being easily crushed in certain pumps such as the gear pump. Other particles and blood mimics no longer used include oil emulsions, air bubbles, milk, silicon carbide and talcum powder (reviewed in Law et al 1989).

One of the more important properties is the backscatter power, as the measurement of Doppler sensitivity is of interest. Figure A2.4 shows the change in backscatter power with time of the nylon-based blood mimic of Table A2.3. The backscatter power at time zero and the time to plateau power are dependent on the preparation of the blood mimic, although the final plateau backscatter power is similar in each case. Nylon particles have a tendency to coalesce, in much the same way that red blood cells do at low shear

Table A2.3 Requirements of a standard blood-mimicking fluid, and reported properties of two blood mimics

Physical property	IEC (draft standard)	Boote and Zagzebski (1988)	Ramnarine et al (1999)
Scatterer concentration (mm^{-3})	> 1000 in sample volume	17	300 000
Scatterer size (µm)	–	30	5
Density (kg m^{-3})	1050 ± 40	1043	1037
Viscosity (mPa s)	4 ± 0.4	–	4.1
Backscatter (f^4 m^{-1} sr^{-1})	$1-10 \times 10^{-31}$	Blood equivalent	Blood equivalent
Acoustic velocity (ms^{-1})	1570 ± 30	1546	1548
Attenuation (dB cm^{-1} MHz^{-1})	< 0.1	0.1 dB cm^{-1} at 1 MHz	0.05

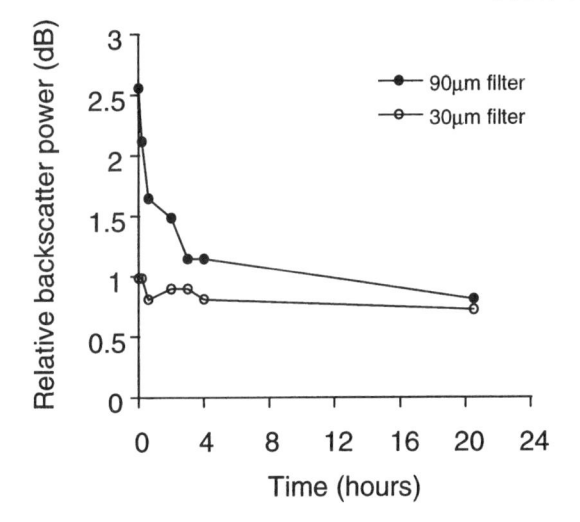

Figure A2.4 Backscatter Doppler power versus time for the blood mimic of Ramnarine et al (1998), demonstrating the effect of the use of different filters during preparation; the power falls over a period of 3 hours if the blood mimic is filtered using a 90 μm filter, but remains constant if a 30 μm filter is used

rates. The interpretation of the data presented in Fig. A2.4 is that it is the amount of clumping which is technique-dependent, so that, for example, use of the finer mesh filter (30 μm instead of 90 μm) removes a greater fraction of the clumps.

Rayleigh scattering is associated with scattering from small particles, typically where the ratio of diameter to wavelength is less than 10%. The features of Rayleigh scattering are that the backscatter is uniform in all directions, the backscatter intensity is dependent on (diameter)^6 and (frequency)^4. The nylon particles in the blood mimic of Ramnarine et al (1998) have the same diameter as the effective diameter of red cells, so it would be anticipated that Rayleigh scattering would continue to be present at higher frequencies (> 10 MHz). At 4 MHz the ratio of diameter to wavelength for 20–70 μm Sephadex particles is 5–18% which is not small. However, it has been shown that at this frequency the statistics of the Doppler spectrum are the same as for human blood (Hoskins et al 1990), so it must be assumed that scattering is still in the Rayleigh region, although this is likely to break down at higher frequencies.

The backscatter power is related to the difference in density and in hardness between the particle and surrounding fluid, and to the sixth power of the particle diameter. The volume concentration of red cells in human blood is in the range 40–45%, compared to 1–2% for blood mimics with blood-equivalent backscatter, due to the increased hardness and diameter of particles used in blood mimics. More complex behaviour of backscatter power will occur as the volume concentration increases, and if there is any kind of interaction between the particles, such as clumping and orientation of clumps with respect to flow direction. The only particle suspension for which these issues have been studied in detail is blood, as discussed in Section 7.2. The peak backscatter for blood occurs around 12–26%, dependent on flow conditions. For Sephadex, a peak backscatter at about 30% volume concentration was found (Hoskins et al 1990).

A2.2.4 Vessel

A large number of different materials have been described for use as the vessel in flow phantoms (reviewed in Law et al 1989). Mismatch in the acoustic velocity between the vessel and the surrounding tissue or blood mimic will produce reflection and refraction of the ultrasound beam, with distortion of the received Doppler signal. This is further exacerbated if there are differences in density and attenuation coefficient (Law et al 1989, Thompson et al 1990, Thompson and Aldis 1996, Tortoli et al 1997). The use of stiff materials, such as glass, acrylic or heatshrink, has the attraction that they are geometrically stable and can be cast or machined to form geometries of interest, such as a stenosis model. It has been recognised that these materials do not have appropriate physical properties and they are no longer used. Table A2.4 shows the IEC requirements for the acoustic properties of the tissue mimic, along with values from various tubing materials. It can be seen that there is no single material with all the correct properties. Figure A2.5 shows the distortion that will occur when different thin-walled vessels are used. Figure A2.6 shows the effect of different wall thicknesses for C-flex. The lack of a suitable vessel

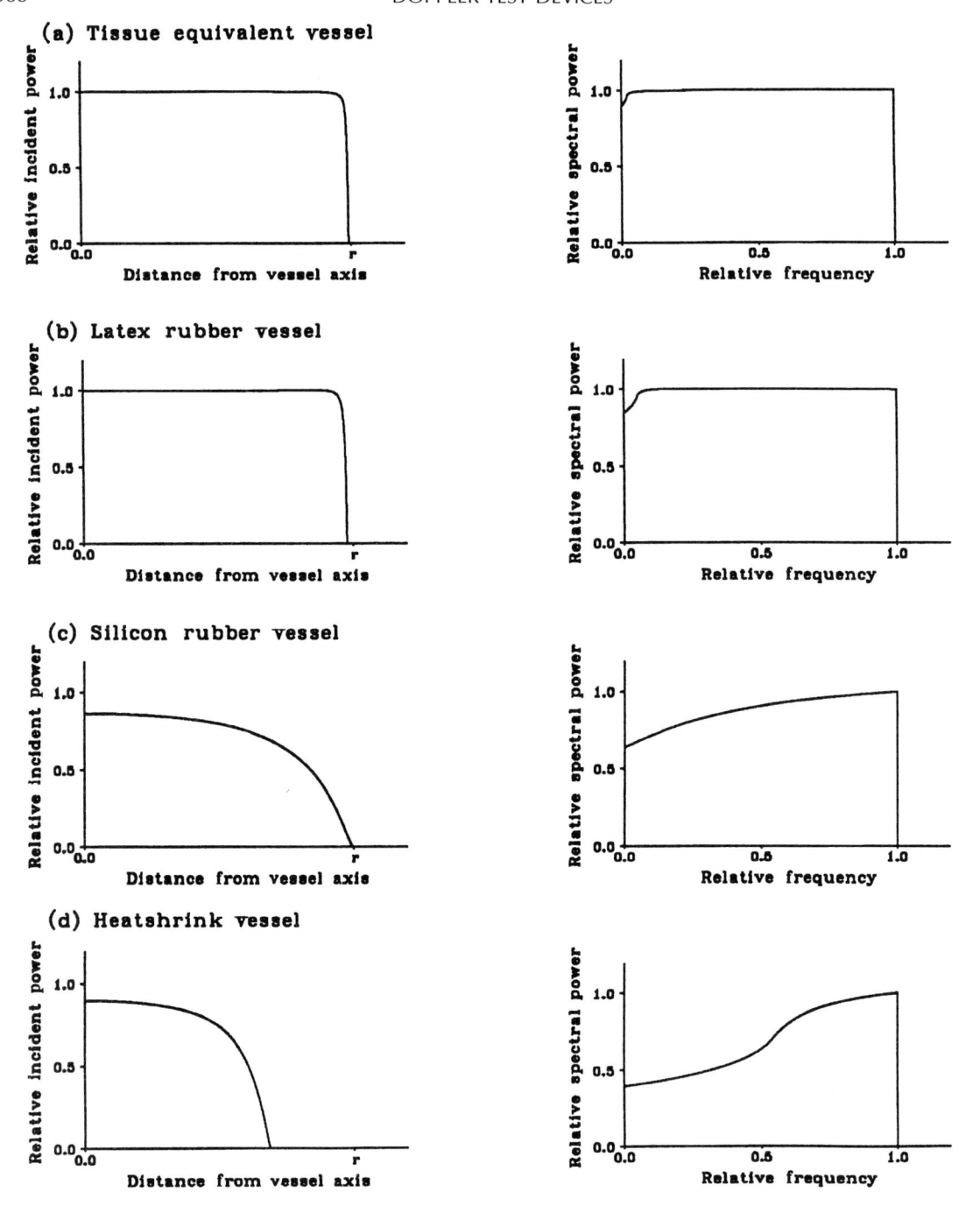

Figure A2.5 Effect of differences in the acoustic properties of different vessels calculated using a thin-wall model under conditions of steady flow and uniform insonation at a beam-vessel angle of 60°. The effect of spectral broadening is ignored. The figure shows the relative sensitivity as a function of radial distance, and the corresponding normalised spectral power as function of normalised Doppler frequency. Below, c_t (ms^{-1}) is the acoustic speed in the tissue mimic, c_w (ms^{-1}) is the acoustic speed in the vessel wall, and c_b (ms^{-1}) is the acoustic speed in the blood mimic: (a) tissue equivalent model, $c_t = 1540$, $c_w = 1570$, $c_b = 1590$; (b) latex rubber vessel $c_t = 1540$, $c_w = 1566$, $c_b = 1540$; (c) silicone rubber vessel $c_t = 1540$, $c_w = 1005$, $c_b = 1540$; (d) heatshrink vessel $c_t = 1540$, $c_w = 1979$, $c_b = 1540$ (reproduced by permission of the Institute of Physics and Engineering in Medicine from Hoskins 1994c)

Table A2.4 Requirements of a standard tissue mimic, and the acoustic properties of various tubes (reproduced by permission from Hoskins et al 1994c)

Physical quantity	Tissue mimic	Latex rubber	C-flex	Silicone rubber	Heatshrink
Density (kg m^{-3})	1000	921	886	1140	960
Velocity (ms^{-1})	1540	1564	1553	1005	1979
Attenuation (dB cm^{-1} MHz^{-1})	0.5 (2.5*)	26*	28*	23*	31*

*dB cm^{-1} at 5 MHz.

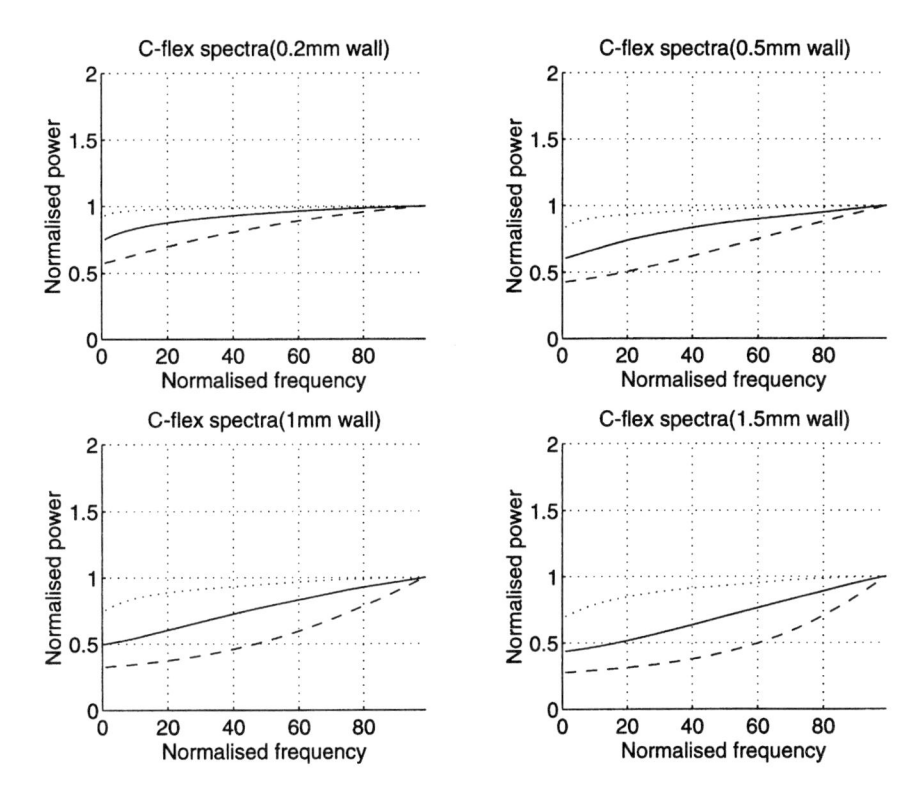

Figure A2.6 Normalised spectral power versus normalised Doppler shift frequency for steady flow in C-flex of different wall thicknesses under conditions of uniform insonation. ($\cdots$ 2 MHz, —— 5MHz, – – – 10 MHz) (reproduced by permission from Fish and Steel 1996)

material has led to the use of 'wall-less' phantoms, in which no vessel is used. Instead, the tissue mimic is cast around rods of known diameter which are removed when casting is complete.

Distortion of the Doppler spectrum arises as a result of differences in the angle of the vessel surface with respect to the beam, and in path length traversed by the beam. These effects are especially marked at the lateral edge of the vessel. Limited tissue equivalence holds under two conditions. The first case is concerned with measurements of maximum frequency shift. When there is alignment of the beam and vessel axis, Fig. A2.6 shows that there is primarily loss of low-frequency amplitude; however, the maximum frequency end is relatively undistorted. It is then possible to perform experiments concerned with the maximum frequency, provided that the vessel wall is straight and the beam and vessel axis are aligned. The second case involves the whole Doppler spectrum. The vessel surface must be sufficiently flat over the whole width of the beam. This requires, in addition to beam and vessel axis alignment, that the beam width is small compared to the vessel diameter. Provided that these conditions are met, the use of tubes such as C-flex or latex rubber, whose acoustic properties are very similar to tissue, is justified. The substantial difference in acoustic properties between acrylic or plexiglas and the ideal means that this assumption is unlikely to be justified for these materials.

A stenosis model is of interest in the development and validation of velocity-based techniques for estimation of the degree of arterial stenosis. Early models were made from acrylic (Morin et al 1987, Douville et al 1983), plexiglas (Fei et al 1988) or heatshrink (McDicken 1986). It has been seen that these materials do not have appropriate acoustic properties, and the resulting spectral distortion is large. An intravascular approach can be used to acquire spectral data from the stenotic region. A wall-less technique may be used, in which the tissue mimic is cast around two rods whose ends are shaped to form the stenosis (Guo and Fenster 1996); however, the absorption of water will distort the stenosis geometry and places a limit on the time over which the model can be used. A stable stenosis model was described by Hoskins (1997), which involved the gluing of stenotic inserts within a latex tube. The inserts were cast from silicone rubber solution, which has a low acoustic velocity of about 1000 ms^{-1}. This allowed the production of asymmetric stenoses in which the insert was located on the back wall of the tube.

A2.2.5 Solid Tissue-mimicking Materials

Solutions of the correct acoustic velocity of 1540 ms^{-1} may be obtained using a 9% glycerol solution or a 10% ethanol solution. It is also possible to use a combination of ethanol, polyethylene glycol and water to obtain a solution with both the desired density and acoustic velocity (Ophir and Jaeger 1992). The simplest attenuative material with approximately tissue-equivalent attenuation is the reticulated foam described by Lerski et al (1982), which is immersed in one of the above fluids. Measurements on different samples show that the attenuation coefficient does not have an exact linear dependence with frequency (Hoskins et al 1994c).

Solid tissue mimics which attempt to match attenuation, velocity and density of tissue may be divided into water-based gels and polymer based materials. Water-based gels are composed of agar (Burlew et al 1980, Rickey et al 1995) or of gelatine (Madsen et al 1982), with particles such as cellulose or graphite added to provide attenuation and to give acoustic speckle on the B-scan image. These materials may be made with acoustic properties corresponding very closely to human tissue (Table A2.5). It has been reported for gelatin-based tissue mimics that greater temperature stability can be obtained by the addition of preservatives, such as gluteraldehyde or formadehyde, without significantly altering the acoustic properties. These may be added to the molten gel (Madsen 1986), or to the set gel. Ryan and Foster (1997) described a vessel phantom suitable for intravascular ultrasound experiments containing two different tissue types whose elastic modulus is matched to normal and hardened arteries. Current polymer materials are based on polyurethane. It is difficult to obtain both the correct acoustic

Table A2.5 Requirements of a standard tissue mimic, and properties of various published tissue mimics

Physical quantity	IEC standard	Madsen et al (1982)	Burlew et al (1980)	Rickey et al (1995)	Teirlinck et al (1998)
Gel base	–	Gelatin	Agar	Agar	Agar
Density (kg m^{-3})	1040 ± 100	980	1060	–	1049
Velocity (m s^{-1})	1540 ± 15	1539	1539	1535	1551
Attenuation (dB cm^{-1} MHz^{-1})	0.5 ± 0.05	0.51	0.85	0.8	0.52

velocity and attenuation at any given frequency; usually one is correct and the other too high. Gel-based tissue-equivalent materials are not stable with time; they lose water if exposed to air, and absorb water if immersed (Fig. A2.7). Polyurethane tissue mimics are more stable, hence they have been commonly used in commercial wall-less systems.

Human tissue has approximately a linear dependence of attenuation coefficient on frequency, whereas for rubbers the dependence is nearer a square law. The distortion of the RF spectrum which this produces compared with human tissue would seem to militate against their use in quantitative experiments involving flow phantoms.

A2.2.6 Cardiac Phantoms

An anthropomorphic cardiac ultrasound phantom has been described by Smith and Rinaldi (1989). This consists of a left ventricle fabricated from polyurethane housed in a larger rigid-walled tank. The fluid in the tank is pumped by a piston pump causing change in volume of the ventricle in a cyclic fashion, with entry and ejection of fluid through prosthetic mitral and aortic valves. Realistic Doppler waveforms were obtained. The design of the left ventricle was later altered to incorporate a polyurethane sponge left ventricle, which gave more realistic B-scan images (Smith et al 1994), and then to replace the artificial left ventricle with an excised porcine heart (Smith et al 1995), which was suitable for B-scan imaging but not flow studies.

A2.2.7 Perfusion Phantoms

When measurements on multiple vessels are of interest then the single-vessel flow phantom is no longer an appropriate model. The requirement of multiple vessels usually refers to Doppler ultrasound measurement of perfusion in microvessels (arterioles, capillaries, venules) whose diameter is less than about 200 μm, and for which the velocity is less than about $3\ mm\ s^{-1}$. Extraction of the Doppler signal is then more difficult than from flow in large vessels, as the velocities are low and the large intensity clutter component from tissue is in the same sample volume as the small intensity signal from blood. Contrast agents and second harmonic imaging have been used, respectively, to increase the Doppler signal strength from the blood, and to reduce the clutter signal strength. Although the exact requirements of a perfusion phantom have not been specified, relevant factors would be the diameter, number, orientation and arrangement of the vessels, the backscatter from the tissue mimic and the blood mimic, and the requirement for non-Newtonian behaviour of the blood mimic.

Perfusion phantoms based on a dialysis filter have been used by Hindle and Perkins (1994) and by Bührer et al (1996). Commercial filters consist of a 2–3 cm bundle of individual tubes made of cellulose, typically with 200 μm diameter and 6–8 μm wall thickness.

A2.3 OTHER MOVING TARGET DEVICES

A2.3.1 String Phantoms

String phantoms have been described by Walker et al (1982), Philips et al (1990), Russell et al (1993) and Lange and Loupas (1996). The main features of these are all similar and are shown in Fig. A2.8. The motor is attached to the drive wheel, and there are one or more other free wheels around which the filament passes. The key element of the phantom is

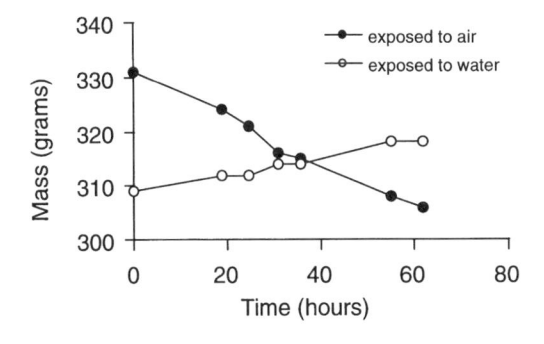

Figure A2.7 Change in weight of a 500 ml sample of tissue mimic formulated according to the recipe of Rickey et al (1995). The sample loses weight if exposed to air and gains weight on exposure to water

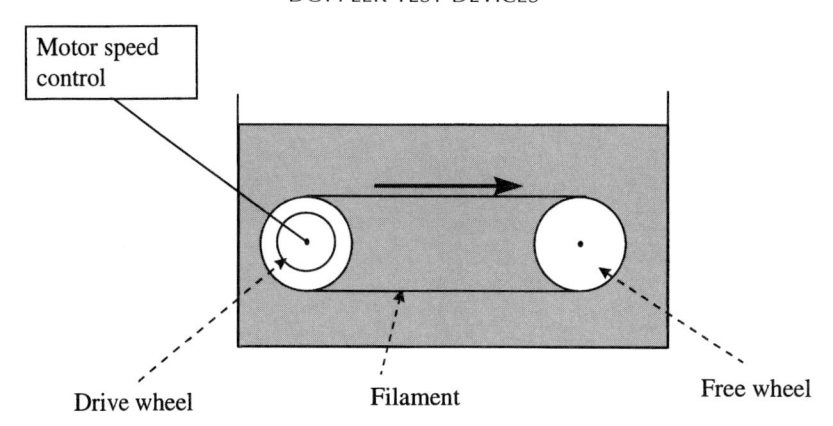

Figure A2.8 Elements of a string phantom

the filament itself. Choice of filament is very important. Braided filaments, such as cotton and silk produce strong backscatter along particular directions (Cathignol et al 1994, Hoskins 1994a). This is due to constructive interference of ultrasound wavelets scattered from the repeating spiral structure of the filament. This is similar to Bragg scattering from a series of point reflectors, in which there is constructive interference at those angles for which the difference in path length from adjacent scatterers is an exact number of wavelengths. This behaviour is contrasted to a non-braided filament, such as O-ring rubber, in which there is no strong angular dependence (Fig. A2.9). The Doppler spectra from a braided and a non-braided filament are shown in Fig. A2.10. There is smooth variation in the brightness levels for the spectra from O-ring rubber; however, there is an intense band of frequencies for the braided filament which corresponds to the peak in angular backscatter.

The phenomenon of angular dependence of backscatter is mainly of concern when the string phantom is used to estimate the error in maximum velocity estimation (Hoskins 1994b). It is known that geometric spectral broadening produces a large degree of overestimation of maximum velocity (Daigle et al 1990, Hoskins et al 1991, Thrush and Evans 1995, Hoskins 1996). However, the presence of the intense band in the spectrum from the braided filament may obscure the true maximum Doppler frequency, leading to an incorrect assessment of the maximum velocity error (Fig. A2.11).

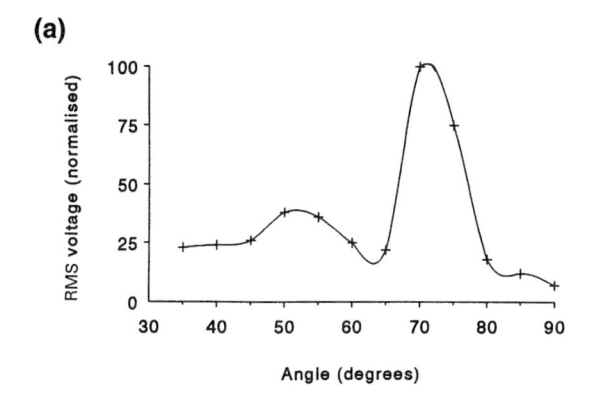

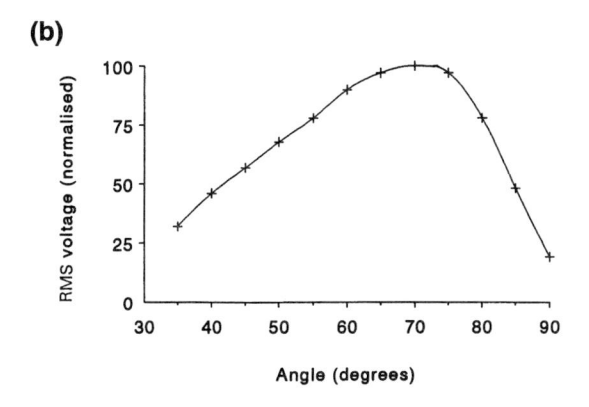

Figure A2.9 Backscattered Doppler amplitude as a function of beam-string angle: (a) surgical silk M2.0; (b) O-ring rubber (reproduced by permission of Elsevier Science, from Hoskins 1994a, *Ultrasound in Medicine and Biology*, © World Federation of Ultrasound in Medicine and Biology)

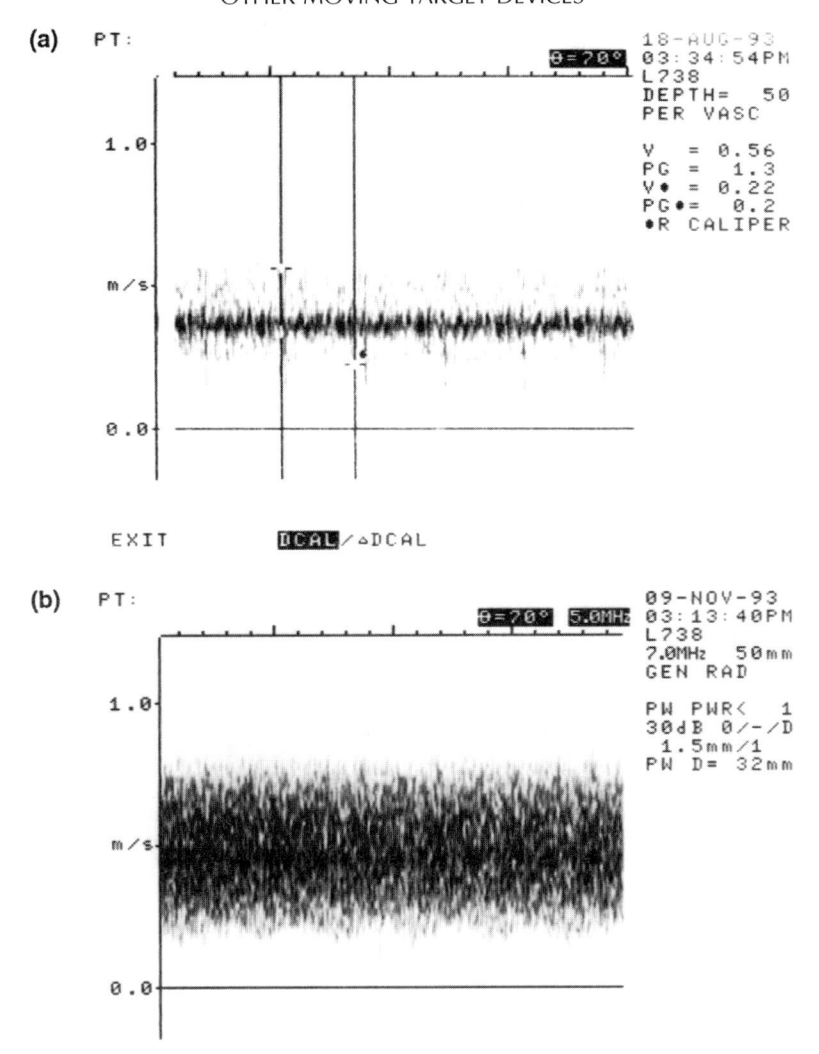

Figure A2.10 Doppler spectra obtained using a 7.5 MHz linear array: (a) surgical silk M2.0; (b) O-ring rubber

For potential use in measurements of sensitivity the change in backscatter from the filament is of interest. The backscatter from braided silk and cotton fall with time, probably because the trapped air is dissolving, whereas that from O-ring rubber is virtually constant over a 24 hour period (Fig. A2.12).

A2.3.2 Belt and Rotating Phantoms

A belt phantom has been described by Rickey et al (1992) in which the moving target is a layer of reticulated foam stitched onto a rubber belt (Fig. A2.13). This provides a moving volume of material which can be used to assess the errors in maximum velocity estimation using spectral Doppler. Its main advantage over the string phantom is that a 2-D colour image may be acquired in which all of the target is moving at the same speed and in the same direction. Transfer of the colour image to the computer enables the precision and accuracy of the colour velocity estimator to be measured.

A follow-up paper showed the device in a modified form (Rickey and Fenster 1996) to

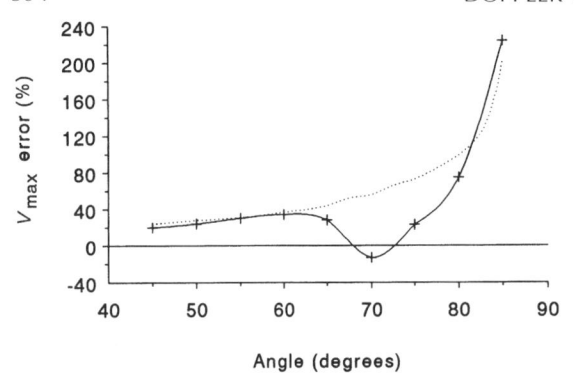

Figure A2.11 Estimated maximum velocity error as a function of beam-string angle, obtained using a 7.5 MHz linear array: solid line, surgical silk M2.0; dotted line, O-ring rubber (reproduced by permission of Elsevier Science, from Hoskins 1994b, *Ultrasound in Medicine and Biology*, © World Federation of Ultrasound in Medicine and Biology)

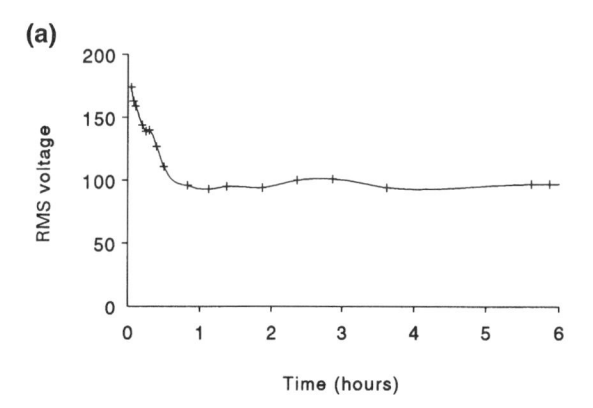

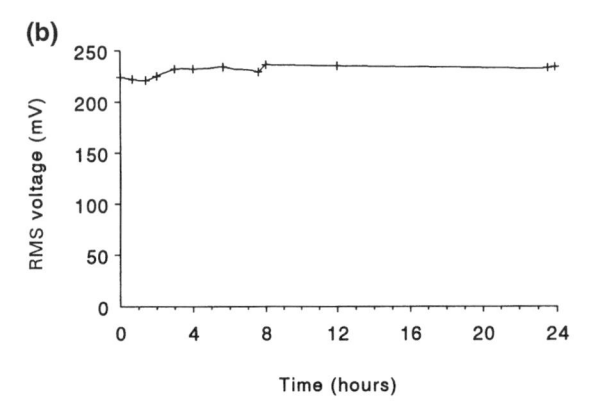

Figure A2.12 Backscattered Doppler amplitude as a function of time: (a) surgical silk M2.0; (b) O-ring rubber (reproduced by permission of Elsevier Science, from Hoskins 1994a, *Ultrasound in Medicine and Biology*, © World Federation of Ultrasound in Medicine and Biology)

enable the investigation of the effect of clutter on the detected Doppler signal from a blood mimic. This used a partially reflective acoustic mirror to produce two sample volumes, enabling combination of Doppler signals from two sources. A clutter signal was obtained by placing one sample volume within a belt of reticulated foam, which was stationary or driven at low velocity. There were two options for the flow signal from the second sample volume. This could be obtained from a second belt moving at higher velocities, with the use of attenuators to adjust the detected ultrasound intensity to be of a similar size to that from blood. Alternatively, a flow phantom could be used.

The rotating phantom is a variant of the belt phantom, in which a circular disc of reticulated foam is rotated around a central axis (McDicken et al 1983). This is technically easier to build than the belt phantom. For a particular rotational speed, the velocities at all points in the phantom can be calculated and used to validate novel velocity measurement techniques such as colour vector Doppler (Hoskins et al 1994a) or Doppler tissue imaging (Fleming et al 1994).

Nowicki et al (1996) have described a simple phantom which allows acquisition of simulated Doppler tissue images from the left ventricle. It consists of a cylinder of sponge fixed to thin rubber. The thin rubber forms a water-filled chamber whose volume may be altered in a cyclical manner by pumping.

A2.3.3 Vibrating Targets

Hoeks et al (1984) described a system for the measurement of the lateral and axial sensitivity plots of the sample volume based on the use of a vibrating ball-bearing. The ball was 0.8 mm in diameter and suspended by wires of 0.01 mm diameter. Vibrations of the order of 1 μm were produced using a loudspeaker attached to the suspending wires.

A vibrating disc was described by Wang et al (1992) which consisted of a diffusely scattering circular plate, 15 cm in diameter, which was made to vibrate at a set audio frequency. The amplitude of oscillation was about 10 μm. A measure of the Doppler sensitivity was derived by assessing the plate drive voltage for which the Doppler signal was just visible on the Doppler spectrum.

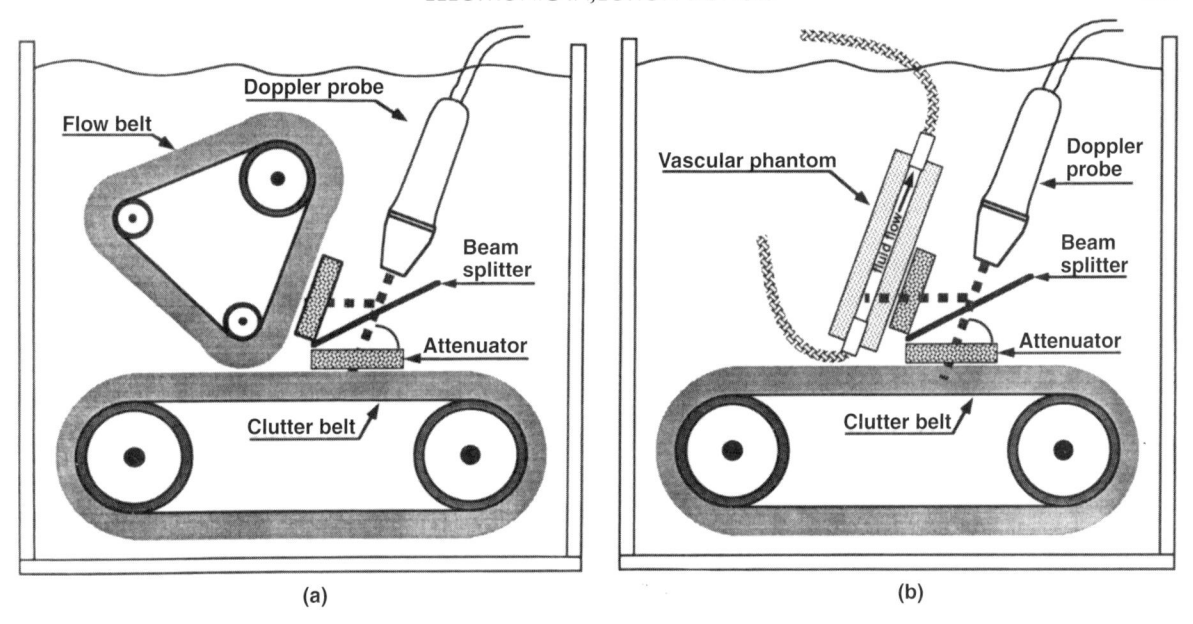

Figure A2.13 Elements of the clutter phantom (reproduced by permission of Elsevier Science, from Rickey and Fenster 1996, *Ultrasound in Medicine and Biology*, © World Federation of Ultrasound in Medicine and Biology)

The main use of a vibrating target has been the assessment of sensitivity of fetal heart rate Doppler detectors. The early work in the area used a spherical target (Ide 1976, JIS-T 1984); however, it has been recognised that the response of a small sphere is highly dependent on frequency. Preston and Bond (1997) studied the frequency dependence of several small targets of different composition and shape. The target of choice was a tungsten carbide rod with a diameter of 1.6 mm tapered to a cylindrical flat end with diameter 0.4–0.6 mm. The reflection loss of this target varied smoothly with frequency and was predictable using a simple theory based on a plane disc reflector, making it a suitable material to meet the requirements of sensitivity measurements described in the IEC standard on fetal Doppler assessment (IEC 1995).

Phillips et al (1997) have described an oscillating thin film test object for making spatial measurements in colour flow systems. A number of precisely deposited subresolvable scatterers are arranged in patterns (vertical and horizontal lines) on a thin film. The film is supported on a frame, which is attached to a modified loudspeaker assembly and aligned to lie in the plane of the ultrasound scan. The loudspeaker is driven by a function generator at a constant frequency of 220 Hz and at a constant amplitude. The images thus generated on the colour flow system are captured digitally for off-line analysis.

A2.4 ELECTRONIC INJECTION DEVICES

In these devices the Doppler signal is synthesised electronically and then injected into the Doppler system. Two realisations of this approach have been described (Fig. A2.14), referred to here as direct injection and acoustic injection.

A2.4.1 Direct Injection

In this approach the signal is directly injected at some point in the signal processing chain. The injected signal could be a synthesised RF signal injected prior to the Doppler detector, or it could consist of two audio channels simulating separated forward and reverse audio signals. The most accessible point for this direct electronic injection is the audio inputs used for playback of the

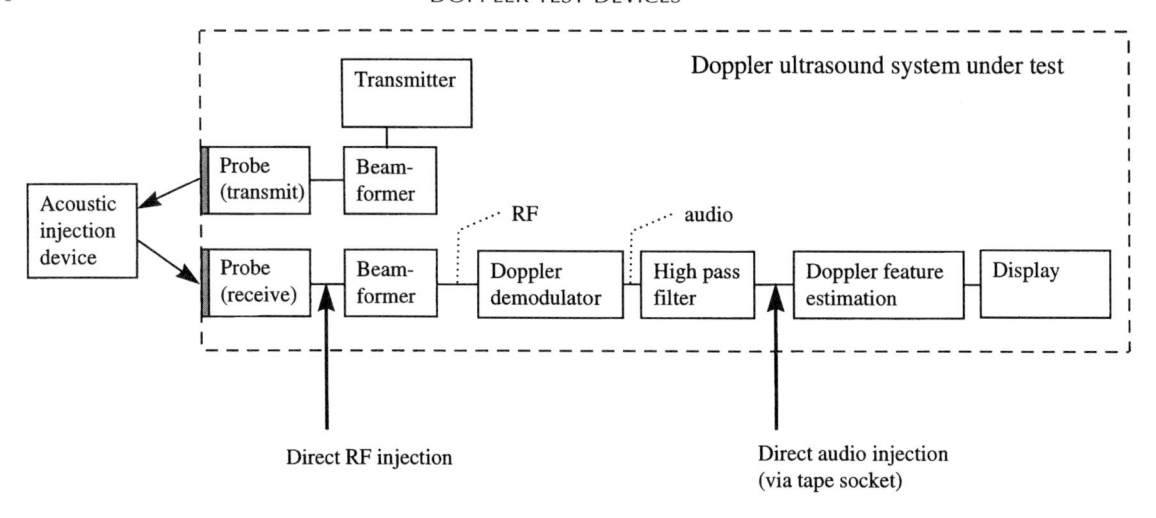

Figure A2.14 Approaches to electronic injection systems; injection can be performed at several points in the system

Doppler signal recorded onto audio or video cassette. The recorded signal is usually a separated forward and reverse flow signal which has already passed through the high pass filter, and the playback audio input signal returns at the same point in the signal processing chain. Direct electronic injection has been described by Sheldon and Duggan (1987), who used a digital white noise source subjected to a voltage-controlled low-pass filter, enabling the production of spectral waveforms with varying degrees of spectral broadening. Bastos and Fish (1991) used a computer to generate the control voltages for the filter and hence produce more complex and physiologically realistic Doppler waveforms.

A2.4.2 Acoustic Injection

In this approach an acoustic signal is produced by a separate transducer which is then detected by the Doppler ultrasound instrument under test. The three approaches to this are described below.

A2.4.2.1 Transmission of Test Signal Without Pulse Capture

By simply transmitting either a continuous wave or pulsed wave signal at a centre frequency which lies within the pass band of the transducer of the system under test, it is possible to produce a Doppler signal. For the detected Doppler signal to be stable, the RF signal generator must be stable to within a few Hz. This method, although unsophisticated, does provide a usable test Doppler signal.

A2.4.2.2 Pulse Capture and Transmission

This technique was originally described by Evans et al (1989), and the essential elements are shown in Fig. A2.15. The pulse transmitted from the scanner under test is captured, mixed with a synthesised audio signal, and the pulse transmitted back to the transducer of the system under test. In its original form, the injected audio signal was a double sideband, hence producing forward and reverse Doppler signals which were identical. A single sideband system has been described by Wallace et al (1993) (Fig. A2.16). The pulse capture and transmission technique is less suitable for colour flow imaging, as only a single colour line at the depth corresponding to the face of the transmission transducer of the injection system is produced.

A2.4.2.3 Pulse Detection and Transmission

An instrument for producing a region of colour, rather than a single line, has been described by Li

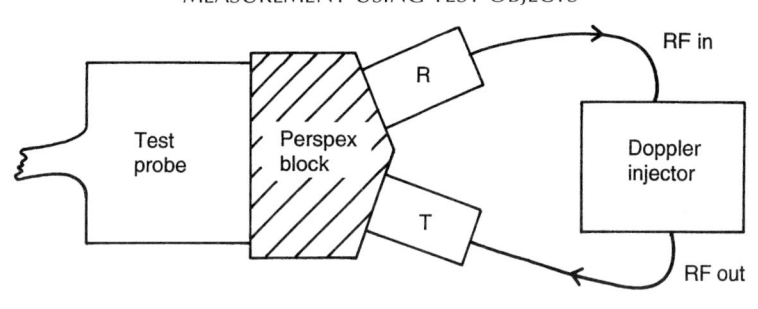

Figure A2.15 Components of the system described by Evans et al (1989)

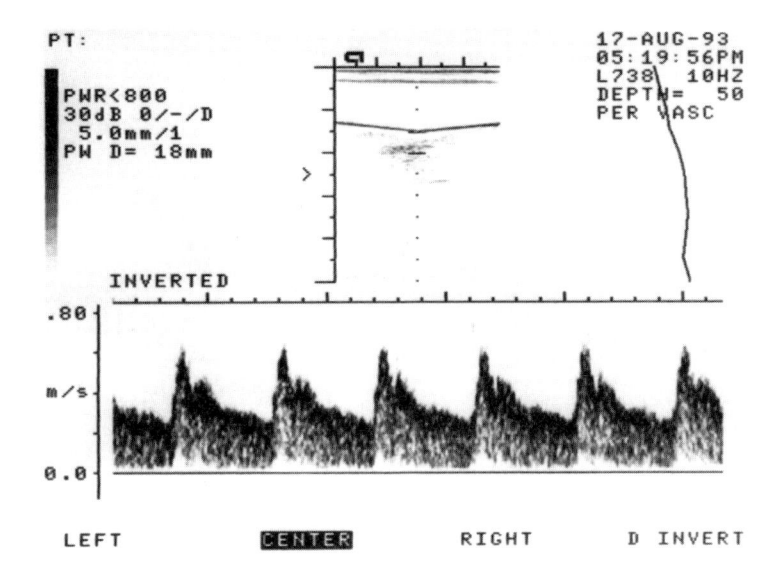

Figure A2.16 Doppler spectra obtained using a single sideband system

et al (1998). This detects the production of an ultrasound pulse from the scanner under test. A longer duration pulse at approximately the same frequency is then produced to which a synthesised audio signal has been added. This produces a colour box on the colour flow image, whose position and length may be controlled by adjusting the onset of transmission and the duration of the synthesised pulse.

A2.5 MEASUREMENT USING TEST OBJECTS

Several types of measurement can be made on Doppler test devices. It is not the purpose of this Appendix to discuss and justify in detail all the different quantities which it would be desirable to measure: for this the reader is referred to other texts (AIUM 1993, Hoskins et al 1994b, IEC 1993, 1995, JIS-T 1984). The application of existing test objects to measurement of several key quantities is discussed below.

A2.5.1 Sample Volume Characteristics

Measurements of the 3-D sample volume shape and size are ideally performed using a point source, hence the vibrating ball-bearing is the most appropriate test tool (Hoeks et al 1984). Measurements may also be made using a line source to give an indication of the dimensions of the sample volume in orthogonal planes. Baker and Yates (1973) used an underwater jet of suspended particles as a line of velocity. The particles passed through distilled water and were captured by an exhaust pipe which ensured minimum mixing of the jet fluid and the containing water. Goldstein (1991) and Hames et al (1991) used a string phantom as a line source. In both cases lateral plots of the sample volume were produced.

Registration of the true sample volume position with the position indicated on the B-scan image is best performed using a thin target, such as the string phantom (Hoskins 1994b). This measurement may also be performed using the electronic injection system (Lunt and Anderson 1993).

A2.5.2 Filter Characteristics

The ability of the Doppler ultrasound device to detect low velocities is related to the design characteristics of the wall thump filter for spectral Doppler and the clutter filter for colour flow. Measurement of the filter characteristics would ideally involve injection of single frequencies prior to the filter, with measurement of signal intensity post-filter (assuming that the filter algorithms are non-adaptive). This is possible on some systems which incorporate analogue circuit design, and for which the circuit diagrams are made available by the manufacturer. For systems in which signal processing is performed digitally, then this is not possible without assistance from the manufacturer. Measurement of the display amplitude may be used as a substitute for the filter output. This would involve image intensity measurements using the ultrasound machine measurement package, if available, or if this was not available, then off-line analysis could be used involving image transfer to computer. If electronic injection cannot be performed, a string phantom can be used; this has a narrow range of frequencies, which could substitute for the single frequency approach. The quality of the measured filter characteristic deteriorates the further the measurement process gets from the ideal.

The effect of the high pass filter may be measured in terms of the minimum velocity that may be detected. The performance of a Doppler system at the low velocity end is related to the velocity of the blood or blood mimic, the design of the high pass filter (wall thump or clutter), the intensity of the Doppler signal from blood or blood mimic and that of the clutter signal. Assessment of minimum detectable velocity requires the use of a tissue-equivalent phantom such as the flow phantom or the modified belt phantom of Rickey and Fenster (1996).

A2.5.3 Velocity Measurement

The precision and accuracy of velocity is of interest in spectral Doppler, where measurements are used for estimation of the degree of arterial stenosis, and in colour Doppler, where higher accuracy is associated with colour images with less noise.

In principle, the flow phantom may be used to assess the accuracy of velocity estimation. The mean velocity V_{mean} and maximum velocity V_{max} of the blood mimic may be estimated from the measured flow rate Q and cross sectional area A, by assuming that the velocity profile is parabolic, in which case $V_{mean} = Q/A$ and $V_{max} = 2Q/A$. This may sometimes be a reasonable assumption to make. Validation of this method using optical particle tracking is possible, provided that the blood mimic is optically transparent; however, many blood mimics are not transparent, which leaves the validity of this method in question. Extreme care must be taken in phantom design to avoid distortion of the Doppler spectrum through mismatch in the acoustic properties of the phantom components, as discussed above. The string or belt phantom may be used to assess velocity measurement errors in spectral Doppler (Fig. A2.17). It is possible to independently estimate string velocity using measurements of string circumference and the time for one revolution of the filament. Modern linear array systems all overestimate maximum velocity as a result of geometric spectral broadening, as has been noted earlier.

Assessment of the variability of maximum velocity estimation under conditions found in clinical practice requires the use of a stenosis model. Figure A2.18 shows that there is a large variation in estimated maximum velocity with beam vessel angle.

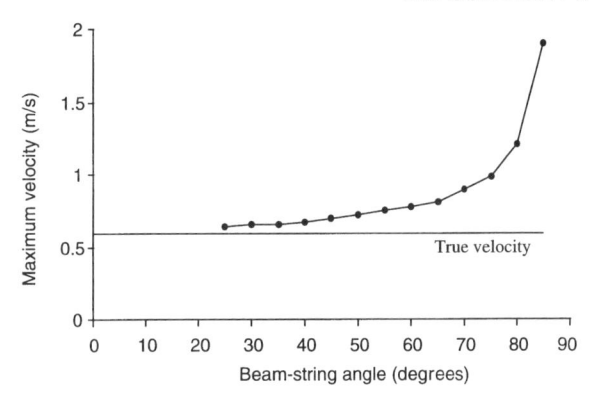

Figure A2.17 Variation of estimated maximum velocity with beam–string angle

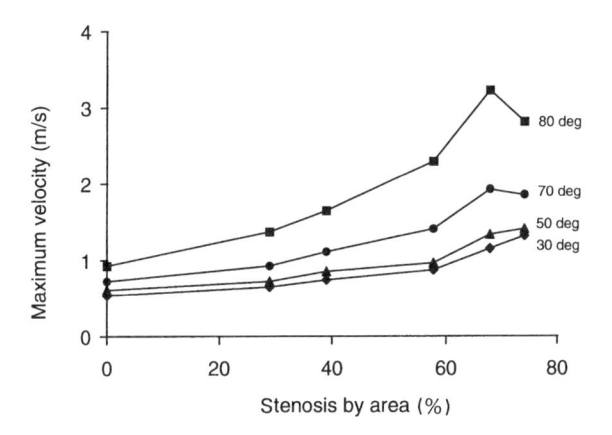

Figure A2.18 Variation of estimated maximum velocity with degree of stenosis and beam–vessel angle

The belt phantom can be used to assess the precision and accuracy of estimated velocity in the colour flow image (Rickey and Fenster 1996) (Fig. A2.19). This requires transfer of the colour image to computer, decoding of the colour image to mean velocity, and calculation of the mean and standard deviation of velocity values.

A2.5.4 Waveform Indices

Physiological Doppler waveforms can be simulated very satisfactorily using the flow phantom (Fig. A2.2). These may be used test the variability of Doppler measurement packages which measure indices, such as resistance index or pulsatility index. Realistic-looking waveforms may also be obtained from the string phantom if the degree of geometric spectral broadening is sufficient (Fig. A2.20), but this approach will never be a substitute for the flow phantom, as there is only a single velocity present within the sample volume at any one time. Electronic injection systems can be used to inject Doppler waveforms acquired onto audio tape from another test device, such as the flow phantom. It is difficult to know whether this is useful, as the saved Doppler waveform will have features of the ultrasound machine used to acquire the data, such as the effect of the high pass filter.

A2.5.5 Sensitivity

Any measure of sensitivity requires a Doppler source that has reproducible, stable backscatter characteristics. This has meant that to date there has not been a good way of measuring Doppler sensitivity. It has been recently shown that it is possible to prepare a blood mimic which has stable backscatter characteristics (Ramnarine et al 1999). Figure A2.21 (see Plate XII) shows measurements of penetration depth made using the blood mimic. A wall-less phantom has been used in which a 4 mm diameter vessel passes diagonally through tissue.

It is possible that measurements of relative sensitivity, adequate for monitoring drift in performance, may be made using other devices, such as the string phantom and disc phantom. The data to support the stability of the backscatter from these sources is not available at the time of writing.

A2.5.6 Image Resolution

The spatial resolution of an imaging device may usually be expressed in terms of the spatial extent of a point source, or the minimum separation for which two targets may be distinguished. Moving point sources, produced by placing acoustic absorber round a vessel, have very small signal intensity. The use of two parallel line sources has been described by Novario et al (1994) using a flow phantom, and by Lange and Loupas (1996) using a

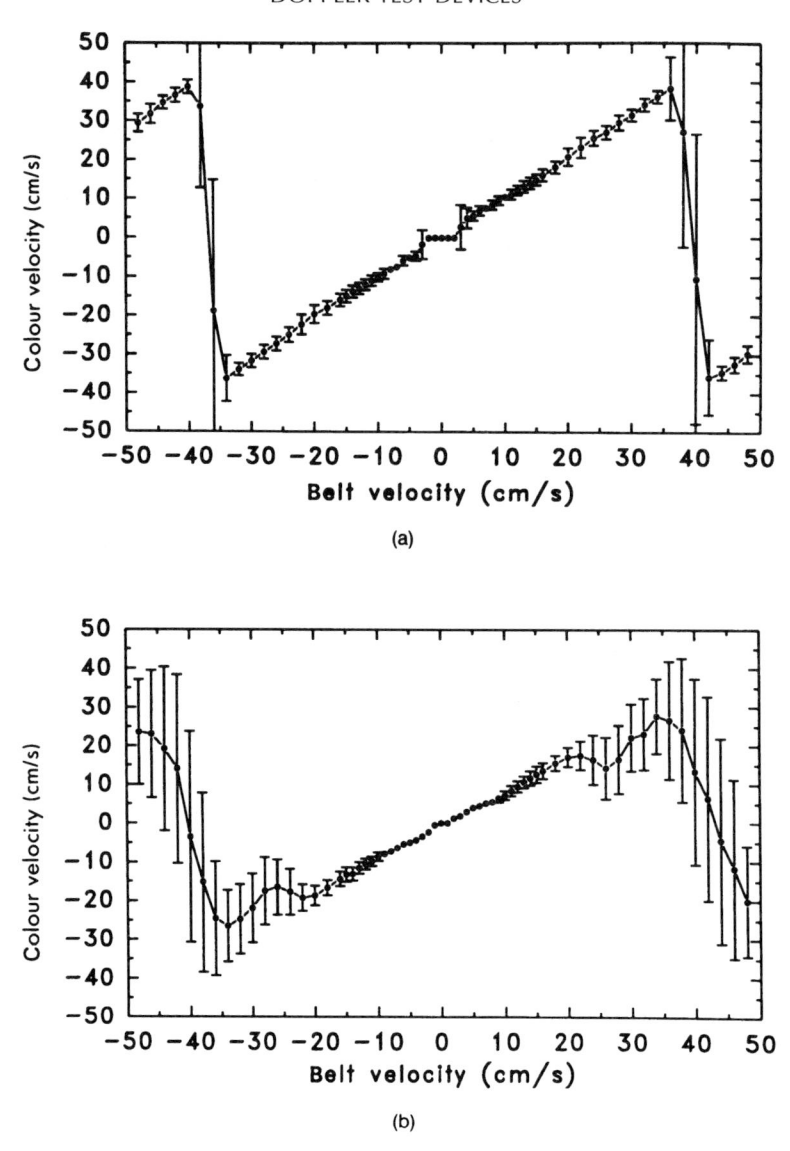

Figure A2.19 Graphs of mean velocity versus true velocity for two-colour flow scanners (reproduced by permission of Elsevier Science, from Rickey et al 1992, *Ultrasound in Medicine and Biology*, © World Federation of Ultrasound in Medicine and Biology)

string phantom. By adjustment of the spacing of the line-sources an indication of the spatial resolution of the colour flow scanner is obtained.

An alternative method was described by Li et al (1997). They used an approach adopted commonly for resolution measurements in diagnostic radiology. A grid was made, consisting of consecutive high- and low-attenuation regions, which when wrapped around the tube of a flow phantom gave consecutive regions of Doppler source and no Doppler source. This produced distinct separated regions of colour on the colour flow image. If the grid spacing was reduced below a critical level, then the colour regions merged together, which was interpreted as being at the resolution limit of the scanner.

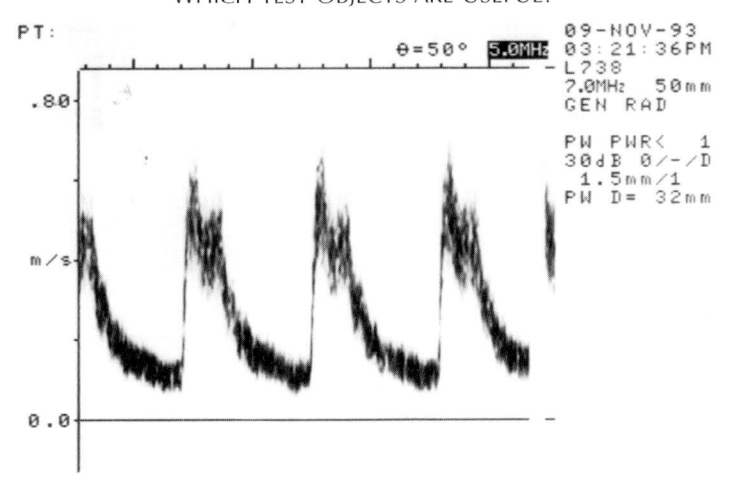

Figure A2.20 Doppler waveforms obtained from the string phantom (reproduced by permission of Elsevier Science, from Hoskins 1994b, *Ultrasound in Medicine and Biology*, © World Federation of Ultrasound in Medicine and Biology)

A method of measuring spatial resolution using an oscillating thin film test object proposed by Phillips et al (1997) has already been described in Section A2.3.3.

A2.6 WHICH TEST OBJECTS ARE USEFUL?

At the time of writing the state-of-the-art tissue-equivalent flow phantom may be obtained using a wall-less design with appropriate choice of tissue mimic and of blood mimic. Remaining problems are concerned with the design of a wall-less tissue mimic resistant to water absorption, and with the design of flow phantom components suitable for high-frequency applications and for applications in very small vessels. The large size of flow phantoms, required because of entry length considerations, means that in general they are large and heavy, restricting them to mainly static laboratory use rather than as portable quality assurance devices. The design area which has received least attention, and which remains the most problematic, is air bubble contamination of the blood mimic. Stringent measures must be taken to degas the blood mimic, by boiling and cooling of water if possible or by exposure to vacuum. Air bubbles may be sucked in at joints in the circuit, and may be produced by cavitation in certain pumps in which large negative pressures are produced, such as the piston or gear pump. The highly tissue-equivalent nature of the flow phantom means that it will always have a place in *in vitro* scientific development work.

Other moving target systems are simpler in design, and with this there is a more limited tissue equivalence. Of these, the string phantom has been most extensively investigated in the literature, and the uses and limitations of the target material are well characterised. The portability of the string phantom makes it ideally suited for quality assurance.

Electronic injection systems are a significantly different class of instruments compared with the moving target instruments above. They have a number of attractive features. There are no moving parts, they are small in scale as there is no large test tank, they are not messy as there is no bulk quantity of fluid. In principle, it is possible to inject signals for which the spectral content (both RF and audio) is known to a high degree of accuracy. The inherent stability of the electronic signal can be very high, which is a significant consideration when attempting to measure sensitivity. Against these are several fundamental limitations. Only aspects of the Doppler system after the point of injection can be tested. No account is taken of the ultrasound beam shape and sample volume characteristics. The Doppler signal from moving targets is strongly influenced by the degree of geometric spectral broadening present, and is critically dependent on the shape, size and location of the sample volume within the velocity field. Many of the quantities it would be desirable to test are influenced by these effects, including all indices related to measured Doppler frequency and measured velocity. Modern ultrasound scanners have sophisticated complex pulse-firing regimes and have detailed control of the RF pulse spectrum. It then becomes increasingly difficult to design an electronic injection system matched to an individual scanner, both in terms of obtaining the required knowledge from the manufacturer and in producing a system which will work with a range of scanners. The lack of a moving target means that it is necessary to prove that measurements made using electronic injection systems are correlated with similar measurements made using a tissue-equivalent test device, such as the flow phantom; this information is not available to date. The electronic injection system may be more appropriate for determination of the electronic properties of the system, such as the high pass filter characteristics; however, this would probably be best performed by direct injection at the appropriate point in the signal processing chain. It is possible that the limitations of the electronic injection approach are so great that they outweigh any potentially useful features.

A2.7 CONCLUSION

It is clear that the highly tissue-equivalent nature of the flow phantom means that it will always have a place in *in vitro* scientific development work. For quality assurance the weight and complexity of the flow phantom makes it unattractive, compared to the string phantom, which is light and which may be used to make a range of measurements. The role of electronic injection systems in the testing of Doppler ultrasound systems is currently far from clear.

A2.8 REFERENCES

AIUM Standards Committee (1993) Performance criteria and measurements for Doppler ultrasound devices. American Institute of Ultrasound in Medicine, Laurel, MD

Baker DW, Yates WG (1973) Technique for studying the sample volume of ultrasonic Doppler devices. Med Biol Eng II:766–770

Bastos CAC, Fish PJ (1991) A Doppler signal simulator. Clin Phys Physiol Meas 12:177–183

Boote EJ, Zagzebski JA (1988) Performance tests of Doppler ultrasound equipment with a tissue and blood mimicking phantom. J Ultrasound Med 7:137–147

Bührer A, Moser UT, Schumacher PM, Pasch T, Anliker M (1996) Subtraction procedure for the registration of tissue perfusion with Doppler ultrasound. Ultrasound Med Biol 22:651–658

Burlew MM, Madsen EL, Zabzebski JA, Banjovc RA, Sum SW (1980) A new tissue equivalent material. Radiology 134:517–520

Cathignol D, Dickerson K, Newhouse VL, Faure P, Chapelon J-Y (1994) On the spectral properties of Doppler thread phantoms. Ultrasound Med Biol 20:601–610

Daigle RJ, Stavros AT, Lee RM (1990) Overestimation of velocity and frequency values by multielement linear array Dopplers. J Vasc Tech 14:206–213

Douville Y, Johnston KW, Kassam M, Zuech P, Cobbold RSC, Jares A (1983) An *in vitro* model and its application for the study of carotid Doppler spectral broadening. Ultrasound Med Biol 9:346–356

Duck FA (1990) Acoustic Properties of Tissue: A Comprehensive Reference Book. Academic Press, London

Evans JA, Price R, Luhana F (1989) A novel testing device for Doppler ultrasound equipment. Phys Med Biol 34:1701–1708

Fei DY, Billan C, Rittgers SE (1988) Flow dynamics in a stenosed carotid artery bifurcation model. Part 1. Basic velocity measurements. Ultrasound Med Biol 14:21–31

Fish P, Steel R (1996) EC project 'Validation of a flow Doppler test object for diagnostic ultrasound scanners'. Summary report on phase 1. University of Wales, Bangor

Fleming AD, McDicken WN, Sutherland GR, Hoskins PR (1994) Assessment of colour Doppler tissue imaging using test-phantoms. Ultrasound Med Biol 20:937–951

Frayne R, Gowman LM, Rickey DW, Holdsworth DW, Picot PA, Drangova M, Chu KC, Caldwell CB, Fenster A, Rutt BK (1993) A geometrically accurate vascular phantom for comparative studies of X-ray, ultrasound, and magnetic resonance vascular imaging: construction and geometrical verification. Med Phys 20:415–425

Goldstein A (1991) Performance tests of Doppler ultrasound equipment with a string phantom. J Ultrasound Med 10:125–139

Guo Z, Fenster A (1996) Three-dimensional power Doppler imaging: a phantom study to quantify vessel stenosis. Ultrasound Med Biol 22:1059–1069

Hames TK, Nelligan BJ, Nelson RJ, Gazzard VM, Roberts J (1991) The resolution of transcranial Doppler scanning: a method for *in vitro* evaluation. Clin Phys Physiol Meas 12:157–161

Hein IA, O'Brien WD (1992) A flexible blood flow phantom capable of independently producing constant and pulsatile flow with a predictable spatial flow profile for ultrasound flow measurement validation. IEEE Trans Biomed Eng 39:1111–1122

Hindle AJ, Perkins AC (1994) A perfusion phantom for the evaluation of ultrasound contrast agents. Ultrasound Med Biol 20:309–314

Hoeks APG, Ruissen CJ, Hick P, Reneman RS (1984) Methods to evaluate the sample volume of pulsed Doppler systems. Ultrasound Med Biol 10:427–434

Holdsworth DW, Rickey DW, Drangova M, Miller DJM, Fenster A (1991) Computer-controlled positive displacement pump for physiological flow simulation. Med Biol Eng Comput 29:565–570

Hoskins PR (1994a) Choice of moving target for a string phantom: I. Measurement of filament backscatter characteristics. Ultrasound Med Biol 20:773–780

Hoskins PR (1994b) Choice of moving target for a string phantom: II. On the performance testing of Doppler ultrasound systems. Ultrasound Med Biol 20:781–789

Hoskins PR (1994c) Review of the design and use of flow phantoms. In: Hoskins PR, Sherriff SB, Evans JA (eds) Testing Doppler Ultrasound Equipment, pp. 12–29. Institute of Physical Sciences in Medicine, York

Hoskins PR (1996) Accuracy of maximum velocity estimates made using Doppler ultrasound systems. Br J Radiol 69:172–177

Hoskins PR (1997) Peak velocity estimation in arterial stenosis models using colour vector Doppler. Ultrasound Med Biol 23:889–897

Hoskins PR, Anderson T, McDicken WN (1989) A computer controlled flow phantom for generation of physiological Doppler waveforms. Phys Med Biol 34:1709–1717

Hoskins PR, Loupas T, McDicken WN (1990) A comparison of the Doppler spectra from human blood and artificial blood used in a flow phantom. Ultrasound Med Biol 16:141–147

Hoskins PR, Li SF, McDicken WN (1991) Velocity estimation using duplex scanners. Ultrasound Med Biol 17:L195-L199

Hoskins PR, Fleming A, Stonebridge P, Allan PL, Cameron D (1994a) Scan-plane vector maps and secondary flow motions in arteries. Eur J Ultrasound 1:159–169

Hoskins PR, Sherriff SB, Evans JA (eds) (1994b) Testing of Doppler Ultrasound Equipment. Institute of Physical Sciences in Medicine, York

Ide M (1976) Steel ball method for measurement of overall sensitivity of ultrasonic diagnostic equipment. Jpn J Med Ultrasonics 3:45–52

IEC (1993) IEC 1206: Ultrasonics—Continuous-wave Doppler Systems—Test Procedures. International Electrotechnical Commission, Geneva

IEC (1995) IEC 1266: Ultrasonics—Hand-held Probe Doppler Fetal Heartbeat Detectors—Performance Requirements and Methods of Measurement and Reporting. International Electrotechnical Commission, Geneva

IEC (draft standard) (in preparation): Ultrasonics—Flow Measurement Systems—Flow Test Object. International Electrotechnical Commission, Geneva, Switzerland

Issartier P, Siouffi M, Pelissier R (1978) Simulation of blood flow by a hydrodynamic generator. Med Prog Technol 6:39–40

Janssens JL, Raman ER, Vanhuyse VJ (1989) A pressure-controlled fluid flow generator for arterial blood-flow simulation. J Med Eng Technol 13:104–108

JIS-T (1984) Japan Industrial Standard 1506: Ultrasonic Doppler fetal diagnostic equipment

Ku DN, Giddens DP, Phillips DJ, Strandness DE (1985) Hemodynamics of the normal human carotid bifurcation: *in vitro* and *in vivo* studies. Ultrasound Med Biol 11:13–26

Lange GJ, Loupas T (1996) Spectral and color Doppler sonographic applications of a new test object with adjustable moving target spacing. J Ultrasound Med 15:775–784

Law YF, Cobbold RSC, Johnstone RW, Bascom PAJ (1987) Computer-controlled pulsatile pump system for physiological flow simulation. Med Biol Eng Comput 25:590–595

Law YF, Johnston KW, Routh HF, Cobbold RSC (1989) On the design and evaluation of a steady flow model for Doppler ultrasound studies. Ultrasound Med Biol 15:505–516

Lerski RA, Duggan TC, Christie J (1982) A simple tissue-like ultrasound phantom material. Br J Radiol 55:156–157

Li S, Hoskins PR, McDicken WN (1997) Rapid measurement of the spatial resolution of colour flow scanners. Ultrasound Med Biol 23:591–596

Li SF, Hoskins PR, Anderson T, McDicken WN (1998) An acoustic injection system for colour flow imaging systems. Ultrasound Med Biol 24:161–164

Lowe GDO (ed) (1988) Clinical Blood Rheology, vol 1. CRC Press, Boca Raton, FL

Lunt MJ, Anderson R (1993) Measurements of Doppler gate length using signal re-injection. Phys Med Biol 38:1631–1636

Madsen EL (1986) Ultrasonically soft tissue mimicking materials and phantoms. pp. 165–181. In: Greenleaf JF (ed) Tissue Characterisation with Ultrasound, vol 1—Methods. CRC Press, Boca Raton, FL

Madsen EL, Zagzebski JA, Frank GR (1982) Oil in gelatin dispersions for use as ultrasonically tissue mimicking materials. Ultrasound Med Biol 8:277–287

McCarty K, Locke DJ (1986) Test objects for the assessment of the performance of Doppler shift flowmeters. pp. 94–106. In: Evans JA (ed) Physics in Medical Ultrasound. Institute of Physical Sciences in Medicine, London

McDicken WN (1986) A versatile test-object for the calibration of ultrasonic Doppler flow instruments. Ultrasound Med Biol 12:245–249

McDicken WN, Morrison DC, Smith DSA (1983) A moving tissue-equivalent phantom for ultrasonic real-time scanning and Doppler techniques. Ultrasound Med Biol 9:L455-L459

Michie DD, Fried WI (1973) An *in vitro* test medium for evaluation clinical Doppler ultrasonic flow systems. J Clin Ultrasound 1:130–133

Morin JF, Johnston KW, Law YF (1987) *In vitro* study of continuous wave Doppler spectral changes resulting from stenoses and bulbs. Ultrasound Med Biol 13:5-13

Novario R, Goddi A, Crespi A, Conte L (1994) A new phantom for quality assurance of color-coded ultrasound flow equipment. Physica Medica 10:101–106

Nowicki A, Olszewski R, Etienne P, Karlowicz P, Adamus J (1996) Assessment of wall velocity gradient imaging using a test phantom. Ultrasound Med Biol 22:1255–1260

Oates CP (1991) Towards an ideal blood analogue for Doppler ultrasound phantoms. Phys Med Biol 36:1433–1442

Ophir J, Jaeger P (1982) A ternary solution for independent acoustic impedance and speed of sound matching to biological tissues. Ultrason Imaging 4:163–170

Pedley TJ (1980) The Fluid Mechanics of Large Blood Vessels. Cambridge University Press, Cambridge

Peterson JN (1984) Digitally controlled system for reproducing blood flow waveforms *in vitro*. Med Biol Eng Comput 22:277–280

Philips DJ, Hossack J, Beach KW, Strandness DE (1990) Testing ultrasonic pulsed Doppler instruments with a physiologic string phantom. J Ultrasound Med 9:426–436

Phillips D, McAleavey S, Parker KJ (1997) Color Doppler spatial resolution measurements with an oscillating thin film test object. In: Schneider SC, Levy M, McAvoy BR (eds) Proceedings of the 1997 IEEE Ultrasonics Symposium, pp. 1517–1520. IEEE, Piscataway, NJ

Preston RC, Bond AD (1997) An experimental study of the reflection from spherical and flat ended cylindrical targets suitable for fetal Doppler performance assessment. Ultrasound Med Biol 23:117–128

Ramnarine KV, Nassiri DK, Hoskins PR, Lubbers J (1998) Validation of a new blood mimicking fluid for use in Doppler flow test objects. Ultrasound Med Biol 24:451–459

Ramnarine KV, Hoskins PR, Routh HF, Davidson F (1999) Doppler backscatter properties of a blood-mimicking fluid for Doppler performance assessment. Ultrasound Med Biol 25:105–110

Rickey DW, Fenster A (1996) A Doppler ultrasound clutter phantom. Ultrasound Med Biol 22:747–766

Rickey DW, Rankin R, Fenster A (1992) A velocity evaluation phantom for colour and pulsed Doppler instruments. Ultrasound Med Biol 18:479–494

Rickey DW, Picot PA, Christopher DA, Fenster A (1995) A wall-less vessel phantom for Doppler ultrasound studies. Ultrasound Med Biol 21:1163–1176

Russell SV, McHugh D, Moreman BR (1993) A programmable Doppler string test object. Phys Med Biol 38:1623–1630

Ryan LK, Foster FS (1997) Tissue equivalent vessel phantoms for intravascular ultrasound. Ultrasound Med Biol 23:261–273

Selfridge AR (1985) Approximate material properties in isotropic materials. IEEE Trans Sonics Ultrason SU-32:381–394

Sheldon CD, Duggen TC (1987) Low-cost Doppler signal simulator. Med Biol Eng Comput 25:226–228

Shortland AP, Cochrane T (1989) Doppler spectral waveform generation *in vitro*: an aid to diagnosis of vascular disease. Ultrasound Med Biol 15:737–748

Smith SW, Rinaldi JE (1989) Anthropomorphic cardiac ultrasound phantom. IEEE Trans Biomed Eng 36:1055–1058

Smith SW, Combs MP, Adams DB, Kissol JA (1994) Improved cardiac anthropomorphic phantom. J Ultrasound Med 13:601–605

Smith SW, Lopath PD, Adams DB, Walcott GP (1995) Cardiac ultrasound phantom using a porcine heart model. Ultrasound Med Biol 21:693–697

Teirlinck CJPM, Bezemer RA, Kollman C, Lubbers J, Hoskins PR, Fish P, Fredfeldt K-E, Schaarschmidt UG (1998) Development of an example flow test object and comparison of five of these test objects in various laboratories. Ultrasonics 36:653–660

Thompson RS, Aldis GK (1996) Effect of a cylindrical refracting interface on ultrasound intensity and the CW Doppler spectrum. IEEE Trans Biomed Eng 43:451–459

Thompson RS, Aldis GK, Linnett IW (1990) Doppler ultrasound spectral power density distribution: measurement artefacts in steady flow. Med Biol Eng Comput 28:60–66

Thrush AJ, Evans DH (1995) Intrinsic spectral broadening: a potential cause of misdiagnosis of carotid artery disease. J Vas Invest 1:187–192

Tortoli P, Berti P, Guidi F, Thompson RS, Aldis GK (1997) Flow imaging with pulsed Doppler ultrasound: refraction artefacts and dual mode propagation. In: Schneider SC, Levy M, McAvoy BR (eds) Proceedings of the 1997 IEEE Ultrasonics Symposium, pp. 1269–1272. IEEE, Piscataway, NJ

Walker AR, Phillips DJ, Powers JE (1982) Evaluating Doppler devices using a moving string test target. J Clin Ultrasound 10:25–30

Wallace JJA, Martin K, Whittingham TA (1993) An experimental single-sideband acoustical re-injection test method for Doppler systems. Physiol Meas 14:479–484

Wang KY, Bone SN, Hossack JM (1992) A tool for evaluating Doppler sensitivity. J Vasc Tech 16:87–94

Wemeck MM, Jones NB, Morgan J (1984) Flexible hydraulic simulator for cardiovascular studies. Med Biol Eng Comput 22:86–89

Appendix 3

Recording and Reproduction of Doppler Signals and Colour Doppler Images

Contributed by T Anderson

Medical Physics Department, Royal Infirmary, Edinburgh EH3 9YW, UK

A3.1 INTRODUCTION

There are now a multitude of options available for recording and reproducing Doppler audio signals and colour Doppler images. The purpose of this appendix is to summarise some of these options to assist the reader in making an appropriate choice for his or her particular application. To this end the types of devices available for each application area are discussed and basic specifications provided for typical units. Mention is also made of their limitations where applicable. Some information regarding relatively new and untried recording technologies (for this application area) has also been included in an attempt to extend the usefulness of this survey. Ignoring costs, the final choice of recording device whether for audio signals or colour images, is dependent on the application and on the use to which the recordings will be put. Careful analysis of the requirements for each application will pay dividends.

The following points, although not exhaustive can be used prior to purchase as the basis for a checklist to assist in reaching a final decision on which device or devices will best meet a particular requirement.

General points

- How many recordings will be required for each study? (As this number increases, so the need for a convenient recording/processing/analysis system increases.)
- What ease of access to individual recordings is required? (Random or sequential access.)
- Will a combination of audio and video recordings be required?
- How will the results be analysed?
- How will the results of this analysis be recorded or stored?

Doppler signals

- What is the likely frequency content of the signals?
- How is the directional information encoded?
- What duration of audio recordings will be required: short (a few seconds); long (a few minutes); or extended periods (tens of minutes)?
- What aspect of the recorded signal will form the basis of further analysis?
- Will an analogue tape-recording be sufficient or will a digital recording be of advantage?

Images

- What type of images are to be recorded: grey-scale or colour?
- What is the likely number of images involved: single images; a series of images; or cine loops?
- What quality of recording will be required?
- Will image processing be involved?
- Will a hard copy of images be required? If so, of what size?

Miscellaneous points

- Will results be required in 'real time'?
- What recording and hard copy devices come with or are already connected to the ultrasound machine?

- To what extent can or will analysis be carried out on the ultrasound machine?
- Will copies be required?
- Will backup be necessary?
- Will distribution of recordings be required?

With a clear understanding of what is to be recorded and why, the next step is to decide on the type of device or devices required. Once this decision has been made it is then possible to select the specific model from those available from the various manufacturers or suppliers.

The following sections briefly describe the characteristics of the Doppler audio signal and then a number of recording and hard copy options available for both audio and video data, together with pointers towards their applications and limitations. Some basic specifications are also given as a rough guide to the capabilities of devices. However, the manufacturers' specification or data sheets should be consulted for full performance details.

A3.2 CHARACTERISTICS OF DOPPLER AUDIO SIGNALS

It is a fortunate coincidence that most ultrasonic Doppler signals found in medical applications lie within the range of 100 to 10 kHz. It is fortunate firstly because since this corresponds to a large degree to the range of human hearing (normally regarded as being approximately 20 Hz to 20 kHz), Doppler signals may be easily appreciated and even partially interpreted simply by listening to them. More fortunate, perhaps, is that since the Doppler shift range corresponds to that of human hearing, there is a mass market for devices that can process, record and play back such signals cheaply and fairly accurately. Nevertheless, there is need for some caution when recording Doppler signals with standard audio equipment, and a number of issues need to be kept in mind. Of particular concern are the frequency range, frequency response and the way in which directional information is encoded.

In terms of frequency range, as mentioned above, most Doppler audio signals fall well within the nominal range of standard audio equipment; however, this is not the case when particularly high transmitted frequencies are used in situations where relatively rapid velocities may be encountered. One example of this is the use of catheter-tipped transducers which are inserted directly into the artery, and which may use carrier frequencies of 20–30 MHz or even higher. Such catheter-tipped probes also often point directly upstream or downstream, which exacerbates the problem because the Doppler angle is approximately 0°, and therefore the cosine term in the standard Doppler equation is approximately unity. Under such circumstances the Doppler shift may easily exceed 30 kHz (Nakatani et al 1992), which means that standard audio frequency recorders are unsuitable and that special strategies have to be adopted to deal with such signals (Moraes et al 1995).

With regard to frequency response, it is important to bear in mind that the ear is essentially a logarithmic sensor which has to deal with a vast dynamic range of signal intensities, and that it is relatively insensitive to small perturbations in the frequency response of instruments like tape recorders. Because of this, frequency responses which might be regarded as 'flat' in domestic hi-fi terms cannot be regarded as flat if detailed analysis of a Doppler signal is required. To give an example, high quality analogue audio cassette recorders may have a frequency response quoted as $30-16$ kHz ± 3.5 dB, which means that although the response is regarded as relatively flat over this range, the playback amplitude may vary by up to $\pm 50\%$ in linear terms. Clearly, the significance of such a variation in a recorded Doppler signal would depend on what information is required from it, and in some applications (perhaps waveform analysis) might not be important, whilst in others (perhaps flow measurement or embolus detection) might be critical. Fortunately, as will be seen, digital recorders tend to perform significantly better in this respect.

Finally, it is a straightforward matter to record directional Doppler signals if the forward and reverse channels are completely separate (although it is of course important that the two recording channels are correctly balanced, which is not always easy because often the recording levels for the two channels are independent, and for the reasons mentioned in the last paragraph exact balance is not of paramount importance in domestic situations). However, if the signals are a phase quadrature pair (see Chapter 6), then there may be considerable difficulties. The reason for this is that since the directional information is encoded in the phase relationship between the direct and quadrature channel, any slight relative phase shift between the two channels causes considerable cross-talk problems. This issue has been discussed in relation to analogue tape recorders by Smallwood (1985), and in relation to digital audio tape recorders by Bush and Evans (1993), and both studies showed the instruments tested to be unsuitable for storing quadrature signals. Thus if a Doppler device only presents the audio output as a quadrature pair, conversion to a directional format should take place before recording. Coghlan and Taylor (1978) and Aydin et al (1994), respectively, have described analogue and digital techniques for achieving this.

A3.3 RECORDING DOPPLER AUDIO SIGNALS

There are essentially only two methods of recording audio signals, one based on analogue techniques, the other on digital. Whilst analogue audio signals are almost exclusively stored on tape, digital information may be stored using a number of different technologies.

A3.3.1 Analogue Recording Systems

Whilst it is possible to record short periods of audio signal direct to a semiconductor device, magnetic tape remains the only practical medium on which to record analogue signals of any significant duration.

A3.3.1.1 Audio Cassette Deck

Where the recording of audio Doppler signals alone is required, using an audio cassette recorder (ACR) has attractions due to the low cost and simplicity of such devices. However, logging the details of each examination must rely on the use of the tape indicator, a method which is prone to errors. The use of voice dubbing for record labelling purposes is also a problem, since few of these devices include a microphone input. They are also limited due to the sequential nature of the recordings, particularly as the number of recordings per tape increases. Locating specific records can be made easier if blank sections are recorded between each record. This allows the use of the 'Skip Search' feature to locate the beginning of each record.

ACRs are not suitable for applications where it is important to retain the exact spectral content of the Doppler signal (see Section A3.2) but nevertheless are probably suitable in some less demanding situations. Under such circumstances a mid-range machine from a well-known manufacturer is likely to be adequate, and the final selection is likely to be based on available features, such as searching facilities and general ease of use. Note that noise reduction systems such as 'Dolby' can alter the spectral content of the audio signal, and are probably best disabled when recording Doppler signals.

Typical specifications:
Channels	2
Frequency response	Normal tape 20 Hz–18 kHz (± 3.5 dB) Chrome 20 Hz–19 kHz (± 3.5 dB) Metal 20 Hz–20 kHz (± 3.5 dB)
Signal-to-noise ratio	70 dB max
Total harmonic distortion	0.8%
Record access	Digital electronic position indication and skip control

If recording of images is also required, then a video cassette recorder may be able to satisfy both requirements since it is able to record both images and audio Doppler signals.

A3.3.2 Digital Recording Systems

Digital recording of audio signals offers distinct advantages over analogue systems such as:

- Generally higher quality recordings.
- No degeneration with storage.

- Random access to recordings.
- Simplified editing.
- Increased options for analysis.
- Potentially lossless copying of information (valuable for back-up or distribution).
- Possibility of direct digital transfer of information from the Doppler instrument to a computer system.

A wide choice of methods exists for digitising audio Doppler signals:

- Digital audio tape.
- Personal computer based system.
- Sound card.
- Compact disk-recordable.
- Analogue/digital capture card.
- CD recorder.
- Mini-disk recorder.
- Digital versatile disks.

Each of these methods has the potential to offer superior quality recording when compared to ACRs and hi-fi VCRs.

A3.3.2.1 Digital Audio Tape

Digital audio tape (DAT) recorders available from specialist hi-fi dealers have a much higher specification than analogue devices. They also have additional facilities such as the ability to record 'sub-code signals' which may be used to identify the beginning of recordings and/or the date and time at which the recordings are made. Although not random access devices, their ×200–×400 fast forward and rewind allows fast access to individual records with a maximum seek time of 27 seconds on a 2 hour tape.

DAT machine performance is on a par with CDs with a wide frequency response and a much flatter performance curve than an ACR. DAT recorders are now often considered the instruments of choice for capturing Doppler audio signals where faithful reproduction is required. Bush and Evans (1993) have reviewed the merits and demerits of DAT as a method of storing Doppler ultrasound audio signals.

Typical specifications:
Channels	2
Recording time	2 hours
Quantisation	16 bits/sample
Sampling rate	48 kSps
Frequency response	10 Hz–20 kHz (±0.5 dB)
Dynamic range	92 dB
SNR ratio	92 dB
Wow and flutter	Unmeasurable

A3.3.2.2 PC-based Systems

Virtually all modern PCs can be configured to operate as a high specification data capture system using standard plug-in cards and data storage devices.

Sound cards. Although developed for multi-media use, sound or audio cards have wider applications. In effect, a sound card comprises a very powerful audio digitiser and playback system able to digitise two channels at up to 55 kSps with a dynamic range of around 85 dB. Considering the specifications, they are available at incredibly low cost. The digitised signals can be recorded direct to the hard disk of the host PC on a continuous basis. The recording time is limited by the space available on the system's hard disk but equates to over 1 hour recording in 1 GB of memory. Data thus collected can be easily manipulated and analysed using readily available software or specific analysis methods, developed as required. With speakers connected the card is also able to reproduce the captured signals as audio by means of the on-board digital to analogue converters and amplifiers.

Typical specifications:
Channels	2
Recording time	Dependent on available storage space
Sampling rate	Variable to 55 kHz
Quantisation	16 bits/sample
Frequency response	15 Hz–20 kHz
Dynamic range	90 dB

CD-recordable/CD-rewritable. Storing analogue audio data in digital format over the course of even a small study requires enormous amounts of storage space. At a rate of approximately 1 GB per hour, even the largest hard disk drive would soon be full. With the advent of the Compact Disk-Recordable (CD-R), however, storing large amounts of data in a convenient, robust, lossless, random access, low cost form presents no difficulty. CD-Rom drives are now a standard feature of most PCs but these are only able to read pre-recorded CDs. To be able to record data onto a CD, a CD-Recorder drive is required. This type of drive is able to use write once type CD recordable media, which is ideal for distribution and archiving. Greater flexibility is offered by CD-Recordable/ReWritable drives (CD-R/RW), which are able to read, write and erase data when used with the appropriate CD media. The CDs produced by these drives may of course be read by a CD-Rom drive. Although robust, manufacturers recommend that CDs should be stored in the dark at less than 40% relative humidity and at a temperature of less than 25°C. Recordable CDs are less robust than pre-recorded CDs.

Typical specifications:
- Interface: IDE/SCSI/parallel port
- Data transfer rate: Read ×6, write ×2 (compared to the standard CD transfer rate of 150 kB/sec)
- Capacity: 650 MB

Analogue-to-digital data capture card. Analogue-to-digital data capture cards represent an alternative method of constructing a PC-based data capture system to that based on sound cards. These cards are not required to meet specific standard specifications and can offer significantly higher digitisation rates, which may be of value for some applications. This type of card, designed for the scientific and industrial signal processing market, carries a significantly higher price tag (×100). A wide range of configurations are available, some with dedicated processing functions, others with signal processing microprocessors. In general, additional software and programming effort is required to develop a specific application. Captured data is often stored direct to the PC system memory or to hard disk, but some store to on-board memory, which can be limiting.

Typical specification:
- Channels: Two simultaneous channels (note: some of these types of card multiplex between channels, cause undesirable time skewing of samples)
- Sample rate: 2 MSps (higher rate cards are available)
- Quantisation: 14 bits/sample

A3.3.2.3 CD Recorder

Complete recording systems based on CD-R technology are available. These devices offer lossless recording to the highest specification. Media costs are similar to ACR tapes. The machines offer direct digital links to and from computer systems as well as analogue audio input and output.

Typical specifications:
- Channels: 2
- Frequency response: 2 Hz–20 kHz
- Quantisation: 16 bits/sample
- SNR: 92 dB
- Channel separation: 100 dB
- Total harmonic distortion: 0.004%
- Wow and flutter: Unmeasurable
- Digital interface: Optical

A3.3.2.4 MiniDisk Systems

MiniDisk systems are magneto optical disk-based devices for the recording of audio signals in digital form. These machines, first developed by Sony in Japan in 1992, are now becoming widely available, models being produced by all the major Japanese electronics companies. With a 74 minute record time, these devices offer in a single unit the ability to record dual channel digital audio on removable media with random access to individual records and the capability of direct data transfer to a PC. At first sight these devices would seem ideal; however, to obtain the 74 minute record time a data compression technique, based on a psycho-acoustic model, is used. This approach relies on frequency-masking effects and hearing thresholds. Whilst achieving a compression rate of 5:1, with little or no audible effect, significant changes in spectral content of the recorded signal are the likely result.

Typical specifications:
- Channels: 2
- Recording time: 74 minutes

Sampling frequency	44.1 kHz
Frequency range	5 Hz–20 kHz
Dynamic range	105 dB
Wow and flutter	Unmeasurable
Compression system	ATRAC (Adaptive TRansform Acoustic Coding)

A3.3.2.5 *Digital Versatile Disk-recordable*

Digital Versatile Disk (DVD) has been developed primarily as a means of providing studio quality interactive audio and video to the consumer. Each DVD has the potential to store four 135 minute movies, two per side. The versatility comes from the fact that DVDs will also be available in a read/write form known as DVD-R. At the time of writing, however, the consortium developing this aspect of the DVD standard has fragmented into a number of camps. The dispute is centred on the storage capacity to be provided (2.6 GB, 3 GB or 5 GB). However, because of the higher storage capacity compared to CD-Rom, backward compatibility with CD-Rom, potentially faster data access rate and enhanced audio/video capabilities, DVD-R may be worth considering at a future date.

Typical specifications:
  Not practical at this time due uncertainty over final standards.

A3.4 RECORDING COLOUR DOPPLER IMAGES

Single image digital capture devices have been available for some years but analogue video cassette recorders have until fairly recently been the only practical option for recording video images over any significant period of time. The most widely used format for VCRs in the medical field is Super-VHS. However, digital video recorders (DVR) are now available, offering superior recording, quality frame freeze and enhanced editing capabilities. Models are available which can be connected to a PC for both control purposes and digital image transfer in both directions.

The selection of a system for recording colour Doppler images largely depends on why the recording has to be made in the first place. If for future reference only, then a hard copy or 'video still recorder' may be ideal. If a sequence of images is required of more than a second or two, then a VCR of some kind is likely to be required. If there is information in the image that needs to be analysed, then a DVR may be the best solution since no other digitiser is involved other than in the DVR, thus preserving video image quality.

As with audio signals, a wide range of options are available for handling video images:

Recording video images

- Video cassette recorder
- Digital video recorder
- Video image processing workstation
- Ultrasonic image capture analysis systems

Frame/image capture

- Multimedia frame grabber
- High performance video acquisition card
- Digital still image recorders (based on MiniDisks)

Hard copy

- Colour video printer

A3.4.1 Recording Video Images

A3.4.1.1 *Video Cassette Recorder*

Recording real-time images for any significant period of time requires the use of a video cassette recorder (VCR). The most widely available analogue format that gives good quality reproduction is Super VHS (S-VHS), a development of the previous VHS standard. In addition to higher quality image reproduction, S-VHS recorders record audio signals to hi-fi standards (stereo). They also have a lower specification 'normal' audio channel, which can be 'dubbed' or recorded onto or over at a later date (on some VCRs, using this channel overrides one of the hi-fi tracks). A good machine will have a digital image store for frame freeze use and possibly time-based correction features to minimise timing errors common to all analogue VCRs. If the application requires image processing, then an analogue VCR may not be the best method of collecting the images.

Typical specification:
Video	PAL, NTSC, S-Video
Search	Shuttle $1/30-\times 10$ normal speed
Recording time	120 minutes with T120 tape
(normal audio)	
Channels	1
Frequency response	50 Hz–12 kHz
SNR	43 dB
(hi-fi audio)	
Channels	2
Frequency response	20 Hz–20 kHz
Dynamic range	90 dB

A3.4.1.2 Digital Video Recorder

The digital video recorder (DVR) boasts all the usual features of a VCR but with a number of advanced capabilities, including direct image transfer to a PC and non-linear editing. DVR systems are significantly smaller and lighter than the typical VCR. Data compression is used to reduce the data flow rate to manageable levels for the $\frac{1}{4}$ inch tape cassette, nevertheless the horizontal resolution is superior to that of S-Video. This type of device features playing time of up to 4.5 hours, computer control by means of an RS232C serial link, and wire remote control. A high speed serial data bus (IEEE 1394) is also a common feature of these machines, which enables the fast transfer of audio and video signals in a digital format.

Typical specifications:
Video	PAL, NTSC, S-Video
Sampling frequency	Y: 13.5 MHz
	C: 3.75 MHz
Quantisation	8 bits/sample
Error correction	Reed–Solomon code
Audio channels	2
Sampling frequency	48 kHz
Quantisation	16 bits/sample
Frequency response	20 Hz–20 kHz
Cross-talk at 1 kHz	−80 dB
Wow and flutter	Below measurable limits

A3.4.1.3 Video Image Processing Workstation

Some UNIX workstations may be configured for image capture and processing, for example adding a SIRIUS Video I/O board to a Silicon Graphics ONYX forms a system capable of real-time digital video capture and manipulation. While detailed description and specifications are beyond the scope of this text, more information can be obtained by consulting the relevant manufacturers or their web site.

A3.4.1.4 Ultrasonic Image Capture and Analysis Systems

Complete video image capture and analysis systems are available targeted at specific application areas, particularly cardiology. Such systems are able to capture a number of cardiac cycles, which may then be reviewed and stored as desired. They may also provide analysis packages for specific examinations.

A3.4.2 Frame/Image Grabbers

A3.4.2.1 Multimedia Frame Grabber

Frame grabbers, in the form of PC cards or parallel-port devices, are widely available. Typical applications for these devices are in the creation of teaching materials and multimedia applications. The use of this type of card should present no difficulties, since at least a minimal set of frame capture and manipulation programs are normally provided as part of the package.

Typical specifications:
Low cost video capture card
Capture modes	24/16/8 bit colour
Video sources	PAL, NTSC
Image capture	760 kB VRAM buffer
Compression	MPEG

Processing	On PC
Intended applications areas	Capturing single frame or motion video from a VCR or camcorders (usually at reduced frame rates), surveillance and video conferencing

A3.4.2.2 High Performance Video Acquisition Card

High performance image acquisition cards intended for scientific or industrial applications are available from a number of manufacturers. These cards are capable of capturing full motion video using any of the common video standards as input. They feature onboard image processing hardware, large buffer memories, high data transfer rates and extensive software libraries. Usually basic application software is supplied but for any kind of sophisticated application, programming effort is likely to be required.

Typical specification:
Video sources	PAL, NTSC, S-Video
Capture mode	16 bit monochrome or YUV 4:2:2 colour
Image capture	Direct to PC bus (70 MB/second sustained transfer rate)
Processing	Onboard i960 Processor with up to 128 MB DRAM Hardware region of interest support External trigger support

A3.4.2.3 Digital Still Image Recorder

A convenient method of collecting single video frames is the Digital Still Image recorder (Sony DRK-700). Based on the MiniDisk (MD) configured for data storage, 100–1000 images can be stored on a single disk, depending on the level of compression used (1/1, 1/4, 1/14). Captured frames can be transferred to a PC for display and further processing using an MD Data drive if required. Again, it should be remembered that compression may result in information loss, so care should be taken in the use of this device when recording images with a view to further processing.

Typical specification:
Input/Output	RGB, Y/C, composite video or digital (SCSI)
External control	Foot pedal, remote control, RS-232
Recording time	2–12 seconds, depending on compression

A3.4.3 Hard Copy

A3.4.3.1 Colour Video Printer

The most convenient method of obtaining high quality hard copy of video images is by means of a colour video printer. Based on dye sublimation thermal transfer, with the dye supplied in an ink film cassette and transferred to cut sheet paper, these devices are able to provide excellent high quality prints. Among the issues to be considered are the number of colours required in the image (usually 256 or 16 million), the range of features required, and the size of paper printout. Printouts can be up to A4 in size, depending on the printer chosen. Printout size does, of course, have running cost implications. Most machines are able to print a number of images on each sheet. These may be selected images or, on some machines, may be a time series captured by means of a strobe option. Typically, up to 16 images may be captured in this way, with the time interval set as required. The final selection of a printer should be made based on an examination of print quality, not just specifications. Test prints should preferably be made using the actual ultrasound machine or workstation with which it is intended to be used.

Typical specification:
Input signals	PAL, NTSC, S-Video, analogue RGB, Digital I/O
Picture size	Up to A4
Print speed	<75 seconds (dependent on paper size)
Gradation	256 gradations or 256 levels each for yellow, magenta and cyan

Where only very small numbers of prints are involved, particularly if prints of the highest quality are not essential, the expense of a professional quality video printer may not be justifiable. Much lower cost video printers, intended to satisfy the demands of the multimedia market, are available. Although these devices are not constructed to the same high standards as their

Table A3.1 Specification terms

Bandwidth	The term 'bandwidth' refers to the range of frequencies a device is able to handle, for example amplify or record. It is usually expressed as the frequency range over which the output signal level will not vary by more than a specified amount, generally 3 dB
Compression	Signal or image compression is a process whereby the amount of data required to store, transmit or reproduce the original signal or image is reduced. Typically, lossless compression can achieve 2:1 data reduction, compression much beyond this is likely to result in a loss of information
Image resolution	The resolution of an image is determined by many factors. Resolution is often expressed in terms of the number of pixels or discrete points used to construct the image
MSps	Million samples per second—The rate at which an analogue signal is converted into its digital representation
Quantisation	Quantisation is the process of dividing an analogue input signal into a number of discrete values by means of an analogue-to-digital converter (A/D). The higher the number of bits used to represent the analogue signal, the more accurate the resultant digital representation becomes. Using an 8 bit A/D converter provides 256 possible values (2^8) and 16 bits 65536 (2^{16}). The number of bits used is constrained by costs and/or noise factors
SCSI	Small computer system interface—An interfacing standard that enables the connection of devices to a computer system for the purpose of data transfer
SNR	Signal-to-noise ratio—all analogue systems suffer from noise to some degree. The SNR of a system is the ratio of the true signal (peak height) to the system noise. This ratio is most often expressed in dB
Total harmonic distortion (THD)	Harmonic distortion is a measure of the degradation undergone by a signal passing through a process. This distortion introduces additional frequency components or harmonics. The level of signal due to these harmonics is expressed as a percentage of the input test signal
Wow and flutter	Wow and flutter are measures of the variations in the speed at which, for example, tape passes the record/play back head. The wow (low) and flutter (high) frequency variation are caused by mechanical imperfections. They are measured against the desired speed and expressed in percentage terms. Digital systems include correction for minor speed variations

Table A3.2 Video signals and standards

Chrominance	The colour information in a video signal
Component video	Each primary colour is represented by a separate video signal, e.g. one line each for the red, green and blue colour signals (RGB)
Composite video	A composite video signal is a video signal in which the luminance and chrominance signals are combined into a single carrier wave
Compression	PAL and NTSC video standards were both devised to compress full bandwidth RGB signals into a single 5.5 or 4.2 MHz channel, respectively
Luminance	Brightness
NTSC	NTSC refers to the video encoding standard used principally in the USA and stands for National Television System Committee, the organisation that devised the standard. The format consists of 525 lines per frame which make up the vertical resolution. With a frame rate of 30 frames per second, each frame consists of two interlaced fields
PAL	PAL is the video encoding standard used in the UK and stands for Phase Alteration by Line. The format consists of 625 horizontal lines per frame which make up the vertical resolution. With a frame rate of 25 frames per second, each frame consists of two interlaced fields
S-Video	S-Video is not a video standard as such, but is a means of delivery. Luminance and chrominance are supplied as separate signals, resulting in higher resolution
YC	Commonly used shorthand to refer to composite video signal composed of luminance (Y) and chrominance (C)
YUV	Commonly used shorthand to refer to luminance and colour difference signals

professional counterparts, the quality of the image produced is reasonable, since the same printing technology is used. For most of these printers the image must first be available on a PC, although JVC produce one model which includes a frame grabber. This unit is therefore able to capture and print images direct from a video source. Details of these devices and occasional reviews may be found in home computer magazines.

A3.5 TERMINOLOGY OF SPECIFICATIONS

Selecting an instrument based on its specification is no simple task, particularly since manufacturers are prone to specify their equipment so as to show it in the most favourable light. The same manufacturer may even specify similar devices in a different way, making comparisons even more difficult. To aid in the interpretation of manufacturers' specifications, a list of some common terms and their meanings are given in Tables A3.1 and A3.2.

A3.6 SUMMARY

There are many options available for recording and reproducing Doppler audio and video signals, and the best choice in any situation will depend on a number of factors. Once having chosen a suitable modality, there is no substitute for hands-on evaluation of a device and, where possible, direct side-by-side comparisons of candidate instruments.

Full information on the instrumentation discussed can be obtained direct from the equipment manufacturers or suppliers; alternatively, a great deal of information is available via the Internet. A number of relevant addresses or URLs (Universal Resource Locations) are listed in Table A3.3. It should be remembered that these URLs can change without warning, although the new location or an alternative may be found by conducting a search using the appropriate key words.

A3.7 REFERENCES

Aydin N, Fan L, Evans DH (1994) Quadrature to directional format conversion of Doppler signals using digital methods. Physiol Meas 15:181–199

Bush G, Evans DH (1993) Digital audio tape as a method of storing Doppler ultrasound signals. Physiol Meas 14:381–386

Coughlan BA, Taylor MG (1978) On mehods for preprocessing direction Doppler signals to allow display of directional blood-velocity waveforms by spectrum analysers. Med Biol Eng Comput 16:549–553

Moraes R, Evans DH, deBono DP (1995) A microcomputer-based system for coronary Doppler studies. Physiol Meas 16:287–294

Nakatani S, Yamnagishi M, Tamai J, Takaki H, Haze K, Miyatake K (1992) Quantitative assessment of coronary artery stenosis by intravascular Doppler catheter technique: application of the continuity equation. Circulation 85:1786–1791

Smallwood RH (1985) Recording Doppler blood flow signals on magnetic tape. Clin Phys Physiol Meas 6:357–359

Table A3.3 Manufacturers' web sites

Equipment	URL
Audio visual equipment	
Hitachi	www.hitachi.com
JVC	www.jvc.com
Panasonic	www.panasonic.com
Sony	www.sony.com
PC audio/video cards	
Creative Labs	www.creative.com
Hauppauge	www.hauppauge.com
Signal/image processing cards	
Blue Wave Systems	www.bluews.com
Data translation	www.datx.com
Datel	www.datel.com
Keithley	www.keithley.com
National Instruments	www.ni.com
Signatec	www.signatec.com

A3.8 FURTHER READING

Pan D (1995) A tutorial on MPEG/Audio Compression. IEEE Multimedia 12:60–74

Steinmetz R, Nahrstedt K (1995) Multimedia: Computing, Communications and Applications (Innovative Technology). Prentice-Hall, London

Tekalp MA (1995) Digital Video Processing. Prentice-Hall, London

Index

Note: page numbers in *italics* refer to figures and tables, those in Roman numerals refer to plates

A-mode technique, tissue motion imaging 344, 346
A/B ratio 205, *206*
A/C ratio 206
absorption 37–9
accuracy performance indicator 220
acoustic imaging, underwater 34
acoustic impedance 36
acoustic injection 396, *397*
acoustic pressure, safety 369
acoustic ranging 340
acoustic shock generation 142
adaptive threshold method of maximum frequency envelope extraction 183, 184
adaptive-Q distribution 166–7
aggregation half-time 128
ALARA principle 371
albumin, air-filled microbubbles 313
alias rejection methods *see* anti-aliasing techniques
aliasing II, 50, 242
  2-D Fourier transform 156
  autocorrelation estimator 279–80
  CFI systems 64, 231–2, 256
  Doppler power mapping 66
  extended autocorrelation estimator 262
  phase domain algorithms 67
  power Doppler imaging 232
  velocity-matched spectrum analysis 159–60
  wide-band/time domain techniques 279–80
  see also anti-aliasing techniques
American Institute of Ultrasound in Medicine (AIUM) 369
  safety statement 375
amplifier, sample-and-hold for PW system 106–7
amplitude, attenuation coefficient 38
amplitude-modulated sine wave 136
analogue envelope detectors 114–15
analogue frequency processors 114–15
analogue recording systems 407
analogue-to-digital converters (ADCs) 116

analogue-to-digital data capture card 409
anemometry, thin-film technique 19
angle estimation techniques
  angle of insonation 304
  blood flow measurement 303–5
  definition of normal 303–4
  dual transducer system 304–5
  imaging 303
  triple transducer system 304–5
angle of propagation 37
anti-aliasing techniques 347–50
  dual frequency technique 349
  extended autocorrelation method 261–2, 350
  frequency tracking 348–9
  non-equally spaced pulse transmission frequencies 348
  velocity-matched spectrum analysis 350
aorta, human 20–1, *22*
aortic arch, velocity measurement 43
aortic blood flow, fetal 200
aortic stenosis
  contour/area ratio 216
  modified threshold-crossing method 181
  pressure drop 18
aortic vasculature, fetal X
apodisation 78, 79
array transducers 76–81
  2-D 80–1
  3-D scans 80–1
  annular 81
  apodisation 78, 79
  beam shape modelling 81
  digital beam forming 78–9
  digital control 77
  piezoelectric elements 77, 78, 80
  reception zones 79
  slow flow detection 77
  types 77
artefacts 241–3
  aliasing 242
  beam/vessel angle VI, 243
  filtering 242
  flash artefact 243

  flow direction sensing 242
  grating-lobes 243
  large signal distortion 242
  machine defects 243
  noise VI, 241
  side-lobes 243
  speckle *169*, 242
  spectral broadening 241–2
arterial disease 239
  waveform shape 200
arteries
  bifurcations 239
  blood flow 5
  branches 19–20, 21
  flow change I, 60
  pulsations 15–16
  side-branches 20
  symmetric bifurcation 19–20
  wall intrinsic distensibility 15
Arts and Roevros circuit 114, 115
atherogenesis, flow patterns 21
atheroma, flow disturbance 17
atherosclerosis, wall shear 7
attack-sustain filter 254
attenuation 37–9, 134
  coefficient 38, 39
  CW systems 174
  frequency-dependent 145, 174–8, 177, 185–6
  maximum frequency envelope extraction 184, 185–6
  mean frequency processors 173, *174*
  parabolic velocity profile *135*, 173, *174*
  pulse centre frequency 175–7
  PW system 144–5, 174, 175
  zero-crossing processors 188
attenuation-compensated volume flowmeter 298–300
  intravascular volumetric flowmetry technique 300
  ultrasound beam production 299–300
audio cassette deck 407
audio Doppler signals
  analogue recording systems 407
  analogue-to-digital data capture card 409

audio Doppler signals (*continued*)
  characteristics 406–7
  compact disk-recordable
    systems 408–9
  digital recording systems 407–10
  digital versatile disk-recordable
    systems 410
  frequency range 406
  MiniDisk systems 409–10
  PC-based systems 408–9
  phase quadrature pair 407
  recording 407–10
  sound cards 408
  see also quadrature phase detection
audio tape, digital 408
auto-covariance function 335
autoclaving 93
autocorrelation estimator 63, 250–5,
    278–80
  2-D 253, 259–61, 279
  aliasing 279–80
  bias 253
  complex 2-D 259
  extended method 261–2, 350
  number of operations 277–8
  packet length 251, 253
  PRF 253
  with sub-sampling 258–9
autoregressive (AR) coefficient 161
autoregressive (AR) estimators, low
    order 262–3
autoregressive (AR) models 161,
    162–3
  order selection 163–4
autoregressive moving average (ARMA)
  all-pole model 162
  all-zero model 162
  models 161–2
  spectral estimation 161–5
axial velocity profile, cardiac cycle 57

B-flow 350
B-mode imaging
  calibration 240
  cardiac phantoms 391
  colour Doppler imaging 239
  contrast agent interactions 315
  duplex imaging 233
  EFSUMB safety statement 376
  grey-shade harmonic 232, 233
  harmonic V, *234*, *235*, 237
  harmonic colour Doppler 232
  pulse inversion *236*, 237
  pulse-echo 236
  real-time 344
  tissue motion imaging 344, 346
  transient *235*, 237
  triplex imaging 233
  ultrasonic 231
  WFUMB safety statement 374–5

B-mode scanner for colour Doppler
    imaging 231–2
back pressure 217
backscatter
  angular dependence phenomenon 392
  cross-section 35
  blood 320–1
  power and blood-mimicking
    fluids 386–7
  string phantoms 393
Bartlett's procedure 152, 155
  spectral variance 146, *151*, 152
Bayes method of classification 221–2
beam
  CW 84, *85*
  digital forming 78
  multi-beam systems 85–6
  steering 78, 79
  stepping 78, 79
  uniform 81
beam/vessel angle VI, 243
belt phantom 393–4, *395*, 398, 399, *400*
Bernoulli's equation 17
  non-steady flow 18
Bessel distribution function 166
bioeffects of ultrasound 359
bit-pattern correlation algorithms 336
block-matching techniques, speckle
    tracking 334
blood
  acoustic properties of constituents 121
  attenuation coefficient 38
  backscatter coefficient 122
  backscatter cross-sections 320–1
  fluids mimicking 386–7
  formed elements 120–1
  Rayleigh scatterer 321
  rouleau-suppressed 126
  scattering cross-section 36
  shear rates 125–6
  stasis 30–1
  ultrasonic target 120–31
  ultrasound scattering 122
    angular dependence 124
    cyclic changes in power in pulsatile
      flow 128–9
    erythrocyte aggregation 125–6
    flow conditions 124–30
    Reynolds number 126–7
    statistical properties 130–1
    turbulence 126–7
    velocity measurement 129–30
blood flow 5
  3-D Doppler imaging 339
  classes (BFCs) 218
  Doppler shift frequency 323
  pulsatile 128–9
  velocity values 348
blood flow measurement
  angle estimation techniques 303–5

angle measurement 291–3
assumed velocity profile method 300
attenuation-compensated
    method 298–300
C-mode Doppler techniques 300–3
  correction factor 302
  multiplanar transducer 302–3
  partial volume compensation 302
contrast agents 303
detected fractional moving blood
    volume 306–7
diameter estimation techniques 305–6
Doppler vector tomography 306
duplex scanners 288–93
errors 307
first moment followers 303
flow vectors 292
intravascular 338
mean Doppler shift 289
plane of scan 292
practical limitations 293
pulse-echo system 288
quantitative 81
velocity of ultrasound in soft
    tissue 292–3
vessel area measurement 289–91
volumetric 288
blood flow measurement with multigate
    systems
  1-D profiles 293–6
  2-D profiles 296–8
  angle measurement 294–5, 296, 297
    accuracy 298
  finite size of volumes 295, 296
  flow velocity 296–7
  radius measurement 294
  ultrasound beam position 296
  velocity measurement 295
  velocity profile symmetry 295–6
  vessel size limitations 298
  volumetric flow 297
blood velocity 15
  IVUS X, 236
  parallel to axis of vessel 294
  profile 5
blood vessels
  attenuation 134
  damage from ultrasound 367
  flow phantom 387, *388*, 389–90
  image production 55
  parabolic velocity profile 132, 133
  power spectrum 132–3
  rectangular beam 132
  uniform insonation 131
  velocity profiles 132–4
  velocity radial distribution 132
blood volume, detected fractional
    moving 306–7
blood-mimicking fluids
  backscatter power 386–7

INDEX

materials 386
   Rayleigh scattering 387
boundary layer thickness 8
broadening, non-stationarity 153–4
broadening function
   Hanning window 154, *155*
   root mean square 153, *154*
Brownian motion 135
bubbles
   backscatter cross-sections 320
   resonance 31, 312
   scattering 311–12
   stabilisation 312–13
   see also contrast agents; microbubbles
bulk modulus 28
butterfly search method 270–1

C-mode Doppler techniques *58*
   blood flow measurement 300–3
Capon method 162
cardiac cycle
   adaptive threshold method 183
   cyclic changes 128–9
   signal-to-noise ratio 187–8
   spectral data analysis 167, 168
   velocity profile 12, 59
      axial *57*
      human carotid bifurcation 21, *23*
   vessel wall motion tracking 306
cardiac output, attenuation-compensated volume flowmeter 300
cardiac phantoms 391
carotid artery, common
   colour-coded Doppler M-mode image I, *60*
   velocity profile 11, *14*
   velocity waveform *57*
carotid artery, internal 169
carotid bifurcation
   3-D Doppler imaging XI
   C-scan *58*
   colour-coded Doppler image I, *58*
   PW Doppler imaging *59*
   velocity profile 20, *23*
carotid circulation, flow patterns 21
carrier frequencies, simultaneous transmission 349
carrier solutions for contrast agents 368
catheter devices 83–4
catheter Doppler imaging 234, 236
catheter transducer *82*
cavitation
   contrast agents 365, 368
   moving blood 367
   threshold 364
   transient threshold 366
   ultrasound 364, 365
   insonation of contrast agents 368
centroid detector
   root *f* power-spectrum 170

single-correlation 169–70
cerebral artery, middle 43
cerebral vessels
   contrast agents VII
   embolic signals 322
   embolus detection 321–2
   velocity measurement 43
Choi–Williams class distribution 165
chromosome damage, ultrasound 366
circle of Willis
   contrast agents VII
   low-amplitude linear scattered echo images VII, 236
classification in pattern
   recognition 200–1, 219–22
   Bayes method 221–2
   multivariate discriminant analysis 221
   nearest neighbour (*NN*) algorithms 221
   ROC curve analysis 220–1
   sensitivity 219–21
   specificity 219–21
   two or more dimensions 221–2
clutter phantom 395
   see also belt phantom
clutter rejection, CFI systems 246–50
coded signal velocity detecting systems 51–4
Cohen joint time–frequency distributions 165–7
coincident bit count 336
colour Doppler images/imaging *see* colour flow imaging
colour Doppler M-model II, 229, 233
heart II, XI
colour flow imaging (CFI) II–III, 59–67, 229, 231–2
   aliasing 64, 231–2, 256
      see also anti-aliasing techniques
   anatomical information 63
   angle independent 331–3
      linear array electronic split 333
      single linear array transducer IX, 332
      spatial filtering 333
      sub-aperture technique 333
      two-component method 332
      velocity components 332–3
      velocity vector reconstruction 333
   applications 60
   autocorrelator 63
   basic operation 61–4
   beam/vessel angle II, VI, 231, 243
   calibration 240
   clutter rejection 62, 66, 246–50
   colour vector information encoding 63–4, 65
   contrast resolution 240
   conventional velocity map 66

delay line cancellers (DLCs) 62, 65
digital still image recorder 412
digital video recorders 410, 411
Doppler bandwidth threshold 255
Doppler power mapping *see* power Doppler imaging
Doppler signal magnitude threshold 255
FFTs 62
filtering 242
filters 246–50
flash artefact 243
flow mapping area 64
fluid jet quantification 215–16
frame/image capture 410, 411–412
grey-scale imaging 61–2, 63
grey-scale information 255
hard copy 410, 412–14
high performance video acquisition card 412
machine defects 243
maximum echo intensity threshold 255
mean angular frequency 63
measurement from images/spectra 240
multigate measurements 293
multi-mode scanning 239
narrow-band systems 245–56
new technologies 66–7
output digitisation 62
performance 239–41
phase domain 61
   algorithms 63
   methods 65
   systems 64
phase-shift estimation algorithms 62–4, 65, 250–3, 256–63
post processing 254–5
power Doppler *see* power Doppler imaging
power resolution 240
priority encoding/artefact suppression algorithms 255
pulse-echo B-scan system I, *60*
quantitative flow measurement 240-1
real-time two-dimensional 59–66
recording 405–6, 406–7, 410–14
recording video images 410–11
reproducing 405–6
resolution 90, 239–40
sample time 60–1
scanning technique 239
signal paths 61
signal processing 245
spatial resolution 239–40
system general layout 245–6
temporal resolution 240

colour flow imaging (CFI) (*continued*)
  time domain correlation methods 64–5, 263–78
  triplex imaging 233
  ultrasonic A-line frequency 65
  ultrasonic image capture and analysis systems 411
  ultrasonic output 363
  ultrasonic signal information extraction 61
  use 239–41
  velocity estimation 66–7
  velocity information 63
    encoding in colour 64
  velocity resolution 240
  velocity threshold 255
  video cassette recorder 410–11
  video image processing workstation 411
  wall thump filters 62, 66, 246–50
  zero-crossing detector 257
colour flow mapping *see* colour flow imaging
colour video printer 412–14
colour-coded continuous wave Doppler imaging 1, 58
colour-coding schemes 237
compact disk-recordable systems 408–9
complex Fourier transform (CFT) 112, *113*, 114
complex linear regression estimator 263
compressibility, distribution 35
computational fluid dynamics (CFD) 22–4
  code coupling with high-level solid mechanics 24
cone-kernel distribution 166
constant flow ratio 205
constrictions, haemodynamic effects 16
continuous wave (CW) beam 84, *85*
  intensity 361
    pressure amplitude 360–1
    sister chromatid exchange 366–7
continuous wave (CW) Doppler imaging 58–9
  colour-coded I, 58
  long scanning time limitation 58–9
continuous wave (CW) Doppler spectra 131–9, *140*, 141–2
  attenuation 134
  filtering 141
  non-uniform insonation 131–4
  non-uniform target distribution 131
  spectral analysis limitations 141–2
  see also intrinsic spectral broadening (ISB)
continuous wave (CW) Doppler units 2
  ultrasonic output 362
continuous wave (CW) fields 84, *85*

continuous wave (CW) systems 28, 97–105
  attenuation 174
  demodulator 100–5
  filtering 105, 141
  ISB 137, 139, *140*
  master oscillator 97, 101, 102
  probe 97–9
  quadrature phase detection 103–5
  receiver 99–100, *101*
  sample volume *87*, 88
  scattering 174
  simple devices 1
  tissue motion imaging 344–6
  transducers 137
  transmitter 97, *98*, *99*
continuous wave (CW) velocity detecting systems 43–5
  demodulated signals 50
  Doppler shift frequency 50–1
contour length 216
contrast agents 311–13, *314*, 315–17
  applications 316–17
  blood flow measurement 303
  bubble size range 318
  carrier solution 368
  cavitation centres 365
  disintegration under ultrasonic pulse influence 317
  imaging techniques 315–16
  intravenous *314*
  lung transit 318
  microbubble interaction with ultrasound 313, 315
  perfusion phantoms 391
  perfusion techniques 317–19
  quantification of passage 317–19
  safety 367–9
  shadowing 318–19
  speckle tracking 311
  time–intensity curves 318
  toxicology 368
  types 311–13
  ultrasound insonation 368–9
contrast Doppler imaging 236–7
contrast resolution, colour Doppler imaging 240
coronary artery blood flow measurement 303
coronary circulation 21
correlation angle estimator *see* autocorrelation estimator
correlation coefficient, maximum of echoes 275
correlation function
  displacement 337
  normalised 335
correlation interpolation
  algorithm 267–8
  method 265

correlation maximum amplitude 261–2
correlation phase estimator *see* autocorrelation estimator
covariance estimator *see* autocorrelation estimator
Cramér–Rao Lower Bound 276, 277
cross-correlation based algorithms 265–7
  analysis of performance 271–3, *273*, 275–7
  analysis of variance of peak location 276
  correlation interpolation algorithm 267–8
  maximum likelihood estimators 268–71
  minimum error derivation 276
  number of operations 277–8
  weighted average technique 266
  windowing 275
cross-correlation estimator 262, 278, 279
  aliasing 279
cross-correlation function
  normalised 272
  time delay *273–4*
cross-correlation method 263–4
  analysis of performance 271–3, *273*, 275–7
  implementation 277–8
cross-correlation model (CCM) 267–8
cross-covariance function 335
cubic spline interpolation, speckle tracking techniques 336
cuff pressure measuring instruments, ultrasonic output 363
curve broadening index 208–9
curve matching 218–19
curved array transducers 80, *82*
curved tubes 18, *19*

damping 31
damping factor, two site measurement 214
DC removal 250
decomposition, singular value 250
decorrelation signal 277, 337
decorrelation-based techniques X, 234, 236, 337–9
  angle calculation between beam and direction of flow 338–9
  correlation map 338, *339*
  intravascular blood flow measurement 338
  volumetric blood flow calculation 338
delay line cancellers (DLCs) 62, 65
demodulation 44
  heterodyne 101–3
  process and signal shift 150
  quadrature phase detection 103–5
  techniques *117*

demodulator
  CW systems 100–5
  PW system 105, 106
density distribution 35
derated (*in situ*) intensity 369–70
dialysis filter 391
diameter estimation techniques 305–6
diffraction 29–30
digital audio tape 408
digital recording systems 407–10
digital still image recorder 412
digital versatile disk-recordable
    systems 410
digital video recorders 410, 411
direction of arrival methods 280–2
  eigenvectors 281
  minimum variance (MV) estimate 280
  MUSIC algorithm 280–2
  spectral resolution 281
direction detection
  complex Fourier transform 112, *113*, 114
  phasing-filter technique 108, *109*, 110, 112
  switched channel processing 107–8
  Weaver receiver technique 110, *111*, 112
directional filter 168–9
directional power Doppler imaging 229
discrete Fourier transform (DFT) 151
discriminant analysis, multivariate 221
displacement correlation function 337
displacement vector, speckle tracking
    techniques 334
distortion, non-linear 31, *33*, 34
Doppler angle 292–3
  blood flow measurement with
    multigate systems 294–8
  small 323
Doppler artefacts 230–1
Doppler bandwidth
  invariance theorem 137–8
  measurement 323, 324, 325
  threshold for CFI systems 255
Doppler effect 1, 39–40
Doppler equation, speed of sound 28
Doppler images 57, 60
  enhancement and contrast agent
    interactions 315
Doppler imaging, 3-D 339–43
  3-Scape™ X, 340–1
  acoustic ranging 340
  B-mode 339
  blood flow imaging 339
  Doppler mode 339
  electronic scanning 342
  free-hand scanning 340–1
  image acquisition techniques 340–2
  image display techniques 342–3
  image reconstruction techniques 342

magnetic position sensing 340
mechanical scanning 341–2
multiplane with textural mapping XI, 343
multiplane viewing X, 342–3
position sensing 340
real-time 339, 342
volume rendering techniques XI, 343
Doppler imaging
  ultrasonic output 363
  WFUMB safety statement 375
Doppler instruments, single-gated 51
Doppler methods, tissue motion
    imaging 346–7
Doppler mode, EFSUMB safety
    statement 376
Doppler power
  mapping III, 65–6
  pulsatile blood flow 128–9
  spectrum 169
  first moment followers 189–90
Doppler principle, medical
    application 2, *3*
Doppler sample volumes *87*, 88–90
Doppler shift 2, 40
  colour-coding I, 58
  signal 2
  spectrum and envelope signals 150
  ultrasound pulse 177–8
Doppler shift frequency 16, 119
  blood flow 323
  instantaneous maximum 300
  low 238
  maximum 185, 324
  pictorial record 45
  PW velocity detecting systems 49
  real-time spectral analysis 45, *46*
  sonogram 45, *46*
  spectrum 135
  time-varying trace 45
  zero-crossing technique 45
Doppler signals
  high-amplitude low-frequency 15
  magnitude threshold in CFI
    systems 255
  recording/reproducing 405–6
  spectral estimation 152, 159–60
  velocity-matched spectrum
    analysis 159–60
  see also audio Doppler signals
Doppler spectral envelope area 168
Doppler spectrum
  estimated 153–4
  order selection for AR models 163–4
Doppler systems
  biological damage potential 55
  C-mode 57
  colour flow mapping 57
  complex 116–17
  duplex scanners 54–6

frequency-modulated 53–4
multichannel 56
phase-shift 51, 54
profile detecting systems 56–7
pseudo-random signal 52–3
random signal 52–3
range gates 56
SNR ratio 51
time-shift 51, 54
velocity detecting 43–54
velocity imaging systems 58–67
Doppler techniques, scattering 35–6
Doppler test devices
  electronic injection systems 401–2
  filter characteristics 398
  image resolution 399–401
  sample volume
    characteristics 397–8
  sensitivity XII, 399
  velocity measurement 398–9
  waveform indices *384*, 399, *401*
Doppler tissue imaging IV, 229, 233
  myocardium XII, 345
  tissue motion imaging IV, XI, XII, 344–7
  ultrasonic output 363
Doppler units
  non-directional 44
  ultrasonic output 362–3
Doppler vector tomography 306
Doppler velocimeter 150
Doppler wires 83
double window modified trimmed mean
    (DWMTM) filter 168
duplex scanners, blood flow
    measurement 288–93
duplex systems 54–6, 233
  angle-measuring cursor 56
  blood vessel image production 55
  flow measurement 56
  operation 55–6
  pulse-echo B-scan system 54
  ultrasonic output 363
  velocity measurement 55–6

early diastolic notch 201, 206
  pulsatility index 206–7
echo amplitude 37
echo cancellers 246–7, *248*
echo images, low-amplitude linear
    scattered VII, 236
echo imaging signals 35
echo intensity threshold in CFI
    systems 255
eigenvectors 213
  direction of arrival methods 281
elastic modulus, taper 20
elastic tubes, pulsatile flow 14–16
electric field distortion constant 72
electro-mechanical coupling coefficient 72

electronic injection devices   395–7, 401
  acoustic injection   396, *397*
electronic scanning   342
emboli detection   VII, 319–23
embolic signals   319
  characteristics   322
  Fourier transform techniques   323
  power   319–22
  processing   VIII, 322–3
  spectral analysis techniques   323
embolus-to-blood ratio (*EBR*)   319–22
energy, rate of flow   29
ensemble averaging   167–8
entrance effects
  human arteries   21
  pulsatile flow   14
  turbulent flow   9
envelope averaging techniques   190–1
envelope extraction   169, 179, 187, 189–90
  technique choice   191–2
  see also maximum frequency envelope extraction; mean frequency processors; zero-crossing processors
envelope signals   169
  Doppler shift spectrum   150
erythrocytes   30, 121
  aggregation   125–6, 128, 129
  backscatter coefficient   122, 123
  clumps   122, 125
  turbulence   126–7
estimated Doppler spectrum, non-stationarity   153–4
ethylene oxide   93
European Federation of Societies for Ultrasound in Medicine and Biology (EFSUMB)   369, 375–6
exponential initialisation   249
extended autocorrelation estimator   261–2, 350
  aliasing   262
extreme angle rays   136

fan scanning   341–2
far field   84
fast Fourier transform (FFT)   62, 142
  analysers   154–5
  AR-based technique comparisons   163, *164*
  PW system   145–6
  spectral estimation techniques   150, 151–5
  spectral variance   142
  see also complex Fourier transform (CFT); Fourier analysis; Fourier transform
feature extraction   200, 201–19
  A/B ratio   205, *206*

A/C ratio   206
back pressure   217
backward selection   211
constant flow ratio   205
contour/area ratio   216
contour length   216
curve broadening index   208–9
curve matching   218–19
D/S ratio   205
damping factor   214
discriminating features   211
field profile index (FPI)   216
fluid jet quantification   215–16
forward selection   211
frequency contours   218
frequency ratio   215
height–width index   *204*, 207
high resistance index   205
indices derived from multiple samples   215–17
Laplace transform   211–12
maximum frequency   202
multiple features   209, 211
path length index   208
Pourcelot's resistance index   204–5
pre-processing   201
principal component analysis   212–14
pulsatility index   203–4, 214
reactive hyperaemia test   217
relative flow index   208
reverse area index (RAI)   216
S/D ratio   205
S/N ratio   206
selection of candidate features   211
simple single-site indices
  normalised frequency   202–7
  that include time   207–9
spectral broadening indices   209
subjective interpretation   202
systolic decay time index   207, *208*
systolic width   209
trans-systolic time   209
transit time   214
  ratio   *214*, 215
two site measurement   214–15
vascularity measures   216–17
velocity acceleration/deceleration   207
velocity gradient index (VGI)   216
femoral artery
  common   11, *14*
  superficial   21
fetal heart rate Doppler, vibrating targets   395
fetal monitoring systems
  EFSUMB safety statement   376
  transducers   76
  ultrasonic output   363
field profile index (FPI)   216
fields

CW   84, 84–5, *85*, *86*
PW   84–5, *86*
filter characteristics of Doppler test device   398
filter function, frequency-dependent   176
filtering   168–9, 242
filtering 2-D   167
filtering
  CW system   141
  median temporal   254
  PW system   145
  singular value decomposition   250
  vessel wall thump   145
filters
  analogue low pass   154
  CFI systems   246–50
  post-processing   254
  directional   168–9
  DWMTM   168
  finite impulse response   246–7, *248*, 253
  infinite impulse response   247–9, 253
  Lee's   168
  low-pass   116
  maximum frequency envelope extraction   186
  mean frequency processors   178–9
  PW system   107
  regression   249–50
  zero-crossing processors   187–8,
finite impulse response (FIR) filters   246–7, *248*
  autocorrelation algorithm   253
first moment followers   189–90, 303
first order autoregressive estimator *see* autocorrelation algorithm
fixed attenuation model   370
fixed target cancellers *see* delay line cancellers (DLCs)
flash artefact   243
flash echo imaging   316
flow   6
  direction sensing   242
  disturbance with atheroma   17
  oscillation amplitude   10
  pressure drop   7
    across stenosis   17–18
  separation with arterial branches   20
  stream line convergence   16
  through orifice   17
  see also laminar flow; parabolic flow; pulsatile flow; sinusoidal flow; turbulent flow; volumetric flow
flow phantom   382–4, *385*, 386–7, *388*, 389–91
  blood-mimicking fluids   386–7
  cardiac   391
  design   383
  gel-based   390, 391
  materials   383, 390

INDEX

solid tissue-mimicking 390–1
perfusion 391
polyurethane-based 390, 391
portable 383
pumps 383–4, *385*
reproducible flow waveforms 382, 383, *384, 385*
solid tissue mimics 390–1
stenosis 390
tissue equivalence 389–90, 391, 401
vessel 387, *388*, 389–90
wall thickness 387, *388*, 389
wall-less 389, 390, 399, 401
see also belt phantom; rotating phantom; string phantom
flow velocity *see* velocity
flow waveforms 9–10
reproducible 382, 383, *384, 385*
fluid dynamics
computational 22–4
model 349
fluid jet quantification, colour flow imaging 215–16
force balance 360
Fourier analysis 9–10
Fourier transform
1-D 157, 158
2-D 155–60, 259
aliasing 156
analyser 151–5
early diastolic notch pulsatility index 206
embolic signals 323
multifrequency Doppler 157–8
pulse complex envelope 158
random signal 154–5
spectral broadening 156
velocity waveform 203
velocity-matched spectrum analysis 158–60, *161*
see also complex Fourier transform (CFT); fast Fourier transform (FFT)
fractional broadening 209
frame averaging 238
frame grabber, multimedia 411–12
frame rates 231
frame/image capture 410, 411–12
Fraunhofer field 84, *85*
free-hand scanning 340–1
frequency 27
contours 218
dependence 39
envelope 201
tracking 348–9
two site measurement of ratio 215
frequency-modulated Doppler systems 53–4
Fresnel field 84, *85*

gas bubbles *see* bubbles; contrast agents; microbubbles
Gaussian noise
non-white 130
white 182
Gaussian random process 130
band-limited 151–2
Gaussian windows 153
gel-based materials 390, 391
geometric changes 16–20
geometric method of maximum frequency envelope extraction 182–3
geometrical broadening 135–6
grating-lobes 243
grid slopes interpolation, speckle tracking techniques 336
growth retardation, ultrasound 367

haematocrit 121, 122, 127
ultrasound backscatter 123–4
haemolysis, ultrasound insonation of contrast agents 368
Hanning window
broadening function 154, *155*
fast Fourier transform analysers 154
hard copy of colour Doppler images 410, 412–14
colour video printer 412–14
harmonic colour Doppler imaging 229, 232–3
harmonic contrast imaging 232, 315
harmonic frequencies 34
harmonic images 236–7
harmonic power Doppler imaging 229, 232–3
harmonic signals, contrast agents 315
harmonic tissue imaging 315
heart
blood flow II, 60
colour Doppler M-mode II, XI
see also myocardium
heart muscle, velocity gradients XI, 233
heart valves, stenosed 17, 18
heating effect of ultrasound 363–4, 365
height–width index *204*, 207
heterodyne demodulation 101–3
high pass filters (HPFs) 116
high resistance index 205
high-amplitude non-linear scattered echo images 236–7
Hilbert transform
operator 271
wide-band digital 110
Homogeneous model 369, 370
hot film anemometry 191
hybrid method 181–2
hydrophone 90–2, 365
materials 91
signal 92
ultrasonic intensity measurement 29

hyperaemia, reactive test 217
image persistence 238
image resolution of Doppler test device 399–401
imagers 237–9
cine-loop 238
colour-coding 237
discrimination 238
display 238
frame averaging 238
image persistence 238
interpolation 238
mobile units 239
priority encoding 238
real-time zoom 238
recording techniques 238
scanning strategy 237–8
sensitivity control 237
slow flow detection 238
infinite impulse response (IIR) filters 247–9
autocorrelation algorithm 253
initialisation 249
inlet length 8
inlet phenomena 21
insonation 368–9
angle 43, 304
contrast agents 368
non-uniform 131–4, 142, 184, 188, 289
mean frequency processors 170–1, *172*, 173
uniform 131
instantaneous frequency detector (IFD) 258
instantaneous maximum Doppler shift frequency 300
instrument specifications *413*, 414
intensity of ultrasound 28–9
attenuation coefficient 38
measurement 29
output 29
intensity-weighted mean bin number (*IWMB*) 170
intensity-weighted mean frequency (IWMF) 169–70, 171, *172*, 173
filters 179
signal-to-noise ratio 179
statistical bias 173
interference 30
intermittent, harmonic and power Doppler imaging (IHPD) 316
intermittent imaging 315–16
internal scanners 81, *82*
interpolation 238, 336
intravascular ultrasound (IVUS)
catheters 337
grey-scale image X, 338
scanners X, 234
transducers 83–4

intravascular volumetric flowmetry technique 300
intrinsic distensibility, arterial wall 15
intrinsic spectral broadening (ISB) 134–9, *140*, 141, 241–2
   error source 139, 141
   extent 136–9, *140*, 141
   geometrical 135–6
   maximum frequency envelope extraction 185, *186*
   PW system 142–4
   shape 136–9, *140*, 141
   transit time 135, 136
   use 141
   zero-crossing processors 188

Japanese Society of Ultrasonics 369
jet area 216
jet energy, summed 215
jet size 215, 216
jitter error magnitude 276, 277
jugular vein I, 60

Kay spectrographs 179

lamina, velocity 6
laminar flow 6
   breakdown 17
   projections 17
   steady 7–8
   taper 20
Laplace transform analysis 211–12
laser Doppler 1
layered model 370
lead zirconate titanate (PZT) 71, 73, 75–6, 91
Lee's filter 168
left ventricular outflow tract (LVOT) 168
linear array transducers 79–80, *82*
linear scanning 341
liver metastases, stimulated acoustic emission imaging V, 237
loss of correlation (LOC) imaging 316
low-amplitude linear scattered echo images VII, 236
low-pass filters (LPF) 116

M-mode technique
   colour Doppler imaging 239
   Doppler tissue imaging XII, 345, 346
   EFSUMB safety statement 376
   pulse-echo I, 60, 345
   real-time 344
   tissue motion imaging 344, 346
machine defects 243
magnetic position sensing 340
maximum frequency, composite 179
maximum frequency envelope extraction 179–84

10% max method 183
adaptive threshold method 183, 184
attenuation 184
   frequency-dependent 185–6
filters 186
geometric method 182–3
hybrid method 181–2
intrinsic spectral broadening 185, *186*
mean frequency relationship 184
modified geometric method *182*, 183
modified threshold-crossing method 181
non-uniform insonation 184
percentile method 180
performance of methods 183–4
scattering 185–6
signal-to-noise ratio 186–7
simple threshold method 180
threshold-crossing method 180–1
windows 180, 181
maximum frequency processors 179–87, 191
maximum likelihood estimators 268–71
   butterfly search method 270–1
   velocity estimation 269
   wide-band point 269
   wide-band range-spread 269
maximum likelihood (ML) method 162–3
mean frequency
   analogue techniques 169–70
   intensity-weighted (IWMF) 169–70, 171, *172*, 173
mean frequency processors 169–71, *172*, 173–9, 191
   attenuation 173, *174*
   filter effects 178–9
   intensity weighted 114–15
   intrinsic spectral broadening 174
   IWMF 179
   non-uniform insonation 170–1, *172*, 173
   PIWMF 179
   signal-to-noise ratio 179
   statistical bias 173
   variance 173
mean velocity, PW systems 175
measured embolic power ratio (*MEPR*) 319, 322, 323
mechanical index (*MI*) 370–1, 373, 374
mechanical scanning 341–2
median frequency followers 190
median temporal filtering 254
microbubbles
   air-filled albumin 313
   air-filled sorbitol 313
   contrast agents 31
   interaction with ultrasound 313, 315

destruction 369
encapsulated 311–13
free unencapsulated 315–16
thin-walled 365
toxicology 368
MiniDisk systems 409–10
minimum variance (MV)
   estimate 280, 281
   method 162
mobile units 239
mode frequency followers 190
Moens–Korteweg equation 15, 16
moving average (MA) coefficient 161
moving mirror transducer *82*
multi-beam systems 85–6
multifrequency Doppler 155
   2-D Fourier transform 157–8
   power spectral estimate 158
   radio frequency 158
multigate pulsed Doppler systems 116–17
multimedia frame grabber 411–12
multiplanar transducer 302–3
multiplane viewing, 3-D Doppler imaging X, 342–3
multiple signal classification (MUSIC) algorithm 280–2
myocardium
   Doppler tissue imaging XII, 345
   power Doppler tissue imaging 345–6

narrow-band methods 256–63
Navier–Stokes equation 22
near field 84, *85*
nearest neighbour (*NN*) algorithms 221
Newtonian fluids
   steady flow 6–9
   viscosity 5
noise
   artefacts in Doppler techniques VI, 241
   random transmission 347–8
non-linear propagation 31–4, *32*, *33*, 142
non-stationary broadening 166
normalised mean power in coloured pixels 217
normalised power-weighted pixel density 217
nylon particles, blood-mimicking fluids 386, 387
Nyquist frequency 49
Nyquist limit 262
   carrier frequencies 349
   frequency tracking beyond 348–9
   pulse inversion Doppler 316

oesophageal probe, diameter estimation techniques 306
one-bit signals, cross-correlation based algorithms 266

oscillating transducer 82
oscillator
 heterodyne 101, 102
 master 97, 101, 102
Output Display Standard (ODS) 371, 373

packing factor 123
parabolic flow 6, 7
parabolic velocity profile 6, 7, 8, 132, 133
 attenuation 135, 173, 174
 distortion 18
 mean velocity overestimation 170
parametric spectral estimation techniques 160–5
particle velocity, speed of sound 32, 34
path length index 208
patient exposure to ultrasound 376–7
pattern recognition 200–1
 classification 200–1, 219–22
 human brain 202
 pre-processing 201
 subjective interpretation 202
 transduction 200, 201
 see also feature extraction
percentile method 180
Percus–Yevick packing theory for pair-correlated hard spheres 123
perfusion phantoms 391
perfusion techniques 317–19
periodogram of stochastic signal 151, 153
phase 27
 detector 257
phase-only poly-pulse-pair processing 262
phase-shift 51, 54
 averaging of velocity waveforms 191
phase-shift estimation
 algorithms 65
 clutter filtering 278
 filters 278
phase-shift estimation methods 256–63, 350
 autocorrelation estimator with sub-sampling 258–9
 complex linear regression estimator 263
 extended autocorrelation estimator 261–2
 instantaneous frequency detector (IFD) 258
 phase detector 257
 zero-crossing detector 256–7
phased array transducers 80, 82
phasing-filter technique 108, 109, 110, 112
piezoelectric ceramics 71, 73
piezoelectric crystals

acoustic match of probe with body 99
 crosstalk 98
 CW system probe 97–8
piezoelectric effect 72
piezoelectric elements 74
 array transducers 77, 78, 80
 hydrophone 90–1
 polymer 71, 73, 91
piezoelectric lead zirconate titanate (PZT) 71, 73, 75, 76
 hydrophones 91
Poiseuille flow 6–7
Poiseuille's law 7, 10, 17
polyurethane-based materials 390, 391
polyvinylidene difluoride (PVDF) 71, 73
 hydrophones 91
position and intensity weighted mean frequency (PIWMF) 170–1, 172, 173
 filters 179
 signal-to-noise ratio 179
 statistical bias 173
post-processing, CFI systems 254–5
Pourcelot's resistance index 204–5
power
 normalised mean in coloured pixels 217
 ultrasonic beam 28, 29, 360
power Doppler imaging III, 65–6, 129, 130, 141, 232
 3-D imaging 232
 aliasing 232
 beam/vessel angle 243
 CFI system 255–6
 contrast agent VII, 311
 direction of flow information IV, 232
 directional 229
 flash artefact 243
 image persistence 238
 noise VI, 241
power Doppler tissue imaging, myocardium 345–6
power spectral density (PSD) 152, 161, 162
power spectrum 119–20
 blood vessels 132–3
 distortion 120
 filtering 141
 velocity profile 119
power-weighted pixel density 217
pre-enactment safety standard 370
pre-multiplication operator 271
predictive value
 negative 220
 positive 219
pregnancy, EFSUMB safety statement 376
pressure
 amplitude 360–1
 change and blood velocity 5

 gradient 9–10
 pressure drop 17
 flow 7
principal component analysis 212–14
 eigenvalues 213
 sample mean record calculation 212–13
priority encoding/artefact suppression algorithms, CFI systems 238, 255
probes
 CW systems 97–9
 PW system 106
profile detecting systems 56–7
projection initialisation 249
projections
 haemodynamic effects 16
 laminar flow 17
pseudo-random noise transmission 347–8
pseudo-random signal Doppler systems 52–3
pseudo-Wigner distribution 165–6
pulsatile flow 6
 elastic tubes 14–16
 entrance effects 14
 Fourier components 13, 14
 rigid tubes 9–14, 13
 velocity profile 12, 13, 14
 velocity signals 12
 viscoelastic tubes 14–16
pulsatility index 203–4
 two site measurement 214
pulse
 capture and transmission 396
 distorted 31
 Doppler shift 177–8
 random staggering 52
 reflected 37
 transmission frequencies 348
 wave velocity 15
pulse centre, frequency 175–7
pulse inversion Doppler imaging 236, 237, 316
pulse pressure amplitude 365
pulse repetition frequency (PRF) 46–7
 autocorrelation algorithm 253
 generator 47
 increase in PW velocity detecting systems 50
 PW system 107
pulse-echo systems 55
 blood flow measurement 288
 tissue motion imaging 60, 346–7
pulsed Doppler units 2
pulsed wave (PW) beam 84–5, 86
 intensity 361
 pressure amplitude 360–1
 sister chromatid exchange 366–7
pulsed wave (PW) Doppler 59

single-element transducer device 74, 75–6
spectra 131
tissue motion imaging 344–5
unit ultrasonic output 362
pulsed wave (PW) fields 84–5, *86*
pulsed wave (PW) systems 1, 28, 43, 45–51, 105–7
  attenuation 144–5, 174, 175
  control logic 107
  demodulated signals 50
  demodulation 105, 106
  depth of examination 46–7
  Doppler shift frequency 49
  fast Fourier transform 145–6
  filtering 107, 145
  FM system comparative performance 54
  frequency dependent scattering 144
  ISB 142–4
  maximum velocity
    detection 46–7
    limit 48–9
  mean velocity 175
  non-linear propagation 142
  non-uniform insonation 142
  operation 47
  PRF 107
    generator 47
    increase 50
  principal sample volume 47
  probe 106
  pulse length 48
  range ambiguity 49–50
  range cell 47–8
  range gate 48, 49–50
  range-velocity limit 50
  receiver 106
  reference oscillator 49
  sample volume 88–9
    length 143–4
  sample and hold amplifier 106–7
  scattering 174, 175
  signals 47
  spectral analysis limitations 145–6
  tissue motion imaging 346
  transmission gate 47
  transmitter 105
pumps, flow phantom 383–4, *385*

quadrature diffractive transducer 326
quadrature phase detection 103–5
  signals 107, *108*, 110, 114, *115*

radial dilatation 15–16
radiation pressure balance 29
radio frequency (RF)
  amplifier 97
  complex cross-correlation model (C3M) 268

fluctuations 259
multifrequency Doppler 158
pulsed Doppler unit *156*
random variations 278
received pulse 156
two-dimensional autocorrelation estimator 259, 261
radio frequency (RF) signal
  CFI system 245–6
  cross-correlation method 264
  decorrelation 337
  digitisation 277
random noise transmission 347–8
random signal Doppler systems 52–3
range–velocity ambiguity 51
rational transfer model 161
Rayleigh scattering 35, 121, 127
  blood 321
  blood-mimicking fluids 387
  frequency-dependent 175, *176*
reactive hyperaemia test 217
real-time spectrum analysis 2
real-time zoom 238
receiver
  CW systems 99–100, *101*
  PW system 106
receiver operating characteristic (ROC) curve 220–1
reception zone
  CW fields and beams 84
  multi-beam systems 85, *86*
  plotting 92
recording
  colour Doppler images 405–6, 410–14
  Doppler signals 405–10
  video images 410–11
red cells
  aggregates 122, 125
  capillaries 30
  see also erythrocytes
reflection 36–7
refraction 28, 36–7
regression filters 249–50
  adaptive 250
relative Doppler bandwidth invariance theorem 325
relative flow index 208
relaxation frequency 38
renal vasculature, power Doppler image III
reproduction of Doppler signals 405–6
resistance index 204–5, 206
resolution 231
resonance
  gas bubble scattering 312
  small bubbles in blood 31
reverse area index (RAI) 216
Reynolds number (*Re*) 8
  arterial branches 19, 20

blood ultrasound scattering 126–7
  critical 8
  human aorta 20
  taper 20
rigid tubes
  pulsatile flow 9–12, *13*, 14
  steady flow 6–9
'root *f*' followers 114, 115
root mean square (RMS)
  followers 187–9
  frequency 189
root-MUSIC algorithm 281–2
rotating phantom 394
rotating transducer *82*
rotational scanning 342
rouleaux 122, 125, 126
  formation 128, 129
Rschevkin formula 124

S-Dopp 259
S/D ratio 205
S/N ratio 206
safety 359
  acoustic pressure 369
  contrast agents 367–9
  derated (*in situ*) intensity 369–70
  machine outputs 369
  mechanical index (*MI*) 370–1, 373, 374
  pre-enactment standard 370
  standards 369–71, *372*, 373–4
  thermal index (*TI*) 370–1, *372*, 373–4
safety statements 374–6
  AIUM 375
  EFSUMB 375–6
  WFUMB 374–5
sample volume
  characteristics of Doppler test device 397–8
  CW system *87*, 88
  length for PW systems 143–4
  PW system 88–9
3-Scape™ X, 340–1
scattering 28, 30, 34–6
  coefficient 39
  cross-section 311–12
  CW systems 174
  differential cross-section 35–6
  frequency-dependent 174–8
    maximum frequency envelope extraction 185–6
  PW systems 174, 175
  signature 264
  speckle pattern 236
schlieren method of field visualisation 92–3
second harmonic imaging 67
  perfusion phantoms 391
  see also harmonic images

Sephadex, blood-mimicking fluids 386, 387
shadowing, contrast agents 318–19
shear modulus 28
shear rate 7
 backscattered power 130
 blood 125–6
 erythrocyte aggregation 125
shear stress 5
 vessel wall 7
shear waves 27
shock discontinuity 32
shock wave 31
 formation 34
side-lobes 243
signal
 decorrelation 277
 distortion 242
 instantaneous mean frequency 349
 processing for colour flow imaging 245
 shift in demodulation process 150
 vector 256
signal location estimators 115
 first moment followers 189–90
 median/mode frequency followers 190
signal processors 150
 see also spectral estimation techniques
signal-to-noise ratio 51
 autocorrelation algorithm 250, 253
 cardiac cycle 187–8
 cross-correlation based algorithms 265, 266
 cross-correlation function 275, 277
 envelope averaging techniques 190–1
 finite impulse response filters 247
 instantaneous frequency detector 258
 low order autoregressive (AR) estimators 262, 263
 maximum frequency envelope extraction 186–7
 mean frequency processors 179
 phase detector 257
 zero-crossing processors 188–9
simple threshold method for maximum frequency envelope extraction 180
single sideband detection (SSB) 100–1
sinusoidal burst spectrum 136
sinusoidal flow 10–11
sinusoidal wave evolution *31*
sister chromatid exchange 366–7
slow flow detection 238
Snell's law 37, 293
soft tissues, speed of ultrasound 28
solid mechanics, high-level code 24
sonoelasticity 233, 344
sonogram post-processing 167–9
 2-D filtering 167, 168–9
 ensemble averaging 167–8

sonoscintigraphy 316
sorbitol, air-filled microbubbles 313
sound cards 408
sound, speed 27–8, 39
 particle velocity 32, 34
sound, velocity (c) 2
spatial resolution, colour Doppler imaging 90, 239–40
speckle *169*, 242
speckle tracking techniques 333–6
 2-D block matching 334
 2-D correlation method 335
 2-D Fourier transforms 336
 B-scan images 336
 bit-pattern correlation algorithms 336
 block-matching techniques 334
 contrast agent 311
 cubic spline interpolation 336
 displacement vector 334
 grid slopes interpolation 336
 interpolation methods 336
 multi-lag processing 336
 multi-level block-matching algorithm 334
 sum-absolute-difference method IX, 335–6
 time domain methods 334
 tissue motion imaging 346–7
 vector Doppler techniques 331
spectral analysis *see* spectral estimation
spectral broadening 138, 139, *140*
 Fourier transform 156
 indices 209, *210*
spectral Doppler
 duplex imaging 233
 filtering 242
 machine defects 243
 post-processing 254–5
 signals VII, 311
 triplex imaging 233
spectral estimation techniques 150–69
 Doppler bandwidth measurement 325
 embolic signals 323
 fast Fourier transform (FFT) 150, 151–5
 Fourier transform analyser 151–5
 limitations 141–2
 procedure 160–1
 real-time 45, *46*
 time interval histogram (TIH) 150–1
 velocity matched 156, 158–60
spectral estimation techniques, parametric
 AR 161, 162–3
 ARMA 161–5
 maximum likelihood (ML) method 162–3
spectral sweep, threshold-crossing method 180–1
spectral variance

Bartlett's averaging procedure 146, *151*, 152
fast Fourier transform (FFT) 142
spectral width 137–8
spectral-MUSIC algorithm 282
speed, mean in coloured pixels 217
speed and power-weighted normalised pixel density 217
speed-weighted pixel density 217
standing waves 30–1
stationary echo cancellers *see* delay line cancellers (DLCs)
stenosis 16
 3-D colour flow XI
 angle independent colour flow imaging IX, 332
 backscattered power 129
 blood backscattering 127
 Doppler signal 131
 flow patterns 16
 flow phantom 390
 frequency ratio 215
 pressure drop derivation 17, 18
 vector Doppler techniques VIII, 329
 velocity detection 202
 see also aortic stenosis
sterilisation, transducers 93
stimulated acoustic emission (SAE) imaging V, 232–3, 237, 316
streaming, ultrasound 364, 365
string phantom 391–3, *394*, 398, *399*, *401*
stroke rate, Doppler power 129
sub-sample volume Doppler processing 253
sum-absolute-difference algorithm IX, 335–6
sum-of-approximate-match 336
summed jet energy 215
superposition principle 30
surface waves 27
sweep generator, reference signal 54
switched channel processing 107–8
systolic decay time index 207, *208*
systolic width 209

tapers 20
target velocity (v) 2
Taylor expansion 32
temporal filtering, non-linear 254
temporal resolution, colour Doppler imaging 240
ten percent max method of maximum frequency envelope extraction 183
test devices 382
 moving target 381
 tissue equivalence 382
testicular varicocele X
thermal effects, WFUMB safety statement 374–5

thermal index (*TI*) 370–1, *372*, 373–4
thermal relaxation 38
three-dimensional Doppler
    imaging X–XI, 233–4, *343*
    colour Doppler imaging 239
    complex vascular beds 234
    real time 234
threshold-crossing method for maximum
    frequency envelope
    extraction 180–1
time domain correlation
    flowmeter 272–3
    methods for CFI systems 64–5
time interval histogram (TIH) 150–1
time shift 51, 54
time shift estimation 263–74, *273*,
    275–80
    cross-correlation approach 263–4
    cross-correlation based
        algorithms 265–7
    techniques 350
time-average volumetric flow through
    vessel 289
time-averaged Doppler frequency,
    maximum/mean 184
time–frequency distributions 165–7
time–frequency representation
    (*TFR*) 218
time–motion pulse-echo scans 60,
    346–7
tissue characterisation parameters 39
tissue elasticity 233
tissue equivalence, flow
    phantom 389–90
tissue layers, non-parallel 293
tissue mimic, acoustic properties 387,
    389
tissue motion imaging 343–7
    A-mode technique 344, 346
    B-mode technique 344, 346
    CW Doppler 344–5, 346
    Doppler tissue imaging IV, XI, XII,
        344, 345–7
    M-mode technique 344, 346
    pulse-echo methods 346–7
    PW Doppler 344–5, 346
    real-time B-mode and M-mode 344
    speckle tracking techniques 346–7
tissue vascularity 317
trans-cranial Doppler 321–2
trans-systolic time 209
transducers 71–81, *82*, 83–4
    angle estimation techniques 304–5
    array 76–81
    commercial 72
    curved array 80
    custom-made 72
    damage 94
    damping 74
    design 31, 71–2
    double-element 76
    elements 71
    fetal monitoring systems 76
    frequency bandwidth response
        74–5
    function 73–5
    heating and WFUMB safety
        statement 375
    internal 81, *82*
    linear array 79–80
    manipulation for best signal 132
    materials 71, 73
    models 74
    phased array 80
    piezoelectric effect 72
    piezoelectric element 74
    power received by 124
    PZT/polymer composite 75, *76*
    Q factor 75
    quadrature diffractive 326
    rotating *82*
    single-element *74*, 75–6
    sterilisation 93
    structure 73–5
    transmission fields 71, *72*
    transoesophageal 80
    types 72, *73*
    ultrasonic pulse 75
    waveplate matching 73–4
transduction, pattern recognition 200,
    201
transient cavitation threshold 366
transient response imaging 316
transit time
    broadening 135, 136, 138–9
    ratio *214*, 215
    two site measurement 214, 215
transmitter
    CW systems 97, *98*, *99*
    PW system 105
transoesophageal imaging 80
    heart 81
transverse Doppler 138, 323–5
    studies 141
    theory 324–5
    vector Doppler technique
        combination 329–30
triplex imaging V, 233
true negative/positive fractions 219
tumour vascularity
    3-D Doppler imaging 399
    contrast imaging 316
turbulence
    blood ultrasound scattering 126–7
    spectral broadening for
        quantification 167–8
turbulent flow 8–9
    average velocity profile 8
    entrance effects 9
    pressure gradient 8–9
two-bit signals, cross-correlation based
    algorithms 266

ultrasonic beam 84–6
    angle of propagation 37
    distortion 37
    finite width 291
    shape 92
ultrasonic field 84–6
    schlieren method of visualisation 92–3
ultrasonic field measurement 90–3,
    359–61
    instruments 359
    intensity 361
    power 360
    pressure amplitude 360–1
ultrasonic output of Doppler units 362–3
ultrasonic phenomena 34–9
    see also absorption; attenuation;
        reflection; refraction; scattering
ultrasonic signal
    characteristics 406–7
    colour flow image formation 61
    reception 230
    transmission 230
ultrasonic signature of target 136
ultrasonic waves 28–32, *33*, 34
    amplitude 31
    non-linear propagation 31–2, *33*, 34,
        142
ultrasound 1–2
    bioeffects 359, 364, 365–7
    blood vessel damage 367
    cavitation 364, 365
        moving blood 367
    CFI systems 230–1
    chromosome damage 366
    Doppler 1, *3*
        devices 2
    epidemiological surveys 367
    frequencies 1–2
    frequency-dependent scattering 144
    growth retardation 367
    heating 363–4, 365
    imaging devices 2
    microbubble contrast agent
        interaction 313, 315
    non-linear propagation 31–2, *33*, 34,
        142
    patient exposure 376–7
    physical effects 363–5
        diagnostic/therapeutic levels 365
    safety 359
    sister chromatid exchange 366–7
    streaming 364, 365
uniform velocity flow, ISB 137

vascularity measures 216–17
vastus medialis muscle, Doppler
    M-mode XII, 346

vector Doppler techniques 325–33
　2-D speckle tracking techniques 331
　2-D velocity profiles 327
　3-D 326–7, 330
　　flow velocity vector 331
　　velocity vector 328
　angle independent colour flow
　　imaging 331–3
　beam/vessel angle 243
　correlation-based techniques 330–1
　duplex scanner with position location
　　system 329
　individual sample volumes 326–31
　multiple crossed-beam
　　ultrasound 328
　probe arrangements 328
　stenosis VIII, 329
　transducers pointing in non-coplanar
　　directions 328
　transmitting and receiving transducer
　　arrangements 328, 329
　transverse Doppler
　　combination 329–30
　triple transducer design 328–9
　two-transducer system 327
velocimeter 150
velocity
　acceleration/deceleration 207
　calibration for colour Doppler
　　imaging 240
　dispersion 272
　distribution 119–20
　　radial in blood vessels 132
　　within beam 209
　estimator 62
　flow through vessel 289
　map 66
　measurement of Doppler test
　　device 398–9
　resolution in colour Doppler
　　imaging 240
　threshold for CFI systems 255
　transverse 272
　transverse component estimation 337
　ultrasound in soft tissue 292–3
　variance estimates 278–9
　see also parabolic velocity profile
velocity detecting systems 43–54
　coded signal 51–4
　continuous wave 43–5
　pulsed wave 43, 45–51
velocity estimate
　2-D autocorrelation 260–1
　autocorrelation estimator with
　　sub-sampling 258
　maximum likelihood estimators 269
velocity gradient index (VGI) 216
velocity imaging systems 58–67

continuous wave Doppler
　imaging 58–9
pulsed wave Doppler imaging 59
real-time two-dimensional colour flow
　imaging 59–66
velocity profile 10–11, 132–4
　assumed in blood flow
　　measurement 300
　carotid bifurcation 20, *23*
　curved tubes 18, *19*
　human arteries 20–1, *22*, *23*
　IWMF 169–70, 171, *172*, 173
　mean component 170
　measurement 53–4
　partial sampling 134
　PIWMF 171, *172*, 173
　taper 20
velocity waveform
　common carotid artery *57*
　Fourier transform 203
　phase-shift averaging 191
　see also feature extraction
velocity-matched spectral analysis 155,
　156, 158–60, *161*, 350
venous flow noise 186
venous graft, 3-D Doppler imaging
　343
vessel area measurement 289–91
　errors 290, 291
vessel diameter measurement 290–1
　calliper setting errors 291
　ultrasound imaging 294
vessel insonation, non-uniform 289
vessel shape 290
vessel size change 290
vessel wall
　distensibility 15
　thickness in flow phantom 387, *388*,
　　389
vessel wall thump 15
　elimination 186
　filtering 66, 145
vibrating targets 394–5
video acquisition card, high
　performance 412
video cassette recorder 410–11
video image processing workstation 411
video signals/standards *413*
video two-bit correlation 336
viscoelastic tubes, pulsatile
　flow 14–16
viscosity 5–6
　absolute 5
　dynamic 5
　kinematic 5, 10
volume rendering techniques XI, 343
volumetric flow 7, 200
　measurement 288

vortex detection, pseudo-Wigner
　distribution 166
wall thump *see* vessel wall thump
water-bath transducer *82*
wave scattering by particles 121
waveform
　distortion 31, *33*
　indices of Doppler test device *384*,
　　399, *401*
　matching technique 218–19
　pulsatility 204
　resistance index 205, 206
　shape 200
　see also feature extraction
wavelet transform-based cross-correlation
　(WTCC) 266–7
waveplate matching 73–4
Weaver receiver technique 110, *111*, 112
weighted average technique 266
wide-band methods 263–74, *273*, 275–80
　correlation interpolation
　　algorithm 267
　cross-correlation 263–7, 271–4, *273*,
　　275–8
　maximum likelihood
　　estimators 268–71
　wide-band/time domain
　　techniques 278–80
Wiener–Khinchin theorem 251
Wigner–Ville class distribution 165
windowing 153
　cross-correlation based
　　algorithms 275
Womersley's parameters 10–11
Womersley's theory 381
World Federation for Ultrasound in
　Medicine and Biology
　(WFUMB) 374–5

Young's modulus of elasticity 15

zero-crossing detector (ZCD) 45, 116,
　256–7
zero-crossing processors 187–9
　attenuation 188
　filters 188
　non-uniform insonation 188
　output filters 187–8
　relationship to mean frequency 188
　root mean square (RMS)
　　frequency 189
　SET-RESET system *188*, 189
　signal-to-noise ratio 188–9
　spectral broadening 188
　statistical noise 187–8
zoom Wigner transform 165

Index compiled by Jill C. Halliday

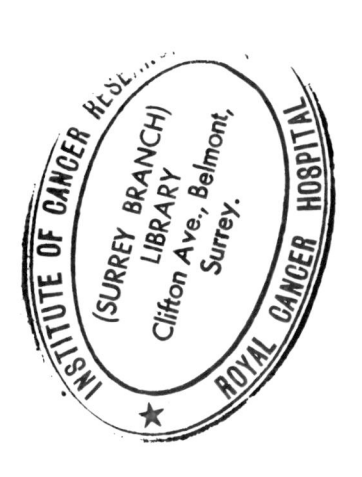